CERAMBYCIDAE
OF THE WORLD
Biology and Pest Management

CONTEMPORARY TOPICS in ENTOMOLOGY SERIES

THOMAS A. MILLER Editor

CERAMBYCIDAE
OF THE WORLD
Biology and Pest Management

edited by **Qiao Wang**

CRC Press
Taylor & Francis Group
Boca Raton London New York

CRC Press is an imprint of the
Taylor & Francis Group, an **informa** business

CRC Press
Taylor & Francis Group
6000 Broken Sound Parkway NW, Suite 300
Boca Raton, FL 33487-2742

First issued in paperback 2020

© 2017 by Taylor & Francis Group, LLC
CRC Press is an imprint of Taylor & Francis Group, an Informa business

No claim to original U.S. Government works

ISBN 13: 978-0-367-57397-3 (pbk)
ISBN 13: 978-1-4822-1990-6 (hbk)

Library of Congress Cataloging-in-Publication Data

Names: Wang, Qiao, 1960 November 16- , editor.
Title: Cerambycidae of the world : biology and pest management / [edited by]
Qiao Wang.
Description: Boca Raton : Taylor & Francis, 2017. | Series: Contemporary
topics in entomology series
Identifiers: LCCN 2016028247 | ISBN 9781482219906 (hardback : alk. paper) |
ISBN 9781315313252 (e-book) | ISBN 9781315313238 (e-book) | ISBN
9781315313245 (e-book) | ISBN 9781315313221 (e-book)
Subjects: LCSH: Cerambycidae. | Insect pests--Control.
Classification: LCC QL596.C4 C465 2017 | DDC 595.76/48--dc23
LC record available at https://lccn.loc.gov/2016028247

Visit the Taylor & Francis Web site at
http://www.taylorandfrancis.com

and the CRC Press Web site at
http://www.crcpress.com

The editor dedicates this book to the late Professor Shu-nan Chiang (1914–2013),

a cerambycidist, who led him to the world of Cerambycidae.

Contents

Preface

There are more than 36,000 described species in the family Cerambycidae worldwide. Although only a small proportion of these species are pests in agriculture, forestry, or horticulture, their economic impact is enormous, costing billions of dollars in production losses, damage to landscapes, and management expenditures. A number of cerambycid species are important pests of various field, vine, and tree crops as well as forest and urban trees in their native regions. However, with the substantial increase of international trade in recent decades, many cerambycid species have become established outside their natural range of distribution, with the potential for causing enormous damage in these novel habitats. So far, no comprehensive work dealing with all aspects of cerambycid biology and management from a global viewpoint has been published.

This volume attempts to address that void by covering the entire spectrum from cerambycid classification, biology, ecology, plant disease transmission to biological, cultural, and chemical control tactics, to the world's major agricultural and tree pests, invasive pests, and biosecurity measures. It is intended to provide an entrance to the scientific literature on Cerambycidae for scientists in research institutions, primary industries, and universities, and an essential reference for quarantine officers in governmental departments charged with detection, exclusion, and control of cerambycids throughout the world. It is hoped that this book will serve as a valuable reference work for many years to come.

This book is divided into 13 chapters, each of which covers a particular topic consisting of our current knowledge and the gaps to be filled. Hundreds of examples, graphs, and photos are presented. The book begins with an introductory chapter dealing with morphology of adults and immature stages, the current classification system, the identification of adults and immatures to subfamily, and biology, global diversity, and distribution of subfamilies. Chapter 2 discusses the types of habitats commonly occupied by cerambycids; oviposition, fecundity, and egg development; voltinism, overwintering, quiescence, and diapause; adult dispersal and longevity; and population dynamics in relation to environmental conditions. Chapter 3 focuses on adult and larval feeding habits and wood digestion; flight, pollination, and plant disease transmission in relation to adult feeding; larval host plant range and conditions, parts and tissue utilized, and voltinism in relation to development and nutrition. Chapter 4 discusses adult phenology and diet in association with host and mate location, mating and oviposition behavior, larval development, and reproductive strategies. Historically, it was thought that cerambycids did not use semiochemicals to mediate reproductive behaviors, but research over the past 15 years suggests that this was erroneous, and that semiochemical use is very common if not ubiquitous within the family. Thus, Chapter 5 summarizes recent research on cerambycid pheromones and their chemistry, the role of plant volatiles as pheromone synergists, mechanisms for maintaining reproductive isolation, and applications of pheromones and kairomones in pest management and detection of invasive pest species. Chapter 6 describes the biology and control of cerambycids as vectors of pathogens (nematodes) of the pine wilt disease and plant–beetle–nematode interactions. Chapter 7 presents a thorough review of laboratory rearing and handling of both cerambycid adults and immature stages with artificial and natural diets. These are followed by three chapters on pest control tactics. Chapter 8 describes natural enemies in relation to cerambycid life history, taxonomic range of natural enemies, impact of natural enemies on cerambycid population dynamics, biological control approaches, and case studies. Using a number of examples, Chapter 9 covers cultural control measures, including mechanical and sanitary techniques, irrigation, plant density management, adjusting planting and harvest times, physical barriers, traps, crop rotation and intercropping, plant resistance, and pest management in relation to climate change. Chapter 10 discusses chemical control of cerambycid pests and provides a number of examples, covering the main classes of chemical insecticides and their field applications, including field sprays, bark treatment, trunk injection and insertion, and soil and root treatment. Chapter 11 presents 43 selected cerambycid species that illustrate the wide range of life history strategies found among cerambycids infesting forest and urban trees throughout the world; information is provided on the identification of adults, native and introduced geographic range, larval hosts, life history, economic impact, and control options. In Chapter 12, 90 cerambycid species of economic importance in field crops, tree crops, and vine crops from around the world are discussed along with their adult diagnoses, native and introduced geographic range, damage, biology, and management measures. Chapter 13 deals with invasive cerambycid pests and biosecurity measures, providing detailed information on interceptions and pathways of invasive pests, inspection and detection methods, pest risk assessments, eradication programs, and establishment and outbreaks of nonnative species.

Because of the nature of multiauthored contributions, it has not been possible to keep strict uniformity in all chapters. The editor has tried, however, to adopt a uniform nomenclature for all cerambycid species throughout the book. This has not been easy because the taxonomy of the Cerambycidae is still in flux, and the recent synonymizations of several species are reflected in a few chapters. Although chapters are logically linked, each represents an independent topic. Therefore, to keep the integrity of each chapter, there is some overlap in subject matter in a few chapters.

The editor is indebted to many people for their advice during the preparation of this book, in particular to Dr. J. Sulzycki, Dr. M. C. Thomas, Dr. T. A. Miller, J. J. Jurgensen, Jennifer Blaise, and to all of the book's contributors. This volume could not have been completed without the generosity of numerous photographers, reviewers, and copyright holders, whose help is gratefully acknowledged in individual chapters. I thank all of my family for their love and support, which have kept me going.

Qiao Wang
Massey University

More than 200 full color illustrations, which will be useful for identification purposes, are available from the CRC Press website under the Downloads tab: https://www.crcpress.com/Cerambycidae-of-the-World-Biology-and-Pest-Management/Wang/p/book/9781482219906

Editor

Qiao Wang, PhD, is a professor of entomology at the Institute of Agriculture and Environment, Massey University, Palmerston North, New Zealand. He earned his MSc under Professor Shu-nan Chiang from Southwest Agricultural University, Chongqing, China, PhD under Professors Ian W.B. Thornton and Tim R. New from La Trobe University, Melbourne, Australia, and postdoctoral experience under Professor Jocelyn G. Millar from the University of California, Riverside, before joining Massey University. He has studied cerambycid beetles since 1982. His research team currently focuses on plant protection, insect behavior, biological control, and evolutionary biology. Dr. Wang's experience in Australia, China, New Zealand, and the United States is reflected in his more than 300 publications; work with more than 70 postgraduate, postdoctoral, and visiting scientists from around the world; service on editorial boards of a number of international journals and international expert panels; and chairmanship of international conference sessions. Dr. Wang was awarded the 2012 Distinguished Scientist Award by the Entomological Society of America for his outstanding contributions to entomological science during his career.

Contributors

Süleyman Akbulut
Department of Forest Engineering
Faculty of Forestry
Düzce University
Düzce, Turkey

Dominic Eyre
Animal and Plant Health
Defra, Department for Environment,
    Food & Rural Affairs
Sand Hutton, York, UK

Robert A. Haack
Northern Research Station
USDA Forest Service
U.S. Department of Agriculture
Lansing, Michigan

Lawrence M. Hanks
Department of Entomology
University of Illinois at Urbana-Champaign
Urbana, Illinois

Melody A. Keena
Northern Research Station
USDA Forest Service
U.S. Department of Agriculture
Hamden, Connecticut

Marc J. Linit
College of Agriculture
    Food and Natural Resources
University of Missouri
Columbia, Missouri

Jocelyn G. Millar
Department of Entomology
University of California
Riverside, California

Marcela L. Monné
Departamento de Entomologia
Museu Nacional, Universidade Federal do
    Rio de Janeiro
Rio de Janeiro, Brasil

Miguel A. Monné
Departamento de Entomologia
Museu Nacional, Universidade Federal do
    Rio de Janeiro
Rio de Janeiro, Brasil

Timothy D. Paine
Department of Entomology
University of California
Riverside, California

Katsumi Togashi
Department of Forest Science
Graduate School of Agricultural and
    Life Sciences
The University of Tokyo
Tokyo, Japan

Qiao Wang
Institute of Agriculture and Environment
Massey University
Palmerston North, New Zealand

1

General Morphology, Classification, and Biology of Cerambycidae

Marcela L. Monné and Miguel A. Monné
Universidade Federal do Rio de Janeiro
Rio de Janeiro, Brasil

Qiao Wang
Massey University
Palmerston North, New Zealand

CONTENTS

1.1 Introduction

Cerambycidae Latreille, 1802, commonly known as longicorns, longhorns, longicorn beetles, longhorned beetles, longhorned borers, round-headed borers, timber beetles, or sawyer beetles, are among the most diverse and economically important families of Coleoptera. Taxonomic interest in the family has been fairly consistent for the past century, but the description of new taxa has accelerated in recent decades. The number of described cerambycid species in the world is about 36,300 in more than 5,300 genera (Tavakilian 2015). The adult body length ranges from less than 2 mm in *Cyrtinus pygmaeus* (Haldeman) (Linsley 1961) to greater than 170 mm in *Titanus giganteus* (L.) (Williams 2001). Cerambycids are widely distributed around the world—from sea level to 4,200 m above—wherever their host plants are found. Distribution and generic diversity of the world's cerambycid subfamilies and tribes are shown in Table 1.1.

The longicorn adults are free-living beetles that may or may not need to feed. They can live for a few days to a few months depending on whether they feed (Hanks 1999; Wang 2008). Cerambycids usually reproduce sexually but, in very rare cases—such as in some species of *Kurarus* Gressitt (Cerambycinae) (Goh 1977) and *Cortodera* Mulsant (Lepturinae) (Švácha and Lawrence 2014), they can reproduce parthenogenetically. Švácha and Lawrence (2014) suggested that at least in *Cortodera*, parthenogenesis probably is of recent origin because the female has a distinct spermatheca with a spermathecal gland. Mate location depends on the occurrence and status of larval hosts, adult food sources, and/or pheromones. Hanks (1999) predicted that the absence of feeding in the adult stage of many species is associated with the production of long-range pheromones, but the current knowledge shows that the use of volatile pheromones is widespread in cerambycids (see Chapter 5). The females lay their eggs on or near their hosts. The larvae of most cerambycid species feed on woody plants, but some select herbaceous hosts. The vast majority of species at the larval stage are living and feeding inside the plants although small minorities are free-living in soil and feed on plant roots.

Many cerambycid larvae are dead plant feeders and play a major role in recycling dead plants; others attack living plants of different health states, ranging from stressed to healthy plants. To date, there are about 200 cerambycid species worldwide that have some economic impact on agriculture, forestry, and horticulture, causing billions of dollars of damage in production losses, environmental disasters, and management costs. They may damage plants by direct feeding and/or transmission of plant diseases.

TABLE 1.1

Distribution and Generic Diversity of Cerambycid Subfamilies and Tribes

Subfamilies and Tribes	Biogeographic Regions	No. Genera
Cerambycinae Latreille, 1802	**All biogeographic regions**	**1,757**
Acangassuini Galileo & Martins, 2001	Neotropical	1
Achrysonini Lacordaire, 1868	All biogeographic regions	20
Agallissini Le Conte, 1873	Neotropical	3
Alanizini Di Iorio, 2003	Neotropical	1
Anaglyptini Lacordaire, 1868	All biogeographic regions	12
Aphanasiini Lacordaire, 1868	Afrotropical and Australian	6
Aphneopini Lacordaire, 1868	Australian	5
Auxesini Lepesme & Breuning, 1952	Afrotropical	8
Basipterini Fragoso, Monné & Campos Seabra, 1987	Neotropical	2
Bimiini Lacordaire, 1868	Australian and Neotropical	7
Bothriospilini Lane, 1950	Neotropical	11
Brachypteromatini Sama, 2008	Palaearctic	1
Callichromatini Swainson, 1840	All biogeographic regions	178
Callidiini Kirby, 1837	All biogeographic regions	38
Callidiopini Lacordaire, 1868	All biogeographic regions	62
Cerambycini Latreille, 1802	All biogeographic regions	99
Certallini Fairmaire, 1864	Palaearctic, Afrotropical, and Australian	9
Chlidonini Waterhouse, 1879	Afrotropical (Madagascar)	2
Cleomenini Lacordaire, 1868	Afrotropical and Oriental	23
Clytini Mulsant, 1839	All biogeographic regions	83
Compsocerini Thomson, 1864	All biogeographic regions	33
Coptommatini Lacordaire, 1869	Australian	1
Curiini LeConte, 1873	Neotropical	1
Deilini Fairmaire, 1864	Palaearctic and Australian	3
Dejanirini Lacordaire, 1868	Oriental	2
Diorini Lane, 1950	Neotropical	1
Distichocerini Pascoe, 1867	Australian	2
Dodecosini Aurivillius, 1912	Neotropical	4
Dryobiini Arnett, 1962	Nearctic and Neotropical	3
Eburiini Blanchard, 1845	Neotropical	23
Ectenessini Martins, 1998	Neotropical	12
Elaphidiini Thomson, 1864	Nearctic and Neotropical	91
Eligmodermini Lacordaire, 1868	Neotropical	5
Erlandiini Aurivillius, 1912	Neotropical	1
Eroschemini Lacordaire, 1868	Australian	2
Eumichthini Linsley, 1940	Nearctic	2
Gahaniini Quentin & Villiers, 1969	Afrotropical	1
Glaucytini Lacordaire, 1868	Oriental and Australian	18
Graciliini Mulsant, 1839	All biogeographic regions	22
Hesperophanini Mulsant, 1839	All biogeographic regions	85
Hesthesini Pascoe, 1867	Australian	1
Heteropsini Lacordaire, 1868	Neotropical and Australian	29
Hexoplini Martins, 2006	Neotropical	22
Holopleurini Chemsak & Linsley, 1974	Nearctic	1
Hyboderini Linsley, 1940	Nearctic and Neotropical	4
Hylotrupini Zagajkevich, 1991	Palaearctic	1
Ideratini Martins & Napp, 2009	Neotropical	1

(Continued)

TABLE 1.1 *(Continued)*

Distribution and Generic Diversity of Cerambycid Subfamilies and Tribes

Subfamilies and Tribes	Biogeographic Regions	No. Genera
Lissonotini Swainson, 1840	Neotropical	1
Luscosmodicini Martins, 2003	Neotropical	1
Lygrini Sama, 2008	Afrotropical	1
Macronini Lacordaire, 1868	Australian	4
Megacoelini Quentin & Villiers, 1969	Afrotropical	2
Methiini Thomson, 1860	Oriental, Afrotropical, and Neotropical	19
Molorchini Gistel, 1848	All biogeographic regions	26
Mythodini Lacordaire, 1868	Oriental	4
Necydalopsini Lacordaire, 1868	Neotropical	12
Neocorini Martins, 2005	Neotropical	7
Neoibidionini Monné, 2012	Neotropical	55
Neostenini Lacordaire, 1868	Australian	4
Obriini Pascoe, 1871	All biogeographic regions	43
Ochyrini Pascoe, 1871	Australian	1
Oedenoderini Aurivillius, 1912	Afrotropical	1
Oemini Lacordaire, 1868	All biogeographic regions	101
Opsimini LeConte, 1873	Nearctic and Palaearctic	3
Oxycoleini Martins & Galileo, 2003	Neotropical	2
Paraholopterini Martins, 1997	Neotropical	1
Phalotini Lacordaire, 1868	Australian	4
Phlyctaenodini Lacordaire, 1868	Australian and Neotropical	17
Phoracanthini Newman, 1840	Australian	22
Phyllarthriini Lepesme & Breuning, 1956	Afrotropical	4
Piesarthriini McKeown, 1947	Australian	4
Piezocerini Lacordaire, 1868	Neotropical	19
Platyarthrini Bates, 1870	Neotropical	1
Plectogasterini Quentin & Villiers, 1969	Afrotropical	8
Plectromerini Nearns & Braham, 2008	Neotropical	1
Pleiarthrocerini Lane, 1950	Neotropical	1
Plesioclytini Wappes & Skelley, 2015	Nearctic	1
Proholopterini Monné, 2012	Neotropical	3
Protaxini Gahan, 1906	Oriental	1
Prothemini Lacordaire, 1868	Oriental	3
Psebiini Lacordaire, 1868	Afrotropical and Neotropical	24
Pseudocephalini Aurivillius, 1912	Australian and Neotropical	4
Pseudolepturini Thomson, 1861	Oriental	6
Psilomorphini Lacordaire, 1868	Australian	3
Pteroplatini Thomson, 1861	Afrotropical and Neotropical	10
Rhagiomorphini Newman, 1841	Australian	4
Rhinotragini Thomson, 1861	Neotropical	82
Rhopalophorini Blanchard, 1845	Nearctic, Neotropical, and Australian	29
Sestyrini Lacordaire, 1868	Oriental	2
Smodicini Lacordaire, 1868	Afrotropical, Nearctic, and Neotropical	8
Spintheriini Lacordaire, 1869	Australian	2
Stenhomalini Miroshnikov, 1989	Oriental	2
Stenoderini Pascoe, 1867	Australian and Oriental	10
Stenopterini Gistel, 1848	Palaearctic and Oriental	14
Strongylurini Lacordaire, 1868	Australian	6

(Continued)

TABLE 1.1 *(Continued)*

Distribution and Generic Diversity of Cerambycid Subfamilies and Tribes

Subfamilies and Tribes	Biogeographic Regions	No. Genera
Tessarommatini Lacordaire, 1868	Australian	1
Thraniini Gahan, 1906	Oriental	3
Thyrsiini Marinoni & Napp, 1984	Neotropical	1
Tillomorphini Lacordaire, 1868	Nearctic, Neotropical, Oriental, and Australian	31
Torneutini Thomson, 1861	Neotropical	16
Trachyderini Dupont, 1836	All biogeographic regions	154
Tragocerini Pascoe, 1867	Australian	1
Trichomesiini Aurivillius, 1912	Australian	1
Trigonarthrini Villiers, 1984	Afrotropical	2
Tropocalymmatini Lacordaire, 1868	Australian	1
Typhocesini Lacordaire, 1868	Australian	4
Unxiini Napp, 2007	Neotropical	8
Uracanthini Blanchard, 1853	Australian	6
Vesperellini Sama, 2008	Palaearctic	1
Xystrocerini Blanchard, 1845	Afrotropical and Australian	2
Dorcasominae Lacordaire, 1868	**Afrotropical, Oriental, and Palaearctic**	**95**
Apatophyseini Lacordaire, 1869	Afrotropical, Oriental, and Palaearctic	90
Dorcasomini Lacordaire, 1868	Afrotropical, Oriental, and Palaearctic	5
Lamiinae Latreille, 1825	**All biogeographic regions**	**2,964**
Acanthocinini Blanchard, 1845	All biogeographic regions	386
Acanthoderini Thomson, 1860	All biogeographic regions	66
Acmocerini Thomson, 1864	Afrotropical	6
Acridocephalini Dillon & Dillon, 1959	Afrotropical	1
Acrocinini Swainson, 1840	Neotropical	1
Aderpasini Breuning & Teocchi, 1978	Afrotropical	1
Aerenicini Lacordaire, 1872	Neotropical	26
Agapanthiini Mulsant, 1839	All biogeographic regions	84
Amphoecini Breuning, 1951	Australian	2
Ancitini Aurivillius, 1917	Australian	1
Ancylonotini Lacordaire, 1869	Afrotropical, Oriental, and Palaearctic	36
Anisocerini Thomson, 1860	Neotropical	26
Apomecynini Thomson, 1860	All biogeographic regions	240
Astathini Thomson, 1864	Australian, Afrotropical, Oriental, and Palaearctic	23
Batocerini Thomson, 1864	Australian, Oriental, and Palaearctic	10
Calliini Thomson, 1864	Neotropical	40
Ceroplesini Thomson, 1860	Afrotropical, Oriental, and Palaearctic	88
Cloniocerini Lacordaire, 1872	Afrotropical	1
Colobotheini Thomson, 1860	Neotropical	12
Compsosomatini Thomson, 1867	Neotropical	13
Cyrtinini Thomson, 1864	Australian and Neotropical	16
Desmiphorini Thomson, 1860	All biogeographic regions	319
Dorcadionini Swainson, 1840	Palaearctic and Oriental	14
Dorcaschematini Thomson, 1860	Oriental and Australian	9
Elytracanthinini Bousquet, 2009	Neotropical	1
Enicodini Thomson, 1864	Australian and Oriental	27
Eupromerini Galileo & Martins, 1995	Neotropical	5
Forsteriini Tippmann, 1960	Neotropical	16
Gnomini Thomson, 1860	Australian, Oriental, and Palaearctic	4

(Continued)

TABLE 1.1 *(Continued)*

Distribution and Generic Diversity of Cerambycid Subfamilies and Tribes

Subfamilies and Tribes	Biogeographic Regions	No. Genera
Gyaritini Breuning, 1950	Australian and Oriental	14
Heliolini Breuning, 1951	Australian	1
Hemilophini Thomson, 1868	Neotropical and Nearctic	127
Homonoeini Thomson, 1864	Australian, Oriental, and Palaearctic	22
Hyborhabdini Aurivillius, 1911	Oriental	1
Lamiini Latreille, 1825	Afrotropical, Australian, Oriental, and Palaearctic	48
Laticraniini Lane, 1959	Neotropical	2
Mauesiini Lane, 1956	Neotropical	4
Megabasini Thomson, 1860	Neotropical	1
Mesosini Mulsant, 1839	All biogeographic regions	99
Microcymaturini Breuning & Teocchi, 1985	Afrotropical	3
Moneilemini Thomson, 1864	Nearctic and Neotropical	1
Monochamini Gistel, 1848	All biogeographic regions	263
Morimonellini Lobanov, Danilevsky & Murzin, 1981	Palaearctic	1
Morimopsini Lacordaire, 1869	All regions except Nearctic	47
Nyctimeniini Gressitt, 1951	Australian and Oriental	1
Obereini Thomson, 1864	All regions except Neotropical	3
Oculariini Breuning, 1950	Afrotropical	2
Onciderini Thomson, 1860	Neotropical and Nearctic	81
Oncideropsidini Aurivillius, 1922	Oriental	1
Onocephalini Thomson, 1860	Neotropical	3
Onychogleneini Aurivillius, 1923	Oriental	1
Parmenini Mulsant, 1839	All biogeographic regions	87
Petrognathini Blanchard, 1845	Afrotropical and Oriental	10
Phacellini Lacordaire, 1872	Neotropical	7
Phantasini Kolbe, 1897	Afrotropical	3
Phrynetini Thomson, 1864	Afrotropical, Oriental, and Palaearctic	14
Phymasternini Teocchi, 1989	Afrotropical	1
Phytoeciini Mulsant, 1839	All biogeographic regions	32
Pogonocherini Mulsant, 1839	All biogeographic regions	33
Polyrhaphidini Thomson, 1860	Afrotropical and Neotropical	2
Pretiliini Martins & Galileo, 1990	Neotropical	1
Proctocerini Aurivillius, 1922	Afrotropical	1
Prosopocerini Thomson, 1864	Afrotropical	18
Pteropliini Thomson, 1860	All biogeographic regions	256
Saperdini Mulsant, 1839	All regions except Neotropical	154
Stenobiini Breuning, 1950	Afrotropical	7
Sternotomini Thomson, 1860	Afrotropical	20
Tapeinini Thomson, 1857	Neotropical and Oriental	2
Tetraopini Thomson, 1860	Nearctic and Neotropical	3
Tetraulaxini Breuning & Teocchi, 1977	Afrotropical	2
Tetropini Portevin, 1927	Palaearctic	2
Theocrini Lacordaire, 1872	Afrotropical	8
Tmesisternini Blanchard, 1853	Australian and Oriental	12
Tragocephalini Thomson, 1857	Afrotropical	63
Xenicotelini Matsushita, 1933	Oriental	1
Xenofreini Aurivillius, 1923	Neotropical	3
Xenoleini Lacordaire, 1872	Australian, Oriental, and Palaearctic	3

(Continued)

TABLE 1.1 *(Continued)*

Distribution and Generic Diversity of Cerambycid Subfamilies and Tribes

Subfamilies and Tribes	Biogeographic Regions	No. Genera
Xylorhizini Lacordaire, 1872	Afrotropical, Australian, Oriental, and Palaearctic	10
Zygocerini Thomson, 1864	Australian and Oriental	9
Lepturinae Latreille, 1802	**All biogeographic regions**	**210**
Desmocerini Blanchard, 1845	Nearctic	1
Encyclopini LeConte, 1873	Nearctic and Palaearctic	2
Lepturini Latreille, 1802	All biogeographic regions	140
Oxymirini Danilevsky, 1997	Palaearctic	1
Rhagiini Kirby, 1837	All biogeographic regions	53
Rhamnusiini Sama, 2009	Palaearctic and Oriental	2
Sachalinobiini Danilevsky, 2010	Nearctic and Palaearctic	1
Teledapini Pascoe, 1871	Oriental	3
Xylosteini Reitter, 1913	Palaearctic and Oriental	7
Necydalinae Latreille, 1825	**Nearctic, Palaearctic, and Oriental**	**2**
Necydalini Latreille, 1825	Nearctic, Palaearctic, and Oriental	2
Parandrinae Blanchard, 1845	**All biogeographic regions**	**19**
Erichsoniini Thomson, 1861	Neotropical	1
Parandrini Blanchard, 1845	All biogeographic regions	18
Prioninae Latreille, 1802	**All biogeographic regions**	**302**
Acanthophorini Thomson, 1864	Afrotropical	7
Aegosomatini Thomson, 1861	Afrotropical, Oriental, and Australian	20
Anacolini Thomson, 1857	Afrotropical, Oriental, and Neotropical	33
Cacoscelini Thomson, 1861	Afrotropical and Australian	5
Callipogonini Thomson, 1861	Afrotropical, Palaearctic, and Neotropical	17
Calocomini Galileo & Martins, 1993	Neotropical	1
Cantharocnemini Thomson, 1861	Afrotropical and Australian	6
Closterini, Lacordaire, 1868	Afrotropical, Australian, and Oriental	8
Ergatini Fairmaire, 1864	Afrotropical, Palaearctic, and Nearctic	5
Eurypodini Gahan, 1906	Palaearctic and Oriental	4
Hopliderini Thomson, 1864	Afrotropical	5
Macrodontiini Thomson, 1861	Neotropical	5
Macrotomini Thomson, 1861	All biogeographic regions	78
Mallaspini Thomson, 1861	Neotropical	10
Mallodonini Thomson, 1861	Afrotropical, Oriental, Nearctic, and Neotropical	10
Meroscelisini Thomson, 1861	Afrotropical, Australian, and Neotropical	21
Prionini Latreille, 1802	All biogeographic regions	50
Remphanini Lacordaire, 1868	Oriental	6
Solenopterini Lacordaire, 1868	Neotropical	7
Tereticini Lameere, 1913	Afrotropical and Australian	3
Vesperoctenini Vives, 2005	Neotropical	1
Spondylidinae Audinet-Serville, 1832	**All biogeographic regions**	**32**
Anisarthrini Mamaev & Danilevsky, 1973	Palaearctic	4
Asemini Thomson, 1861	All biogeographic regions	12
Atimiini LeConte, 1873	Nearctic, Neotropical, and Palaearctic	3
Saphanini Gistel 1848	Afrotropical and Nearctic	10
Spondylidini Audinet-Serville, 1832	Neotropical, Nearctic, and Palaearctic	3

With the increase of international trade in recent years, many cerambycid species have been intercepted; some have become established outside their natural distribution range, causing serious problems globally (Haack et al. 2010; see Chapter 13).

Linsley (1961, 1962a) and Wang (2008) summarize the general morphology and biology of the Cerambycidae. More recently, Švácha and Lawrence (2014) have made a very detailed treatment of the morphology and a general account of the ecology of the Cerambycidae. Ślipiński and Escalona (2013) gave a good introduction to the morphology and ecology of Australian cerambycids. In this chapter, we summarize the current knowledge about this family, including the definition and morphology, and a brief introduction to the taxonomy, distribution, and general biology at the subfamily level. We aim to provide readers with a fundamental knowledge of cerambycids as well as a guide for those who may wish to consult specific chapters in this book where detailed treatments of Cerambycid biology and pest management are discussed.

1.2 Definition and Morphology of the Family Cerambycidae

1.2.1 Definition

Traditionally, the family Cerambycidae had wider scope, including nine subfamilies: Anoplodermatinae, Aseminae, Cerambycinae, Lamiinae, Lepturinae, Parandrinae, Philinae, Prioninae, and Spondylidinae (Napp 1994). In the current classification system (Bouchard et al. 2011; Monné 2012; Švácha and Lawrence 2014), Oxypeltinae, Vesperinae, and Disteniinae are considered independent families. We use the new system in this book and discuss eight subfamilies: Cerambycinae, Dorcasominae, Lamiinae, Lepturinae, Necydalinae, Parandrinae, Prioninae, and Spondylidinae. Table 1.1 summarizes the distribution and generic diversity of cerambycid subfamilies and tribes.

1.2.2 General Morphology

The general morphology of Cerambycidae is extracted from Ślipiński and Escalona (2013) and Švácha and Lawrence (2014).

1.2.2.1 Adult

1.2.2.1.1 Diagnosis

General external morphology of cerambycid adults is illustrated in Figures 1.1 and 1.2. Antennae usually filiform, elongate, and 11-segmented, rarely serrate and >12-segmented, usually inserted on pronounced tubercles; eyes usually emarginate; prothorax without pleural sutures; tibia with two distinct tibial spurs; tarsi usually pseudotetramerous with fourth tarsomere usually minute and concealed by third tarsomere; elytra usually covering abdomen; hind wings with a spur on radio-medial crossvein; abdomen usually with five visible sternites, fifth sternite entire.

1.2.2.1.2 Description

1.2.2.1.2.1 Head The head is prognathous and more or less horizontal in the Parandrinae (Figures 1.3 and 1.4). It is produced anteriorly to form a short to moderately long muzzle in some Lepturinae (Figures 1.5 and 1.6), Dorcasominae, and Cerambycinae, inclined anteriorly in the Spondylidinae, and is vertical or retracted, with the genal line directed posteriorly, in the Lamiinae (Figure 1.7). The eyes are entire in the Parandrinae (Figure 1.3), most Lepturinae, and some Prioninae; feebly emarginate in the Spondylidinae (Figure 1.8) and most Prioninae (Figure 1.9); emarginate to entire in the Dorcasominae; and usually are deeply emarginate and reniform in the Cerambycinae (Figure 1.10) and Lamiinae (Figure 1.7); although occasionally they are divided—as in *Tetraopes* Schönherr—or lacking the upper lobe—as in *Tillomorpha* Blanchard. The facets of the eyes are large and coarse in the Parandrinae, most Prioninae, and some Asemini and Cerambycinae; usually, they are finer in the Lepturinae, Lamiinae, and more specialized Cerambycinae.

The antennae usually have 11 antennomeres (Figures 1.1 and 1.2) that are inserted near the base of the mandibles in the Parandrinae (Figure 1.3), Prioninae (Figure 1.9), and in some Spondylidinae;

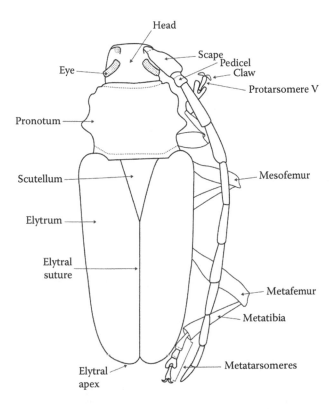

FIGURE 1.1 General morphology, dorsal view of *Trachyderes succinctus* (L.) (Cerambycinae).

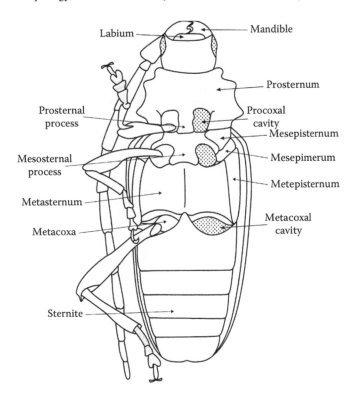

FIGURE 1.2 General morphology, ventral view of *Trachyderes succinctus* (L.) (Cerambycinae).

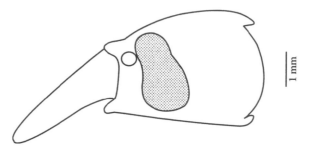

FIGURE 1.3 Head, lateral view of *Parandra (Parandra) glabra* (De Geer) (Parandrinae). (Reprinted with permission from C. J. B. Carvalho, editor. Napp, D. S., *Rev. Bras. Entomol.*, 38, 265–419, 1994.)

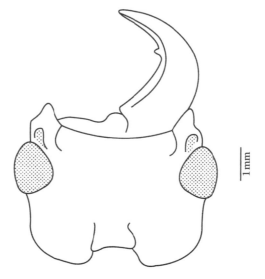

FIGURE 1.4 Head, dorsal view of *Parandra (Parandra) glabra* (De Geer) (Parandrinae). (Reprinted with permission from C. J. B. Carvalho, editor. Napp, D. S., *Rev. Bras. Entomol.*, 38, 265–419, 1994.)

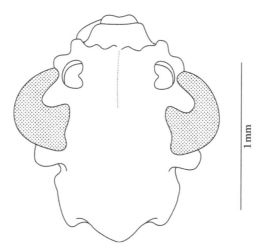

FIGURE 1.5 Head, dorsal view of *Leptura rubra* L. (Lepturinae). (Reprinted with permission from C. J. B. Carvalho, editor. Napp, D. S., *Rev. Bras. Entomol.*, 38, 265–419, 1994.)

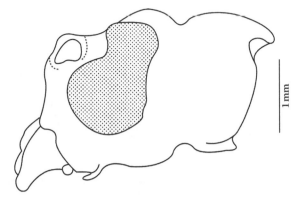

FIGURE 1.6 Head, lateral view of *Leptura rubra* L. (Lepturinae). (Reprinted with permission from C. J. B. Carvalho, editor. Napp, D. S., *Rev. Bras. Entomol.,* 38, 265–419, 1994.)

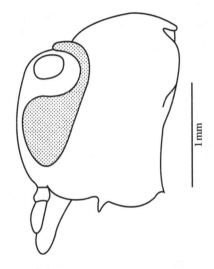

FIGURE 1.7 Head, lateral view of *Estola obscura* Thomson (Lamiinae). (Reprinted with permission from C. J. B. Carvalho, editor. Napp, D. S., *Rev. Bras. Entomol.,* 38, 265–419, 1994.)

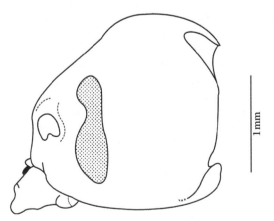

FIGURE 1.8 Head, lateral view of *Asemum striatum* (L.) (Spondylidinae). (Reprinted with permission from C. J. B. Carvalho, editor. Napp, D. S., *Rev. Bras. Entomol.,* 38, 265–419, 1994.)

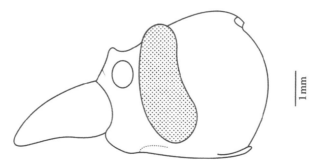

FIGURE 1.9   Head, lateral view of *Mallodon spinibarbis* (L.) (Prioninae). (Reprinted with permission from C. J. B. Carvalho, editor. Napp, D. S., *Rev. Bras. Entomol.*, 38, 265–419, 1994.)

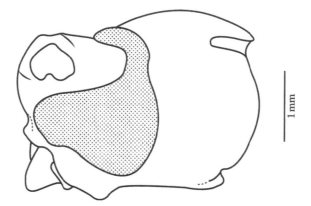

FIGURE 1.10   Head, lateral view of *Achryson surinamum* (L.) (Cerambycinae). (Reprinted with permission from C. J. B. Carvalho, editor. Napp, D. S., *Rev. Bras. Entomol.*, 38, 265–419, 1994.)

are near the eyes in the Asemini and Lepturinae (Figure 1.6); and are usually more or less embraced by the eyes in the Cerambycinae (Figure 1.10) and Lamiinae (Figure 1.7). In some diurnal Lamiinae (e.g., *Octotapnia* Galileo & Martins and *Pseudotapnia* Chemsak & Linsley) and Prioninae, the antennae may have fewer antennomeres. In some Lamiinae (e.g., *Paratenthras* Monné), the first three antennomeres are long, whereas the remaining flagella are reduced and sometimes moniliform. The number of antennomeres may be 12 in a number of unrelated groups and more than 12 in a few Cerambycinae and Prioninae (up to 30 in some species of *Prionus* Müller). The antennal structure is similar between sexes in the Parandrinae, Spondylidinae, and Lepturinae, and strikingly dissimilar in many Prioninae and in most Cerambycinae and Lamiinae. In the Parandrinae and Spondylidinae, differentiation of antennomeres is not well marked; the scape is short, the second antennomere is not greatly reduced in size, half as long as, or subequal to the third antennomere, and the segments that follow are subequal in length. In the remaining subfamilies, the scape is usually more elongate, the second segment is greatly reduced, and the following antennomeres are unequal in length— with the third usually greatly elongated and those that follow diminishing to the ultimate antenno- mere. The antennal segments are glabrous in the Parandrinae, Prioninae, and Spondylidinae, and are pubescent in other subfamilies.

The labrum is fused with the epistoma in the Parandrinae and Prioninae but free in other subfami- lies. The mandibles are acute in all of the Cerambycidae; large and often toothed in the Parandrinae (Figure 1.4) and Prioninae (Figure 1.11); long, slender, and untoothed in the Spondylidinae; shorter in most other groups; and are provided with a dense fringe of hairs in the inner margin of the Dorcasominae and Lepturinae. The maxillae are typically bilobed; the inner lobe is obsolete in the Parandrinae (Figure 1.12) and Prioninae (Figure 1.13). The ultimate segment of the palpi (both

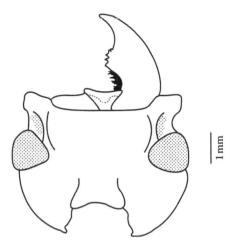

FIGURE 1.11 Head, dorsal view of *Mallodon spinibarbis* (L.) (Prioninae). (Reprinted with permission from C. J. B. Carvalho, editor. Napp, D. S., *Rev. Bras. Entomol.*, 38, 265–419, 1994.)

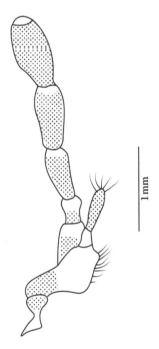

FIGURE 1.12 Maxilla, ventral view of *Parandra (Parandra) glabra* (De Geer) (Parandrinae). (Reprinted with permission from C. J. B. Carvalho, editor. Napp, D. S., *Rev. Bras. Entomol.*, 38, 265–419, 1994.)

maxillary and labial) is pointed at the apex in the Lamiinae (Figures 1.14 and 1.15) and truncated (Figures 1.16 through 1.19) in other subfamilies. The submentum projects between the bases of the maxillae in the Lepturinae; is short in many Cerambycinae; and is absent in the Parandrinae, Prioninae, and Spondylidinae. The mentum is distinctly transverse in the Parandrinae (Figure 1.20), Prioninae, and Spondylidinae, and trapezoidal in the Lepturinae, Cerambycinae (Figure 1.18), and Lamiinae (Figure 1.14). The ligula is corneous in the Parandrinae and Spondylidinae, and membranous or coriaceous in the Lepturinae, Cerambycinae (Figure 1.18) (except Oemini and Methini), and Lamiinae (Figure 1.14).

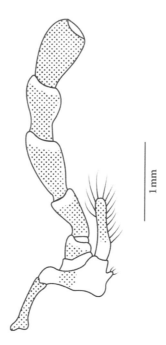

1 mm

FIGURE 1.13 Maxilla, ventral view of *Mallodon spinibarbis* (L.) (Prioninae). (Reprinted with permission from C. J. B. Carvalho, editor. Napp, D. S., *Rev. Bras. Entomol.,* 38, 265–419, 1994.)

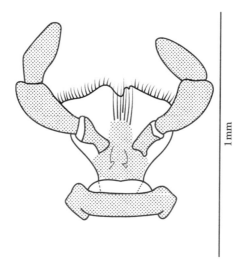

1 mm

FIGURE 1.14 Labium, ventral view of *Estola obscura* Thomson (Lamiinae). (Reprinted with permission from C. J. B. Carvalho, editor. Napp, D. S., *Rev. Bras. Entomol.,* 38, 265–419, 1994.)

1.2.2.1.2.2 Thorax   The prothorax bears lateral carinae in the Parandrinae (Figure 1.21) and Prioninae (Figures 1.22 and 1.23), which are lacking in other subfamilies (Figures 1.24 through 1.26). The procoxae are strongly transverse in the Parandrinae and Prioninae, less so in some Spondylidinae—such as Asemini, subconical in the rest Spondylidinae, conical in the Lepturinae, and usually rounded in the Cerambycinae and Lamiinae. The procoxal cavities are closed behind in some Parandrinae, in some Spondylidinae, and in most Lamiinae (Figure 1.26); wide open in the Prioninae (Figure 1.23), Asemini, and most Lepturinae (Figure 1.25); and open or closed in the Cerambycinae. The scutellum is visible, sometimes well developed (Figure 1.1) and usually is not abruptly elevated, anteriorly flat, or separated from

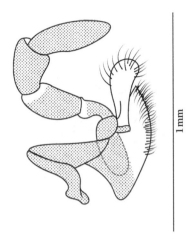

FIGURE 1.15 Maxilla, ventral view of *Estola obscura* Thomson (Lamiinae). (Reprinted with permission from C. J. B. Carvalho, editor. Napp, D. S., *Rev. Bras. Entomol.*, 38, 265–419, 1994.)

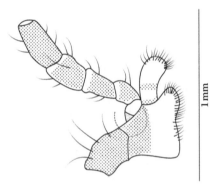

FIGURE 1.16 Maxilla, ventral view of *Asemum striatum* (L.) (Spondylidinae). (Reprinted with permission from C. J. B. Carvalho, editor. Napp, D. S., *Rev. Bras. Entomol.*, 38, 265–419, 1994.)

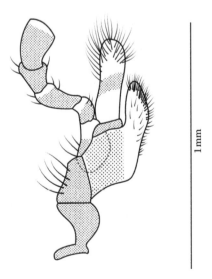

FIGURE 1.17 Maxilla, ventral view of *Necydalis major* L. (Necydalinae). (Reprinted with permission from C. J. B. Carvalho, editor. Napp, D. S., *Rev. Bras. Entomol.*, 38, 265–419, 1994.)

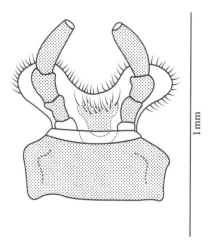

FIGURE 1.18 Labium, ventral view of *Rhopalophora collaris* (Germar) (Cerambycinae). (Reprinted with permission from C. J. B. Carvalho, editor. Napp, D. S., *Rev. Bras. Entomol.*, 38, 265–419, 1994.)

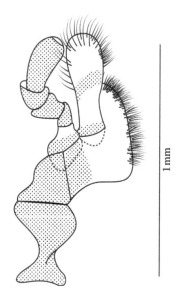

FIGURE 1.19 Maxilla, ventral view of *Trachyderes succinctus* (L.) (Cerambycinae). (Reprinted with permission from C. J. B. Carvalho, editor. Napp, D. S., *Rev. Bras. Entomol.*, 38, 265–419, 1994.)

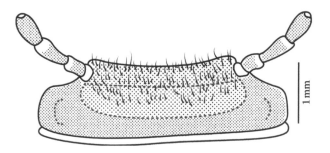

FIGURE 1.20 Labium, ventral view of *Parandra (Parandra) glabra* (De Geer) (Parandrinae). (Reprinted with permission from C. J. B. Carvalho, editor. Napp, D. S., *Rev. Bras. Entomol.*, 38, 265–419, 1994.)

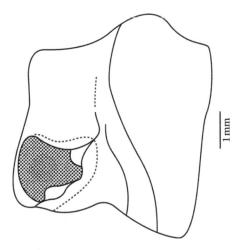

FIGURE 1.21 Prothorax, lateral view of *Parandra (Parandra) glabra* (De Geer) (Parandrinae). (Reprinted with permission from C. J. B. Carvalho, editor. Napp, D. S., *Rev. Bras. Entomol.,* 38, 265–419, 1994.)

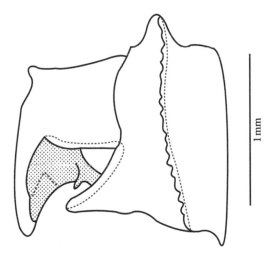

FIGURE 1.22 Prothorax, lateral view of *Mallodon spinibarbis* (L.) (Prioninae). (Reprinted with permission from C. J. B. Carvalho, editor. Napp, D. S., *Rev. Bras. Entomol.,* 38, 265–419, 1994.)

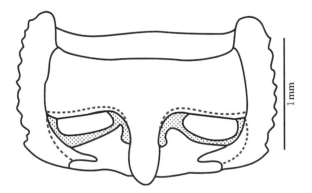

FIGURE 1.23 Prosternum, ventral view of *Mallodon spinibarbis* (L.) (Prioninae). (Reprinted with permission from C. J. B. Carvalho, editor. Napp, D. S., *Rev. Bras. Entomol.,* 38, 265–419, 1994.)

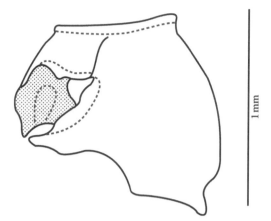

FIGURE 1.24 Prothorax, lateral view of *Leptura rubra* L. (Lepturinae). (Reprinted with permission from C. J. B. Carvalho, editor. Napp, D. S., *Rev. Bras. Entomol.*, 38, 265–419, 1994.)

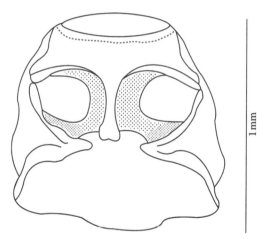

FIGURE 1.25 Prosternum, ventral view of *Leptura rubra* L. (Lepturinae). (Reprinted with permission from C. J. B. Carvalho, editor. Napp, D. S., *Rev. Bras. Entomol.*, 38, 265–419, 1994.)

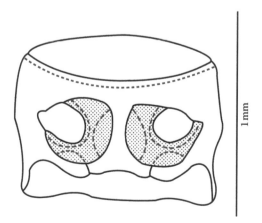

FIGURE 1.26 Prosternum, ventral view of *Adesmus hemispilus* Germar (Lamiinae). (Reprinted with permission from C. J. B. Carvalho, editor. Napp, D. S., *Rev. Bras. Entomol.*, 38, 265–419, 1994.)

mesoscutum by an impression. The mesonotum lacks a stridulatory plate in the Parandrinae (Figure 1.27), Prioninae, and some Spondylidinae; has a divided stridulatory plate in the Dorcasominae (if present), Asemini (Spondylidinae) (Figure 1.28), and Lepturinae (Figure 1.29), and an undivided stridulatory plate in the Necydalinae (Figure 1.30), Cerambycinae (Figure 1.31), and Lamiinae (Figure 1.32).

Legs mostly are cursorial (Figure 1.1) and usually moderately long in most longicorn beetles but can be very long in some species such as males of lamiine *Gerania* (Audinet-Serville); fore legs are enlarged in some (particularly males) Prioninae and Lamiinae and extremely long in the lamiine *Acrocinus* Illiger (fore femora in large males can be as long as body), where they reportedly are used for traversing tree branches; hind legs may be enlarged, such as metafemora in the male cerambycine *Utopia* Thomson, or plate-like tibial extensions in some Cerambycinae, but are never adapted for jumping. The tibia usually has two spurs at the terminal end (Figure 1.33). The legs exhibit an oblique groove along the inner side of the protibiae and a notch or groove on the outer face of the mesotibiae in the Lamiinae; these grooves and notches are lacking in other subfamilies. The tarsi are distinctly pentamerous without pubescent ventral

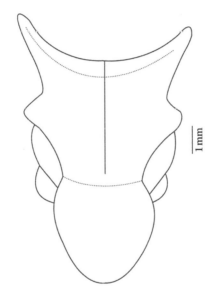

FIGURE 1.27 Mesonotum, dorsal view of *Parandra (Parandra) glabra* (De Geer) (Parandrinae). (Reprinted with permission from C. J. B. Carvalho, editor. Napp, D. S., *Rev. Bras. Entomol.*, 38, 265–419, 1994.)

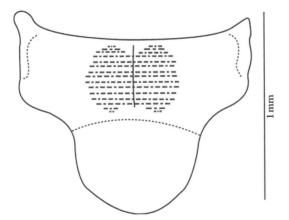

FIGURE 1.28 Mesonotum, dorsal view of *Asemum striatum* (L.) (Spondylidinae). (Reprinted with permission from C. J. B. Carvalho, editor. Napp, D. S., *Rev. Bras. Entomol.*, 38, 265–419, 1994.)

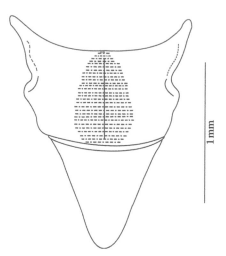

FIGURE 1.29 Mesonotum, dorsal view of *Leptura rubra* L. (Lepturinae). (Reprinted with permission from C. J. B. Carvalho, editor. Napp, D. S., *Rev. Bras. Entomol.,* 38, 265–419, 1994.)

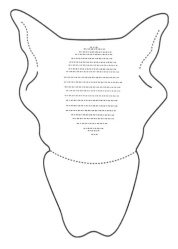

FIGURE 1.30 Mesonotum, dorsal view of *Necydalis major* L. (Necydalinae). (Reprinted with permission from C. J. B. Carvalho, editor. Napp, D. S., *Rev. Bras. Entomol.,* 38, 265–419, 1994.)

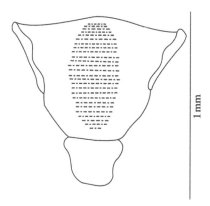

FIGURE 1.31 Mesonotum, dorsal view of *Rhopalophora collaris* (Germar) (Cerambycinae). (Reprinted with permission from C. J. B. Carvalho, editor. Napp, D. S., *Rev. Bras. Entomol.,* 38, 265–419, 1994.)

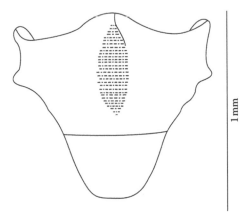

FIGURE 1.32 Mesonotum, dorsal view of *Adesmus hemispilus* Germar (Lamiinae). (Reprinted with permission from C. J. B. Carvalho, editor. Napp, D. S., *Rev. Bras. Entomol.,* 38, 265–419, 1994.)

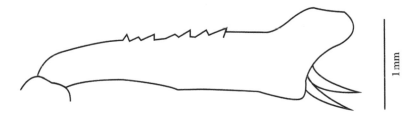

FIGURE 1.33 Protibia, lateral view of *Spondylis buprestoides* (L.) (Spondylidinae). (Reprinted with permission from C. J. B. Carvalho, editor. Napp, D. S., *Rev. Bras. Entomol.,* 38, 265–419, 1994.)

pads in the Parandrinae (Figure 1.34), although they are pseudotetramerous with ventral pads in the Prioninae, Lepturinae, Cerambycinae, Spondylidinae (Figure 1.35), and Lamiinae. The third tarsomere is simple in the Parandrinae but dilated in the remaining subfamilies (Figure 1.35). The tarsal claws are appendiculate or cleft in the most specialized Lamiinae but simple in all other subfamilies.

The hind wing usually has a moderately to very long apical field (though this is short in some very large forms, such as *Titanus* Audinet-Serville) with two more or less complete radial extensions converging and then diverging to form a scissor-like figure, with a dark sclerite apicad of radial cell and a subtriangular sclerite crossing r4. The radial cell often is well developed and more or less elongate (although sometimes it is short and broad or lacking basal limit). Cross-vein r3 is slightly to strongly oblique and sometimes is absent. The basal portion of radius posterior (RP) is long to very short and not surpassing r4. The medial field usually has four or five free veins (sometimes with three or, rarely, fewer) and always lacks medial fleck. The wedge cell is well developed in almost all Prioninae and some Lepturinae and Spondylidinae and is absent in all other subfamilies. If the elytra are shortened in macropterous forms, the hind wings are exposed (often giving the beetles a hymenopteran appearance) and their apex is then sometimes not folded (all Necydalinae, Figures 1.36 and 1.37). The hind wings are highly reduced or disappear in numerous Cerambycinae (such as males of *Torneutes* Reich), Lamiinae (usually both sexes) and Prioninae (more often only females), Lepturinae, and Spondylidinae (both sexes of *Michthisoma* LeConte); in some taxa, the beetles apparently are flightless even if wings are present.

1.2.2.1.2.3 Abdomen   The abdomen usually has five free, visible ventrites (belonging to segments III–VII; sternites 1 and 2 form the posterointernal wall of metacoxal acetabula) (Figure 1.2), with the first usually not much longer than the second; rarely, it is almost as long as the remaining combined (females of the cerambycine Obriini). The intercoxal process is acute to broadly rounded or angulate—or absent, with the medial part of reduced sternum II visible between the hind coxae (Necydalinae and some slender wasp-mimicking

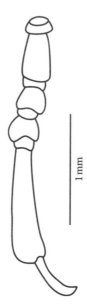

FIGURE 1.34   Metatarsus, lateral view of *Parandra (Parandra) glabra* (De Geer) (Parandrinae). (Reprinted with permission from C. J. B. Carvalho, editor. Napp, D. S., *Rev. Bras. Entomol.*, 38, 265–419, 1994.)

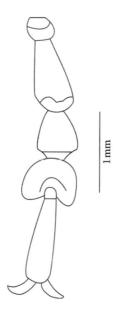

FIGURE 1.35   Metatarsus, dorsal view of *Spondylis buprestoides* (L.) (Spondylidinae). (Reprinted with permission from C. J. B. Carvalho, editor. Napp, D. S., *Rev. Bras. Entomol.*, 38, 265–419, 1994.)

Cerambycinae and the telescoped, with segment II forming a petiolus-like basal piece). The abdominal tergites 1–6 are semisclerotized. Functional spiracles are present on each side of abdominal segments I–VII (the first pair is very large, particularly in flying forms), and spiracles VIII are rudimentary and closed.

1.2.2.1.2.4 External Morphology of Terminalia   Male terminalia (Figures 1.38 through 1.40) consist of three abdominal segments. The anterior edge of sternite VIII (Figure 1.38) usually bears a median strut (that is rudimentary or absent in some taxa); the anterior edge of sternite IX has spiculum gastrale; tergites IX and X are fused together and usually membranous. The anterior edge of tegmen (Figure 1.39)

FIGURE 1.36 *Necydalis major* L. (Necydalinae).

FIGURE 1.37 *Ulochaetes leoninus* LeConte (Necydalinae).

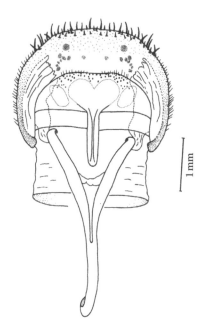

FIGURE 1.38 Male sternite VIII of *Hedypathes betulinus* (Klug) (Lamiinae). (Reprinted with permission from C. J. B. Carvalho, editor. Galileo, M. H. M., et al., *Rev. Bras. Entomol.,* 37, 705–715, 1993.)

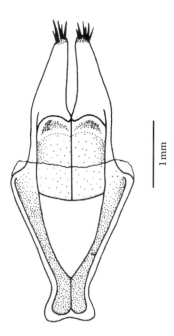

FIGURE 1.39 Tegmen of *Hedypathes betulinus* (Klug) (Lamiinae). (Reprinted with permission from C. J. B. Carvalho, editor. Galileo, M. H. M., et al., *Rev. Bras. Entomol.,* 37, 705–715, 1993.)

has a single or no strut; the parameres usually are fused to phallobase and free from one another, but they are more or less completely fused in some Cerambycinae (such as the Molorchini–Obriini complex— very short in some and nearly absent in the Neotropical Ectenessini). The aedeagus is cucujiform and symmetrical, but usually is rotated to one side in the abdominal cavity when at rest. Surrounding struc- tures therefore may not be entirely symmetrical. The anterior edge of the aedeagus (Figure 1.40) almost

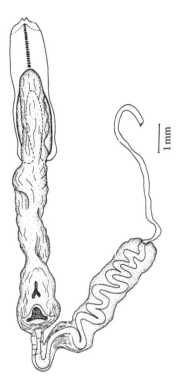

FIGURE 1.40   Aedeagus and internal sac of *Hedypathes betulinus* (Klug) (Lamiinae). (Reprinted with permission from C. J. B. Carvalho, editor. Galileo, M. H. M., et al., *Rev. Bras. Entomol.*, 37, 705–715, 1993.)

always has paired struts; the internal sac (endophallus) (Figure 1.40) may bear distinctive sclerotized structures such as asperities, paired or unpaired sclerites, or longitudinal sclerotized rods.

The terminalia of female cerambycids (Figures 1.41 and 1.42) follow the same structural plan as the male, with sclerites sequentially distributed along a more or less membranous tube, which is kept invaginated at rest. Segment VIII is entirely contained within the partially double-walled segment VII (where the first invagination occurs). Sternite VIII (Figure 1.41) bears a long ventral apodeme that is closely related with the ovipositor length among different taxa. Sternite and tergite VIII are mostly partly desclerotized and tend to fuse into a tube enclosing the "anus–ovipositor" complex, sometimes protruding from the abdomen, either "naked" or protected by posterior sternal and tergal projections of segment VII (e.g., some Acanthocinini of Lamiinae). The ovipositor (Figure 1.42) usually is long but secondarily shortened and modified—particularly in some Cerambycinae (such as Trachyderini) ovipositing on the host surface; the paraprocts are short without baculi (in all Lamiinae) and flexible with subapical styli; in some groups, the apex of the ovipositor is sclerotized with styli being lateral or laterodorsal and often reduced or virtually inbuilt in coxitis (this type occurs in several subfamilies depending on biology but is common in the Prioninae and universal in the Parandrinae). One or two pairs of glandular integumental invaginations often are present at the base of the ovipositor on both sides of anus.

1.2.2.1.2.5 Reproductive System   Testes (Figure 1.43) consist of one to several pairs of testicular lobes, with each lobe having a number of radially arranged testicular follicles. The basal parts of vasa deferentia may be broadened into seminal vesicles. Usually, there are two pairs (or at least one pair) of accessory glands at or before the fusion of vasa deferentia. Ducts are more or less completely paired (mostly up to paired gonopores on the internal sac) in Lamiini and several related tribes of Lamiinae. Primary gonopore seldom projects into a long sclerotized flagellum.

Ovaries (Figure 1.44) are paired, each with a variable number (up to several tens) of ovarioles. There is a single more or less sclerotized spermatheca of simple shape (often an elongate, curved capsule bridged

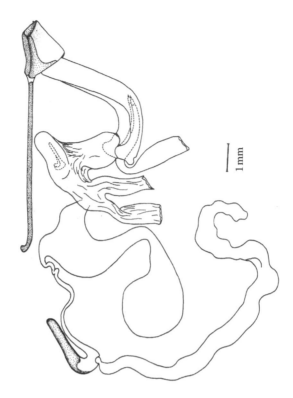

FIGURE 1.41 Female terminalia of *Hedypathes betulinus* (Klug) (Lamiinae). (Reprinted with permission from C. J. B. Carvalho, editor. Galileo, M. H. M., et al., *Rev. Bras. Entomol.,* 37, 705–715, 1993.)

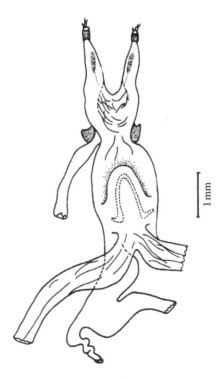

FIGURE 1.42 Ovipositor of *Hedypathes betulinus* (Klug) (Lamiinae). (Reprinted with permission from C. J. B. Carvalho, editor. Galileo, M. H. M., et al., *Rev. Bras. Entomol.,* 37, 705–715, 1993.)

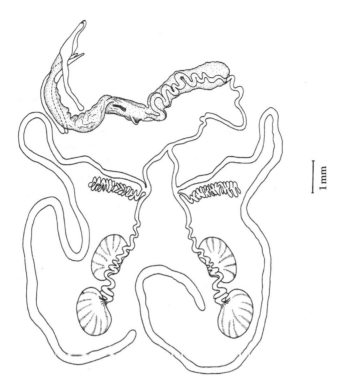

FIGURE 1.43 Male reproductive system of *Hedypathes betulinus* (Klug) (Lamiinae). (Reprinted with permission from C. J. B. Carvalho, editor. Galileo, M. H. M., et al., *Rev. Bras. Entomol.,* 37, 705–715, 1993.)

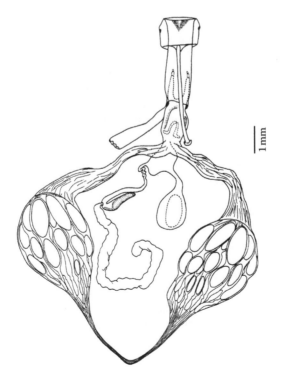

FIGURE 1.44 Female reproductive system of *Hedypathes betulinus* (Klug) (Lamiinae). (Reprinted with permission from C. J. B. Carvalho, editor. Galileo, M. H. M., et al., *Rev. Bras. Entomol.,* 37, 705–715, 1993.)

by spermathecal compressor) with a distinct, sometimes very large, spermathecal gland attached to it. A bursa copulatrix usually is present in the form of a soft diverticulum of various sizes and locations; the spermathecal duct arises on its base.

1.2.2.1.2.6 Digestive System   The adult gut may be primitively rather reduced. In prionines and some cerambycines, the midgut is short and reduced with the posterior midgut being extremely reduced and thread-like (slightly less so in some floricolous Clytini of the Cerambycinae), suggesting adult aphagy. However, the gut is long and well developed in the Lamiinae whose adults feed extensively.

The digestive tube (Figure 1.45) usually does not have a distinct crop/proventriculus. The anterior midgut in some taxa (Necydalinae, Spondylidinae, and most Lepturinae) produces morphologically distinct mycetomes in the form of gut wall diverticula whose cells harbor intracellular yeast-like symbionts. The posterior midgut often bears numerous small, scattered evaginated crypts. Six cryptonephridial Malpighian tubules enter the gut separately in two clusters of three.

1.2.2.2 Immature Stages

The following morphological descriptions are based on Butovitsch (1939), Duffy (1953, 1957, 1960, 1968), Gardiner (1966), and Švácha and Lawrence (2014).

1.2.2.2.1 Eggs

The eggs (Figure 1.46) are elongate, ovoid, or fusiform and often have thin, flexible chorion so that their shape can adapt to the tight spaces in which they usually are laid. A female can lay a dozen to several hundred eggs in her lifetime. They usually hatch in a few days to a few weeks after oviposition, depending on species and climates. In some lamiine species, the larvae may overwinter within the chorion, particularly if the eggs are laid late in the season.

1.2.2.2.2 Larvae

The larvae are soft-bodied, eucephalic, oligopodous to apodous, prognathous, more or less elongate, and subcylindrical to dorsoventrally depressed (Figures 1.47 through 1.58). Their body shape and mechanics largely depend on hemolymph pressure.

The cranium (particularly its anterior part, which supports mouthparts, often called the "mouth-frame") may be strongly sclerotized and pigmented, whereas the body generally is soft and white to yellow (Figures 1.51 through 1.54). In rare cases, the body can be grayish with some prothoracic regions and the abdominal end sclerotized and pigmented. The skin of the prothorax is not attached to the submentum. The ventral mouthparts are protracted; the mandibles (Figure 1.59) lack a molar tooth or other appendage; the labium bears a setose ligula, and gula and hypostoma are present. The abdomen, at least dorsally, has more or less retractile and often characteristically sculptured, protuberances called ambulatory ampullae, providing support in galleries; abdominal segments 1 to 6 or 7 have dorsal ampullae. The spiracular system is peripneustic, with one pair of functional spiracles on the mesothorax (Figure 1.60) and one pair on each of eight abdominal segments. The digestive system is similar to that of adults.

1.2.2.2.3 Pupae

The pupae (Figure 1.61a and b) are similar to adults in size, shape, and proportions of cephalic and thoracic appendages. Secondary sexual differences in adults generally are evident in the pupae. They are adecticous, exarate, and generally soft and pale (except for some special structures like spines or gin traps), with a strongly ventrally bent head so that mouthparts point caudally (except for some Prioninae). The body usually is waxy or milky white to testaceous, often with scattered setae or spinose areas or combinations of both.

The antennae extend at least as far as the mesothorax but generally to the abdominal segments, where they are nearly always curved downward beneath the body. The elytra are always glabrous (except Acanthocinini). The abdomen usually has nine movable segments, with the tenth (and occasionally

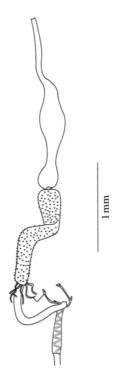

FIGURE 1.45   Digestive system of *Oxymerus luteus luteus* (Voet) (Cerambycinae). (Reprinted with permission from W. F. de Azevedo, Jr., editor. Moura, L. A., and A. F. Franceschini, *Biociências,* 2, 135–143, 1994.)

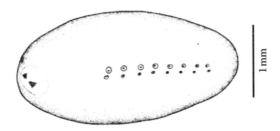

FIGURE 1.46   Egg of *Hedypathes betulinus* (Klug) (Lamiinae). (Reprinted with permission from C. J. B. Carvalho, editor. Galileo, M. H. M., et al., *Rev. Bras. Entomol.,* 37, 705–715, 1993.)

the ninth) being telescoped within the preceding segments. Abdominal segments 7 and 8 usually are more elongate than the preceding ones but sometimes considerably produced. The abdomen has five to seven pairs of functional spiracles. Segment 9 often ends in a vertical or horizontal spine or process or with a pair of incurved or outwardly curved urogomphi. Some prionines (tribes Callipogonini and Macrotomini) have paired paramedian gin-traps. The legs often have subapical setae on the femora and sometimes one or two setae on the tarsi.

1.3   Key to Subfamilies of the Family Cerambycidae

The key to adults is based on the work of Linsley (1962b), Ślipiński and Escalona (2013), and Švácha and Lawrence (2014). The key to larvae is based on Duffy (1953, 1957, 1960, 1968), Švácha and Danilevsky (1987, 1988, 1989), and Švácha and Lawrence (2014).

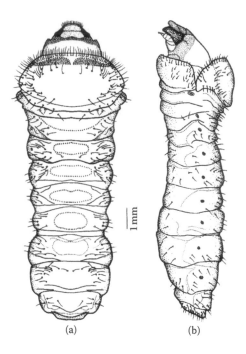

(a)   (b)

FIGURE 1.47 Larvae of *Acanthoderes (Psapharochrus) melanosticta* White (Lamiinae), dorsal view (a) and lateral view (b). (Reprinted with permission from C. J. B. Carvalho editor. Mermudes, J. R. M., and M. L. Monné, *Rev. Bras. Entomol.,* 45, 331–334, 2001.)

FIGURE 1.48 Larva of *Parandra* sp. (Parandrinae), dorsal view. (Reprinted from Costa, C., et al., *Larvas de Coleoptera do Brasil,* Museu de Zoologia, Universidade de São Paulo, São Paulo, 1988. With permission.)

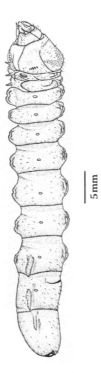

FIGURE 1.49 Larva of *Parandra* sp. (Parandrinae), lateral view. (Reprinted from Costa, C., et al., *Larvas de Coleoptera do Brasil,* Museu de Zoologia, Universidade de São Paulo, São Paulo, 1988. With permission.)

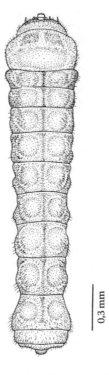

FIGURE 1.50 Larva of *Acyphoderes aurulenta* (Kirby) (Cerambycinae), dorsal view. (Reprinted from Costa, C., et al., *Larvas de Coleoptera do Brasil,* Museu de Zoologia, Universidade de São Paulo, São Paulo, 1988. With permission.)

FIGURE 1.51   Larva of *Tsivoka simplicicollis* (Gahan) (Dorcasominae), laterodorsal view. (Reprinted with permission from Petr Švácha, owner. Švácha, P., and J. F. Lawrence, 2.1 Vesperidae Mulsant, 1839; 2.2 Oxypeltidae Lacordaire, 1868; 2.3 Disteniidae J. Thomson, 1861; 2.4 Cerambycidae Latreille, 1802, In *Handbook of zoology, Arthropoda: Insecta; Coleoptera, beetles, Volume 3: Morphology and systematics (Phytophaga)*, eds. R. A. B. Leschen and R. G. Beutel, Walter de Gruyter, Berlin, 2014, 16–177.)

FIGURE 1.52   Larva of *Tsivoka simplicicollis* (Gahan) (Dorcasominae), ventral view. (Reprinted with permission from Petr Švácha, owner. Švácha, P., and J. F. Lawrence, 2.1 Vesperidae Mulsant, 1839; 2.2 Oxypeltidae Lacordaire, 1868; 2.3 Disteniidae J. Thomson, 1861; 2.4 Cerambycidae Latreille, 1802, In *Handbook of zoology, Arthropoda: Insecta; Coleoptera, beetles, Volume 3: Morphology and systematics (Phytophaga)*, eds. R. A. B. Leschen and R. G. Beutel, Walter de Gruyter, Berlin, 2014, 16–177.)

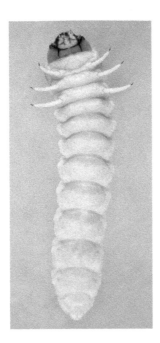

FIGURE 1.53 Larva of *Judolia sexmaculata* (L.) (Lepturinae), ventral view. (Reprinted with permission from Petr Švácha, owner. Švácha, P., and J. F. Lawrence, 2.1 Vesperidae Mulsant, 1839; 2.2 Oxypeltidae Lacordaire, 1868; 2.3 Disteniidae J. Thomson, 1861; 2.4 Cerambycidae Latreille, 1802, In *Handbook of zoology, Arthropoda: Insecta; Coleoptera, beetles, Volume 3: Morphology and systematics (Phytophaga)*, eds. R. A. B. Leschen and R. G. Beutel, Walter de Gruyter, Berlin, 2014, 16–177.)

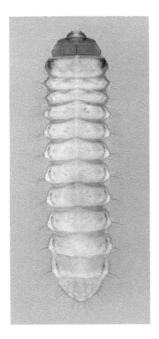

FIGURE 1.54 Larva of *Dinoptera collaris* (L.) (Lepturinae), dorsal view. (Reprinted with permission from Petr Švácha, owner. Švácha, P., and J. F. Lawrence, 2.1 Vesperidae Mulsant, 1839; 2.2 Oxypeltidae Lacordaire, 1868; 2.3 Disteniidae J. Thomson, 1861; 2.4 Cerambycidae Latreille, 1802, In *Handbook of zoology, Arthropoda: Insecta; Coleoptera, beetles, Volume 3: Morphology and systematics (Phytophaga)*, eds. R. A. B. Leschen and R. G. Beutel, Walter de Gruyter, Berlin, 2014, 16–177.)

FIGURE 1.55 Larva of *Prionus coriarius* (L.) (Prioninae), lateral view. (Reprinted with permission from Petr Švácha, owner. Švácha, P., and J. F. Lawrence, 2.1 Vesperidae Mulsant, 1839; 2.2 Oxypeltidae Lacordaire, 1868; 2.3 Disteniidae J. Thomson, 1861; 2.4 Cerambycidae Latreille, 1802, In *Handbook of zoology, Arthropoda: Insecta; Coleoptera, beetles, Volume 3: Morphology and systematics (Phytophaga)*, eds. R. A. B. Leschen and R. G. Beutel, Walter de Gruyter, Berlin, 2014, 16–177.)

FIGURE 1.56 Larva of *Prionus coriarius* (L.) (Prioninae), ventral view. (Reprinted with permission from Petr Švácha, owner. Švácha, P., and J. F. Lawrence, 2.1 Vesperidae Mulsant, 1839; 2.2 Oxypeltidae Lacordaire, 1868; 2.3 Disteniidae J. Thomson, 1861; 2.4 Cerambycidae Latreille, 1802, In *Handbook of zoology, Arthropoda: Insecta; Coleoptera, beetles, Volume 3: Morphology and systematics (Phytophaga)*, eds. R. A. B. Leschen and R. G. Beutel, Walter de Gruyter, Berlin, 2014, 16–177.)

FIGURE 1.57 Larva of *Atimia okayamensis* Hayashi (Spondylidinae), lateral view. (Reprinted with permission from Petr Švácha, owner. Švácha, P., and J. F. Lawrence, 2.1 Vesperidae Mulsant, 1839; 2.2 Oxypeltidae Lacordaire, 1868; 2.3 Disteniidae J. Thomson, 1861; 2.4 Cerambycidae Latreille, 1802, In *Handbook of zoology, Arthropoda: Insecta; Coleoptera, beetles, Volume 3: Morphology and systematics (Phytophaga)*, eds. R. A. B. Leschen and R. G. Beutel, Walter de Gruyter, Berlin, 2014, 16–177.)

FIGURE 1.58 Larva of *Arhopalus rusticus* (L.) (Spondylidinae), ventral view. (Reprinted with permission from Petr Švácha, owner. Švácha, P., and J. F. Lawrence, 2.1 Vesperidae Mulsant, 1839; 2.2 Oxypeltidae Lacordaire, 1868; 2.3 Disteniidae J. Thomson, 1861; 2.4 Cerambycidae Latreille, 1802, In *Handbook of zoology, Arthropoda: Insecta; Coleoptera, beetles, Volume 3: Morphology and systematics (Phytophaga)*, eds. R. A. B. Leschen and R. G. Beutel, Walter de Gruyter, Berlin, 2014, 16–177.)

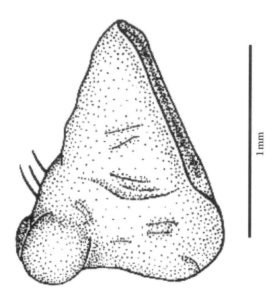

FIGURE 1.59 Mandible of the larva of *Acyphoderes aurulenta* (Kirby) (Cerambycinae), dorsal view. (Reprinted from Costa, C., et al., *Larvas de Coleoptera do Brasil,* Museu de Zoologia, Universidade de São Paulo, São Paulo, 1988. With permission.)

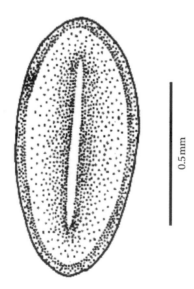

FIGURE 1.60 Thoracic spiracle of the larva of *Acyphoderes aurulenta* (Kirby) (Cerambycinae). (Reprinted from Costa, C., et al., *Larvas de Coleoptera do Brasil,* Museu de Zoologia, Universidade de São Paulo, São Paulo, 1988. With permission.)

1.3.1 Adults

1. Tarsi distinctly pentamerous (Figure 1.34); lateral pronotal carinae entire and simple (Figure 1.21) .... **Parandrinae**

   a. Tarsi pseudotetramerous (Figure 1.35); lateral pronotal carinae absent (Figure 1.24), or present, often dentate or spinose (Figures 1.22 and 1.23) 2

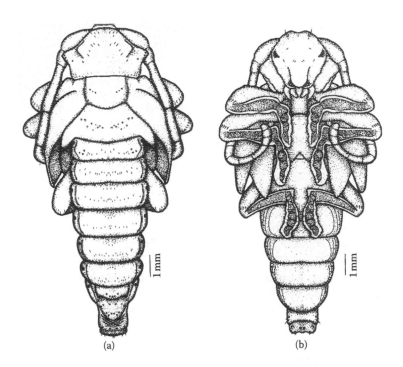

(a)   (b)

FIGURE 1.61 Pupae of *Acanthoderes (Psapharochrus) melanosticta* (White) (Lamiinae), dorsal view (a) and ventral view (b). (Reprinted with permission from C. J. B. Carvalho, editor. Mermudes, J. R. M., and M. L. Monné, *Rev. Bras. Entomol.,* 45, 331–334, 2001.)

2. Head vertical or retracted (Figure 1.7), genal margin always directed posteriorly; protibiae with mesial sinus; mesotibiae usually notched or grooved externally; last segment of maxillary palpi usually pointed at apex (Figures 1.14 and 1.15) .... **Lamiinae**

   a. Head usually prognathous or weakly deflexed (Figure 1.10), genal margin never directed posteriorly; protibiae without mesial sinus; mesotibiae never notched or grooved externally; last segment of maxillary palpi obtuse or truncate at apex (Figures 1.13 and 1.16 through 1.19) 3

3. Pronotum with elevated, often dentate or spinose lateral carinae (Figures 1.22 and 1.23); labrum fused with epistoma; inner lobe of maxillae obsolete (Figure 1.13); procoxae strongly transverse (Figure 1.23); mesoscutum without a striated stridulatory area .... **Prioninae**

   a. Pronotum without distinct lateral carinae; labrum free; inner lobe of maxillae usually well developed (Figures 1.16, 1.17, and 1.19); procoxae rarely transverse, usually rounded; mesoscutum with a finely striated stridulatory area (Figures 1.28 through 1.30) or without a stridulatory area 4

4. Mesoscutum without a median endocarina (Figure 1.30); elytra shortened, covering only the pterothorax, exposed hind wings with unfolded apex (Figures 1.36 and 1.37) ..... **Necydalinae**

   a. Mesoscutum with a median endocarina (Figures 1.28 and 1.29); elytra usually well developed, hind wings folded at apex 5

5. Head with region behind eyes usually having prominent temples followed by a constricted neck (Figures 1.5 and 1.6); procoxae conical, prominent, and strongly projecting below prosternal process .... **Lepturinae**

   a. Head may be gradually narrowing to abruptly constricted behind eyes, without prominent temples, followed by a constricted neck; procoxae of variable shape usually subglobular, seldom strongly projecting below prosternal process 6

6. Mandibula with incisor edge without fringe of hairs; hind wing with edge cell .... **Spondylidinae**

   a. Mandibula with incisor edge usually with fringe of long hairs; hind wing without edge cell 7

7. Mesonotum with an undivided stridulatory plate (Figure 1.31); notosternal suture rarely complete, usually indistinct or incomplete anteriorly or absent; empodium usually small or indistinct .... **Cerambycinae**

   a. Mesonotum with a divided stridulatory plate; notosternal suture may be relatively distinct and complete; empodium indistinct .... **Dorcasominae**

1.3.2 Larvae

1. Clypeus (Figure 1.62) very narrow, with only slender basal arms reaching to mandibular articulations; these arms may be more or less sclerotized and fused with epistomal margin. Mandibular apex and dorsal angle lacking; mandibles (Figure 1.59) short and apically rounded, spoon-like .... **Cerambycinae** (Figure 1.50)

   a. Clypeus more or less trapezoidal, filling entire space between dorsal mandibular articulations. Mandibles with distinct apex and more or less distinct dorsal angle 2

2. Legs absent, or present with only two minute segments visible under high magnification. Cardo extremely reduced, labiomaxillary base firmly attached to cranium along whole width; maxillae movable only from stipes .... **Lamiinae** (Figure 1.47a and b)

   a. Distinct four-segmented legs (Figure 1.65) present though may be strongly reduced and inconspicuous. Free movable cardo present 3

3. Main antennal sensillum flat, rarely convex, never conical 4

   a. Main antennal sensillum prominent and conical 5

4. Basal half of pronotum more or less roughly asperate. Labrum cordate (Figures 1.63 and 1.64), very long. Epistomal, frontal, and postcondylar carinae absent ..... **Parandrinae** (Figures 1.48 and 1.49)

   a. Body without coarse asperities. Labrum never as long as in *Parandra*. Distinct epistomal, frontal, and postcondylar carinae often present .... **Prioninae** (Figures 1.55 and 1.56)

5. Pretarsus without setae. Abdominal epipleurum protuberant on segments 7–9. Lateral furrows of pronotum long and distinct 6

   a. Pretarsus with distinct setae. Abdominal epipleurum protuberant on at least segments 6–9. Lateral furrows of pronotum rarely long and distinct 7

6. Large postnotal fold present. Urogomphi absent. Dorsal ampullae with one lateral impression on each side .... **Dorcasominae** (Figures 1.51 and 1.52)

   a. Postnotum absent. Urogomphi present or absent. Dorsal ampullae with two lateral impressions on each side .... **Spondylidinae** (Figures 1.57 and 1.58)

7. Dorsal ampullae with two broadly separate lateral impressions on each side (Figure 1.66) Prothoracic lateropresternum largely microspiculate .... **Necydalinae**

   a. Dorsal ampullae with one lateral impression. Prothoracic lateropresternum microspiculate at most along anterior margin .... **Lepturinae** (Figures 1.53 and 1.54)

1.4 Diagnosis, Biodiversity, Distribution, and Biology of Subfamilies

Phylogenetic relationships within the Cerambycidae are still highly controversial (e.g., Wang and Chiang 1991; Napp 1994; Švácha and Lawrence 2014). As a result, we order subfamilies alphabetically in this section. Morphological features of subfamilies mainly are extracted from Linsley (1962a, 1962b, 1963, 1964), Linsley and Chemsak (1972, 1984, 1995), Chemsak (1996), Ślipiński and Escalona (2013), and Švácha and Lawrence (2014).

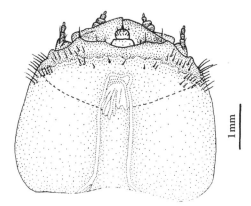

FIGURE 1.62   Head of the larva of *Acyphoderes aurulenta* (Kirby) (Cerambycinae), dorsal view. (Reprinted from Costa, C., et al., *Larvas de Coleoptera do Brasil,* Museu de Zoologia, Universidade de São Paulo, São Paulo, 1988. With permission.)

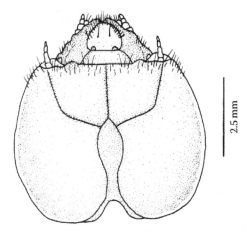

FIGURE 1.63   Head of the larva of *Parandra* sp. (Parandrinae), dorsal view. (Reprinted from Costa, C., et al., *Larvas de Coleoptera do Brasil,* Museu de Zoologia, Universidade de São Paulo, São Paulo, 1988. With permission.)

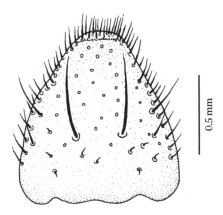

FIGURE 1.64   Labrum of the larva of *Parandra* sp. (Parandrinae), dorsal view. (Reprinted from Costa, C., et al., *Larvas de Coleoptera do Brasil,* Museu de Zoologia, Universidade de São Paulo, São Paulo, 1988. With permission.)

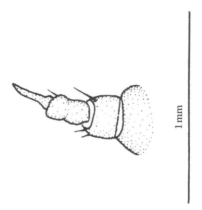

1 mm

FIGURE 1.65 Proleg of the larva of *Parandra* sp. (Parandrinae), dorsal view. (Reprinted from Costa, C., et al., *Larvas de Coleoptera do Brasil*, Museu de Zoologia, Universidade de São Paulo, São Paulo, 1988. With permission.)

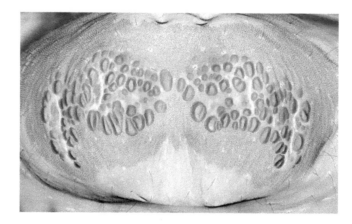

FIGURE 1.66 Larva of *Necydalis major* L. (Necydalinae), detail of ventral ambulatory ampulla 4. (Reprinted with permission from Petr Švácha, owner. Švácha, P., and J. F. Lawrence, 2.1 Vesperidae Mulsant, 1839; 2.2 Oxypeltidae Lacordaire, 1868; 2.3 Disteniidae J. Thomson, 1861; 2.4 Cerambycidae Latreille, 1802, In *Handbook of zoology, Arthropoda: Insecta; Coleoptera, beetles, Volume 3: Morphology and systematics (Phytophaga)*, eds. R. A. B. Leschen and R. G. Beutel, Walter de Gruyter, Berlin, 2014, 16–177.)

1.4.1 Subfamily Cerambycinae Latreille, 1802

1.4.1.1 Diagnosis

Small to large, elongate beetles (Figures 1.67 through 1.76). Head (Figure 1.10) subvertical, scarcely narrowed behind eyes; antennae inserted high on frons between eyes, usually very elongate, second antennomere short; eyes reniform, usually embracing antennal insertion; mandibles acute without molar plate; incisor edge with or without pubescent fringe; maxillae and labium variable, lacinia usually well developed (Figure 1.19); mentum usually trapezoidal (Figure 1.18); submentum sometimes produced between bases of maxillae as a short process. Pronotum without lateral margin; procoxae rarely prominent, usually rounded, cavities variable; notosternal suture rarely complete, usually indistinct or incomplete anteriorly, or absent; mesoscutum with a median endocarina; stridulatory plate, when present, undivided (Figure 1.31). Elytra usually not abbreviated; hind wings without closed cell in anal sector, radial cell closed. Legs moderately long; protibiae without mesial sinus; tarsi pseudotetramerous, padded beneath, third tarsomere dilated, bilobed concealing minute fourth tarsomere, empodium small or absent.

FIGURE 1.67 *Coccoderus sexmaculatus* Buquet (Cerambycinae).

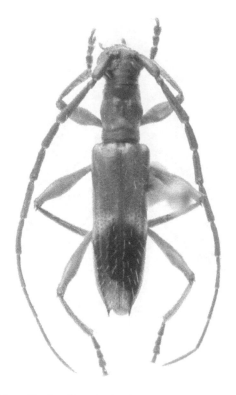

FIGURE 1.68 *Compsibidion divisum* Martins (Cerambycinae).

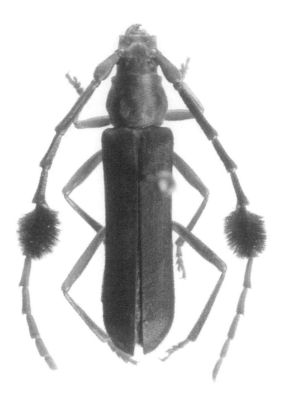

FIGURE 1.69 *Compsocerus deceptor* Napp (Cerambycinae).

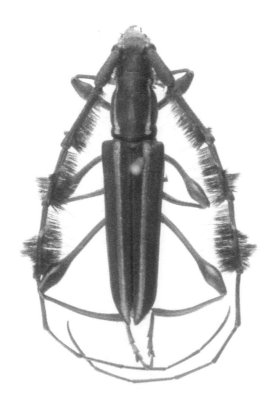

FIGURE 1.70 *Disaulax hirsuticornis* (Kirby) (Cerambycinae).

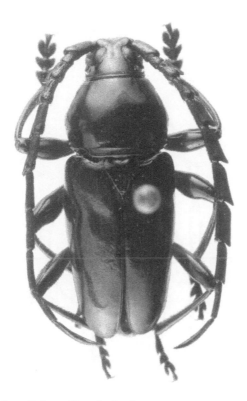

FIGURE 1.71 *Lissonotus spadiceus* Dalman (Cerambycinae).

FIGURE 1.72 *Megacyllene (Megacyllene) patruelis* (Chevrolat) (Cerambycinae).

FIGURE 1.73 *Mionochroma chloe* (Gounelle) (Cerambycinae).

FIGURE 1.74 *Neoregostoma coccineum* (Gory) (Cerambycinae).

FIGURE 1.75 *Pronuba decora* Thomson (Cerambycinae).

FIGURE 1.76 *Psygmatocerus wagleri* Perty (Cerambycinae).

1.4.1.2 Comments

Due to the diversity of forms, the subfamily is one of the most difficult to define, with uncertain limits and relationships (Napp, 1994). Two groups of genera previously included in the Necydalinae are recognized as incertae sedis of Cerambycinae by Švácha and Lawrence (2014): (1) *Atelopteryx* Lacordaire, *Callisphyris* Newman, *Hephaestion* Newman, *Parahephaestion* Melzer, *Planopus* Bosq and *Stenorhopalus* Blanchard; and (2) *Cauarana* Lane, *Mendesina* Lane, *Rhathymoscelis* Thomson and *Hephaestioides* Zajciw.

1.4.1.3 Diversity and Distribution

This is the second largest subfamily in Cerambycidae, with 1,757 genera and more than 11,200 species in the world (Bouchard et al. 2011; Monné 2012; Tavakilian 2015) (Table 1.1). The Cerambycinae are widely distributed in all biogeographic regions. In the Australian, Nearctic, and southern Neotropical regions, the Cerambycinae are the most speciose subfamily compared to other subfamilies (e.g., Forchhammer and Wang 1987; Švácha and Lawrence 2014).

1.4.1.4 Biology

Adults are extremely diverse, from dark nocturnal forms to brightly colored mimetic diurnal species (Švácha and Lawrence 2014) (Figures 1.67 through 1.76). Linsley (1962b) has attempted to divide this subfamily into two groups based on adult activity patterns. Adults of many species from the Callidiopini, Gracilliini, Opsimini, Methiini, and so on are active during the night, for example, *Oemona hirta* (F.) (Wang et al. 1998). Some adults may be crepuscular (vespertine) such as *Nadezhdiella cantori* (Hope) (Wang et al. 2002), *Phoracantha semipunctata* (F.), and *P. recurva* Newman (Wang at el. 2008). Nocturnal or crepuscular adults may or may not need to feed depending on species. Adults of most species appear to be diurnal, such as species from the Aphneopini, Callidiini, Cleomenini, Clytini, Molorchini, and so forth. Many diurnally active adults visit flowers and feed on pollens and nectar, such as *Zorion guttigerum* Westwood (Wang 2002). Some cerambycine adults feed on tree foliage, such as *Lissonotus spadiceus* Dalman (Figure 1.71) (M. L. Monné, personal observation) and *Xylotrechus pyrrhoderus* Bates (Guo 1999); a few in Trachyderini feed on fruit. Adults can live for a week to a few months depending on whether they feed. Male-produced long-range sex and aggregation pheromones have been identified in many cerambycine species (see Chapter 5). Mating may occur on larval hosts or adult feeding sites (see Chapter 4). Depending on the nature of larval feeding biology, adults may be attracted to larval hosts of certain conditions for oviposition (Hanks 1999; see Chapter 3). They then lay their eggs on the surface of larval host plants or in crevices and wounds of bark or under loose bark. Each female can lay dozens to hundreds of eggs in her lifetime.

The distinctive cerambycine larval mouthparts are well suited for solid hosts; most larvae do not occur in soft rotten wood or in soil, and species feeding in soft herbs are rare (Švácha and Lawrence 2014). Although larvae of many species feed on dead (not rotten) plants, most species probably attack living plants that may be perfectly healthy, weakened, or stressed (Hanks 1999). The larval host range of cerambycine species can be from oligophagous to highly polyphagous across both angiosperms and gymnosperms (see Chapters 11 and 12). Larvae bore in branches and stems of host plants and sometimes enter roots, such as *O. hirta* (Wang et al. 1998). Mature larvae usually pupate in their host plants. The life cycle usually lasts one to four years. Many cerambycine species are important pests of trees and logs.

1.4.2 Subfamily Dorcasominae Lacordaire, 1868

1.4.2.1 Diagnosis

Small to moderately large, usually elongate beetles with tapering or subparallel elytra and often long cursorial legs; no strongly depressed forms. Head prognathous or distinctly rostrate, usually constricted immediately behind eyes but never with prominent temples followed by a constricted neck; antennal insertions of variable position but at least slightly away from mandibular condyle, antennal sockets usually facing laterally or laterodorsally; eyes moderate to very large, emarginate to entire; mandibles never enlarged, apex unidentate, inner margin usually with a distinct fringe of hairs; maxillae and labium well developed, lacinia distinct, submentum with a very short to long intermaxillary process, ligula usually large, membranous and emarginate or bilobed, terminal segments of both palps usually more or less truncate. Pronotum without lateral carina, often with a pair of lateral tubercles or spines; procoxal cavities closed internally and at least narrowly open posteriorly, prosternal process usually narrow but complete; notosternal suture may be relatively distinct and complete; mesonotum usually with a divided stridulatory plate; mesocoxal cavities open laterally. Elytra in some taxa strongly narrowed and separate or also shortened posteriorly, partly exposing hind wings yet almost always distinctly surpassing posterior pterothoracic margin; hind wing with radial cell closed proximally but without edge cell. Legs short to moderately long; procoxae transverse to subglobular, prominent, projecting at least slightly below prosternal process; tarsi pseudotetramerous and padded beneath, tarsomere 5 in males of some taxa remarkably broadened distally, claws free, divaricate to moderately divergent. Empodium indistinct.

1.4.2.2 Comments

Some authors treat this group of cerambycids as two separate subfamilies, Dorcasominae and Apatophyseinae, each with one tribe (e.g., Danilevsky 1979; Tavakilian 2015). However, most authors accept that these two subfamilies should be two tribes, Apatophyseini and Dorcasomini, under the subfamily Dorcasominae (e.g., Švácha and Danilevsky 1987, 1989; Özdikmen 2008; Švácha and Lawrence 2014; Adlbauer et al. 2015; Vives 2015). We adopt the latter opinion in this chapter.

1.4.2.3 Diversity and Distribution

There are about 340 described species in 95 genera and two tribes occurring in the Oriental, southern Palaearctic, and Afrotropical regions (Švácha and Lawrence 2014; Adlbauer et al. 2015; Tavakilian 2015; Vives 2015) (Table 1.1).

1.4.2.4 Biology

Similar to the Lepturinae, adults of many dorcasomine species are diurnal with some apatophyseine species being floricolous (Švácha and Lawrence 2014). Most adults may be nocturnally active and hide under the bark of trees or between dead logs and the ground. For example, the nocturnal *Apterotoxitiades vivesi* Adlbauer adults are found under one- to two-year-old pine logs lying on the ground adjacent to grassland (Adlbauer et al. 2015). It is not clear whether adults feed and how they reproduce. No long-range pheromones have been found.

Švácha and Lawrence (2014) summarized the known biology of this subfamily. Larvae of *Dorcasomus gigas* Aurivillius make wide galleries along the center of stems and branches of living trees and pupate in the host plant. Larvae of *Apatophysis* Chevrolat develop in dead or moribund underground parts of trees and shrubs and in dry, often treeless, habitats with large perennial herbs. Adlbauer et al. (2015) speculate that *A. vivesi* larvae may feed on grass roots. Undescribed larvae of many Madagascan and one South African (*Otteissa* Pascoe) genera were found in dead, often rotting, wood, mostly above the ground; but some species are subterranean (and larvae also tend to lose stemmata). They are found less frequently in relatively fresh dead branches where larvae usually feed subcortically; unidentified dorcasomine larvae were also found in the outer bark layer of large living broad-leaved trees. Mature larvae of nearly all known species leave the host material and pupate in soil. The life-cycle length is unknown for this subfamily.

1.4.3 Subfamily Lamiinae Latreille, 1825

1.4.3.1 Diagnosis

Small to large, elongate to robust beetles (Figures 1.77 through 1.91). Head (Figure 1.7) vertical in front or retracted and hypognathous, genal line directly posterior; antennae inserted high on frons between eyes;

FIGURE 1.77   *Demophoo hammatus* (Chabrillac) (Lamiinae).

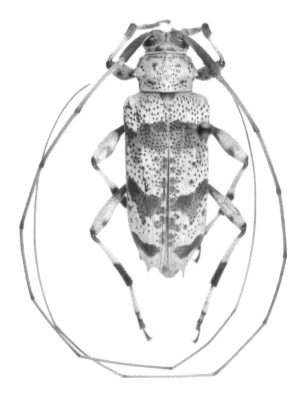

FIGURE 1.78   *Hamastatus conspectus* Monné (Lamiinae).

FIGURE 1.79 *Hydraschema fabulosa* Thomson (Lamiinae).

FIGURE 1.80 *Hylettus stigmosus* Monné (Lamiinae).

FIGURE 1.81   *Lycaneptia nigrobasalis* Tippmann (Lamiinae).

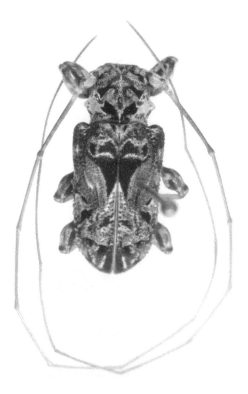

FIGURE 1.82   *Macronemus filicornis* (Thomson) (Lamiinae).

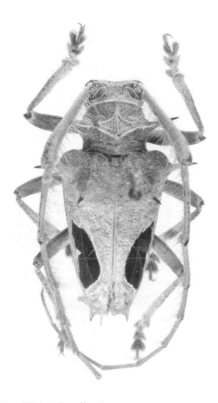

FIGURE 1.83 *Megabasis speculifera* (Kirby) (Lamiinae).

FIGURE 1.84 *Onychocerus albitarsis* Pascoe (Lamiinae).

FIGURE 1.85   *Pertyia sericea* (Perty) (Lamiinae).

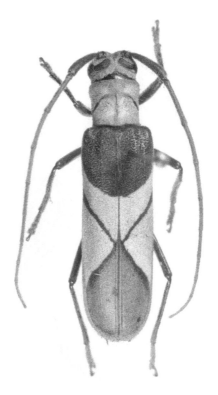

FIGURE 1.86   *Melzerella lutzi* Lima (Lamiinae).

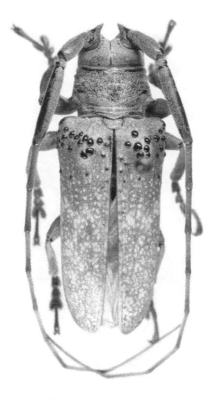

FIGURE 1.87   *Oncideres ulcerosa* (Germar) (Lamiinae).

FIGURE 1.88   *Onocephala obliquata* Lacordaire (Lamiinae).

FIGURE 1.89 *Psapharochrus luctuosus* (Bates) (Lamiinae).

FIGURE 1.90 *Scleronotus scabrosus* Thomson (Lamiinae).

FIGURE 1.91   *Trachysomus fragifer* (Kirby) (Lamiinae).

second antennomere small; eyes finely or coarsely faceted, emarginate, frequently divided; labrum free; mandibles acute without a molar plate and incisor edge without fringe of hairs; maxillae and labium well-developed; palpi with last segment pointed at apex (Figures 1.14 and 1.15); ligula membranous; mentum trapezoidal; gula usually with a short intermaxillary mentigerous process. Pronotum without a lateral margin; procoxae globose or subconical, cavities usually angulated externally (Figure 1.26); mesoscutum usually with a complete endocarina and an undivided stridulatory plate (Figure 1.32); meso-coxal cavities usually open to epimera. Legs usually short; protibiae with an oblique internal mesial sinus; mesotibiae usually with an oblique external sinus, tibial spurs short; tarsi pseudotetramerous, third tarsomere dilated, concealing the minute fourth tarsomere; tarsal claws divaricate, divergent, appendiculate, or bifid; empodium absent.

1.4.3.2 Comments

The subfamily Lamiinae had always been considered a very well-defined and monophyletic group (Napp 1994; Švácha and Lawrence 2014). Audinet-Serville (1832) was the first to propose a classification of the Cerambycidae, which included three "tribus": Prioniens, Cerambycins, and Lamiares. Audinet-Serville (1834) divided Lamiares by the body form into "Déprimés" and "Convexes," describing numerous genera and species. This classification generally was accepted in Europe by Bates (1861, 1862, 1863, 1864, 1865, 1866), Pascoe (1864–1869), and Thomson (1860, 1861, 1864).

LeConte and Horn (1883) recognized the Lamiinae by the prothorax not being margined, the palpi with the last joint cylindrical and pointed, and frequently the front tibiae obliquely sulcate on the inner side. They arranged the tribes represented in North America into nine series (including "Lamioides") and several subdivisions.

Linsley and Chemsak (1984) were unable to resolve contradictions and inconsistencies in the recent usage of Lamiinae tribal classification, mainly in North America. The classification proposed by Linsley and Chemsak represents a compromise between that of LeConte and Horn and more recent work, thus retaining as many of the familiar names as possible of those currently in use by researchers of North American longicorn fauna.

1.4.3.3 Diversity and Distribution

This is the largest subfamily, with more than 21,000 species in 2,964 genera. It has representatives in all biogeographic regions and is particularly diverse in the tropics and subtropics (Table 1.1). Although the Lamiinae are the species-richest subfamily in most regions, they are outnumbered by the Cerambycinae in the Australian, Nearctic, and Neotropical regions (Forchhammer and Wang 1987).

1.4.3.4 Biology

A number of authors have summarized the general biology of the Lamiinae (e.g., Butovitsch 1939; Linsley 1961; Linsley and Chemsak 1984, 1995; Wang and Chiang 1991; Hanks 1999; Wang 2008; Ślipiński and Escalona 2013; Švácha and Lawrence 2014). Adults of most lamiine species probably are nocturnal or crepuscular in habit, based on evidence from morphological (such as coarsely faceted eyes) and biological studies. Many are diurnal, and a few are flower visitors. In almost all species of known biology, adults feed before sexual maturation; for example, many species feed for an average of seven days before mating (Hanks 1999). Most adults feed on leaves and stems of their larval hosts (Hanks 1999), such as *Paraglenea fortunei* Saunders (Wang et al. 1990), *Phytoecia rufiventris* Gautier (Wang et al. 1992), and *Glenea cantor* (F.) (Lu et al. 2011). Others feed on conifer needles and cones, such as *Acanthocinus* and *Monochamus* species, and a few on pollen, stamens, nectar, fungus, or fermented tree oozes (Butovitsch 1939). Due to the nature of adult feeding, active adult life span is relatively long, ranging from weeks to months. Flightlessness is not infrequent and obviously of multiple origin and apparently always occurs in both sexes. So far, only male-produced long-range aggregation pheromones have been identified for lamiines (see Chapter 5).

Lamiine adults usually prepare the oviposition sites in the bark of stems or branches using mandibles to cut slits where eggs are deposited. Some genera even girdle the plants before making oviposition slits. For example, the parent beetles of *Oncideres* Lepeletier and Serville and their relatives completely girdle twigs and branches and then lay eggs beyond the girdle, thus providing freshly killed tissue in which their larvae can develop (e.g., Linsley and Chemsak 1984; Cordeiro et al. 2010). Species from *Phytoecia* Dejean also girdle the twigs or shoots, but they lay eggs slightly below the girdled point, and neonate larvae bore away from the killed part of the twigs or shoots (e.g., Wang et al. 1992). Females never oviposit on barkless wood.

The larvae of most lamiines feed on living or recently dead plants, including both angiosperms and gymnosperms. Most are woodborers, such as *Anoplophora*, *Apriona*, and *Monochamus* species, and some are herbaceous plant feeders, such as *Dorcadion*, *Tetraopes*, and *Phytoecia* species. Most species feed on plant tissues above the ground level, and some consume those in the soil including roots. Lamiine larvae are very rarely found in strongly rotten wood (*Rhodopina* is an exception) and in dry, hard, long-dead wood, including seasoned construction timber. Most species probably have one generation a year although some need two years, and rarely three years, to complete a life cycle. Development may be very rapid for some tropical species; for example, *Steirastoma breve* (Sulzer) can complete four generations a year (Kliejunas et al. 2001) and *G. cantor* five generations a year (Lu et al. 2011). Except for terricolous larvae, pupation almost always occurs in the host. Many lamiine species are among the world's most important pests (see Chapters 11 and 12).

1.4.4 Subfamily Lepturinae Latreille, 1802

1.4.4.1 Diagnosis

Small to moderate, elongate, or robust beetles (Figures 1.92 and 1.93). Head oblique, rarely subvertical, gradually or suddenly narrowed behind the eyes (Figures 1.5 and 1.6); antennae inserted before or between the eyes, distinctly separated from bases of mandibles, moderately long, second antennomere short, transverse; eyes oval or emarginate, often entire, finely, or coarsely faceted; mandibles usually

FIGURE 1.92 *Euryptera unilineaticollis* Fuchs (Lepturinae).

FIGURE 1.93 *Strangalia melanura* (Redtenbacher) (Lepturinae).

flat, acute, usually with a molar tooth and a dense pubescence fringe; maxillae and labium well developed; palpi with last segment truncate at apex; ligula emarginate, frequently bilobed, membranous, or coriaceous; mentum flat, usually trapezoidal; gula produced between bases of maxillae as a distinct mentigerous process. Pronotum without lateral carinae (Figure 1.24); procoxae prominent, conical, cavities angulated externally, usually open posteriorly (Figure 1.25); mesoscutum usually with a complete median endocarina and a divided stridulatory plate (Figure 1.29); mesocoxal cavities open to epimera; metepisterna broad, usually narrowed posteriorly. Elytra usually narrowed posteriorly, rarely abbreviated, apices variable; posterior wings folded at apex, with or without a closed cell in the anal sector, radial cell closed. Legs slender; protibiae without a mesial sinus; tarsi pseudotetramerous, padded beneath (pads sometimes lacking on first tarsomere); claws divaricate to moderately divergent; empodium variable.

1.4.4.2 Comments

Audinet-Serville (1835) established the fourth "tribu" Lepturétes under the Cerambycidae with some characters that allow us to partially identify the subfamily: eyes rounded, scarcely reniforms; antennae inserted near the anterior margin of the eyes; head projected or suddenly constricted behind the eyes.

In their revision on North American Lepturinae, Linsley and Chemsak (1976) did not solve the taxonomic problems that had arisen since revisions by Swaine and Hopping (1928), Hopping (1937), and Hopping and Hopping (1947). The groupings are still not very clear because some groups are quite diverse in morphological and biological characters and their phylogenetic relationships are not solved.

1.4.4.3 Diversity and Distribution

The Lepturinae have more than 1,500 species in 210 genera and nine tribes in the world (Table 1.1). The subfamily is dominant in the Northern Hemisphere, particularly in the Holarctic, with a few species expanding to Afrotropical, Australian, Neotropical, and Oriental regions (Monné and Monné 2008; Švácha and Lawrence 2014).

1.4.4.4 Biology

Adults often are diurnally active flower visitors with head and mouthpart morphology of many taxa strongly adapted to pollen and nectar feeding (Linsley 1961; Linsley and Chemsak 1972; Gosling and Gosling 1976; Taki et al. 2013; Švácha and Lawrence 2014). They visit flowers of shrubs, trees, and even herbs depending on the availability (Gosling and Gosling 1976). Some species may feed on fungus spores (Linsley 1961). For example, pollens and fungus spores are found in guts of three Japanese *Rhagium* species (Kanda 1980). However, adults of most species from the Xylosteini appear to be nocturnally active (Linsley and Chemsak 1972). Some crepuscular or nocturnal adults may also visit flowers and feed on pollen. For example, pollen is found on the body of *Enoploderes sanguineum* Faldermann (Danilevsky and Miroshnikov 1981) and *Centrodera spurca* (LeConte) (Leech 1963). Adults live for a few weeks to a few months. Mating usually occurs on flowers. Although female-produced sex pheromones were evidenced earlier (Linsley 1961), they have not been identified until recently. For example, these have been identified for several species in the genera *Ortholeptura* Casey (Ray et al. 2011) and *Desmocerus* Serville (Ray et al. 2014). Millar and Hanks provide details about pheromones in this subfamily in Chapter 5. Females usually lay their eggs on or in the larval food material without special preparation of the oviposition site. However, females of some root feeders oviposit in soil or on the ground, and first instar larvae dig into the soil and search for the roots.

Linsley (1961) and Švácha and Lawrence (2014) summarized the biology of immature stages. Larvae of most species probably are dead wood feeders. Subcortical larval feeding and strongly flattened larval forms are widespread in Rhagiini but rare (Lepturini) or unknown in other tribes. Some dead wood feeders are associated with specific fungi; for example, *Pseudovadonia livida* (F.)

larvae tunnel in humus with mycclium of the fungus *Marasmius oreades* (Bolton) Fr. Larvae of *Encyclops* Newman and some *Pidonia* species bore in thick outer bark of living trees. In rare cases, larvae of some species from *Pseudogaurotina* Plavilstshikov and *Desmocerus* Serville feed within living tissues of woody plants. A few groups develop in or on the underground parts of living herbs, such as *Brachyta* Fairmaire, *Brachysomida* Casey, *Vadonia* Mulsant, many *Cortodera*, and some *Typocerus*; other *Cortodera* feed in wood fragments or conifer cones buried in humus. Larvae of many species may bore into the roots and, in specialized root feeders such as *Pachyta* Dejean, *Stenocorus* Geoffroy, and *Akimerus* Serville, the larvae start feeding in thinner distal roots and proceed toward the thicker proximal ones. Larvae of *Pidonia* Mulsant also are frequently subterranean, and related taxa (*Pseudosieversia* Pic, *Macropidonia* Pic) appear to be at least partly terricolous, feeding on the roots externally (Cherepanov 1979). Depending on the species, mature larvae may construct pupal chambers in the host plant or soil. The life cycle lasts from several months to several years.

1.4.5 Subfamily Necydalinae Latreille, 1825

1.4.5.1 Diagnosis

Small to moderate, usually elongate (Figures 1.36 and 1.37) but occasionally robust beetles. Head constricted behind eyes; frons large, vertical; antennae inserted high on frons; eyes finely faceted, deeply emarginate; mandibles small, robust, acute, densely fringed along inner margin; mentum short, transverse; ligula membranous, bilobed; palpi short. Pronotum with sides convex or tuberculate, constricted anteriorly and posteriorly; notosternal suture fine or incomplete; prosternum short, intercoxal process narrow; procoxae large, conical, nearly contiguous, cavities open to epimera; mesoscutum without median endocarina but with an undivided stridulatory plate (Figure 1.30); metepisterna broad, attenuated posteriorly. Elytra (Figures 1.36 and 1.37) short, scarcely longer than thorax, exposing abdomen, apices dehiscent; hind wings lying flat on abdomen, not folded at apex, without an anal cell. Legs slender; tibial spurs short; first metatarsomere without a pubescent sole. Empodium variable, in some cases distinct and multisetose. Abdomen with tergite corneous; sternites subequal in length, fifth sternite broadly emarginate in male.

1.4.5.2 Comments

The Necydalinae often have been classified as a tribe of the Lepturinae (Linsley 1940). Adult synapomorphies may include the mandibular molar plate, which is not known in other cerambycid subfamilies (but occurs in many Disteniidae and Oxypeltidae) and seems to be best developed in some derived floricolous Lepturinae (whereas it is less distinct in some presumed basal forms and in the Necydalinae).

1.4.5.3 Diversity and Distribution

This subfamily has about 70 species in two genera (Table 1.1): *Necydalis* L. and *Ulochaethes* LeConte. The former is distributed in the Nearctic, Palaearctic, and Oriental regions, although the latter is monotypic and only occurs in the western Nearctic. All other genera classified in the Necydalinae or Necydalini should be placed in the Cerambycinae (Švácha and Lawrence 2014).

1.4.5.4 Biology

Adults are predominantly diurnal and good fliers (Linsley 1940, 1961; Zolotov 2001). They mimic hymenopterans morphologically and behaviorally with *Necydalis* (Figure 1.36) resembling large wasps and *Ulochaetes* (Figure 1.37) similar to bumblebees (Linsley 1940, 1961; Švácha and Lawrence 2014). Some species of *Necydalis* visit flowers. Adults usually lay their eggs at the base of standing dead trees or stumps. Adults may feed on flowers or pollens but are relatively short-lived, probably less than a month (Linsley 1961).

Larvae of most species feed on the heartwood of dying or recently killed trees, with *Ulochaetes* attacking only conifers and most *Necydalis* mainly feeding on angiosperms and only some on conifers (Linsley 1961; Švácha and Lawrence 2014). *Ulochaetes* larvae bore in the base and roots of trees. Most *Necydalis* larvae may bore inside the trunk and stumps and occasionally in the roots. The life cycle normally lasts one year but, in some species, there may be two generations a year.

1.4.6 Subfamily Parandrinae Blanchard, 1845

1.4.6.1 Diagnosis

Moderate to large and usually robust beetles (Figure 1.94). Head (Figures 1.3 and 1.4) oblique to horizontal; antennae inserted in front of eyes, near base of mandibles, not extending beyond base of pronotum, similar in both sexes, glabrous, sensory areas deeply impressed; second antennomere at least one-half as long as third, remaining antennomeres subequal in length; eyes small, vertical, convex, feebly emarginate, coarsely faceted; labrum triangular fused with epistoma; mandibles dentate, molar tooth and pubescent fringe lacking; lobes of maxillae small, inner one obsolete (Figure 1.12); palpi with last segment truncate at apex; ligula very transverse, corneous; mentum transverse (Figure 1.20); submentum without intermaxillary process. Pronotum with an elevated, entire, and simple carina at each side (Figure 1.21); procoxae transverse; cavities open or closed posteriorly; mesoscutum without stridulatory plate (Figure 1.27); mesocoxal cavities open externally; metepisterna narrow, parallel-sided. Elytra parallel-sided, apices rounded; hind wings without closed cell in anal sector, radial cell open. Legs short; femora and tibiae compressed; procoxae without mesial sinus; tarsi (Figure 1.34) slender, five-segmented, without pubescent ventral pads, third tarsomere entire or feebly emarginate, not dilated, empodium small, slender, occasionally reduced or obsolete, usually uni- or bisetose.

FIGURE 1.94   *Erichsonia dentifrons* Westwood (Parandrinae).

1.4.6.2 Comments

Taxonomy of this small subfamily has been progressing rapidly in recent decades (Monné 2006; Švácha and Lawrence 2014). For example, there have been several revisions on the Afrotropical (Bouyer et al. 2012), Australian and Oriental (Santos-Silva et al. 2010; Komiya and Santos-Silva 2011), and Neotropical (Santos-Silva 2002, 2015) genera. Furthermore, Biffi and Fuhrmann (2013) gave a detailed treatment of the larvae and pupae of a Neotropical species, *Parandra longicollis* Thomson, generally discuss parandrine immature stages, and provide a key to parandrine larvae of the world.

1.4.6.3 Diversity and Distribution

Worldwide, the Parandrinae have about 120 species in 19 genera and two tribes (Table 1.1). The Erichsoniini contains only one genus and one species, *Erichsonia dentifrons* Westwood (Figure 1.94), and occurs only in Mexico and Central America. The Parandrini has representatives in all biogeographic regions, with six genera in the Oriental and Australian, five in the Neotropical, one in the Nearctic, one in Hawaii, one in the Palaearctic, one in the Afrotropical, and several species of uncertain generic position in the Afrotropical. Most parandrine species are distributed in warmer regions, and only a few occur in temperate zones.

1.4.6.4 Biology

The biology of the Parandrinae is still poorly understood. So far, nothing is known about the biology of the tribe Erichsoniini. Based on literature, Švácha and Lawrence (2014) summarized the general biology of the Parandrini. Adults are mostly nocturnal and hide in tree hollows, cracks, and under the loose bark during the day. Females lay eggs on the wood, with several generations often developing within the same dead wood. The life cycle usually lasts more than a year. For example, *Parandra brunnea* (F.) takes two to three years to complete growth and development (Gahan 1911). Lingafelter (1998) described mating behavioral interactions between sexes of a parandrine species, *P. (P.) glabra* (De Geer), for the first time. No pheromones have been found in this subfamily.

Larvae usually develop in dead moist logs, timber poles close to the ground surface, or in dead wood of living trees; sometimes they even develop in closed (healed over) hollows in which the adults may reproduce without leaving the tree (Linsley 1962a). Some species can feed on both dead and living plants. For example, the larvae of *P. brunnea* feed in the decayed heartwood and moist structural timber of many hardwood species and can also cause severe injury to living trees where the heartwood has been exposed (Gosling 1973). Most species are highly polyphagous, occasionally even attacking both gymnosperms and angiosperms with a preference for angiosperms (Linsley 1962a; Monné 2002). However, some southern species appear to be associated with, or at least frequently develop in, the gymnosperm tree genus *Araucaria* (Mecke et al. 2000; Mecke 2002). Mature larvae usually construct pupal cells inside the wood and pupate there.

1.4.7 Subfamily Prioninae Latreille, 1802

1.4.7.1 Diagnosis

Moderate to very large, usually robust beetles (Figures 1.95 and 1.96). Head (Figures 1.9 and 1.11) oblique or subvertical; antennae inserted near base of mandibles, variable in length but always surpassing base of pronotum, antennomeres glabrous, frequently produced or serrate in male; eyes large, usually coarsely faceted; labrum fused with epistoma; mandibles large, molar tooth and pubescent fringe lacking; maxillae with inner lobes obsolete (Figure 1.13); palpi with last segment truncate at apex; ligula transverse, corneous; mentum transverse; submentum without intermaxillary process. Pronotum with an elevated, often spinose or dentate, carina (Figures 1.22 and 1.23) at each side; cavities open behind; mesoscutum without a stridulatory plate; mesocoxal cavities usually open externally;

FIGURE 1.95   *Chariea birai* Monné and Monné (Prioninae).

FIGURE 1.96   *Poecilopyrodes pictus* (Perty) (Prioninae).

metepisterna variable. Elytra with variable apices; hind wings with a closed cell in anal and in radial sectors. Legs moderately robust; procoxae transverse; protibiae without mesial sinus; tarsi pseudote-tramerous, padded beneath, third tarsomere dilated, concealing minute fourth tarsomere. Empodium from prominent and multisetose to indistinct.

1.4.7.2 Comments

Prioninae are remarkable by almost complete absence of very small forms—except the Neotropical genus *Chariea*. For example, *C. birai* Monné & Monné (Figure 1.95) is only 4.2 mm long, which is the smallest species of Prioninae (Monné and Monné 2015). This subfamily contains some of the largest known beetles (*Titanus*, *Xixuthrus*, and *Macrodontia* with specimens reaching 150–175 mm). Extreme sexual dimorphism in size, general form, antennal morphology, and color is common in some species, such as those in the tribe Anacolini.

1.4.7.3 Diversity and Distribution

About 1,100 species in 302 genera and 21 tribes have been described for this subfamily worldwide (Švácha and Lawrence 2014; Tavakilian 2015). The subfamily occurs in all biogeographic regions but predominantly in the tropical and subtropical regions (Table 1.1) such as the warmer areas of the Afrotropical, Australian, Oriental, and Neotropical. Relatively few species are distributed in temperate areas of the Nearctic and Palaearctic regions.

1.4.7.4 Biology

Švácha and Lawrence (2014) summarized some aspects of the general biology of this subfamily. Typically, adults are crepuscular or nocturnal and correspondingly sober, dull, or dark in color. In the few known diurnal species that exclusively occur in the tropics, adults can be brightly colored or metallic. Flightlessness or brachyptery is relatively common, particularly in dry regions; but except for some Neotropical long-legged "pedestrian" species, this usually is restricted to females. Some winged females apparently cannot fly until they lay a portion of their eggs. Adults usually do not feed, or at most feed on imbibed fluids. Most adults live for a few days, but some can survive for a few months. Mating usually occurs on the emergence day or the next. Long-range female-produced sex pheromones have been identified for *Prionus californicus* Motschulsky (Rodstein et al. 2011) and several *Tragosoma* species (Ray et al. 2012), all of which are from the Nearctic. Although female-produced sex pheromones are evidenced in a New Zealand species, *Prionoplus reticularis* White (Edwards 1961), these have never been identified. In Chapter 5, Millar and Hanks provide a detailed treatment of pheromones in this subfamily. Females lay eggs immediately or one day after mating; in many species, they can lay hundreds of eggs. For example, the female of the Oriental species *Dorysthenes granulosus* (Thomson) can lay up to 783 eggs in her lifetime (Zeng and Huang 1981).

Larvae of most species develop in dead plants, but those of some species feed on living plants. Dead plant feeders usually are highly polyphagous, while those that feed on living plants can be either polyphagous or oligophagous. In several tribes, such as Prionini and Cantharocnemini, larvae develop more or less exclusively underground, and some can even move through soil and feed on roots of trees or herbs externally and then internally. Some species can bore from the main roots to the stems up to 60 cm above the ground surface. Females of such species usually oviposit in soil near the host plants, for example, *P. californicus* (Linsley 1962a; Alston et al. 2014) and the Oriental species *D. hugelii* (Redtenbacher) (Duffy 1968; Sharma and Khajuria 2005; Singh et al. 2010) and *D. granulosus* (Zeng and Huang 1981). Mature larvae usually pupate in the pupal cells in soil. However, most prionines lay eggs directly on the hosts. For example, the females of the Afrotropical species *Macrotoma palmata* (F.) lay their eggs in the cracks and crevices in the bark of the trunks and main branches of healthy trees (Duffy 1957; Tawfik et al. 2014). The larvae bore into the bark and then into the wood. For example, *P. reticularis* bore inside logs, stumps, dead parts of living trees, and untreated sawn timber (Edwards 1961). Mature larvae pupate in the wood. However, the larvae of some trunk borer species may exit from the wood when mature and enter soil for pupation—for example, the Neotropical species *Macrodontia cervicornis* (L.) (Paprzycki 1942).

Almost all species of known biology from this subfamily take more than a year to complete their life cycles. For example, the growth and development of immature stages require three to five years in *P. californicus* (Linsley 1962a; Alston et al. 2014) and three to four years in *D. hugelii* (Duffy 1968) and *M. palmata* (Duffy 1957; Tawfik et al. 2014). A shorter life history has been recorded (about two years) in *D. granulosus* (Zeng and Huang 1981; Yu et al. 2012).

1.4.8 Subfamily Spondylidinae Audinet-Serville, 1832

1.4.8.1 Diagnosis

Moderate in size and elongate, sometimes robust in form (Figures 1.97 and 1.98). Head (Figure 1.8) large, oblique, may be constricted behind eyes but without prominent temples; mouthparts moderately to strongly oblique; frontoclypeal suture complete or somewhat obliterated medially, postclypeus strongly transverse to shortly triangular, protentorial pits mostly distinct, sublateral to dorsal/frontal; anteclypeus small, very short in Spondylidini; labrum separate but often short and transverse; antennal sockets broadly separate, relatively distant from mandibular condyle, and facing laterally in Anisarthrini, Saphanini, and Atimiini, closer to condyle and facing slightly anteriorly in Asemini and Spondylidini; eyes huge to very small, more or less emarginate, may be strongly constricted or completely divided; in the saphanine branch, eyes may be slightly protruding between mandibular articulation and antennal socket. Pronotum without elevated carina, sides entire; prosternal process present, procoxal cavities of variable shape (but lateral procoxa and trochantin at least partly exposed), closed internally, open or closed posteriorly; meso-scutum with a medial endocarina, and stridulatory plate (if present) divided (Figure 1.28); mesocoxal cavities open in most species or narrowly closed laterally in Atimiini. Elytra parallel-sided, apices rounded, unarmed; hind wings with edge cell. Legs short; femora and tibiae compressed; protibiae without mesial sinus, apex with a broad terminal lamella (Figure 1.33); tarsi pseudotetramerous (Figure 1.35), claws divaricate, empodium small and bisetose to indistinct; legs modified in Spondylidini (short and stout with slightly compressed dentate tibiae, somewhat reduced tarsal pads and enlarged fourth tarsomere).

FIGURE 1.97   *Asemum striatum* (L.) (Spondylidinae).

FIGURE 1.98 *Neospondylis mexicanus* (Bates) (Spondylidinae).

1.4.8.2 Comments

The relationships of the Spondylidinae with other subfamilies have been controversial. For example, the subfamily was considered more closely related with the Prioninae by Audinet-Serville (1832), although it was treated as a closer relative of the Anoplodermatinae by Thomson (1860). Based on larval morphology, Craighead (1923), Duffy (1960), and Švácha and Danilevsky (1987) included *Spondylis* in the Aseminae. In the current system, the tribes Asemini, Atimiini, and Spondylidini are included in the Spondylidinae.

1.4.8.3 Diversity and Distribution

The subfamily consists of about 160 species in 32 genera and 5 tribes and is distributed in all biogeographic regions (Table 1.1). However, Švácha and Lawrence (2014) listed only 20 genera. They show that the spondylidine branch (Spondylidini and Asemini) generally is Holarctic, with some Neotropical and Oriental representatives. Several *Arhopalus* species have been introduced to the Australian (Wang and Leschen 2003), Afrotropical (Adlbauer 2001), and Neotropical (López et al. 2008). In the saphanine branch, Anisarthrini primarily occur in the Palaearctic and Afrotropical, Saphanini mainly in the Palaearctic and Nearctic, and Atimiini in the Palaearctic, Nearctic, and Oriental.

1.4.8.4 Biology

Adults are predominantly crepuscular and nocturnal, usually somber-colored, nonfeeding and short-lived (Švácha and Lawrence 2014). Some species of *Tetropium* (tribe Asemini) are diurnal in habit and brightly colored. The Saphanini are mostly flightless; for example, *Saphanus* is macropterous, but at least females of some, if not all, populations do not fly and *Drymochares* and *Michthisoma* are micropterous. Male-produced long-range sex pheromones in *Tetropium fuscum* (F.) and *T. cinnamopterum* Kirb have been identified (Silk et al. 2007). Lemay et al. (2010) described calling behavior by *T. fuscum* males, during which time sex pheromones are released. After females are attracted to calling males, the latter use female-produced contact sex pheromones for sex recognition and mating (Silk et al. 2011). Details about pheromones in this subfamily are given in Chapter 5. Some *Arhopalus* species are attracted to newly burned pine forest for oviposition (Suckling et al. 2001).

Larvae from Saphanini and Anisarthrini usually feed on dead angiosperms, but those of most species from the remaining three tribes live on dead conifers. According to Švácha and Lawrence (2014), Asemini and Spondylidini adults usually do not oviposit on barkless wood. Larvae from *Tetropium* are completely subcortical, sometimes in freshly dead or live trees (Gosling 1973). Species of *Nothorhina* and *T. aquilonium* develop exclusively within the bark of large, standing, living trees (Heliövaara et al. 2004). Larvae of many Asemini feed on dead or stressed coniferous trees (Hanks 1999) and may bore into underground parts of the host trees. Many species from the Spondylidini are specialized root feeders, working from distal roots toward the tree base so that mature larvae may reach it and adults may emerge from stem or stump bases aboveground. Female *Spondylis* dig into the soil and oviposit directly on the root bark (Cherepanov 1979). All taxa pupate in the food material. The life cycle of this subfamily lasts one to several years, depending on species and climate.

ACKNOWLEDGMENTS

We thank Eugenio H. Nearns for reviewing an earlier version of the manuscript, Luiza Silvério da Cruz for assistance in illustrations, and Cleide Costa, Petr Švácha, Julia Reindlmeier, Claudio J. B. Carvalho, Carlos Lamas, and Walter F. de Azevedo Jr. for permission to use their illustrations.

REFERENCES

Adlbauer, K. 2001. *Katalog und Fotoatlas der Bockkafer Namibias (Cerambycidae)*. Hradec Kralove: Taita Publishers.

Adlbauer, K., A. Bjørnstad, and R. Perissinotto. 2015. Description of a new species of *Apterotoxitiades* Adlbauer, 2008 (Cerambycidae, Dorcasominae, Apatophyseini) and the female of *A. vivesi* Adlbauer, 2008, with notes on the biology of the genus. *ZooKeys* 482: 9–19.

Alston, D. G., S. A. Steffan, and M. Pace. 2014. Prionus root borer (*Prionus californicus*). Utah Pests Fact Sheet. http://utahpests.usu.edu/ipm/htm/fruits/fruit-insect-disease/prionus-borers10 (accessed November 15, 2015).

Audinet-Serville, J. G. 1832. Nouvelle classification de la famille des longicornes. *Annales de la Société Entomologique de France* 1: 118–201.

Audinet-Serville, J. G. 1834. Nouvelle classification de la famille des longicornes (suite). *Annales de la Société Entomologique de France* 3: 5–110.

Audinet-Serville, J. G. 1835. Nouvelle classification de la famille des longicornes (suite et fin). *Annales de la Société Entomologique de France* 4: 197–228.

Bates, H. W. 1861. Contributions to an insect fauna of the Amazon Valley. Coleoptera: Longicornes. *The Annals and Magazine of Natural History* 8: 40–52, 147–152, 212–219, 471–478.

Bates, H. W. 1862. Contributions to an insect fauna of the Amazon Valley. Coleoptera: Longicornes. *The Annals and Magazine of Natural History* 9: 117–124, 396–405, 446–458.

Bates, H. W. 1863. Contributions to an insect fauna of the Amazon Valley. Coleoptera: Longicornes. *The Annals and Magazine of Natural History* 12: 100–109, 275–288, 367–381.

Bates, H. W. 1864. Contributions to an insect fauna of the Amazon Valley. Coleoptera: Longicornes. *The Annals and Magazine of Natural History* 13: 43–56, 144–164; 14, 11–24.

Bates, H. W. 1865. Contributions to an insect fauna of the Amazon Valley. Coleoptera: Longicornes. *The Annals and Magazine of Natural History* 15: 213–225, 382–394; 16, 101–113, 167–182, 308–314.

Bates, H. W. 1866. Contributions to an insect fauna of the Amazon Valley. Coleoptera: Longicornes. *The Annals and Magazine of Natural History* 17: 31–42, 191–202, 288–303, 367–373, 425–435.

Biffi, G., and J. Fuhrmann. 2013. Immatures of *Parandra* (*Tavandra*) *longicollis* Thomson, 1861 and comments on the larvae of Parandrinae (Coleoptera: Cerambycidae). *Insecta Mundi* 323: 1–14.

Bouchard, P., Y. Bousquet, A. E. Davies, et al. 2011. Family-group names in Coleoptera (Insecta). *ZooKeys* 88: 1–972.

Bouyer, T., A. Drumont, and A. Santos-Silva. 2012. Revision of African Parandrinae (Coleoptera, Cerambycidae). *Insecta Mundi* 241: 1–85.

Butovitsch, V. 1939. Zur Kenntnis de Paarung, Eiablage und Ernöhrung der Cerambyciden. *Entomologisk Tidskrift* 60: 206–258.

Chemsak, J. A. 1996. *Illustrated revision of the Cerambycidae of North America. Parandrinae, Spondylidinae, Aseminae, Prioninae*. Burbank, CA: Wolfsgarden Books.

Cherepanov, A. I. 1979. *Cerambycidae of Northern Asia (Prioninae, Disteniinae, Lepturinae, Aseminae)*. Novosibirsk: Nauka [in Russian].

Cordeiro, G., N. Anjos, P. G. Lemes, and C. A. R. Matrangolo. 2010. Ocorrencia de *Oncideres dejeanii* Thomson (Cerambycidae) em *Pyrus pyrifolia* (Rosaceae), em Minas Gerais. *Pesquisa Florestal Brasileira* 30: 153–156.

Craighead, F. C. 1923. North American cerambycid larvae. *Bulletin of the Canada Department of Agriculture* 27: 1–239.

Danilevsky, M. L. 1979. Description of the female, pupa and larva of *Apatophysis pavlovskii* Plav., and discussion of systematic position of the genus *Apatophysis* Chevr. (Coleoptera, Cerambycidae). *Entomologicheskoe, Obozrenie* 58: 821–828 [in Russian].

Danilevsky, M. L., and A. I. Miroshnikov. 1981. New data about biology of *Enoploderes sanguineum* Fald. and *Isotomus comptus* Mannh. (Coleoptera, Cerambycidae) with descriptions of their larvae. *Biologicheskie Nauki* 9: 50–53 [in Russian].

Duffy, E. A. J. 1953. *A monograph of the immature stages of British and imported timber beetles (Cerambycidae)*. London: British Museum (Natural History).

Duffy, E. A. J. 1957. *A monograph of the immature stages of African timber beetles (Cerambycidae)*. London: British Museum (Natural History).

Duffy, E. A. J. 1960. *A monograph of the immature stages of Neotropical timber beetles (Cerambycidae)*. London: British Museum (Natural History).

Duffy, E. A. J. 1968. *A monograph of the immature stages of Oriental timber beetles (Cerambycidae)*. London: British Museum (Natural History).

Edwards, J. S. 1961. Observations on the ecology and behaviour of the huhu beetle *Prionoplus reticularis* White (Coleoptera: Cerambycidae). *Transactions of the Royal Society of New Zealand* 88: 733–741.

Forchhammer, P., and Q. Wang. 1987. An analysis of the subfamily distribution and composition of the longicorn beetles (Coleoptera: Cerambycidae) in the provinces of China. *Journal of Biogeography* 14: 583–593.

Gahan, A. B. 1911. Some notes on *Parandra brunnea* Fabr. *Journal of Economic Entomology* 4: 299–301.

Gardiner, L. M. 1966. Egg bursters and hatching in the Cerambycidae (Coleoptera). *Canadian Journal of Zoology* 44: 199–212.

Goh, T. 1977. A study on thelytokous parthenogenesis of *Kurarua rhopalophoroides* Hayashi (Col., Cerambycidae). *Elytra* 5: 13–16.

Gosling, D. C. L. 1973. An annotated list of the Cerambycidae of Michigan (Coleoptera). Part I: Introduction and the subfamilies Parandrinae, Prioninae, Spondylinae, Aseminae, and Cerambycinae. *The Great Lakes Entomologist* 6: 65–84.

Gosling, D. C. L., and N. M. Gosling. 1976. An annotated list of the Cerambycidae of Michigan (Coleoptera). Part II: The subfamilies Lepturinae and Lamiinae. *The Great Lakes Entomologist* 10: 1–37.

Guo, J. F. 1999. Control of grape tiger longicorn. *China Rural Science & Technology* (12): 16–16 [in Chinese].

Haack, R. A., F. Hérard, J. H. Sun, and J. J. Turgeon. 2010. Managing invasive populations of Asian long-horned beetle and citrus longhorned beetle: A worldwide perspective. *Annual Review of Entomology* 55: 521–46.

Hanks, L. M. 1999. Influence of the larval host plant on reproductive strategies of cerambycid beetles. *Annual Review of Entomology* 44: 483–505.

Heliövaara, K., I. Mannerkoski, and J. Siitonen. 2004. *Longhorn beetles of Finland (Coleoptera, Cerambycidae)*. Helsinki: Tremex Press [in Finnish with English abstract].

Hopping, R. 1937. The Lepturini of America, north of Mexico. *Bulletin of the National Museum of Canada* 85: 1–42.

Hopping, R., and G. R. Hopping. 1947. The Lepturini of America, north of Mexico. Part III: *Cortodera*. *Scientific Agriculture* 27: 220–236.

Kanda, E. 1980. Hind gut contents in adult beetles of the genus *Rhagium* (Coleoptera: Cerambycidae: Lepturinae). *New Entomologist* 29: 27–32.

Kliejunas, J. T., B. M. Tkacz, H. H. Burdsall Jr., et al. 2001. Pest risk assessment of the importation into the United States of unprocessed Eucalyptus logs and chips from South America. *USDA Forest Service General Technical Report FPL-GTR-124*: 1–144.

Komiya, Z., and A. Santos-Silva. 2011. Two new species of *Stenandra* Lameere, 1912 (Coleoptera, Cerambycidae, Parandrinae) from the Oriental Region. *ZooKeys* 103: 41–47.

LeConte, J. L., and G. H. Horn. 1883. Classification of the Coleoptera of North America. Prepared for the Smithsonian Institution. *Smithsonian Miscellaneous Collections* 26(507): 1–567.

Leech, H. B. 1963. *Centrodera spurca* (LeConte) and two new species resembling it, with biological and other notes (Coleoptera: Cerambycidae). *Proceedings of the California Academy of Sciences (Series 4)* 32: 149–218.

Lemay, M. A., P. J. Silk, and J. Sweeney. 2010. Calling behavior of *Tetropium fuscum* (Coleoptera: Cerambycidae: Spondylidinae). *Canadian Entomologist* 142: 256–260.

Lingafelter, S. W. 1998. Observations of interactive behavior in *Parandra glabra* (Coleoptera: Cerambycidae). *Entomological News* 109: 75–80.

Linsley, E.G. 1940. A revision of the North American Necydalini (Coleoptera, Cerambycidae). *Annals of the Entomological Society of America* 33: 269–281.

Linsley, E. G. 1961. *The Cerambycidae of North America. Part I: Introduction.* Berkeley, CA: University of California Press.

Linsley, E. G. 1962a. *The Cerambycidae of North America. Part II: Taxonomy and classification of the Parandrinae, Prioninae, Spondylinae and Aseminae.* Berkeley, CA: University of California Press.

Linsley, E. G. 1962b. *The Cerambycidae of North America. Part III: Taxonomy and classification of the subfamily Cerambycinae, tribes Opsimini through Megaderini.* Berkeley, CA: University of California Press.

Linsley, E. G. 1963. *The Cerambycidae of North America. Part IV: Taxonomy and classification of the subfamily Cerambycinae, tribes Elaphidionini through Rhinotragini.* Berkeley, CA: University of California Press.

Linsley, E. G. 1964. *The Cerambycidae of North America. Part V: Taxonomy and classification of the subfamily Cerambycinae, tribes Callichromini through Ancylocerini.* Berkeley, CA: University of California Press.

Linsley, E. G., and J. A. Chemsak. 1972. *The Cerambycidae of North America. Part VI, No. 1: Taxonomy and classification of the subfamily Lepturinae.* Berkeley, CA: University of California Press.

Linsley, E.G., and J. A. Chemsak. 1976. *The Cerambycidae of North America. Part VI, No. 2: Taxonomy and classification of the subfamily Lepturinae.* Berkeley, CA: University of California Press.

Linsley, E. G., and J. A. Chemsak. 1984. *The Cerambycidae of North America. Part VII, No. 1: Taxonomy and classification of the subfamily Lamiinae, tribes Parmenini through Acanthoderini.* Berkeley, CA: University of California Press.

Linsley, E. G., and J. A. Chemsak. 1995. *The Cerambycidae of North America. Part VII, No. 2: Taxonomy and classification of the subfamily Lamiinae, tribes Acanthocinini through Hemilophini.* Berkeley, CA: University of California Press.

López, A., J. Garcéa, M. Demaestri, O. Di Iorio, and R. Magris. 2008. The genus *Arhopalus* Serville, 1834 (Insecta: Coleoptera: Cerambycidae: Aseminae) in association to *Sirex noctilio* in Argentina. *Boletin de Sanidad Vegetal Plagas* 34: 529–531.

Lu, W., Q. Wang, M. Y. Tian, J. Xu, and A. Z. Qin. 2011. Phenology and laboratory rearing procedures of an Asian longicorn beetle, *Glenea cantor* (Coleoptera: Cerambycidae: Lamiinae). *Journal of Economic Entomology* 104: 509–516.

Mecke, R., 2002. *Insetos do Pinheiro brasileiro—Insekten der brasilianischen Araukarie—Insects of the Brasilian Pine.* Tübingen: Attempto Service GmbH.

Mecke, R., M. H. M. Galileo, and W. Engels. 2000. Insetos e ácaros associados à *Araucaria angustifolia* (Araucariaceae, Coniferae) no sul do Brasil. *Iheringia, Zoologia* 88: 165–172.

Monné, M. A. 2002. Catalogue of the Neotropical Cerambycidae (Coleoptera) with known host plant—Part V: Subfamilies Prioninae, Parandrinae, Oxypeltinae, Anoplodermatinae, Aseminae and Lepturinae. *Publicacoes Avulsas do Museu Nacional* 96: 3–70.

Monné, M. A. 2006. Catalogue of the Cerambycidae (Coleoptera) of the Neotropical Region. Part III: Subfamilies Parandrinae, Prioninae, Anoplodermatinae, Aseminae, Spondylidinae, Lepturinae, Oxypeltinae, and addenda to the Cerambycinae and Lamiinae. *Zootaxa* 1212: 1–244.

Monné, M. A. 2012. Catalogue of the type-species of the genera of the Cerambycidae, Disteniidae, Oxypeltidae and Vesperidae (Coleoptera) of the Neotropical Region. *Zootaxa* 3213: 1–183.

Monné, M. L., and M. A. Monné. 2008. The tribe Lepturini in South America (Coleoptera: Cerambycidae: Lepturinae). *Zootaxa* 1858: 37–52.

Monné, M. L., and M. A. Monné. 2015. A new species of *Chariea* Audinet-Serville, 1832 (Coleoptera, Cerambycidae, Prioninae). *Arquivos de Zoologia* 46: 79–81.

Napp, D. S. 1994. Phylogenetic relationships among the subfamilies of Cerambycidae (Coleoptera, Chrysomeloidea). *Revista Brasileira de Entomologia* 38: 265–419.

Özdikmen, H. 2008. A nomenclatural act: Some nomenclatural changes on Palearctic longhorned beetles (Coleoptera: Cerambycidae). *Munis Entomology and Zoology* 3: 707–715.

Paprzycki, P. 1942. Datos para la captura y crianza del más grande de los cerambícidos *"Macrodontus cervicornis"* en la selva peruana. *Boletín del Museo de Historia Natural Javier Prado* 6: 349–351.

Pascoe, F. P. 1864–1869. Longicornia Malayana: A descriptive catalogue of the species of the three longicorn families Lamiidae, Cerambycidae and Prionidae collected by Mr. A. R. Wallace in the Malay Archipelago. *The Transactions of the Entomological Society of London* 3(3): 1–689.

Ray, A. M., R. A. Arnold, I. Swift, et al. 2014. (R)-Desmolactone is a sex pheromone or sex attractant for the endangered valley elderberry longhorn beetle *Desmocerus californicus dimorphus* and several congeners (Cerambycidae: Lepturinae). *PLos One* 9: e115498.

Ray, A. M., J. D. Barbour, J. S. McElfresh, et al. 2012. 2,3-Hexanediols as sex attractants and a female-produced sex pheromone for cerambycid beetles in the prionine genus *Tragosoma. Journal of Chemical Ecology* 38: 1151–1158.

Ray, A. M., A. Zunic, R. L. Alten, J. S. McElfresh, L. M. Hanks, and J. G. Millar. 2011. cis-Vaccenyl acetate, a female-produced sex pheromone component of *Ortholeptura valida*, A longhorned beetle in the subfamily Lepturinae. *Journal of Chemical Ecology* 37: 173–178.

Rodstein, J., J. G. Millar, J. D. Barbour, et al. 2011. Determination of the relative and absolute configurations of the female-produced sex pheromone of the cerambycid beetle *Prionus californicus. Journal of Chemical Ecology* 37: 114–124.

Santos-Silva, A. 2002. Notas e descricoes em Parandrini (Coleoptera, Cerambycidae, Parandrinae). *Iheringia (Serie Zoologia)* 92: 29–52.

Santos-Silva, A. 2015. A new species of *Parandra* (*Parandra*) Latreille from Peru (Coleoptera, Cerambycidae, Parandrinae). *Insecta Mundi* 405: 1–5.

Santos-Silva, A., D. Heffern, and K. Matsuda. 2010. Revision of Hawaiian, Australasian, Oriental, and Japanese Parandrinae (Coleoptera, Cerambycidae). *Insecta Mundi* 130: 1–120.

Sharma, J. P., and D. R. Khajuria. 2005. Distribution and activity of grubs and adults of apple root borer *Dorysthenes hugelii* Redt. *Acta Horticulturae* 696: 387–393.

Silk, P. J., J. Sweeney, J. P. Wu, J. Price, J. M. Gutowski, and E. G. Kettela. 2007. Evidence for a male-produced pheromone in *Tetropium fuscum* (F.) and *Tetropium cinnamopterum* (Kirby) (Coleoptera: Cerambycidae). *Naturwissenschaften* 94: 697–701.

Silk, P. J., J. Sweeney, J. P. Wu, S. Sopow, P. D. Mayo, and D. Magee. 2011. Contact sex pheromones identified for two species of longhorned beetles (Coleoptera: Cerambycidae) *Tetropium fuscum* and *T. cinnamopterum* in the subfamily Spondylidinae. *Environmental Entomology* 40: 714–726.

Singh, M., J. P. Sharma, and D. R. Khajuria. 2010. Impact of meteorological factors on the population dynamics of the apple root borer, *Dorysthenes hugelii* (Redt.), adults in Kullu valley of Himachal Pradesh. *Pest Management and Economic Zoology* 18: 134–139.

Ślipiński, A., and H. E. Escalona. 2013. *Australian longhorn beetles (Coleoptera: Cerambycidae), Vol. 1: Introduction and subfamily Lamiinae*. Melbourne: CSIRO Publishing.

Suckling, D. M., A. R. Gibb, J. M. Daly, D. Chen, and E. G. Brockerhoff. 2001. Behavioral and electrophysiological responses of *Arhopalus tristis* to burnt pine and other stimuli. *Journal of Chemical Ecology* 27: 1091–1104.

Švácha, P., and M. L. Danilevsky. 1987. Cerambycoid larvae of Europe and Soviet Union (Coleoptera, Cerambycoidea). *Part I: Acta Universitatis Carolinae, Biologica* 30: 1–176.

Švácha, P., and M. L. Danilevsky. 1988. Cerambycoid larvae of Europe and Soviet Union (Coleoptera, Cerambycoidea). *Part II: Acta Universitatis Carolinae, Biologica* 31: 121–284.

Švácha, P., and M. L. Danilevsky. 1989. Cerambycoid larvae of Europe and Soviet Union (Coleoptera, Cerambycoidea). *Part III: Acta Universitatis Carolinae, Biologica* 32: 1–205.

Švácha, P., and J. F. Lawrence. 2014. 2.1 Vesperidae Mulsant, 1839; 2.2 Oxypeltidae Lacordaire, 1868; 2.3 Disteniidae J. Thomson, 1861; 2.4 Cerambycidae Latreille, 1802; In *Handbook of zoology, Arthropoda: Insecta; Coleoptera, beetles, Volume. 3: Morphology and systematics (Phytophaga)*, eds. R. A. B. Leschen, and R. G. Beutel, pp. 16–177. Berlin: Walter de Gruyter.

Swaine, J. M., and R. Hopping. 1928. The Lepturini of America north of Mexico. Part I. *Bulletin of the Canadian National Museum* 52: 1–97.

Taki, H., H. Makihara, T. Matsumura, et al. 2013. Evaluation of secondary forests as alternative habitats to primary forests for flower-visiting insects. *Journal of Insect Conservation* 17: 549–556.

Tavakilian, G. 2015. Base de données Titan sur les Cerambycidés ou Longicornes. Paris: Institut de Recherche pour le Développement. http://lis-02.snv.jussieu.fr/titan/ (maintained by H. Chevillotte; accessed February 11, 2016)

Tawfik, H. M., W. A. Shehata, F. N. Nasr, and F. F. Abd-Allah. 2014. Population dynamics of *Macrotoma palmata* F. (Col.: Cerambycidae) on casuarina trees in Alexandria, Egypt. Alex. *Journal of Agricultural Research* 59: 141–146.

Thomson, J. 1860. *Essai d'une classification de la famille des cérambycides et matériaux pour servir a une monographie de cette famille.* Paris: Bouchard-Huzard.

Thomson, J. 1861. *Essai d'une classification de la famille des cérambycides et matériaux pour servir a une monographie de cette famille.* Paris: Bouchard-Huzard.

Thomson, J. 1864. Systema cerambycidarum ou exposé de tous les genres compris dans la famille des cérambycides et familles limitrophes. *Mémoires de la Société Royale des Sciences de Liège* 19: 1–352.

Vives, E. 2015. Revision of the genus *Trypogeus* Lacordaire, 1869 (Cerambycidae, Dorcasominae). *ZooKeys* 502: 39–60.

Wang, Q. 2002. Sexual selection of *Zorion guttigerum* Westwood (Coleoptera: Cerambycidae) in relation to body size and color. *Journal of Insect Behavior* 15: 675–687.

Wang, Q. 2008. Longicorn, longhorned, or round-headed beetles (Coleoptera: Cerambycidae). In *Encyclopedia of entomology (2nd ed.)*, ed. J. Capinera, 2227–2232. Dordrecht, The Netherlands: Springer.

Wang, Q., and S. N. Chiang. 1991. The evolution in the higher taxa of the Cerambycidae (Coleoptera). *Entomotaxonomia* 8: 93–114.

Wang, Q., and R. A. B. Leschen. 2003. Identification and distribution of *Arhopalus* species (Coleoptera: Cerambycidae: Aseminae) in Australia and New Zealand. *New Zealand Entomologist* 26: 53–59.

Wang, Q., J. G. Millar, D. A. Reed, et al. 2008. Development of a strategy for selective collection of a parasitoid attacking one member of a large herbivore guild. *Journal of Economic Entomology* 101: 1771–1778.

Wang, Q., G. L. Shi, and L. K. Davis. 1998. Reproductive potential and daily reproductive rhythms of *Oemona hirta* (Coleoptera: Cerambycidae). *Journal of Economic Entomology* 91: 1360–1365.

Wang, Q., X. Z. Xiong, and J. S. Li. 1992. Observations on oviposition and adult feeding behavior of *Phytoecia rufiventris* Gautier (Coleoptera: Cerambycidae). *The Coleopterits Bulletin* 46: 290–295.

Wang, Q., W. Y. Zeng, L. Y. Chen, J. S. Li, and X. M. Yin. 2002. Circadian reproductive rhythms, pair-bonding, and evidence for sex specific pheromones in *Nadezhdiella cantori* (Coleoptera: Cerambycidae). *Journal of Insect Behavior* 15: 527–539.

Wang, Q., W. Y. Zeng, and J. S. Li. 1990. Reproductive behavior of *Paraglenea fortunei* Saunders (Coleoptera: Cerambycidae). *Annals of the Entomological Society of America* 83: 860–866.

Williams, D. M. 2001. Chapter 30: Largest. In *The University of Florida Book of Insect Records*, ed. T. J. Walker. Gainesville: University of Florida. http://entomology.ifas.ufl.edu/walker/ufbir/chapters/chapter_30.shtml (accessed August 13, 2015).

Yu, Y. H., Chen, G. J., D. W. Wei, X. R. Zeng, and T. Zeng. 2012. Division of larval instars of *Dorysthenes granulosus* based on Crosby growth rule. *Journal of Southern Agriculture* 43: 1485–1489 [in Chinese with English abstract].

Zeng, C. F., and Q. S. Huang. 1981. Preliminary study on occurrence and control of *Dorysthenes granulosus*. *Entomological Knowledge* (1): 18–20 [in Chinese].

Zolotov, V. V. 2001. The *Necydalis major* determines the obstacle size both visually and tactilly. *Vestnik Zoologii* 35: 93–96.

2

Life History and Population Dynamics of Cerambycids

Robert A. Haack
USDA Forest Service
Lansing, Michigan

Melody A. Keena
USDA Forest Service
Hamden, Connecticut

Dominic Eyre
Defra, Department for Environment, Food & Rural Affairs
Sand Hutton, York, UK

CONTENTS

2.1 Introduction

The Cerambycidae comprise a large and diverse family of beetles with more than 36,000 species recognized worldwide (see Chapter 1). Cerambycids vary greatly in adult body length, from as short as 1.5 mm long in the Caribbean twig-boring lamiine *Decarthria stephensi* Hope (Villiers 1980; Peck 2011) to as long as 167 mm in the prionine *Titanus giganteus* (L.) (Bleuzen 1994), the larvae of which likely develop in decaying wood in South American rain forests. Cerambycids are native to all continents with the exception of Antarctica and can be found from sea level [e.g., the cerambycine *Ceresium olidum* (Fairmaire) in the Society Islands and Fiji; Blair 1934] to alpine sites as high as 4200 m (e.g., the cerambycine *Molorchus relictus* Niisato in China [Niisato 1996; Pesarini and Sabbadini 1997] and the lamiine *Lophopoeum forsteri* Tippmann in Bolivia [Tippmann 1960]). In this chapter, we will discuss the types of habitats commonly occupied by cerambycids, the development of the immature stages, diapause, adult dispersal and longevity, and population dynamics.

2.2 Larval and Adult Feeding Habits

Nearly all cerambycids are phytophagous as both adults and larvae. One exception is exemplified by adults in the cerambycine genus *Elytroleptus* that mimic and prey on adult lycid beetles (Eisner et al. 2008; Grzymala and Miller 2013). Likewise, although cerambycid larvae generally are considered phytophagous, facultative inter- and intraspecific predation has been observed when larvae encounter other individuals while constructing galleries within their host plants (Dodds et al. 2001; Ware and Stephen 2006; Schoeller et al. 2012).

A detailed account of the feeding biology of cerambycids is presented in Chapter 3 of this book; therefore, only a brief summary will be presented here. Cerambycids utilize a wide variety of plant species as larval hosts, including both monocots and dicots, and many species are pests of forest and urban trees (see Chapter 11) and crops (see Chapter 12). Worldwide, the vast majority of cerambycids develop in woody plants, especially trees. Almost every plant part is consumed by at least a few cerambycid species, with the vast majority developing in the stems, branches, and roots. Moreover, nearly every plant tissue is consumed by cerambycid larvae, with some species feeding mostly in the bark, some mostly in sapwood, others mostly in heartwood, and still others mostly in pith. Most cerambycid larvae develop within the tissues of their host plant, but there are some soil-dwelling species that feed externally on plant roots. With respect to host condition, some species develop in living hosts, although others are found in recently dead or even well-decayed hosts (Hanks 1999). In addition, some species prefer to infest dry wood (Hickin 1975). As for the larval host range, some cerambycids are highly monophagous, feeding on a single species or genus of plants, although others develop on several genera within a single family, and still others are highly polyphagous, developing in several plant families (Craighead 1923; Duffy 1953; Linsley 1959, 1961). There also is great variation in the types of food consumed by adult cerambycids, with some apparently not feeding at all, while others feed on flowers, bark, foliage, cones, sap, fruit, roots, and fungi (Trägårdh 1930; Butovitsch 1939; Duffy 1953; Linsley 1959, 1961; also see Chapter 3).

2.3 Oviposition, Fecundity, and Egg Development

Cerambycids oviposit on, in, or near their larval host plants (Trägårdh 1930; Butovitsch 1939; Duffy 1953; Linsley 1959, 1961). The behaviors displayed during oviposition also vary greatly among species. For example, in many Lepturinae and Prioninae, females simply push their ovipositor into the substrate when laying eggs, which is often soft and partially decayed wood or soil near the base of the larval host plant. Some cerambycids oviposit on the outer surface of their hosts (e.g., some Cerambycinae and Spondylidinae), while most non-Lamiinae cerambycids that infest woody plants lay eggs under bark scales or in bark crevices (Linsley 1959). In the Lamiinae, however, females typically prepare the oviposition site with their mandibles by chewing a slit or pit through the outer plant tissues into which they oviposit (Figures 2.1 and 2.2), while some lamiines use the tip of their abdomen to enlarge the oviposition site after first using their mouthparts (Linsley 1959).

Members of several cerambycid subfamilies oviposit in the entrance holes, exit holes, and other gallery structures created by various bark- and wood-boring insects (Linsley 1959). For example, Youngs (1897) reported on the lepturine *Anthophylax attenuates* (Haldeman) ovipositing in the galleries of the ptinid beetle *Ptilinus ruficornis* Say, apparently by inserting their ovipositor into the exit holes. Similarly, some members of the lamiine genus *Acanthocinus* construct their oviposition pits over bark beetle (Scolytinae) entrance holes and ventilation holes (i.e., the holes constructed by bark beetles along the length of their egg galleries that extend into the outer bark). For example, 56% of the oviposition pits made by *A. aedilis* (L.) in Europe and 99% of the oviposition pits made by *A. nodosus* (F.) in the southern United States were centered over bark beetle entrance holes and ventilation holes on infested pine trees (Schroeder 1997; Dodds et al. 2002).

Relatively few cerambycids oviposit directly on wood that is bark free (Duffy 1953; Linsley 1961). The cerambycine *Hylotrupes bajulus* (L.) is one exception in that females lay batches of eggs directly in cracks and crevices on exposed wood, primarily the sapwood of various conifers (Duffy 1953).

Some adult females girdle the host tissue with their mouthparts prior to oviposition, especially twigs and small-diameter branches. Twig girdling is common among the two lamiine genera *Oberea* and

FIGURE 2.1 Adult female of the lamiine *Anoplophora glabripennis* (Motschulsky) ovipositing on a branch of Norway maple (*Acer platanoides* L.) in the United States. (Courtesy of Melody Keena [Bugwood image 5431704].)

FIGURE 2.2 Adults and oviposition pits of the lamiine *Anoplophora glabripennis* on maple (*Acer*) in New York. (Courtesy of Kenneth Law [Bugwood image 0949056].)

Oncideres, with *Oberea* females tending to oviposit below the girdle (i.e., toward the trunk, proximal) while *Oncideres* females oviposit above the girdle (i.e., toward the branch tip, distal). Solomon (1995) provided details on the girdling behavior of several woody dicot-infesting, North American species, including *Oberea bimaculata* (Olivier), *O. ocellata* Haldeman, *O. ruficollis* (Fabricius), *O. tripunctata* (Swederus), and *Oncideres cingulata* (Say), *O. pustulata* LeConte, *O. rhodosticta* Bates, and *O. quercus* Skinner. Typically, for the *Oberea* and *Oncideres* species listed here, the *Oberea* adult females first chew two rings of punctures around the stem or twig about 13–25 mm apart and then chew a pit between the two rings and oviposit a single egg with the resulting larva tunneling downward (Figure 2.3). By contrast, *Oncideres* females girdle small to large branches (usually 5–20 mm, but up to 65 mm in *O. pustulata*) by chewing a ring around the circumference of the branch deep into the xylem tissue (Figure 2.4). *Oncideres* females usually girdle first and then oviposit in the branch, laying eggs singly in bark slits but often multiple eggs per branch (Rogers 1977a; Solomon 1995). The girdling of a branch appears to benefit the cerambycid larvae by reducing host defenses and elevating the nutritional quality of the host tissues (Forcella 1982; Rice 1995; Hanks 1999).

Fecundity in the Cerambycidae varies considerably among species from tens of eggs per female to several hundreds. Duffy (1953) warned, however, that field estimates often underestimate true total fecundity because eggs of cerambycids are often well concealed, making accurate counts difficult. A few examples of average lifetime fecundity based on laboratory studies are 80 eggs for the cerambycine *Megacyllene robinae* (Galford 1984), 119 eggs for the cerambycine *Enaphalodes rufulus* (Haldeman) (Donley 1978), 133 eggs for the lamiine *Glenea cantor* (F.) (Lu et al. 2013), 159 eggs for the spondylidine *Arhopalus ferus* (Mulsant) (Hosking and Bain 1977), 161 eggs for the cerambycine *H. bajulus* (Cannon and Robinson 1982), 165 eggs for the lamiine *Neoptychodes trilineatus* (L.) (Horton 1917), 200–451 eggs for the lamiine *Monochamus carolinensis* (Olivier) (Walsh and Linit 1985; Zhang and Linit 1998), 250–350 eggs for the prionine *Prionoplus reticularis* White (Rogers et al. 2002), and 581 eggs for *M. alternatus* Hope (Zhang and Linit 1998). In a study using 15 field-collected prionine *Prionus laticollis* (Drury) adult females, Farrar and Kerr (1968) reported that on average they laid 388 eggs but retained an average of 255 eggs, with one female having a total of 1,211 eggs (laid and unlaid).

Eggs usually are shades of white to yellow when first deposited and vary in length from under 0.5 mm to more than 1 cm in the prionine *T. giganteus* (Duffy 1953). Cerambycid eggs typically are elongate, usually being at least twice as long as broad (Butovitsch 1939; Figure 2.5). For example, based on data

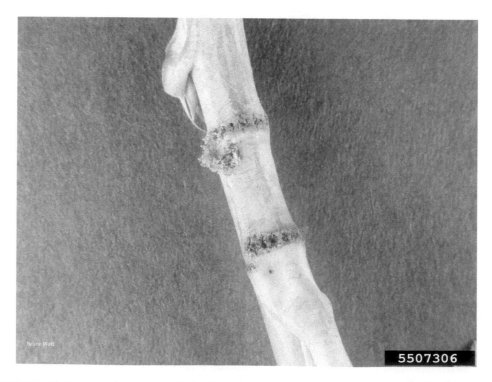

FIGURE 2.3 Typical oviposition damage by the lamiine *Oberea perspicillata* Haldeman on its larval host *Rubus* in which the adult female makes two rows of punctures that encircle the stem and then lays an egg in the stem between the punctures. Larval frass is being extruded from the oviposition hole. (Courtesy of Bruce Watt, photo taken in Maine, USA. [Bugwood image 5507306].)

FIGURE 2.4 Typical oviposition damage by an adult female of the lamiine *Oncideres cingulata* (Say) on one of its larval hosts, *Carya*, in the eastern United States. Females first girdle the twigs or small branches by chewing a ring around the twig's circumference and then laying one or more eggs in the distal portion of the branch. The girdled branches eventually fall to the ground and the larvae develop within them. (Photo from the USDA Cooperative Extension Slide Series [Bugwood image 1435156].)

FIGURE 2.5 Egg of the spondylidine *Tetropium fuscum* (Fabricius) with the larva visible inside. (Courtesy of Jessica Price, photo taken in Canada [Bugwood image 5331003].)

in Duffy (1953) and Rogers et al. (2002), eggs average 1.1 mm long and 0.3 mm wide at their greatest width for the cerambycine *Molorchus minor* (L.), and similarly 1.5 and 0.5 mm for the parandrine *Neandra* (= *Parandra*) *brunnea*, 2.4 and 0.4 mm for the prionine *P. reticularis*, 3.1 and 1.5 mm for the prionine *Ergates spiculatus* (LeConte), and 4.1 and 0.5 mm for the prionine *Prionus coriarius* (L.).

Eggs usually are laid singly or occasionally in small clusters. For example, Walsh and Linit (1985) examined 652 oviposition pits constructed by the lamiine *M. carolinensis* and found that 89 pits had no eggs, 559 had one egg, and 4 had two eggs. Similarly, eggs mostly were laid singly by the branch-borer cerambycine *Osphranteria coerulescens* Redtenbacher but occasionally in groups of two to three (Sharifi et al. 1970). By contrast, in the cerambycine *Phoracantha semipunctata* (F.), eggs commonly are deposited under bark in groups of 3–30 eggs (Scriven et al. 1986).

Most reviews state that cerambycid eggs usually hatch in a few days or up to four to five weeks at times, with two weeks being average (Craighead 1923; Butovitsch 1939; Duffy 1953; Linsley 1961). Data on the timing of egg hatch from several field and laboratory studies are listed in Table 2.1. For the field studies presented, the shortest incubation period was three to seven days for the lamiine *Nealcidion deletum* (Bates) in Guyana (Cleare 1931), and one of the longest was 22–24 days for the lamiine *Apriona germari* Hope in India (Hussain and Buhroo 2012). For the laboratory studies, which were conducted at temperatures ranging from 15°C to 34°C, the shortest average time to egg hatch was four days for the lamiines *Dectes texanus* LeConte at 27°C (Hatchett et al. 1973), *Phytoecia rufiventris* Gautier at 30°C (Shintani 2011), and *M. carolinensis* at 34°C (Pershing and Linit 1986). By contrast, some of the longest average times to egg hatch were 54 days for the lamiine *Anoplophora glabripennis* (Motschulsky) at 15°C (Keena 2006) and 55 days for the prionine *P. laticollis* at 16°C (Farrar and Kerr 1968). In addition, Keena (2006) reported that the average time to egg hatch for *A. glabripennis* was 25 days at 20°C, 15 days at 25°C, and 13 days at 30°C (Table 2.1).

2.4 Larval Development and Voltinism

During eclosion, cerambycid larvae rupture the egg chorion with the use of cephalic, thoracic, or abdominal spines (so-called egg burster spines) or with their mandibles (Duffy 1949, 1953; Linsley 1961; Gardiner 1966). After eclosion, most cerambycid larvae quickly tunnel into the host tissues or soil where they feed and develop over the next several months to years. As a result of these long generation times and the extensive feeding that usually occurs within the host plant, the larval stage of cerambycids is by far the most injurious life stage to the host plant. Cerambycid larvae generally are elongate and subcylindrical, with protracted mouthparts and poorly developed thoracic legs. However, some lepturine larvae have well-developed thoracic legs. For example, larvae of the European lepturine *Dinoptera collaris* (L.) are very mobile, live under loose bark and in insect galleries where they feed on frass of other borers, and can walk over the soil among logs and stumps (Duffy 1953; Bílý and Mehl 1989). Similarly, larvae of the *Quercus*-infesting lepturine *Leptura pacifica* (Linsley) have been reported to feed on frass of other cerambycids (Skiles et al. 1978).

Cerambycid larval galleries tend to be oval in cross-section with a meandering configuration when tunneling occurs in the cambial region but usually are round in cross-section and straighter when tunneling occurs in wood (Craighead 1923; Duffy 1953). The frass of cerambycid larvae, especially when feeding in woody tissue, often is granular with coarsely shredded or fibrous pieces of wood present (Hay 1968; Solomon 1977). The granular portion of the frass is material that passed through the digestive tract of the larva, whereas the shreds of wood are pieces torn off by the larva's mouthparts but not consumed (Craighead 1923; Solomon 1977).

Usually one to three years are required for most cerambycids to complete a single generation. However, some cerambycids apparently are able to complete two generations (bivoltine) per year (Matsumoto et al. 2000; Pershing and Linit 1986; Watari et al. 2002; Logarzo and Gandolfo 2005) or perhaps even three or more (multivoltine). For example, in southern China, Lu et al. (2011) demonstrated that the lamiine *G. cantor* can complete a generation in about 70 days or five generations per year. Similarly, Swezey (1950), working in Hawaii, reared adults of four cerambycids species within four months after live branches of a breadfruit tree [*Artocarpus altilis* (Parkinson) Fosberg] were cut and allowed to undergo

TABLE 2.1

Summary Data for Egg Development Time under Field or Laboratory Conditions for Selected Cerambycids

Species	Subfamily[a]	Study Location	Host Genera in Study	Study Conditions	Egg Hatch (Days)		References
					Mean	Range	
Aeolesthes holosericea Fabricius	Cer	India	Hardwoods	Field	–	7–12	Khan and Khan 1942
Alcidion cereicola Fisher	Lam	Argentina	*Harrisia*	Field	–	5–9	McFadyen and Fidalgo 1976
Anoplophora chinensis (Forster)	Lam	Japan	*Citrus*	20°C	14	–	Adachi 1994
				25°C	10	–	
				30°C	8	–	
Anoplophora glabripennis (Motschulsky)	Lam	USA	*Acer*	15°C	54	37–84	Keena 2006
				20°C	25	16–59	
				25°C	15	8–38	
				30°C	13	8–27	
Apriona germari Hope	Lam	India	*Morus*	Field	–	22–24	Hussain and Buhroo 2012
				25°C	18	–	Yoon and Mah 1999
Calchaenesthes pistacivora Holzschuh	Cer	Iran	*Pistacia*	Field	–	10–14	Rad 2006
Chion cinctus (Drury)	Lam	USA	*Carya*	Field	–	7–9	Hovey 1941
Colobothea distincta Pascoe	Lam	Costa Rica	*Theobroma*	Field	–	8–12	Lara and Shenefelt 1966
Dectes texanus LeConte	Lam	USA.	*Glycine*	27°C	4	3–5	Hatchett et al. 1973
Eupromus ruber (Dalman)	Lam	Japan	*Persea*	Field	–	7–10	Banno and Yamagami 1991
Megacyllene mellyi (Chevrolat)	Cer	Brazil	*Baccharis*	Field	14	–	McFadyen 1983
Megacyllene robiniae (Forster)	Cer	USA	*Robinia*	27°C	7	4–11	Wollerman et al. 1969
Monochamus carolinensis (Olivier)	Lam	USA	*Pinus*	18°C	14	–	Pershing and Linit 1986
				26°C	6	–	
				34°C	4	–	
Neoptychodes trilineatus (L.)	Lam	USA	*Ficus*	Field	6	3–8	Horton 1917
Nealcidion deletum (Bates)	Lam	Guyana	*Solanum*	Field	–	3–7	Cleare 1931
Oberea schaumi LeConte	Lam	USA	*Populus*	Field	–	14–15	Nord et al. 1972
Oemona hirta (F.)	Cer	New Zealand	*Citrus*	23°C	9	–	Wang et al. 1998

(Continued)

TABLE 2.1 (Continued)

Summary Data for Egg Development Time under Field or Laboratory Conditions for Selected Cerambycids

Species	Subfamily[a]	Study Location	Host Genera in Study	Study Conditions	Egg Hatch (Days) Mean	Egg Hatch (Days) Range	References
Osphranteria coerulescens Redtenbacher	Cer	Iran	*Prunus*	30°C	9	7–11	Sharifi et al. 1970
Phoracantha semipunctata (F.)	Cer	USA	*Eucalyptus*	20°C	5	–	Hanks et al. 1993
Phoracantha semipunctata	Cer	Tunisia	*Eucalyptus*	26°C	–	6–7	Chararas 1969
Phytoecia rufiventris Gautier	Lam	Japan	*Chrysanthemum*	20°C	10	–	Shintani 2011
				25°C	5	–	
				30°C	4	–	
Prionoplus reticularis White	Prio	New Zealand	*Pinus*	Field	–	16–25	Rogers et al. 2002
Prionus laticollis (Drury)	Prio	USA	*Malus*	27°C	19	–	Farrar and Kerr 1968
				16°C	55	–	
Saperda populnea (L.)	Lam	Turkey	*Populus*	Field	–	11–14	Tozlu et al. 2010
Saperda populnea	Lam	Korea	*Populus*	25°C	–	8–11	Park and Paik 1986
Semanotus litigiosus (Casey)	Cer	USA	*Abies*	Field	–	10–30	Wickman 1968
Stromatium longicorne (Newman)	Cer	China	Hardwoods	Field	–	10–15	Shi et al. 1982
Stromatium longicorne	Cer	Japan	Hardwoods	Field	–	8–12	Yashiro 1940
Xylotrechus colonus (Fab.)	Cer	Canada	*Betula*	Field	21	–	Gardiner 1960
Xylotrechus quadripes Chev.	Cer	India	*Coffea*	Field	–	5–6	Seetharama et al. 2005
Xylotrechus quadripes	Cer	Thailand	*Coffea*	29–31°C	5	3–9	Visitpanich 1994

[a] Cer = Cerambycinae, Lam = Lamiinae, Prio = Prioninae.

natural infestation for two months before being caged. In the tropical climate of Zambia, Löyttyniemi (1983) reported that *P. semipunctata*, an introduced cerambycine pest of eucalypts, could complete two to three generations per year. In Southern California, however, Bybee et al. (2004b) found that *P. semipunctata* generally is univoltine, but that *P. recurva* Newman could complete one and a partial second generation each year. Bybee et al. (2004b) suggested that this difference in generation time may partially explain how *P. recurva* is displacing *P. semipunctata* in California even though *P. semipunctata* was first reported in California in 1984 and *P. recurva* not until 1995. However, Luhring et al. (2004) suggested that differential susceptibility to natural enemies may also be an important factor leading to *P. recurva*'s displacement of *P. semipunctata*. In the case of the spondylidine *Tetropium gabrieli* Weise, Duffy (1953) presented data showing that this beetle usually completes one generation per year in the United Kingdom, but two generations can be completed there during very warm summers. By contrast, larval development can be greatly protracted in dry wood such as flooring, molding, and furniture (Duffy 1953; Hickin 1975). For example, an adult beetle of the cerambycine *Eburia quadrigeminata* (Say) emerged from a bookcase that was constructed more than 40 years earlier (Jaques 1918).

2.5 Pupal Development and Adult Emergence

Typically, pupation occurs at the end of the larval feeding galleries between the bark and wood; within the bark, sapwood, or heartwood of woody plants; inside the stems or roots of herbaceous plants; or in the soil (Craighead 1923; Duffy 1953). There are two general types of pupal chambers that are commonly referred to as cells and cocoons. Cells are chambers constructed near the terminal end of the larval gallery in which the pupa is in direct contact with the host tissues (Figure 2.6), while cocoons are chambers formed in the soil or in wood in which larvae first line the inner walls with a calcareous or gum-like secretion produced by the larvae (Duffy 1953; Linsley 1961; Figure 2.7). Many larvae do not pack their gallery with frass as they tunnel will isolate themselves in the gallery prior to pupation by plugging

FIGURE 2.6 Pupa of the lamiine *Anoplophora glabripennis* within the pupal chamber constructed at the end of the larval gallery in *Acer* in the eastern United States. Note how the larva plugged the gallery with wood shavings prior to pupating. (Courtesy of Melody Keena [Bugwood image 5431706].)

FIGURE 2.7 Larva of the prionine *Prionus coriarius* (Linnaeus) in an earthen cocoon that it constructed in the soil in which to pupate. (Courtesy of Gyorgy Csoka, photo taken in Hungary [Bugwood image 1231074].)

the gallery with a wad of shredded plant tissue (Craighead 1923). Depending on the species and where pupation occurs within the gallery, larvae will plug one or both sides of the gallery around themselves (Linsley 1961). These wads of plant tissue are thought to provide limited protection against natural enemies and possibly aid in regulating humidity (Duffy 1953). Many larvae extend their galleries to near the outer surface of the host plant and then return deeper within the gallery to pupate and, in so doing, reduce the amount of tunneling required to exit the host after they transform into adults (Linsley 1961).

The following discussion on the types of pupation cells and cocoons constructed by cerambycid larvae is based primarily on Craighead (1923), Duffy (1953), and Linsley (1961). Some species of *Cyrtinus, Leiopus, Poecilium,* and *Rhagium* construct shallow elliptical cells between the bark and wood, often plugging the galleries with wood fibers or surrounding the cell with shredded wood. For species that commonly pupate in the bark, they will instead pupate in wood if the bark is relatively thin or has fallen away. Several species that pupate in wood (e.g., species of *Aeolesthes, Aromia, Callidium, Cerambyx, Enaphalodes, Molorchus,* and *Saperda*) turn around at the end of the gallery prior to pupation so that once transformed the new adult can use the same gallery to exit, and moreover, several of these species (especially Cerambycini) plug the gallery entrance with a calcareous secretion prior to pupation. In others (e.g., *Apriona, Goes,* and *Monochamus*), the larvae construct pupation cells at the end of their galleries near the outer sapwood but do not turn; thus, the new adults must extend the gallery to exit the host. In many cerambycids that tunnel in the stems of herbaceous plants and small-diameter woody stems and twigs (e.g., *Agapanthia, Oberea,* and *Phytoecia*), the larvae plug one or both ends of the gallery around themselves with shredded plant tissue, and later, after pupating, the new adults chew through the wall of the stem or twig to exit. Many root-feeding cerambycids as well as some wood-feeding lepturines that exit their host and fall to the ground to pupate prepare earthen cocoons in the soil for pupation by hollowing out a chamber and often cementing the inner soil particles with a secretion produced by larvae (e.g., *Acmaeops, Anthophylax, Gaurotes, Judolia, Pachyta, Prionus,* and *Tetraopes*). Some cerambycids (e.g., *Plocaederus* and *Xystrocera*) construct calcareous cocoons in wood by coating the inner walls of the chamber with a calcium carbonate solution that is regurgitated by the larvae.

Prior to pupation, cerambycid larvae stop feeding, become quiescent, and contract in body length (Duffy 1953). Most cerambycids are oriented head-up for those that pupate within the host plant or horizontal for those that pupate in the soil (Duffy 1953). When pupation occurs within the host plant, stout spines usually are present on the abdominal tergites of the pupa, and these structures likely aid in anchoring the pupa within the gallery (Duffy 1953). Such spines are lacking in species that pupate horizontally in the soil (Craighead 1923). Although most pupae are inactive, pupae of the lamiine *Agapanthia villosoviridescens* (DeGeer) can use their spines to quickly ascend and descend within the larval gallery that is found in stems of its herbaceous hosts (Duffy 1953).

The duration of the pupal period varies considerably among cerambycids and is greatly affected by ambient temperature. Duffy (1953) stated that the pupal stage usually takes three to four weeks to complete, but some species require up to six weeks. By contrast, Linsley (1961) stated that the pupal stage lasts 7–10 days in most cases but up to a month can be required. Several examples of the time required to complete the pupal stage are presented in Table 2.2. For the field studies listed, the range of time required to complete the pupal stage varies from as little as six to eight days, in the lamiine *Nealcidion deletum* in Guyana (Cleare 1931), to 25–35 days in the cerambycine *Xylotrechus quadripes* Chevrolat in Thailand (Visitpanich 1994). In laboratory studies conducted at constant temperatures, the mean time to complete the pupal stage varies from 47 days at 15°C to 12 days at 30°C for the lamiine *A. glabripennis* (Keena and Moore 2010) and from 15 days at 22°C to 8 days at 34°C for the lamiine *M. carolinensis* (Pershing and Linit 1986). In the species where data are available (Table 2.2), males generally have a shorter pupal period than females.

After completion of the pupal stage, many physiological changes occur within newly eclosed adults, including sclerotization of the exoskeleton (Neville 1983). This process may take several days, and once complete, the new adult will initiate emergence, which often takes several more days, especially for adults that must chew through wood and bark to exit the host plant. For example, newly eclosed *A. glabripennis* adults took an average of seven days before starting to tunnel out of the wood and another five days, on average, to tunnel through the wood and emerge at 20°C and, similarly, five plus four days at 25°C and four plus four days at 30°C (Sánchez and Keena 2013). Adult cerambycids construct exit holes that are broadly oval to circular (Figure 2.8).

2.6 Overwintering, Quiescence, and Diapause

During periods of adverse environmental conditions, insects as well as invertebrates in general become dormant for varying periods of time. There are two general types of dormancy: quiescence and diapause. Quiescence is controlled exogenously (e.g., low ambient temperatures), whereas diapause is controlled endogenously (e.g., hormonal changes within the insect). For many cerambycids, the lower threshold temperature for development is about 10–12°C (Pershing and Linit 1986; Keena 2006; Naves and de Sousa 2009; Keena and Moore 2010; García-Ruiz et al. 2011). Therefore, in theory, when cerambycids experience these or lower temperatures, they would become quiescent and not develop further until the threshold temperature is again exceeded.

It appears that some cerambycids undergo true diapause although others simply become quiescent when temperatures drop below the developmental threshold. For example, while developing rearing methods for several cerambycid species, Gardiner (1970) reported that the final larval instars of the lamiine *Graphisurus fasciatus* (DeGeer) appeared to undergo true diapause and required about one month of cold treatment to break diapause. Similarly, when developing rearing methods for cerambycid pests of sunflowers, it was discovered that the lamiine *Ataxia hubbardi* Fisher undergoes facultative diapause (= quiescence), although the lamiine *Mecas cana saturnina* (LeConte) [reported as *M. inornata* Say; Linsley and Chemsak (1995)] undergoes obligatory diapause (Rogers 1977b; Rogers and Serda 1979). For some members of the lamiine genus *Monochamus*, an obligatory diapause has been reported for *M. alternatus* (Togashi 1991), *M. galloprovincialis* (Olivier) (Naves et al. 2007; Koutroumpa et al. 2008), and *M. saltuarius* (Gebler) (Togashi et al. 1994), but not for *M. carolinensis*, which can complete two generations per year (Pershing and Linit 1986).

Various combinations of temperature and photoperiod have been investigated to explore the conditions most favorable for diapause induction (Shintani et al. 1996; Asano et al. 2004) and termination (Togashi 1987, 1991; Jikumaru and Togashi 1996; Kitajima and Igarashi 1997; Esaki 2001; Rogers et al. 2002; Asano et al. 2004; Naves et al. 2007). By contrast, in southern China, where ambient temperatures

TABLE 2.2

Summary Data for Pupal Development Time under Field or Laboratory Conditions for Selected Cerambycids

Species	Subfamily[a]	Country	Host Genus	Rearing Temp. or Conditions	Pupation Time (Days) Mean	Range	References
Acalolepta vastator (Newman)	Lam	Australia	*Vitis*	Field, male	20	–	Goodwin and Pettit 1994
				Field, female	22	–	
Alcidion cereicola Fisher	Lam	Argentina	*Harrisia*	Field	–	10–11	McFadyen and Fidalgo 1976
Anoplophora chinensis	Lam	Japan	*Citrus*	20°C	24	–	Adachi 1994
Anoplophora glabripennis (Motschulsky)	Lam	USA	*Acer*	15°C	47	–	Keena and Moore 2010
				20°C	26	–	
				25°C	18	–	
				30°C	12	–	
Anoplophora macularia (Thomson)	Lam	Taiwan	*Citrus* (diet)	25°C	–	10–26	Lee and Lo 1998
Apriona germari Hope	Lam	India	*Morus*	Field	28	–	Hussain and Buhroo 2012
				25°C, male	18	–	Yoon and Mah 1999
				25°C, female	19	–	
Calchaenesthes pistacivora Holzschuh	Cer	Iran	*Pistacia*	Field	45	–	Rad 2006
Dectes texanus LeConte	Lam	USA	*Glycine*	27°C	10	7–15	Hatchett et al. 1973
Megacyllene robiniae (Forster)	Cer	USA	*Robinia*	27°C	13	5–17	Wollerman et al. 1969
				27°C, male	11	7–14	Galford 1984
				27°C, female	12	8–15	
Monochamus carolinensis (Olivier)	Lam	USA	*Pinus*	22°C	15	–	Pershing and Linit 1986
				26°C	9	–	
				34°C	8	–	
Nealcidion deletum (Bates)	Lam	Guyana	*Solanum*	Field	–	6–8	Cleare 1931
Neoptychodes trilineatus (L.)	Lam	USA	*Ficus*	Field	24	5–73	Horton 1917

(Continued)

TABLE 2.2 (Continued)

Summary Data for Pupal Development Time under Field or Laboratory Conditions for Selected Cerambycids

Species	Subfamily[a]	Country	Host Genus	Rearing Temp. or Conditions	Pupation Time (Days) Mean	Range	References
Oemona hirta (F.)	Cer	New Zealand	*Populus*	23°C	15–19	–	Wang et al. 2002
Osphranteria coerulescens Redtenbacher	Cer	Iran	*Prunus*	30°C	15	12–21	Sharifi et al. 1970
Phytoecia rufiventris Gautier	Lam	Japan	*Chrysanthemum*	20°C	22	–	Shintani 2011
				25°C	15	–	
				30°C	11	–	
Prionoplus reticularis White	Prio	New Zealand	*Pinus*	Field	25	20–24	Morgan 1960; Rogers et al. 2002
Prionus laticollis (Drury)	Prio	USA	*Malus*	18°C	25	–	Benham 1969
Saperda populnea (L.)	Lam	Korea	*Populus*	25°C	11	–	Park and Paik 1986
Semanotus litigiosus (Casey)	Cer	USA	*Abies*	Field	–	14–28	Wickman 1968
Stromatium longicorne (Newman)	Cer	China	Hardwoods	Field	–	15–18	Shi et al. 1982
Xylotrechus arvicola (Olivier)	Cer	Spain	*Vitis*	24°C, male	16	–	García-Ruiz et al. 2012
				24°C, female	18	–	
Xylotrechus quadripes Chevrolat	Cer	India	*Coffea*	Field	–	25–35	Seetharama et al. 2005
Xylotrechus quadripes	Cer	Thailand	*Coffea*	29–31°C	11	9–15	Visitpanich 1994

[a] Cer = Cerambycinae, Lam = Lamiinae, Prio = Prioninae.

FIGURE 2.8 Exit hole made by an adult of the lamiine *Saperda populnea* (L.) upon emergence from its host, *Populus*. (Courtesy of Gyorgy Csoka, photo taken in Hungary [Bugwood image 1141004].)

often support insect development year-round, Lu et al. (2011) found no evidence of diapause in the multivoltine lamiine *G. cantor*.

The vast majority of cerambycids overwinter in the larval stage. For example, in field studies conducted during a U.S. eradication program for the lamiine *A. glabripennis* in Illinois, of 569 *A. glabripennis* life stages recovered during dissections of infested trees in winter and early spring, 542 individuals were live larvae (95%) and 27 appeared to be viable eggs (5%) (Haack et al. 2006). Similarly, at an *A. glabripennis* outbreak site in northeastern Italy, most individuals overwintered as mature larvae in xylem, although some apparently living eggs and young larvae were found in the phloem (Faccoli et al. 2015). Although the overwintering eggs and young larvae were likely the result of late-season oviposition, many had died—possibly reflecting lower winter temperatures in the phloem compared to the xylem and lower fat stores of young larvae compared to mature larvae (Faccoli et al. 2015). Similarly, of 78 North American tree- and shrub-infesting cerambycids for which Solomon (1995) listed the overwintering life stage, 76 overwintered as larvae, although 2 species were said to overwinter as either larvae or pupae, both being cerambycines with two-year life cycles: *Anelaphus parallelus* (Newman), a twig pruner primarily of *Quercus*, and *Xylocrius agassizi* (LeConte), a root borer on *Ribes*.

A few cerambycids have been reported to overwinter as adults. Generally, these are individuals that pupate in late summer or autumn, and the adults then remain within the pupal cell until the following spring. This behavior has been reported for some species of the cerambycine genus *Cerambyx*; the lamiine genera *Mesosa, Phytoecia*, and *Pogonocherus*; and the lepturine genus *Rhagium* (Duffy 1953; Linsley and Chemsak 1972; Bílý and Mehl 1989; Bense 1995). Another example is the cerambycine *Aeolesthes holosericea* Fabricius, which either pupates in late summer and then overwinters as an adult within the pupal chamber or overwinters as a larva and then pupates and emerges as an adult the following year (Gupta and Tara 2013). In addition, adults of a few cerambycids do overwinter outside their pupal cells (Linsley 1936, 1961). For example, adults of the lamiine *Psenocerus supernotatus* (Say) have been found overwintering in the outer folds of the cocoons of the large saturniid moth *Hyalophora cecropia* (L.) (Lepidoptera: Saturniidae) in Pennsylvania, in the United States (Hamilton 1884).

2.7 Dispersal

Although some cerambycids are flightless, the majority are capable of flight (Figure 2.9). Duffy (1953) states that, in the United Kingdom, most cerambycids are day flyers (diurnal), while in the tropics most fly at dusk or dawn (crepuscular) or at night (nocturnal). In general, species that are diurnal fliers are faster and more agile in flight than nocturnal fliers (Linsley 1959). Many of the nocturnal species are attracted to lights. Most cerambycids demonstrate slow, directed flight, usually in the direction of food plants, larval host plants, or potential mates. At the subfamily level, the Lamiinae and Prioninae tend to be nocturnal while the Lepturinae typically are diurnal. Adults of many Lepturinae are strong and agile flyers, which is beneficial given that these adults usually feed and mate on flowers that are different plant species than the plants used as larval hosts (Hanks 1999). Although many cerambycids are slow flyers, especially many Prioninae, the lepturine *Judolia cerambyciformis* (Schrank) can hover and fly up and down over flowers, and the necydaline *Ulochaetes leoninus* LeConte can fly like a *Bombus* bumblebee (Hymenoptera: Apidae) (Duffy 1953). Lamiine adults usually feed daily and mate, feed, and oviposit on the same host plant and therefore commonly fly between the crown foliage where they feed and the branches and stems where they oviposit (Craighead 1923; Linsley 1961; Hanks 1999).

Dispersal in the Cerambycidae has been studied primarily in species that are forest pests as well as a few rare species. These types of studies are important—especially for introduced (= alien or exotic) species where quarantine zones need to be established—and there is a need to set survey boundaries and model potential pest spread (Kobayashi 1984; Takasu et al. 2000; Smith et al. 2001; Haack et al. 2010a; Hernández et al. 2011; Akbulut and Stamps 2012; also see Chapter 13). In other studies, researchers explored various attributes of the beetles themselves that would help support longer flight as well as attributes of the hosts that would increase attraction. For example, Hanks et al. (1998) reported that, in the cerambycine *P. semipunctata*, larger individuals tended to disperse further than smaller adults. Similarly, in field studies on the cerambycine *Semanotus japonicus* Lacordaire, Ito (1999) reported that adults preferentially landed on larger-diameter trees. In another study on *S. japonicus*, Shibata (1989) predicted that the average dispersal distance would increase throughout the period of adult emergence given that ambient temperatures would be near the flight threshold early in the flight season but that later in the flight season, the air temperatures would be well above the threshold.

The distances that cerambycid adults can fly have been estimated in laboratory studies using flight mills and in field studies using mark–recapture techniques (Table 2.3). The maximum distances recorded

FIGURE 2.9 Adult beetle of the lamiine *Anoplophora glabripennis* preparing for flight. (Courtesy of Roger Zerillo, photo taken in USA.)

TABLE 2.3

Summary Data for Selected Cerambycids with Published Dispersal Information

Species	Subfamily[a]	Country	Host Plant	Study Details	Findings	References
Anoplophora chinensis (Forster)	Lam	Japan	*Citrus*	Mark recapture in citrus orchard	Proof of immigration into orchard	Komazaki and Sakagami 1989
Anoplophora glabripennis (Motschulsky)	Lam	Korea	*Salix*	Used harmonic radar for 14 days	Mean dispersal 14 m (max 92 m) in 14 days	Williams et al. 2004
Anoplophora glabripennis	Lam	China	*Populus*	Mark–recapture in poplar stand	Mean dispersal 42 m (max 214 m) in 19 days	Zhou et al. 1984
Anoplophora glabripennis	Lam	China	?	Mark–recapture	Mean dispersal of 106 m over 20–28 days	Wen et al. 1998
Anoplophora glabripennis	Lam	China	*Populus*	Mark–recapture	Mean dispersal 266 m (max 1442 m) in 9 wks	Smith et al. 2001
Anoplophora glabripennis	Lam	China	*Populus, Salix, Ulmus*	Mark–recapture, season-long study	98% of adults dispersed under 920 m; max dispersal was 2,394 m for a male and 2,644 m for a female	Smith et al. 2004
Hirticlytus comosus (Matsushita)	Cer	Japan	*Podocarpus*	Tethered flight in laboratory	Mean estimated flight 122 m, max = 1,170 m	Sato 2005
Monochamus alternatus Hope	Lam	Japan	*Pinus*	Mark recapture in young pine stand	Early season mean dispersal was 19 m for males (max 55 m) and 23 m for females (max 59 m)	Shibata 1986a
Monochamus alternatus	Lam	Japan	*Pinus*	Circumstantial evidence	Two reports cited stating that islands 3.3 km from nearest outbreak became infested	Kobayashi et al. 1984; Togashi 1990a
Monochamus carolinensis (Olivier)	Lam	U.S.	*Pinus*	Flight mill study	Mean flight was 2.2 km in 2 hr; max = 10.3 km in 115 min	Akbulut and Linit 1999

(Continued)

TABLE 2.3 (Continued)

Summary Data for Selected Cerambycids with Published Dispersal Information

Species	Subfamily[a]	Country	Host plant	Study details	Findings	References
Monochamus galloprovincialis (Olivier)	Lam	France	*Pinus*	Flight mill study with adults flown once weekly, for up to 2 hr, until death	Average total distance flown was 15.6 km for males, 16.3 km for females, with a max. of 62.7 km	David et al. 2014
Monochamus galloprovincialis	Lam	Spain	*Pinus*	Mark recapture	One adult was captured in the most distant trap at 8.3 km	Gallego et al. 2012
Monochamus galloprovincialis	Lam	Spain	*Pinus*	Mark recapture	Season-long study, with several flying >3 km, and one flew 7.1 km	Hernández et al. 2011
Monochamus galloprovincialis	Lam	Spain	*Pinus*	Mark recapture	Some adults flew 13.6 to 22.1 km	Mas et al. 2013
Phoracantha semipunctata (F.)	Cer	South Africa	*Eucalyptus*	Circumstantial evidence	An isolated outbreak was 14 km from any other known source	Drinkwater 1975
Phoracantha semipunctata	Cer	Spain	*Eucalyptus*	Circumstantial evidence	Two isolated outbreaks were about 2 and 5 km from any known source	Martínez Egea 1982
Rosalia alpina (L.)	Cer	Czech Republic	*Fagus*	Mark recapture	Furthest dispersal detected: 1.6 km in 11 days	Drag et al. 2011
Semanotus japonicas Lacordaire	Cer	Japan	*Cryptomeria*	Mark recapture	Seasonal mean dispersal was 9 m for males (max 80 m) and 16 m for females (max 150 m)	Shibata 1986b
Tetraopes tetrophthalmus (Forster)	Lam	USA	*Asclepias*	Mark recapture	Average dispersal over 10 days was less than 40 m	McCauley et al. 1981

[a] Cer = Cerambycinae, Lam = Lamiinae.

in field studies can be influenced by the number of insects released, the number of traps deployed, the trapping distances used, and the length of time over which the study is conducted. Some of the extreme dispersal distances recorded based on mark–recapture studies were 2.6 km for the lamiine *A. glabripennis* (Smith et al. 2004) and 22.1 km for the lamiine *M. galloprovincialis* (Mas et al. 2013). For the flight-mill studies, the estimated maximum distances flown were 1.2 km for the cerambycine *Hirticlytus comosus* (Matsushita) (Sato 2005), 10.3 km for *M. carolinensis* (Akbulut and Linit 1999), and 62.7 km for *M. galloprovincialis* (David et al. 2014).

There are also several estimates of cerambycid dispersal based on circumstantial evidence (Table 2.3). For example, in Japan, *Pinus* trees infected with pine wilt disease, which is caused by an exotic xylem-invading nematode that is vectored by *M. alternatus* adults, were found on isolated islands that were approximately 3.3 km from the nearest disease centers. Therefore, *M. alternatus* adults were assumed to have flown that distance over water and carried the nematodes (Kobayashi et al. 1984; Togashi 1990a). Similarly, based on the nearest known outbreaks to newly discovered infestations, cerambycine *P. semipunctata* adults were assumed to have dispersed at least 5 km at a site in Spain (Martínez Egea 1982) and 14 km in South Africa (Drinkwater 1975). In Nova Scotia, Canada, where the European spondylidine *Tetropium fuscum* (F.) was introduced around 1990, new infestations have extended about 80 km from the original site of introduction after 20 years of spread (Rhainds et al. 2011).

Dispersal has also been studied in a few flightless cerambycids, such as the European lamiine *Dorcadion fuliginator* (L.), an endangered grass-feeding species (Baur et al. 2005). At a study site in Central Europe where several isolated grassland patches were surrounded by agricultural fields and human settlements, several beetles were observed to move 20–100 m, with one male moving a maximum of 218 m in 12 days. Similarly, the cactus-feeding lamiine *Moneilema* species, which are native to western North America, are also flightless, and several species occur in isolated patches along mountain slopes and therefore would likely have restricted capacity for dispersal (Lingafelter 2003; Smith and Farrell 2005).

In addition to natural dispersal, several cerambycids have been moved outside their native range as a result of inadvertent, human-assisted transport, including trade or movement of live plants, solid wood packaging materials, and firewood (Haack 2006; Cocquempot and Lindelöw 2010; Haack et al. 2010a, 2010b; Hu et al. 2013; Haack et al. 2014; Rassati et al. 2016; also see Chapter 13). For example, *A. chinensis* (Forster) has been moved primarily from its native range in Asia to other countries in live trees, including both nursery stock and bonsai plants, while *A. glabripennis* has been moved primarily in wood packaging materials such as pallets and crating (Haack et al. 2010a). During inspections of wood packaging materials entering U.S. ports of entry, cerambycids were second only to scolytines in being the most common group of wood borers encountered, representing about 20–25% of wood borers intercepted (Haack 2006; Haack et al. 2014). Moreover, during a survey conducted in Michigan of firewood transported in vehicles by the public, live bark- and wood-infesting insects were found in 23% of the individual firewood pieces, with most of the live borers encountered being cerambycids (Haack et al. 2010b).

2.8 Adult Longevity

Duffy (1953) and Linsley (1959) reported that cerambycid adults generally live from several days to several months, with females usually living longer than males in any given species. Longevity likely is linked to adult feeding habits and, given that adults of some subfamilies seldom feed (e.g., Prioninae) while others feed almost daily (e.g., Lamiinae and Lepturinae), it would not be surprising that on average the prionines would tend to have shorter adult life spans than the lamiines and lepturines. Nevertheless, Craighead (1923) reported that some prionine adults have been kept in captivity for 30–40 days without feeding. By contrast, Beeson and Bhatia (1939) reported that adults of the lamiine *Batocera rufomaculata* (De Geer) can live up to eight months. Duffy (1953) suggested that some of the longest-lived adults would likely be those that pupate in late summer, eclose, and then overwinter as adults within their pupal cells and thereby be in the adult stage for at least seven months. As mentioned in the earlier overwintering discussion (Section 2.6), these species would include cerambycines (*Cerambyx*), lamiines (*Mesosa, Phytoecia,* and *Pogonocherus*), and lepturines (*Rhagium*).

Longevity data from field and laboratory studies for a number of cerambycids are presented in Table 2.4. Mean longevity values varied from about one to four months in the field studies and from less than one month to more than seven months in the laboratory studies. The greatest mean values (more than 200 days) were reported for the cerambycine *P. recurva* at 20°C (Bybee et al. 2004a) and the lamiine *M. alternatus* at 28°C (Zhang and Linit 1998; Table 2.4).

Adult longevity can be significantly impacted by the ambient temperature as well as by larval and adult food sources (Table 2.4). For example, Keena (2006) reported that longevity for the lamiine *A. glabripennis* peaked at 10°C and declined at lower and higher temperatures, while Smith et al. (2002) found that *A. glabripennis* adults lived longer when fed on *Acer* twigs compared with *Salix* twigs. In the case of the lamiine *M. galloprovincialis*, Akbulut et al. (2007) reported that adults lived longer when (as larvae) they developed in logs that were cut in spring or fall as compared with logs cut in summer, possibly reflecting seasonal variation in wood moisture content.

2.9 Population Dynamics in Relation to Environmental Conditions

Many factors can affect the population dynamics of bark- and wood-infesting insects, such as availability and susceptibility of host plants, intra- and interspecific competition, parasitization, predation, and climatic factors such as temperature and rainfall (Coulson 1979). As detailed by Hanks (1999), cerambycids often are very selective about the condition of the host plant chosen for oviposition, with some favoring apparently healthy hosts, others weakened or severely stressed hosts, and still others hosts that are dead. There are many physical and environmental stressors that can affect individual plants and move the individual plants along a continuum from healthy to stressed to dead. Some of the physical factors that can weaken a plant include soil nutrient levels, soil pH, and soil compaction. Similarly, some of the environmental stressors that alter plant resistance to insects include air pollution (Alstad et al. 1982), defoliation (Kulman 1971), drought (Mattson and Haack 1987; Wallner 1987), fire (McCullough et al. 1998), and ice and wind damage (Gandhi et al. 2007; Schowalter 2012). Besides lowering a tree's resistance to insect infestation, environmental stressors can interact in complicated ways to affect not only the host plant but also the herbivore and the herbivore's natural enemies. In the case of drought, for example, Mattson and Haack (1987) contend that drought-stressed plants are more attractive and more susceptible to colonizing herbivores; the plant tissues of stressed plants are nutritionally superior; and the elevated plant and ambient temperatures during drought favor herbivores over their natural enemies as well as favor the herbivore's detoxification system, immune system, and symbiotic microorganisms.

In life-table studies of various cerambycids, the highest levels of mortality usually occurred during the larval stage—often the early larval stages. This relationship has been reported for the cerambycines *P. semipunctata* (Powell 1982) and *Styloxus bicolor* (Champlain & Knull) (Itami and Craig 1989) as well as for the lamiines *A. glabripennis* (Zhao et al. 1993), *M. galloprovincialis* (Koutroumpa et al. 2008), *Oberea schaumii* LeConte (Grimble and Knight 1971), and *Saperda inornata* Say (Grimble and Knight 1970). Although most researchers reported that mortality usually was highest among early larval instars, Rogers (1977a) reported that most mortality in the twig-girdling lamiine *O. cingulata* occurred in the egg stage, whereas Togashi (1990b) reported that the highest mortality for *M. alternatus* occurred among late larval instars, often when they were in their pupal cells, and that insect predators were the leading mortality agents. Similarly, in studies on the cerambycine *E. rufulus*, which has a two-year life cycle, Haavik et al. (2012) reported that the highest mortality occurs in the second summer of larval development when larvae tunnel from the cambial region into the sapwood. Researchers have also reported that larval survivorship increases with log diameter (Akbulut et al. 2004; Koutroumpa et al. 2008) and that the first larva to colonize a particular area of a log tends to have a higher probability of surviving encounters with other larvae that are tunneling nearby, especially when the neighboring larva is younger (Anbutsu and Togashi 1997). For twig- and branch-infesting cerambycids, premature branch breakage and subsequent early drying of the host tissues can lead to high larval mortality (Itami and Craig 1989). Shibata (2000)

TABLE 2.4

Summary Data for Adult Longevity under Field or Laboratory Conditions for Selected Cerambycids

Species	Subfamily[a]	Country	Hosts Genera in Study	Study Details	Longevity (Days)		References
					Mean	Range	
Acalolepta vastator (Newman)	Lam	Australia	Vitis	Field, male	42	20–103	Goodwin and Pettit 1994
				Female	47	22–131	
Anoplophora chinensis	Lam	Japan	Citrus	Females caged at ambient conditions	78	47–109	Adachi 1988
Anoplophora glabripennis (Motschulsky)	Lam	USA	Acer	25°C, male	99–106	–	Keena 2002
				Female	73–88	–	
Anoplophora glabripennis	Lam	USA	Acer, Salix	22–25°C, adult food			Smith et al. 2002
				Acer platanoides	104	44–131	
				Acer rubrum	97	30–137	
				Salix nigra	83	58–107	
Anoplophora glabripennis	Lam	USA	Acer	−1°C (♂, ♀)	19, 21	–	Keena 2006
				5°C (♂, ♀)	42, 44	–	
				10°C (♂, ♀)	145, 136	–	
				15°C (♂, ♀)	102, 76	–	
				20°C (♂, ♀)	128, 85	–	
				25°C (♂, ♀)	98, 79	–	
				30°C (♂, ♀)	57, 56	–	
				35°C (♂, ♀)	19, 21	–	
Apriona garmari Hope	Lam	Korea	Morus	25°C (♂, ♀)	44, 41	–	Yoon and Mah 1999
Ataxia hubbardi Fisher	Lam	USA	Glycine	26°C (♂, ♀)	87, 67	–	Rogers and Serda 1979
Callidiellum rufipenne (Motschulsky)	Cer	Japan	Cryptomeria	Ambient (♂, ♀)	18, 17	–	Shibata 1994
Enaphalodes rufulus (Haldeman)	Cer	USA	Quercus	20°C	21	–	Galford 1985
Dorcadion fulginator (L.)	Lam	Border of Switzerland, Germany and France	Bromus erectus and others	Field mark–recapture study, longevity estimated from results	11	–	Baur et al. 2005
Glenea cantor (F.)	Lam	China	Bombax	25°C (♂, ♀)	47, 72	–	Lu et al. 2011
Megacyllene robiniae (Forster)	Cer	USA	Robinia	27°C	34	14–55	Wollerman et al. 1969
Monochamus alternatus Hope	Lam	USA	Pinus	28°C, mated unmated	180	–	Zhang and Linit 1998
					207	–	
Monochamus carolinensis (Olivier)	Lam	USA	Pinus	28°C, mated unmated	173	–	Zhang and Linit 1998
					103	–	

(Continued)

TABLE 2.4 (Continued)

Summary Data for Adult Longevity under Field or Laboratory Conditions for Selected Cerambycids

Species	Subfamily[a]	Country	Hosts Genera in Study	Study Details	Longevity (Days) Mean	Longevity (Days) Range	References
Monochamus galloprovincialis (Olivier)	Lam	Turkey	*Pinus*	24–26°C, larval food			Akbulut et al. 2007
				Spring-cut logs	41	–	
				Summer-cut logs	18	–	
				Fall-cut logs	39	–	
Monochamus galloprovincialis	Lam	Portugal	*Pinus*	25°C, males	61	5–128	Naves et al. 2006
				25°C, females	64	3–125	
Monochamus galloprovincialis	Lam	France	*Pinus*	23°C, females		75–113	Koutroumpa et al. 2008
Monochamus leuconotus (Pascoe)	Lam	South Africa	*Coffea*	Field, male	112	–	Schoeman et al. 1998
				Field, female	122	–	
Monochamus saltuarius (Gebler)	Lam	Korea	*Pinus*	23–27°C			Yoon et al. 2011
				Fed current year twigs (♂, ♀)	63, 58	–	
				Fed 1-yr-old twigs	46, 42	–	
				Fed 2-yr-old twigs	40, 36	–	
Neoptychodes trilineatus (L.)	Lam	USA	*Ficus*	Field	115	75–213	Horton 1917
Oemona hirta (F.)	Cer	New Zealand	*Populus*	Lab (♂, ♀)	30–50, 36–52	–	Wang et al. 2002
				Field (♂, ♀)	52, 33	–	
Phoracantha recurva Newman, *Phoracantha semipunctata* (F.)	Cer	USA	*Eucalyptus*	10°C	~60, 65[b]	–	Bybee et al. 2004a
				15°C	~160, 130	–	
				20°C	~220, 120	–	
				25°C	~130, 100	–	
Semanotus japonicus Lacordaire	Cer	Japan	*Cryptomeria*	20–22°C (♀)	19–20	–	Shibata 1995
Xylotrechus arvicola (Olivier)	Cer	Spain	*Vitis*	24°C, larval collection site: Field	♀ 24	–	García-Ruiz et al. 2012
				Artificial diet	♀ 37	–	
Xylotrechus pyrrhoderus Bates	Cer	Japan	*Vitis*	25°C, males	18	–	Iwabuchi 1988
				25°C, females	20	–	
Xylotrechus quadripes Chevrolat	Cer	Thailand	*Coffea*	Field, male	24	7–46	Visitpanich 1994
				Field, female	29	81–53	

[a] Cer = Cerambycinae, Lam = Lamiinae.

[b] First value is for *P. recurva*; second value is for *P. semipunctata*.

concluded that high resin flow in the conifer *Cryptomeria japonica* D. Don was responsible for much of the early larval mortality in the cerambycine *S. japonicus* based on experiments where newly hatched larvae were introduced into host plants that had undergone various degrees of pruning and girdling to alter resin flow.

Population outbreaks have been reported for a number of cerambycid species. Some outbreaks have been studied at the stand level, such as a single outbreak of *S. japonicus* that occurred over a 10-year study period in Japan (Ito and Kobayashi 1991). By contrast, during the late 1990s and early 2000s, there was a regional outbreak in the United States of *E. rufulus* in the *Quercus*-dominated forests of the Ozark Mountains in Arkansas (Stephen et al. 2001; Riggins et al. 2009). Populations of this cerambycid, which has a highly synchronous two-year life cycle, peaked in 1999–2003 and then collapsed during the 2003–2007 generations (Riggins et al. 2009; Haavik et al. 2010). For example, population estimates of *E. rufulus* on a per tree basis peaked at an average of 174 borers per tree in 2001 and then fell to an average of 32 in 2003, 2 in 2005, and 1 in 2007 (Stephen et al. 2001; Riggins et al. 2009). No specific stress event was ever identified as the causal agent for triggering the *E. rufulus* outbreak, but rather the outbreak appears linked to an abundance of even-aged, overmature, and densely stocked forests in the Ozark Mountains (Riggins et al. 2009).

Another example of long-term population fluctuations in a cerambycid was described by Haack (2012) for the cerambycine *Anelaphus parallelus* (Newman) in Michigan. This *Quercus*-infesting, twig-pruning beetle has a highly synchronous two-year life cycle in Michigan with adults emerging primarily in odd-numbered years and twigs falling to the ground in even-numbered years (Gosling 1978, 1981; Figure 2.10). From 1990 to 2011, observations were made at nearly one- to two-week intervals on the occurrence of fallen twigs along the same 1-km-long section of trail in a natural hardwood forest (Haack 2012). However, accurate counts of the number of fallen shoots that were infested by *A. parallelus* did

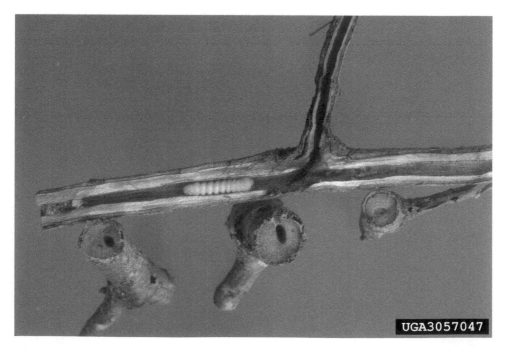

FIGURE 2.10 Larva, larval gallery, and typical pruning damage made by larvae of the cerambycine *Anelaphus parallelus* (Newman) on its most common host, *Quercus*, in the eastern United States. This species has a two-year life cycle in which a single egg is laid on a small twig with the resulting larva tunneling downward to the adjoining larger branch during its first summer. In the second summer, the larva extends the gallery and eventually consumes a disc of wood, leaving only the bark intact, and plugs the feeding hole with wood fibers. Later, the "pruned" branch breaks and falls to the ground; therein, the larva completes development and emerges as an adult the following spring. (Courtesy of James Solomon (Bugwood image 3057047].)

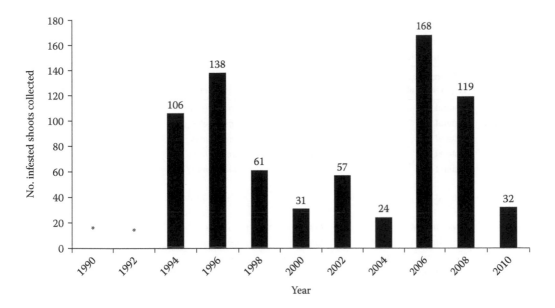

FIGURE 2.11 Number of *Anelaphus parallelus*-infested twigs collected along the same 1-km-long forest trail near Dansville, Michigan, at annual intervals from 1990 to 2011 (Haack 2012; R. A. Haack, unpublished data). Accurate counts were not made in 1990 or 1992, although newly fallen infested twigs were present in both years (signified as *). No newly fallen, infested twigs were found during any of the odd-numbered years from 1991 to 2011. *Anelaphus parallelus* has a two-year life cycle in Michigan, with twigs falling to the ground (after being partially severed by the larvae) in even-numbered years and the adults emerging from the fallen twigs in odd-numbered years.

not begin until 1994. Based on the data shown in Figure 2.11 (R. A. Haack, unpublished data), it is clear that *A. parallelus* has a two-year life cycle in Michigan and that local populations can vary widely from generation to generation as indicated by as many as 168 infested twigs being counted in 2006 to as few as 32 twigs in 2010 along the same transect.

As mentioned earlier, many species of cerambycids commonly are associated with stressed trees. Several examples of situations where cerambycids infested trees that had been impacted by air pollution, defoliation, drought, fire, ice storms, or wind storms are listed in Table 2.5. In many of these reports, bark beetles (Scolytinae) infested the same trees either before or concurrently with the cerambycids—see Zabecki (1988) for an air pollution example; Basham and Belyea (1960) and Mamaev (1990) for defoliation; Wermelinger et al. (2008) for drought; Kimmey and Furniss (1943), Gardiner (1957), Zhang et al. (1993), and Saint-Germain et al. (2004) for fire; Ryall and Smith (2001) for ice storms; and Connola et al. (1953), Gardiner (1975), and Gandhi et al. (2009) for wind storms. By contrast, in the air pollution gradient study reported by Haack (1996) in the central United States, no major *Quercus*-infesting bark beetle species were found in living *Quercus* trees infested with cerambycids in the cerambycine genus *Enaphalodes* and the lamiine genus *Goes*, both of which infest living oak (*Quercus*) trees (Solomon 1995). This was not unexpected given that there are no major tree-killing bark beetles that infest stressed oaks in North America (Solomon 1995)—unlike the situation in Europe with the oak-infesting bark beetle *Scolytus intricatus* (Ratzeburg) (Yates 1984).

2.10 Final Note

Given the thousands of cerambycid species found worldwide, it is not surprising that these insects successfully occupy a diverse array of habitats and are capable of developing within a wide assortment of plant species and their parts and tissues. Nevertheless, basic life history information, including larval host records, still is lacking for most cerambycids therefore demanding the continued study of this family of fascinating and economically important beetles.

TABLE 2.5

Examples of where Tree-Infesting Cerambycids Populations Increased after Various Stress Events that were either Natural or Induced Experimentally

Stress Event	Natural or Experimental	Country	Major Tree Genera	Major Cerambycid Genera	References
Air pollution	Natural	Poland	*Abies*	*Clytus, Obrium*	Zabecki 1988
Air pollution	Natural	USA	*Quercus*	*Enaphalodes, Goes*	Haack 1996
Air pollution	Natural	Russia	*Abies, Pinus*	*Monochamus*	Isaev et al. 1988
Defoliation	Natural	Russia	*Larix*	*Acanthocinus, Tetropium*	Mamaev 1990
Defoliation	Natural	Canada	*Abies*	*Monochamus, Tetropium*	Basham and Belyea 1960
Drought	Natural	Switzerland	*Pinus*	*Acanthocinus*	Wermelinger et al. 2008
Drought	Experimental	Mexico	*Prosopis*	*Oncideres*	Martínez et al. 2009
Drought	Experimental	USA	*Eucalyptus*	*Phoracantha*	Hanks et al. 1999
Drought	Experimental	Portugal	*Eucalyptus*	*Phoracantha*	Caldeira et al. 2002
Fire	Natural	Canada	*Pinus*	*Acanthocinus, Acmaeops, Asemum, Monochamus, Rhagium, Xylotrechus*	Gardiner 1957
Fire	Natural	USA	*Abies, Larix, Picea, Pinus*	*Acanthocinus, Anoplodera, Astylopsis, Callidium, Monochamus, Pogonocherus, Rhagium*	Parmelee 1941
Fire	Natural	USA	*Pinus*	*Acanthocinus, Monochamus, Rhagium, Stictoleptura*	Costello et al. 2011, 2013
Fire	Natural	USA	*Pseudotsuga*	*Arhopalus, Asemum, Ergates, Neoclytus, Xylotrechus*	Kimmey and Furniss 1943
Fire	Natural	Canada	*Picea*	*Monochamus*	Saint-Germain et al. 2004; Cobb et al. 2010
Fire	Natural	China	*Larix, Pinus*	*Monochamus*	Zhang et al. 1993
Ice storm	Natural	Canada	*Pinus*	*Acanthocinus, Monochamus, Rhagium*	Ryall and Smith 2001; Ryall 2003
Wind storm	Natural	USA	*Picea, Pinus*	*Callidium, Monochamus, Tetropium*	Connola et al. 1953
Wind storm	Natural	Canada	*Picea, Pinus*	*Monochamus, Tetropium*	Gardiner 1975
Wind storm	Natural	USA	*Pinus*	*Monochamus*	Gandhi et al. 2009
Wind storm	Natural	USA	*Pinus*	*Monochamus*	Webb 1909

ACKNOWLEDGMENTS

The authors thank Toby R. Petrice and Vicente Sánchez for comments on an earlier draft of this chapter and the many photographers who supplied images to the Bugwood website at the University of Georgia.

REFERENCES

Adachi, I. 1988. Reproductive biology of the white-spotted longicorn beetle, *Anoplophora malasiaca* Thomson (Coleoptera: Cerambycidae), in citrus trees. *Applied Entomology and Zoology* 23: 256–264.

Adachi, I. 1994. Development and life cycle of *Anoplophora malasiaca* (Thomson) (Coleoptera: Cerambycidae) on citrus trees under fluctuating and constant temperature regimes. *Applied Entomology and Zoology* 29: 485–497.

Akbulut, S., A. Keten, I. Baysal, and B. Yuksel. 2007. The effect of log seasonality on the reproductive potential of *Monochamus galloprovincialis* Olivier (Coleoptera: Cerambycidae) reared in black pine logs under laboratory conditions. *Turkish Journal of Agriculture and Forestry* 31: 413–422.

Akbulut, S., and M. J. Linit. 1999. Flight performance of *Monochamus carolinensis* (Coleoptera: Cerambycidae) with respect to nematode phoresis and beetle characteristics. *Environmental Entomology* 28: 1014–1020.

Akbulut, S., and W. T. Stamps. 2012. Insect vectors of the pinewood nematode: A review of the biology and ecology of *Monochamus* species. *Forest Pathology* 42: 89–99.

Akbulut, S., W. T. Stamps, and M. J. Linit. 2004. Population dynamics of *Monochamus carolinensis* (Col., Cerambycidae) under laboratory conditions. *Journal of Applied Entomology* 128: 17–21.

Alstad, D. N., G. F. Edmunds, Jr., and L. H. Weinstein. 1982. Effects of air pollutants on insect populations. *Annual Review of Entomology* 27: 369–384.

Anbutsu, H., and K. Togashi. 1997. Effects of spatio-temporal intervals between newly-hatched larvae on larval survival and development in *Monochamus alternatus* (Coleoptera: Cerambycidae). *Researches on Population Ecology* 39: 181–189.

Asano, W., F. N. Munyiri, Y. Shintani, and Y. Ishikawa. 2004. Interactive effects of photoperiod and temperature on diapause induction and termination in the yellowspotted longicorn beetle, *Psacothea hilaris*. *Physiological Entomology* 29: 458–463.

Banno, H., and A. Yamagami. 1991. Life cycle and larval survival rate of the redspotted longicorn beetle, *Eupromus ruber* (Dalman) (Coleoptera: Cerambycidae). *Applied Entomology and Zoology* 26: 195–204.

Basham, J. T., and R. M. Belyea. 1960. Death and deterioration of balsam fir weakened by spruce budworm defoliation in Ontario, Part III: The deterioration of dead trees. *Forest Science* 6: 78–96.

Baur, B., A. Coray, N. Minoretti, and S. Zschokke. 2005. Dispersal of the endangered flightless beetle *Dorcadion fuliginator* (Coleoptera: Cerambycidae) in spatially realistic landscapes. *Biological Conservation* 124: 49–61.

Beeson, C. F. C., and B. M. Bhatia. 1939. On the biology of the Cerambycidae (Coleopt.). *Indian Forest Records (new series): Entomology* 5: 1–235.

Benham, G. S. 1969. The pupa of *Prionus laticollis* (Coleoptera: Cerambycidae). *Annals of the Entomological Society of America* 62: 1331–1335.

Bense, U. 1995. *Longhorn beetles: Illustrated key to the Cerambycidae and Vesperidae of Europe*. Weikersheim: Margraf Verlag.

Bílý, S., and O. Mehl. 1989. Longhorn beetles (Coleoptera, Cerambycidae) of Fennoscandia and Denmark. *Fauna Entomologica Scandinavica* 22: 1–204.

Blair, K. G. 1934. Cerambycidae from the Society Islands. *B. P. Bishop Museum Bulletin* 113: 127–129.

Bleuzen, P. 1994. *The Beetles of the World, Vol. 21: Macrodontini & Prionini 1*. Venette: Sciences Naturelles.

Butovitsch, V. 1939. Zur Kenntnis der Paarung, Eiablage und Ernährung der Cerambyciden. *Entomologisk Tidskrift* 60: 206–258.

Bybee, L. F., J. G. Millar, T. D. Paine, and K. Campbell. 2004a. Effects of temperature on fecundity and longevity of *Phoracantha recurva* and *P. semipunctata* (Coleoptera: Cerambycidae). *Environmental Entomology* 33: 138–146.

Bybee, L. F., J. G. Millar, T. D. Paine, K. Campbell, and C. C. Hanlon. 2004b. Seasonal development of *Phoracantha recurva* and *P. semipunctata* (Coleoptera: Cerambycidae) in southern California. *Environmental Entomology* 33: 1232–1241.

Caldeira, M. C., V. Fernandéz, J. Tomé, and J. S. Pereira. 2002. Positive effect of drought on longicorn borer larval survival and growth on eucalyptus trunks. *Annals of Forest Science* 59: 99–106.

Cannon, K. F., and W. H. Robinson. 1982. An artificial diet for the laboratory rearing of the old house borer, *Hylotrupes bajulus* (Coleoptera, Cerambycidae). *Canadian Entomologist* 114: 739–742.

Chararas, C. 1969. Biologie et écologie de *Phoracantha semipunctata* F. (Coléoptère Cerambycidae xylophage) ravageur des Eucalyptus en Tunisie, et méthodes de protection des peuplements. *Annales de l'Institut National de Recherches Forestières de Tunisie* 2: 1–37.

Cleare, L. D. 1931. The egg-plant stem borer, *Alcidion deletum* Bates (Col. Cerambycidae). *The Agricultural Journal of British Guiana* 4: 82–90.

Cobb, T. P., K. D. Hannam, B. E. Kishchuk, D. W. Langor, S. A. Quideauand, and J. R. Spence. 2010. Wood-feeding beetles and soil nutrient cycling in burned forests: Implications of post-fire salvage logging. *Agricultural and Forest Entomology* 12: 9–18.

Cocquempot, C., and A. Lindelöw. 2010. Longhorn beetles (Coleoptera, Cerambycidae). *BioRisk* 4: 193–218.

Connola, D. P., C. J. Yops, J. A. Wilcox, and D. L. Collins. 1953. Survey and control studies of beetles attacking windthrown trees in the Adirondacks. *Journal of Economic Entomology* 46: 249–254.

Costello, S. L., W. R. Jacobi, and J. F. Negrón. 2013. Emergence of Buprestidae, Cerambycidae, and Scolytinae (Coleoptera) from mountain pine beetle-killed and fire-killed ponderosa pines in the Black Hills, South Dakota, USA. *The Coleopterists Bulletin* 67: 149–154.

Costello, S. L., J. F. Negron, and W. R. Jacobi. 2011. Wood-boring insect abundance in fire-injured ponderosa pine. *Agricultural and Forest Entomology* 13: 373–381.

Coulson, R. N. 1979. Population dynamics of bark beetles. *Annual Review of Entomology* 24: 417–447.

Craighead, F. C. 1923. North American cerambycid-larvae. *Bulletin of the Canada Department of Agriculture* 27: 1–239.

David, G., B. Giffard, D. Piou, and H. Jactel. 2014. Dispersal capacity of *Monochamus galloprovincialis*, the European vector of the pine wood nematode, on flight mills. *Journal of Applied Entomology* 138: 566–576.

Dodds, K. J., C. Graber, and F. M. Stephen. 2001. Facultative intraguild predation by larval Cerambycidae (Coleoptera) on bark beetle larvae (Coleoptera: Scolytidae). *Environmental Entomology* 30: 17–22.

Dodds, K. J., C. Graber, and F. M. Stephen. 2002. Oviposition biology of *Acanthocinus nodosus* (Coleoptera: Cerambycidae) in *Pinus taeda*. *Florida Entomologist* 85: 452–457.

Donley, D. E. 1978. Oviposition by the red oak borer, *Enaphalodes rufulus* Coleoptera: Cerambycidae. *Annals of the Entomological Society of America* 71: 496–498.

Drag, L., D. Hauck, P. Pokluda, K. Zimmermann, and L. Cizek. 2011. Demography and dispersal ability of a threatened saproxylic beetle: A mark-recapture study of the *Rosalia* longicorn (*Rosalia alpina*). *PLoS ONE* 6(6): e21345. doi: 10.1371/journal.pone.0021345

Drinkwater, T. W. 1975. The present pest status of eucalyptus borers *Phoracantha* spp. in South Africa. In *Proceedings of the First Congress of the Entomological Society of Southern Africa*, eds. H. J. R. Durr, J. H. Giliomee, and S. Neser, 119–129. Pretoria: Entomological Society of Southern Africa.

Duffy, E. A. J. 1949. A contribution towards the biology of *Aromia moschata* L., the musk beetle. *Proceedings and Transactions of the South London Entomological and Natural History Society* 1947–1948: 82–110.

Duffy, E. A. J. 1953. *A monograph of the immature stages of African timber beetles (Cerambycidae)*. London: British Museum (Natural History).

Eisner, T., F. C. Schroeder, N. Snyder, et al. 2008. Defensive chemistry of lycid beetles and of mimetic cerambycid beetles that feed on them. *Chemoecology* 18: 109–119.

Esaki, K. 2001. Artificial diet rearing and termination of larval diapause in the mulberry longicorn beetle, *Apriona japonica* Thomson (Coleoptera: Cerambycidae). *Japanese Journal of Applied Entomology and Zoology* 45: 149–151.

Faccoli, M., R. Favaro, M. T. Smith, and J. Wu. 2015. Life history of the Asian longhorn beetle *Anoplophora glabripennis* (Coleoptera: Cerambycidae) in southern Europe. *Agricultural and Forest Entomology*. 17: 188–196.

Farrar, R. J., and T. W. Kerr. 1968. A preliminary study of the life history of the broad-necked root borer in Rhode Island. *Journal of Economic Entomology* 61: 563–564.

Forcella, F. 1982. Why twig-girdling beetles girdle twigs. *Naturwissenschaften* 69: 398–400.

Galford, J. R. 1984. *The locust borer*. Washington, DC: U.S. Department of Agriculture, Forest Service, Forest Insect & Disease Leaflet 71.

Galford, J. R. 1985. *Enaphalodes rufulus*. In *Handbook of Insect Rearing, Vol. 1*, eds. P. Singh, and R. F. Moore, 255–264. New York: Elsevier.

Gallego, D., F. J. Sanchez-Garcia, H. Mas, M. T. Campo, and Y. J. L. Lencina. 2012. Estudio de la capacidad de vuelo a larga distancia de *Monochamus galloprovincialis* (Olivier 1795). (Coleoptera: Cerambycidae) en un mosaico agro-forestal. *Boletín de Sanidad Vegetal Plagas* 38: 109–123.

Gandhi, K. J. K., D. W. Gilmore, S. A. Katovich, W. J. Mattson, J. R. Spence, and S. J. Seybold. 2007. Physical effects of weather events on the abundance and diversity of insects in North American forests. *Environmental Review* 15: 113–152.

Gandhi, K. J. K., D. W. Gilmore, R. A. Haack, et al. 2009. Application of semiochemicals to assess the biodiversity of subcortical insects following an ecosystem disturbance in a sub-boreal forest. *Journal of Chemical Ecology* 35: 1384–1410.

García-Ruiz, E., V. Marco, and I. Pérez-Moreno. 2011. Effects of variable and constant temperatures on the embryonic development and survival of a new grape pest, *Xylotrechus arvicola* (Coleoptera: Cerambycidae). *Environmental Entomology* 40: 939–947.

García-Ruiz, E., V. Marco, and I. Perez-Moreno. 2012. Laboratory rearing and life history of an emerging grape pest, *Xylotrechus arvicola* (Coleoptera: Cerambycidae). *Bulletin of Entomological Research* 102: 89–96.

Gardiner, K. M. 1957. Deterioration of firekilled pine in Ontario and the causal wood-boring beetles. *Canadian Entomologist* 89: 241–263.

Gardiner, L. M. 1960. Descriptions of immature forms and biology of *Xylotrechus colonus* (Fab.) (Coleoptera: Cerambycidae). *Canadian Entomologist* 92: 820–825.

Gardiner, L. M. 1966. Egg bursters and hatching in the Cerambycidae (Coleoptera). *Canadian Journal of Zoology* 44: 199–212.

Gardiner, L. M. 1970. Rearing wood-boring beetles (Cerambycidae) on artificial diet. *Canadian Entomologist* 102: 113–117.

Gardiner, L. M. 1975. Insect attack and value loss in wind-damaged spruce and jack pine stands in northern Ontario. *Canadian Journal of Forest Research* 5: 387–398.

Goodwin, S., and M. A. Pettit. 1994. *Acalolepta vastator* (Newman) (Coleoptera: Cerambycidae) infesting grapevines in the Hunter Valley, New South Wales. Biology and ecology. *Journal of the Australian Entomological Society* 33: 391–397.

Gosling, D. C. L. 1978. Observations on the biology of the oak twig pruner, *Elaphidionoides parallelus*, (Coloeoptera: Cerambycidae) in Michigan. *The Great Lakes Entomologist* 11: 1–10.

Gosling, D. C. L. 1981. Correct identity of the oak twig pruner (Coloeoptera: Cerambycidae). *The Great Lakes Entomologist* 14: 179–180.

Grimble, D. G., and F. B. Knight. 1970. Life tables and mortality factors for *Saperda inornata* (Coleoptera: Cerambycidae). *Annals of the Entomological Society of America* 63: 1309–1319.

Grimble, D. G., and F. B. Knight. 1971. Mortality factors for *Oberea schaumii* (Coleoptera: Cerambycidae). *Annals of the Entomological Society of America* 64: 1417–1420.

Grzymala, T. L., and K. B. Miller. 2013. Taxonomic revision and phylogenetic analysis of the genus *Elytroleptus* Dugés (Coleoptera: Cerambycidae: Cerambycinae: Trachyderini). *Zootaxa* 3659: 1–62.

Gupta, R., and J. S. Tara. 2013. First record on the biology of *Aeolesthes holosericea* Fabricius, 1787 (Coleoptera: Cerambycidae), an important pest on apple plantations (*Malus domestica* Borkh.) in India. *Munis Entomology & Zoology* 8: 243–251.

Haack, R. A. 1996. Patterns of forest invertebrates along an acid deposition gradient in the midwestern United States. In *Air pollution & multiple stresses, Proceedings of the IUFRO 2.05 Conference, 7–9 September 1994, Fredericton, New Brunswick*, eds. R. Cox, K. Percy, K. Jensen, and C. Simpson, 245–257. Fredericton: Canadian Forest Service.

Haack, R. A. 2006. Exotic bark- and wood-boring Coleoptera in the United States: Recent establishments and interceptions. *Canadian Journal of Forest Research* 36: 269–288.

Haack, R. A. 2012. Seasonality of oak twig pruner shoot fall: A long-term dog walking study. *Newsletter of the Michigan Entomological Society* 57: 24.

Haack, R. A., L. S. Bauer, R.-T. Gao, et al. 2006. *Anoplophora glabripennis* within-tree distribution, seasonal development, and host suitability in China and Chicago. *The Great Lakes Entomologist* 39: 169–183.

Haack, R. A., F. Hérard, J. Sun, and J. J. Turgeon. 2010a. Managing invasive populations of Asian long-horned beetle and citrus longhorned beetle: A worldwide perspective. *Annual Review of Entomology* 55: 521–546.

Haack, R. A., T. R. Petrice, and A. C. Wiedenhoeft. 2010b. Incidence of bark- and wood-boring insects in firewood: A survey at Michigan's Mackinac Bridge. *Journal of Economic Entomology* 103: 1682–1692.

Haack, R. A., K. O. Britton, E. G. Brockerhoff, et al. 2014. Effectiveness of the international phytosanitary standard ISPM No. 15 on reducing wood borer infestation rates in wood packaging material entering the United States. *PLoS ONE* 9(5): e96611. doi: 10.1371/journal.pone.0096611

Haavik, L. J., M. K. Fierke, and F. M. Stephen. 2010. Factors affecting suitability of *Quercus rubra* as hosts for *Enaphalodes rufulus* (Coleoptera: Cerambycidae). *Environmental Entomology* 39: 520–527.

Haavik, L. J., D. J. Crook, M. K. Fierke, L. D. Galligan, and F. M. Stephen. 2012. Partial life tables from three generations of *Enaphalodes rufulus* (Coleoptera: Cerambycidae). *Environmental Entomology* 41: 1311–1321.

Hamilton, J. 1884. Notes on a few species of Coleoptera which are confused in many collections, and on some introduced European species. *Canadian Entomologist* 16: 35–38.

Hanks, L. M. 1999. Influence of the larval host plant on reproductive strategies of cerambycid beetles. *Annual Review of Entomology* 44: 483–505.

Hanks, L. M., J. S. Mcelfresh, J. G. Millar, and T. D. Paine. 1993. *Phoracantha semipunctata* (Coleoptera: Cerambycidae), a serious pest of eucalyptus in California: Biology and laboratory-rearing procedures. *Annals of the Entomological Society of America* 86: 96–102.

Hanks, L. M., J. G. Millar, and T. D. Paine. 1998. Dispersal of the eucalyptus longhorned borer (Coleoptera: Cerambycidae) in urban landscapes. *Environmental Entomology* 27: 1418–1424.

Hanks, L. M., T. D. Paine, J. G. Millar, C. D. Campbell, and U. K. Schuch. 1999. Water relations of host trees and resistance to the phloem-boring beetle *Phoracantha semipunctata* F. (Coleoptera: Cerambycidae). *Oecologia* 119: 400–407.

Hatchett, J. H., R. D. Jackson, R. M. Barry, and E. C. Houser. 1973. Rearing a weed cerambycid, *Dectes texanus*, on an artificial medium, with notes on biology. *Annals of the Entomological Society of America* 66: 519–522.

Hay, C. J. 1968. Frass of some wood-boring insects in living oak (Coleoptera: Cerambycidae: Lepidoptera: Cossidae and Aegeriidae). *Annals of the Entomological Society of America* 61: 255–258.

Hernández, R., A. Ortiz, V. Pérez, J. M. Gil, and G. Sánchez. 2011. *Monochamus galloprovincialis* (Olivier, 1975) (Coleoptera: Cerambycidae), comportamiento y distancias de vuelo. *Boletín de Sanidad Vegetal Plagas* 37: 79–96.

Hickin, N. E. 1975. *The insect factor in wood decay* (3rd ed.). London: Hutchinson.

Horton, J. R. 1917. Three-lined fig-tree borer. *Journal of Agricultural Research* 11: 371–382.

Hosking, G. P., and J. Bain. 1977. *Arhopalus ferus* (Coleoptera: Cerambycidae): Its biology in New Zealand. *New Zealand Journal of Forestry Science* 7: 3–15.

Hovey, C. L. 1941. Studies on *Chion cinctus* (Drury) (Coleoptera, Cerambycidae) in Oklahoma. *Proceedings of the Oklahoma Academy of Science* 21: 23–24.

Hu, S.-J., T. Ning, D.-Y. Fu, et al. 2013. Dispersal of the Japanese pine sawyer, *Monochamus alternatus* (Coleoptera: Cerambycidae), in mainland China as inferred from molecular data and associations to indices of human activity. *PLoS ONE* 8(2): e57568. doi: 10.1371/journal.pone.0057568

Hussain, A., and A. A. Buhroo. 2012. On the biology of *Apriona germari* Hope (Coleoptera: Cerambycidae) infesting mulberry plants in Jammu and Kashmir, India. *Nature and Science* 10: 24–35.

Isaev, A. S., A. S. Rozhkov, and V. V. Kiselev. 1988. *Fir sawyer beetle* Monochamus urussovi *(Fisch.)*. Novosibirsk: Nauka Publishing House.

Itami, J. K., and T. P. Craig. 1989. Life History of *Styloxus bicolor* (Coleoptera: Cerambycidae) on *Juniperus monosperma* in northern Arizona. *Annals of the Entomological Society of America* 82: 582–587.

Ito, K. 1999. Differential host residence of adult cryptomeria bark borer, *Semanotus japonicus* Lacordaire (Coleoptera: Cerambycidae), in relation to tree size of Japanese cedar, *Cryptomeria japonica* D. Don. *Journal of Forest Research* 4: 151–156.

Ito, K., and K. Kobayashi. 1991. An outbreak of the cryptomeria bark borer, *Semanotus japonicus* Lacordaire (Coleoptera: Cerambycidae) in a young Japanese cedar (*Cryptomeria japonica* D. Don) plantation. I. Annual fluctuations in adult population size and impact on host trees. *Applied Entomology and Zoology* 26: 63–70.

Iwabuchi, K. 1988. Mating behavior of *Xylotrechus pyrrhoderus* Bates (Coleoptera: Cerambycidae). IV. Mating frequency, fecundity, fertility and longevity. *Applied Entomology and Zoology* 23: 127–134.

Jaques, H. E. 1918. A long-lifed woodboring beetle. *Proceedings of the Iowa Academy of Science* 25: 175.

Jikumaru, S., and K. Togashi. 1996. Effect of temperature on the post-diapause development of *Monochamus saltuarius* (Gebler) (Coleoptera: Cerambycidae). *Applied Entomology and Zoology* 31: 145–148.

Keena, M. A. 2002. *Anoplophora glabripennis* (Coleoptera: Cerambycidae) fecundity and longevity under laboratory conditions: Comparison of populations from New York and Illinois on *Acer saccharum*. *Environmental Entomology* 31: 490–498.

Keena, M. A. 2006. Effects of temperature on *Anoplophora glabripennis* (Coleoptera: Cerambycidae) adult survival, reproduction, and egg hatch. *Environmental Entomology* 35: 912–921.

Keena, M. A., and P. M. Moore. 2010. Effects of temperature on *Anoplophora glabripennis* (Coleoptera: Cerambycidae) larvae and pupae. *Environmental Entomology* 39: 1323–1335.

Khan, A. R., and A. W. Khan. 1942. Bionomics and control of *Aeolesthes holosericea* F. (Cerambycidae: Coleoptera). *Proceedings of the Indian Academy of Sciences, Section B* 15: 181–185.

Kimmey, J. W, and R. L. Furniss. 1943. *Deterioration of fire-killed Douglas-fir*. Washington, DC: U.S. Department of Agriculture Technical Bulletin 851.

Kitajima, H., and M. Igarashi. 1997. Rearing of the cryptomeria bark borer, *Semanotus japonicas* (Coleoptera: Cerambycidae) larvae on bolts of the Japanese cedar, *Cryptomeria japonica* and termination of adult diapause by the low temperature treatments. *Journal of the Japanese Forestry Society* 79: 186–190.

Kobayashi, F., A. Yamane, and T. Ikeda. 1984. The Japanese pine sawyer beetle as the vector of pine wilt disease. *Annual Review of Entomology* 29: 115–135.

Komazaki, S., and Y. Sakagami. 1989. Capture-recapture study on the adult population of the white spotted longicorn beetle, *Anoplophora malasiaca* (Thomson) (Coleoptera: Cerambycidae), in a citrus orchard. *Applied Entomology and Zoology* 24: 78–84.

Koutroumpa, F. A., B. Vincent, G. Roux-Morabito, C. Martin, and F. Lieutier. 2008. Fecundity and larval development of *Monochamus galloprovincialis*. *Annals of Forest Science* 65: 707. doi: 10.1051/forest:2008056

Kulman, H. M. 1971. Effects of insect defoliation on growth and mortality of trees. *Annual Review of Entomology* 16: 289–324.

Lara, E. F., and R. D. Shenefelt. 1966. *Colobothea distincta* (Coleoptera: Cerambycidae) on cacao: Notes on its morphology and biology. *Annals of the Entomological Society of America* 59: 453–458.

Lee, C. Y., and K. C. Lo. 1998. Rearing of *Anoplophora macularia* (Thomson) (Coleoptera: Cerambycidae) on artificial diets. *Applied Entomology and Zoology* 33: 105–109.

Lingafelter, S. W. 2003. New host and elevation records for *Moneilema appressum* LeConte (Coleoptera: Cerambycidae: Lamiinae). *Journal of the New York Entomological Society* 111: 57–60.

Linsley, E. G. 1936. Hibernation in the Cerambycidae. *Pan-Pacific Entomologist* 12: 119.

Linsley, E. G. 1959. Ecology of Cerambycidae. *Annual Review of Entomology* 4: 99–138.

Linsley, E. G. 1961. The Cerambycidae of North America. Part I: Introduction. *University of California Publications in Entomology* 18: 1–135.

Linsley, E. G., and J. A. Chemsak. 1972. Cerambycidae of North America. Part VI, No. 1: Taxonomy and classification of the subfamily Lepturinae. *University of California Publications in Entomology* 69: 1–138.

Linsley, E. G., and J. A. Chemsak. 1995. The Cerambycidae of North America. Part VII, No. 2: Taxonomy and classification of the subfamily Lamiinae, tribes Acanthocinini through Hemilophini. *University of California Publications in Entomology* 114: 1–292.

Logarzo, G. A., and D. E. Gandolfo. 2005. Análisis de voltinismo y la diapausa en poblaciones de *Apagomerella versicolor* (Coleoptera: Cerambycidae) en el gradiente latitudinal de su distribución en la Argentina. *Revista de la Sociedad Entomológica Argentina* 64: 143–146.

Löyttyniemi, K. 1983. Flight pattern and voltinism of *Phoracantha* beetles (Coleoptera: Cerambycidae) in a semihumid tropical climate in Zambia. *Annales Entomologici Fennici* 49: 49–53.

Lu, W., Q. Wang, M.-Y. Tian, J. Xu, and A.-Z. Qin. 2011. Phenology and laboratory rearing procedures of an Asian longicorn beetle, *Glenea cantor* (Coleoptera: Cerambycidae: Lamiinae). *Journal of Economic Entomology* 104: 509–516.

Lu, W., Q. Wang, M.-Y. Tian, et al. 2013. Reproductive traits of *Glenea cantor* (Coleoptera: Cerambycidae: Lamiinae). *Journal of Economic Entomology* 106: 215–220.

Luhring, K. A., J. G. Millar, T. D. Paine, D. Reed, and H. Christiansen. 2004. Ovipositional preferences and progeny development of the egg parasitoid *Avetianella longoi*: Factors mediating replacement of one species by a congener in a shared habitat. *Biological Control* 30: 382–391.

Mamaev, Y. B. 1990. [Outbreaks of stem pests in larch forests of the Tuva ASSR damaged by *Dendrolimus sibiricus.*] *Izvestiya Vysshikh Uchebnykh Zavedenii—Lesnoi Zhurnal* 2: 16–19 [in Russian].

Martínez, A. J., J. López-Portillo, A. Eben, and J. Golubov. 2009. Cerambycid girdling and water stress modify mesquite architecture and reproduction. *Population Ecology* 51: 533–541.

Martínez Egea, J. M. 1982. *Phoracantha semipunctata* Fab. en el suroeste español: Resumen de la campaña de colocacion de arboles cebo. *Boletin de la Estacion Central de Ecologia* 11: 57–69.

Mas, H., R. Hernandez, M. Villaroya, et al. 2013. Comportamiento de dispersión y capacidad de vuelo a larga distancia de *Monochamus galloprovincialis* (Olivier 1795). 6 Congreso forestal español. 6CFE01-393. http://www.congresoforestal.es/actas/doc/6CFE/6CFE01-393.pdf (accessed January 27, 2016).

Matsumoto, K., R. S. B. Irianto, and H. Kitajima. 2000. Biology of the Japanese green-lined Albizzia longicorn, *Xystrocera globosa* (Coleoptera: Cerambycidae). *Entomological Science* 3: 33–42.

Mattson, W. J., and R. A. Haack. 1987. The role of drought in outbreaks of plant-eating insects. *BioScience* 37: 110–118.

McCauley, D. E., J. R. Ott, A. Stine, and S. McGrath. 1981. Limited dispersal and its effect on population structure in the milkweed beetle *Tetraopes tetraophthalmus*. *Oecologia* 51: 145–150.

McCullough, D. G., R. A. Werner, and D. Neumann. 1998. Fire and insects in northern boreal ecosystems of North America. *Annual Review of Entomology* 43: 107–127.

McFadyen, P. J. 1983. Host specificity and biology of *Megacyllene mellyi* [Col.: Cerambycidae] introduced into Australia for the biological control of *Baccharis halimifolia* [Compositae]. *Entomophaga* 28: 65–71.

McFadyen, R. E., and A. P. Fidalgo. 1976. Investigations on *Alcidion cereicola* [Col.: Cerambycidae] a potential agent for the biological control of *Eriocereus martinii* [Cactaceae] in Australia. *Entomophaga* 21: 103–111.

Morgan, F. D. 1960. The comparative biologies of certain New Zealand Cerambycidae. *New Zealand Entomologist* 2: 26–34.

Naves, P., E. de Sousa, and J. A. Quartau. 2006. Reproductive traits of *Monochamus galloprovincialis* (Coleoptera: Cerambycidae) under laboratory conditions. *Bulletin of Entomological Research* 96: 289–294.

Naves, P., E. de Sousa, and J. A. Quartau. 2007. Winter dormancy of the pine sawyer *Monochamus galloprovincialis* (Col., Cerambycidae) in Portugal. *Journal of Applied Entomology* 131: 669–673.

Naves, P. M., and E. M. de Sousa. 2009. Threshold temperatures and degree-day estimates for development of post-dormancy larvae of *Monochamus galloprovincialis* (Coleoptera: Cerambycidae). *Journal of Pest Science* 82: 1–6.

Neville, A. C. 1983. Daily cuticular growth layers and the teneral adults stage in insects: A review. *Journal of Insect Physiology* 29: 211–219.

Niisato, T. 1996. Occurrence of an archaic Molorchine beetle (Coleoptera, Cerambycidae) in western Sichuan, southwest China. *Elytra* (Tokyo) 24: 375–381.

Nord, J. C., D. G. Grimble, and F. B. Knight. 1972. Biology of *Oberea schaumii* (Coleoptera: Cerambycidae) in trembling aspen, *Populus tremuloides*. *Annals of the Entomological Society of America* 65: 114–119.

Park, K. T., and H. R. Paik. 1986. Seasonal fluctuation, reproduction, development and damaging behavior of *Compsidia populnea* L. (Coleoptera: Cerambycidae) on *Populus alba* X *glandulosa*. *Korean Journal of Plant Protection* 24: 195–201.

Parmelee, F. T. 1941. Longhorned and flat headed borers attacking fire killed coniferous timber in Michigan. *Journal of Economic Entomology* 34: 377–380.

Peck, S. B. 2011. The beetles of Martinique, Lesser Antilles (Insecta: Coleoptera); diversity and distributions. *Insecta Mundi* 0178: 1–57.

Pershing, J. C., and M. J. Linit. 1986. Development and seasonal occurrence of *Monochamus carolinensis* (Coleoptera: Cerambycidae) in Missouri. *Environmental Entomology* 15: 251–253.

Pesarini, C., and A. Sabbadini. 1997. Notes on new or poorly known species of Asian Cerambycidae (Insecta, Coleoptera) II Naturalista Valtellinese. *Atti del Museo Civico di Storia Naturale di Morbegno* 7: 95–129.

Powell, W. 1982. Age-specific life-table data for the Eucalyptus boring beetle, *Phoracantha semipunctata* (F.) (Coleoptera: Cerambycidae), in Malawi. *Bulletin of Entomological Research* 72: 645–653.

Rad, H. H. 2006. Study on the biology and distribution of long-horned beetles *Calchaenesthes pistacivora* n. sp. (Col.: Cerambycidae): A new pistachio and wild pistachio pest in Kerman Province. *ISHS Acta Horticulturae* 726: 425–430.

Rassati, D., F. Lieutier, and M. Faccoli. 2016. Alien wood-boring beetles in Mediterranean regions. In *Insects and diseases of Mediterranean forest systems*, eds. T. D. Paine, and F. Lieutier, 293–327. Cham, Switzerland: Springer International Publishing.

Rhainds, M., W. E. Mackinnon, K. B. Porter, J. D. Sweeney, and P. J. Silk. 2011. Evidence for limited spatial spread in an exotic longhorn beetle, *Tetropium fuscum* (Coleoptera: Cerambycidae). *Journal of Economic Entomology* 104: 1928–1933.

Rice, M. E. 1995. Branch girdling by *Oncideres cingulata* (Coleoptera: Cerambycidae) and relative host quality of persimmon, hickory and elm. *Annals of the Entomological Society of America* 88: 451–455.

Riggins, J. J., L. D. Galligan, and F. M. Stephen. 2009. Rise and fall of red oak borer (Coleoptera: Cerambycidae) in the Ozark Mountains of Arkansas, USA. *Florida Entomologist* 92: 426–433.

Rogers, C. E. 1977a. Bionomics of *Oncideres cingulata* (Coleoptera: Cerambycidae) on mesquite. *Journal of the Kansas Entomological Society* 50: 222–228.

Rogers, C. E. 1977b. Cerambycid pests of sunflower: Distribution and behavior in the southern plains. *Environmental Entomology* 6: 833–838.

Rogers, C. E., and J. G. Serda. 1979. Rearing and biology of *Ataxia hubbardi* and *Mecas inornata* (Coleoptera: Cerambycidae), girdling pests of sunflower. *Journal of the Kansas Entomological Society* 52: 546–549.

Rogers, D. J., S. E. Lewthwaite, and P. R. Dentener. 2002. Rearing huhu beetle larvae, *Prionoplus reticularis* (Coleoptera: Cerambycidae) on artificial diet. *New Zealand Journal of Zoology* 29: 303–310.

Ryall, K. L. 2003. Response of the pine engraver beetle, *Ips pini* (Coleoptera: Scolytidae) and associated natural enemies to increased resource availability following a major ice storm disturbance. Ph.D. Thesis, University of Toronto, Toronto, Ontario, Canada.

Ryall, K. L., and S. M. Smith. 2001. Bark and wood-boring beetle response in red pine (*Pinus resinosa* Ait.) plantations damaged by the 1998 ice storm: Preliminary observations. *Forestry Chronicle* 77: 657–660.

Saint-Germain, M., P. Drapeau, and C. Hébert. 2004. Xylophagous insect species composition and patterns of substratum use on fire-killed black spruce in central Quebec. *Canadian Journal of Forest Research* 34: 677–685.

Sánchez, V., and M. A. Keena. 2013. Development of the teneral adult *Anoplophora glabripennis* (Coleoptera: Cerambycidae): Time to initiate and completely bore out of maple wood. *Environmental Entomology* 42: 1–6.

Sato, Y. 2005. Flight ability of the podocarp bark borer, *Hirticlytus comosus* (Matsushita) (Coleoptera: Cerambycidae). *Journal of the Japanese Forestry Society* 87: 247–250.

Schoeller, E. N., C. Husseneder, and J. D. Allison. 2012. Molecular evidence of facultative intraguild predation by *Monochamus titillator* larvae (Coleoptera: Cerambycidae) on members of the southern pine beetle guild. *Naturwissenschaften* 99: 913–924.

Schoeman, P. S., H. V. Hamburg, and B. P. Pasques. 1998. The morphology and phenology of the white coffee stem borer, *Monochamus leuconotus* (Pascoe) (Coleoptera: Cerambycidae), a pest of Arabica coffee. *African Entomology* 6: 83–89.

Schowalter, T. D. 2012. Insect responses to major landscape-level disturbance. *Annual Review of Entomology* 57: 1–20.

Schroeder, L. M. 1997. Oviposition behavior and reproductive success of the cerambycid *Acanthocinus aedilis* in the presence and absence of the bark beetle *Tomicus piniperda*. *Entomologia Experimentalis et Applicata* 82: 9–17.

Scriven, G. T., E. L. Reeves, and R. F. Luck. 1986. Beetle from Australia threatens eucalyptus. *California Agriculture* 40 (7–8): 4–6.

Seetharama, H. G., V. Vasudev, P. K. V. Kumar, and K. Sreedharan. 2005. Biology of coffee white stem borer *Xylotrechus quadripes* Chev. (Coleoptera: Cerambycidae). *Journal of Coffee Research* 33: 98–107.

Sharifi, S., I. Javadi, and J. A. Chemsak. 1970. Biology of the Rosaceae branch borer, *Osphranteria coerulescens* (Coleoptera: Cerambycidae). *Annals of the Entomological Society of America* 63: 1515–1520.

Shi, Z. H., K. G. Cen, and S. Q. Tan. 1982. Studies on *Stromatium longicorne* (Newman) (Coleoptera: Cerambycidae). *Acta Entomologica Sinica* 25: 35–41.

Shibata, E. 1986a. Dispersal movement of the adult Japanese pine sawyer, *Monochamus alternatus* Hope (Coleoptera: Cerambycidae) in a young pine forest. *Applied Entomology and Zoology* 21: 184–186.

Shibata, E. 1986b. Adult populations of the sugi bark borer, *Semanotus japonicus* Lacordaire (Coleoptera: Cerambycidae), in Japanese cedar stands: Population parameters, dispersal, and spatial distribution. *Researches on Population Ecology* 28: 253–266.

Shibata, E. 1989. The influence of temperature upon the activity of the adult sugi bark borer, *Semanotus japonicus* Lacordaire (Coleoptera: Cerambycidae). *Applied Entomology and Zoology* 24: 321–325.

Shibata, E. 1994. Population studies of *Callidiellum rufipenne* (Coleoptera: Cerambycidae) on Japanese cedar logs. *Annals of the Entomological Society of America* 87: 836–841.

Shibata, E. 1995. Reproductive strategy of the Sugi bark borer, *Semanotus japonicus* (Coleoptera: Cerambycidae) on Japanese cedar, *Cryptomeria japonica*. *Researches on Population Ecology* 37: 229–237.

Shibata, E. 2000. Bark borer *Semanotus japonicus* (Col., Cerambycidae) utilization of Japanese cedar *Cryptomeria japonica*: A delicate balance between a primary and secondary insect. *Journal of Applied Entomology* 124: 279–285.

Shintani, Y. 2011. Quantitative short-day photoperiodic response in larval development and its adaptive significance in an adult-overwintering cerambycid beetle, *Phytoecia rufiventris*. *Journal of Insect Physiology* 57: 1053–1059.

Shintani, Y., Y. Ishikawa, and S. Tatsuki. 1996. Larval diapause in the yellow-spotted longicorn beetle, *Psacothea hilaris* (Pascoe) (Coleoptera: Cerambycidae). *Applied Entomology and Zoology* 31: 489–494.

Skiles, D. D., F. T. Hovore, and E. F. Giesbert. 1978. Biology of *Leptura pacifica* (Linsley). *Coleopterists Bulletin* 32: 107–112.

Smith, C. I., and B. D. Farrell. 2005. Phylogeography of the longhorn cactus beetle *Moneilema appressum* LeConte (Coleoptera: Cerambycidae): Was the differentiation of the Madrean sky islands driven by Pleistocene climate changes? *Molecular Ecology* 14: 3049–3065.

Smith, M. T., J. Bancroft, G. Li, R Gao, and S. Teale. 2001. Dispersal of *Anoplophora glabripennis* (Cerambycidae). *Environmental Entomology* 30: 1036–1040.

Smith, M. T., J. Bancroft, and J. Tropp. 2002. Age-specific fecundity of *Anoplophora glabripennis* (Coleoptera: Cerambycidae) on three tree species infested in the United States. *Environmental Entomology* 31: 76–83.

Smith, M. T., P. C. Tobin, J. Bancroft, G. H. Li, and R. T. Gao. 2004. Dispersal and spatiotemporal dynamics of Asian longhorned beetle (Coleoptera: Cerambycidae) in China. *Environmental Entomology* 33: 435–442.

Solomon, J. D. 1977. Frass characteristics for identifying insect borers (Lepidoptera: Cossidae and Sesiidae; Coleoptera: Cerambycidae) in living hardwoods. *Canadian Entomologist* 109: 295–303.

Solomon, J. D. 1995. *Guide to insect borers in North American broadleaf trees and shrubs.* Washington, DC: U.S. Department of Agriculture, Forest Service, Agriculture Handbook AH-706.

Stephen, F. M., V. B. Salisbury, and F. L. Oliveria. 2001. Red oak borer, *Enaphalodes rufulus* (Coleoptera: Cerambycidae), in the Ozark Mountains of Arkansas, U.S.A.: An unexpected and remarkable forest disturbance. *Integrated Pest Management Reviews* 6: 247–252.

Swezey, O. H. 1950. Notes on the life cycle of certain introduced cerambycid beetles. *Proceedings of the Hawaiian Entomological Society* 14: 187–188.

Takasu, F., N. Yamamoto, K. Kawasaki, K. Togashi, Y. Kishi, and N. Shigesada. 2000. Modeling the expansion of an introduced tree disease. *Biological Invasions* 2: 141–150.

Tippmann, F. F. 1960. Studien über neotropische Longicornier III (Coleoptera, Cerambycidae). *Koleopterologische Rundschau* 37–38: 82–217.

Togashi, K. 1987. Diapause termination in the adult cryptomeria bark borer, *Semanotus japonicus* (Coleoptera,Cerambycidae). *Kontyû* 55: 169–175.

Togashi, K. 1990a. A field experiment on dispersal of newly emerged adults of *Monochamus alternatus* (Coleoptera: Cerambycidae). *Researches on Population Ecology* 32: 1–13.

Togashi, K. 1990b. Life table for *Monochamus alternatus* (Coleoptera: Cerambycidae) within dead trees of *Pinus thunbergii*. *Japanese Journal of Entomology* 58: 217–230.

Togashi, K. 1991. Larval diapause termination of *Monochamus alternatus* Hope (Coleoptera: Cerambycidae) under natural conditions. *Applied Entomology and Zoology* 26: 381–386.

Togashi, K., S. Jikumaru, A. Taketsune, and F. Takahashi. 1994. Termination of larval diapause in *Monochamus saltuarius* (Coleoptera: Cerambycidae) under natural conditions. *Journal of the Japanese Forestry Society* 76: 30–34.

Tozlu, G., S. Çoruh, and H. Özbek. 2010. Biology and damage of *Saperda populnea* (L.) (Coleoptera: Cerambycidae) in Aras valley (Kars and Erzurum provinces), Turkey. *Anadolu Tarim Bilimleri Dergisi* 25: 151–158.

Trägårdh, I. 1930. Some aspects in the biology of longicorn beetles. *Bulletin of Entomological Research.* 21: 1–8.

Villiers, A. 1980. Coléoptères Cerambycidae des Antilles Françaises. III. Lamiinae. *Annales de la Société Entomologique de France* 16: 541–598.

Visitpanich, J. 1994. The biology and survival rate of the coffee stem borer, *Xylotrechus quadripes* Chevrolat (Coleoptera: Cerambycidae) in northern Thailand. *Japanese Journal of Entomology* 62: 731–745.

Wallner, W. E. 1987. Factors affecting insect population dynamics: Differences between outbreak and non-outbreak species. *Annual Review of Entomology* 32: 317–340.

Walsh, K. D., and M. J. Linit. 1985. Oviposition biology of the pine sawyer, *Monochamus carolinensis* (Coleoptera: Cerambycidae). *Annals of the Entomological Society of America* 78: 81–85.

Wang, Q., G. Shi, and L. K. Davis. 1998. Reproductive potential and daily reproductive rhythms of *Oemona hirta* (Coleoptera: Cerambycidae). *Journal of Economic Entomology* 91: 1360–1365.

Wang, Q., G. Shi, D. Song, D. J. Rogers, L. K. Davis, and X. Chen. 2002. Development, survival, body weight, longevity, and reproductive potential of *Oemona hirta* (Coleoptera: Cerambycidae) under different rearing conditions. *Journal of Economic Entomology* 95: 563–569.

Ware, V. L., and F. M. Stephen. 2006. Facultative intraguild predation of red oak borer larvae (Coleoptera: Cerambycidae). *Environmental Entomology* 35: 443–447.

Watari, Y., T. Yamanaka, W. Asano, and Y. Ishikawa. 2002. Prediction of the life cycle of the west Japan type yellow-spotted longicorn beetle, *Psacothea hilaris* (Coleoptera: Cerambycidae) by numerical simulations. *Applied Entomology and Zoology* 37: 559–569.

Webb, J. L. 1909. Some insects injurious to forests: The southern pine sawyer. U.S. Department of Agriculture, Bureau of Entomology, Bulletin 58, Part IV, 41–56. Washington, DC: U.S. Government Printing Office.

Wen, J., Y. Li, N. Xia, and Y. Luo. 1998. Dispersal pattern of adult *Anoplophora glabripennis* on poplars. *Acta Ecologica Sinica* 18: 269–277.

Wermelinger, B., A. Rigling, D. Schneidermathis, and M. Dobbertin. 2008. Assessing the role of bark- and wood-boring insects in the decline of Scots pine (*Pinus sylvestris*) in the Swiss Rhone valley. *Ecological Entomology* 33: 239–249.

Wickman, B. E. 1968. The biology of the fir tree borer, *Semanotus litigiosus* (Coleoptera: Cerambycidae), in California. *Canadian Entomologist* 100: 208–220.

Williams, D. W., G.-H. Li, and R.-T. Gao. 2004. Tracking movements of individual *Anoplophora glabripennis* (Coleoptera: Cerambycidae) adults: Application of harmonic radar. *Environmental Entomology* 33: 644–649.

Wollerman, E. H., C. Adams, and G. C. Heaton. 1969. Continuous laboratory culture of the locust borer, *Megacyllene robiniae*. *Annals of the Entomological Society of America* 62: 647–649.

Yashiro, H. 1940. On the life-history of *Stromatium longicorne* Newman. *Bulletin of Okinawa Forestry Society* 1: 1–10.

Yates, M. G. 1984. The biology of the oak bark beetle, *Scolytus intricatus* (Ratzeburg) (Coleoptera: Scolytidae), in southern England. *Bulletin of Entomological Research* 74: 569–579.

Yoon, C., Y. H. Shin, J. O. Yang, J. H. Han, and G. H. Kim. 2011. *Pinus koraiensis* twigs affect *Monochamus saltuarius* (Coleoptera: Cerambycidae) longevity and reproduction. *Journal of Asia-Pacific Entomology* 14: 327–333.

Yoon, H., and Y. Mah. 1999. Life cycle of the mulberry longicorn beetle, *Apriona germari* Hope on an artificial diet. *Journal of Asia-Pacific Entomology* 2: 169–173.

Youngs, D. B. 1897. Oviposition of *Anthophilax attenuates*. *Entomological News* 8: 192.

Zabecki, W. 1988. Role of cambio- and xylophagous insects in the process of decline of silver fir stands affected by the industrial air pollution in Ojcow National Park. *Acta Agraria et Silvestria. Series Silvestris* 27: 17–30 [in Polish].

Zhang, Q.-H., J. A. Byers, and X.-D. Zhang. 1993. Influence of bark thickness, trunk diameter and height on reproduction of the longhorned beetle, *Monochamus sutor* (Coleoptera: Cerambycidae) in burned larch and pine. *Journal of Applied Entomology* 115: 145–154.

Zhang, X., and M. J. Linit. 1998. Comparison of oviposition and longevity of *Monochamus alternatus* and *M. carolinensis* (Coleoptera: Cerambycidae) under laboratory conditions. *Environmental Entomology* 27: 885–891.

Zhao, R.-L., Z.-S. Lu, X.-H. Lu, and X.-Y. Wu. 1993. Life table study of *Anoplophora glabripennis* (Coleoptera: Cerambycidae) natural populations. *Journal of Beijing Forestry University* 15(4): 125–129.

Zhou, J.-X., K.-B. Zhang, and Y.-Z. Lu. 1984. Study on adult activity and behavioral mechanism of *Anoplophora nobilis* Ganglbauer. *Scientia Silvae Sinica* 20: 372–379.

3

Feeding Biology of Cerambycids

Robert A. Haack
USDA Forest Service
Lansing, Michigan

CONTENTS

3.1 Introduction

There are more than 36,000 species of Cerambycidae recognized throughout the world (see Chapter 1), occurring on all continents except Antarctica (Linsley 1959). Given such numbers, it is not surprising that cerambycids display great diversity in their feeding habits. Both adults and larvae are almost exclusively phytophagous. Some adults appear not to feed at all, while others feed daily. Larvae primarily utilize woody host plants, but some species develop within herbaceous plants. Cerambycid larvae infest nearly all plant parts, especially stems, branches, and roots, as well as feed on nearly all plant tissues, especially bark, cambium, and wood. As expected in such a large insect family, some cerambycids are strictly monophagous while others are highly polyphagous. Similarly, some cerambycids infest live, healthy plants while others develop in dead plants; likewise, some species prefer moist wood, while others prefer dry wood. Cerambycid larvae are able to digest woody tissues with the aid of enzymes that they sometimes secrete themselves or that they obtain from symbionts. Many details on the feeding biology of cerambycids will be provided in this chapter, including the types of food consumed by adults and larvae, the common parts of plants that larvae infest and the tissues they consume, and aspects of wood digestion.

3.2 Adult Feeding Habits

Although few detailed studies have been conducted on the feeding habits of cerambycid adults, some general trends are apparent at the subfamily level (Duffy 1953). For example, parandrine, prionine, and spondylidine adults appear not to feed at all (Benham and Farrar 1976; Bense 1995; Švácha and Lawrence 2014). By contrast, it appears that all Lamiinae as well as most Lepturinae feed as adults. Although in the Cerambycinae, some adults are known to feed, while others do not—such as *Hylotrupes* and *Stromatium* (Duffy 1953).

3.2.1 Types of Adult Food

Butovitsch (1939) categorized the general types of food consumed by adult cerambycids as flowers, bark, foliage, cones, sap, fruit, roots, and fungi. Most lepturine adults, as well as many cerambycine (e.g., *Batyle, Euderces, Megacyllene,* and *Molorchus*) and a few lamiine (e.g., *Phytoecia* and *Tetrops*) adults, visit flowers and feed on pollen and nectar (Duffy 1953; Linsley 1959; see Figure 3.1). The bark and stem feeders are almost entirely lamiines (e.g., *Acanthocinus, Lamia,* and *Monochamus*; Duffy 1953). Leaves are consumed primarily by lamiines (e.g., *Batocera, Oberea,* and *Saperda*) as well as needles and developing cones (e.g., *Monochamus*; Butovitsch 1939; Duffy 1953). Adults of some lamiine species feed on both bark and foliage (e.g., *Goes, Monochamus, Plectrodera,* and *Saperda*; Webb 1909; Brooks 1919; Nord et al. 1972; Solomon 1974, 1980; see Figure 3.2). Various cerambycine (e.g., *Cerambyx* and *Hoplocerambyx*) and lamiine (e.g., *Moneilema*) adults feed on fruit and sap exudates (Duffy 1953). Roots of grasses are fed on by both the larvae and, at times, the adults of the soil-dwelling lamiine genus *Dorcadion* (Duffy 1953; Linsley 1959). Only a few adult cerambycids are known to feed on fungi, such as members of the lamiine genus *Leiopus* (Craighead 1923; Duffy 1953; Michalcewicz 2002).

FIGURE 3.1 *Megacyllene robiniae* (Forster) adult feeding on pollen of goldenrod (*Solidago*) flowers. The larval hosts of this North American cerambycine are locust trees in the genus *Robinia*. (Courtesy of David Cappaert [Bugwood image 2106090].)

FIGURE 3.2 An example of maturation feeding by the Asian lamiine *Anoplophora glabripennis* (Motschulsky) on maple (*Acer*) branches. (Courtesy of Dean Morewood [Bugwood image 1193003].)

3.2.2 Food and Adult Reproduction

As mentioned earlier, many cerambycids do not feed as adults and are typically capable of reproducing soon after emergence. For example, the cerambycine *Xylotrechus pyrrhoderus* Bates was capable of flight, responding to pheromones, mating, and egg laying at the time of emergence from its host plant without any additional feeding (Iwabuchi 1982). However, for species that feed after emergence, such as the Lamiinae, adults typically feed for one to three weeks before becoming sexually mature (Alya and Hain 1985; Hanks 1999), a period of time referred to as maturation feeding (Edwards 1961; Slipinski and Escalona 2013). For example, in the lamiine *Anoplophora glabripennis* (Motschulsky), adult females become sexually mature about 10 days after emergence, and maturation feeding is required for ovary development (Li and Liu 1997). Similarly, Keena (2002) and Smith et al. (2002) reported that the mean time from adult emergence to first oviposition in *A. glabripennis* varied from 9 to 17 days. Members of the lamiine genus *Monochamus* often feed for 7–12 days after emergence before becoming sexually mature (Akbulut and Stamps 2012). Similarly, in the lamiine *Glenea cantor* (F.), Lu et al. (2013) noted that adults required an average of five to seven days of feeding before mating and about another week of feeding before initiating oviposition.

The type of food consumed by cerambycid adults can influence their longevity and fecundity. In *A. glabripennis*, for example, the species of tree selected as the source of twigs to feed the adults influences their fecundity (Smith et al. 2002; Hajek and Kalb 2007). Similarly, for the cerambycine borers *Phoracantha recurva* Newman and *Phoracantha semipunctata* F., which as larvae develop in *Eucalyptus* trees and as adults feed on *Eucalyptus* pollen, adult longevity and fecundity increased when adults were maintained on a diet rich in *Eucalyptus* pollen compared with other pollen sources (Millar et al. 2003).

3.2.3 Food and Adult Flight, Pollination, and Disease Transmission

Many of the feeding habits of adult cerambycids have a direct influence on their dispersal behavior as well as on their role in pollination and disease transmission. For example, in the flower-feeding cerambycids, such as the Lepturinae, the flowers visited by adults typically are on different plant species than the larval hosts. Therefore, these adults need to disperse at least twice after emerging—first to flowers where they feed and often mate and then to the larval host plants to oviposit (Duffy 1953; Linsley 1959; Bílý and Mehl 1989; Bense 1995; Hanks 1999). By contrast, cerambycid adults that feed on bark and foliage usually feed on the same species of plants that serve as the larval hosts (Duffy 1953; Solomon 1995; Hanks 1999); therefore, adult dispersal may be minimal if the original host plant is still suitable for oviposition (see Chapter 2). For example, in *A. glabripennis*, which typically is univoltine, the same tree is often reinfested for several years until it dies, although a small portion of the progeny may disperse more widely (Haack et al. 2006, 2010). Many of the flower-feeding cerambycids pollinate their food plants as they feed on pollen and nectar (Willemstein 1987; Gutowski 1990; Hawkeswood and Turner 2007). With respect to disease transmission, the fact that adults of the pine-infesting lamiine genus *Monochamus* conduct maturation feeding on the bark, twigs, and foliage of pines (*Pinus*) enables them to be efficient vectors of the pinewood nematode, *Bursaphelenchus xylophilus* (Steiner & Buhrer) Nickle, given that the nematodes depart the adult's body during feeding and enter the trees through the feeding wounds created in the bark (Linit 1990; Akbulut and Stamps 2012; see Chapter 6).

3.2.4 Predatory Cerambycids

Although nearly all cerambycid adults that are known to feed are phytophagous, there is at least one cerambycine genus (*Elytroleptus*) where the species are carnivorous, preying on adult lycid beetles (net-winged beetles). Most *Elytroleptus* species are native to Mexico and the southwestern United States (Linsley 1962). In general, lycid adults are protected chemically from predation, and they often have aposematic coloration and form dense aggregations on plants (Eisner et al. 2008; Grzymala and Miller 2013). *Elytroleptus* adults mimic the appearance of these lycid beetles, allowing them to join their aggregations and prey on them (Linsley et al. 1961; Eisner et al. 1962; Selander et al. 1963).

3.3 Larval Feeding Habits

Cerambycid larvae are phytophagous (Linsley 1959; Hanks 1999; Slipinski and Escalona 2013), although facultative inter- and intraspecific predation has been observed when larvae encounter other individuals within the host plant as they construct their galleries (Togashi 1990; Victorsson and Wikars 1996; Dodds et al. 2001; Ware and Stephen 2006; Schoeller et al. 2012). Cerambycid larvae feed on a wide diversity of plant species, plant parts, plant tissues, as well as on plants in various conditions from living and healthy to dead and decaying. Before looking at these trends in host usage, readers need to be aware that information on larval hosts is best known for the economically important species but usually is incomplete for most cerambycids or entirely lacking for others. Moreover, larval host information is at times incorrect, often as a result of inaccurate plant or insect identification, changes in taxonomic status of a species or species complex, in situations when the plant on which an adult beetle is collected is assumed to be the larval host, or when such information is not presented clearly in the literature or on museum specimen labels. Nevertheless, for those world regions where the larval hosts are relatively well known for the local cerambycids, the clear trend is for most species to develop in woody plants, especially conifers and hardwood (broadleaf) trees (see Section 3.3.1).

3.3.1 Larval Host Plants

Information in Table 3.1 shows the number of cerambycid species that develop in various groupings of host plants for Montana (89% of species with known hosts; Hart et al. 2013) and Florida (77%; Thomas et al. 2005) in the United States; as well as the geographic regions of Fennoscandia (comprising Norway, Sweden, Finland, and a small part of neighboring Russia) and Denmark (100%; Bílý and Mehl 1989); Israel (91%; Sama et al. 2010); and Korea (57%; Lim et al. 2014). The checklist provided in Hart et al. (2013) for the cerambycids of Montana was supplemented with host data from the "Montana Wood Boring Insect Project" database (http://www.mtent.org/Cerambycidae.html) and, occasionally, from

TABLE 3.1

Percent of Cerambycids with Known Larval Hosts that Feed as Larvae in Various Types of Host Plants in Five World Regions

Larval Hosts[a]	Percent of Cerambycids that Develop in Different Hose Groupings[b]				
	Montana	Florida	Fennoscandia and Denmark	Israel	Korea
C	45.2	16.2	30.1	5.3	15.5
H	19.3	69.3	44.7	41.1	49.7
S	5.2	3.0	0	3.2	2.2
V	0.7	1.7	0	1.0	3.9
HC	5.9	5.6	14.6	5.3	8.8
HS	7.4	0.6	5.7	6.3	9.9
HCS	0.7	0	0.8	1.0	1.1
HCSV	0.7	0	0	0	1.1
HV	0	1.1	0	2.1	2.8
HSV	3.7	0	0.8	0	0.6
P	10.4	1.7	3.2	32.6	1.1
Cactus	0.7	0	0	0	0
Palm	0	1.1	0	2.1	0
HP	0	0	0	0	1.1
VP	0	0	0	0	0.6
HVB	0	0	0	0	0.6
HB	0	0	0	0	1.1
No. spp. With host data	135	179	123	95	181
Total No. spp.	152	233	123	104	318

Source: Data were based on Hart et al. (2013), http://www.mtent.org/Cerambycidae.html, and occasionally on other sources (see Section 3.3.1) for Montana, United States; Thomas et al. (2005) for Florida, United States; Bílý and Mehl (1989) for Fennoscandia and Denmark; Sama et al. (2010) for Israel; and Lim et al. (2014) for Korea.

[a] Larval host categories: B = bamboo; C = conifers; H = hardwood trees; S = woody shrubs; P = herbaceous plants; V = woody vines. Categories with more than one letter or plant group represent cerambycids that utilize plant genera in each of the listed plant groups.

[b] Percentage values based on only those cerambycid species with known larval hosts.

Linsley (1962a, 1962b, 1963, 1964) and Linsley and Chemsak (1972, 1976, 1984, 1995). These five world regions were selected because they represent different parts of the Northern Hemisphere where there is a good knowledge of the local larval host plants. In each of these world regions, assuming the information is accurate and complete, the majority of the cerambycid species develop strictly in trees, shrubs, and woody vines, with these woody plants constituting the larval hosts of about 89% of the cerambycids in Montana, 98% in Florida, 97% in Fennoscandia/Denmark, 65% in Israel, and 96% in Korea (Table 3.1). Coniferous trees are the most commonly utilized group of host plants for the cerambycids of Montana, although hardwood trees are the most common larval hosts in the other four world regions (Table 3.1). The dominance of woody plants serving as larval hosts likely is the general pattern for cerambycids worldwide. For example, recent larval host records for 180 South American cerambycids indicated that about 92% of the species listed developed strictly in trees, shrubs, and woody vines (Machado et al. 2012). Likewise, in Hawaii, Gressitt and Davis (1972) reported that nearly all of the 120 endemic cerambycids developed in trees and shrubs.

Herbaceous plants that traditionally are considered nonwoody, as well as cacti (Cactaceae) and various monocots that have some "woody" parts, occasionally are used by cerambycids as larval hosts—for example, some agave (Asparagaceae, formerly in Liliaceae), orchids (Orchidaceae), palms (Arecaceae), and yucca (Asparagaceae). About 11% of the cerambycids in Montana develop in herbaceous plants and cacti, 3% of Florida species develop in herbaceous plants and palms, 3% of Fennoscandian species develop in herbaceous plants, 35% of Israeli species develop in herbaceous plants and palms, and 5% of Korean species develop in herbaceous plants and bamboo (Table 3.1). Both annual and perennial herbaceous plants are used as hosts by certain cerambycids (Linsley 1959).

Several examples of cerambycids that feed on herbaceous plants, cacti, and woody monocots are listed in Table 3.2. In general, larvae of these cerambycids feed inside the roots and stems of their hosts, but some larvae live in the soil and feed externally on host tissues. For example, larvae of the prionine *Prionus emarginatus* Say and the lamiine *Dorcadion pseudopreissi* Breuning feed externally on the roots of grasses (Poaceae) (Craighead 1923; Gwynne and Hostetler 1978; Kumral et al. 2012), and larvae of the lepturine *Pseudovadonia livida* (F.) feed on decaying roots and stalks of grasses as well as fungal mycelium (Burakowski 1979; Bense 1995).

Larvae of many cerambycids are economic pests on herbaceous plants in various parts of the world. For example, the lamiine *Apomecyna binubila* Pascoe is a pest of melons (*Cucurbita* sp.) in Africa (Pollard 1954); the lamiine *Dectes texanus* LeConte is a pest of soybeans [*Glycine max* (L.) Merr.] in the United States (Tindall et al. 2010); the prionine *Dorysthenes buqueti* (Guérin-Méneville) is a pest of sugarcane (*Saccharum officinarum* L.) in Asia (Sugar Research Australia 2013); the lamiine *Acalolepta mixta* (Hope) and the cerambycine *Xylotrechus arvicola* (Olivier) are pests of grapes (*Vitis*) in Australia and Spain, respectively (Goodwin and Pettit 1994; García-Ruiz et al. 2012); the lamiine *Agapanthia cardui* L. is a pest of artichokes (*Cynara scolymus* L.) in the Mediterranean region (Baragaño Galán et al. 1981); and the cerambycine *Plagionotus floralis* (Pallas) is a pest of alfalfa (*Medicago sativa* L.) in Europe (Toshova et al. 2010). Likewise, a few of the cerambycids that develop in woody monocots are economic pests, such as the cerambycines *Jebusaea hammerschmidti* Reiche, a pest of date palms (*Phoenix dactilifera* L.) in the Middle East (Giblin-Davis 2001), and *Chlorophorus annularis* F., a pest of bamboo (*Bambusa* sp.) in Asia (Barak et al. 2009). With respect to the cactus-feeding cerambycids, a few species are of concern in the United States because they infest rare species of cacti that are protected under the U.S. Federal Endangered Species Act. For example, the lamiines *Moneilema armatum* LeConte and *Moneilema semipunctatum* LeConte infest at least five species of endangered and threatened cacti in the western United States (Kass 2001; Ferguson and Williamson 2009; USDA 2013; Figure 3.3).

3.3.2 Plant Parts Utilized by Larvae

Cerambycids develop in nearly all parts of their host plants but mainly in the stems, branches, and roots. Some species develop primarily in twigs (Figure 3.4), such as the lamiine *Oberea tripunctata* (Swederus) and the cerambycine *Tessaropa tenuipes* (Haldeman) in North America (Linsley 1962b, Solomon 1995) and the lamiine *Pogonocherus hispidus* (L.) in Europe (Bílý and Mehl 1989; Bense 1995). Some species oviposit primarily along the lower trunk of their hosts, such as the cerambycines *Enaphalodes rufulus* (Haldeman) (Donley and Rast 1984) and *Megacyllene robiniae* (Forster) (Harman and Harman 1990) in the United States. Still others oviposit predominantly at the base of trees such as the prionines *Mallodon* (= *Stenodontes*) *dasystomus* (Say) in the United States (Linsley 1962a, Solomon 1995) and the prionine *Prionus coriarius* (L.) in Europe (Bílý and Mehl 1989; Bense 1995).

To further illustrate the utilization of various tree parts among different cerambycids, several species are listed in Table 3.3 that develop primarily in the twigs, branches, trunks, and roots of oak (*Quercus*) and pine trees in the United States and northern Europe. The typical range in adult body length is given for each of the species listed in Table 3.3 and, assuming these sizes are typical of cerambycids that infest these different parts of a tree, there is an apparent trend where twig- and branch-infesting cerambycids generally are smaller than root- and trunk-infesting cerambycids. Such a pattern in beetle size is logical given the differences in physical size of these plant parts and the faster decay rates of twigs and branches compared with trunks and stumps (Cornelissen et al. 2012) and thus reflect the constraints that would be placed on potential beetle size and voltinism in twig- versus trunk-infesting cerambycids, for example. Similar lists of cerambycids that infest different parts of trees could be developed for many other tree genera throughout the world.

Although most cerambycid larvae develop in the roots, stems, and branches of their host plants, a few develop in other plant parts such as seeds, pods, fruits, and cones (Table 3.4 and Figure 3.5).

TABLE 3.2

Examples of Cerambycids that Develop in Nonwoody Plants

Species	Sub-Family[a]	Common Host Genera in Nature	Host Plant Family	Plant Part Infested	Source
Agapanthia villosoviridescens (DeGeer)	Lam	*Cirsium, Angelica, Carduus, Senecio*	Apiaceae, Asteraceae	Stem	Bílý and Mehl 1989
Ataxia hubbardi Fisher	Lam	*Ambrosia, Erigeron Helianthus, Vernonia, Heracleum*	Asteraceae	Root, stem	Rogers 1977; Schwitzgebel and Wilbur 1942; Twinn and Harding 1999
Brachysomida californica (LeConte)	Lep	*Lomatium*	Apiaceae	Root	Swift 2008
Chlorophorus annularis (Fabricius)	Cer	*Phyllostachys, Sasa*	Poaceae	Stem	Friedman et al. 2008; Lim et al. 2014
Coenopoeus palmeri (LeConte)	Lam	*Opuntia*	Cactaceae	Stem, branch	Raske 1972
Cortodera flavimana (Waldl)	Lep	*Ranunculus*	Ranunculaceae	Root	Özdikmen 2003
Dectes texanus LeConte	Lam	*Ambrosia, Anoda, Glycine, Xanthium*	Asteraceae, Fabaceae, Malvaceae	Stem	Tindall et al. 2010
Diaxenes dendrobii Gahan	Lam	*Coelogyne, Dendrobium, Laelia, Odontoglossum*	Orchidaceae	Stem	MacDougall 1900
Hemierana marginata (Fabricius)	Lam	*Ambrosia, Erigeron, Vernonia*	Asteraceae	Stem	Schwitzgebel and Wilbur 1942
Hippopsis lemniscata (Fabricius)	Lam	*Ambrosia, Vernonia, Xanthium*	Asteraceae	Stem	Piper 1977; Rogers 1977
Jebusaea hammerschmidti Reiche	Cer	*Phoenix*	Arecaceae	Stem	Giblin-Davis 2001
Mecas cana saturnina (LeConte)[b]	Lam	*Ambrosia, Helianthus, Iva*	Asteraceae	Root, stem	Rogers 1977
Moneilema appressum LeConte	Lam	*Echinocereus, Opuntia*	Cactaceae	Root, stem	Lingafelter 2003
Moneilema armatum LeConte	Lam	*Astrophytum, Opuntia*	Cactaceae	Root, stem	Ferguson and Williamson 2009
Nealcidion cereicola (Fisher)	Lam	*Cereus, Cleistocactus, Echinopsis, Monvillea, Stetsonia*	Cactaceae	Stem, branch	Machado et al. 2012; McFadyen and Fidalgo 1976
Phytoecia cylindrica (L)	Lam	*Anthriscus, Daucus*	Apiaceae	Stem	Bílýand Mehl 1989; Twinn and Harding 1999
Prionus emarginatus Say	Prio	*Grasses*	Poaceae	root	Craighead 1923; Gwynne and Hostetler 1978
Tetraopes tetraophthalmus (Forster)	Lam	*Asclepias*	Asclepidacae	Root	Matter 2001
Tragidion agave Swift & Ray	Cer	*Agave*	Asparagaceae	Flower stalk	Chemsak and Powell 1966; Swift and Ray 2008
Tragidion armatum LeConte	Cer	*Agave, Yucca*	Asparagaceae	Stem	Craighead 1923; Linsley 1962a; Waring and Smith 1987
Zagymnus clerinus (LeConte)	Cer	*Chamaerops*	Arecaceae	Leaf stem	Beutenmuller 1896; Blatchley 1928

[a] Cer = Cerambycinae; Lam = Lamiinae; Lep = Lepturinae; Prio = Prioninae.

[b] In the original publication, *Mecas cana saturnina* (LeConte) was reported as *Mecas inornata* Say; see Linsley and Chemsak (1995).

FIGURE 3.3 Adult lamiine *Moneilema armatum* LeConte, a flightless cerambycid that feeds on and develops in cacti in the western United States. (Courtesy of Whitney Cranshaw [Bugwood image 5393466].)

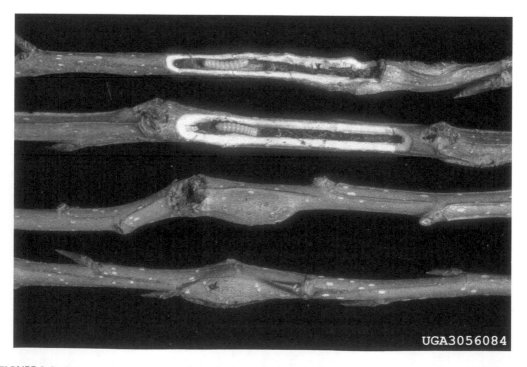

FIGURE 3.4 Larvae, galls, and larval galleries of the lamiine *Oberea delongi* Knull on its larval host, eastern cotton-wood (*Populus deltoides* Bartram ex Marsh.), in the eastern United States. (Courtesy of James Solomon [Bugwood image 3056084].)

TABLE 3.3

Examples of Cerambycids that Develop in Various Parts of Oak (*Quercus*) and Pine (*Pinus*) Trees Growing in the United States and in Fennoscandia and Denmark[a]

Tree Part	Species, Subfamily,[b] and Adult Length (mm)[c]					
	United States			Fennoscandia and Denmark		
Oak-Infesting Cerambycids						
Twig	*Anelaphus parallelus* (Newman)	Cer	10–15	*Anaesthetis testacea* (Fabricius)	Lam	5–10
Twig	*Psyrassa unicolor* (Randall)	Cer	9–15	*Phymatodes pusillus* (Fabricius)	Cer	5–10
Branch	*Goes debilis* LeConte	Lam	10–16	*Pyrrhidium sanguineum* (L.)	Cer	6–15
Branch	*Purpuricenus axillaris* Haldeman	Cer	12–29	*Xylotrechus antilope* (Schönherr)	Cer	7–14
Trunk	*Enaphalodes rufulus* (Haldeman)	Cer	23–33	*Cerambyx cerdo* L.	Lam	24–53
Trunk	*Goes tigrinus* (De Geer)	Lam	22–38	*Rhagium sycophanta* (Schrank)	Lep	17–26
Root	*Archodontes melanopus melanopus* (L.)	Prio	33–57	*Prionus coriarius* (L.)	Prio	19–45
Root	*Prionus imbricornis* (L.)	Prio	31–42	*Stenocorus meridianus* (L.)	Lep	15–25
Pine-Infesting Cerambycids						
Twig and branch	*Phymatodes hirtellus* (LeConte)	Cer	5–7	*Pogonocherus decoratus* Fairmaire	Lam	4–7
Branch	*Haplidus testaceus* LeConte	Cer	8–16	*Molorchus minor* (L.)	Cer	6–16
Branch	*Neoclytus muricatulus* Kirby	Cer	5–9	*Pogonocherus fasciculatus* (DeGeer)	Lam	5–8
Trunk	*Callidium antennatum* Newman	Cer	9–15	*Ergates faber* (L.)	Prio	23–60
Trunk	*Monochamus titillator* (Fabricius)	Lam	17–31	*Monochamus sutor* (L.)	Lam	15–25
Root	*Typocerus zebra* (Olivier)	Lep	10–16	*Judolia sexmaculata* (L.)	Lep	8–14
Root	*Ulochaetes leoninus* LeConte	Nec	20–30	*Pachyta quadrimaculata* (L.)	Lep	11–20

[a] Craighead (1923) and Solomon (1995) for the United States; Bílý and Mehl (1989) for Fennoscandia and Denmark.

[b] Cer = Cerambycinae; Lam = Lamiinae; Lep = Lepturinae; Nec = Necydalinae (formerly a tribe in Lepturinae); Prio = Prioninae.

[c] Adult length data from Bílý and Mehl 1989; Furniss and Carolin 1977; Linsley 1962b, 1964; Solomon 1995; Yanega 1996.

Seed-infesting cerambycids have been recorded from trees in the families Rhizophoraceae and Sapindaceae, pod-infesting species from both herbaceous and woody Leguminosae, and cone-infesting species from conifers in the Cupressaceae and Pinaceae (Table 3.4). All the dicot-infesting cerambycids listed in Table 3.4 are members of the subfamily Lamiinae, while the conifer-infesting species represent the subfamilies Cerambycinae, Lepturinae, and Spondylidinae. It is surprising that so few cerambycids have evolved to develop in seeds, pods, fruits, and cones—or perhaps many more await discovery. The cerambycids that develop inside seeds and fruit generally are small in size. For example, adults of the seed-infesting lamiine *Ataxia falli* Breuning are 12–16 mm in length (Linsley and Chemsak 1984), and adults of the fruit-infesting lamiine *Leptostylus gibbulosus* Bates are 8–11 mm in length (Linsley and Chemsak 1995; Table 3.4).

Leaves are seldom mined by cerambycid larvae, but there are a few exceptions. The lamiine *Microlamia pygmaea* Bates, for example, is a small (adults are about 2–4 mm long) cerambycid in New Zealand that develops in fallen twigs and dead leaves of kauri trees, *Agathis australis* (D. Don) Lindley (Martin 2000). Another example is the cerambycine *Jebusaea hammerschmidti* Reiche (syn. *Pseudophilus testaceus* Gah.), a large cerambycid (21–40 mm) native to the Middle East that, as larvae, first mines the

TABLE 3.4

Records of Cerambycid Species that Feed on Seeds, Pods, and Cones

Species	Sub-Family[a]	Plant Host and Plant Part Infested			Country or Location of Study	Source
		Genus or Species	Plant Family	Part		
Ataxia sulcata Fall (= *Ataxia falli* Breuning)	Lam	*Rhizophora mangle*	Rhizophoraceae	Seed	United States (Florida)	Craighead 1923
Chlorophorus strobilicola Champion	Cer	*Pinus roxburghii*	Pinaceae	Cone	India	Champion 1919; Pande and Bhandari 2006
Cortodera femorata (Fabricius)	Lep	*Picea* *Pinus*	Pinaceae Pinaceae	Cone	Serbia, Latvia	Pil and Stojanovic 2005; Barševskis and Savenkov 2013
Enaretta castelnaudii Thomson	Lam.	*Acacia*	Leguminosae	Pod	Africa	Schabel 2006
Leptostylus gibbulosus Bates	Lam	*Sapindus saponaria*	Sapindaceae	Fruit	United States (Texas), Mexico, Colombia	Vogt 1949; Romero Nápoles et al. 2007; Hernandez-Jaramillo et al. 2012
Leptostylus gundlachi Fisher [= *Leptostylopsis gundlachi* (Fisher), and *Styloleptus gundlachi* (Fisher)]	Lam	*Erythrina fusca*	Leguminosae	Pod	Puerto Rico	Wolcott 1948
Leptostyulus spermovoratis Chemsak	Lam	*Diospyros*	Ebenaceae	Fruit	Costa Rica	Chemsak 1972
Leptostylus terracolor Horn, [= *Leptostylopsis terraecolor* (Horn)]	Lam	*Rhizophora mangle*	Rhizophoraceae	Seed	United States (Florida)	Craighead 1923
Lepturges guadeloupensis Fleutiaux & Salle [= *Urgleptes guadeloupensis* (Fleutiaux & Salle)]	Lam	*Acacia farnesiana*	Leguminosae	Pod	Puerto Rico	Wolcott 1948
Lepturges spermophagus Fisher [= *Atrypanius spermophagus* (Fisher)]	Lam	*Vigna*	Leguminosae	Pod	Mexico	Fisher 1917
Lophopoeum timbouvae Lameere (= *Baryssinus leguminicola* Linell)	Lam	*Enterolobium, Gleditsia, Inga, Prosopis, Tamarindus*	Leguminosae	Pod	Argentina, Brazil, Paraguay	Duffy 1960
Paratimia conicola Fisher	Spo	*Pinus attenuate, Pinus contorta ssp. bolanderi*	Pinaceae	Cone	United States (California, Oregon)	Craighead 1923; Linsley 1962a
Phymatodes nitidus LeConte	Cer	*Sequoia sempervirens, Sequoiadendron giganteum*	Cupressaceae	Cone	United States (California)	Keen 1958; Stecker 1980
Xylotrechus schaefferi Schott	Cer	*Pinus banksiana, Pinus rigida*	Pinaceae	Cone	United States (New York)	Hoebeke and Huether 1990

[a] Cer = Cerambycinae; Lam = Lamiinae; Lep = Lepturinae; Spo = Spondylidinae.

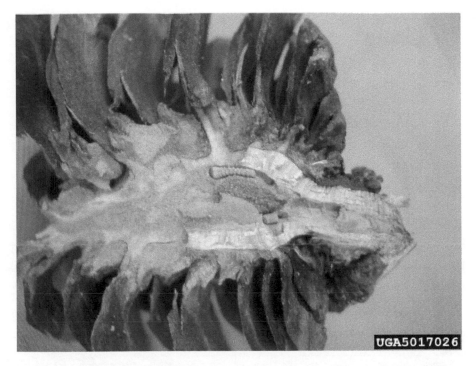

FIGURE 3.5 Larvae of the cerambycine *Chlorophorus strobilicola* Champion construct galleries in pine (*Pinus*) cones. (Courtesy of the Pennsylvania Department of Conservation and Natural Resources [Bugwood image 5017026].)

leaf petioles of date palms (*Phoenix dactylifera* L.) before entering the main stem of the plant where they complete larval development and pupate (Carpenter and Elmer 1978; Giblin-Davis 2001).

3.3.3 Host Tissues Utilized

The host tissues in the major roots, trunks, and branches of woody plants form distinctive bands, starting first (from the outside) with the outer bark and then the inner bark, cambium, sapwood, and heartwood. Cerambycid larvae have evolved to feed on all of these tissues, with some developing almost entirely in a single tissue and others feeding on several tissues. For example, the lepturine *Encyclops caeruleus* (Say) and the cerambycine *Microclytus gazellula* (Haldeman) develop mainly in the outer bark of various North American hardwoods; the lamiine *Acanthocinus* species develop almost entirely in the inner and outer bark of conifers (Figure 3.6); and the cerambycine *Eburia quadrigeminata* (Say) and the lepturine *Pyrotrichus vitticollis* LeConte develop primarily in the heartwood of hardwoods (Craighead 1923; Hardy 1944; Linsley 1962b; Baker 1972; Dodds et al. 2002; Yuan et al. 2008). In most cases when larvae utilize multiple tissues, they complete early larval development in the cambial region and then tunnel deeper into the sapwood—and possibly the heartwood—during late larval development. There are exceptions to this rule, such as the hardwood trunk-infesting species of the lamiine genus *Goes* and the cerambycine *Neoclytus caprea* Say where the newly hatched larvae enter the sapwood with little feeding in the cambial region (Craighead 1923; Solomon 1995). In this chapter, the term cambial region includes the inner bark, cambium, and outer sapwood.

Several hardwood-infesting cerambycids that are native to eastern North America are listed in Table 3.5. The species are grouped by the host tissues typically consumed during larval development, starting with species that feed primarily on the bark and ending with species that feed primarily on wood. Host condition can influence which host tissues are utilized by cerambycid larvae. For example, in the *Ulmus*-infesting cerambycine *Physocnemum brevilineum* (Say), larvae typically feed and later construct pupal cells in the outer bark when developing in living trees, but in cut logs, the larvae develop in the cambial region and construct pupal chambers in the outer sapwood (Haliburton 1951).

FIGURE 3.6 Larvae of the lamiine *Acanthocinus aedilis* (L.) construct galleries primarily in the outer bark of their coniferous hosts. (Courtesy of Valentyna Meshkova [Bugwood image 5425792].)

3.3.4 Host Range

The larval host range of cerambycids varies from some that feed on a single species or genus of plants (monophagous) to cerambycids that develop on multiple plant species, either all within a single plant family (oligophagous) or multiple families (polyphagous). For the cerambycids of Montana (Hart et al. 2013, supplemented with data from http://www.mtent.org/Cerambycidae.html; Linsley 1962a, 1962b, 1963, 1964; Linsley and Chemsak 1972, 1976, 1984, 1995), Fennoscandia and Denmark (Bílý and Mehl 1989), Israel (Sama et al. 2010), and Korea (Lim et al. 2014), for which host genera were provided, the average number of plant genera used as larval hosts was 3.9 genera per cerambycid species (range 1–26 genera) in Montana, 4.4 genera (range 1–16) in Fennoscandia and Denmark, 3.3 genera (1–19) in Israel, and 3.5 genera (1–27) in Korea. These numbers would likely be higher if complete larval host data were known for all cerambycids.

Examples of monophagous North American species would include the cerambycine *Megacyllene robiniae* that develops in *Robinia* trees (Solomon 1995; Figure 3.7) and all species of the lepturine genus *Desmocerus* that develop in species of *Sambucus* (Burke 1921; Figure 3.8). Similarly, in Europe, the lamiine *Saperda punctata* (L.) mostly infests *Ulmus*, and the cerambycine *Xylotrechus antilope* (Schönherr) mostly infests *Quercus* (Bense 1995). One North American example of a polyphagous species is the cerambycine *Neoclytus acuminatus* (Fabricius), which has developed in at least 26 genera of hardwood trees (Linsley 1964; Solomon 1995; Hart et al. 2013) and has also become established in Europe (Cocquempot and Lindelöw 2010). European examples of polyphagous species include the lepturine *Rhagium bifasciatum* Fabricius, which develops in at least 16 genera of conifers and hardwoods (Bílý and Mehl 1989), and the cerambycine *Penichroa fasciata* (Stephens), which develops in at least 19 genera of conifers and hardwoods (Sama et al. 2010).

Several cerambycids have been introduced to new countries as biological control agents for both herbaceous and woody plants because of their high host specificity. For example, the lamiine *Apagomerella versicolor* (Boheman) from Argentina is a potential biological control agent of *Xanthium* species in

TABLE 3.5

Typical Generation Time for Selected Cerambycids Native to the Northeastern United States that Infest Hardwood Trees, Organized by the Host Tissues Consumed by the Larvae, Starting from the Outer Bark and Moving Inward to the Heartwood

Species	Sub-Family[a]	Common Larval Hosts[b]	Adult Length (mm)	Larval Host Tissues[c]	Generation Time (yr)	Tree Parts Infested
Enaphalodes cortiphagus (Craighead)	Cer	*Quercus*	16–30	OB, CAM	3	Trunk
Parelaphidion incertum (Newman)	Cer	*Morus, Quercus*	9–17	OB, CAM	2–3	Trunk
Physocnemum brevilineum (Say)	Cer	*Ulmus*	9–20	OB, CAM, SW	1–2	Trunk, branch
Strophiona nitens (Forster)	Lep	*Castanea, Quercus*	10–15	OB, SW	2	Trunk, branch
Saperda discoidea (Fabricius)	Lam	*Carya, Juglans*	10–11	CAM	1	Trunk
Saperda tridentata Olivier	Lam	*Ulmus*	9–17	CAM	1	Trunk, branch
Enaphalodes rufulus (Haldeman)	Cer	*Quercus*	23–33	CAM, SW	2	Trunk, branch
Glycobius speciosus (Say)	Cer	*Acer*	22–27	CAM, SW	2	Trunk, branch
Neoclytus acuminatus (Fabricius)	Cer	*Fraxinus, Quercus*	4–18	CAM, SW	1	Trunk, branch
Tylonotus bimaculatus Haldeman	Cer	*Fraxinus, Ligustrum*	10–18	CAM, SW	2	Trunk
Xylotrechus quadrimaculatus (Haldeman)	Cer	*Betula, Fagus*	8–16	CAM, SW	1	Branch
Dorcaschema alternatum (Say)	Lam	*Morus, Maclura*	7–16	CAM, SW	1–2	Trunk, branch
Dorcaschema wildii Uhler	Lam	*Morus, Maclura*	16–22	CAM, SW	1–2	Trunk, branch
Plectrodera scalator (Fabricius)	Lam	*Populus, Salix*	25–40	CAM, SW	1–2	Root
Saperda calcarata Say	Lam	*Populus*	20–30	CAM, SW	2–3	Trunk
Saperda cretata Newman	Lam	*Malus, Crateagus*	10–20	CAM, SW	2–3	Trunk, branch
Saperda fayi Bland	Lam	*Crateagus*	12–13	CAM, SW	2	Branch
Saperda inornata Say	Lam	*Populus, Salix*	8–13	CAM, SW	2	Trunk
Megacyllene robiniae (Forster)	Cer	*Robinia*	12–19	CAM, SW, HW	1	Trunk, branch
Aegomorphus morrisi (Uhler)	Lam	*Nyssa*	20–26	CAM, SW, HW	2	Trunk
Saperda candida Fabricius	Lam	*Cydonia, Malus*	13–25	CAM, SW, HW	2–4	Trunk
Saperda vestita Say	Lam	*Tilia*	12–21	CAM, SW, HW	3	Trunk
Purpuricenus axillaris Haldeman	Cer	*Quercus, Castanea*	12–29	SW	2	Branch
Goes debelis LeConte	Lam	*Quercus*	10–16	SW	3–4	Branch
Goes pulcher (Haldeman)	Lam	*Carya, Juglans*	17–25	SW	3–5	Trunk
Goes pulverulentus (Haldeman)	Lam	*Fagus, Quercus*	18–25	SW	3–5	Trunk, branch

(Continued)

TABLE 3.5 *(Continued)*

Typical Generation Time for Selected Cerambycids Native to the Northeastern United States that Infest Hardwood Trees, Organized by the Host Tissues Consumed by the Larvae, Starting from the Outer Bark and Moving Inward to the Heartwood

Species	Sub-Family[a]	Common Larval Hosts[b]	Adult Length (mm)	Larval Host Tissues[c]	Generation Time (yr)	Tree Parts Infested
Goes tesselatus (Haldeman)	Lam	*Quercus*	20–27	SW	3–5	Trunk
Goes tigrinus (De Geer)	Lam	*Quercus*	22–38	SW	3–4	Trunk
Oberea ruficollis (Fabricius)	Lam	*Sassafras*	17	SW	2–3	Trunk
Oberea schaumii LeConte	Lam	*Populus*	12–16	SW	2–3	Branch
Desmocerus palliatus (Forster)	Prio	*Sambucus*	18–27	SW	2–3	Trunk
Dryobius sexnotatus (Linsley)	Cer	*Acer, Tilia*	20–26	SW, HW	2–3	Trunk
Xylotrechus aceris Fisher	Cer	*Acer*	10–14	SW, HW	2	Trunk, branch
Neandra brunnea (Fabricius)	Par	*Juglans, Carya*	8–12	SW, HW	3–4	Trunk
Prionus imbricornis (L.)	Prio	*Quercus, Castanea*	31–42	SW, HW	3–5	Root

Source: Data based almost entirely on Solomon, J. D., *Guide to insect borers in North American broadleaf trees and shrubs.* USDA Forest Service, Washington, DC, 1995.

[a] Cer = Cerambycinae; Lam = Lamiinae; Lep = Lepturinae; Par = Parandrinae; Prio = Prioninae.

[b] Primary larval hosts listed in Solomon (1995).

[c] Host tissues include OB = outer bark; CAM = cambial region, including inner bark and outer sapwood; SW = sapwood, and HW = heartwood.

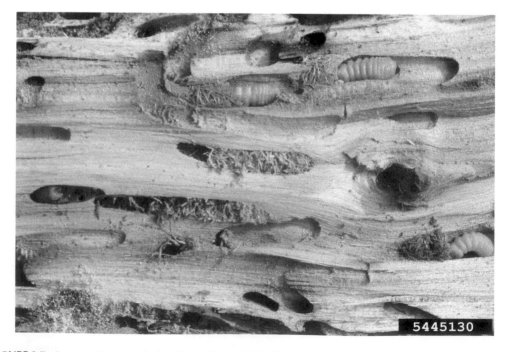

FIGURE 3.7 Larvae of the cerambycine *Megacyllene robiniae* (Forster) construct galleries throughout the sapwood and heartwood of their typical larval host, black locust (*Robinia pseudoacacia* L.). (Courtesy of Whitney Cranshaw [Bugwood image 5445130].)

FIGURE 3.8 Pupae of the lepturine *Desmocerus palliatus* (Forster) in the larval galleries that extend downward in the stems to the base of the plant—primarily in the pith of its host elder (*Sambucus*). (Courtesy of James Solomon [Bugwood image 3066074].)

the United States (Logarzo et al. 2002); the cerambycine *Megacyllene mellyi* (Chevrolat) from South America was introduced into Australia to control *Baccharis halimifolia* L. (McFadyen 1983); the lamiine *Nealcidion cereicola* (Fisher) from South America was introduced into Australia for biological control of *Harrisia* (= *Eriocereus*) *martinii* (Labouret) Britton (McFadyen and Fidalgo 1976); the lamiine *Oberea erythrocephala* (Schrank) from Europe was introduced into North America to control *Euphorbia esula* L. (Hansen et al. 1997); and the lamiines *Aerenicopsis championi* Bates and *Plagiohammus spinipennis* (Thoms.) from Mexico were introduced into Hawaii to control *Lantana camara* L. (Davis et al. 1993).

Typically, species that develop in healthy live plants tend to be monophagous or oligophagous, while those that develop in dead or decaying hosts tend to be polyphagous (Hanks 1999). However, given that there are thousands of cerambycid species worldwide, there are always exceptions. For example, the two Asian lamiines, *Anoplophora chinensis* (Forster) and *A. glabripennis* (Motschulsky), can develop in dozens of genera of apparently healthy hardwood trees and eventually kill them, which is the key reason why active eradication programs have been initiated in Europe and North America—where these two beetles have been introduced (MacLeod et al. 2002; Haack et al. 2010; Turgeon et al. 2015; Rassati et al. 2016; see Chapter 13).

3.3.5 Host Condition

Cerambycid larvae develop in host plants that vary in condition from healthy to dead and from moist to dry. Hanks (1999) noted the following general trends in the condition of the host plants selected for oviposition by adult females in several of the cerambycid subfamilies: Lepturinae, Prioninae, and Spondylidinae often develop in dead and decaying wood; Lamiinae usually develop in living and weakened hosts but seldom in dead hosts, whereas Cerambycinae develop in living, dying, and dead hosts. Many cerambycids that develop in living trees—but not all (see as follows) can complete development even in dead wood, especially when moisture levels are maintained at suitable levels. Although most cerambycids show an ovipositional preference for hosts in a particular condition, this relationship can be altered when the insect encounters new hosts. For example, in its native range of Europe, the spondylidine *Tetropium fuscum* F. tends to infest stressed or recently cut *Picea* trees, whereas in Canada, where it was introduced, it infests apparently healthy *Picea* trees (Flaherty et al. 2011). Another widespread European species, the cerambycine *Xylotrechus arvicola*, whose larvae generally feed on dead and decaying wood of hardwood trees, has recently been found to infest and damage living grape stems, branches, and roots (García-Ruiz et al. 2012).

Examples of cerambycid genera whose species are commonly associated with living hosts include *Anoplophora, Enaphalodes, Goes, Lamia, Megacyllene, Oberea, Oncideres, Plectrodera*, and *Saperda* (Figure 3.9). Similarly, examples of cerambycids that typically infest dead hosts are members of the genera *Arhopalus, Ergates, Parandra*, and *Rhagium* (Craighead 1923; Linsley 1959; Bílý and Mehl 1989; Solomon 1995). The requirement for living hosts appears to be particularly strong in species of the lamiine genera *Goes* and *Saperda*, given that their larvae seldom complete development when an infested tree is cut (Craighead 1923; Linsley 1959). Similarly, cerambycids that develop in the roots and stems of herbaceous plants usually select living hosts for oviposition (Piper 1978; Bense 1995; Rejzek et al. 2001), but some, such as the lamiines *Lepromoris gibba* (Brulle) (Duffy 1953) and *Parmena pubescens* (Dalman) (Duffy 1957), infest dead stalks of herbaceous Euphorbiaceae, perhaps because less milky sap is present in the dead tissue (Duffy 1953). When considering wood moisture content, several species of *Mallodon, Rhagium*, and *Rutpela* (=*Stragalic*) generally favor moist, decaying wood, while many species of *Chlorophorus, Gracilia, Hylotrupes*, and *Stromatium* develop in dry wood (Craighead 1923; Duffy 1953; Linsley 1959; Bense 1995). As logs decompose from the time of initial death or cutting to wood in advanced stages of decay, there are successional changes in the wood borer community structure, including cerambycids, providing further evidence that cerambycids vary in their preferences for hosts of a particular condition (Blackman and Stage 1924; Graham 1925; Savely 1939; Parmelee 1941; Howden and Vogt 1951; Haack et al. 1983; Khan 1985; Harmon et al. 1986; Hanula 1996; Saint-Germain et al. 2007; Costello et al. 2013; Lee et al. 2014; Ulyshen 2016).

FIGURE 3.9 Larva, larval galleries, pupa, and adult of the lamiine *Saperda calcarata* on its larval host, eastern cottonwood (*Populus deltoides* Bartram ex Marsh.), in the eastern United States. (Courtesy of James Solomon [Bugwood image 0284067].)

3.4 Larval Development, Nutrition, and Voltinism

As mentioned in Sections 3.3.2 and 3.3.3, cerambycid larvae feed and develop in almost every major plant part (roots, stems, branches, fruit, and seeds) and plant tissue (outer bark, inner bark, cambium, sapwood, and heartwood). These plant tissues differ greatly in thickness, nutrient levels, amount of living cells, degree of lignification (toughness), and their functional role within plants (Kramer and Kozlowski 1979; Haack and Slansky 1987; Pallardy 2008) and therefore can have a strong influence on cerambycid development rates and voltinism.

A brief discussion follows on the major physical and nutritional differences among these tree tissues based largely on the work by Kramer and Kozlowski (1979), Haack and Slansky (1987), and Pallardy (2008). The outer bark is largely dead, dense, corky tissue that functions in protecting the underlying tissues and reducing water loss. Outer bark generally is low in water and nutrients. By contrast, the inner bark (often called phloem) largely is soft living tissue, consisting of thin-walled cells. The inner bark is rich in water and nutrients and is the major tissue for the transport of photosynthates. The vascular cambium (or simply the cambium) is the ring of living cells that produces phloem cells to the outside and xylem (wood) cells to the inside. Because the cambium consists of living cells and is meristematically active, it likely has the highest water and nutrient levels of any major tissue within a woody plant. The sapwood consists mostly of dead, highly lignified cells that function in water and mineral transport. About 5–35% of the sapwood is living parenchyma cells, depending on the tree species (Panshin and de Zeeuw 1980). The water content of sapwood generally is high, being similar to inner bark, but the nutrient levels are usually much lower than the cambium or inner bark. The most nutritious portion of the sapwood is the outer portion closest to the cambium. Parenchyma cells die during the transition from sapwood to heartwood, and thus heartwood consists of dead tissue (Spicer 2005). Heartwood, in comparison with sapwood, generally is similar in density and lower in nutrients but is higher in secondary compounds. The water content of heartwood usually is similar to that of sapwood in hardwoods but is lower in conifers (Peck 1959; Skarr 1972).

Considering the overall nutritional quality of these tissues, cambium would rank highest, followed by inner bark, then sapwood, with heartwood and outer bark being lowest. Of course, the volume of each of these tissues available for larval consumption also varies considerably. For example, in the trunk of a large tree, the cambium would provide the smallest volume of these major tissues, followed by the inner bark, and likely then the outer bark, sapwood, and heartwood. However, great variation can occur in the thickness of these tissues, depending on plant age and diameter, as well as among and within plant families, genera, and species (Wilkins 1991; Sellin 1994; Pallardy 2008). Another tissue that varies dramatically in width is the pith, which commonly is found at the center of young stems and branches and initially consists of soft, spongy parenchyma cells. The pith usually is very distinct in young branches and stems of woody plants, but it becomes crushed and difficult to discern in older stem sections. In species of elderberry (*Sambucus*), which usually grow as shrubs or small trees, the pith is relatively wide and solid and is the primary tissue consumed by larvae of the North American lepturine *Desmocerus* species, especially the early instars (Burke 1921; Solomon 1995; Figure 3.8).

Many factors can influence cerambycid development time, including nutritional quality of the host tissues, host condition, and ambient temperatures. As detailed by Haack and Slansky (1987) for tree-infesting, temperate-zone cerambycids, development usually occurs in one to two years for larvae that feed primarily in the nutrient-rich cambial region, in two to three years when development occurs in both the cambial region and sapwood, and in three years or longer when most development occurs in the sapwood and heartwood. To demonstrate this pattern with cerambycids, several species native to the eastern United States are listed in Table 3.5 based on life-history data presented in Solomon (1995). These cerambycids are grouped by the host tissues commonly consumed by larvae of each species, starting with species that feed primarily on bark and ending with species that feed primarily on wood. An attempt was made to select species that were broadly similar in adult size so that voltinism patterns could be compared more on the basis of variation in nutritional quality of the host tissues consumed rather than on variation in final body size. Overall, for the species listed in Table 3.5, cerambycids that developed primarily in the outer and inner bark of tree trunks often required two to three years to complete one generation.

Those that developed primarily in the cambial region typically were univoltine. For species that developed in the cambial region as well as the sapwood, one to two years usually were required to complete development, although some species needed two to three years. For species that developed almost entirely in the sapwood or sapwood and heartwood, two to three years to as many as three to five years usually were required to complete one generation (Table 3.5).

There are exceptions to the aforementioned voltinism patterns. For example, the North American cerambycine *Megacyllene robiniae* is univoltine on its host (*Robinia*), although the larvae spend much of their life tunneling and feeding in the sapwood and heartwood (Galford 1984; Harman and Harman 1990). Similarly, the Asian lamiine *A. glabripennis*, which develops in dozens of species of living hardwood trees, generally is univoltine, although it too feeds extensively in sapwood and heartwood (Haack et al. 2010). The high microbial diversity in the guts of *A. glabripennis* larvae likely improves the digestion and assimilation of woody tissues and thereby may allow for their relatively short generation time (Geib et al. 2008, 2009a, 2009b; Scully et al. 2014).

Although most wood-infesting cerambycids complete development within a few years, there are numerous records of adults emerging from various wood products many years, or even decades, after the product was constructed. In these cases, oviposition is presumed to have occurred in the forest or lumberyard prior to milling the logs rather than sometime after construction of the wood products. This scenario can be assumed to have happened for most of the cerambycids reared from wood products given that these species require bark for oviposition, none of which usually is present on the final constructed product. Several examples of prolonged cerambycid development are shown in Table 3.6, with the longest being for the cerambycine *Eburia quadrigeminata* that emerged from a bookcase that was constructed more than 40 years earlier. More examples of protracted cerambycid development are given in Packard (1881) and Duffy (1953), and similar examples exist for several species of wood-boring Buprestidae (Coleoptera) (Spencer 1930; Linsley 1943; Smith 1962a, 1962b), with the longest records exceeding 50 years for *Buprestis aurulenta* L. (Smith 1962a). Although such records are exceptions, these data provide evidence that certain buprestids and cerambycids have the longest generation times of all insects.

Over the years, some authors have questioned the validity of these records of prolonged development, suggesting that oviposition could have occurred on the actual wood products sometime after final construction

TABLE 3.6

Summary Data for Cerambycids that Exhibited Prolonged Larval Development in Various Wood Items

Species	Sub-Family[a]	Country (State or Province)[b]	Wood Item	Min. Age (yr)[d]	References
Anaglyptus mysticus (L.)	Cer	The United Kingdom	*Fagus* drawer[c]	13	Hickin 1947
Phymatodes dimidiatus (Kirby)	Cer	Canada (BC)	*Abies* rafters	6	Spencer 1930
Eburia quadrigeminata (Say)	Cer	United States (IN)	*Acer* flooring	14	Webster 1889
Eburia quadrigeminata	Cer	United States (IA)	*Betula* bookcase	40	Jaques 1918
Eburia quadrigeminata	Cer	United States (IN)	*Fraxinus* door sill	19	McNeil 1886
Eburia quadrigeminata	Cer	United States (IN)	Bedstead (*Quercus*)[c]	20	Troop 1915
Eburia quadrigeminata	Cer	The United Kingdom	*Quercus* wardrobe	19	Hickin 1951
Ergates faber (L.)	Prio	The United Kingdom	*Pinus* pier	20	Fraser 1948
Hylotrupes bajulus (L.)	Cer	Canada	Dry wood in attic	12–15	Campbell et al. 1989
Hylotrupes bajulus	Cer	The United Kingdom	*Pinus* cupboard	17	Bayford 1938

[a] Cer = Cerambycinae; Prio = Prioninae.
[b] BC = British Columbia; IA = Iowa; IN = Indiana.
[c] Likely host, but not confirmed.
[d] Values represent the likely number of years since construction of the wood items prior to adult emergence and thus the estimated minimum number of years required to complete development.

(Spencer 1930; Beer 1949). However, some of the records are of buprestids and cerambycids that were native to the country where the furniture or wood product was constructed but that were not native to the country where the beetles emerged (Linsley 1943; Smith 1962a). Given that moisture content and nutrient levels decline as wood dries (Haack and Slansky 1987), it is not surprising that larval development would be prolonged when larvae are present in finished wood products that are relatively dry and kept indoors.

3.5 Wood Digestion

The woody environment is fibrous, tough, and nutritionally poor and therefore presents many challenges to wood-boring insects as they tunnel and feed in wood (Haack and Slansky 1987). For example, the nitrogen content of wood typically ranges from only 0.03% to 0.1% on a dry weight basis (Cowling and Merrill 1966). Considering the tree trunk in cross-section, the nitrogen content of sapwood generally decreases from the annual rings nearest the cambium inward to the sapwood–heartwood interface and then stays relatively constant throughout the heartwood until rising somewhat again near the pith at the center of the trunk or branch (Merrill and Cowling 1966). The toughness of wood results from the highly polymerized cell walls, which provide rigidity to woody plants and consist primarily of cellulose microfibrils, hemicellulose, and lignin (Gilbert 2010). Cellulose consists of thousands of glucose molecules, linked end-to-end in long straight chains that bind with other cellulose molecules to form microfibrils. By contrast, hemicellulose is a branched chain of mostly five carbon sugars that help link cellulose and lignin in each major layer of the cell wall. Lignin is an aromatic three-dimensional polymer that acts to cement the microfibrils together and gives woody plants their rigidity (Rowell et al. 2005). In North American conifers, for example, 40–45% of wood on a dry weight basis is cellulose, 7–14% is hemicellulose, and 26–34% is lignin. By contrast, in North American hardwoods, 38–49% of wood is cellulose, 19–26% is hemicellulose, and 23–30% is lignin (Rowell et al. 2005).

For more than two centuries, biologists have been interested in understanding how wood-boring insects are able to develop and survive in such a harsh environment as sapwood and heartwood. Mansour and Mansour-Bek (1934) and Parkin (1940) reviewed the early research in this field, including a discussion of the researchers involved, the test insects used, and their general findings and interpretations. A variety of techniques were utilized in these early studies but most involved comparing the chemical constituents of larval frass with the wood being consumed to detect differences as well as testing extracts from the borer's gut for their ability to enzymatically degrade specific wood constituents (Parkin 1940). As a result of these early studies, researchers developed a basic understanding that symbiotic microorganisms were involved in wood digestion in insects through enzymatic activity, including bacteria, fungi, and protozoans (Mansour and Mansour-Bek 1934; Parkin 1940; Graham 1967; Breznak 1982; Breznak and Brune 1994). More recently, using modern molecular and biochemical techniques, many more details have been elucidated about the symbionts, enzymes, and genes involved in wood degradation (Sugimura et al. 2003; Lee et al. 2004; Geib et al. 2008; Zhou et al. 2009; Watanabe and Tokuda 2010; Calderón-Cortés et al. 2012; Scully et al. 2013; Brune and Dietrich 2015). Moreover, it is recognized that wood-degrading enzymes in cerambycids are produced both exogenously (symbiont dependent) (Delalibera et al. 2005; Park et al. 2007; Zhou et al. 2009; Geib et al. 2009b) and endogenously (symbiont independent) (Scrivener et al. 1997; Lee et al. 2005; Wei et al. 2006; Calderón-Cortés et al. 2012).

In the early 1900s, several cerambycids were known to harbor yeast-like fungi as endosymbionts in their midgut intestinal walls (Graham 1967). For example, Schomann (1937) reported that fungal endosymbionts were common in conifer-infesting cerambycid larvae but were rare in hardwood-infesting cerambycids. More recently, Grünwald et al. (2010) described several strains of ascomycetous yeasts in the guts of conifer-infesting cerambycids. In addition, Geib et al. (2008) demonstrated that certain gut fungi in *A. glabripennis* aided in lignin degradation. Transfer of symbiotic fungi between generations is accomplished during oviposition when fungi are deposited externally on the egg surface, with the new larvae becoming inoculated as they chew through the egg chorion (Schomann 1937; Graham 1967). In addition to fungal endosymbionts, several cerambycids utilize cellulolytic enzymes from fungi that they ingest while tunneling in wood (i.e., so-called acquired digestive enzymes). Cerambycids became the focus of this line of research in the 1980s, using species such as the conifer-infesting lamiine

Monochamus marmorator Kirby and the hardwood-infesting lamiine *Saperda calcarata* Say (Martin 1983; Kukor and Martin 1986a, 1986b; Martin 1992).

The role of bacteria in the digestion of cellulose, hemicellulose, and other polysaccharides in cerambycid larvae has been recognized for decades (Mansour and Mansour-Bek 1934; Parkin 1940). The bacterial diversity in the guts of cerambycid larvae has been elucidated in several species recently, with several classes of bacteria identified such as Actinobacteria and Gammaproteobacteria (Delalibera et al. 2005; Heo et al. 2006; Schloss et al. 2006; Park et al. 2007; Mazza et al. 2014). It is important to note that the bacterial community in the insect's gut is highly variable and can be influenced by the host plant. For example, Gieb et al. (2009a) demonstrated that the community of gut bacteria in *A. glabripennis* was most diverse when larvae fed in their preferred host trees (*Acer*) but much less diverse when they fed in nonpreferred hosts (*Pyrus*). Moreover, Schloss et al. (2006) reported that bacterial diversity was relatively high in the guts of *A. glabripennis* larvae, which have a broad host range, whereas bacterial diversity was relatively low in *Saperda vestita* Say, a lamiine with a narrow host range (primarily *Tilia*). It would be interesting to determine if bacterial diversity typically is greater in the gut tract of polyphagous cerambycid larvae compared to monophagous species. As for the transfer of bacteria between generations of cerambycids, it likely occurs during oviposition with bacteria being placed on the outside or inside of eggs or deposited near the oviposition site (Gieb et al. 2009b).

3.6 Summary and Future Directions

The thousands of cerambycid species found worldwide display great diversity in their feeding habits, including a wide variety of host plants, plant parts, and tissues consumed. The great success that cerambycids have had in exploiting the woody environment is related to their ability to enzymatically degrade many wood constituents through symbionts and endogenously produced enzymes. Knowledge of the feeding habits of adult and larval cerambycids can be used in developing integrated pest management programs for pest species. For example, knowing when and where adults feed is useful when scheduling detection surveys and pesticide applications. Similarly, knowing in which plant tissues the larvae feed and tunnel allows managers to judge the value, for example, of using systemic insecticides for their control, knowing that larvae that feed in or close to the cambial region would be much more susceptible to systemic insecticides than those that feed deep within the sapwood or heartwood (Poland et al. 2006). Although much has been learned about the feeding biology of the world's cerambycids, there are still many gaps in our basic understanding of their life history, larval and adult host plants, digestive symbionts, and wood-degrading enzymes; therefore, much more research remains to be conducted on these topics.

ACKNOWLEDGMENTS

The author thanks Toby Petrice for comments on an earlier draft of this chapter and the many photographers who supplied images to the Bugwood website at the University of Georgia.

REFERENCES

Akbulut, S., and W. T. Stamps. 2012. Insect vectors of the pinewood nematode: a review of the biology and ecology of *Monochamus* species. *Forest Pathology* 42:89–99.

Alya, A. B., and F. P. Hain. 1985. Life histories of *Monochamus carolinensis* and *M. titillator* (Coleoptera: Cerambycidae) in the Piedmont of North Carolina. *Journal of Entomological Science* 20:390–397.

Baker, W. L. 1972. *Eastern forest insects*. Washington, DC: USDA Forest Service, Miscellaneous Publication No. 1175.

Baragaño Galán, J. R., A. Notario Gómez, and C. Sa Montero. 1981. *Agapanthia asphodeli* Latreille (Col.: Cerambycidae): cría artificial y estudio cariológico. *Boletín de Sanidad Vegetal - Plagas* 7:161–167.

Barak, A. V., W. D. Yang, D. J. Yu, et al. 2009. Methyl bromide as a quarantine treatment for *Chlorophorus annularis* F. (bamboo borer) (Coleoptera: Cerambycidae) in raw bamboo poles. *Journal of Economic Entomology* 102:913–920.

Barševskis, A., and N. Savenkov. 2013. Contribution to the knowledge of long-horned beetles (Coleoptera: Cerambycidae) in Latvia. *Baltic Journal of Coleopterology* 13:91–102.

Bayford, E. G. 1938. Retarded development: *Hylotrupes bajulus* L. *The Naturalist, London* 980:254–256.

Beer, F. M. 1949. The rearing of Buprestidae and delayed emergence of their larvae. *The Coleopterists Bulletin* 3:81–84.

Benham, G. S., and R. J. Farrar. 1976. Notes on the biology of *Prionus laticollis* (Coleoptera: Cerambycidae). *Canadian Entomologist* 108:569–576.

Bense, U. 1995. *Longhorn beetles. Illustrated key to the Cerambycidae and Vesperidae of Europe.* Weikersheim: Margraf Verlag.

Beutenmuller, W. 1896. Food-habits of North American Cerambycidae. *Journal of the New York Entomological Society* 4:73–81.

Bílý, S., and O. Mehl. 1989. Longhorn beetles (Coleoptera, Cerambycidae) of Fennoscandia and Denmark. *Fauna Entomologica Scandinavica* 22:1–204.

Blackman, M. W., and H. H. Stage. 1924. On the succession of insects living in the bark and wood of dying, dead, and decaying hickory. *New York State College of Forestry, Syracuse, New York, Technical Publication* 17:3–269.

Blatchley, W. S. 1928. Notes on some Florida Coleoptera with descriptions of new species. *Canadian Entomologist* 60:60–73.

Breznak, J. A. 1982. Intestinal microbiota of termites and other xylophagous insects. *Annual Review of Microbiology* 36:323–343.

Breznak, J. A., and A. Brune. 1994. Role of microorganisms in the digestion of lignocellulose by termites. *Annual Review of Entomology* 39:453–487.

Brooks, F. E. 1919. *The roundheaded apple-tree borer.* Washington, DC: USDA Farmers Bulletin 675.

Brune, A., and C. Dietrich. 2015. The gut microbiota of termites: digesting the diversity in the light of ecology and evolution. *Annual Review of Microbiology* 69:145–166.

Burakowski, B. 1979. Immature stages and bionomics of *Vadonia livida* (F.) Coleoptera, Cerambycidae. *Annales Zoologici (Warsaw)* 352:25–42.

Burke, H. E. 1921. Biological notes on *Desmocerus*, a genus of roundhead borers, the species of which infests various elders. *Journal of Economic Entomology* 14:450–452.

Butovitsch, V. 1939. Zur Kenntnis der Paarung, Eiablage und Ernährung der Cerambyciden. *Entomologisk Tidskrift* 60:206–258.

Calderón-Cortés, N., M. Quesada, H. Watanabe, H. Cano-Camacho, and K. Oyama. 2012. Endogenous plant cell wall digestion: a key mechanism in insect evolution. *Annual Review of Ecology Evolution and Systematics* 43:45–71.

Campbell, J. M., M. J. Sarazin, and D. B. Lyons. 1989. *Canadian beetles (Coleoptera) injurious to crops, ornamentals, stored products and buildings.* Ottawa: Agriculture Canada, Publication 1826.

Carpenter, J. B., and H. S. Elmer. 1978. *Pests and diseases of the date palm.* Washington, DC: U.S. Department of Agriculture, Agriculture Handbook No. 527.

Champion, H. G. 1919. A cerambycid infesting pine cones in India, *Chlorophorus strobilicola*, n. sp. *Entomologist's Monthly Magazine* 58:219–224.

Chemsak, J. A. 1972: A new seed inhabiting cerambycid from Costa Rica (Coleoptera). *The Pan-Pacific Entomologist* 48:150–152.

Chemsak, J. A., and J. Powell. 1966. Studies on the bionomics of *Tragidion armatum* LeConte. *The Pan-Pacific Entomologist* 42:36–47.

Cocquempot, C., and A. Lindelöw. 2010. Longhorn beetles (Coleoptera, Cerambycidae). *BioRisk* 4:193–218. doi: 10.3897/biorisk.4.56

Cornelissen, J. H. C., U. Sass-Klaassen, L. Poorter, et al. 2012. Controls on coarse wood decay in temperate tree species: birth of the LOGLIFE experiment. *Ambio* 41(Supplement 3):231–245

Costello, S. L., W. R. Jacobi, and J. F. Negrón. 2013. Emergence of Buprestidae, Cerambycidae, and Scolytinae (Coleoptera) from mountain pine beetle-killed and fire-killed ponderosa pines in the Black Hills, South Dakota, USA. *The Coleopterists Bulletin* 67:149–154.

Cowling, E. B., and W. Merrill. 1966. Nitrogen in wood and its role in wood deterioration. *Canadian Journal of Botany* 44:1539–1554.

Craighead, F. C. 1923. North American cerambycid-larvae. *Bulletin of the Canada Department of Agriculture* 27:1–239.

Davis, C. J., E. Yoshioka, and D. Kageler. 1993. Biological control of *Lantana*, prickly pear, and *Hamakua pamakane* in Hawaii: a review and update. In *Alien plant invasions in native ecosystems of Hawaii*, ed. C. P. Stone, C. W. Smith, and J. T. Tunison, 411–431. Honolulu: University of Hawaii Press.

Delalibera, I., J. Handelsman, and K. F. Raffa. 2005. Contrasts in cellulolytic activities of gut microorganisms between the wood borer, *Saperda vestita* (Coleoptera: Cerambycidae), and the bark beetles, *Ips pini* and *Dendroctonus frontalis* (Coleoptera: Curculionidae). *Environmental Entomology* 34:541–547.

Dodds, K. J., C. Graeber, and F. M. Stephen. 2001. Facultative intraguild predation by larval Cerambycidae (Coleoptera) on bark beetle larvae (Coleoptera: Scolytidae). *Environmental Entomology* 30:17–22.

Dodds K. J., C. Graber, and F. M. Stephen. 2002. Oviposition biology of *Acanthocinus nodosus* (Coleoptera: Cerambycidae) in *Pinus taeda*. *Florida Entomologist* 85:452–457.

Donley, D. E., and E. Rast. 1984. Vertical distribution of the red oak borer, *Enaphalodes rufulus* (Coleoptera: Cerambycidae), in red oak. *Environmental Entomology* 13:41–44.

Duffy, E. A. J. 1953. *A monograph of the immature stages of African timber beetles (Cerambycidae)*. London: British Museum (Natural History).

Duffy, E. A. J. 1957. *A monograph of the immature stages of British and imported timber beetles (Cerambycidae)*. London: British Museum (Natural History).

Duffy, E. A. J. 1960. *A monograph of the immature stages of Neotropical timber beetles (Cerambycidae)*. London: British Museum (Natural History).

Edwards, J. S. 1961. On the reproduction of *Prionoplus reticularis* (Coleoptera, Cerambycidae), with general remarks on reproduction in the Cerambycidae. *Quarterly Journal of Microscopical Science* 102:519–529.

Eisner, T., F. C. Kafatos, and E. G. Linsley. 1962. Lycid predation by mimetic adult Cerambycidae (Coleoptera). *Evolution* 16:316–324.

Eisner, T., F. C. Schroeder, N. Snyder, et al. 2008. Defensive chemistry of lycid beetles and of mimetic cerambycid beetles that feed on them. *Chemoecology* 18:109–119.

Ferguson, A. W., and P. S. Williamson. 2009. A new host plant record, the endangered star cactus, *Astrophytum asterias* (Zuccarini) Lemaire, for *Moneilema armatum* LeConte (Coleoptera: Cerambycidae: Lamiinae). *The Coleopterists Bulletin* 63:218–220.

Fisher, W. S. 1917. A new species of longhorn beetle infesting cowpeas from Mexico. *Proceedings of the Entomological Society of Washington* 19:173–174.

Flaherty, L., J. D. Sweeney, D. Pureswaran, and D. T. Quiring. 2011. Influence of host tree condition on the performance of *Tetropium fuscum* (Coleoptera: Cerambycidae). *Environmental Entomology* 40:1200–1209.

Fraser, M. G. 1948. *Ergates faber* F. (Col. Prionidae) in Lancashire. *Entomologists Monthly Magazine* 84:47.

Friedman, A. L. L., O. Rittner, and V. I. Chikatunov. 2008. Five new invasive species of longhorn beetles (Coleoptera: Cerambycidae) in Israel. *Phytoparasitica* 36:242–246.

Furniss, R. L. and V. M. Carolin. 1977. *Western forest insects*. Washington, DC: USDA Forest Service, Miscellaneous Publication No. 1339.

Galford, J. R. 1984. *The locust borer*. Washington, DC: USDA Forest Service, Forest Insect & Disease Leaflet 71.

García-Ruiz, E., V. Marco, and I. Perez-Moreno. 2012. Laboratory rearing and life history of an emerging grape pest, *Xylotrechus arvicola* (Coleoptera: Cerambycidae). *Bulletin of Entomological Research* 102:89–96.

Geib, S. M., T. R. Filley, P. G. Hatcher, et al. 2008. Lignin degradation in wood-feeding insects. *Proceedings of the National Academy of Sciences of the United States of America* 105:12932–12937.

Geib, S. M., M. D. M. Jimenez-Gasco, J. E. Carlson, M. Tien, and K. Hoover. 2009a. Effect of host tree species on cellulase activity and bacterial community composition in the gut of larval Asian longhorned beetle. *Environmental Entomology* 38:686–699.

Geib, S. M., M. D. M. Jimenez-Gasco, J. E. Carlson, M. Tien, R. Jabbour, and K. Hoover. 2009b. Microbial community profiling to investigate transmission of bacteria between life stages of the woodboring beetle, *Anoplophora glabripennis*. *Microbial Ecology* 58:199–211.

Giblin-Davis, R. M. 2001. Borers of palms. In *Insects on palms*, ed. W. Howard, D. Moore, R. M. Giblin-Davis, and R. G. Abad, 267–304. Wallingford, CT: CABI Publishing.

Gilbert, H. J. 2010. The biochemistry and structural biology of plant cell wall deconstruction. *Plant Physiology* 153:444–455.

Goodwin, S., and M. A. Pettit. 1994. *Acalolepta vastator* (Newman) (Coleoptera: Cerambycidae) infesting grapevines in the Hunter Valley, New-South-Wales. 2. Biology and ecology. *Journal of the Australian Entomological Society* 33:391–397.

Graham, S. A. 1925. The felled tree trunk as an ecological unit. *Ecology* 6:397–411.

Graham, K. 1967. Fungal-insect mutualism in trees and timber. *Annual Review of Entomology.* 12:105–126.

Gressitt, J. L., and C. J. Davis. 1972. *Seasonal occurrence and host-lists of Hawaiian Cerambycidae.* Honolulu: Island Ecosystems IRP, U.S. International Biological Program, International Biological Program Technical Report No. 5.

Grünwald, S., M. Pilhofer, and W. Höll. 2010. Microbial associations in gut systems of wood- and bark-inhabiting longhorned beetles (Coleoptera: Cerambycidae). *Systematic and Applied Microbiology* 33:25–34.

Grzymala, T. L., and K. B. Miller. 2013. Taxonomic revision and phylogenetic analysis of the genus *Elytroleptus* Dugés (Coleoptera: Cerambycidae: Cerambycinae: Trachyderini). *Zootaxa* 3659:1–62.

Gutowski, J. M. 1990. Pollination of the orchid *Dactylorhiza fuchsii* by longhorn beetles in primeval forests of Northeastern Poland. *Biological Conservation* 51:287–297.

Gwynne, D. T., and B. B. Hostetler. 1978. Mass emergence of *Prionus emarginatus* (Say) (Coleoptera Cerambycidae). *The Coleopterists Bulletin* 32:347–348.

Haack, R. A., and F. Slansky. 1987. Nutritional ecology of wood-feeding Coleoptera, Lepidoptera, and Hymenoptera. In *Nutritional ecology of insects, mites, spiders and related invertebrates*, ed. F. Slansky Jr., and J. G. Rodriguez, 449–486. New York: John Wiley.

Haack, R. A., D. M. Benjamin, and K. D. Haack. 1983. Buprestidae, Cerambycidae, and Scolytidae associated with successive stages of *Agrilus bilineatus* (Coleoptera: Buprestidae) infestations of oaks in Wisconsin. *The Great Lakes Entomologist* 16:47–55.

Haack, R. A., L. S. Bauer, R.-T. Gao, et al. 2006. *Anoplophora glabripennis* within-tree distribution, seasonal development, and host suitability in China and Chicago. *The Great Lakes Entomologist* 39: 169–183.

Haack, R. A., F. Hérard, J. Sun, and J. J. Turgeon. 2010. Managing invasive populations of Asian longhorned beetle and citrus longhorned beetle: a worldwide perspective. *Annual Review of Entomology* 55:521–546.

Hajek, A. E., and D. M. Kalb. 2007. Suitability of *Acer saccharum* and *Acer pensylvanicum* (Aceraceae) for rearing *Anoplophora glabripennis* (Coleoptera: Cerambycidae). *Canadian Entomologist* 139:751–755.

Haliburton, W. 1951. On the habits of the elm bark borer *Physocnemum brevilineum* (Say); (Coleoptera: Cerambycidae). *Canadian Entomologist* 83:36–38.

Hanks, L. M. 1999. Influence of the larval host plant on reproductive strategies of cerambycid beetles. *Annual Review of Entomology* 44:483–505.

Hansen, R. W., R. D. Richard, P. E. Parker, and L. E. Wendel. 1997. Distribution of biological control agents of leafy spurge (*Euphorbia esula* L.) in the United States: 1988–1996. *Biological Control* 10:129–142.

Hanula, J. L. 1996. Relationship of wood-feeding insects and coarse woody debris. In *Proceedings of the Workshop on Coarse Woody Debris in Southern Forests: Effects on Biodiversity*, ed. J. W. McMinn, and D. A. Crossley Jr., 55–81. Asheville, NC: USDA Forest Service, Southern Research Station, General Technical Report SE-94.

Hardy, G. A. 1944. Further notes on the Cerambycidae of Vancouver Island (Coleoptera). *Proceedings of the Entomological Society of British Columbia* 41:15–18.

Harman, D. M., and A. L. Harman. 1990. Height distribution of locust borer attacks (Coleoptera: Cerambycidae) in black locust. *Environmental Entomology* 19:501–504.

Harmon, M. E., J. F. Franklin, F. J. Swanson, et al. 1986. Ecology of coarse woody debris in temperate ecosystems. *Advances in Ecological Research* 15:133–302.

Hart, C. J., J. S. Cope, and M. A. Ivie. 2013. A checklist of the Cerambycidae (Coleoptera) of Montana, USA, with distribution maps. *The Coleopterists Bulletin* 67:133–148.

Hawkeswood, T. J., and J. R. Turner. 2007. Record of pollination of *Lomatia silaifolia* (Sm.) R.Br. (Proteaceae) by the longicorn beetle *Uracanthus triangularis* (Hope, 1833) (Coleoptera: Cerambycidae). *Calodema Supplementary Paper* 53:1–3.

Heo, S., J. Kwak, H. W. Oh, et al. 2006. Characterization of an extracellular xylanase in *Paenibacillus* sp. HY-8 isolated from an herbivorous longicorn beetle. *Journal of Microbiology and Biotechnology* 16:1753–1759.

Hernandez-Jaramillo, A., O. P. Pinzón, and A. Parrado-Rosselli. 2012. Seed predation of *Sapindus saponaria* L. by *Leptostylus gibbulosus* Bates (Coleoptera: Cerambycidae) and its effects on germination. *Colombia Forestal* 15:247–260.

Hickin, N. E. 1947. The length of larval stage of *Anaclyptus* (*Clytus*) *mysticus* L. (Col., Cerambycidae). *Entomologists Monthly Magazine* 83:224.

Hickin, N. E. 1951. Delayed emergence of *Eburia quadrigeminata* Say (Col., Cerambycidae). *Entomologists Monthly Magazine* 87:51.

Hoebeke, E. R., and J. P. Huether. 1990. Biology and recognition of *Xylotrechus schaefferi* Schott, an enigmatic longhorn in northcentral and eastern North America, with a description of the larva (Coleoptera: Cerambycidae). *Journal of the New York Entomological Society* 98:441–449.

Howden, H. F., and G. B. Vogt. 1951. Insect communities of standing dead pine (*Pinus virginiana* Mill.). *Annals of the Entomological Society of America* 44:581–595.

Iwabuchi K. 1982. Mating behavior of *Xylotrechus pyrrhoderus* Bates (Coleoptera: Cerambycidae). I. Behavioral sequences and existence of the male pheromone. *Applied Entomology and Zoology* 17:494–500.

Jaques, H. E. 1918. A long-lifed woodboring beetle. *Proceedings of the Iowa Academy of Science* 25:175.

Kass, R. J. 2001. Mortality of the endangered Wright fishhook cactus (*Sclerocactus wrightiae*) by an *Opuntia*-borer beetle (Cerambycidae: *Moneilema semipunctatum*). *Western North American Naturalist* 61:495–497.

Keen, F. P. 1958. *Cone and seed insects of western forest trees*. Washington, DC: USDA Forest Service, Technical Bulletin 1169.

Keena, M. A. 2002. *Anoplophora glabripennis* (Coleoptera: Cerambycidae) fecundity and longevity under laboratory conditions: comparison of populations from New York and Illinois on *Acer saccharum*. *Environmental Entomology* 31:490–498.

Khan, T. N. 1985. Community and succession of the round-head borers (Coleoptera: Cerambycidae) infesting the felled logs of White Dhup, *Canarium euphyllum* Kurz. *Proceedings of the Indian Academy of Sciences (Animal Sciences)* 94:435–441.

Kramer, P. J., and T. T. Kozlowski. 1979. *Physiology of woody plants*. New York: Academic Press.

Kukor, J. J., and M. M. Martin.1986a. Cellulose digestion in *Monochamus marmorator* Kby. (Coleoptera: Cerambycidae): role of acquired fungal enzymes. *Journal of Chemical Ecology* 12:1057–1070.

Kukor, J. J., and M. M. Martin. 1986b. The transformation of *Saperda calcarata* (Coleoptera: Cerambycidae) into a cellulose digester through the inclusion of fungal enzymes in its diet. *Oecologia* 71:138–141.

Kumral, N. A., U. Bilgili, and E. Acikgoz. 2012. *Dorcadion pseudopreissi* (Coleoptera: Cerambycidae), a new turf pest in Turkey, the bio-ecology, population fluctuation and damage on different turf species. *Türkiye Entomoloji Dergisi* 36:123–133 [in Turkish].

Lee, S. J., S. R. Kim, H. J. Yoon, et al. 2004. cDNA cloning, expression, and enzymatic activity of a cellulase from the mulberry longicorn beetle, *Apriona germari*. *Comparative Biochemistry and Physiology Part B: Biochemistry and Molecular Biology* 139:107–116.

Lee, S. J., K. S. Lee, S. R. Kim, et al. 2005. A novel cellulase gene from the mulberry longicorn beetle, *Apriona germari*, gene structure, expression and enzymatic activity. *Comparative Biochemistry and Physiology Part B: Biochemistry and Molecular Biology* 140:551–560

Lee, S.-I., J. R. Spence, and D. W. Langor. 2014. Succession of saproxylic beetles associated with decomposition of boreal white spruce logs. *Agricultural and Forest Entomology* 16:391–405.

Li, D.-J., and Y.-N. Liu. 1997. Correlations between sexual development, age, maturation feeding, and mating of adult *Anoplophora glabripennis* Motsch. (Coleoptera: Cerambycidae). *Journal of Northwest Forestry College* 12(4): 19–23.

Lim, J., S.-Y. Jung, J.-S. Lim, et al. 2014. A review of host plants of Cerambycidae (Coleoptera: Chrysomeloidea) with new host records for fourteen cerambycids, including the Asian longhorn beetle (*Anoplophora glabripennis* Motschulsky), in Korea. *Korean Journal of Applied Entomology* 53:111–133.

Lingafelter, S. W. 2003. New host and elevation records for *Moneilema appressum* LeConte (Coleoptera: Cerambycidae: Lamiinae). *Journal of the New York Entomological Society* 111:57–60.

Linit, M. J. 1990. Transmission of pinewood nematode through feeding wounds of *Monochamus carolinensis* (Coleoptera: Cerambycidae). *Journal of Nematology* 22:231–236.

Linsley, E. G. 1943. Delayed emergence of *Buprestis aurulenta* from structural timbers. *Journal of Economic Entomology* 36:348–349.

Linsley, E. G. 1959. Ecology of Cerambycidae. *Annual Review of Entomology* 4:99–138.

Linsley, E. G. 1962a. The Cerambycidae of North America. Part II. Taxonomy and classification of the Parandrinae, Prioninae, Spondylinae and Aseminae. *University of California Publications in Entomology* 19:1–102.

Linsley, E. G. 1962b. The Cerambycidae of North America. Part III. Taxonomy and classification of the sub-family Cerambycinae, tribes Opsimini through Megaderini. *University of California Publications in Entomology* 20:1–188.

Linsley, E. G. 1963. The Cerambycidae of North America. Part IV. Taxonomy and classification of the sub-family Cerambycinae, tribes Elaphidionini through Rhinotragini. *University of California Publications in Entomology* 21:1–165.

Linsley, E. G. 1964. The Cerambycidae of North America. Part V. Taxonomy and classification of the subfamily Cerambycinae, tribes Callichromini through Ancylocerini. *University of California Publications in Entomology* 22:1–197.

Linsley, E. G., and J. A. Chemsak. 1972. Cerambycidae of North America, Part VI, No. 1. Taxonomy and classification of the subfamily Lepturinae. *University of California Publications in Entomology* 69:1–138.

Linsley, E. G., and J. A. Chemsak. 1976. Cerambycidae of North America, Part VI, No. 2. Taxonomy and classification of the subfamily Lepturinae. *University of California Publications in Entomology* 80:1–186.

Linsley, E. G., and J. A. Chemsak. 1984. The Cerambycidae of North America, Part VII, No. 1: Taxonomy and classification of the subfamily Lamiinae, tribes Parmenini through Acanthoderini. *University of California Publications in Entomology* 102:1–258.

Linsley, E. G., and J. A. Chemsak. 1995. The Cerambycidae of North America, Part VII, No. 2: Taxonomy and classification of the subfamily Lamiinae, tribes Acanthocinini through Hemilophini. *University of California Publications in Entomology* 114:1–292.

Linsley, E. G., T. Eisner, and A. B. Klots. 1961. Mimetic assemblages of sibling species of lycid beetles. *Evolution* 15:15–29.

Logarzo, G., D. Gandolfo, and H. Cordo. 2002. Biology of *Apagomerella versicolor* (Boheman) (Coleoptera: Cerambycidae) in Argentina, a candidate for biological control of cocklebur (*Xanthium* spp.). *Biological Control* 25:22–29.

Lu, W, Q. Wang, M. Y. Tian, et al. 2013. Reproductive traits of *Glenea cantor* (Coleoptera: Cerambycidae: Lamiinae). *Journal of Economic Entomology* 106:215–220.

MacDougall, R. S. 1900. The Life-history and habits of *Diaxenes dendrobii*, Gahan, with notes on prevention and remedy. *Notes from the Royal Botanic Garden, Edinburgh* 1:1–12.

Machado, V. S., J. P. Botero, A. Carelli, M. Cupello, H. Y. Quintino, and V. P. Simões Marianna. 2012. Host plants of Cerambycidae and Vesperidae (Coleoptera, Chrysomeloidea) from South America. *Revista Brasileira de Entomologia* 56:186–198.

MacLeod, A., H. F. Evans, and R. H. A. Baker. 2002. An analysis of pest risk from an Asian longhorn beetle (*Anoplophora glabripennis*) to hardwood trees in the European community. *Crop Protection* 21:635–645.

Mansour, K., and J. J. Mansour-Bek. 1934. The digestion of wood by insects and the supposed role of micro-organisms. *Biological Reviews* 9:363–382.

Martin, M. M. 1983. Cellulose digestion in insects. *Comparative Biochemistry and Physiology Part A: Physiology* 75:313–324.

Martin, M. M. 1992. The evolution of insect- fungus associations: from contact to stable symbiosis. *American Zoologist* 32:593– 605.

Martin, N. A. 2000. A longicorn leaf miner, *Microlamia pygmaea* (Coleoptera: Cerambycidae: Lamiinae) found in New Zealand. *New Zealand Entomologist* 23:86.

Matter, S. F. 2001. Effects of above and below ground herbivory by *Tetraopes tetraophthalmus* (Coleoptera: Cerambycidae) on the growth and reproduction of *Asclepias syriaca* (Asclepidacae). *Environmental Entomology* 30:333–338.

Mazza, G., B. Chouaia, G. C. Lozzia, and M. Montagna. 2014. The bacterial community associated to an Italian population of *Psacothea hilaris*: a preliminary study. *Bulletin of Insectology* 67:281–285.

McFadyen, P. J. 1983. Host specificity and biology of *Megacyllene mellyi* [Col.: Cerambycidae] introduced into Australia for the biological control of *Baccharis halimifolia* [Compositae]. *Entomophaga* 28:65–71.

McFadyen, R. E., and A. P. Fidalgo. 1976. Investigations on *Alcidion cereicola* [Col.: Cerambycidae] a potential agent for the biological control of *Eriocereus martinii* [Cactaceae] in Australia. *Entomophaga* 21:103–111.

McNeil, J. 1886. A remarkable case of longevity in a longicorn beetle (*Eburia quadrigeminata*). *American Naturalist* 20:1055–1057.

Merrill, W., and E. B. Cowling. 1966. Role of nitrogen in wood deterioration: amounts and distribution of nitrogen in tree stems. *Canadian Journal of Botany* 44:1555–1580.

Michalcewicz, J. 2002. Feeding of adults of *Leiopus nebulosus* (L.) (Coleoptera: Cerambycidae) on fruit-bodies of the fungus *Diatrype bullata* (Hoffm.: Fr.) Tul. (Ascomycotina: Sphaeriales). *Wiadomosci Entomologiczne* 21:19–22.

Millar, J. G., T. D. Paine, A. L. Joyce, and L. M. Hanks. 2003. The effects of *Eucalyptus* pollen on longevity and fecundity of *Eucalyptus* longhorned borers (Coleoptera: Cerambycidae). *Journal of Economic Entomology* 96:370–376.

Nord, J. C., D. G. Grimble, and F. B. Knight. 1972. Biology of *Saperda inornata* (Coleoptera: Cerambycidae) in trembling aspen, *Populus tremuloides*. *Annals of the Entomological Society of America* 65:127–135.

Özdikmen H. 2003. The genus *Cortodera* Mulsant, 1863 (Cerambycidae: Coleoptera) in Turkey. *Phytoparasitica* 31:433–441.

Pallardy, S. G. 2008. *Physiology of woody plants*, 3rd ed. London: Academic Press.

Pande, S., and R. S. Bhandari. 2006. Bio-ecology of a major cone and seed insect of chir pine, *Chlorophorus strobilicola* Champ. (Coleoptera: Cerambycidae). *Indian Journal of Forestry* 29:253–256.

Panshin, A. J., and C. de Zeeuw. 1980. *Textbook of wood technology: structure, identification, properties, and uses of the commercial woods of the United States and Canada*. New York: McGraw-Hill.

Park, D. S., H. W. Oh, W. J. Jeong, H. Kim, H. Y. Park, and K. S. Bae. 2007. A culture-based study of the bacterial communities within the guts on nine longicorn beetle species and their exo-enzyme producing properties for degrading xylan and pectin. *Journal of Microbiology* 45:394–401.

Parkin, E. A., 1940. The digestive enzymes of some wood-boring beetle larvae. *Journal of Experimental Biology* 17:364–377.

Parmelee, F. T. 1941. Longhorned and flat headed borers attacking fire killed coniferous timber in Michigan. *Journal of Economic Entomology* 34:377–380.

Peck, E.C. 1959. *The sap or moisture in wood*. Madison, WI: USDA Forest Service, Forest Products Laboratory, Report No.768.

Pil, N., and D. Stojanović. 2005. New longhorn beetles (Coleoptera: Cerambycidae) from Serbia. *Archives of Biological Sciences (Belgrade)* 57:27–28.

Piper, G. L. 1977. Biology and habits of *Hippopsis lemniscata* (Coleoptera: Cerambycidae). *The Coleopterists Bulletin* 31:299–306.

Piper, G. L. 1978. Biology and immature stages of *Dectes sayi* Dillon and Dillon (Coleoptera: Cerambycidae). *The Coleopterists Bulletin* 32:273–278.

Poland, T. M., R. A. Haack, T. R. Petrice, D. L. Miller, L. S. Bauer, and R. Gao. 2006. Field evaluations of systemic insecticides for control of *Anoplophora glabripennis* (Coleoptera: Cerambycidae) in China. *Journal of Economic Entomology* 99:383–392.

Pollard, D. G. 1954. The melon stem-borer, *Apomecyna binubila* Pascoe (Coleoptera: Lamiinae) in the Sudan. *Bulletin of Entomological Research* 45:553–561.

Raske, A. G. 1972. Immature stages, genitalia, and notes on the biology of *Coenopoeus palmeri* (Coleoptera: Cerambycidae). *Canadian Entomologist* 104:121–128.

Rassati, D., F. Lieutier, and M. Faccoli. 2016. Alien wood-boring beetles in Mediterranean regions. In *Insects and diseases of Mediterranean forest systems*, ed. T. D. Paine, and F. Lieutier, 293–327. Cham, Switzerland: Springer International Publishing.

Rejzek, M., G. Sama, and G. Alziar. 2001. Host plants of several herb-feeding Cerambycidae mainly from east Mediterranean region (Coleoptera: Cerambycidae). *Biocosme Mésogéen, Nice* 17:263–294.

Rogers, C. E. 1977. Cerambycid pests of sunflower: distribution and behavior in the southern plains. *Environmental Entomology* 6:833–838.

Romero Nápoles, J., J. A. Chemsak, and C. Rodriguez Hernández. 2007. Some notes on natural history and distribution of *Leptostylus gibbulosus* Bates, 1874 (Coleoptera: Cerambycidae). *Acta Zoológica Mexicana* 23:171–173.

Rowell, R. M., R. Pettersen, J. S. Han, J. S. Rowell, and M. A. Tshabalala. 2005. Cell wall chemistry. In *Handbook of wood chemistry and wood composites*, ed. R. M. Rowell, 35–74. Boca Raton, FL: CRC Press.

Saint-Germain, M., P. Drapeau, and C. M. Buddle. 2007. Host-use patterns of saproxylic phloeophagous and xylophagous Coleoptera adults and larvae along the decay gradient in standing dead black spruce and aspen. *Ecography* 30:737–748.

Sama, G., J. Buse, E. Orbach, A. L. L. Friedman, O. Rittner, and V. Chikatunov. 2010. A new catalogue of the Cerambycidae (Coleoptera) of Israel with notes on their distribution and host plants. *Munis Entomology & Zoology* 5:1–51.

Savely, H. E. 1939. Ecological relations of certain animals in dead pine and oak logs. *Ecological Monographs* 9:323–385.

Schabel, H. G. 2006 *Forest entomology in East Africa*. Dordrecht, The Netherlands: Springer.

Schloss, P. D., I. Delalibera, J. Handelsman, and K. F. Raffa. 2006. Bacteria associated with the guts of two woodboring beetles: *Anoplophora glabripennis* and *Saperda vestita* (Cerambycidae). *Environmental Entomology* 35:625–629.

Schoeller, E. N., C. Husseneder, and J. D. Allison. 2012. Molecular evidence of facultative intraguild predation by *Monochamus titillator* larvae (Coleoptera: Cerambycidae) on members of the southern pine beetle guild. *Naturwissenschaften* 99:913–924.

Schomann, H. 1937. Die Symbiose der Bockkäfer. *Zeitschrift für Morphologie und Ökologie der Tiere* 32:542–612.

Scrivener, A. M., H. Watanabe, and H. Noda. 1997. Diet and carbohydrate digestion in the yellowspotted longicorn beetle *Psacothea hilaris*. *Journal of Insect Physiology* 43:1039–1052.

Schwitzgebel, R. B., and D. A. Wilbur. 1942. Coleoptera associated with ironweed, *Vernonia interior* Small in Kansas. *Journal of the Kansas Entomological Society* 15:37–44.

Scully, E. D., S. M. Geib, K. Hoover, et al. 2013. Metagenomic profiling reveals lignocellulose degrading system in a microbial community associated with a wood-feeding beetle. *PLoS One* 8(9):e73827.

Scully, E. D., S. M. Geib, J. E. Carlson, M. Tien, D. McKenna, and K. Hoover. 2014. Functional genomics and microbiome profiling of the Asian longhorned beetle (*Anoplophora glabripennis*) reveal insights into the digestive physiology and nutritional ecology of wood feeding beetles. *BMC Genomics* 15:1096. http://www.biomedcentral.com/1471-2164/15/1096/

Selander, R. B., J. L. Miller, and J. M. Mathieu. 1963. Mimetic associations of lycid and cerambycid beetles. *Journal of the Kansas Entomological Society* 36:45–52.

Sellin, A. 1994. Sapwood–heartwood proportion related to tree diameter, age, and growth rate in *Picea abies*. *Canadian Journal of Forest Research* 24:1022–1028.

Skaar, C. 1972. *Water in wood*. Syracuse, NY: Syracuse University Press.

Slipinski, S. A., and H. E. Escalona. 2013. *Australian longhorn beetles (Coleoptera: Cerambycidae) Volume 1, Introduction and subfamily Lamiinae*. Collingwood: CSIRO Publishing.

Smith, D. N. 1962a. Prolonged larval development in *Buprestis aurulenta* L. (Coleoptera: Buprestidae). A review with new cases. *Canadian Entomologist* 94:586–593.

Smith, D. N. 1962b. A note on the longevity and behavior of adult golden buprestids, *Buprestis aurulenta* L. (Coleoptera: Buprestidae) under artificial conditions. *Canadian Entomologist* 94:672.

Smith, M. T., J. Bancroft, and J. Tropp. 2002. Age-specific fecundity of *Anoplophora glabripennis* (Coleoptera: Cerambycidae) on three tree species infested in the United States. *Environmental Entomology* 31:76–83.

Solomon, J. D. 1974. Biology and damage of the hickory borer, *Goes pulcher*, in hickory and pecan. *Annals of the Entomological Society of America* 67:257–260.

Solomon, J. D. 1980. *Cottonwood borer (Plectrodera scalator)—a guide to its biology, damage, and control*. New Orleans, LA: USDA Forest Service, Southern Forest Experiment Station, Research Paper, SO-157.

Solomon, J. D. 1995. *Guide to insect borers in North American broadleaf trees and shrubs*. Washington, DC: USDA Forest Service, Agriculture Handbook AH-706.

Spencer, G. J. 1930. Insects emerging from prepared timber in buildings. *Proceedings of the Entomological Society of British Colombia* 27:6–10.

Spicer, R. 2005. Senescence in secondary xylem: heartwood formation as an active developmental program. In *Vascular transport in plants*, ed. N. M. Holbrook, and M. Zwieniecki, 457–475. Oxford: Elsevier/ Academic Press.

Stecker, R. E. 1980. The role of insects in giant sequoia reproduction. In *Giant sequoia ecology, fire and reproduction*, ed. H. T. Harvey, H. S. Shellhammer, and R. E. Stecker, 83–100. Washington, DC: National Park Service, U.S. Department of the Interior, Scientific Monograph Series 12. http://www.nps.gov/ history/history/online_books/science/12/chap7.htm

Sugar Research Australia. 2013. Dossier on *Dorysthenes buqueti* as a pest of sugarcane. http://www.sugarresearch.com.au/icms_docs/163515_Sugarcane_longhorn_stemborer_Dorysthenes_buqueti_Dossier.pdf (accessed December 29, 2014)

Sugimura, M., H. Watanabe, N. Lo, and H. Saito. 2003. Purification, characterization, cDNA cloning and nucleotide sequencing of a cellulase from the yellow-spotted longicorn beetle, *Psacothea hilaris*. *European Journal of Biochemistry* 270:3455–3460.

Švácha, P., and J. F. Lawrence. 2014: 2.4 Cerambycidae Latreille, 1802. In *Handbook of zoology, Arthropoda: Insecta; Coleoptera, beetles, Volume 3: Morphology and systematics (Phytophaga)*, ed. R. A. B. Leschen, and R. G. Beutel, 77–177. Berlin: Walter de Gruyter.

Swift, I. 2008. Ecological and biogeographical observations on Cerambycidae (Coleoptera) from California, USA. *Insecta Mundi* 26:1–7.

Swift, I., and A. M. Ray. 2008. A review of the genus *Tragidion* Audinet-Serville, 1834 (Coleoptera: Cerambycidae: Cerambycinae: Trachyderini). *Zootaxa* 1892:1–25.

Thomas, M. C., S. Hill, R. F. Morris, and E. Nearns. 2005. The Cerambycidae of Florida. http://www.fsca-dpi. org/Coleoptera/Mike/FloridaCerambycids/openingpage.htm (accessed December 29, 2014).

Tindall, K. V., S. Stewart, F. Musser, et al. 2010. Distribution of the long-horned beetle, *Dectes texanus*, in soybeans of Missouri, western Tennessee, Mississippi, and Arkansas. *Journal of Insect Science* 10:178. http://144.92.199.79/10.178/

Togashi, K. 1990. Life table for *Monochamus alternatus* (Coleoptera: Cerambycidae) within dead trees of *Pinus thunbergii*. *Japanese Journal of Entomology* 58:217–230.

Toshova, T. B., D. I. Atanasova, M. Tóth, and M. A. Subchev. 2010. Seasonal activity of *Plagionotus* (*Echinocerus*) *floralis* (Pallas) (Coleoptera: Cerambycidae, Cerambycinae) adults in Bulgaria established by attractant baited fluorescent yellow funnel traps. *Acta Phytopathologica et Entomologica Hungarica* 45:391–399.

Troop, J. 1915. Cerambycid in bedstead (Col.). *Entomological News* 26:281.

Turgeon, J. J., M. Orr, C. Grant, Y. Wu, and B. Gasman. 2015. Decade-old satellite infestation of *Anoplophora glabripennis* Motschulsky (Coleoptera: Cerambycidae) found in Ontario, Canada outside regulated area of founder population. *The Coleopterists Bulletin* 69:674–678.

Twinn, P. F. G., and P. T. Harding. 1999. *Provisional atlas of the longhorn beetles (Coleoptera, Cerambycidae) of Britain*. Cambridgeshire, UK: Biological Records Centre, Institute of Terrestrial Ecology.

Ulyshen, M. D. 2016. Wood decomposition as influenced by invertebrates. *Biological Review* 91:70–85.

USDA (United States Department of Agriculture). 2013. *Threatened, endangered, and candidate plant species of Utah*. Boise, ID: USDA Natural Resource Conservation Service, Technical Note Plant Materials 52.

Victorsson, J., and L. Wikars. 1996. Sound production and cannibalism in larvae of the pine-sawyer beetle *Monochamus sutor* L. (Coleoptera: Cerambycidae). *Entomologisk Tidskrift* 117:29–33.

Vogt, G. B. 1949. Notes on Cerambycidae from the Lower Rio Grande Valley, Texas (Coleoptera). *The Pan-Pacific Entomologist* 25:175–184.

Ware, V. L., and F. M. Stephen. 2006. Facultative intraguild predation of red oak borer larvae (Coleoptera: Cerambycidae). *Environmental Entomology* 35:443–447.

Waring, G. L., and R. L. Smith. 1987. Patterns of faunal succession in *Agave palmeri*. *The Southwestern Naturalist* 32:489–497.

Watanabe, H., and G. Tokuda. 2010. Cellulolytic systems in insects. *Annual Review of Entomology* 55:609–632.

Webb, J. L. 1909. *Some insects injurious to forests: the southern pine sawyer*. Washington, DC: USDA Bureau of Entomology, Bulletin 58, Part IV, 41–56.

Webster, F. M. 1889. Notes upon the longevity of the early stages of *Eburia quadrimaculata* Say. *Insect Life* 1:339.

Wei, Y. D., K. S. Lee, Z. Z. Gui, et al. 2006. Molecular cloning, expression, and enzymatic activity of a novel endogenous cellulase from the mulberry longicorn beetle, *Apriona germari*. *Comparative Biochemistry and Physiology Part B: Biochemistry and Molecular Biology* 145:220–229.

Wilkins A. P. 1991. Sapwood, heartwood and bark thickness of silviculturally treated *Eucalyptus grandis*. *Wood Science and Technology* 25:415–423.

Wolcott, G. N. 1948. The insects of Puerto Rico. Coleoptera. *The Journal of Agriculture of the University of Puerto Rico* 32:225–416.

Yuan, F., Y.-Q. Luo, J. Shi, et al. 2008. Biological characteristics of *Acanthocinus carinulatus*, a new record insect pest in Aershan, Inner Mongolia. *Forestry Studies in China* 10:14–18.

Zhou, J., H. Huang, K. Meng, et al. 2009. Molecular and biochemical characterization of a novel xylanase from the symbiotic *Sphingobacterium* sp. TN19. *Applied Microbiology and Biotechnology* 85:323–333.

4

Reproductive Biology of Cerambycids

Lawrence M. Hanks
University of Illinois at Urbana-Champaign
Urbana, Illinois

Qiao Wang
Massey University
Palmerston North, New Zealand

CONTENTS

4.1 Introduction

There is extensive literature on the geographic ranges of cerambycid species and the host associations of their larvae, much of it published by naturalists (e.g., references for volumes indexed by Linsley and Chemsak 1997). Much less is known, however, about the behavior of the adult beetles. The earliest research on cerambycid species was often the most thorough because researchers at that time were free to devote their full attention to pest species for years on end. For example, Atkinson (1926) summarized years of research on the cerambycine *Hoplocerambyx spinicornis* Newman, an important pest of sal trees (*Shorea robusta* Gaertn. f.; Dipsocarpaceae) in India, during which time he had studied sensory cues involved in locating and assessing the girth of larval hosts (the basis for oviposition preference), the resistance response of the host, and the behavior and life history of the larvae; the work included many beautiful hand-drawn illustrations. These early publications are still valuable because they describe behaviors and summarize natural histories that are common among cerambycids.

Little is known of the biology of many cerambycid species because their adults are rarely encountered, especially species that are active at night or that are confined to the forest canopy (e.g., Vance et al. 2003). The volume of literature devoted to the biology of a species is usually a measure of its economic importance as a pest and of how long it has been considered a pest. Much of the published research concerns species in the largest subfamilies, the Lamiinae and Cerambycinae, simply because of the great number of these species whose larvae damage and kill woody plants. The smaller subfamilies Prioninae and Spondylidinae also have important pest species that have been intensively studied (e.g., Duffy 1946;

Solomon 1995). On the other hand, less is known of the behavior of species in the Lepturinae and Parandrinae because their larvae usually develop within dead and decaying hosts and so generally are not considered pests (e.g., Solomon 1995; Evans et al. 2004).

A few cerambycid species have been the focus of purely basic research. The red milkweed beetle, *Tetraopes tetrophthalmus* (Forster), has served as a model organism in studies of behavior and sexual selection—in part because it is common and abundant, and the aposematically colored adults are diurnal and conspicuous (McCauley 1982; Lawrence 1990; Matter 1996). The larvae feed on roots of milkweed (*Asclepias* species; Apocyanaceae), and so usually are not considered pests. Other species that have been studied as model organisms for sexual selection are the sexually dimorphic species *Trachyderes mandibularis* Dupont (Goldsmith and Alcock 1993) and the harlequin beetle, *Acrocinus longimanus* (L.) (Zeh et al. 1992).

A comparative study of reproductive behaviors of cerambycid species is often hindered by inconsistencies in experimental conditions, population densities, and seasonality. For example, local population density, and sex ratio within populations, may strongly influence the duration of copulation and pair bonding, the impact of body size on mate choice, and the incidence of aggressive competition among males (Lawrence 1986; McLain and Boromisa 1987). Behaviors of adults may be altered by unnaturally high population densities, such as during an outbreak of an exotic species (e.g., Hanks et al. 1996a). Study of cerambycids in artificial environments, especially when adults are caged, can result in unrealistic assessments of copulation frequency, longevity, lifetime fecundity, and the outcome of aggressive competition among males because the beetles are not subject to the physiological costs of dispersal and are free of their natural enemies, and because females cannot escape from males and males cannot avoid one another (e.g., Pilat 1972; Saliba 1974; Wang et al. 1990; Hanks et al. 1996a; Lu et al. 2013a). Nevertheless, even caged beetles may behave naturally with regard to mate recognition, courtship, fighting behavior of males, copulation, and oviposition (McCauley 1982; Goldsmith et al. 1996; Hanks et al. 1996a; Yang et al. 2007, 2011; Godinez-Aguilar et al. 2009).

In this chapter, we have drawn references from the literature that most thoroughly review earlier research on reproductive behavior of cerambycids or that provide the most detailed descriptions of reproductive behaviors. We emphasize a set of 36 species that represent a broad geographical sampling and a variety of host plant relationships (Table 4.1). These include the following 14 species of the subfamily Cerambycinae: *Cerambyx dux* (Faldermann), *Megacyllene caryae* (Gahan), *Megacyllene robiniae* (Forster), *Nadezhdiella cantori* Hope, *Neoclytus acuminatus acuminatus* (F.), *Oemona hirta* (F.), *Perarthrus linsleyi* (Knull), *Phoracantha semipunctata* (F.), *Plagithmysus bilineatus* Sharp, *Rosalia batesi* Harold, *Semanotus litigiosus* (Casey), *T. mandibularis*, *Xystrocera globosa* (Olivier), and *Zorion guttigerum* (Westwood). The 15 species in the Lamiinae are *Acalolepta luxuriosa* (Bates), *A. longimanus*, *Anoplophora chinensis* (Forster), *Anoplophora glabripennis* (Motschulsky), *Batocera horsfieldi* Hope, *Dectes texanus* LeConte, *Glenea cantor* (F.), *Monochamus alternatus* Hope, *Monochamus s. scutellatus* (Say), *Paraglenea fortunei* Saunders, *Phytoecia rufiventris* Gautier, *Plectrodera scalator* (F.), *Psacothea hilaris* (F.), *T. tetrophthalmus*, and *Tetrasarus plato* Bates. The two species in the Lepturinae are *Desmocerus californicus californicus* Horn and *Rhagium inquisitor inquisitor* (L.). The three species in the Prioninae are *Prionoplus reticularis* White, *Prionus coriarius* (L.), and *Prionus laticollis* (Drury). Finally, the two species in the Spondylidinae are *Arhopalus ferus* (Mulsant) and *Tetropium fuscum* (F.).

In the following sections, we refer to these 36 species as "example species" and cite their names as references, rather than publications (which can be found in Table 4.1), so as to better illustrate taxonomic patterns in behavior. When more than one example species is referenced, they are ordered as in Table 4.1.

4.2 Phenology of Adults

Cerambycids usually are protandrous (Linsley 1959). Species native to temperate regions often are univoltine or semivoltine, the adults being active for as briefly as a few weeks (*M. caryae*, *M. robiniae*, *R. batesi*, *D. texanus*, *M. alternatus*, *P. rufiventris*, *P. scalator*, *T. tetrophthalmus*, *A. ferus*; also see Hanks et al. 2014). Voltinism can vary with latitude, however, such as is the case with *N. a. acuminatus*, which is univoltine in the north-central United States but trivoltine in southern states. Tropical and

TABLE 4.1

Example Species of Cerambycid Beetles Used as References

Subfamily/Species	Geographical Distribution	Primary Host Plants of Larvae (L) and Adults (A)	References
Cerambycinae			
Cerambyx dux (Faldermann)	Palearctic	L: trunks of living rosaceous trees; A: bark, stems, and fruit of larval host species	Jolles 1932; Saliba 1974, 1977
Megacyllene caryae (Gahan)	Eastern North America	L: trunks and branches of dead and dying hickory (*Carya* spp., Juglandaceae); A: pollen of oaks	Dusham 1921; Ginzel and Hanks 2003, 2005; Ginzel et al. 2006; Lacey et al. 2008b; Hanks and Millar 2013; Hanks et al. 2014; adult diet: LMH, unpublished data
Megacyllene robiniae (Forster)	Eastern North America	L: trunks of stressed black locust (*Robinia pseudoacacia* L., Fabaceae); A: pollen of goldenrod (*Solidago altissima* L., Asteraceae)	Galford 1977; Harman and Harman 1987; Solomon 1995; Ginzel and Hanks 2003, 2005; Ginzel et al. 2003; Ray et al. 2009
Nadezhdiella cantori Hope	East Asia	L: trunks of living *Citrus* spp. (Rutaceae); A: nonfeeding	Lieu 1947; Wang et al. 2002; Wang and Zeng 2004
Neoclytus a. acuminatus (F.)	Eastern North America	L: trunks and branches of stressed hardwoods; A: nonfeeding	Waters 1981; Lacey et al. 2004, 2007b, 2008a; Hanks and Millar 2013; Hanks et al. 2014; Hughes et al. 2015
Oemona hirta (F.)	New Zealand	L: branches of living trees and vines of many deciduous genera; A: pollen and nectar	Wang et al. 1998; Wang and Davis 2005
Perarthrus linsleyi (Knull)	Mexico, southwestern North America	L: stems of asteraceous shrubs; A: flowers of *Larrea tridentata* (DC.) Coville (Zygophyllaceae)	Goldsmith 1987b
Phoracantha semipunctata (F.)	Australia, introduced North America, Middle East, Africa	L: trunks and branches of stressed *Eucalyptus* species (Myrtaceae); A: flowers of *Eucalyptus* species	Hanks et al. 1991, 1993, 1995, 1996a, 1996b, 1998; Barata and Araújo 2001; Millar et al. 2003; Bybee et al. 2005; Lopes et al. 2005; Wang et al. 2008
Plagithmysus bilineatus Sharp	Hawaii	L: trunks of stressed and dying *Metrosideros polymorpha* Gaudich. (Myrtaceae); A: unknown	Papp and Samuelson 1981; Stein and Nagata 1986
Rosalia batesi Harold	Asia	L: trunks of dead hardwoods of many genera; A: pollen	Iwata et al. 1998
Semanotus litigiosus (Casey)	North America	L: trunks of damaged and dying conifers; A: nonfeeding	Wickman 1968
Trachyderes mandibularis Dupont	Mexico, Southwestern North America	L: trunks of dead hardwoods; A: tree sap, cactus fruits	Goldsmith 1987a; Goldsmith and Alcock 1993
Xystrocera globosa (Olivier)	Tropical Eurasia, introduced Hawaii, Puerto Rico	L: trunks of weakened and dying leguminous hardwoods of many genera; A: pollen	Khan 1996; Matsumoto et al. 1996
Zorion guttigerum (Westwood)	New Zealand	L: dead twigs of conifers and hardwoods; A: pollen of many plant species	Wang 2002; Wang and Chen 2005

(Continued)

TABLE 4.1 (*Continued*)

Example Species of Cerambycid Beetles Used as References

Subfamily/Species	Geographical distribution	Primary host plants of Larvae (L) and Adults (A)	References
Lamiinae			
Acalolepta luxuriosa Bates	Asia	L: trunks of araliaceous woody plants; A: unknown	Akutsu and Kuboki 1983a, 1983b; Kuboki et al. 1985
Acrocinus longimanus (L.)	Mexico, Central and South America	L: trunks of dying or decaying hardwoods; A: tree sap	Zeh et al. 1992
Anoplophora chinensis (Forster)	East Asia	L: trunks and branches of living hardwood trees of many genera; A: twig bark and foliage of larval hosts	Wang et al. 1996b; Wang 1998; Lingafelter and Hoebeke 2002; Hansen et al. 2015
Anoplophora glabripennis (Motschulsky)	East Asia, introduced North America, Europe	L: trunks and branches of living hardwood trees of many families; A: twig bark and foliage of larval hosts	Lingafelter and Hoebeke 2002; Morewood et al. 2003; Zhang et al. 2003; Hoover et al. 2014; Meng et al. 2015
Batocera horsfieldi Hope	East Asia	L: trunks and branches of living hardwoods of many genera; A: foliage and tender bark of larval hosts	Duffy 1968; Luo et al. 2011; Yang et al. 2011
Dectes texanus LeConte	Eastern North America, Mexico	L: stems of asteraceous perennials, soybean (*Glycine max* [L] Merr.; Fabaceae); A: foliage and tender stems of larval host species	Hatchett et al. 1975; Crook et al. 2004; Michaud and Grant 2005, 2010
Glenea cantor (F.)	China	L: trunks and branches of weakened hardwoods of many families; A: foliage and tender bark of larval host species	Lu et al. 2007, 2011a, 2011b, 2013a, 2013b
Monochamus alternatus Hope	East Asia	L: trunks of stressed conifers; A: tender bark of conifer branchlets	Kobayashi et al. 1984; Fauziah et al. 1987; Anbutsu and Togashi 2002; Zhou and Togashi 2006; Fan et al. 2007a, 2007b; Yang et al. 2007; Teale et al. 2011; Huang et al. 2015
Monochamus s. scutellatus (Say)	North America	L: trunks of dead and dying conifers; A: twig bark and foliage of conifers	Hughes 1979, 1981; Hughes and Hughes 1982, 1985, 1987; Allison and Borden 2001; Peddle et al. 2002; Fierke et al. 2012; Macias-Samano et al. 2012; Breton et al. 2013
Paraglenea fortunei Saunders	East Asia	L: stems and roots of living herbaceous and woody plants; A: twigs and foliage of larval host species	Wang et al. 1990, 1991; Togashi 2007; Tsubaki et al. 2014
Phytoecia rufiventris Gautier	East Asia	L: living stems of asteraceous perennials; A: foliage of larval host species	Wang et al. 1992, 1996a
Plectrodera scalator (F.)	North America	L: crown and tap roots of living *Populus, Salix* (Salicaceae); A: twigs and foliage of larval host species	Milliken 1916; Solomon 1980, Goldsmith et al. 1996; Ginzel and Hanks 2003
Psacothea hilaris (F.)	Eastern Asia	L: trunks of weakened hardwoods; A: foliage of larval host species	Yokoi 1989; Fukaya and Honda 1992, 1995, 1996; Iba 1993; Fukaya 2004

(Continued)

TABLE 4.1 (Continued)

Example Species of Cerambycid Beetles Used as References

Subfamily/Species	Geographical distribution	Primary host plants of Larvae (L) and Adults (A)	References
Tetraopes tetrophthalmus (Forster)	Eastern North America	L: roots and rhizomes of living *Asclepias* species (Apocyanaceae); A: pollinia of larval host species	Chemsak 1963; Pilat 1972; Price and Willson 1976; McCauley 1982; McCauley and Reilly 1984; Lawrence 1986, 1990; McLain and Boromisa 1987; Matter 1996; Reagel et al. 2002; Agrawal 2004
Tetrasarus plato Bates	Mexico, Central America	L: dying branches of *Alchornea latifolia* Sw. (Euphorbiaceae); A: foliage of larval host species	Godinez-Aguilar et al. 2009
Lepturinae			
Desmocerus c. californicus Horn	Coastal California, USA	L: stems of living elderberry trees (*Sambucus* species, Adoxaceae); A: foliage of larval host species	Davis 1931; Barr 1991; Solomon 1995; Collinge et al. 2001; Talley 2007; Ray et al. 2012b
Rhagium inquisitor inquisitor (L.)	Holarctic	L: trunks of recently deceased conifers; A: pollen	Hess 1920; Craighead 1923; Michelsen 1963
Prioninae			
Prionoplus reticularis White	New Zealand	L: trunks of recently dead to decaying conifers and hardwoods; A: nonfeeding	Edwards 1959, 1961a, 1961b
Prionus coriarius (L.)	Europe	L: trunks of dead and decaying hardwoods, or roots of living hardwoods; A: nonfeeding	Duffy 1946; Bense 1995; Barbour et al. 2011
Prionus laticollis (Drury)	North America	L: roots of living hardwoods; A: nonfeeding	Farrar 1967; Benham 1970; Benham and Farrar 1976; Barbour et al. 2011
Spondylidinae			
Arhopalus ferus (Mulsant)	Europe	L: trunks and branches of recently dead to decaying conifers; A: nonfeeding	Duffy 1953, 1968; Wallace 1954; Hosking and Bain 1977; Hosking and Hutcheson 1979; Suckling et al. 2001; Wang and Leschen 2003
Tetropium fuscum F.	Eurasia, introduced North America	L: trunks of stressed spruce trees (*Picea* species; Pinaceae); A: nonfeeding	Schimitschek 1929; Sweeney et al. 2006; Silk et al. 2007, 2011; Flaherty et al. 2013

Source: For additional information about geographical distribution and host plant relationships, see Bense 1995; Duffy 1953, 1960, 1963, 1968; Lingafelter 2007; Linsley 1962a, 1962b; Linsley and Chemsak 1972, 1984, 1995; Löbl and Smetana 2010.

subtropical species may be active for much longer periods, with multiple overlapping generations (*G. cantor*, also see Shah and Vora 1976; Qian 1985; Sontakke 2002), and may even fly year round (*P. bilineatus, X. globosa, A. longimanus*; also see Gressitt and Davis 1972). Regardless of seasonal phenology, the adult beetles may mate and oviposit for only a few hours each day (Linsley 1959). Adult prionines typically are crepuscular or nocturnal (all three species in Table 4.1), whereas lepturines typically are diurnal (both species in Table 4.1). Cerambycines, lamiines, and spondylidines, however, vary from diurnal (*C. dux, M. caryae, M. robiniae, N. a. acuminatus, P. linsleyi, R. batesi, S. litigiosus, X. globosa, T. mandibularis, Z. guttigerum, A. chinensis, B. horsfieldi, D. texanus, G. cantor, M. s. scutellatus, P. fortunei, P. rufiventris, T. tetrophthalmus, T. plato, T. fuscum*) to between crepuscular and nocturnal (i.e., active between sunset and midnight; *N. cantori, P. semipunctata, P. hilaris*), or truly are nocturnal (*O. hirta, A. longimanus, A. glabripennis, M. alternatus, A. ferus*). Diel patterns in flight or mating activity may be unimodal (*N. a. acuminatus, O. hirta, R. batesi, M. s. scutellatus, P. fortunei*) or bimodal (*N. cantori, X. globosa, Z. guttigerum, A. chinensis, M. alternatus*).

4.3 Diet of Adults

The diet of adult cerambycids is an important consideration because of its implications for dispersal and mate location (see Section 4.8). Cerambycid species vary considerably in adult diet and as to whether the adults feed at all (see Chapter 3). Longevity of adults and fecundity are usually enhanced by feeding (*C. dux, P. semipunctata, X. globosa, G. cantor*) and may vary with host plant species (*A. glabripennis, B. horsfieldi, D. texanus*).

Lamiines are synovigenic, the adults eclosing with undeveloped eggs and sperm (Linsley 1959). Adults of both sexes may have to feed for days, or even weeks, before they reach sexual maturity; they commonly feed on foliage and tender bark of the larval host plant species (Table 4.1). The species of host plants that adult lamiines prefer for their own feeding may also be favored for oviposition (*D. texanus*).

Cerambycids other than the lamiines typically are pro-ovigenic, the adults eclosing with eggs and sperm fully developed, or nearly so, and they usually can mate and oviposit within a day after eclosing. This trait has been documented in the cerambycines *C. dux, M. caryae, N. cantori, N. a. acuminatus, P. semipunctata, S. litigiosus*, and *X. globosa*; in the lepturine *R. i. inquisitor*; in the prionines *P. reticularis* and *P. coriarius*; and in the spondylidines *A. ferus* and *T. fuscum*. Adult cerambycines commonly feed on pollen and nectar, plant sap, or tree fruit (Table 4.1). The exceptions among the example species are *N. cantori, N. a. acuminatus*, and *S. litigiosus*, which apparently do not feed as adults.

Prionines typically do not feed as adults and mate soon after emergence (all three species in Table 4.1). The lack of feeding is indicated by the atrophied digestive tract of adult *P. reticularis* and *P. laticollis*. Spondylidines also may not feed as adults, mating and ovipositing soon after emergence (*A. ferus, T. fuscum*; also see Ross and Vanderwal 1969). Adults of most lepturines feed on pollen and nectar (*R. inquisitor*; see Linsley 1959), although adult *D. c. californicus* also feed on foliage of larval host species.

4.4 Location of Host Plants and Mates

Linsley (1959) observed that adults of many cerambycid species mate on the larval host plant and concluded that the processes of finding hosts and finding mates are intimately associated. Research over the last decade has revealed that mate location in many cerambycids is mediated by long-range volatile pheromones of two types: aggregation pheromones, produced by males and attracting both sexes (see Cardé 2014), or sex pheromones, produced by females and attracting only males (see Chapter 5). Aggregation pheromones have been identified now, or are at least suspected (i.e., males produce chemicals that are known pheromones of other cerambycids, or both sexes are attracted to such chemicals even though males have not yet been shown to produce them), for more than 60 species of cerambycines, more than 30 species of lamiines, and four species of spondylidines. Sex pheromones are known to be used by 14 species of prionines and 7 species of lepturines. The consistent reports of aggregation pheromones among so many species of cerambycines, lamiines, and spondylidines, and the reports of sex pheromones only among prionines and lepturines, suggest that pheromone type is consistent within subfamilies.

Among insects that use aggregation pheromones, males usually call from either the larval host plant or that of the adults (Landolt and Phillips 1997), and the same holds true for the cerambycine, lamiine, and spondylidine example species. Because males are first to arrive on the host plant, they must be proficient at locating hosts independently of pheromones. Females also must be capable of locating larval hosts for oviposition, independently of males and their pheromones. It has long been suspected that host location in cerambycids is mediated primarily by plant volatiles (e.g., Beeson 1930), which seems especially likely for nocturnal species. Diurnal species, on the other hand, are more likely to use visual cues to find host plants and so may prefer plants that are especially conspicuous (due to larger size, brighter color, exposure to sunlight) for feeding (*M. caryae, M. robiniae, P. linsleyi, Z. guttigerum, A. chinensis, T. tetrophthalmus, P. rufiventris*) or ovipositing (*S. litigiosus, A. chinensis, M. s. scutellatus*).

Much of the research on host plant attractants of cerambycids has concerned species whose larvae feed in conifers, testing attraction to volatiles that are typical of conifers (especially α-pinene) or that are associated with physiological stress and weakened resistance (especially ethanol; e.g., Fan et al. 2007b; Jurc et al. 2012). Such general plant kairomones attract both sexes of the cerambycine *Semanotus japonicus* Lacordaire (Yatagai et al. 2002), the lamiine *M. alternatus* and its congeners (e.g., Allison et al. 2012), the spondylidine *T. fuscum* and its congeners (Silk et al. 2007), and the lepturine *R. inquisitor* (Table 4.1). Much less is known of the olfactory cues from deciduous hosts, although laboratory experiments have demonstrated that both sexes of some cerambycines and lamiines are attracted by host volatiles (*M. caryae, M. robiniae, P. semipunctata, P. bilineatus, A. longimanus, A. glabripennis, G. cantor*; also see Ginzel and Hanks 2005). Floral volatiles from host plants of adults play a similar role in attracting both sexes of the cerambycine *Anaglyptus subfasciatus* Pic (Nakashima et al. 1994).

The chances of finding a mate may be improved if both sexes aggregate on larval host plants that are in a particular physiological condition (*P. semipunctata, P. bilineatus, G. cantor, M. alternatus, T. fuscum*), or on plants of a certain size (*A. longimanus, A. chinensis, M. s. scutellatus*), or on certain areas of large hosts, such as the lower trunk (*M. s. scutellatus*), or where the bark is exposed to sunlight (*S. litigiosus*). Males and females also may aggregate on certain plants for feeding, such as the tallest perennials for the foliage feeder *P. rufiventris*, tall trees with dark green leaves for the foliage feeder *A. chinensis*, and plants with the largest or greatest number of blooms for the pollen feeders *M. robiniae, P. linsleyi*, and *T. tetrophthalmus*. Aggregation in response to host plant cues may seem sufficient to bring the sexes together for mating, leading some researchers to conclude that a species does not use a volatile pheromone (*N. cantor, P. semipunctata, P. bilineatus, Z. guttigerum, A. longimanus, A. chinensis, A. glabripennis, G. cantor, P. rufiventris, P. hilaris, T. tetrophthalmus*). It should be noted, however, that much of this research was conducted when there was scant evidence for use of volatile pheromones by cerambycids (i.e., prior to 2004; Allison et al. 2004). For example, early research on *M. alternatus* suggested that mate location is mediated entirely by host plant volatiles (Kobayashi et al. 1984), but it was later discovered that males produce the aggregation pheromone monochamol (Teale et al. 2011).

Volatiles from host plants of the larvae or adults may strongly influence attraction of cerambycids to their aggregation pheromones and, again, most of this information comes from research on conifer feeders. For example, attraction of several *Monochamus* species to monochamol is strongly—even critically—synergized by ethanol and various monoterpenes that are at best only weakly attractive when tested alone (*M. alternatus, M. s. scutellatus*; also see Pajares et al. 2010; Allison et al. 2012; Macias-Samano et al. 2012). Attraction to monochamol also may be enhanced by pheromones of bark beetles with or without plant volatiles (*M. s. scutellatus*; also see Pajares et al. 2010, 2013; Macias-Samano et al. 2012; Ryall et al. 2015).

Sensitivity to bark beetle pheromones probably is adaptive because it serves to guide cerambycids to weakened hosts that are vulnerable to colonization and in which the larvae can feed opportunistically on bark beetle larvae (e.g., Schoeller et al. 2012). Volatiles from coniferous hosts also synergize attraction of the spondylidine *Tetropium fuscum* to its pheromone, fuscumol (Silk et al. 2007). Volatiles from deciduous hosts synergized attraction of adult *Anoplophora glabripennis* to an aggregation pheromone in olfactometer experiments, although the overall response was weak (Nehme et al. 2009). Similarly, floral volatiles from the host plants of adult *Anaglyptus subfasciatus* synergize attraction to the aggregation pheromone produced by males (Nakamuta et al. 1997).

Males of the prionine example species apparently require only the sex pheromones of their females to locate mates (also see Cervantes et al. 2006). Because the adult males do not feed (see Section 4.3), they would not seem to benefit from an ability to detect host plant volatiles. A lack of attraction to plant volatiles in males, and the characteristic sedentary nature of females (see Section 4.8), may account for the rarity with which prionines of either sex have been trapped during field bioassays of conifer volatiles in the southeastern United States (Dodds 2011; Miller et al. 2011) and in Europe (Jurc et al. 2012). However, most of the prionine species native to the areas where the bioassays were conducted develop in deciduous trees (see Bense 1995; Lingafelter 2007), and such species usually are not attracted by ethanol or monoterpenes (see Chapter 5). An exception is the conifer-feeding prionine *Prionus pocularis* Dalman of the eastern United States, adult males and females of which are attracted by ethanol and α-pinene (Miller et al. 2015; D. R. Miller, personal communication). Laboratory studies of the prionine *Mallodon dasystomus* (Say) have revealed that both sexes are attracted to volatiles from the deciduous larval hosts (Paschen et al. 2012). These findings suggest that host plant volatiles may play a role in mate location for prionines, such as by drawing males into habitats where receptive females are emerging.

Lepturines also use female-produced sex pheromones; however, they differ from prionines in that the adults usually feed and mate on flowers (see Section 4.3). The larval hosts are usually dead and decaying trees (Linsley 1959), in which case, adult females must regularly fly between their own hosts for feeding and the larval hosts for oviposition. Adults may be attracted to their host plants by floral volatiles (e.g., Sakakibara et al. 1998). *Ortholeptura valida* (LeConte) is unusual for a lepturine in that the adults are nocturnal, which may explain why host plants of adults have never been reported (see Švácha and Lawrence 2014). Females of *O. valida* produce the sex pheromones *cis*-vaccenyl acetate (Ray et al. 2011) and presumably call from their host plant. Mate location in the diurnal lepturine *Desmocerus c. californicus*, and several congeners, is mediated by the sex pheromone *R*-desmolactone (Ray et al. 2012b). Larvae of *D. c. californicus* feed in the pith of living trees, and adults feed and mate on foliage of the same host species (Table 4.1).

A pheromone has not yet been identified for the lepturine *Rhagium i. inquisitor* whose larvae develop in stressed and dying conifers (Table 4.1). The adults mate on flowers from which females presumably call. However, both sexes are known to be attracted by ethanol and α-pinene (Sweeney et al. 2006; Miller et al. 2015; J. Sweeney, D. R. Miller, personal communication). Females probably use host plant volatiles to locate hosts for oviposition. It is not at all clear, however, how males would benefit from attraction to volatiles from larval hosts because it would seem to draw them into habitats where females are ovipositing and not receptive to mating, and where food is not available.

4.5 Recognition of Mates

Once adults of some lamiine species are in proximity, males apparently can recognize females over short distances (no more than a few centimeters). Males of diurnal species may use visual signals for this purpose (*A. chinensis, G. cantor, P. hilaris, T. tetrophthalmus*). Other signals that operate over short distances, regardless of diel phenology, include substrate-borne vibration (*P. fortunei, P. rufiventris*) and volatile short-range pheromones (*G. cantor, M. alternatus, P. fortunei*). Males of *B. horsfieldi* and *M. s. scutellatus* also respond to females over short distances, but the signals they use have yet to be identified.

Male cerambycids ultimately recognize females by contacting them with their antennae, mouthparts, or tarsi, and this apparently stereotypical behavior has been reported for many species across subfamilies (*C. dux, M. robiniae, N. cantori, N. a. acuminatus, O. hirta, P. semipunctata, R. batesi, Z. guttigerum, A. luxuriosa, A. chinensis, B. horsfieldi, D. texanus, G. cantor, M. alternatus, M. s. scutellatus, P. scalator, P. hilaris, T. tetrophthalmus, T. plato, P. laticollis, T. fuscum*; also see Michelsen 1963). Mate recognition is mediated by species- and sex-specific contact pheromones in the cuticular wax layer of females. Females walking on the host plant may deposit nonvolatile compounds that serve as a chemical trail that guides males to females (*M. robiniae, N. cantori, A. chinensis, A. glabripennis, G. cantor*).

Recognition of conspecific males also may be mediated by contact chemoreception (*P. semipunctata, M. s. scutellatus, P. fortunei, T. plato*).

Contact pheromones may be entirely sufficient for mate recognition, with males not requiring any tactile, visual, or behavioral signals from females (*N. cantori, P. semipunctata, A. luxuriosa, D. texanus, T. plato, T. fuscum*). Thus, males often will attempt to mate with dead females (*M. caryae, M. robiniae, N. cantori, P. semipunctata, A. luxuriosa, B. horsfieldi, D. texanus, G. cantor, P scalator, P. hilaris, T. fuscum*) and even with crude models such as glass rods coated with solvent extracts of cuticular waxes of females or with synthesized contact pheromones (*N. cantori, A. luxuriosa, A. chinensis, A. glabripennis, B. horsfieldi, G. cantor*). Although a solvent extract of females applied to models of various shapes and sizes elicited mating behavior in males of *P. hilaris*, they could only successfully mount models that at least conformed to the general shape and dimensions of a female (Table 4.1). However, male *P. rufiventris* apparently also require a tactile signal to recognize females.

Adult cerambycids of the major subfamilies (excluding parandrines and prionines) have stridulatory organs on the pronotum and mesonotum with which they can produce a squeaking sound (Finn et al. 1972; Švácha and Lawrence 2014). Some species of prionines stridulate by rubbing their ridged hind femora against elytral margins (*P. coriarius*; also see Švácha and Lawrence 2014). Both sexes can stridulate and appear to do so when disturbed, which may serve as a startle tactic against vertebrate predators. Many authors have concluded that stridulation does not serve as a signal for mate location or recognition (*A. luxuriosa, D. texanus, P. fortunei, P. rufiventris, P. hilaris, T. tetrophthalmus*). Males often stridulate when interacting with one another (*R. batesi, A. chinensis, M. s. scutellatus, P. fortunei, P. rufiventris, T. tetrophthalmus, T. plato*), which may serve to intimidate rivals. It is also common for both sexes to stridulate just before and during copulation (*P. linsleyi, A. luxuriosa, A. chinensis, M. s. scutellatus, P. fortunei, P. rufiventris, P. coriarius, A. ferus*).

4.6 Copulation

Female cerambycids of most species must be fertilized to produce viable eggs, and their longevity and lifetime fecundity are usually positively correlated with body size (*O. hirta, X. globosa, A. glabripennis, D. texanus, G. cantor, P. rufiventris, T. tetrophthalmus*). Both sexes typically mate repeatedly with individual mates and with multiple partners, which maximizes lifetime fecundity for some species (*X. globosa, A. luxuriosa*) but has no such effect for other species (*G. cantor, T. tetrophthalmus*). Polyandry is detrimental to female reproductive success in the cerambycine *P. semipunctata*, possibly due to the energetic cost of repeated harassment by males or because fecundity is maximized at some intermediate number of matings (see Arnqvist and Nillson 2000). Parthenogenetic reproduction is apparently rare in cerambycids, having been reported for only a few species of cerambycines (*Kurarus* Gressitt species; Goh 1977) and lepturines (*Cortodera* Mulsant species; Švácha and Lawrence 2014).

Copulation behavior has been thoroughly characterized for many species of cerambycids and, in general, seems to be quite consistent across subfamilies in its basic components (species in Table 4.1, Michelsen 1963). That is: (1) the male approaches the female from behind; (2) he mounts by grasping her prothorax, abdomen, or elytra with his fore and mid legs—his hind legs remaining on the substrate; (3) he engages his genitalia with hers, rears back, and withdraws the aedeagus and ovipositor; (4) he inserts his genitalia (i.e., the reversed endophallus) into hers; and (5) upon completion of copulation, he withdraws his genitalia. (For detailed analyses of behavior, anatomy, and function, see Edwards 1961a; Michelsen 1963; Pilat 1972; Kobayashi et al. 2003; Lu et al. 2013b.) Nevertheless, even closely related species may vary in the details of their mating behaviors (e.g., Michelsen 1963).

Females of some cerambycid species may appear entirely passive as regards copulation, mating with any male they encounter, which suggests that there are no courtship or precopulatory behaviors (*M. caryae, M. robiniae, N. cantori, P. linsleyi, S. litigiosus, T. mandibularis, A. longimanus, M. s. scutellatus, P. scalator*; also see Michelsen 1963). However, females of many species are usually refractory, avoiding males by running, flying, or dropping from the host plant, or discouraging their advances by bucking, shaking, kicking, stridulating, or positioning their body such that their genitalia are inaccessible (*C. dux, O. hirta, P. semipunctata, Z. guttigerum, A. luxuriosa, A. chinensis, B. horsfieldi, D. texanus,*

G. cantor, M. alternatus, M. s. scutellatus, P. fortunei, P. rufiventris, P. hilaris, T. tetrophthalmus, T. plato, R. i. inquisitor; also see Michelsen 1963, 1966). The same behaviors may serve to dislodge males during and after copulation. Females of some prionine species may even attack and injure court- ing males (*P. laticollis*).

Male cerambycids may respond to uncooperative females by cornering or penning them, prevent- ing their escape, and by clinging to them, biting and pulling their antennae, and even amputating their appendages (*S. litigiosus, A. luxuriosa, P. fortunei, P. rufiventris, P. hilaris, T. plato, P. coriarius*; also see Michelsen 1963). Another stereotypical behavior is for males to apply their mouthparts to the pro- notum or elytra of the female before, during, and after copulation, which apparently calms the female or stimulates her to mate (reviewed by Michelsen 1963, 1966). This behavior, variously described as lick- ing, stroking, tapping, scraping, massaging, or biting, has been reported for species in all of the major subfamilies (*C. dux, O. hirta, P. semipunctata, R. batesi, A. luxuriosa, A. longimanus, A. chinensis, B. horsfieldi, D. texanus, M. alternatus, M. s. scutellatus, P. rufiventris, P. hilaris, T. plato, R. i. inquisitor, P. reticularis*). While a male is so engaged, the repeated deflection of his prothorax may result in stridulation (Michelsen 1963). Michelsen (1957) simulated this calming effect with the lep- turine *Rhagium bifasciatum* F. by rubbing the dorsum of females with a soft brush. Similarly, agitated adult *P. semipunctata* of both sexes can be placated by scratching their elytra with a toothpick (Hanks et al. 1996b). Males also may subdue females by using their antennae or legs to stroke, rub, tap, or bat the head, antennae, or prothorax of the female (*C. dux, P. linsleyi, P. semipunctata, S. litigiosus, A. chinensis, P. fortunei, P. hilaris, T. tetrophthalmus*). A male may attempt multiple calming strategies on one female, even simultaneously (Michelsen 1966).

4.7 Larval Host Plants, Oviposition Behavior, and Larval Development

Larvae of most cerambycid species develop in woody plants and often are restricted to either the roots, crown, trunk, branches, or branchlets (reviewed by Linsley 1959; Švácha and Lawrence 2014; see Chapter 2). Some species develop in stems of herbaceous plants or in stems or small branches of woody plants, usually tunneling through the pith (*Z. guttigerum, D. texanus, P. fortunei, P. rufiventris, T. tetrophthalmus, D. c. californicus*). Some species require living and healthy hosts, while others can thrive only within hosts that are stressed or damaged (Table 4.1). Species whose larvae feed on dead hosts may require that they be freshly dead; other species require seasoned dead hosts or hosts in some particular state of decay that harbor specific types of fungi that render the wood palatable and that pro- vide essential nutrients (*A. ferus*; Hanks 1999; also see Saint-Germain et al. 2010; Filipiak and Weiner 2014). Larvae are usually limited to feeding within particular tissues of woody plants, depending on the condition of the host. For example, larvae of species that feed in stressed, moribund, or freshly dead hosts usually feed, at least initially, on subcortical tissues, which have the highest nutrient content (*M. caryae, P. semipunctata, P. bilineatus, S. litigiosus, X. globosa, G. cantor, M. alternatus, M. s. scutellatus, P. hilaris, R. inquisitor, T. fuscum*: also see Hanks 1999). The nutritional quality of subcortical tissues steadily declines after the death of the host and, for that reason, larvae that develop in long-dead hosts usually feed within the xylem (*R. batesi, A. ferus*). For species whose hosts range from dying to decay- ing, the larvae may feed subcortically at first but later move to the sapwood as the host declines in qual- ity (*P. reticularis, A. ferus*). For species that develop in woody plants that are alive (although possibly stressed), the larvae may feed subcortically (*A. dux, N. cantori, O. hirta, A. glabripennis, A. chinensis, P. scalator*) or within the xylem (*M. robiniae, N. a. acuminatus, B. horsfieldi*).

Lamiines differ from other cerambycids in that females prepare the host for oviposition by chewing a niche into the bark of woody hosts, or into the stems of herbaceous plants, into which they then insert one or a few eggs (see Chapter 2). Females of some lamiine species also manipulate the physiology of their hosts by chewing through the vascular tissues and girdling branches, thereby rendering part of a resis- tant host suitable for larval development (e.g., Solomon and Payne 1986; Lemes et al. 2015). *Tetraopes tetrophthalmus* is unusual in that females oviposit onto stems of grasses, the neonates later falling to the ground and tunneling through the soil to the roots of their milkweed hosts. Females of species in other subfamilies typically deposit their eggs (singly or in batches) into adventitious cracks in bark, or under loose bark, which they find by exploring the host with their antennae or probing with the ovipositor (nearly all the cerambycines, prionines, lepturines, and spondylidines in Table 4.1).

Because cerambycid larvae usually are legless and incapable of moving among hosts (except larvae of root-feeding species such as *Prionus coriarius*, which can move through soil), the choice of larval host depends entirely on the mother. Thus, females are under strong selection to oviposit into the appropriate parts of plants that are of the correct species and that are in suitable condition for larvae development. For species that mate on larval hosts, the males also must be proficient at locating suitable hosts. Host range is quite variable within the family, with a few species specialized on a single host species (*M. robiniae, N. cantori, P. bilineatus*), some species limited to hosts of one genus (*P. semipunctata, T. tetrophthalmus, D. c. californicus*), but many species being highly polyphagous (Table 4.1; also see Chapter 3). The apparent polyphagy of species that feed in decaying hosts may result from the broad host ranges of their associated fungi, rather than from adaptation to the nutritional or resistance traits of the host plant species (Švácha and Lawrence 2014).

It has been proposed that females of polyphagous cerambycid species prefer to oviposit into plants of their natal host species, the so-called Hopkins' host-selection principle (see Barron 2001), but this hypothesis has not gained support (*C. dux, D. texanus*; also see Duffy 1953). There is, however, ample evidence that females effectively choose host plants that are suitable for their larvae, and that oviposition preference is positively associated with larval performance (Thompson 1988). For example, females presented with a variety of plants in field or laboratory experiments usually oviposit on the plant species or plant materials that represent the highest quality hosts for their larvae (*P. semipunctata, A. glabripennis, D. texanus, M. alternatus M. s. scutellatus*; also see Harley and Kunimoto 1969; Boldt 1987; Mazaheri et al. 2011). Nevertheless, they sometimes may choose plants of inferior species (*M. alternatus*). Females also may prefer to oviposit on the most suitable host plants within a species (*P. semipunctata, P. bilineatus, D. texanus, G. cantor, M. s. scutellatus, T. fuscum*) and/or the parts of individual hosts that best support larval development (*G. cantor, M. s. scutellatus*). Oviposition preference may be associated with the girth of hosts (*S. litigiosus, A. longimanus, A. chinensis, A. glabripennis, M. s. scutellatus, P. scalator, D. c. californicus*), their physiological condition (*P. bilineatus, X. globosa, M. alternatus, A. ferus, T. fuscum*), damage by fire (*A. ferus*), or exposure of the host to solar radiation (*S. litigiosus, X. globosa, A. longimanus*).

Given that adult female cerambycids choose hosts for their offspring, host range could be determined entirely by oviposition preference and so may not represent the full range of species that could be used by larvae. Females of some species will oviposit on filter paper and other inappropriate substrates when caged in the absence of suitable hosts (*C. dux, P. semipunctata, S. litigiosus, T. fuscum*), suggesting an urgency in releasing eggs. Thus, potential host range could be assessed by forcing females to oviposit into various natural and unnatural hosts or by manually transferring neonates to such hosts (e.g., Hanks et al. 1995). When larvae have been introduced into unnatural hosts by these methods, their survivorship is poor, development time is prolonged, and they develop into small, probably less fit adults (*P. semipunctata, M. alternatus*; also see Harley and Kunimoto 1969; Boldt 1987). These findings suggest that the full range of plant species that are suitable for larval development can be estimated quite reliably from oviposition preference in nature and from host records from rearing beetles.

To a limited extent, cerambycid larvae may compensate for poor host choice by their mother by migrating to higher quality tissues as they feed within the host (Saint-Germain et al. 2010). In most cases, however, larvae probably are incapable of moving beyond the part of the host that they colonized as neonates. Cerambycid larvae colonizing woody hosts may curtail host defenses by first feeding subcortically and perpendicularly to the vascular tissues, encircling and girdling large branches, roots, or even the entire trunk (*P. semipunctata, P. bilineatus, P. scalator, P. laticollis*, LMH, personal observation; also see Summerland 1932; Tapley 1960). Similarly, larvae of branch-pruning species disarm their host plant by girdling from within (e.g., Solomon and Payne 1986). Nutritional quality of hosts typically varies across species (*P. semipunctata, D. texanus*) and with physiological condition (*P. semipunctata*). Moreover, quality may degrade after the death of the host as well as due to colonization by other species of wood-boring insects. Larvae in poor quality hosts may nevertheless survive, but develop into adults of small size (Hanks 1999). Variability in host quality for wood borers in general, and their inability to alter their nutritional environment, are thought to be responsible for broad variation in adult body size within species when compared to other types of phytophagous insects (Andersen and Nilssen 1983; Walczyńska et al. 2010). Variation in adult body size has important implications for reproductive behavior and fitness of cerambycids, as discussed in Section 4.8.

Cerambycid females may use other strategies to facilitate colonization and survival of their offspring. For example, ovipositing females of some lamiine species will avoid hosts that already have been colonized by larvae, which may serve to minimize competition (*P. fortunei*). This behavior may be mediated by a marking pheromone that is deposited during oviposition (*A. glabripennis, M. alternatus, M. s. scutellatus, P. rufiventris*). Marking pheromones may be present in the gelatinous or resinous substance deposited by female lamiines after ovipositing that seals the egg niche (*P. scalator, P. hilaris*). On the other hand, females of *G. cantor* apparently deposit a chemical while ovipositing that encourages other females to oviposit—the advantage being that host plant defenses can be overwhelmed by greater numbers of larvae.

4.8 Mating Strategy

As seems to be common among insects (Thornhill and Alcock 1983), female cerambycids allocate their energy to finding suitable hosts for their larvae, whereas males devote their time to searching for females and mating. In general, adult male cerambycids are much more active than females, flying more often and for greater distances and actively searching for females on host plants (*M. caryae, N. cantori, O. hirta, P. linsleyi, P. semipunctata, A. luxuriosa, A. chinensis, A. glabripennis, B. horsfieldi, T. tetrophthalmus, T. plato*). Males commonly compete aggressively for mates, with females playing no obvious role in the outcome.

Emlen and Oring (1977) proposed that mating strategies of animals evolve in response to selective pressures related to the spatial distribution and abundance of the resources that are required by adult females and to the ability of individual males to monopolize these resources. Among cerambycids, mating strategy also is almost certainly influenced by whether location of mates is mediated by male-produced aggregation pheromones or female-produced sex pheromones. For many species of cerambycids, both sexes aggregate on either the host plant of the larvae or that of the adults, and aggregation sets the stage for competition among males and for mating. It should be no surprise that these aggregating species are in the Cerambycinae, Lamiinae, and Spondylidinae, which are the subfamilies known to use aggregation pheromones (see Chapter 5). In some cases, however, aggregation apparently is mediated entirely by attraction to host plant volatiles, as already discussed. Aggregation behavior is thought to be adaptive if it improves the efficiency of finding mates and/or in utilizing resources (Wertheim et al. 2005). Thus, attraction to an aggregation pheromone not only would, of course, provide females with a means of finding mates but also may expedite their search for food (if males call from the host plant of adults) or hosts for oviposition (if males call from the larval host).

Reproductive behaviors of a few species of cerambycines and lamiines are consistent with resource defense polygyny, a strategy that is thought to be adaptive if individual males are capable of monopolizing a limiting resource that is required by females and is discretely and patchily distributed (Thornhill and Alcock 1983). Males compete for possession of the resource, and the dominant male mates with the arriving females. Thus, a characteristic of this strategy is prolonged battles among males in the absence of females.

The resource defense strategy is used by the cerambycine *Trachyderes mandibularis*, adults of which feed on plant sap and cactus fruits, with the males fighting to control sites where sap is oozing and individual cactus fruits. Males fight primarily with their mandibles, and selection has resulted in sexual dimorphism in mandible size and allometric development in males (i.e., larger males having disproportionately large mandibles). That strategy is also used by the tropical lamiine *A. longimanus*, the harlequin beetle, males of which compete to control fallen and decaying trees that are the larval hosts. Entire trees might seem a resource too large to be defended by individual males, but it may be possible because *A. longimanus* is among the largest of beetles (exceeding 7 cm in body length; Duffy 1960). Males compete with an intimidation display involving exaggerated movements of the head, antennae, elytra, and abdomen. Dominance is associated with the length of the forelegs, which develop allometrically and are disproportionately long in large males. Males apparently can gauge the leg lengths of one another during head-to-head confrontations, with smaller males tending to withdraw. Interactions among males may escalate to sparring with the forelegs and biting, each male attempting to overturn the other or throw him from the host, sometimes resulting in mutilation (Zeh et al. 1992).

Resource defense polygyny has also been proposed for the lamiine *Monochamus s. scutellatus*, adults of which aggregate on fallen conifers (a patchy and unpredictable resource). Males produce an aggregation pheromone that presumably mediates aggregation on larval hosts. They compete aggressively for areas of the trunk that are favored by females for oviposition (where trunk circumference is greatest). Larval hosts are often small enough that they can be monopolized by individual males. Males fight by lashing with their antennae, and the disproportionately longer antennae of large males apparently allow them to intimidate or overpower smaller rivals. More evenly matched males fight by locking mandibles and biting. Dominant males control the resource and mate with arriving females, and females may later disperse alone to seek other hosts for oviposition.

Resource defense polygyny has also been proposed for the lamiine *Tetraopes tetrophthalmus* and the cerambycine *Zorion guttigerum*, adults of which feed and mate on inflorescences—the critical resource exploited by males. Males of both species fight to control particular plants, and larger males have a competitive advantage. In *T. tetrophthalmus*, fighting primarily takes the form of head-butting, locking mandibles, and pushing, while male *Z. guttigerum* fight by lashing with their antennae. Individual males of both species may try multiple tactics against one rival. The host plants of *T. tetrophthalmus* and *Z. guttigerum* seem not to conform to the limiting and patchy distribution of resources that is characteristic of the resource defense strategy—their floral hosts are commonly abundant over large areas. If these species use aggregation pheromones, however, males calling from host plants could attract females and other males, with the males competing to control the plant and mate with females.

Among species using the resource defense strategy, males mate with females on first contact (all five species); although when presented a choice, both sexes may prefer larger mates (*Z. guttigerum, M. s. scutellatus, T. tetrophthalmus*). After mating, males usually remain with the females in a half mount position for a period of time, copulating at intervals. If the critical resource is the larval host, males may attend females as they oviposit (*A. longimanus, M. s. scutellatus*); but if it is the adult host, males attend females as they feed (*T. mandibularis, Z. guttigerum, T. tetrophthalmus*). Females of the flower-feeding species eventually disperse alone to oviposit (*Z. guttigerum, T. tetrophthalmus*). Although the same milkweed plant may serve as a host for both adults and larvae of *T. tetrophthalmus*, the females leave the plant alone to oviposit into the stems of nearby grasses.

Postcopulatory mate guarding provides males with the opportunity to copulate repeatedly with the same female, but it also prevents her from copulating with other males, which could result in sperm displacement. Last-male sperm precedence may assure paternity of eggs laid by the female during the pair bond (*M. s. scutellatus, T. tetrophthalmus*). Thus, male *M. s. scutellatus* benefit by immediately and forcefully mating with females before permitting them to oviposit. As already mentioned, larger males of all five species usually dominate smaller males in aggressive contests and so have a reproductive advantage (all five species), but dominance may not necessarily translate into greater mating success (*Z. guttigerum, T. mandibularis*). Smaller males may compensate for their disadvantages in aggressive contests by being more agile and more efficient in finding females (*T. tetrophthalmus*), or they may adopt alternative strategies, such as intercepting females as they arrive, or inhabiting less desirable sites that nevertheless attract some females (*T. mandibularis, A. longimanus, M. s. scutellatus*). Smaller males may benefit from aggregating (and from attraction by pheromones of conspecific males) if it allows them to mate without the cost of producing pheromones themselves (see Landolt and Phillips 1997).

Female defense polygyny is another mating strategy that is associated with aggregation, and apparently is much more common among cerambycids than resource defense polygyny. In this case, females again require some critical resource that is both limiting and patchily distributed, but the resource is too large to be monopolized by individual males (Thornhill and Alcock 1983). Females are the limiting resource for males, and males compete by guarding individual females and preventing their mating with other males. Although female defense polygyny is rarely specified in the literature on reproductive behavior of cerambycids, it appears to be common among species that use aggregation pheromones, including most of the example species in the Cerambycinae, Lamiinae, and Spondylidinae (excluding the five species already mentioned that use the resource defense strategy; Table 4.1). Males mate with females on first contact, attend them while they oviposit (if on the larval host) or while they feed (if on the adult host), repeat copulation at intervals, and fight with other males to maintain control of the female. Thus, the behavior of males is quite similar to that of species using the resource defense strategy, suggesting that a species could shift between strategies, depending only on the dimensions of the critical resource. If males are naturally pugnacious and intolerant of other males, they may drive other males from a small resource (e.g., a fallen branch or an isolated blooming plant) that they could then control; but they may be incapable of defending a larger resource (a fallen tree or large patch of blooming plants) and so have no option but to guard individual females.

There is some evidence of strategy switching in studies of *Monochamus s. scutellatus* and its congeners. As already mentioned, adults of this species show behaviors consistent with resource defense polygyny, the key criterion being that females prefer to oviposit on particular portions of larval hosts that otherwise would be too large to be defended by a single male. Other species of *Monochamus* are similar to *M. s. scutellatus* in their natural history and behavior (larval hosts are stressed and dying conifers, and males produce aggregation pheromones) but appear to use the female defense strategy (*M. alternatus*; also see Zhang et al. 1993; Naves et al. 2008; see Chapter 5). Whether males show behaviors consistent with female defense or resource defense could be determined simply by the nature of available hosts, which would make it impossible to assign any one strategy to the species. Another example may be *P. scalator*, the mating behavior of which seems not to conform to any of the preconceived strategies (Goldsmith et al. 1996).

Adults of the cerambycine *Phoracantha semipunctata* show stereotypical behaviors that are associated with the female defense strategy among many of the example species. The adults feed on pollen of living eucalypts but aggregate and mate on stressed trees and fallen branches of eucalypts, which are the larval hosts. Aggregation apparently is cued by volatiles emanating from larval hosts rather than being mediated by an aggregation pheromone. Adult males and females arrive independently on larval hosts shortly after sunset, the males constantly running in search of females, while females typically wander slowly as they search for oviposition sites. After mating, the males guard females as they oviposit and may be challenged repeatedly by other males that attempt to supplant them. Among other example species, aggregation is probably mediated by aggregation pheromones produced by males (*M. caryae, N. a. acuminatus, A. chinensis, M. alternatus,* and *T. fuscum*).

For species using female defense polygyny, aggregation may be adaptive for males if their cumulative output of pheromones improves mating opportunity. There are two lines of evidence to suggest that high release rates of pheromones are necessary to attract female cerambycines: (1) individual males commonly produce relatively large quantities of pheromones for insects of their size (exceeding 30 μg/h; Iwabuchi et al. 1987; Hall et al. 2006; Lacey et al. 2007a) and (2) high release rates of synthetic pheromones are necessary to attract females (as well as other males) to traps (10–20 μg/h; LMH, unpublished data). Males usually attempt to mate with any and every female they encounter (*P. semipunctata, P. fortunei, P. rufiventris, P. scalator, P. hilaris*). As would be expected with this strategy, and in direct contrast with resource defense polygyny, solitary males usually ignore one another (*M. robiniae, N. cantori, P. semipunctata, A. chinensis*) because the sole impetus for fighting is the presence of a female (but see Goldsmith et al. 1996). A male paired with a female is challenged by rivals, and their fighting tactics are similar to those of species that use the resource defense strategy—often biting and/or pulling with the mandibles (*M. caryae, M. robiniae, P. semipunctata, R. batesi, A. luxuriosa, A. chinensis, P. hilaris, P. scalator*). Males may even amputate appendages of their rivals (*O. hirta, A. luxuriosa, T. plato, P. reticularis, P. coriarius*), but this degree of escalation could be an artifact of confining beetles during observational studies (*C. dux, T. plato, A. ferus*). Other agonistic tactics of males include pushing, lashing or batting with the antennae, head butting, kicking, and wedging themselves between the mated pair (*C. dux, M. robiniae, N. cantori, P. semipunctata, R. batesi, X. globosa, Z. guttigerum, A. luxuriosa, A. chinensis, B. horsfieldi, M. alternatus*). Challenging males may try several different tactics in rapid succession while attempting to displace a paired rival (*M. robiniae, P. semipunctata, A. luxuriosa, A. longimanus, A. chinensis*).

Among species using the female defense strategy, larger males may have the advantage in both defending females and displacing paired males (*N. cantori, P. semipunctata, R. batesi, P. scalator*), but again, dominance may not translate into greater mating success (*P. scalator*). Laboratory studies of *M. robiniae* have suggested that large body size does not confer an advantage in aggressive competition for females. Small males of *A. chinensis* and *P. hilaris* compensate for their size to some extent by locating mates more efficiently than larger males, although the females of the latter species are more likely to reject smaller males. After copulating, the male usually remains with the female in a half-mount position as she resumes ovipositing or feeding—a form of mate guarding (*M. caryae, N. cantori, N. a. acuminatus, O. hirta, P. semipunctata, R. batesi, S. litigiosus, X. globosa, A. luxuriosa, A. chinensis, B. horsfieldi, M. alternatus, P. scalaris, P. hilaris, T. plato*). Males of *P. hilaris* apparently are capable of extracting sperms of earlier rivals from the reproductive tract of females (Yokoi 1990). Solitary females may slowly wander in search of oviposition sites, and conspecific females usually ignore or avoid one another (*C. dux, N. cantori, P. semipunctata, A. chinensis, P. fortunei, P. rufiventris, T. plato*).

A third mating strategy, scramble competition, has been proposed for the cerambycine *Perarthrus linsleyi*, adults of which feed and mate on flowers of the creosote bush, *Larrea tridentata* (DC.) Coville (Zygophyllaceae). The creosote bush is the dominant woody plant over broad areas of the warm desert regions of the southwestern United States (Baldwin et al. 2002). The larvae of *P. linsleyi* develop in asteraceous shrubs such as brittlebush, *Encelia farinosa* A. Gray ex Torr., which is also abundant and broadly distributed (Baldwin et al. 2002). Goldsmith (1987b) proposed that adult females of *P. linsleyi* become unpredictably distributed as they disperse through populations of creosote bushes, and fitness in males is measured by their ability to find females, rather than by dominance in aggressive competition. It should be noted, however, that this mechanism of mate location is not in accordance with the likely pheromone chemistry of *P. linsleyi*. Being a cerambycine, mate location presumably is mediated by an aggregation pheromone produced by males. Although a pheromone has not yet been identified for this species, males have conspicuous gland pores on the prothorax, which are absent in females, that are associated with pheromone production in other species of cerambycines (LMH, unpublished data; see Chapter 5). If males can induce aggregation by producing a pheromone, there would be no need for them to search for females, which explains why males were no more likely than females to disperse long distances (Goldsmith 1987b). Moreover, the female defense strategy accounts for the behavior of adult *P. linsleyi* during a dry year in which fewer bushes flowered and flower densities were lower: beetles aggregated on plants with the greatest densities of blooms, but males did not compete aggressively for control of plants.

Resource and female defense polygyny would not seem to be available to prionines and lepturines because the sex pheromones of their females do not induce aggregation of both sexes. Quite unlike the aggregation pheromones, sex pheromones are produced in small quantities (a few nanograms/hour; Rodstein et al. 2009; Ray et al. 2012a). Males therefore must be highly sensitive to pheromones to locate single females, as is exemplified by the prionine *Prionus californicus* Motschulsky. Males of that species are attracted by nanogram quantities of pheromones (Rodstein et al. 2011), from as far as 0.5 km away (Maki et al. 2011), and arrive in great numbers at traps baited with a single virgin female (Cervantes et al. 2006). Because prionines do not feed as adults (see Section 4.3), the males apparently have only one goal: to find females and copulate. Attraction of males to individual females, independent of host plants, seems most consistent with scramble competition polygyny (Thornhill and Alcock 1983). In this case, the fitness of males is tested both by their chemosensory abilities in locating females and by aggressive competition with other males that are drawn to the same female. After mating, males of *P. laticollis* usually remain on the back of females as they oviposit, which would discourage females from mating with other males.

The female-produced sex pheromones of prionines seem to have predisposed them to reproductive behaviors quite different from those of species using aggregation pheromones. For example, newly emerged females need not disperse in search of food (because they do not feed) or to find males (because they can attract males with a pheromone), which may account for their characteristic sedentary nature: female prionines often are disinclined to fly or even are incapable of flight (*P. reticularis, P. laticollis*; also see Summerland 1932; Hovore 1981; Barbour et al. 2006; Nilsson 2015). Of course, this sedentary nature could only be adaptive if newly emerged females would be assured of finding suitable larval hosts in their immediate vicinity. Such is likely the case with species whose larvae feed on the roots of living woody plants or in decomposing logs—hosts that can support multiple generations of larvae (Table 4.1; also see Solomon 1995; Wikars 2003). The characteristic polyphagy of prionines (Linsley 1959) also improves the chances that females will find larval hosts without dispersing long distances.

The use of sex pheromones by lepturines might be expected to select for mating strategies similar to those of prionines, but they differ from prionines in one important aspect: the adults feed. Females therefore must travel regularly between their host plants and those of the larvae. For example, vagility in females of *Ortholeptura valida* is consistent with their need to alternately seek floral hosts to feed and the decaying larval hosts (see Section 4.4; also see Ray et al. 2011). On the other hand, females of the lepturine *Desmocerus c. californicus* are so sedentary that populations remain contagiously distributed in the face of habitat loss, raising concern that the species is prone to extinction (Talley 2007; Holyoak et al. 2010). The sedentary nature of female *D. c. californicus* is associated with life history traits similar to those of prionines. This would seem to minimize the need to disperse under natural circumstances as follows (Table 4.1): (1) unlike most lepturines, the larvae feed within living trees, and individual elderberry trees can support multiple generations of larvae; (2) adults can feed on the foliage of their natal tree; (3) females can oviposit into their natal tree; and (4) females attract males with the sex pheromone *R*-desmolactone.

4.9 Conclusion

Certain behavioral traits related to reproduction appear to be universal in the Cerambycidae, such as the contact pheromones of females that mediate mate recognition (reviewed by Ginzel 2010). Early researchers noted that male cerambycids recognize females only after contacting them with their antennae (e.g., Heintze 1925), and the same behavior has since been reported for many and diverse cerambycid species (Section 4.5). Copulation behavior also varies little within the family, with the male mounting the female from behind, clinging to her with his first and second pairs of legs, using his mouthparts to placate her, and then extracting the ovipositor (Section 4.6). Nevertheless, other reproductive traits seem to vary among cerambycids along subfamily lines. For example, the required maturation feeding in adult lamiines contrasts with the complete lack of feeding in the adult stage of prionines. Lepturine adults are usually diurnal, whereas prionines typically are nocturnal. In particular, the use of different pheromone types appears to break out along subfamily lines, with aggregation pheromones used by cerambycines, lamiines, and spondylidines, and sex pheromones used by prionines and lepturines.

The apparent immutability of certain traits at the level of the subfamily gives rise to suites of associated traits that show the same taxonomic affinities. For example, the more vertically directed mandibles of lamiines allow females to chew into host plants upon which they stand, such as when preparing their characteristic oviposition niche (see Section 4.3). This behavior apparently is not possible for the other types of cerambycids due to their prognathous mouthparts (Trägårdh 1930; Ślipiński and Escalona 2013). The oviposition niche conceals the egg from natural enemies, provides a suitable microclimate, and serves as a support against which neonates can brace themselves while boring into the wood. The vertical mouthparts of lamiines also make it possible for females to girdle small branches of woody plants and stems of herbaceous plants (*P. rufiventris*), providing the larvae with a freshly killed (or at least greatly compromised) section of a living host that they otherwise would not be able to colonize (e.g., see Stride and Warwick 1962; Asogwa et al. 2011; Lemes et al. 2015).

Having to prepare an egg niche would seem to be a disadvantage for lamiines because it requires much more time than it would take to oviposit into a bark crack as do females of most species in the other subfamilies (see Section 4.7). However, female lamiines may have an advantage in that they need not search for oviposition sites once they are on the larval host but rather can create them *de novo* anywhere. Thus, their oviposition rate (eggs per unit time) may approach that of other types of cerambycids. For example, females of the lamiine *Monochamus s. scutellatus* can complete an egg niche in approximately 1 minute (Hughes 1979), while females of the cerambycine *Cerambyx dux* may require 30 minutes on average to locate a bark crack suitable for oviposition (Saliba 1974). Because *C. dux* oviposits eggs in batches, however, the oviposition rate of the two species may be comparable.

Girdling a branch certainly involves a major energetic investment by females because it may require hours or even days to complete (Stride and Warwick 1962; Solomon 1995) and may support development of only one or a few offspring. (See Herrick 1902 for an amusing account of the arduous process of girdling by a female of *Oncideres cingulata texana* Herrick.) Nevertheless, there may be no incentive to expedite oviposition for these species because the larval hosts do not vary over time in quality (assuming they remain alive). This relatively leisurely lifestyle seems suited to lamiines because the adults must feed for at least a few days before they can mate (see Section 4.3), they commonly feed on host plants of the larvae (minimizing the need to disperse), they spend a considerable portion of their adult lives feeding, and feeding extends their longevity and fecundity (see Linsley 1959; Švácha and Lawrence 2014). Lamiines in general enjoy a relatively long life as adults (several weeks to months) when food is available (Linsley 1959; Švácha and Lawrence 2014). For example, adult *Anoplophora glabripennis* of both sexes are characteristically sedentary, which has been attributed to the fact that the larvae develop in living hosts that can support multiple generations; therefore, adults can feed, mate, and oviposit on their natal host (Table 4.1). The same is true of *Tetraopes tetrophthalmus*, the adults of which feed and mate on inflorescences of milkweed plants that also can serve as larval hosts (Table 4.1). Adult *T. tetrophthalmus* move among patches of milkweeds infrequently, with the males being much more likely to disperse long distances than females (reviewed by Matter 1996).

In direct contrast to these lamiines is the cerambycine *Semanotus litigiosus*, larvae of which develop in damaged trees and in fallen branches of conifers. In spring, this is the first cerambycid to colonize coniferous host material (downed and damaged trees, and fallen branches) that has accumulated during winter. The adults overwinter and emerge so early in the season that snow may still be on the ground. They do not feed, mate soon after emerging, and females then immediately begin depositing clusters of eggs in bark cracks. The larvae quickly colonize the subcortical zone, and their feeding degrades the quality of those tissues for later arriving xylophagous species. This highly competitive lifestyle, in which reproductive success depends on rapid colonization of larval hosts, would seem unattainable for lamiines because of their required period of maturation feeding. Thus, the lamiine *Monochamus scutellatus oregonensis* LeConte avoids direct competition with *S. litigiosus* by emerging later in the season and by colonizing hosts of smaller diameter than those already occupied by larvae of *S. litigiosus* (Wickman 1968).

Most other reproductive traits of cerambycids appear to be quite variable. For example, even congeners may differ as to whether adults mate on host plants of the larvae or those of adults and in the condition of the host plant that their larvae require (i.e., whether hosts are living, stressed, dying, dead, or decaying). For example, adults of *Megacyllene robiniae* emerge in late fall and aggregate, feed, and mate on flowers of goldenrod. Females oviposit in living, but stressed, black locust trees, and the larvae damage, but rarely kill, their hosts. In contrast, adults of the congener *M. caryae* emerge in very early spring and feed on the pollen of oaks but aggregate, mate, and oviposit on larval hosts (dead and dying hickories). Both species appear to use the female defense strategy—the males fighting for control of females—consistent with the inability of individual males to monopolize the expanses of goldenrod (for *M. robiniae*) and tree trunks (for *M. caryae*) where mating takes place.

Any review of reproductive behavior for an insect family as large and diverse as the Cerambycidae must speak in broad generalities, glossing over many exceptions, and inevitably raising further questions about the evolution of behavior. Indeed, not all "longhorned borers" have exceptionally long antennae (see Švácha and Lawrence 2014), and larvae of some species are not even wood borers, including the common and familiar *Tetraopes tetrophthalmus* (larvae feeding on roots of milkweed) and the important pest *Dectes texanus* (larvae feeding within stems of herbaceous plants). One lingering question is the adaptive advantage of aggregating on the larval host: does it serve merely to bring the sexes together for mating, is reproductive success higher if multiple females oviposit into the same host, do all the males call, is fitness in males correlated with pheromone release rate, or are there sneaky males that exploit the pheromones of conspecifics to find hosts and intercept females? Is the apparent mating strategy of a species an inherited trait, or is it malleable and shaped by the variable qualities of host plants. The recent advances in identification of cerambycid pheromones raise questions about their role in reproduction for many of the species in Table 4.1 whose behaviors were characterized previously. For example, how might an aggregation pheromones in *Tetraopes tetrophthalmus* inform differences between the sexes in dispersal behavior and the influence of habitat patch characteristics? Why do species of *Monochamus* vary so markedly in the degree to which plant volatiles influence attraction to their aggregation pheromones (see Chapter 5)? Why have species such as the cerambycine *Phoracantha semipunctata* evolved to exploit host plant volatiles as aggregation cues rather than use aggregation pheromones as do so many other cerambycines? These questions illustrate how little we know about the biology and ecology of the Cerambycidae and, particularly, the role of pheromones in reproduction.

ACKNOWLEDGMENTS

The authors thank J. G. Millar, M. D. Ginzel, A. M. Ray, and R. F. Mitchell for their helpful comments on an early draft of the manuscript. We appreciate funding support from the Alphawood Foundation of Chicago (to LMH).

REFERENCES

Agrawal, A. A. 2004. Resistance and susceptibility of milkweed: Competition, root herbivory, and plant genetic variation. *Ecology* 85: 2118–2133.

Akutsu, K., and M. Kuboki. 1983a. Mating behavior of the udo longicorn beetle, *Acalolepta luxuriosa* Bates (Coleoptera: Cerambycidae). *Japanese Journal of Applied Entomology and Zoology* 27: 189–196.

Akutsu, K., and M. Kuboki. 1983b. Analysis of mating behavior of udo longicorn beetle, *Acalolepta luxuriosa* Bates (Coleoptera: Cerambycidae). *Japanese Journal of Applied Entomology and Zoology* 27: 247–251.

Allison, J. D., and J. H. Borden. 2001. Observations on the behavior of *Monochamus scutellatus* (Coleoptera: Cerambycidae) in northern British Columbia. *Journal of the Entomological Society of British Columbia* 98: 195–200.

Allison, J. D., J. H. Borden, and S. J. Seybold. 2004. A review of the chemical ecology of the Cerambycidae (Coleoptera). *Chemoecology* 14: 123–150.

Allison, J. D., J. L. McKenney, J. G. Millar, et al. 2012. Response of the woodborers *Monochamus carolinensis* and *Monochamus titillator* (Coleoptera: Cerambycidae) to known cerambycid pheromones in the presence and absence of the host plant volatile α-pinene. *Environmental Entomology* 41: 1587–1596.

Anbutsu, H., and K. Togashi. 2002. Oviposition deterrence associated with larval frass of the Japanese pine sawyer, *Monochamus alternatus* (Coleoptera: Cerambycidae). *Journal of Insect Physiology* 48: 459–465.

Andersen, J., and A. C. Nilssen. 1983. Intrapopulation size variation of free-living and tree-boring Coleoptera. *The Canadian Entomologist* 115: 1453–1464.

Arnqvist, G., and T. Nilsson. 2000. The evolution of polyandry: Multiple mating and female fitness in insects. *Animal Behaviour* 60: 145–164.

Asogwa, E. U., T. C. N. Ndubuaku, and A. T. Hassan. 2011. Distribution and damage characteristics of *Analeptes trifasciata* Fabricius 1775 (Coleoptera: Cerambycidae) on cashew (*Anacardium occidentale* Linnaeus 1753) in Nigeria. *Agriculture and Biology Journal of North America* 2: 421–431.

Atkinson, D. J. 1926. *Hoplocerambyx spinicornis*—An important pest of sal. *Indian Forest Bulletin* 70: 1–14.

Baldwin, B. G., S. Boyd, B. Ertter, et al. 2002. *The Jepson desert manual: Vascular plants of southeastern California.* Berkeley: University of California Press.

Barata, E. N., and J. Araújo. 2001. Olfactory orientation responses of the eucalyptus woodborer, *Phoracantha semipunctata*, to host plant in a wind tunnel. *Physiological Entomology* 26: 26–37.

Barbour, J. D., D. E. Cervantes, E. S. Lacey, et al. 2006. Calling behavior in the primitive longhorned beetle *Prionus californicus* Mots. *Journal of Insect Behavior* 19: 623–629.

Barbour, J. D., J. G. Millar, J. Rodstein, et al. 2011. Synthetic 3,5-dimethyldodecanoic acid serves as a general attractant for multiple species of *Prionus* (Coleoptera: Cerambycidae). *Annals of the Entomological Society of America* 104: 588–593.

Barr, C. B. 1991. *The distribution, habitat, and status of the valley elderberry longhorn beetle,* Desmocerus californicus dimorphus *Fisher (Insecta: Coleoptera: Cerambycidae).* Sacramento: U.S. Fish and Wildlife Service.

Barron, A. B. 2001. The life and death of Hopkins' host-selection principle. *Journal of Insect Behavior* 14: 725–737.

Beeson, C. F. C. 1930. Sense of smell of longicorn beetles. *Nature* 126: 12.

Benham, G. S. 1970. Gross morphology of the digestive tract of *Prionus laticollis* (Coleoptera: Cerambycidae). *Annals of the Entomological Society of America* 63: 1413–1419.

Benham, G. S., and R. J. Farrar. 1976. Notes on the biology of *Prionus laticollis* (Coleoptera: Cerambycidae). *The Canadian Entomologist* 108: 569–576.

Bense, U. 1995. *Longhorn beetles: Illustrated key to the Cerambycidae and Vesperidae of Europe.* Weikersheim: Margraf Verlag.

Boldt, P. E. 1987. Host specificity and laboratory rearing studies of *Megacyllene mellyi* (Coleoptera: Cerambycidae): A potential biological control agent of *Baccharis neglecta* Britt. (Asteraceae). *Proceedings of the Entomological Society of Washington* 89: 665–672.

Breton, Y., C. Hébert, J. Ibarzabal, et al. 2013. Host tree species and burn treatment as determinants of preference and suitability for *Monochamus scutellatus scutellatus* (Coleoptera: Cerambycidae). *Environmental Entomology* 42: 270–276.

Bybee, L. F., J. G. Millar, T. D. Paine, et al. 2005. Effects of single versus multiple mates: Monogamy results in increased fecundity for the beetle *Phoracantha semipunctata*. *Journal of Insect Behavior* 18: 513–527.

Cardé, R. T. 2014. Defining attraction and aggregation pheromones: Teleological versus functional perspectives. *Journal of Chemical Ecology* 40: 519–520.

Cervantes, D. E., L. M. Hanks, E. S. Lacey, et al. 2006. First documentation of a volatile sex pheromone in a longhorned beetle (Coleoptera: Cerambycidae) of the primitive subfamily Prioninae. *Annals of the Entomological Society of America* 99: 718–722.

Chemsak, J. A. 1963. Taxonomy and bionomics of the genus *Tetraopes* (Coleoptera: Cerambycidae). *University of California Publications in Entomology* 30: 1–90.

Collinge, S. K., M. Holyoak, C. B. Barr, et al. 2001. Riparian habitat fragmentation and population persistence of the threatened valley elderberry longhorn beetle in central California. *Biological Conservation* 100: 103–113.

Craighead, F. C. 1923. *North American cerambycid larvae: A classification and the biology of North American cerambycid larvae.* Ottawa, Canada: Dominion of Canada Department of Agriculture Bulletin No. 27 New Series (Technical).

Crook, D. J., J. A. Hopper, S. B. Ramaswamy, et al. 2004. Courtship behavior of the soybean stem borer *Dectes texanus texanus* (Coleoptera: Cerambycidae): Evidence for a female contact sex pheromone. *Annals of the Entomological Society of America* 97: 600–604.

Davis, A. C. 1931. California collecting notes. II. *Bulletin of the Brooklyn Entomological Society* 26: 187–188.

Dodds, K. J. 2011. Effects of habitat type and trap placement on captures of bark (Coleoptera: Scolytidae) and longhorned (Coleoptera: Cerambycidae) beetles in semiochemical-baited traps. *Journal of Economic Entomology* 104: 879–888.

Duffy, E. A. J. 1946. A contribution toward the biology of *Prionus coriarius* L. (Coleoptera, Cerambycidae). *Transactions of the Royal Entomological Society of London* 97: 419–442.

Duffy, E. A. J. 1953. *A monograph of the immature stages of British and imported timber beetles.* London: British Museum of Natural History.

Duffy, E. A. J. 1960. *A monograph of the immature stages of Neotropical timber beetles (Cerambycidae).* London: British Museum of Natural History.

Duffy, E. A. J. 1963. *A monograph of the immature stages of Australasian timber beetles (Cerambycidae).* London: British Museum of Natural History.

Duffy, E. A. J. 1968. *A monograph of the immature stages of Oriental timber beetles (Cerambycidae).* London: British Museum of Natural History.

Dusham, E. J. 1921. The painted hickory borer. *Bulletin of the Cornell University Agricultural Experiment Station* 407: 175–203.

Edwards, J. S. 1959. Host range in *Prionoplus reticularis* White. *Transactions of the Royal Society of New Zealand* 87: 315–318.

Edwards, J. S. 1961a. On the reproduction of *Prionoplus reticularis* (Coleoptera, Cerambycidae), with general remarks on reproduction in the Cerambycidae. *Quarterly Journal of Microscopical Science* 102: 519–529.

Edwards, J. S. 1961b. Observations on the ecology and behaviour of the huhu beetle, *Prionoplus reticularis* White. (Col. Ceramb.). *Transactions of the Royal Society of New Zealand* 88: 733–741.

Emlen, S. T., and L. W. Oring. 1977. Ecology, sexual selection, and the evolution of mating systems. *Science* 197: 215–223.

Evans, H. F., L. G. Moraal, and J. A. Pajares. 2004. Biology, ecology and economic importance of Buprestidae and Cerambycidae. In *Bark and wood boring insects in living trees in Europe, a synthesis*, eds. F. Lieutier, K. R. Day, A. Battisti, et al., 447–474. Dordrecht: Kluwer Academic Publishers.

Fan, J., L. Kang, and J. Sun. 2007a. Role of host volatiles in mate location by the Japanese pine sawyer, *Monochamus alternatus* Hope (Coleoptera: Cerambycidae). *Environmental Entomology* 36: 58–63.

Fan, J., J. Sun, and J. Shi. 2007b. Attraction of the Japanese pine sawyer, *Monochamus alternatus*, to volatiles from stressed host in China. *Annals of Forest Science* 64: 67–71.

Farrar, R. J. 1967. A study of the life history and control of *Prionus laticollis* (Drury) in Rhode Island. MSc Thesis, University of Rhode Island, Kingston.

Fauziah, B. A., T. Hidaka, and K. Tabata. 1987. The reproductive behavior of *Monochamus alternatus* Hope (Coleoptera: Cerambycidae). *Applied Entomology and Zoology* 22: 272–285.

Fierke, M. K., D. D. Skabeikis, J. G. Millar, et al. 2012. Identification of a male-produced aggregation pheromone for *Monochamus scutellatus scutellatus* and an attractant for the congener *Monochamus notatus* (Coleoptera: Cerambycidae). *Journal of Economic Entomology* 105: 2029–2034.

Filipiak, M., and J. Weiner. 2014. How to make a beetle out of wood: Multi-elemental stoichiometry of wood decay, xylophagy and fungivory. *PLoS One* 9: e115104.

Finn, W. E., V. C. Mastro, and T. L. Payne. 1972. Stridulatory apparatus and analysis of the acoustics of four species of the subfamily Lamiinae (Coleoptera: Cerambycidae). *Annals of the Entomological Society of America* 65: 644–647.

Flaherty, L., D. Quiring, D. Pureswaran, et al. 2013. Preference of an exotic wood borer for stressed trees is more attributable to pre-alighting than post-alighting behaviour. *Ecological Entomology* 38: 546–552.

Fukaya, M. 2004. Effects of male body size on mating activity and female mate refusal in the yellow-spotted longicorn beetle, *Psacathea hilaris* (Pascoe) (Coleoptera: Cerambycidae): Are small males inferior in mating? *Applied Entomology and Zoology* 39: 603–609.

Fukaya, M., and H. Honda. 1992. Reproductive biology of the yellow-spotted longicorn beetle, *Psacothea hilaris* (Pascoe) (Coleoptera: Cerambycidae). I. Male mating behaviors and female sex pheromones. *Applied Entomology and Zoology* 27: 89–97.

Fukaya, M., and H. Honda. 1995. Reproductive biology of the yellow-spotted longicorn beetle, *Psacothea hilaris* (Pascoe) (Coleoptera: Cerambycidae). II. Evidence for two female pheromone components with different functions. *Applied Entomology and Zoology* 30: 467–470.

Fukaya, M., and H. Honda. 1996. Reproductive biology of the yellow-spotted longicorn beetle, *Psacothea hilaris* (Pascoe) (Coleoptera: Cerambycidae). IV. Effects of shape and size of female models on male mating behaviors. *Applied Entomology and Zoology* 31: 51–58.

Galford, J. R. 1977. Evidence for a pheromone in the locust borer. *United States Department of Agriculture Forest Service Research Note* NE-240: 1–3.

Ginzel, M. D. 2010. Hydrocarbons as contact pheromones of longhorned beetles (Coleoptera: Cerambycidae). In *Insect hydrocarbons: Biology, biochemistry and chemical ecology*, eds. J. G. Blomquist and A. Bagnères, 375–389. Cambridge: Cambridge University Press.

Ginzel, M. D., and L. M. Hanks. 2003. Contact pheromones as mate recognition cues of four species of longhorned beetles (Coleoptera: Cerambycidae). *Journal of Insect Behavior* 16: 181–187.

Ginzel, M. D., and L. M. Hanks. 2005. Role of host plant volatiles in mate location for three species of longhorned beetles. *Journal of Chemical Ecology* 31: 213–217.

Ginzel, M. D., J. G. Millar, and L. M. Hanks. 2003. (Z)-9-Pentacosene—Contact sex pheromone of the locust borer, *Megacyllene robiniae*. *Chemoecology* 13: 135–141.

Ginzel, M. D., J. A. Moreira, A. M. Ray, et al. 2006. (Z)-9-Nonacosene—Major component of the contact sex pheromone of the beetle *Megacyllene caryae*. *Journal of Chemical Ecology* 32: 435–451.

Godinez-Aguilar, J. L., J. E. Macias-Samano, and A. Moron-Rios. 2009. Notes on biology and sexual behavior of *Tetrasarus plato* Bates (Coleoptera: Cerambycidae), a tropical longhorn beetle in coffee plantations in Chiapas, Mexico. *The Coleopterists Bulletin* 63: 311–318.

Goh, T. 1977. A study on thelytokous parthenogenesis of *Kurarua rhopalophoroides* Hayashi (Col., Cerambycidae). *Elytra* 5: 13–16.

Goldsmith, S. K. 1987a. The mating system and alternative reproductive behaviors of *Dendrobias mandibularis* (Coleoptera: Cerambycidae). *Behavioral Ecology and Sociobiology* 20: 111–115.

Goldsmith, S. K. 1987b. Resource distribution and its effect on the mating system of a longhorned beetle, *Perarthrus linsleyi* (Coleoptera: Cerambycidae). *Oecologia* 73: 317–320.

Goldsmith, S. K., and J. Alcock. 1993. The mating chances of small males of the cerambycid beetle *Trachyderes mandibularis* differ in different environments (Coleoptera: Cerambycidae). *Journal of Insect Behavior* 6: 351–360.

Goldsmith, S. K., Z. Stewart, S. Adams, et al. 1996. Body size, male aggression, and male mating success in the cottonwood borer, *Plectrodera scalator* (Coleoptera: Cerambycidae). *Journal of Insect Behavior* 9: 719–727.

Gressitt, J. L., and C. J. Davis. 1972. Seasonal occurrence of the Hawaiian Cerambycidae (Col.). *Proceedings of the Hawaiian Entomological Society* 21: 213–221.

Hall, D. R., A. Cork, S. J. Phythian, et al. 2006. Identification of components of male-produced pheromone of coffee white stem borer, *Xylotrechus quadripes*. *Journal of Chemical Ecology* 32: 195–219.

Hanks, L. M. 1999. Influence of the larval host plant on reproductive strategies of cerambycid beetles. *Annual Review of Entomology* 44: 483–505.

Hanks, L. M., and J. G. Millar. 2013. Field bioassays of cerambycid pheromones reveal widespread parsimony of pheromone structures, enhancement by host plant volatiles, and antagonism by components from heterospecifics. *Chemoecology* 23: 21–44.

Hanks, L. M., J. G. Millar, and T. D. Paine. 1991. Mechanisms of resistance in *Eucalyptus* against larvae of the eucalyptus longhorned borer (Coleoptera: Cerambycidae). *Environmental Entomology* 20: 1583–1588.

Hanks, L. M., J. G. Millar, and T. D. Paine. 1993. Host species preference and larval performance in the wood-boring beetle *Phoracantha semipunctata* F. *Oecologia* 95: 22–29.

Hanks, L. M., J. G. Millar, and T. D. Paine. 1995. Biological constraints on host-range expansion by the wood-boring beetle *Phoracantha semipunctata* (Coleoptera: Cerambycidae). *Annals of the Entomological Society of America* 88: 183–188.

Hanks, L. M., J. G. Millar, and T. D. Paine. 1996a. Body size influences mating success of the eucalyptus long-horned borer (Coleoptera: Cerambycidae). *Journal of Insect Behavior* 9: 369–382.

Hanks, L. M., J. G. Millar, and T. D. Paine. 1996b. Mating behavior of the eucalyptus longhorned borer (Coleoptera: Cerambycidae) and the adaptive significance of long "horns". *Journal of Insect Behavior* 9: 383–393.

Hanks, L. M., J. G. Millar, and T. D. Paine. 1998. Dispersal of the eucalyptus longhorned borer (Coleoptera: Cerambycidae) in urban landscapes. *Environmental Entomology* 27: 1418–1424.

Hanks, L. M., P. F. Reagel, R. F. Mitchell, et al. 2014. Seasonal phenology of the cerambycid beetles of east central Illinois. *Annals of the Entomological Society of America* 107: 211–226.

Hansen, L., T. Xu, J. Wickham, et al. 2015. Identification of a male-produced pheromone component of the citrus longhorned beetle, *Anoplophora chinensis*. *PLoS One* 10: e0134358.

Harley, K. L. S., and R. K. Kunimoto. 1969. Assessment of the suitability of *Plagiohammus spinipennis* (Thoms.) (Col., Cerambycidae) as an agent for control of weeds of the genus *Lantana* (Verbenaceae). II. Host specificity. *Bulletin of Entomological Research* 58: 787–792.

Harman, D. M., and A. L. Harman. 1987. Distribution pattern of adult locust borers, (Coleoptera: Cerambycidae) on nearby goldenrod, *Solidago* spp. (Asteraceae), at a forest-field edge. *Proceedings of the Entomological Society of Washington* 89: 706–710.

Hatchett, J. H., D. M. Daugherty, J. C. Robbins, et al. 1975. Biology in Missouri of *Dectes texanus*, a new pest of soybean. *Annals of the Entomological Society of America* 68: 209–213.

Heintz, A. 1925. Lepturinernas blombesök och sekundära könskaraktärer. *Entomologisk Tidskrift* 46: 21–34.

Herrick, G. W. 1902. Notes on the life-history and habits of *Oncideres texana*. *Journal of the New York Entomological Society* 10: 15–19.

Hess, W. M. 1920. The ribbed pine-borer. *Cornell University Agricultural Experiment Station Memoir* 33: 367–381.

Holyoak, M., T. S. Talley, and S. E. Hogle. 2010. The effectiveness of US mitigation and monitoring practices for the threatened valley elderberry longhorn beetle. *Journal of Insect Conservation* 14: 43–52.

Hoover, K., M. Keena, M. Nehme, et al. 2014. Sex-specific trail pheromone mediates complex mate finding behavior in *Anoplophora glabripennis*. *Journal of Chemical Ecology* 40: 169–180.

Hosking, G. P., and J. Bain. 1977. *Arhopalus ferus* (Coleoptera: Cerambycidae); Its biology in New Zealand. *New Zealand Journal of Forestry Science* 7: 3–15.

Hosking, G. P., and J. A. Hutcheson. 1979. Nutritional basis for feeding zone preference of *Arhopalus ferus* (Coleoptera: Cerambycidae). *New Zealand Journal of Forestry Science* 9: 185–192.

Hovore, F. T. 1981. Two new species of *Prionus Homaesthesis* from the southwestern United States, with notes on other species (Coleoptera: Cerambycidae). *The Coleopterists Bulletin* 35: 453–457.

Huang, J., J. Zhang, M. Li, et al. 2015. Seasonal variations in the incidence of *Monochamus alternatus* adults (Coleoptera: Cerambycidae) and other major Coleoptera: A two-year monitor in the pine forests of Hangzhou, Eastern China. *Scandinavian Journal of Forest Research* 30: 507–515.

Hughes, A. L. 1979. Reproductive behavior and sexual dimorphism in the white-spotted sawyer *Monochamus scutellatus* (Say). *The Coleopterists Bulletin* 33: 45–47.

Hughes, A. L. 1981. Differential male mating success in the white spotted sawyer *Monochamus scutellatus* (Coleoptera: Cerambycidae). *Annals of the Entomological Society of America* 74: 180–184.

Hughes, A. L., and M. K. Hughes. 1982. Male size, mating success, and breeding habitat partitioning in the whitespotted sawyer *Monochamus scutellatus* (Say) (Coleoptera: Cerambycidae). *Oecologia* 55: 258–263.

Hughes, A. L., and M. K. Hughes. 1985. Female choice of mates in a polygynous insect, the whitespotted sawyer *Monochamus scutellatus*. *Behavioral Ecology and Sociobiology* 17: 385–387.

Hughes, A. L., and M. K. Hughes. 1987. Asymmetric contests among sawyer beetles (Cerambycidae: *Monochamus notatus* and *Monochamus scutellatus*). *Canadian Journal of Zoology* 65: 823–827.

Hughes, G. P., J. E. Bello, J. G. Millar, et al. 2015. Determination of the absolute configuration of female-produced contact sex pheromone components of the longhorned beetle, *Neoclytus acuminatus acuminatus* (F.) (Coleoptera: Cerambycidae). *Journal of Chemical Ecology* 41: 1050–1057.

Iba, M. 1993. Studies on the ecology and control of the yellow-spotted longicorn beetle, *Psacothea hilaris* Pascoe (Coleoptera: Cerambycidae), in the mulberry fields. *Bulletin of the National Institute of Sericultural and Entomological Science* 8: 1–119.

Iwabuchi, K., J. Takahashi, and T. Sakai. 1987. Occurrence of 2,3-octanediol and 2-hydroxy-3-octanone, possible male sex pheromone in *Xylotrechus chinensis*. *Applied Entomology and Zoology* 22: 110–111.

Iwata, R., M. Aoki, T. Nozaki, et al. 1998. Some notes on the biology of a hardwood-log-boring beetle, *Rosalia batesi* Harold (Coleoptera: Cerambycidae), with special reference to its occurrences in a building and a suburban lumberyard. *Japanese Journal of Environmental Entomology and Zoology* 9: 83–97.

Jolles, P. 1932. A study of the life-history and control of *Cerambyx dux*, Fald., a pest of certain stone-fruit trees in Palestine. *Bulletin of Entomological Research* 23: 251–256.

Jurc, M., S. Bojovic, M. F. Fernández, et al. 2012. The attraction of cerambycids and other xylophagous beetles, potential vectors of *Bursaphelenchus xylophilus*, to semio-chemicals in Slovenia. *Phytoparasitica* 40: 337–349.

Khan, T. N. 1996. Comparative ecobiology of *Xystrocera globosa* (Olivier) (Coleoptera: Cerambycidae) in the Indian subcontinent. *Journal of Bengal Natural History Society* 15: 8–25.

Kobayashi, F., A. Yamane, and T. Ikeda. 1984. The Japanese pine sawyer beetle as the vector of pine wilt disease. *Annual Review of Entomology* 29: 115–135.

Kobayashi, H., A. Yamane, and R. Iwata. 2003. Mating behavior of the pine sawyer, *Monochamus saltuarius* (Coleoptera: Cerambycidae). *Applied Entomology and Zoology* 38: 141–148.

Kuboki, M., K. Akutsu, A. Sakai, et al. 1985. Bioassay of the sex pheromone of the udo longicorn beetle, *Acalolepta luxuriosa* Bates (Coleoptera: Cerambycidae). *Applied Entomology and Zoology* 20: 88–89.

Lacey, E. S., M. D. Ginzel, J. G. Millar, et al. 2004. Male-produced aggregation pheromone of the cerambycid beetle *Neoclytus acuminatus acuminatus*. *Journal of Chemical Ecology* 30: 1493–1507.

Lacey, E. S., M. D. Ginzel, J. G. Millar, et al. 2008a. 7-Methylheptacosane is a major component of the contact sex pheromone of the cerambycid beetle *Neoclytus acuminatus acuminatus*. *Physiological Entomology* 33: 209–216.

Lacey, E. S., J. A. Moreira, J. G. Millar, et al. 2007a. Male-produced aggregation pheromone of the cerambycid beetle *Neoclytus mucronatus mucronatus*. *Entomologia Experimentalis et Applicata* 122: 171–179.

Lacey, E. S., J. A. Moreira, J. G. Millar, et al. 2008b. A male-produced aggregation pheromone blend consisting of alkanediols, terpenoids, and an aromatic alcohol from the cerambycid beetle *Megacyllene caryae*. *Journal of Chemical Ecology* 34: 408–417.

Lacey, E. S., A. M. Ray, and L. M. Hanks. 2007b. Calling behavior of the cerambycid beetle *Neoclytus acuminatus acuminatus* (F.). *Journal of Insect Behavior* 20: 117–128.

Landolt, P. J., and T. W. Phillips. 1997. Host plant influences on sex pheromone behavior of phytophagous insects. *Annual Review of Entomology* 42: 371–91.

Lawrence, W. S. 1986. Male choice and competition in *Tetraopes tetraophthalmus*: Effects of local sex ratio variation. *Behavioral Ecology and Sociobiology* 18: 289–296.

Lawrence, W. S. 1990. Effects of body size and repeated matings on female milkweed beetle (Coleoptera: Cerambycidae) reproductive success. *Annals of the Entomological Society of America* 83: 1096–1100.

Lemes, P. G., G. Cordeiro, I. R. Jorge, et al. 2015. Cerambycidae and other Coleoptera associated with branches girdled by *Oncideres saga* Dalman (Coleoptera: Cerambycidae: Lamiinae: Onciderini). *The Coleopterists Bulletin* 69: 159–166.

Lieu, K. O. V. 1947. The study of wood borers in China. II. Biology and control of the citrus-trunk-cerambycids, *Nadezhdiella cantori* (Hope) (Coleoptera). *Notes D'Entomologie Chinoise* 11: 70–117.

Lingafelter, S. W. 2007. *Illustrated key to the longhorned woodboring beetles of the eastern United States. Special Publication No. 3.* North Potomac, MD: Coleopterists Society.

Lingafelter, S. W., and E. R. Hoebeke. 2002. *Revision of the genus* Anoplophora *(Coleoptera: Cerambycidae).* Washington, DC: Entomological Society of Washington.

Linsley, E. G. 1959. Ecology of Cerambycidae. *Annual Review of Entomology* 4: 99–138.

Linsley, E. G. 1962a. The Cerambycidae of North America. Part II: Taxonomy and classification of the Parandrinae, Prioninae, Spondylinae, and Aseminae. *University of California Publications in Entomology* 19: 1–102.

Linsley, E. G. 1962b. The Cerambycidae of North America. Part III: Taxonomy and classification of the subfamily Cerambycinae, tribes Opsimini through Megaderini. *University of California Publications in Entomology* 20: 1–188.

Linsley, E. G., and J. A. Chemsak. 1972. Cerambycidae of North America, Part VI, No. 1: Taxonomy and classification of the subfamily Lepturinae. *University of California Publications in Entomology* 69: 1–142.

Linsley, E. G., and J. A. Chemsak. 1995. The Cerambycidae of North America, Part VII, No. 2: Taxonomy and classification of the subfamily Lamiinae, tribes Acanthocinini through Hemilophini. *University of California Publications in Entomology* 114: 1–292.

Linsley, E. G., and J. A. Chemsak. 1997. The Cerambycidae of North America, Part VIII: Bibliography, index, and host plant index. *University of California Publications in Entomology* 117: 1–534.

Löbl, I., and A. Smetana. 2010. *Catalogue of Palaearctic Coleoptera: Chrysomeloidea.* Stenstrup, Denmark: Apollo Books.

Lopes, O., P. C. Marques, and J. Araujo. 2005. The role of antennae in mate recognition in *Phoracantha semipunctata* (Coleoptera: Cerambycidae). *Journal of Insect Behavior* 18: 243–257.

Lu, W., Q. Wang, M. Y. Tian, et al. 2007. Mate location and recognition in *Glenea cantor* (Fabr.) (Coleoptera: Cerambycidae: Lamiinae): Roles of host plant health, female sex pheromone, and vision. *Environmental Entomology* 36: 864–870.

Lu, W., Q. Wang, M. Y. Tian, et al. 2011a. Phenology and laboratory rearing procedures of an Asian longicorn beetle, *Glenea cantor* (Coleoptera: Cerambycidae: Lamiinae). *Journal of Economic Entomology* 104: 509–516.

Lu, W., Q. Wang, M. Y. Tian, et al. 2011b. Host selection and colonization strategies with evidence for a female-produced oviposition attractant in a longhorn beetle. *Environmental Entomology* 40: 1487–1493.

Lu, W., Q. Wang, M. Tian, et al. 2013a. Reproductive traits of *Glenea cantor* (Coleoptera: Cerambycidae: Lamiinae). *Journal of Economic Entomology* 106: 215–220.

Lu, W., Q. Wang, M. Tian, et al. 2013b. Mating behavior and sexual selection in a polygamous beetle. *Current Zoology* 59: 257–264.

Luo, S., P. Zhuge, and M. Wang. 2011. Mating behavior and contact pheromones of *Batocera horsfieldi* (Hope) (Coleoptera: Cerambycidae). *Entomological Science* 14: 359–363.

Macias-Samano, J. E., D. Wakarchuk, J. G. Millar, et al. 2012. 2-Undecyloxy-1-ethanol in combination with other semiochemicals attracts three *Monochamus* species (Coleoptera: Cerambycidae) in British Columbia, Canada. *The Canadian Entomologist* 144: 764–768.

Maki, E. C., J. G. Millar, J. Rodstein, et al. 2011. Evaluation of mass trapping and mating disruption for managing *Prionus californicus* (Coleoptera: Cerambycidae) in hop production yards. *Journal of Economic Entomology* 104: 933–938.

Matsumoto, K., S. Santosa, N. Irianto, et al. 1996. Biology of the green lined albizzia longicorn, *Xystrocera globosa* Olivier (Coleoptera: Cerambycidae), from Sumatra, based on laboratory breeding. *Tropics* 6: 79–89.

Matter, S. F. 1996. Interpatch movement of the red milkweed beetle, *Tetraopes tetraophthalmus*: Individual responses to patch size and isolation. *Oecologia* 105: 447–453.

Mazaheri, A., J. Khajehali, and B. Hatami. 2011. Oviposition preference and larval performance of *Aeolesthes sarta* (Coleoptera: Cerambycidae) in six hardwood tree species. *Journal of Pest Science* 84: 355–361.

McCauley, D. 1982. The behavioural components of sexual selection in the milkweed beetle *Tetraopes tetraophthalmus*. *Animal Behaviour* 30: 23–28.

McCauley, D. E., and L. M. Reilly. 1984. Sperm storage and sperm precedence in the milkweed beetle *Tetraopes tetraophthalmus* (Forster) (Coleoptera: Cerambycidae). *Annals of the Entomological Society of America* 77: 526–530.

McLain, D. K., and R. D. Boromisa. 1987. Male choice, fighting ability, assortative mating and the intensity of sexual selection in the milkweed longhorn beetle, *Tetraopes tetraophthalmus* (Coleoptera, Cerambycidae). *Behavioral Ecology and Sociobiology* 20: 239–246.

Meng, P. S., K. Hoover, and M. A. Keena. 2015. Asian longhorned beetle (Coleoptera: Cerambycidae), an introduced pest of maple and other hardwood trees in North America and Europe. *Journal of Integrated Pest Management* 6: 1–13.

Michaud, J. P., and A. K. Grant. 2005. The biology and behavior of the longhorned beetle, *Dectes texanus* on sunflower and soybean. *Journal of Insect Science* 5: 1–15.

Michaud, J. P., and A. K. Grant. 2010. Variation in fitness of the longhorned beetle, *Dectes texanus*, as a function of host plant. *Journal of Insect Science* 10: Article 206: 1–13.

Michelsen, A. 1957. Undersögelse af *Rhagium mordax* De G. og *bifasciatum* F.'s udbredelsesforhold paa Nordfyn og deres parringsbiologi. *Entomologiske Meddelelser* 28: 77–83.

Michelsen, A. 1963. Observations on the sexual behaviour of some longicorn beetles, subfamily Lepturinae (Coleoptera, Cerambycidae). *Behaviour* 22: 152–166.

Michelsen, A. 1966. On the evolution of tactile stimulatory actions in longhorned beetles (Cerambycidae, Coleoptera). *Zeitschrift für Tierpsychologie* 23: 258–266.

Millar, J. G., et al. 2003. The effects of *Eucalyptus* pollen on longevity and fecundity of *Eucalyptus* longhorned borers (Coleoptera: Cerambycidae). *Journal of Economic Entomology* 96: 370–376.

Miller, D. R., C. Asaro, C. M. Crowe, et al. 2011. Bark beetle pheromones and pine volatiles: Attractant kairomone lure blend for longhorn beetles (Cerambycidae) in pine stands of the southeastern United States. *Journal of Economic Entomology* 104: 1245–1257.

Miller, D. R., C. M. Crowe, K. J. Dodds, et al. 2015. Ipsenol, ipsdienol, ethanol, and α-pinene: Trap lure blend for Cerambycidae and Buprestidae (Coleoptera) in pine forests of eastern North America. *Journal of Economic Entomology* 108: 1838–1851.

Milliken, F. B. 1916. *Bulletin No. 424. The cottonwood borer*. Washington, DC: United States Department of Agriculture.

Morewood, W. D., P. R. Neiner, J. R. McNeil, et al. 2003. Oviposition preference and larval performance of *Anoplophora glabripennis* (Coleoptera: Cerambycidae) in four eastern North American hardwood tree species. *Environmental Entomology* 32: 1028–1034.

Nakamuta, K., W. S. Leal, T. Nakashima, et al. 1997. Increase of trap catches by a combination of male sex pheromones and floral attractant in longhorn beetle, *Anaglyptus subfasciatus*. *Journal of Chemical Ecology* 23: 1635–1640.

Nakashima, T., K. Nakamuta, H. Makihara, et al. 1994. Field response of *Anaglyptus subfasciatus* Pic (Coleoptera: Cerambycidae) to benzyl acetate and structurally related esters. *Applied Entomology and Zoology* 29: 421–425.

Naves, P. M., E. Sousa, and J. M. Rodrigues. 2008. Biology of *Monochamus galloprovincialis* (Coleoptera: Cerambycidae) in the pine wilt disease affected zone, Southern Portugal. *Silva Lusitana* 16: 133–148.

Nehme, M. E., M. A. Keena, A. Zhang, et al. 2009. Attraction of *Anoplophora glabripennis* to male-produced pheromone and plant volatiles. *Environmental Entomology* 38: 1745–1755.

Nilsson, A. 2015. Trapped in the forest: The longhorn beetle *Tragosoma depsarium* L. in south-east Sweden. MS Thesis, Linköping University, Linköping, Sweden.

Pajares, J. A., G. Álvarez, D. R. Hall, et al. 2013. 2-(Undecyloxy)-ethanol is a major component of the male-produced aggregation pheromone of *Monochamus sutor*. *Entomologia Experimentalis et Applicata* 149: 118–127.

Pajares, J. A., G. Álvarez, F. Ibeas, et al. 2010. Identification and field activity of a male-produced aggregation pheromone in the pine sawyer beetle, *Monochamus galloprovincialis*. *Journal of Chemical Ecology* 36: 570–583.

Papp, R. P., and G. A. Samuelson. 1981. Life history and ecology of *Plagithmysus bilineatus*, an endemic Hawaiian borer associated with *Ohia lehua* (Myrtaceae). *Annals of the Entomological Society of America* 74: 387–391.

Paschen, M. A., N. M. Schiff, and M. D. Ginzel. 2012. Role of volatile semiochemicals in the host and mate location behavior of *Mallodon dasystomus* (Coleoptera: Cerambycidae). *Journal of Insect Behavior* 25: 569–577.

Peddle, S., P. De Groot, and S. Smith. 2002. Oviposition behaviour and response of *Monochamus scutellatus* (Coleoptera: Cerambycidae) to conspecific eggs and larvae. *Agricultural and Forest Entomology* 4: 217–222.

Pilat, T. J. 1972. Copulation in two species of *Tetraopes*: Behavior, mechanics and morphology (Coleoptera, Cerambycidae). MSc Thesis, Loyola University, Chicago, IL.

Price, P. W., and M. F. Willson. 1976. Some consequences for a parasitic herbivore, the milkweed longhorn beetle, *Tetraopes tetrophthalmus*, of a host-plant shift from *Asclepias syriaca* to *A. verticillata*. *Oecologia* 25: 331–340.

Qian, T. Y. 1985. A note on gourd stem borer *Apomecyna saltator neveosparsa* infesting Cucurbitaceae (Coleoptera: Cerambycidae). *Wuyi Science Journal* 5: 47–50.

Ray, A. M., J. D. Barbour, J. S. McElfresh, et al. 2012a. 2,3-Hexanediols as sex attractants and a female-produced sex pheromone for cerambycid beetles in the prionine genus *Tragosoma*. *Journal of Chemical Ecology* 38: 1151–1158.

Ray, A. M., M. D. Ginzel, and L. M. Hanks. 2009. Male *Megacyllene robiniae* (Coleoptera: Cerambycidae) use multiple tactics when aggressively competing for mates. *Environmental Entomology* 38: 425–432.

Ray, A. M., I. P. Swift, J. S. McElfresh, et al. 2012b. (*R*)-Desmolactone, a female-produced sex pheromone component of the cerambycid beetle *Desmocerus californicus californicus* (subfamily Lepturinae). *Journal of Chemical Ecology* 38: 157–167.

Ray, A. M., A. Žunič, R. L. Alten, et al. 2011. *cis*-Vaccenyl acetate, a female-produced sex pheromone component of *Ortholeptura valida*, a longhorned beetle in the subfamily Lepturinae. *Journal of Chemical Ecology* 37: 173–178.

Reagel, P. F., M. D. Ginzel, and L. M. Hanks. 2002. Aggregation and mate location in the red milkweed beetle (Coleoptera: Cerambycidae). *Journal of Insect Behavior* 15: 811–830.

Rodstein, J., J. S. McElfresh, J. D. Barbour, et al. 2009. Identification and synthesis of a female-produced sex pheromone for the cerambycid beetle *Prionus californicus*. *Journal of Chemical Ecology* 35: 590–600.

Rodstein, J., J. G. Millar, J. D. Barbour, et al. 2011. Determination of the relative and absolute configurations of the female-produced sex pheromone of the cerambycid beetle *Prionus californicus*. *Journal of Chemical Ecology* 37: 114–124.

Ross, D. A., and H. Vanderwal. 1969. A spruce borer, *Tetropium cinnamopterum* Kirby, in interior British Colombia. *Journal of the Entomological Society of British Columbia* 66: 10–14.

Ryall, K., P. Silk, R. P. Webster, et al. 2015. Further evidence that monochamol is attractive to *Monochamus* (Coleoptera: Cerambycidae) species, with attraction synergised by host plant volatiles and bark beetle (Coleoptera: Curculionidae) pheromones. *The Canadian Entomologist* 147: 564–579.

Saint-Germain, M., C. M. Buddle, and P. Drapeau. 2010. Substrate selection by saprophagous wood-borer larvae within highly variable hosts. *Entomologia Experimentalis et Applicata* 134: 227–233.

Sakakibara, Y., A. Kikuma, R. Iwata, et al. 1998. Performances of four chemicals with floral scents as attractants for longicorn beetles (Coleoptera: Cerambycidae) in a broadleaved forest. *Journal of Forest Resources* 3: 221–224.

Saliba, L. J. 1974. The adult behavior of *Cerambyx dux* Faldermann. *Annals of the Entomological Society of America* 67: 47–50.

Saliba, L. J. 1977. Observations on the biology of *Cerambyx dux* Faldermann in the Maltese Islands. *Bulletin of Entomological Research* 67: 107–117.

Schimitschek, E. 1929. *Tetropium gabrieli* Weise und *Tetropium fuscum* F.—Ein Beitrag zu ihrer lebensgeschichte und lebensgemeinschaft. *Zeitschrift für Angewandte Entomologie* 15: 229–334.

Schoeller, E. N., C. Husseneder, and J. D. Allison. 2012. Molecular evidence of facultative intraguild predation by *Monochamus titillator* larvae (Coleoptera: Cerambycidae) on members of the southern pine beetle guild. *Naturwissenschaften* 99: 913–924.

Shah, A. H., and V. J. Vora. 1976. Biology of the pointed gourd vine borer, *Apomecyna neglecta* Pasc. (Cerambycidae: Coleoptera) in South Gujarat. *Indian Journal of Entomology* 36: 308–311.

Silk, P. J., J. Sweeney, J. Wu, et al. 2007. Evidence for a male-produced pheromone in *Tetropium fuscum* (F.) and *Tetropium cinnamopterum* (Kirby) (Coleoptera: Cerambycidae). *Naturwissenschaften* 94: 697–701.

Silk, P. J., J. Sweeney, J. Wu, et al. 2011. Contact sex pheromones identified for two species of longhorned beetles (Coleoptera: Cerambycidae) *Tetropium fuscum* and *T. cinnamopterum* in the subfamily Spondylidinae. *Environmental Entomology* 40: 714–726.

Slipinski, A., and H. Escalona. 2013. *Australian longhorn beetles (Coleoptera: Cerambycidae). Vol. 1: Introduction and subfamily Lamiinae*. Collingwood: CSIRO Publishing.

Solomon, J. D. 1980. *Research paper SO-157: Cottonwood borer* (Plectrodera scalator)—*A guide to its biology, damage, and control*. New Orleans, LA: United States Department of Agriculture, Forest Service, Southern Forest Experiment Station.

Solomon, J. D. 1995. *Handbook 706. Guide to insect borers in North American broadleaf trees and shrubs*. Washington, DC: United States Department of Agriculture Forest Service.

Solomon, J. D., and J. A. Payne. 1986. *General Technical Report SO-64. A guide to the insect borers, pruners, and girdlers of pecan and hickory*. New Orleans, LA: United States Department of Agriculture, Forest Service, Southern Forest Experiment Station.

Sontakke, B. K. 2002. Biology of the pointed gourd vine borer, *Apomecyna saltator* Fab. (Coleoptera: Cerambycidae) in Western Orissa. *Journal of Applied Zoological Researches* 13: 197–198.

Stein, J. D., and R. F. Nagata. 1986. Response of *Plagithmysus bilineatus* Sharp (Coleoptera: Cerambycidae) to healthy and stressed ohia trees. *Pan-Pacific Entomologist* 62: 344–349.

Stride, G. O., and E. P. Warwick. 1962. Ovipositional girdling in a North American cerambycid beetle, *Mecas saturnina*. *Animal Behaviour* 10: 112–117.

Suckling, D. M., A. R. Gibb, J. M. Daly, et al. 2001. Behavioral and electrophysiological responses of *Arhopalus tristis* to burnt pine and other stimuli. *Journal of Chemical Ecology* 27: 1091–1104.

Summerland, S. A. 1932. The tile-horned prionus as a pest of apple trees. *Journal of Economic Entomology* 25: 1172–1176.

Sweeney, J., J. M. Gutowski, J. Price, et al. 2006. Effect of semiochemical release rate, killing agent, and trap design on detection of *Tetropium fuscum* (F.) and other longhorn beetles (Coleoptera: Cerambycidae). *Environmental Entomology* 35: 645–654.

Švácha, P., and J. F. Lawrence. 2014. Cerambycidae Latreille, 1802. In *Handbook of zoology: Arthropoda: Insecta: Coleoptera, beetles. Vol. 3: Morphology and systematics (Phytophaga)*, eds. R. A. B. Leschen and R. G. Beutel, 77–177. Berlin/Boston: Walter de Gruyter.

Talley, T. S. 2007. Which spatial heterogeneity framework? Consequences for conclusions about patchy population distributions. *Ecology* 88: 1476–1489.

Tapley, R. G. 1960. The white coffee borer, *Anthores leuconotus* Pasc., and its control. *Bulletin of Entomological Research* 51: 279–301.

Teale, S. A., J. D. Wickham, F. Zhang, et al. 2011. A male-produced aggregation pheromone of *Monochamus alternatus* (Coleoptera: Cerambycidae), a major vector of pine wood nematode. *Journal of Economic Entomology* 104: 1592–1598.

Thompson, J. N. 1988. Evolutionary ecology of the relationship between oviposition preference and performance of offspring in phytophagous insects. *Entomologia Experimentalis et Applicata* 47: 3–14.

Thornhill, R., and J. Alcock. 1983. *The evolution of insect mating systems*. Cambridge, MA: Harvard University Press.

Togashi, K. 2007. Lifetime fecundity and female body size in *Paraglenea fortunei* (Coleoptera: Cerambycidae). *Applied Entomology and Zoology* 42: 549–556.

Trägårdh, I. 1930. Some aspects in the biology of longicorn beetles. *Bulletin of Entomological Research* 21: 1–8.

Tsubaki, R., N. Hosoda, H. Kitajima, et al. 2014. Substrate-borne vibrations induce behavioral responses in the leaf-dwelling cerambycid, *Paraglenea fortunei*. *Zoological Science* 31: 789–794.

Vance, C. C., K. R. Kirby, J. R. Malcolm, et al. 2003. Community composition of longhorned beetles (Coleoptera: Cerambycidae) in the canopy and understory of sugar maple and white pine stands in south-central Ontario. *Environmental Entomology* 32: 1066–1074.

Walczyńska, A., M. Dańko, and J. Kozłowski. 2010. The considerable adult size variability in wood feeders is optimal. *Ecological Entomology* 35: 16–24.

Wallace, H. R. 1954. Notes on the biology of *Arhopalus ferus* Mulsant (Coleoptera: Cerambycidae). *Proceedings of the Royal Entomological Society of London, Series A, General Entomology* 29: 99–113.

Wang, Q. 1998. Evidence for a contact female sex pheromone in *Anoplophora chinensis* (Forster) (Coleoptera: Cerambycidae: Lamiinae). *The Coleopterists Bulletin* 52: 363–368.

Wang, Q. 2002. Sexual selection of *Zorion guttigerum* Westwood (Coleoptera: Cerambycidae: Cerambycinae) in relation to body size and color. *Journal of Insect Behavior* 15: 675–687.

Wang, Q., and L. Chen. 2005. Mating behavior of a flower-visiting longhorn beetle *Zorion guttigerum* (Westwood) (Coleoptera: Cerambycidae: Cerambycinae). *Naturwissenschaften* 92: 237–241.

Wang, Q., L. Chen, J. Li, et al. 1996a. Mating behavior of *Phytoecia rufiventris* Gautier (Coleoptera: Cerambycidae). *Journal of Insect Behavior* 9: 47–60.

Wang, Q., L. Chen, W. Zeng, et al. 1996b. Reproductive behaviour of *Anoplophora chinensis* (Forster) (Coleoptera: Cerambycidae: Lamiinae), a serious pest of citrus. *The Entomologist* 115: 40–49.

Wang, Q., and L. Davis. 2005. Mating behavior of *Oemona hirta* (F.) (Coleoptera: Cerambycidae: Cerambycinae) in laboratory conditions. *Journal of Insect Behavior* 18: 187–191.

Wang, Q., and R. A. B. Leschen. 2003. Identification and distribution of *Arhopalus* species (Coleoptera: Cerambycidae: Aseminae) in Australia and New Zealand. *New Zealand Entomologist* 26: 53–59.

Wang, Q., J. Li, W. Zeng, et al. 1991. Sex recognition by males and evidence for a female sex pheromone in *Paraglenea fortunei* (Coleoptera: Cerambycidae). *Annals of the Entomological Society of America* 84: 107–110.

Wang, Q., J. G. Millar, D. A. Reed, et al. 2008. Development of a strategy for selective collection of a parasitoid attacking one member of a large herbivore guild. *Journal of Economic Entomology* 101: 1771–1778.

Wang, Q., G. Shi, and L. K. Davis. 1998. Reproductive potential and daily reproductive rhythms of *Oemona hirta* (Coleoptera: Cerambycidae). *Journal of Economic Entomology* 91: 1360–1365.

Wang, Q., X. Xiong, and J. Li. 1992. Observations on oviposition and adult feeding behavior of *Phytoecia rufiventris* Gautier (Coleoptera: Cerambycidae). *The Coleopterists Bulletin* 46: 290–295.

Wang, Q., and W. Zeng. 2004. Sexual selection and male aggression of *Nadezhdiella cantori* (Coleoptera: Cerambycidae: Cerambycinae) in relation to body size. *Environmental Entomology* 33: 657–661.

Wang, Q., W. Zeng, L. Chen, et al. 2002. Circadian reproductive rhythms, pair-bonding, and evidence for sex-specific pheromones in *Nadezhdiella cantori* (Coleoptera: Cerambycidae). *Journal of Insect Behavior* 15: 527–539.

Wang, Q., W. Zeng, and J. Li. 1990. Reproductive behavior of *Paraglenea fortunei* (Coleoptera: Cerambycidae). *Annals of the Entomological Society of America* 83: 860–866.

Waters, D. J. 1981. Life history of *Neoclytus acuminatus* with notes on other cerambycids associated with dead or dying deciduous trees. Master's Thesis, Auburn University, Auburn, AL.

Wertheim, B., E. J. van Baalen, M. Dicke, et al. 2005. Pheromone-mediated aggregation in nonsocial arthropods: An evolutionary ecological perspective. *Annual Review Entomology* 50: 321–346.

Wickman, B. E. 1968. The biology of the fir tree borer, *Semanotus litigiosus* (Coleoptera: Cerambycidae), in California. *The Canadian Entomologist* 100: 208–220.

Wikars, L. 2003. Raggbocken *(Tragosoma depsarium)* gynnas tillfalligt av hyggen men behover gammelskogen. *Entomologisk Tidskrift* 124: 1–12.

Yang, H., J. Wang, Z. Zhao, et al. 2007. Mating behavior of *Monochamus alternatus* Hope (Coleoptera: Cerambycidae). *Acta Entomologica Sinica* 50: 807–812.

Yang, H., W. Yang, M. F. Yang, et al. 2011. Mating and oviposition behavior of *Batocera horsfieldi*. *Scientia Silvae Sinicae* 47: 88–92.

Yatagai, M., H. Makihara, and K. Oba. 2002. Volatile components of Japanese cedar cultivars as repellents related to resistance to Cryptomeria bark borer. *Journal of Wood Science* 48: 51–55.

Yokoi, N. 1989. Observation on the mating behavior of the yellow-spotted longicorn beetle, *Psacothea hilaris* Pascoe (Coleoptera: Cerambycidae). *Japanese Journal of Applied Entomology and Zoology* 33: 175–179.

Yokoi, N. 1990. The sperm removal behavior of the yellow spotted longicorn beetle *Psacothea hilaris* (Coleoptera: Cerambycidae). *Applied Entomology and Zoology* 25: 383–388.

Zeh, D. W., J. A. Zeh, and G. Tavakilian. 1992. Sexual selection and sexual dimorphism in the harlequin beetle *Acrocinus longimanus*. *Biotropica* 24: 86–96.

Zhang, A., J. E. Oliver, K. Chauhan, et al. 2003. Evidence for contact sex recognition pheromone of the Asian longhorned beetle, *Anoplophora glabripennis* (Coleoptera: Cerambycidae). *Naturwissenschaften* 90: 410–413.

Zhang, Q. H., J. A. Byers, and X. D. Zhang. 1993. Influence of bark thickness, trunk diameter and height on reproduction of the longhorned beetle, *Monochamus sutor* (Col., Cerambycidae) in burned larch and pine. *Journal of Applied Entomology* 115: 145–154.

Zhou, Z., and K. Togashi. 2006. Oviposition and larval performance of *Monochamus alternatus* (Coleoptera: Cerambycidae) on the Japanese cedar *Cryptomeria japonica*. *Journal of Forest Research* 11: 35–40.

5

Chemical Ecology of Cerambycids

Jocelyn G. Millar
University of California
Riverside, California

Lawrence M. Hanks
University of Illinois at Urbana-Champaign
Urbana, Illinois

CONTENTS

5.1   Introduction

Given their popularity with naturalists and collectors, and in some cases their economic importance, it is remarkable how little is known about the basic biology of most cerambycid species. This is particularly true of cerambycid semiochemistry, which remains largely unexplored. In a 1999 review of cerambycid mate location and recognition, Hanks exhaustively reviewed the available data, which suggested that pheromones that act over long distances appeared to be uncommon in the Cerambycidae (Hanks 1999). Similarly, in a 2004 review of cerambycid chemical ecology, Allison et al. (2004) stated that most cerambycids did not use sex or aggregation pheromones. However, studies over the past decade have shown that, if anything, cerambycid species that do not use some form of attractant pheromones actually may be in the minority. Even more surprising is the fact that careful studies of a number of economically important species (e.g., the *Monochamus* and *Megacyllene* species, reviewed in Hanks 1999) had concluded that these species did not use long-range attractant pheromones, whereas we now have abundant evidence—from multiple species in the five major subfamilies—that the use of volatile pheromones is widespread within the family and that these compounds are often powerful attractants of one or both sexes.

There are several possible causes of the initial erroneous dogma that attractant pheromones were rare among the cerambycids. First, in species with relatively long-lived, feeding adults, such as the *Monochamus* species, a sexual maturation period and/or host feeding may be required before the adults produce or respond to pheromones (see Chapter 3). Second, in many cases, the pheromone is strongly synergized by host plant volatiles—in extreme cases to the extent that the pheromone alone is completely unattractive (see next section). Furthermore, a number of species are attracted to host volatiles independently of pheromones, which may have helped to obscure the fact that pheromones also were being used. Third, pheromone-mediated behavior may be much less obvious than in other types of insects. For example, male moths presented with the sex pheromones of females (during their normal activity period) immediately become active, wing-fanning and then initiating upwind flight (e.g., Hagaman and Cardé 1984). In contrast, cerambycids presented with male-produced aggregation pheromones have not been observed to display such overt behaviors. However, among those species that have female-produced sex pheromones, males may indeed activate rapidly when presented with odors from females (e.g., Cervantes et al. 2006). Fourth, cerambycids may only produce and respond to pheromones during specific daily time windows, and in fact, this appears to be one mechanism by which sympatric and seasonally synchronic species, which share one or more pheromone components, are able to maintain reproductive isolation (Mitchell et al. 2015). Fifth, there is growing evidence that adults of at least some cerambycids are highly stratified within forests, with some species rarely leaving the canopy. Thus, such species are unlikely to be captured during field bioassays using traps in the understory (e.g., Graham et al. 2012). Sixth, early investigators may not have appreciated that relatively high doses and high release rates of synthetic pheromones typically are required to attract those species that use male-produced aggregation pheromones. That is, studies have shown that individual males may release many micrograms of pheromones per hour (e.g., Iwabuchi 1986; Iwabuchi et al. 1987; Hall et al. 2006a; Lacey et al. 2007a), which translates into release rates of ~1 mg/day or more from a passive pheromone dispenser that releases pheromones constantly in order to match the release rate from a single male. Release rates lower than this may result in weak or no attraction to the male-produced aggregation pheromones (e.g., Lacey et al. 2004). Finally, it has become clear that use of appropriate trap designs is critically important in trapping cerambycids, with more than 10-fold differences in trap catch among different trap types (e.g., Graham et al. 2010; Allison et al. 2011, 2014). Any one, or any subset of these reasons may have obscured evidence of pheromone-mediated attraction among conspecifics and contributed to the erroneous impression that cerambycids generally did not use long-range sex or aggregation pheromones. As should become clear in the sections that follow, several cumulative factors contributed to the reversal of this dogma and resulted in our current working hypothesis that the use of attractant pheromones is widespread among the Cerambycidae.

5.2 Use of Pheromones in Cerambycid Reproduction

As with most insects, cerambycids use two distinct types of chemical signals to mediate reproductive interactions: volatile attractants that act over a distance to bring conspecifics together and less volatile to nonvolatile cuticular lipids for close-range recognition of sex and species. There is also increasing evidence that male beetles may use nonvolatile compounds deposited by females on substrates on which they have walked as a form of trail pheromone, allowing males to track and locate females.

The earlier literature reviewed for this chapter suggested three general scenarios for bringing the sexes together for mating:

1. One sex produces a pheromone that attracts the other sex, or both sexes, over long distances, with no additive or synergistic effects from host plant odors. As discussed here, earlier reports of this behavior may have been lacking due to inappropriate bioassay conditions, such as insufficient pheromone release rates or inefficient traps. Since then, a large number of species that fit this scenario have been discovered.

2. One sex produces an attractant pheromone, which acts additively or synergistically with host plant volatiles, to attract one or both sexes. It has become clear that this is a common scenario and that different species display a range of responses—from essentially no attraction to pheromones or host volatiles alone but strongly synergistic attraction to blends of the two, to weaker, additive attraction to blends as compared to pheromones alone.

3. Neither sex produces long-range attractant pheromones; instead, males and females are brought together by their mutual attraction to larval host plants (Linsley 1959), with attraction mediated by plant volatiles (e.g., Hanks et al. 1996; Lu et al. 2007; Flaherty et al. 2013). *Phoracantha semipunctata* (F.) may represent an example of this type of strategy because repeated attempts by the authors and others have failed to find any evidence of behaviors mediated by volatile pheromones. Alternatively, both sexes may be drawn by visual or olfactory cues to flowers or other food sources upon which adults feed and so are brought into contact (e.g., Wang et al. 1996; Reagel et al. 2002; Wang and Chen 2005).

These probably cover the main scenarios and strategies for mate location. Nevertheless, it is entirely possible that there are other scenarios still waiting to be discovered, given the great number and diversity of cerambycid species.

To date, attractant pheromones have been identified from five of the eight currently recognized subfamilies within the Cerambycidae (see Švácha and Lawrence 2014). The pheromones are of two types:

1. Male-produced pheromones that attract both sexes. By the standard definition, these would be classified as aggregation pheromones (produced by one or both sexes and attracting both sexes). But, in actuality, it is likely that they primarily serve to bring the sexes together for mating or to enhance mating in other ways, and so it may be more appropriate to refer to them as aggregation-sex pheromones (see discussion in Cardé 2014).

2. Female-produced sex pheromones that attract males exclusively.

Although these two types of pheromones likely serve the same overall purpose of bringing the sexes together for mating, some further discussion of the differences between them is warranted because these differences are almost certainly correlated with the ecology and life history of the producing species. Thus, the male-produced aggregation pheromones are produced in large amounts, sometimes more than 30 µg/h (Iwabuchi 1986; Iwabuchi et al. 1987; Hall et al. 2006a; Lacey et al. 2007a). These amounts far exceed the amount required to be detected by females because, in electroantennogram assays, antennae of females are sensitive to nanogram quantities of the pheromones (e.g., Zhang et al. 2002; Hall et al. 2006a; Pajares et al. 2010). It also is likely that these compounds, as their name implies, are involved in the formation of the conspecific aggregations that have been commonly observed among some species (see Chapter 4). Furthermore, there is some evidence that calling males stimulate other males to call in a type of "chorusing" behavior (Lemay et al. 2010). In combination, these three points suggest that aggregation pheromones also are involved in sexual selection, with males competing to advertise their fitness to females by their production of voluminous (and costly) amounts of pheromone.

In contrast, the female-produced sex pheromones appear to be emitted in much smaller amounts, perhaps on the order of a few nanograms per female per hour (Rodstein et al. 2009; Ray et al. 2012b). Furthermore, in at least some instances, it has been shown that they elicit immediate activation and upwind orientation by males (Cervantes et al. 2006; Rodstein et al. 2011) and that males can detect and respond to pheromone sources over distances of hundreds of meters (Maki et al. 2011a). Among the species that use these sex pheromones, it is likely that scramble competition prevails, with the first male to find a calling female immediately attempting to mate with her, and later arriving males challenging the paired male (see Chapter 4). These species also may be more likely to have nonfeeding adults, so that there is some urgency to find a mate before they run out of energy and die (see Chapter 4).

From among the ~115 species for which pheromones or likely pheromones are known (Table 5.1), without exception to date, the use of these two types of pheromones breaks out along taxonomic lines. Thus, all known attractant pheromones or likely pheromones for species in the subfamilies Prioninae (identified for ~25 species) and Lepturinae (~10 species) are female-produced sex pheromones, whereas

TABLE 5.1

Summary of Published Research from Field Trials on Volatile Pheromones, Attractants, Synergists, and Antagonists of Cerambycid Beetles

Taxonomy	Volatiles emitted (Pher = confirmed attractant)	Sex	Attractant (Attr), enhancer (Enh), antagonist (Antag)	References
CERAMBYCINAE				
Anaglyptini				
Anaglyptus colobotheoides Bates	—	–	Attr: 3-C6-ketol	Sweeney et al. 2014
Anaglyptus subfasciatus Pic	Pher: 3R-C6-, 3R-C8-ketol	M	Attr and Enh: floral volatiles	Nakashima et al. 1994; Leal et al. 1995
Cyrtophorus verrucosus (Olivier)	Pher: 3R-C6-ketol, nonan-2-one	M	Antag: (2,3)-C6-diol, plant volatiles	Hanks and Millar 2013; Mitchell et al. 2013, 2015
Callidiini				
Callidiellum rufipenne (Motschulsky)	Pher: 3R-, 3S-, 2R-, 2S-C6-ketol	M	Enh: 1-(1H-pyrrol-2-yl)-1,2-propanedione	Zou et al. 2016
Hylotrupes bajulus (L.)	Pher: 3R-C6-ketol, (2S,3R)-, (2R,3R)-C6-diol, (2,3)-C6-dione	M	—	Fettköther et al. 1995
Phymatodes aereus (Newman)	Pher: 3R-, 3S-C6-ketol	M	Attr: 3-C6-ketol + 3-C8-ketol; Enh: (2,3)-C6-diol; Antag: 2-methylbutan-1-ol	Hanks and Millar 2013; Mitchell et al. 2015
Phymatodes amoenus (Say)	Pher: 3R-C6-ketol, (R)-2-methylbutan-1-ol	M	Antag: plant volatiles	Hanks and Millar 2013; Mitchell et al. 2015
Phymatodes grandis (= *lecontei*) Casey	Pher: (R)-2-methylbutan-1-ol	M	Attr: 3R-, 3S-C6-ketol	Hanks et al. 2007
Phymatodes lengi Joutel	Pher: 3R-C6-ketol, (R)-2-methylbutan-1-ol	M	—	Hanks et al. 2012; Mitchell et al. 2015
Phymatodes pusillus (F.)	1-hexanol, 1-butanol, 1-octanol	M	—	Schröder 1996
Phymatodes testaceus (L.)	Pher: (R)-2-methylbutan-1-ol	M	Attr: 3-C6-ketol + (2,3)-C6-diol; Antag: plant volatiles	Sweeney et al. 2004; Hanks and Millar 2013
Phymatodes varius (F.)	Pher: 3R-C6-ketol, (R)-2-methylbutan-1-ol	M	Attr: syn-C6-diol; Antag: plant volatiles	Hanks and Millar 2013; Mitchell et al. 2015
Pyrrhidium sanguineum (L.)	3R-C6-ketol, (2R,3R)-, (2S,3R)-C6-diol	M	—	Schröder et al. 1994
Clytini				
Clytus arietis (L.)	3R-, 2S-C8-ketol, (2,3)-C8-diol	M	—	Schröder 1996
Clytus ruricola (Olivier)	—	–	Attr: plant volatiles	Montgomery and Wargo 1983
Curius dentatus Newman	—	–	Attr: anti-C6-diol; Antag: syn-C6 diol	Lacey et al. 2004; Hanks and Millar 2013
Demonax balyi Pascoe	—	–	Attr: 2S-C10-ketol	Hall et al. 2006a
Demonax gracilestriatus Gressitt & Rondon	—	–	Attr: 3-C6-ketol	Wickham et al. 2014

(Continued)

TABLE 5.1 (Continued)

Summary of Published Research from Field Trials on Volatile Pheromones, Attractants, Synergists, and Antagonists of Cerambycid Beetles

Taxonomy	Volatiles emitted (Pher = confirmed attractant)	Sex	Attractant (Attr), enhancer (Enh), antagonist (Antag)	References
Demonax l. literalis Gahan	—	–	Attr: 3-C6-ketol	Wickham et al. 2014
Demonax ordinatus Pascoe	—	–	Attr: *syn*-C8-diol	Wickham et al. 2014
Demonax theresae Pic		–	Attr: *anti*-C8-diol	Wickham et al. 2014
Megacyllene caryae (Gahan)	Pher: (2R,3S)-, (2S,3R)-C6-diol, (S)-(−)-limonene, 2-phenylethanol, (−)-α-terpineol, nerol, neral, geranial	M	Attr: citral, *syn*-C6-diol; Antag: plant volatiles	Lacey et al. 2008b; Hanks and Millar 2013; Handley et al. 2015
Neoclytus a. acuminatus (F.)	Pher: (2S,3S)-C6-diol	M	Attr: plant volatiles; Antag: *anti*-C6-diol, 3-C6-ketol, plant volatiles	Lacey et al. 2004; Hanks et al. 2012; Hanks and Millar 2013
Neoclytus balteatus LeConte	3R-C6-ketol, (2S,3S)-C6-, C8-diol	M		Ray et al. 2015
Neoclytus caprea (Say)	Pher: 3R-C6-ketol	M	Antag: plant volatiles	Hanks and Millar 2013; Ray et al. 2015
Neoclytus conjunctus (LeConte)	Pher: 3R-C6-ketol	M	—	Ray et al. 2015
Neoclytus irroratus (LeConte)	Pher: 3R-C6-ketol	M	Antag: 3S-C6-ketol	Ray et al. 2015
Neoclytus m. modestus Fall	Pher: 3R-C6-ketol	M	Antag: 3S-C6-ketol	Ray et al. 2015
Neoclytus m. mucronatus (F.)	Pher: 3R-C6-ketol	–	Attr: plant volatiles; Antag: (2,3)-C6-diol	Lacey et al. 2007a; Hanks et al. 2012; Miller et al. 2015a, 2015b; Ray et al. 2015
Neoclytus mucronatus vogti Linsley	3R-C6-ketol	M	—	Ray et al. 2015
Neoclytus scutellaris (Olivier)	Pher: 3R-C6-ketol	M	—	Ray et al. 2015
Neoclytus tenuiscriptus Fall	Pher: (2S,3S)-C6-diol	M	Antag: *anti*-C6-diol	Ray et al. 2015
Plagionotus arcuatus L.	3R-, 2-C10-ketol	M	—	Schröder 1996
Plagionotus christophi (Kraatz)	2-C8-ketol, (2,3)-C8-diol	M	—	Iwabuchi 1999
Rhaphuma horsfieldi (White)	—	–	Attr: *syn*-C6-, *syn*-C8-diol	Wickham et al. 2014
Rhaphuma laosica Gressitt & Rondon	—	–	Attr: *anti*-C6-diol	Wickham et al. 2014
Sarosesthes fulminans (F.)	Pher: 3R-C6-ketol, (2S,3R)-C6-diol	M	Attr: *anti*-C6-diol	Lacey et al. 2009; Hanks and Millar 2013
Xylotrechus antilope Schönh.	2S-, 3R-, 3S-C8-ketol	M	—	Schröder 1996
Xylotrechus atronotatus draconiceps Gressitt	—	–	Attr: 3-C6-ketol	Wickham et al. 2014
Xylotrechus chinensis (Chevrolat)	2S-, 3-C8-ketol, (2S,3S)-C8-diol	M	—	Iwabuchi et al. 1987; Kuwahara et al. 1987
Xylotrechus colonus (F.)	Pher: 3R-, 3S-C6-ketol, (2R,3R)-, (2S,3S)-, (2R,3S)-, (2S,3R)-C6-diol	M	Attr: plant volatiles; Antag: (2,3)-C6-diol	Lacey et al. 2009; Hanks et al. 2012; Hanks and Millar 2013, LMH, unpublished data

(Continued)

TABLE 5.1 (Continued)

Summary of Published Research from Field Trials on Volatile Pheromones, Attractants, Synergists, and Antagonists of Cerambycid Beetles

Taxonomy	Volatiles emitted (Pher = confirmed attractant)	Sex	Attractant (Attr), enhancer (Enh), antagonist (Antag)	References
Xylotrechus incurvatus (Chevrolat)	—	—	Attr: 3-C6-ketol	Wickham et al. 2014
Xylotrechus longitarsus Casey	—	—	Attr: plant volatiles	Morewood et al. 2002
Xylotrechus nauticus (Mannerheim)	Pher: 3R-, 3S-C6-ketol	M	—	Hanks et al. 2007
Xylotrechus pyrrhoderus Bates	Pher: (2S,3S)-C8-diol, 2S-C8-ketol	M	Antag: 2R-C8-ketol	Sakai et al. 1984; Iwabuchi et al. 1986; Narai et al. 2015
Xylotrechus quadripes Chevrolat	Pher: 2S-, 3-C10-ketol, 2-C8-ketol, (2S,3S)-C10-diol, (2,3)-C10-dione, 2-phenylethanol	M	—	Jayarama et al. 1998; Hall et al. 2006a
Xylotrechus rufilius Bates	—	—	Attr: 2-C8-ketol; Antag: *syn*-C8-diol	Iwabuchi 1999; Narai et al. 2015
Xylotrechus s. sagittatus (Germar)	—	—	Attr: plant volatiles	Miller et al. 2011, 2015a, 2015b; Hanks and Millar 2013
Xylotrechus undulatus (Say)	—	—	Attr: plant volatiles	Chénier and Philogène 1989
Xylotrechus villioni (Villard)	2-C8-ketone, (2,3)-C8-diol	M		Iwabuchi 1999
Compsocerini				
Rosalia funebris Motschulsky	Pher: (Z)-3-decenyl (E)-2-hexenoate	M	—	Ray et al. 2009a
Elaphidiini				
Anelaphus inflaticollis Chemsak	Pher: 3R-, 2S-C6-ketol, (2R,3R)-, (2R,3S)-C6-diol	M		Ray et al. 2009b
Anelaphus parallelus (Newman)	—	—	Attr: *syn*-C6-diol; Antag: 3-C6-ketol + *anti*-C6-diol	Hanks and Millar 2013
Anelaphus pumilus (Newman)	Pher: 3R-C5-ketol	M		Mitchell et al. 2015
Anelaphus villosus (F.)	—	—	Attr: *syn*-C6-diol, plant volatiles; Antag: plant volatiles	Montgomery and Wargo 1983; Hanks and Millar 2013
Elaphidion mucronatum (Say)	—	—	Attr: plant volatiles	Dunn and Potter 1991
Enaphalodes rufulus (Haldeman)	(2S,3S)-, (2S,3R)-C6-diol	M	—	Dahl 2006
Orwellion gibbulum arizonense (Casey)	3R-C6-ketol, decan-2-one	M	—	Mitchell et al. 2013
Parelaphidion aspersum (Haldeman)	3R-C6-ketol, nonan-2-one	M	—	Mitchell et al. 2013

(Continued)

TABLE 5.1 *(Continued)*

Summary of Published Research from Field Trials on Volatile Pheromones, Attractants, Synergists, and Antagonists of Cerambycid Beetles

Taxonomy	Volatiles emitted (Pher = confirmed attractant)	Sex	Attractant (Attr), enhancer (Enh), antagonist (Antag)	References
Hesperophanini				
Tylonotus bimaculatus Haldeman	Pher: (2S,4E)-2-hydroxy-4-octen-3-one; (3R,4E)-3-hydroxyoct-4-en-2-one, (E)-4-octen-2,3-dione, 2,3-C8-dione	M	–	Zou et al. 2015
Molorchini				
Molorchus minor (L.)	–	–	Attr: 3-C8-ketol + plant volatiles	Sweeney et al. 2014
Molorchus umbellatarum Schreb.	–	–	Attr: *anti*-C8-diol	Imrei et al. 2013
Obriini				
Obrium maculatum (Olivier)	–	–	Attr: fuscumol acetate	Mitchell et al. 2011; Hanks and Millar 2013
Tillomorphini				
Euderces picipes (F.)	3R-C6-ketol	M	Antag: 2-methylbutan-1-ol, nonan-2-one	Mitchell et al. 2015
Trachyderini				
Tragidion armatum brevipenne Linsley	3R-C6-ketol	M	–	Hanks et al. 2007
LAMIINAE				
Acanthocerini				
Acanthocinus aedilis (L.)	–	–	Attr: plant volatiles	Schroeder and Weslien 1994
Acanthocinus nodosus (F.)	–	–	Attr: plant volatile + bark beetle pheromones	Miller et al. 2015a, 2015b
Acanthocinus obliquus (LeConte)	–	–	Attr: plant volatile + bark beetle pheromones	Costello et al. 2008
Acanthocinus obsoletus (Olivier)	–	–	Attr: Plant volatiles and/or bark beetle pheromones	Miller and Asaro 2005; Miller et al. 2015a, 2015b
Acanthocinus princeps (Walker)	–	–	Attr: plant volatile + bark beetle pheromones	Macias-Samano et al. 2012
Acanthocinus spectabilis (LeConte)	–	–	Attr: plant volatile + bark beetle pheromones	Costello et al. 2008
Astyleiopus variegatus (Haldeman)	Pher: *S*-fuscumol, *S*-fuscumol acetate	M	Enh: plant volatiles; Antag: plant volatiles	Mitchell et al. 2011; Hanks et al. 2012; Hanks and Millar 2013; Hughes et al. 2013

(Continued)

TABLE 5.1 (Continued)

Summary of Published Research from Field Trials on Volatile Pheromones, Attractants, Synergists, and Antagonists of Cerambycid Beetles

Taxonomy	Volatiles emitted (Pher = confirmed attractant)	Sex	Attractant (Attr), enhancer (Enh), antagonist (Antag)	References
Astylidius parvus (LeConte)	–	–	Attr: fuscumol, fuscumol + fuscumol acetate; Enh: plant volatiles	Mitchell et al. 2011; Hanks et al. 2012
Astylopsis macula (Say)	–	–	Attr: fuscumol + fuscumol acetate, plant volatiles	Hanks and Millar 2013
Astylopsis sexguttata (Say)	–	–	Attr: plant volatiles and/or bark beetle pheromones; Enh: bark beetle pheromones	Miller et al. 2011, 2015a, 2015b; Hanks and Millar 2013
Graphisurus fasciatus (Degeer)	–	–	Attr: fuscumol + fuscumol acetate, fuscumol acetate, plant volatiles; Enh: plant volatiles; Antag: plant volatiles	Mitchell et al. 2011; Hanks et al. 2012; Hanks and Millar 2013
Leptostylus transversus (Gyllenhal)	–	–	Attr: fuscumol	Mitchell et al. 2011
Lepturges angulatus (LeConte)	–	–	Attr: fuscumol acetate, fuscumol + fuscumol acetate; Enh: plant volatiles	Mitchell et al. 2011; Hanks et al. 2012; Hanks and Millar 2013
Lepturges confluens (Haldeman)	–	–	Attr: fuscumol + fuscumol acetate; Enh: plant volatiles	Hanks and Millar 2013
Acanthoderini				
Acanthoderes quadrigibba (Say)	–	–	Attr: fuscumol + fuscumol acetate; Antag: plant volatiles	Hanks and Millar 2013
Aegomorphus modestus (Gyllenhal)	–	–	Attr: fuscumol acetate, fuscumol + fuscumol acetate; Antag: plant volatiles	Mitchell et al. 2011; Hanks et al. 2012; Hanks and Millar 2013
Hedypathes betulinus (Klug)	Fuscumol acetate, geranylacetone, fuscumol	M	–	Fonseca et al. 2010
Sternidius alpha (Say)	–	–	Attr: plant volatiles, fuscumol, fuscumol acetate; Antag: plant volatiles	Mitchell et al. 2011; Hanks and Millar 2013
Steirastoma breve (Sulzer)	Fuscumol	M	–	Liendo-Barandiaran et al. 2010b
Urgleptes querci (Fitch)	–	–	Attr: plant volatiles	Montgomery and Wargo 1983
Agniini				
Acalolepta formosana (Breuning)	–	–	Attr: monochamol	Wickham et al. 2014
Dorcaschematini				
Dorcaschema alternatum (Say)	–	–	Attr: *anti*-C6-diol; Antag: 3-C6-ketol + *syn*-C6-diol	Hanks and Millar 2013

(Continued)

TABLE 5.1 (Continued)

Summary of Published Research from Field Trials on Volatile Pheromones, Attractants, Synergists, and Antagonists of Cerambycid Beetles

Taxonomy	Volatiles emitted (Pher = confirmed attractant)	Sex	Attractant (Attr), enhancer (Enh), antagonist (Antag)	References
Lamiini				
Pharsalia subgemmata (Thomson)	–	–	Attr: monochamol	Wickham et al. 2014
Pseudomacrochenus antennatus (Gahan)	–	–	Attr: monochamol	Wickham et al. 2014
Xenohammus bimaculatus Schwarzer	–	–	Attr: monochamol	Wickham et al. 2014
Monochamini				
Anoplophora chinensis (Forster)	Pher: 4-(heptyloxy)butan-1-ol	M	–	Hansen et al. 2015
Anoplophora glabripennis (Motschulsky)	Pher: 4-(heptyloxy)butanal, 4-(heptyloxy)butan-1-ol, (*E,E*)-α-farnesene	M	Attr: plant volatiles	Zhang et al. 2002; Nehme et al. 2010, 2014; Crook et al. 2014
Monochamus alternatus Hope	Pher: monochamol	M	Attr and Enh: plant volatiles	Fan et al. 2007a, 2007b; Teale et al. 2011
Monochamus bimaculatus Gahan	–	–	Attr: monochamol	Wickham et al. 2014
Monochamus carolinensis (Olivier)	Pher: monochamol	M	Attr, Enh, and Antag: plant volatiles	Phillips et al. 1988; Allison et al. 2012; Hanks et al. 2012; Hanks and Millar 2013; Ryall et al. 2015
Monochamus clamator (LeConte)	–	–	Attr: monochamol + plant volatiles; Enh: bark beetle pheromones	Miller and Borden 1990; Allison et al. 2001, 2003; Morewood et al. 2003; Costello et al. 2008; Macias-Samano et al. 2012
Monochamus galloprovincialis (Olivier)	Pher: monochamol	M	Attr and Enh: plant volatiles	Ibeas et al. 2007; Pajares et al. 2010; Jurc et al. 2012, Rassati et al. 2012
Monochamus leuconotus LeConte	2-(4-heptyloxy-1-butyloxy)ethan-1-ol	M	–	Hall et al. 2006b; Pajares et al. 2010
Monochamus marmorator Kirby	–	–	Attr: monochamol + plant volatiles	Ryall et al. 2015
Monochamus mutator LeConte	–	–	Attr: monochamol + plant volatiles	Ryall et al. 2015
Monochamus notatus (Drury)	–	–	Attr: monochamol, *syn*-C6-diol + plant volatiles, plant volatiles, bark beetle pheromones; Enh: plant volatiles	Allison et al. 2001, 2012; Fierke et al. 2012; Hanks and Millar 2015; Ryall et al. 2015
Monochamus obtusus Casey	–	–	Attr: monochamol + plant volatiles, plant volatiles, bark beetle pheromones	Allison et al. 2001; Macias-Samano et al. 2012

(Continued)

TABLE 5.1 (Continued)

Summary of Published Research from Field Trials on Volatile Pheromones, Attractants, Synergists, and Antagonists of Cerambycid Beetles

Taxonomy	Volatiles emitted (Pher = confirmed attractant)	Sex	Attractant (Attr), enhancer (Enh), antagonist (Antag)	References
Monochamus saltuarius (Gebler)	–	–	Attr: monochamol	Ryall et al. 2015
Monochamus s. scutellatus (Say)	Pher: monochamol	M	Attr and Enh: plant volatiles, bark beetle pheromones; Antag: bark beetle pheromones	Chénier and Philogène 1989; Allison et al. 2001, 2003; Morewood et al. 2003; Fierke et al. 2012; Macias-Samano et al. 2012; Hanks and Millar 2013; Ryall et al. 2015
Monochamus sutor (L.)	Pher: monochamol	M	Attr: Bark beetle pheromones	Pajares et al. 2013
Monochamus titillator (F.)	–	–	Attr: monochamol + plant volatiles, plant volatiles, bark beetle pheromones	Billings and Cameron 1984; Phillips et al. 1988; Miller and Asaro 2005; Allison et al. 2012
Monochamus urussovii (Fischer)	–	–	Attr: monochamol + plant volatiles, plant volatiles	Sweeney et al. 2004; Ryall et al. 2015
LEPTURINAE				
Desmocerini				
Desmocerus a. aureipennis Chevrolat	Pher: (*R*)-desmolactone	F	–	Ray et al. 2014
Desmocerus a. cribripennis Horn	–	–	Attr: (*R*)-desmolactone	Ray et al. 2014
Desmocerus a. lacustris Linsley and Chemsak	–	–	Attr: (*R*)-desmolactone	Ray et al. 2014
Desmocerus c. californicus Horn	Pher: (*R*)-desmolactone	F	–	Ray et al. 2012
Desmocerus californicus dimorphus Fisher	–	–	Attr: (*R*)-desmolactone; Antag: (*S*)-desmolactone	Ray et al. 2014
Desmocerus palliatus (Forster)	–	–	Attr: (*R*)-desmolactone; Antag: (*S*)-desmolactone	Ray et al. 2014
Lepturini				
Analeptura lineola (Say)	–	–	Attr: plant volatiles	Montgomery and Wargo 1983
Ortholeptura valida (LeConte)	Pher: *cis*-vaccenyl acetate	F	–	Ray et al. 2011
Rhagiini				
Acmaeops proteus (Kirby)	–	–	Attr: monoterpenes, plant volatiles; Enh: bark beetle pheromones	Chénier and Philogène 1989; Costello et al. 2008; Miller et al. 2015a, 2015b

(Continued)

TABLE 5.1 (Continued)

Summary of Published Research from Field Trials on Volatile Pheromones, Attractants, Synergists, and Antagonists of Cerambycid Beetles

Taxonomy	Volatiles emitted (Pher = confirmed attractant)	Sex	Attractant (Attr), enhancer (Enh), antagonist (Antag)	References
Rhagium inquisitor (L.)	–	–	Attr: plant volatiles, bark beetle pheromones	Schroeder and Weslien 1994; Miller et al. 2011; Jurc et al. 2012; Hanks and Millar 2013; Miller et al. 2015a, 2015b
PRIONINAE				
Megopidini				
Megopis costipennis White	Pher: (2*R*,3*S*)-C8 diol	F	Attr: *anti*-C8-diol	Wickham et al. 2014, 2016
Meroscelisini				
Tragosoma depsarium "*harrisi*" LeConte	–	–	Attr: (2*S*,3*R*)-C6-diol; Antag: *syn*-C6-diol	Ray et al. 2012a
Tragosoma depsarium "sp. nov. Laplante"	Pher: (2*R*,3*R*)-C6 diol	F	Antag: *anti*-C6-diol	Ray et al. 2012a
Tragosoma pilosicorne Casey	–	–	Attr: (2*S*,3*R*)-C6-diol	Ray et al. 2012a
Prionini				
Dorythenes granulosus (Thomson)	–	–	Attr: prionic acid	Wickham et al. 2016a
Prionus aztecus Casey	–	–	Attr: prionic acid	Barbour et al. 2011
Prionus californicus Motschulsky	Pher: prionic acid	F	–	Rodstein et al. 2009, 2011
Prionus coriarius (L.)	–	–	Attr: prionic acid	Barbour et al. 2011
Prionus imbricornis (L.)	–	–	Attr: prionic acid	Barbour et al. 2011
Prionus integer LeConte	–	–	Attr: prionic acid	Barbour et al. 2011
Prionus laticollis (Drury)	–	–	Attr: prionic acid	Barbour et al. 2011
Prionus lecontei Lameere	–	–	Attr: prionic acid	Barbour et al. 2011
Prionus linsleyi Hovore	–	–	Attr: prionic acid	Barbour et al. 2011
Prionus pocularis Dalman	–	–	Attr: plant volatiles	Miller et al. 2015a, 2015b

(Continued)

TABLE 5.1 (*Continued*)

Summary of Published Research from Field Trials on Volatile Pheromones, Attractants, Synergists, and Antagonists of Cerambycid Beetles

Taxonomy	Volatiles emitted (Pher = confirmed attractant)	Sex	Attractant (Attr), enhancer (Enh), antagonist (Antag)	References
SPONDYLIDINAE				
Asemini				
Arhopalus rusticus (L.)	–	–	Attr: plant volatiles	Jurc et al. 2012; Miller et al. 2015a, 2015b
Arhopalus ferus (Mulsant)	–	–	Attr: plant volatiles; Antag: plant volatiles	Suckling et al. 2001; Wang and Leschen 2003
Asemum striatum (L.)	–	–	Attr: plant volatiles	Chénier and Philogène 1989; Sweeney et al. 2004; Miller et al. 2011, 2015a, 2015b; Hanks and Millar 2013
Tetropium castaneum (L.)	–	–	Attr: plant volatiles; Enh: fuscumol + fuscumol acetate	Sweeney et al. 2004, 2010, 2014
Tetropium cinnamopterum (Kirby)	Pher: *S*-fuscumol	M	Attr and Enh: plant volatiles	Silk et al. 2007; Sweeney et al. 2010; Hanks and Millar 2013
Tetropium fuscum (F.)	Pher: *S*-fuscumol	M	Attr and Enh: plant volatiles	Sweeney et al. 2004, 2010; Silk et al. 2007
Tetropium schwarzianum Casey	–	–	Attr: fuscumol + fuscumol acetate; Enh: plant volatiles	Hanks and Millar 2013
Spondylidini				
Spondylis buprestoides (L.)	–	–	Attr: plant volatiles	Sweeney et al. 2004; Jurc et al. 2012

Note: The sex that emits volatiles is indicated by M (male) and F (female). Confirmed pheromones (indicated by "Pher") that were produced by males attracted both sexes, whereas those produced by females attracted only males (see text). Compound abbreviations: carbon chainlength indicated by C6, C8, or C10; ketol, 2- or 3-hydroxyalkanone; diol, (2,3)-alkanediol; dione, (2,3)-alkanedione. Blends of chemicals tested in field trials are indicated by "+".

all known attractant pheromones or likely pheromones from species in the subfamilies Cerambycinae, Lamiinae, and Spondylidinae are male-produced aggregation pheromones (Table 5.1). Thus, through evolutionary time, the subfamilies appear to have diverged in their use of different types of pheromones, but the underlying causes for this divergence are not yet clear.

No attractant pheromones are known for any species in the remaining three subfamilies, (Dorcasominae, Necydalinae, Parandrinae), but it should be pointed out that these subfamilies are relatively small. It also must be pointed out that pheromones have been identified for two species, *Migdolus fryanus* Westwood (Leal et al. 1994) and *Vesperus xatarti* Mulsant (Boyer et al. 1997), which formerly were considered in the Cerambycidae, but taxonomic revisions have now placed these species in their own family, the Vesperidae (Švácha and Lawrence 2014). Hence, the pheromones of these two species will not be discussed here.

5.3 Volatile Pheromones from the Various Subfamilies

5.3.1 Subfamily Cerambycinae

The first volatile pheromone found in the Cerambycidae was identified from the Japanese cerambycine *Xylotrechus pyrrhoderus* Bates (Sakai et al. 1984; Iwabuchi 1986). The pheromone is produced by males and consists of two components, (S)-2-hydroxyoctan-3-one and (2S,3S)-2,3-octanediol (Figure 5.1 and Table 5.1). Wind tunnel and field cage experiments showed that the ketol component was somewhat attractive on its own, whereas the diol was not, but it synergized attraction to the ketol (Iwabuchi et al. 1986; Iwabuchi 1988). Furthermore, the "nonnatural" enantiomer (R)-2-hydroxyoctan-3-one inhibited attraction. The same two gross structures were found in the congener *X. chinensis* (Chevrolat), but it was not possible to determine which stereoisomers were produced (Iwabuchi et al. 1987). Over the next 15 years, male-produced pheromones were reported from another four cerambycine species (Table 5.1), including *Pyrrhidium sanguineum* (L.), *Hylotrupes bajulus* (L.), *Anaglyptus subfasciatus* Pic, and *Xylotrechus quadripes* Chevrolat. Remarkably, the structures of the pheromones of all six of these cerambycine species, from different areas of the world, were all very similar, with all of them consisting of unbranched 6-, 8-, or 10-carbon chains with either a ketone and an alcohol function in the two and three positions or alcohols in both positions (Figure 5.1). On this basis, we synthesized and (with the help of numerous collaborators) tested a library of these types of structures in field screening trials in North America and other parts of the world, with remarkable results (see Table 5.1). To summarize a decade of trials, these types of compounds now are known to be male-produced pheromone components for numerous cerambycine species and likely pheromones for many more that have been attracted in significant numbers in field trials but that have not yet been shown to produce the compounds to which they were attracted. The cumulative data have elucidated a number of trends:

1. Chain lengths of the hydroxyketones and 2,3-alkanediols appear to be restricted to 6, 8, or 10 carbons.
2. Although 6-, 8-, and 10-carbon 2,3-alkanediones frequently are found in extracts of headspace volatiles from males, they do not appear to be active components of the pheromones. Rather, they may be biosynthetic intermediates or artifacts from degradation of the corresponding ketols during analysis (Sakai et al. 1984; Leal et al. 1995; Millar et al. 2009).
3. In the hydroxyketones, the ketone can be in the 2- or 3-position. When in the 2-position, for example, 3-hydroxyalkan-2-one, the (R)-configuration is strongly favored, whereas when the ketone is in the 3-position (2-hydroxyalkan-3-one), the (S)-configuration appears most common (Table 5.1). For reasons that are not known, 3-hydroxyalkan-2-ones appear to be much more common than the isomeric 2-hydroxyalkan-3-ones, but this may be due to the fact that more field screening trials have been carried out with the former. Furthermore, beetles generally appear to be insensitive to the enantiomeric purity of the hydroxyketones (but see Iwabuchi et al. 1986 for an example where they are not), whereas there are a number of examples of beetles being inhibited by other structural analogs (Table 5.1).

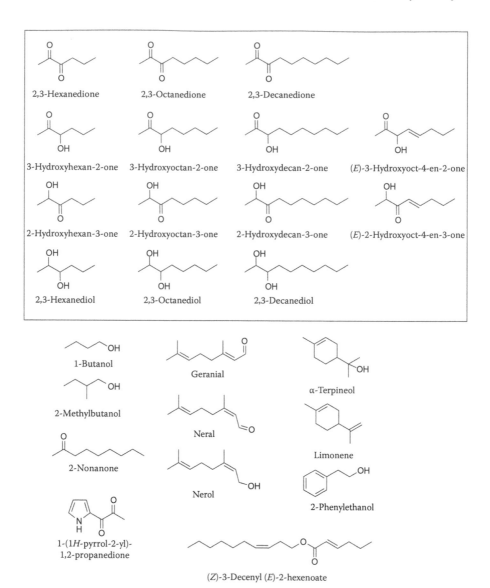

FIGURE 5.1 Pheromone structures and related compounds identified from species in the subfamily Cerambycinae. The structures in the box are the fairly well-defined group of 2,3-alkanediones, 2-hydroxyalkan-3-ones, 3-hydroxylalkan-2-ones, and 2,3-alkanediols. The remaining structures belong to various classes of compounds.

4. The 2,3-alkanediols each have four possible stereoisomers (two diastereomeric pairs of enantiomers). For those species using 2,3-alkanediol pheromone components, there are no examples known of inhibition by the "non-natural" enantiomer, but there are several examples known of species being inhibited by one or both of the diastereomers (Table 5.1).

5. All known examples of volatile pheromones from species in the subfamily Cerambycinae are male-produced, and many of the pheromones appear to be produced from glandular tissue in the prothorax (see following).

To date, the only variation that has appeared on these hydroxyketone and diol structures is the addition of a double bond between carbons 4 and 5, for the cerambycine *Tylonotus bimaculatus* Haldeman, males of which produce (2S,4E)-2-hydroxy-4-octen-3-one and lesser amounts of relatively unstable (3R,4E)-3-hydroxy-4-octen-2-one (Figure 5.1 and Table 5.1).

However, it is rapidly becoming clear that cerambycine species use a wide variety of other compounds as male-produced pheromone components, and these compounds currently defy ready classification into related structural groups (see Figure 5.1 and Table 5.1). For example, several species use 2-methylbutan-1-ol as a pheromone component, either as a single component or in combination with other components such as 3-hydroxyhexan-2-one. Several others produce a novel diketopyrrole structure (1-[1*H*-pyrrol-2-yl]-1,2-propanedione) (Zou et al. 2016). The North American species *Megacyllene caryae* (Gahan) produces a complicated blend consisting of two 2,3-hexanediols, along with several terpenoids and aromatic compounds. Whereas initial field trials suggested that most of the components were necessary to attract beetles, subsequent bioassays over several years have shown that an isomeric blend of geranial and neral (= citral), among the most abundant components in the blend, is highly attractive without the remaining minor components (e.g., Handley et al. 2015). Other compounds appear to be unique; for example, the North American species *Rosalia funebris* Motschulsky uses (Z)-3-decenyl (E)-2-hexenoate as its male-produced pheromone, which has not attracted any other cerambycid species in field trials in North America, Asia, or Europe. Overall though, based on the trends that have emerged so far, it seems safe to say that most of the compounds mentioned earlier in this chapter will turn out to be pheromone components of additional species. Conversely, it is certain that other entirely new cerambycine pheromone structures remain to be discovered.

Two general types of glandular tissues that produce volatile pheromones have been identified within the Cerambycidae. The first type, typified by males of many cerambycine species, consists of sex-specific multicellular glands in the endocuticle that connect via ducts to pores in the prothorax. These were first identified in males of the cerambycine *X. pyrrhoderus* by Iwabuchi (1986), who also noted that the pheromone-producing tissues were undeveloped in newly emerged males but developed and became functional as the males became sexually mature. More recently, Iwabuchi and coworkers published a follow-up study in which the glandular structures were illustrated in much greater detail in *X. pyrrhoderus* and a number of other species (Hoshino et al. 2015). Males of the cerambycines *Neoclytus a. acuminatus* (F.) (Lacey et al. 2007b), *A. subfasciatus* (Nakamuta et al. 1994), and *H. bajulus* (Noldt et al. 1995) have prothoracic pores similar to those of *X. pyrrhoderus*, and these pores also have been associated with production of pheromones. In a more comprehensive examination of males and females of 65 species in 24 tribes within the Cerambycinae, Ray et al. (2006) found male-specific prothoracic gland pores in 49 species across 15 tribes, whereas neither sex of the remaining 16 species appeared to have pores, even though at least one of these species (*R. funebris*) is known to have a male-produced volatile pheromone (Ray et al. 2009a). However, it must be noted that the structure of its pheromone ([Z]-3-decenyl [E]-2-hexenoate) is quite different than those of any other pheromones known from cerambycines (see Figure 5.1), and so the pheromone may be produced by glands elsewhere on the body. Furthermore, Ray et al. (2006) did not find consistent patterns within tribes or even within a genus. For example, males of one species in the genus *Lissonotus* had gland pores, whereas they were absent in a congener. Male-specific gland pores also were found in all nine species in another study of cerambycine species in the tribe Clytini (Li et al. 2013). In a more recent study comparing 12 Asian cerambycine species in the tribes Clytini and Anaglyptini, males of 11 species, including four *Xylotrechus* species, had gland pores, whereas a fifth species, *X. cuneipennis* (Kraatz), did not (Hoshino et al. 2015).

The large number of examples now known suggests that the presence of male-specific pores is a likely indication of aggregation pheromone production within the subfamily Cerambycinae. In fact, 11 species shown by Ray et al. (2006) to have pores were subsequently found to use male-produced pheromones (Lacey et al. 2008b, 2009; Mitchell et al. 2013, 2015; Ray et al. 2015; Zou et al. 2016). However, no conclusions can be drawn about pheromone use in cerambycine species in which the pores are absent because at least one species without pores (*R. funebris*) clearly does use male-produced pheromones. Thus, the prothoracic pores offer a one-sided test: If they are present, the species probably uses a male-produced pheromone, but if absent, it cannot be predicted whether or not pheromones are used.

In addition to having specialized morphological structures for pheromone production, cerambycine species have evolved specific behaviors that aid in the dissemination of volatile pheromones. Thus, calling male *N. a. acuminatus* adopt a "push-up stance" in which the head and thorax are raised, which increases air flow over the pheromone-releasing prothoracic structures (Lacey et al. 2007b). Since then, similar behaviors have been noted in males of a number of other cerambycine species that have prothoracic pores (Lacey et al. 2007a, 2009; Ray et al. 2009b).

5.3.2 Subfamily Lamiinae

To date, all known volatile pheromones of species within the Lamiinae are male-produced aggregation pheromones. Two general motifs have been found—one clearly terpenoid based and the other based on hydroxyethers (Figure 5.2). The first lamiine pheromone was identified from males of *Anoplophora glabripennis* (Motschulsky), the Asian longhorned beetle, and consisted of a blend of 4-(heptyloxy)butanol and the corresponding aldehyde, 4-(heptyloxy)butanal (Zhang et al. 2002). However because of the relatively weak activity of the pheromone in numerous field trials conducted in both the United States (Nehme et al. 2014) and this beetle's native region, China (Nehme et al. 2010; Meng et al. 2014), it remains uncertain whether the pheromone has been completely identified. In fact, a recent paper suggested that (*E,E*)-α-farnesene may comprise a third component of the pheromone but, to date, only laboratory bioassays have been reported (Crook et al. 2014). Combinations of host plant volatiles with the pheromone lure do seem to increase attraction, but the overall levels of attraction remain relatively weak (Nehme et al. 2010; Meng et al. 2014).

Not unexpectedly, males of the sympatric congener *A. chinensis* (Forster) also have been shown to produce 4-(heptyloxy)butanol but not the aldehyde, and in field trials, beetles were attracted to blends

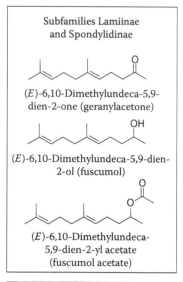

Subfamilies Lamiinae and Spondylidinae

(*E*)-6,10-Dimethylundeca-5,9-dien-2-one (geranylacetone)

(*E*)-6,10-Dimethylundeca-5,9-dien-2-ol (fuscumol)

(*E*)-6,10-Dimethylundeca-5,9-dien-2-yl acetate (fuscumol acetate)

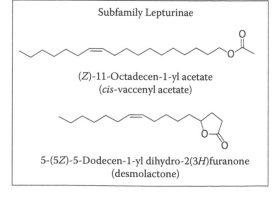

Subfamily Lamiinae

2-(Undecyloxy)ethanol (monochamol)

4-(Heptyloxy)butanol

4-(Heptyloxy)butanal

2-(Heptyloxybutoxy)ethanol

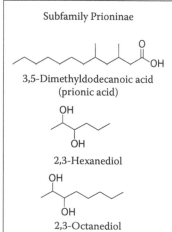

Subfamily Prioninae

3,5-Dimethyldodecanoic acid (prionic acid)

2,3-Hexanediol

2,3-Octanediol

Subfamily Lepturinae

(*Z*)-11-Octadecen-1-yl acetate (*cis*-vaccenyl acetate)

5-(5*Z*)-5-Dodecen-1-yl dihydro-2(3*H*)furanone (desmolactone)

FIGURE 5.2 Pheromone structures identified from species in the subfamilies Lamiinae, Spondylidinae, Prioninae, and Lepturinae. Boxes delineate groups of structures associated with the various subfamilies.

of the pheromone when released with α-pinene and ethanol as crude mimics of host plant volatiles (Hansen et al. 2015).

Another example of the hydroxyether motif was identified in the European species *Monochamus galloprovincialis* (Olivier) (Pajares et al. 2010), males of which produce 2-(undecyloxy)ethanol (given the common name monochamol). This pheromone, in combination with host plant volatiles and bark beetle pheromone synergists, is an excellent attractant for both sexes, and lures have been commercialized for trapping this species, specifically because it is a major vector of pinewood nematode in Europe (Sousa et al. 2001). Monochamol now has been identified as a pheromone component for five additional *Monochamus* species and as a likely pheromone component for another eight (Table 5.1). For some species, the pheromone alone is attractive, whereas with others, the pheromone is synergized by plant volatiles and/or bark beetle pheromones (see Section 5.7). Extending the biosynthetic parsimony further, monochamol has now been identified as a likely pheromone for beetles in four other genera in the same tribe as *Monochamus* (Monochamini), specifically *Acalolepta formosana* (Breuning), *Pharsalia subgemmata* (Thomson), *Pseudomacrochenus antennatus* (Gahan), and *Xenohammus bimaculatus* Schwarzer (Table 5.1). However, it must be noted that even within the genus *Monochamus*, there is some variation in pheromone structures. Specifically, males of the African species *Monochamus leuconotus* (Pascoe) produce 2-(heptyloxybutoxy)ethanol, almost a hybrid structure of the other two (Figure 5.2 and Table 5.1). To date, no field trials have been reported for this species.

Given the similarities among these hydroxyether type pheromones, and the overall high degree of conservation of pheromone structures among related cerambycid taxa, we synthesized and field tested four analogs of these compounds (2-[nonyloxy]ethanol, 4-[nonyloxy]butanol, 2-[heptyloxy]ethanol, and 6-[heptyloxy]hexanol) in the United States, China, and Costa Rica (unpublished data). None of them attracted any cerambycids, suggesting that if there are other pheromones with these types of structures, they likely consist of only a few specific structures. From the three known pheromones of this type (four if you also count 4-[heptyloxy]butanal), one consistent structural feature that may be important is that the nonfunctionalized end of the molecule consists of a fragment with an odd number of carbons (7 or 11), whereas the second part of the chain contains an even number of carbons (2 or 4; Figure 5.2).

The second major pheromone structural type found from lamiine species to date is based on the sesquiterpene degradation product geranylacetone, the corresponding alcohol ([*E*]-6,10-dimethylundeca-5,9-dien-2-ol, originally called fuscol, but now generally referred to as fuscumol), and the acetate ester of the alcohol ([*E*]-6,10-dimethylundeca-5,9-dien-2-yl acetate, or fuscumol acetate; Figure 5.2). The alcohol and acetate can exist in two enantiomeric forms, each of which will likely have different biological activities. This structural motif was first identified in two species in the genus *Tetropium* (subfamily Spondylidinae; Silk et al. 2007; see following text), with fuscumol being named after *T. fuscum* (F.), but fuscumol and/or fuscumol acetate were later shown to attract adults of a number of lamiine species as well (Mitchell et al. 2011). In the Lamiinae, components with this structural motif were first identified from the South American species *Hedypathes betulinus* (Klug), males of which produce fuscumol acetate as a major component, with lesser amounts of fuscumol and geranylacetone (Table 5.1). In laboratory bioassays, fuscumol acetate was not attractive to females as a single component, but blends of fuscumol acetate with the two minor components, or with host plant volatiles, were attractive. Further studies showed that the males produced pure (*R*)-fuscumol acetate but, unexpectedly, a ~4:1 mixture of (*R*)- and (*S*)-fuscumol (Vidal et al. 2010). Similarly, fuscumol has been identified from male-produced volatiles of the South American lamiine *Steirastoma breve* (Sulzer), and females were attracted to extracts of males in laboratory bioassays (Table 5.1). There have been no published reports of field trials for either species to date.

Fuscumol and fuscumol acetate, as single components or blended, have now been shown to attract a number of North American lamiine species (Table 5.1), usually in screening trials that used racemic fuscumol and fuscumol acetate (e.g., Hanks et al. 2012) or even "technical grade" mixtures of the (*E*)- and (*Z*)-isomers of the two compounds (Mitchell et al. 2011; Wong et al. 2012; Hanks and Millar 2013; Hanks et al. 2014). However, for almost all of these species, it is not yet known which enantiomer (or blend of enantiomers) of fuscumol and/or fuscumol acetate is produced. The single exception is *Astyleiopus variegatus* (Haldeman), which has been shown to produce the (*S*)-enantiomer of both compounds (Hughes et al. 2013). The results of field trials to determine whether the "non-natural" (*R*)-enantiomers either inhibit or have no effect on attraction have not yet been reported.

Despite the rapidly increasing number of pheromones known from the Lamiinae, little is known about where these compounds are produced. Prothoracic pores were found in *Hedypathes betulinus* (Klug), males of which produce fuscumol, fuscumol acetate, and geranylacetone, but the pores were present in both sexes (Fonseca et al. 2010). By analyzing extracts of body sections of males, Zarbin et al. (2013) found that pheromones were concentrated in the prothorax; furthermore, isotopically labeled precursors applied to the prothoraces of males were converted to labeled pheromone (Zarbin et al. 2013). Thus, the evidence suggests that pheromone production likely occurs somewhere in the prothorax, even though the role of the identified pores is equivocal. Similarly, males of *M. s. scutellatus* (Say), which produce the pheromone monochamol (Fierke et al. 2012), were found to have prothoracic pores, but no females were examined so it is not known whether the pores are sex-specific (Brodie 2008). Thus, all that can really be said is that the pheromone-producing structures of lamiine species are not yet known.

5.3.3 Subfamily Spondylidinae

Male-produced volatile pheromones, including fuscumol and fuscumol acetate (Figure 5.2), have been identified from two species in this subfamily (Table 5.1). As already mentioned, fuscumol was first identified from the congeners *Tetropium fuscum*, native to Europe but established in North America since about 1999, and the North American *T. c. cinnamopterum* Kirby (Silk et al. 2007; Sweeney et al. 2010). Both species produce and respond to (*S*)-fuscumol, but they respond equally well to racemic fuscumol, indicating that the (*R*)-enantiomer is not inhibitory (Sweeney et al. 2010). Also, the pheromone alone is minimally active for both species, as well as for the European congener *T. castaneum* (L.), but there is strong synergism between the pheromone and host volatiles (Sweeney et al. 2010). Fuscumol and fuscumol acetate also attracted the congener *T. schwarzianum* Casey in field screening trials (Hanks and Millar 2013), providing more evidence that this is likely a conserved and widely shared pheromone motif.

Within the subfamily Spondylidinae, Mayo et al. (2013) reported that neither sex of *T. fuscum* and *T. c. cinnamopterum* has the prothoracic gland pores that have been correlated with pheromone production in cerambycine species. However, isotopically labeled precursors applied to the abdominal sternites were incorporated into the pheromone, leading the authors to hypothesize that pheromone production might be occurring in the midgut, but this remains to be verified. It has also been observed that calling males of *Tetropium* species adopt a "push-up" stance, analogous to that noted with cerambycine species (Section 5.3.1), as a mechanism to more effectively disperse their pheromones (Lemay et al. 2010; Mayo et al. 2013). The presence of calling males also stimulates other males to call, increasing the strength of the signal (Lemay et al. 2010).

5.3.4 Subfamily Prioninae

Švácha and Lawrence (2014) suggested that pheromones of prionine species were likely to be produced by females based on the fact that females of many species are flightless, or at least incapable of flight until they have laid some of their eggs and reduced their mass, whereas males fly readily. In contrast, few cerambycine species have flightless females (Švácha and Lawrence 2014). To date, two disparate types of female-produced sex pheromones have been identified from prionine species (Figure 5.2 and Table 5.1). The first was identified from *Prionus californicus* Motschulsky as 3,5-dimethyldodecanoic acid (prionic acid) (Rodstein et al. 2009), and this compound as a mixture of all four possible stereoisomers subsequently was shown to attract males of at least six *Prionus* species in North America as well as the European *P. coriarius* L. (Barbour et al. 2011). The pheromone appears to be produced from an eversible gland on the dorsum of the ovipositor: a calling female elevates her abdomen, extrudes her long ovipositor, and the gland everts on the dorsal surface (Barbour et al. 2006). Other researchers have reported similar behaviors by females of other prionine species (Rotrou 1936; Linsley 1962; Benham and Farrar 1976; Gwynne and Hostetler 1978; Paschen et al. 2012), suggesting that the eversible gland on the ovipositor may be widespread in some but not all tribes of the Prioninae (see following for examples of other pheromone gland structures).

Extracts of the pheromone glands of female *P. californicus* were found to contain a number of analogs and homologs of prionic acid, some of which also were attractive to males; but prionic acid as a single component was as attractive as any blend, showing that the single compound was both necessary and sufficient for attraction (Maki et al. 2011b). Very recently, males of an Asian prionine species in another genus, *Dorysthenes granulosus* (Thomson), have been shown to be attracted by prionic acid (Table 5.1; Wickham et al. 2016a), indicating that this compound is a likely pheromone for this species as well and that the structure has been conserved across genera of the tribe Prionini on different continents. It has not been reported whether females of *D. granulosus* display the same type of calling behavior, with an eversible gland on the ovipositor. Furthermore, given the high degree of conservation of pheromone structures within closely related taxa of cerambycids, it seems likely that prionic acid, or a closely related compound, will be found to attract males of other *Dorysthenes* species that are important pests in Asia (Hill 2008).

The second type of female-produced sex pheromones are comprised of 2,3-alkanediols, which have been identified within the Prioninae (Figure 5.2 and Table 5.1). Females of the North American *Tragosoma depsarium* "sp. nov. Laplante" (tribe Meroscelisini) produce (2*R*,3*R*)-2,3-hexanediol, and only males were attracted by the synthesized compound in field bioassays (Ray et al. 2012a). During those field trials, males of the congeners *T. depsarium* "*harrisi*" LeConte and *T. pilosicorne* Casey were specifically attracted by the (2*S*,3*R*)-enantiomer, strong evidence that it is likely to be their female-produced sex pheromone. Similarly, females of the Asian prionine *Nepiodes* (formerly *Megopis*) *c. costipennis* White (tribe Megopidini) produce (2*R*,3*S*)-2,3-octanediol, and only males were attracted by the synthetic compound in field bioassays (Wickham et al. 2016b). These 2,3-alkanediols are unrelated to prionic acid biosynthetically and, furthermore, apparently are produced from entirely different glands than prionic acid. Remarkably, females of these diol-producing prionine species have prothoracic gland pores analogous to the pheromone-producing gland pores in male cerambycines (Ray et al. 2012a; Wickham et al. 2016b), some species of which produce exactly the same 2,3-alkanediols as aggregation pheromones. Thus, the compounds and the morphological structures from which they are produced have been conserved across subfamilies, even though the sexes producing the compounds are different, and the contexts in which the pheromones are used also are different.

5.3.5 Subfamily Lepturinae

The first attractant pheromone of a lepturine species was identified in 2011 for the North American species *Ortholeptura valida* (LeConte) (Ray et al. 2011). The compound, (Z)-11-octadecen-1-yl acetate (*cis*-vaccenyl acetate; Figure 5.2), is a female-produced sex pheromone but also is produced by male fruit flies in the genus *Drosophila* as an aggregation pheromone (Bartelt et al. 1985) and as an anti-aphrodisiac that males apply to females (Jallon et al. 1981).

The other sex pheromone structure that has been identified from lepturine species, (4*R*,9*Z*)-hexadec-9-en-4-olide (desmolactone; Figure 5.2), is likely to be biosynthetically related to *cis*-vaccenyl acetate, both probably being derived from fatty acid precursors. Desmolactone was initially described from *Desmocerus californicus californicus* Horn (Ray et al. 2012b) but has since been shown to be a pheromone, or likely pheromone, for several other North American species and subspecies of *Desmocerus*, including the threatened *D. californicus dimorphus* Fisher (Ray et al. 2014; Table 5.1).

Nothing is known of the site of pheromone production in lepturine species. Hoshino et al. (2015) noted that males of the lepturine *Leptura o. ochraceofasciata* (Motschulsky) lacked the prothoracic gland pores that have been found in males of many cerambycine species, but this may be largely irrelevant because the evidence suggests that lepturine species use female-produced sex pheromones rather than male-produced aggregation pheromones.

5.4 Contact Pheromones

As with most other insect species, cerambycids use relatively nonvolatile compounds as close range or contact pheromones for the recognition of species and sex, and specifically, so that males can recognize females for mating. This dependence on contact for mate recognition was reported quite early (e.g., Heintze 1925), and it has since been reported for cerambycid species in all of the major subfamilies (e.g., Bouhelier and Hudault 1936; Michelsen 1963; Farrar 1967; Pilat 1972; Godinez-Aguilar et al. 2009). Thus, upon contacting females with their antennae, males display a sequence of characteristic behavioral steps, including arrestment and orientation toward the female, alignment of the male's body with that of the female and mounting on her back, culminating in coupling of the genitalia and copulation (see Chapter 4). It has been demonstrated repeatedly that the contact pheromone is a subset of the cuticular lipids, by solvent extraction of freshly killed female carcasses. Males readily try to mate with carcasses of females before extraction but ignore the carcasses after the cuticular lipids have been removed (reviewed in Ginzel 2010). The fact that males readily attempt copulation with female carcasses also precludes the possibility of acoustic signals or behavioral responses on the part of the female being critical components in close-range sexual recognition. The cuticular extract can then be transferred onto a previously washed female or male carcass, or a model such as a glass rod, resulting in at least partial restoration of the activity, with males orientating toward and attempting to copulate with the treated models. In fact, successful interspecific copulations can be induced by treating a live female of one species with the cuticular extract of another (e.g., *Phoracantha semipunctata* and *Phoracantha recurva* Newman; JGM, LMH, unpublished data; *Megacyllene* spp.; Galford 1980a). However, components of the active pheromones have been identified in only a few cases, which is due in part to the complexity of the cuticular lipids and the difficulty in purifying or obtaining pure standards of individual compounds for tests of function in bioassays.

In the first study in which a cerambycid contact pheromone was identified, Fukaya and Honda (1992) conducted a sequence of methodical studies with the yellow-spotted longicorn, *Psacothea hilaris* (Pascoe) (Lamiinae), in which they first described the sequence of male mating behaviors, then showed that the pheromone was removed by extraction with organic solvents and could be reapplied to previously extracted carcasses or neutral substrates, and finally showed that antennae, palpi, and tarsi were all involved in various steps of the behavioral sequence. In this and subsequent papers (e.g., Fukaya and Honda 1995; Fukaya et al. 1996), the authors also presented data indicating that the different steps in the behavioral sequence were elicited by different components, with the compound(s) eliciting the initial orientation response being localized on the female's prothorax, whereas the component(s) eliciting abdominal bending/genital coupling by the male were likely distributed over the entire body surface. A major component of the pheromone was identified as the methyl-branched alkene (Z)-21-methylpentatriacont-8-ene, which elicited mating behaviors from males, although at lower levels than the crude extract from females, suggesting that other components might be involved (Fukaya et al. 1996). In particular, the authors noted that this compound did not induce the orientation behavioral step of males upon initial contact with females. Furthermore, because of the presence of the double bond and a chiral center in the molecule, the compound has four possible stereoisomers. Synthesis and bioassay of all four isomers showed that, somewhat surprisingly, all four isomers elicited some degree of response. The (Z)-isomers elicited stronger responses than the "non-natural" (E)-isomers, but there were no significant differences in responses of males to the (R)- or (S)-enantiomers of the (Z)-alkene, suggesting that the males could not discriminate the absolute configuration (Fukaya et al. 1997). Fukaya and Honda (1996) showed that the size and shape of models treated with extracts of females were also important, with the best responses being obtained with models that approximated the size and shape of a female beetle. Somewhat surprisingly, these authors also reported that extracts of males could elicit partial mating responses from other males (Fukaya and Honda 1992, 1996), suggesting that the relevant components were not sex-specific. However, no chromatographic comparisons of males and females were presented, so this finding remains speculative.

Fukaya and coworkers followed this work with a comprehensive study of the contact pheromone of a Japanese population of another lamiine species, *Anoplophora malasiaca* (Thomson). Although this species was synonymized with *A. chinensis* by Lingafelter and Hoebeke (2002), more recent evidence suggests that the two may indeed be separate species (e.g., Muraji et al. 2011). The pheromone was perceived only after the male contacted the female with his antennae. Liquid chromatographic fractionation of cuticular extracts of females showed that the pheromone consisted of both saturated hydrocarbon(s) and more polar compounds (Fukaya et al. 1999, 2000). Although the hydrocarbon fractions of extracts from males and females had some components in common, they were clearly different both qualitatively and quantitatively (Akino et al. 2001). A reconstruction of the eight major alkanes from females, when mixed with a more polar fraction, elicited the full suite of mating behaviors, and of the eight compounds tested, the 9-methylalkanes (9-methylheptacosane and 9-methylnonacosane) and the 15-methylalkanes (15-methylhentriacontane and 15-methyltritriacontane) appeared to be most important in eliciting responses from males. Initial analyses of the more polar fraction identified a blend of four straight-chain ketones (heptacosan-10-one, [Z]-18-heptacosen-10-one, [18Z,21Z]-heptacosa-18,21-dien-10-one, and [18Z,21Z,24Z]-heptacosa-18,21,24-trien-10-one) as additional active components, whereas a fifth component, heptacosan-12-one, appeared to be inhibitory (Yasui et al. 2003). However, an additional subfraction from the polar fraction was required for full activity. Further analyses identified three complex bicyclic lactones, termed gomadalactones A, B, and C, as the missing components (Yasui et al. 2007a). Their absolute configurations were inferred by comparison of their circular dichroism spectra with those of synthesized model compounds containing the bicyclic lactone core (Mori 2007). As with *P. hilaris*, visual cues also were important in directing males toward females at short range (Fukaya et al. 2004, 2005). To our knowledge, this contact pheromone blend consisting of eight hydrocarbons, four ketones, and three lactones is the most complex insect pheromone blend yet identified.

Contact sex pheromone components also have been identified from one other lamiine species, *A. glabripennis*, a congener of *A. malasiaca*. Zhang et al. (2003) identified five alkenes ([Z]-9-tricosene, Z9-C_{23}, subsequent abbreviations follow this pattern, Z9-C_{25}, Z7-C_{25}, Z9-C_{27}, Z7-C_{27}) that were consistently more abundant in extracts of female *A. glabripennis* than in corresponding extracts of males. Individually, the compounds were inactive, but the blend of all five elicited mating behaviors from males analogous to those elicited by crude extracts of females. Subsets of these five compounds were not tested, so it remains unclear as to whether all five components are required for activity.

Within the subfamily Spondylidinae, contact pheromones have been identified from two species in the genus *Tetropium*. Thus, bioassay evidence suggested that (S)-11-methylheptacosane, present in extracts of females in much larger amounts than in extracts of males, is a contact pheromone component of *T. fuscum*, with the synthetic enantiomer eliciting copulation attempts by males (Silk et al. 2011). In contrast, the racemic mixture and the (R)-enantiomer did not elicit mating attempts, indicating that the males could clearly discriminate between the enantiomers and that the (R)-enantiomer was inhibitory. Because of the relatively small amounts available for analysis, and the very small specific rotations of long-chain methyl-branched hydrocarbons, it was not possible to verify that the females produced the (S)-enantiomer by analytical means. Addition of several other methylalkanes that were present in higher proportions in extracts of females than in extracts of males did not increase the activity. However, only the (S)-enantiomers were tested. For the congeneric *T. c. cinnamopterum*, a blend of (S)-11-methylheptacosane with Z9-C_{25} was required to elicit significant levels of arrestment, mounting, and copulation attempts (Silk et al. 2011).

In the subfamily Cerambycinae, contact sex pheromones have been identified from five species. Thus, the contact pheromone of *N. a. acuminatus* consists of a major component, 7Me-C_{27}, and two synergistic minor components, 7Me- and 9Me-C_{25} (Lacey et al. 2008a). In a follow-up study, Hughes et al. (2015) proved that the insect-produced 7Me-C_{25} had the (R)-configuration by isolating the compound, measuring its optical rotation, and comparing the rotation to those of the pure synthesized enantiomers. It was not possible to isolate the other two components to measure their optical rotations, but bioassays with the enantiomers of the two compounds suggested that they also were likely to have the (R)-configuration.

For the cerambycine *Callidiellum rufipenne* (Motschulsky), the synthesized mixture of the four isomers of the female-specific 5,17-dimethylnonacosane proved sufficient to elicit complete mating sequences from males (Rutledge et al. 2009). Similarly, Z9-C_{25}, the most abundant hydrocarbon on the cuticle of female *Megacyllene robiniae* (Forster), elicited the full array of mating responses from males (Ginzel et al. 2003), whereas females of the congeneric *M. caryae* used the analogous Z9-C_{29} as a contact pheromone component (Ginzel et al. 2006). In both of these *Megacyllene* species, there were marked differences in the cuticular hydrocarbon profiles of males and females, suggesting that other components probably are involved in the pheromones as well. Finally, three contact pheromone components, C_{25}, 9Me-C_{25}, and 3Me-C_{25}, were identified from female *Xylotrechus colonus* (F.) (Ginzel et al. 2003), and the absolute configuration of the latter compound was subsequently shown to be (R) by isolation and measurement of its optical rotation (Bello et al. 2013). In fact, a recent study of the absolute configurations of 36 methyl-branched cuticular hydrocarbons from 20 species in 9 insect orders, including three cerambycid species [*Brothylus gemmulatus* LeConte, *Monochamus titillator* (F.), *Monochamus c. clamator* (LeConte)] determined that they all had the (*R*)-configuration (Bello et al. 2015). Thus, it seems that the majority of methyl-branched hydrocarbons from cerambycids are likely to have the (*R*)-configuration. However, in some species, gas chromatographic (GC) analyses have shown the cuticular hydrocarbon profiles of males and females to be very similar, even though males can clearly discriminate between females and other males upon antennal contact with the cuticle (e.g., *P. semipunctata*; Hanks et al. 1996; JGM and LMH, unpublished data). However, the enantiomers of cuticular hydrocarbons cannot be separated chromatographically, and so one possible explanation for this ability to discriminate among what superficially appear to be similar cuticular profiles could be that females produce the (*S*)-enantiomer of one or more of the hydrocarbons to create a sex-specific signal.

Overall, these studies illustrate several important points about cerambycid contact pheromones. First, considerable information about the possible pheromone components can be obtained by simple fractionation steps that break down crude extracts from females into distinct chemical classes, which can then be bioassayed to determine which classes of chemicals are involved (see Bello et al. 2015). Second, the pheromones may be simple, consisting of one or a few compounds, or remarkably complex, as in the case of *A. malasiaca*. In fact, it is likely that the incomplete recovery of the full activity of many insect contact pheromones is due to the reconstructions of the blends with synthetic compounds being incomplete. Stated another way, it is likely that the contact pheromones of at least some insects contain far more compounds than the few that have been identified and associated with activity, particularly given the relatively large number of compounds present in crude cuticular extracts. Third, contact pheromones of females may be entirely sufficient to elicit the complete suite of mating behaviors in males of at least some species (Fukaya and Honda 1996; Wang et al. 2002; Crook et al. 2004; Lu et al. 2007; Luo et al. 2011). In others, physical cues, such as size and shape, or acoustic cues such as substrate-borne vibrations from walking beetles also may be important (e.g., Wang et al. 1990, 1996; Fukaya and Honda 1996; Fukaya et al. 2004, 2005; Wang and Chen 2005; Lu et al. 2007; Tsubaki et al. 2014).

5.5 Trail Pheromones

As described here, contact sex pheromones have been shown to induce a very characteristic sequence of behavioral steps in males whose antennae come in contact with the pheromones, including mounting and copulation attempts. However, there are also a number of cases in which it has been noted that males can detect where females have walked, and they will follow this trail or at least slow down when the trail is perceived. It is not known whether females actively apply compounds to the substrate or whether the trail is deposited simply by compounds rubbing off the cuticle as the female walks. Galford (1977) appeared to have been the first to demonstrate this phenomenon by showing that males of *M. robiniae* remained longer on sticks or filter papers upon which females had walked than on controls. Wang et al. (2002) showed that male *Nadezhdiella cantori* Hope followed trails deposited by walking females and that the walking responses were even stronger to extracts of females that were painted onto a substrate. In the only example to date in which cerambycid trail pheromone components have been identified, Hoover et al. (2014) identified two major components ($2Me$-C_{22} and $Z9$-C_{23}) and two minor components ($Z9$-C_{25} and $Z7$-C_{25}) from residues deposited on filter papers by walking virgin or mated females but not males of *A. glabripennis*. In bioassays, males followed trails of a reconstructed blend of the four components, and virgin but not mated males followed trails comprised of the two major components alone. It should be noted that the three alkene components had been previously identified as contact sex pheromone components for this species (Zhang et al. 2003). Thus, not surprisingly, there appears to be considerable overlap in both the function and the components of female-produced contact sex pheromones and trail pheromones. The authors of the present chapter have heard anecdotal evidence and seen videos of similar instances of male cerambycids tracking the "footprints" of their females, and it seems likely that this will prove to be a common mate-location strategy within the family, given the propensity of males of many species to patrol host trees, sampling the substrate with their antennae, and ready to pounce on a female as soon as they contact her with their antennae (see Chapter 4).

5.6 Mechanisms for Maintaining Reproductive Isolation

From the earlier descriptions of the types of pheromones found within the different subfamilies, it is clear that pheromone structures are often highly conserved within the Cerambycidae, with the same compound(s) being shared by species within the same genus, tribe, subfamily or, in the cases of the 2,3-alkanediols and fuscumol, even among species in different subfamilies (see Table 5.1). Superficially, this might seem to create multiple opportunities for cross-attraction among species, and at first glance, catches in pheromone-baited traps would appear to bear this out, with traps baited with a single compound often catching beetles of multiple species simultaneously (e.g., Hanks and Millar 2012). However, it must be remembered that pheromone lures are passive devices that release pheromones continuously, whereas the available data suggest that different cerambycid species have distinct diel activity patterns, including specific time windows in which they produce and respond to pheromones (see Chapter 4). Thus, catches in traps baited with lures that release pheromones continuously may be deceiving.

Generally speaking, there are multiple, complimentary mechanisms that assure prezygotic reproductive isolation among sympatric cerambycid species, including spatial, temporal, and chemical factors. Thus, species that are geographically separated can obviously use the same pheromone components without interfering with each other. However, even species that are sympatric can remain spatially separated in one or both of two ways. First, because of the frequent strong synergism of pheromones with host plant volatiles, species that share one or more pheromone components may avoid cross-attraction if they infest hosts that have distinctly different odor profiles. For example, adults of species whose larvae develop in oaks may be repelled by blends of their pheromones with conifer host odors (Collignon et al. 2016). Furthermore, adult males may not call unless they are on suitable hosts (e.g., Fonseca et al. 2010). Second, there is growing evidence that different species segregate themselves at different heights within the forest—from ground level to the upper canopy of host trees. Remarkably, this segregation by height can be absolute, with some species only being caught at ground level, others only being caught in the upper canopy, and another subset of species demonstrating no clear preference for one height over the other (Graham et al. 2012; Dodds 2014; Webster et al. 2016).

With regard to temporal factors, most cerambycids have distinct seasonal activity patterns, with the adults of most species of temperate zones being present for only a few weeks per year (see Chapter 4). Thus, the same compound can be used as a pheromone by a succession of sympatric species as long as their seasonal activity periods have minimal overlap, and this has indeed been shown to be the case (Hanks and Millar 2013; Hanks et al. 2014; Handley et al. 2015; Mitchell et al. 2015). In fact, some species have taken this to extremes, such as two *Megacyllene* species native to the eastern United States. Thus, adults of *M. caryae* emerge very early in spring, whereas adults of *M. robiniae* emerge very late in the fall, in both cases avoiding overlap with other cerambycid species with which they share pheromone components (Hanks et al. 2014; JGM and LMH, unpublished data). In the mountains of California, adults of *Semanotus ligneus amplus* (Casey) and *Callidium pseudotsugae* Fisher emerge so early in spring that there is still snow on the ground (I. P. Swift, personal observation). Emerging very early may have additional advantages, such as obtaining first choice of the oviposition sites created by trees and branches that have fallen during the winter.

Even sympatric species that have similar seasonal activity patterns can minimize cross-attraction by producing and responding to pheromones at different times of day. It is well known that cerambycid species are limited in their daily flight periods, being strictly diurnal, crepuscular, or nocturnal (Linsley 1959), and this phenology likely extends to their pheromone-mediated behavior as well. For example, males of *H. betulinus* produced pheromones primarily between four and six hours after the onset of photophase (Fonseca et al. 2010). These types of observational data have now been reinforced with data from traps on timers, which have clearly shown that sympatric species that share pheromone components have largely nonoverlapping diel response windows (Mitchell et al. 2015).

There is a growing body of evidence that, similar to other insect taxa, cerambycid species may use blends of pheromone components that inhibit cross-attraction in both a positive and a negative sense. That is, the pheromone blends of some species consist of two or more components, with each individual component having no activity, or much reduced activity, in comparison to the blend. For example,

Mitchell et al. (2015) reported that two cerambycine species that overlap in seasonal and daily activity periods have very similar pheromones: males of *Euderces pini* (Olivier) produce (*R*)-3-hydroxyhexan-2-one, while the pheromone of male *Phymatodes amoenus* (Say) is composed primarily of the same compound but with (*R*)-2-methylbutanol as a minor component. Nevertheless, the two species remain segregated by that minor component because it is a critical synergist for adult *P. amoenus*, while simultaneously antagonizing attraction of *E. pini*. Similarly, the cerambycines *X. colonus* and *Sarosesthes fulminans* (F.) both have (*R*)-3-hydroxyhexan-2-one as their dominant pheromone component but differ in the stereochemistry of their 2,3-hexanediol minor components (Lacey et al. 2009). Cross-attraction is averted because attraction of each species is synergized by its natural minor component(s) and antagonized by the minor component(s) produced by the other species. In another subfamily, the Prioninae, female *T. depsarium* "nov. sp. Laplante" produce (2*R*,3*R*)-2,3-hexanediol as their sex pheromone, and attraction of males is antagonized by one or both of the diastereomers, which are produced by its sympatric and synchronic congeners (Ray et al. 2012a).

There are also many examples known of species whose responses are inhibited by isomers or other analogs of their pheromone components, even though it is not yet clear which sympatric species might potentially interfere with their pheromone communication channels. This was first observed in the cerambycine *X. pyrrhoderus*, whose pheromone consists of a blend of (*S*)-2-hydroxyoctan-3-one + (2*S*,3*S*)-2,3-octanediol, and that is inhibited by (*R*)-2-hydroxyoctan-3-one (Iwabuchi et al. 1986). Similarly, *H. bajulus* (Reddy et al. 2005) and *Neoclytus m. mucronatus* (F.) (Lacey et al. 2007a) both use (*R*)-3-hydroxyhexan-2-one as a major pheromone component, and both are inhibited by (*S*)-3-hydroxyhexan-2-one. In another example, males of *N. a. acuminatus* produce (2*S*,3*S*)-2,3-hexanediol as their sole pheromone component. This species is not inhibited by the enantiomeric (2*R*,3*R*)-2,3-hexanediol, but it is inhibited by one or both of the diastereomeric (2*R*,3*S*)- or (2*S*,3*R*)-2,3-hexanediols (Lacey et al. 2004), and by (*R*)-3-hydroxyhexan-2-one (Hanks et al. 2012). An analogous situation was found more recently for the western congener *N. tenuiscriptus* Fall, males of which also produce (2*S*,3*S*)-2,3-hexanediol as their sole pheromone component (Ray et al. 2015). Further examples of inhibition by isomers and analogs of a particular species' pheromone can be found in Table 5.1.

The high degree of conservation of pheromone structures, with identical pheromone structures being shared by species on different continents, may have important consequences for invasion biology. That is, a new invader must overcome Allee effects, whereby low population densities reduce population growth due to the difficulty of males and females finding each other in order to reproduce. Below a certain minimum density, populations will become extinct (Courchamp et al. 1999). Thus, any environmental effect that interferes with mate location will reinforce Allee effects, further hindering the chances of an invasive species becoming established. This concept is the foundation of the authors' working hypothesis that an invasive (cerambycid) species whose sex or aggregation pheromone is also used by one or more endemic species in the country being invaded (i.e., where the native species can interfere with mate location by the invasive species in a form of natural mating disruption) will have a reduced chance of becoming established. Conversely, an invasive species that has a pheromone that is not shared by native species (i.e., that has a "pheromone-free space" to invade) will be more likely to establish. In consequence, we suggest that knowledge of the pheromone of an invasive species, and whether or not the pheromone is shared by native species, should be a factor to consider during risk analysis for potentially invasive pest species.

5.7 Exploitation of Plant Volatiles and Bark Beetle Pheromones as Kairomones

In 1930, Beeson observed that adults of the cerambycine *Hoplocerambyx spinicornis* Newman were attracted from at least a half mile away by volatile chemicals emanating from newly felled host trees. Since that time, there has been abundant observational evidence that cerambycids use plant volatiles to locate hosts for their larvae, based on the behavior of adults (e.g., Hughes and Hughes 1982; Kobayashi et al. 1984; Zeh et al. 1992). These observations have been further supported by formal experiments that tested attraction of beetles to host material in the laboratory (e.g., Hanks et al. 1996; Fujiwara-Tsujii et al. 2012) and in the field (e.g., Liendo-Barandiaran et al. 2010a, 2010b; Coyle et al. 2015). In many cases, both sexes are attracted by host plant volatiles (reviewed in Hanks 1999; Allison et al. 2004; also see Chapter 4) and with both sexes being equally capable of discriminating the odors of stressed trees from those of healthy trees that would be less suitable hosts. Plant volatiles therefore serve to bring the sexes of many cerambycid species together on larval hosts, suggesting that host plant location and mate location are intimately associated (Linsley 1959; Hanks 1999). Further evidence for this mode of mate location is provided by observations of the independent arrival of males and females on larval hosts (see Chapter 4).

The oviposition behavior of female cerambycids also may be strongly influenced by plant volatiles. For example, in laboratory bioassays, Li et al. (2007) showed that female *Monochamus alternatus* Hope were not attracted to volatiles from non-host plants and would not oviposit on natural hosts to which such chemicals had been applied. Laboratory studies (e.g., Hanks et al. 1995; Mazaheri et al. 2011) and field studies (e.g., Collignon et al. 2016) designed to evaluate the potential host range of cerambycid species further attest to the ability of females to use chemical cues to distinguish host from nonhost species.

Larvae of many wood-boring insects (bark and ambrosia beetles, buprestids, cerambycids, and wood wasps) require hosts that are in a particular physiological condition, often attacking trees that have been stressed by adverse environmental conditions such as water deficit, fire, or air pollution (Wood 1982; Hanks 1999). However, it can be difficult to gauge the physiological status of plants when they first become attractive to beetles. A tree that appears perfectly healthy to a human observer may in fact be suffering severe stress, and any one measure of vigor, such as leaf water potential or branch growth increment, may be an unreliable indicator of the potential to resist attacks by wood-boring insects (e.g., Flaherty et al. 2011). Researchers often have assumed that trees that already have been colonized by larvae are representative of a vulnerable physiological condition, and the volatiles that these trees produce therefore are likely attractants (e.g., Sweeney et al. 2004). However, the larvae inevitably alter the condition of their host by destroying tissues and disrupting the vascular system, with concomitant changes in the host odor profile. Thus, the opportunity to identify the particular blend of odor constituents that attracted the first adult colonizers to the host tree has already passed by the time developing larvae are present. Even designating a tree as alive or dead may not be straightforward. For example, felling a tree does not immediately kill it because at least some tissues remain alive for some time; a freshly felled tree or fallen branch is not dead but slowly dying. Many cerambycids colonize such hosts, whereas they will not colonize completely dead hosts in which the tissues are hard and dry (Linsley 1959). The ability to discriminate among suitable and unsuitable hosts may be mediated by both chemical cues (e.g., water content, odors, and possibly contact chemical cues) and physical cues (e.g., bark or wood hardness).

Because many cerambycids are attracted to stressed hosts, research on host plant attractants naturally has centered on volatiles that are associated with plant stress (e.g., Fan et al. 2007b). Females of most cerambycid species have quite broad oviposition preferences, attacking plant species in multiple genera and even families (see volumes indexed in Linsley and Chemsak 1997). It therefore seems plausible that the volatile cues used by adults of a given species for host location may be common to all its hosts. Consistent with this hypothesis, ethanol is produced by stressed or diseased plants, or by plants that are long dead and decaying, and ethanol attracts many species of wood-boring insects (reviewed by Allison et al. 2004; Byers 2004; Kelsey et al. 2014). Thus, ethanol can serve as one possible cue to attract insects that require stressed hosts, including freshly fallen trees or branches (reviewed by Joseph et al. 2001), but by itself, it may not be a completely reliable cue because of the wealth of other possible sources of ethanol—for example, from decay of organic matter. This situation may in part explain why ethanol alone is not a general attractant for cerambycids that attack stressed hosts; in field experiments, ethanol lures typically attract only a few cerambycid species in large numbers, with many more species represented by a few specimens that may be random captures (e.g., Montgomery and Wargo 1983; Miller et al. 2015a, 2015b). Species that are not attracted by ethanol alone likely use other types of plant volatiles, alone or in combination with ethanol, as more reliable cues to locate hosts in the particular condition that is most suitable for their larvae. Nevertheless, high-release rate ethanol lures have been used in a nationwide trapping program to monitor for invasive wood-borers in the United States (Rabaglia et al. 2008).

Consistent with attraction to ethanol, some cerambycid species whose larvae develop in deciduous woody plants are attracted to fermenting molasses or sugar solutions, often with other ingredients such as beer, wine, or decaying fruit added (Linsley 1959). Attraction of insects to such baits has been summarized by Champlain and Kirk (1926), Frost and Dietrich (1929), Champlain and Knull (1932), and Galford (1980b). It also has been suggested that fermenting baits might attract cerambycids because the baits simulate the volatiles of fermenting sap on which adult beetles feed rather than volatiles from larval hosts (Champlain and Knull 1932). This hypothesis is supported by the trapping surveys cited here because all of the species attracted were in the Cerambycinae and Lepturinae, the only two subfamilies in which sap feeding is common (Linsley 1959). Conversely, lamiines were poorly represented in these studies, whereas they would have been expected to be attracted if the fermenting baits did indeed emulate the odor of larval hosts.

Much of the research on host attractants of wood-boring insects has centered on bark and ambrosia beetles that attack conifers, largely due to their importance as forest pests. Stressed conifers may produce ethanol as well as increased amounts of a broad spectrum of terpenes (e.g., Fan et al. 2007b). The monoterpene α-pinene is among the dominant volatiles of many conifers as well as being a major component of the distilled essential oils known as turpentines (Kozlowski and Pallardy 1997; Byers and Zhang 2011). Many species of wood-boring insects have long been known to be attracted by α-pinene and turpentines, including cerambycids that feed on conifers (e.g., Gardiner 1957; Byers 2004). In addition to α-pinene, conifer essential oils contain a variety of other monoterpenes and sesquiterpenes, and blends of these compounds have been formulated as attractants for wood-boring insects, including cerambycids (reviewed in Allison et al. 2004; Sweeney et al. 2004; Fan et al. 2007b; Collignon et al. 2016). Reconstruction of these blends can be complicated by the difficulty in identifying the complex structures of some sesquiterpenes, particularly those that exist in two or more stereoisomeric forms and/or that are not commercially available (e.g., see Millar et al. 1986; Barata et al. 2000; Sweeney et al. 2004; Crook et al. 2008).

Terpenes also may play a role in host location and/or recognition for cerambycids that infest deciduous hosts. In detailed studies of the attraction of cerambycids to host volatiles, Yasui and coworkers showed that the cuticle of both sexes of _A. malasiaca_ acquired host plant odors by contact, feeding, or passive adsorption, and that these odors were subsequently used as mate location cues (see Yasui et al. 2007b, 2008; Yasui 2009). Thus, analysis of volatiles from the elytra of adult beetles reared on satsuma mandarin trees (_Citrus unshiu_ Marcovitch) showed them to contain the same sesquiterpenes as were released by the host after feeding or other types of damage. However, _A. malasiaca_ is polyphagous, and further investigations revealed that the attraction of males was affected by the host upon which they had been reared. For example, males reared on satsuma mandarin, species of willow (_Salix_), or blueberry (_Vaccinium_) were most strongly attracted to the volatiles of the host from which they had been reared or collected, and to those odors released from the elytra of females exposed to plants of those same species (Yasui et al. 2011; Fujiwara-Tsujii et al. 2012, 2013). Thus, the host plant volatiles appeared to act in two sequential ways—first as longer-range attractants to bring the sexes together on the same host plants and then, over shorter ranges, by attraction to host plant volatiles being released from the cuticle of conspecifics. At close range, both visual signals and contact sex pheromones were important in the final steps of mate location and recognition (Yasui 2009).

Field studies have revealed that attraction to terpenes is synergized by ethanol for some conifer-feeding cerambycids (e.g., Kobayashi et al. 1984; Chénier and Philogène 1989). For example, Sweeney et al. (2004) found that adults of the spondylidines _T. fuscum_, _T. castaneum_, and _Spondylis buprestoides_ (L.) were attracted to a blend of monoterpenes associated with host plant stress, and attraction was strongly synergized by ethanol. The synergistic combination of monoterpenes (especially α-pinene) and ethanol as an attractant for many cerambycids has resulted in the combination being considered a general "plant volatile" treatment by researchers (e.g., Costello et al. 2008; Campbell and Borden 2009; Miller et al. 2015a, 2015b). However, the two components do not always act synergistically and are by no means a general attractant for all cerambycid species (e.g., Collignon et al. 2016). For example, a fourth spondylidine species captured by Sweeney et al. (2004), _Asemum striatum_ (L.), was attracted by α-pinene alone but not influenced by ethanol. Species of _Monochamus_ also vary as to whether ethanol synergizes attraction to terpenes (e.g., Chénier and Philogène 1989; Allison et al. 2001; Ibeas et al. 2007). This variation among wood-boring insects in their responses to terpenes and ethanol again may be due to interspecific differences in the host condition required for their larvae (see Phillips et al. 1988; Schroeder and Lindelöw 1989) and/or the types of hosts that they attack (e.g., conifers versus deciduous trees).

Some cerambycid species are strongly attracted to smoke volatiles from forest fires, with the dead, dying, or severely stressed trees providing a large but ephemeral resource for the development of their larvae (reviewed by Suckling et al. 2001; Allison et al. 2004). For example, Boulanger et al. (2013) found three species of cerambycids colonizing fire-damaged spruce, including the lamiine *Monochamus s. scutellatus* and the lepturine congeners *Acmaeops pratensis* (Laicharting) and *Acmaeops proteus proteus* (Kirby). Kelsey and Joseph (2003) reported that pines damaged by fire produced ethanol, with emission rates increasing with the severity of tissue damage. In addition to several species of bark and ambrosia beetles, the cerambycid *Spondylis upiformis* (Mannerheim) was attracted to scorched pines, with the numbers attracted increasing with the concentration of ethanol in the sapwood, suggesting that ethanol was at least in part responsible for attracting adult beetles to potential larval hosts. Adults of other pyrophilous species also are attracted to fire-damaged hosts by smoke volatiles. For example, Gardiner (1957) found that smoke from smoldering pine slash attracted the lepturines *A. p. proteus* and *Stictoleptura c. canadensis* (Olivier) (formerly *Anoplodera canadensis*) as well as the spondylidines *T. c. cinnamopterum* and *Asemum striatum* (formerly *A. atrum*). Hovore and Giesbert (1976) reported that *Tragidion annulatum* LeConte were strongly attracted to brushfires and burning vegetation, and this was corroborated by Swift and Ray (2008). Suckling et al. (2001) found that females of the spondylidine *Arhopalus ferus* (Mulsant) preferred to oviposit on pines that had been burned and used smoke to locate burned hosts. To our knowledge, none of the volatile components of smoke that mediate the attraction of pyrophilous species have been identified, although a recent study showed that specific olfactory neurons on the antennae of adults of both sexes of *Monochamus galloprovincialis* were stimulated by typical smoke volatiles (Álvarez et al. 2015).

Because many cerambycid species are known to exploit volatiles from woody plants as host location cues, it should come as no surprise that volatiles from unsuitable hosts also may influence the behavior of adult beetles. For example, Yatagai et al. (2002) found that adults of the cryptomeria bark borer, *Semanotus japonicus* Lecordaire, were repelled by essential oils of *Cryptomeria* species that are resistant to their larvae. Yamasaki et al. (1997) found that attraction of *Monochamus alternatus* to monoterpenes of its pine hosts was antagonized by the sesquiterpene (–)-germacrene D, which characterizes healthy trees that would be unsuitable larval hosts. Moreover, Suckling et al. (2001) found that attraction of *Arhoplaus ferus* to smoke volatiles was effectively blocked by release of green leaf volatiles that presumably contradict the signal that fire has rendered the upwind host vulnerable to attack. A recent study also has found that species that attack deciduous hosts are inhibited by monoterpene blends characteristic of conifers (Collignon et al. 2016).

The odor blends from both coniferous and deciduous woody plants can be complex, consisting as they do of compounds as variable in properties and chemistry as ethylene (a gas at room temperature), through relatively small and volatile compounds such as green leaf volatiles and monoterpenes, to less volatile compounds such as sesquiterpene hydrocarbons and oxygenated sesquiterpenes (e.g., Kimmerer and Kozlowski 1982; Yatagai et al. 2002). The problem of identifying even individual chemicals is compounded for those components that have two or more stereoisomers because each of these is perceived by biological receptors as a distinct entity. Thus, in theory, the problem of reconstructing the specific subset of compounds that comprise the attractant cue for a particular insect species, including the correct relative ratios, can be daunting (e.g., Millar et al. 1986). Consequently, in practice, many researchers have resorted to using only crude approximations of host volatiles as test attractants, such as the α-pinene or turpentine baits mentioned earlier. In part, these crude baits may work reasonably well for polyphagous species, which presumably have flexible host plant attractant "templates" in order to be able to respond to the varying odor profiles of the variety of species that they infest.

Many researchers have used electroantennography (EAG) or, even better, gas chromatography coupled with electroantennographic detection (GC-EAD) to aid in identifying potential host plant attractants (Millar and Haynes 1998). For example, there is no evidence that the cerambycine *Phoracantha semipunctata* uses volatile pheromones for long-range attraction; instead, both sexes are strongly attracted to the odors of fallen trees, cut logs, or dying foliage (Hanks et al. 1996; Barata and Araújo 2001). The beetle's known host range is restricted to a single genus, *Eucalyptus*, but *Eucalyptus* is among the largest plant genera (~750 species; Elliot and Jones 1990), and the phytochemistry of the various species is both variable and complex (e.g., Boland et al. 1991). Barata et al. (2000) conducted GC-EAD studies with *P. semipunctata* and found that the beetle antennae detected 43 host volatiles—about a third of which the authors could not identify. It also must be remembered that the EAD response of an antenna to a particular compound provides no insight into behavioral response. Thus, some of the compounds eliciting antennal responses may mediate behavioral responses other than attraction and may even be repellent. Because of the complexity of the blend and the resulting uncertainty as to which components might form part of the attractant bouquet, to our knowledge it has not been possible to reconstruct a blend of *Eucalyptus* volatiles that is as attractive as crude host material for *P. semipunctata*. Similarly, Fan et al. (2007a, 2007b) conducted GC-EAD studies with *Monochamus alternatus* and found that antennae of males responded most strongly to α-pinene, β-pinene, 3-carene, and terpinolene. Further complicating matters, males were attracted only by α- and β-pinene in olfactometer bioassays, but they were significantly attracted to all four compounds when they were tested as individual components in field bioassays. Females of *M. alternatus* showed much weaker responses to these compounds, suggesting to the authors that the sexes were brought together by males first arriving on larval hosts and then producing a pheromone that attracts beetles of both sexes. The recent identification of a male-produced pheromone for *M. alternatus* (Teale et al. 2011) has added credence to this theory.

As mentioned here, much of the research on cerambycid host plant attractants has focused on species that attack stressed hosts, with few studies of those that attack healthy hosts or those whose larvae feed in totally dead or decaying hosts. Among the cerambycids that attack healthy hosts are those that render only part of a tree highly stressed, irrespective of its original condition, by disrupting the vascular system. Such species include (1) branch-girdling lamiines, females of which chew a trench through the bark encircling the branch and then oviposit distal to the trench so that the developing larvae are isolated from any host plant reaction; and (2) "pruning" cerambycine species whose larvae girdle branches from within and then feed distally to the girdle, so that they again are protected from host responses (Linsley 1959; Solomon 1995). For such beetle species, any tree of the appropriate species may be suitable as a larval host, and the volatiles that adults use to find hosts are likely to be those that are produced by hosts of any physiological condition.

Little is known of host attractants or other cues used by cerambycids whose larvae develop in hosts that are completely dead. The larvae of parandrines and lepturines commonly develop in decaying hosts, while a few species of cerambycines, spondylidines, and necydalines develop in dead and seasoned wood (Švácha and Lawrence 2014). Species that feed in dead hosts are among the most polyphagous of cerambycids, with some species developing in both coniferous and deciduous hosts (Linsley 1959). This polyphagy is likely due to the fact that the larvae of these species derive their nutrition primarily from the tissues of fungi growing in the dead wood rather than from the dead wood itself (see Švácha and Lawrence 2014). One such species in this category is the old house borer, *Hylotrupes bajulus*, whose larvae develop in dead and seasoned wood (Linsley 1959). This species is an important pest because the larvae can damage structural timber (White 1954; Noldt et al. 1995, and references therein). Higgs and Evans (1978) first showed that female *H. bajulus* preferred to oviposit in host material that was already infested, with oviposition being stimulated by (−)-verbenone present in the larval frass and thus probably derived from the host. Subsequent attempts to identify volatile attractants from hosts and frass in olfactometer studies were hindered by the incomplete responses of adults under experimental conditions (e.g., Fettköther et al. 2000). Nevertheless, these studies indicated that females responded from a distance to crude extracts of host material and larval frass as well as to (−)-verbenone and several other monoterpenes from pines. Males, on the other hand, also responded to the crude extracts but differed from females somewhat in their response to the various monoterpenes—for example, by not responding to (−)-verbenone. A later study by Reddy et al. (2005) revealed that host monoterpenes synergized attraction of females to synthetic blends of the primary components of the male-produced pheromone blend ([*R*]-3-hydroxyhexan-2-one + 1-butanol; Table 5.1).

It also has been determined that cerambycids that share hosts with bark and ambrosia beetles can exploit the pheromones of the latter as kairomones (reviewed by Ibeas et al. 2007). This chemical eavesdropping is adaptive because it directs adult cerambycids to trees that are the highest quality hosts for their larvae—for example, those whose defenses have been compromised by bark beetle attack and that can provide their larvae with added nutrition from facultatively preying upon bark beetle larvae (e.g., Billings and Cameron 1984; Allison et al. 2001, 2003; Dodds et al. 2001). There is a wealth of literature on pheromones of scolytine bark beetles, with many different structures identified (reviewed in Francke and Dettner 2005). Among the most common—and most studied—pheromone components are ipsenol and ipsdienol (e.g., Wood 1982; Seybold et al. 1995, and references that follow). As with host plant volatiles, cerambycid species vary in how they respond to the particular bark beetle pheromones that they exploit as kairomones. Thus, no particular bark beetle compound can be considered a truly general attractant. For example, pheromone components of various bark beetle species may act as attractants, antagonists, or be entirely neutral for sympatric species of *Monochamus* (reviewed by Ibeas et al. 2007). To further complicate matters, the sexes may differ in their responses to plant volatiles and other kairomones. For example, Costello et al. (2008) found that males of *M. c. clamator* were most strongly attracted to host plant volatiles alone, whereas ipsdienol acted synergistically with host volatiles to attract females. The response of a cerambycid species to pheromone components of a given bark beetle species likely depends on whether they share hosts (i.e., the cerambycid depends on bark beetles to weaken host defenses) and on their coevolutionary history (i.e., whether they evolved in sympatry).

Host plant volatiles may strongly influence the response of cerambycids to bark beetle pheromones—for example, by providing "context" to the pheromone signal, indicating that bark beetles are not only present but are colonizing a vulnerable plant. Cerambycid species vary in how they are influenced by the host plant volatiles—again probably due to their ecological relationships with the bark beetles and their host plants. For example, an extensive trapping survey conducted by Miller et al. (2015a, 2015b) revealed that the cerambycid *Astylopsis sexguttata* (Say) was significantly attracted by the blend of ipsenol and ipsdienol, and attraction was synergized by ethanol plus α-pinene. However, adults of *Acanthocinus obsoletus* (Olivier) were attracted only by the combination of bark beetle pheromones and the plant volatiles but not to either tested separately. Additional species that were attracted by plant volatiles alone were not influenced by bark beetle pheromones.

Little is known of the sensory cues used by adult cerambycids to locate the host plants upon which they feed as opposed to the larval hosts upon which they oviposit. In some groups, adults and larvae exploit the same host plants, whereas in others, adults and larvae feed on quite different hosts (see Linsley 1959; also see Chapter 4). For example, adult lamiines typically feed on foliage or tender bark tissues of their larval host species. In contrast, adult lepturines and some groups of cerambycines feed on pollen and nectar from flowers, whereas their larvae develop in dying and dead wood. Other cerambycine adults feed on plant sap or do not feed at all as adults. For flower feeders, mutual attraction of both sexes to plants may be sufficient to bring the sexes together, analogous to species that are attracted to larval hosts for mating. For example, males and females of the cerambycine *Zorion guttigerum* (Westwood) are independently attracted to flowers, and so it has been suggested that they may not require a long-range pheromone (Wang and Chen 2005). Adults of the cerambycine *A. subfasciatus* feed on nectar and pollen; in a field bioassay, Ikeda et al. (1993) found that the adults were attracted to floral volatiles produced by the host, including benzyl acetate, phenylethyl propionate, and linalool. In a screening trial of floral volatiles in Japan, Sakakibara et al. (1998) found that two lepturine species were significantly attracted to benzyl alcohol, another lepturine species to methyl benzoate, and a fourth lepturine and one cerambycine species to methyl phenylacetate. Many other species, including another 20 lepturines, showed no significant treatment effect, most likely because the selected test chemicals were dissimilar to the odor profiles of their floral hosts. Overall, this study highlights the fact that, whereas odors of hosts fed on by adults may be useful for sampling particular species, these odors will not attract a representative sample of the cerambycid community.

The recent advances described here showing that volatile pheromones are common, and perhaps almost ubiquitous, among cerambycids have revealed another role for plant volatiles as pheromone synergists. The accumulated literature shows that, for some groups of cerambycids, pheromones alone are sufficient to strongly attract adults and thus may be effective as trap lures for a variety of species that differ in their larval hosts (e.g., Hanks and Millar 2013). Conversely, plant volatiles are critical synergists of pheromones for other cerambycid species. For example, adults of *Tetropium* species (Spondylidinae) are not attracted to the male-produced pheromone fuscumol but are strongly attracted to traps baited with a combination of the pheromone, ethanol, and a synthetic blend of monoterpenes mimicking volatiles of the spruce host plants (Sweeney et al. 2004; Silk et al. 2007). The blend of ethanol and the monoterpenes was attractive in its own right, but attraction was strongly synergized by fuscumol.

In an experiment that tested attraction of cerambycids to a blend of synthetic pheromones from multiple species, Hanks et al. (2012) found that attraction to pheromones was synergized by ethanol but not affected by α-pinene for the cerambycine *Neoclytus m. mucronatus* and the lamiines *Astyleiopus variegatus* and *Astylidius parvus* (LeConte). In contrast, for the lamiine *Graphisurus fasciatus* (Degeer), ethanol synergized and α-pinene antagonized attraction to the pheromone. The fact that α-pinene was neutral or even repellent to these species is consistent with their larval hosts being deciduous trees (Lingafelter 2007) that do not release large amounts of α-pinene. For the conifer-feeding species *Monochamus carolinensis* Olivier, however, neither the host volatile α-pinene nor the male-produced pheromone monochamol was attractive alone, but they strongly synergized each other. Allison et al. (2012) confirmed this strongly synergistic interaction for *M. carolinensis*.

Monochamol is now known to be a pheromone or likely pheromone for more than a dozen *Monochamus* species (Table 5.1). Among these species, the activity of monochamol varies from being attractive as a single component, through facultative synergism by host plant volatiles and/or bark beetle phero-mones, to obligate synergism between monochamol and host plant volatiles. For example, monochamol alone attracts the Eurasian *M. galloprovincialis*, *M. saltuarius* (Gebler), *M. sutor* (L.), and *M. urussovii* (Fischer), and the North American *M. notatus* (Drury) and *M. s. scutellatus*, and attraction may be synergized by host plant volatiles and/or bark beetle pheromones (Pajares et al. 2010, 2013; Fierke et al. 2012; Ryall et al. 2015). On the contrary, the Asian *M. alternatus* and North American *M. titillator* are attracted only by the combination of monochamol and plant volatiles (α-pinene, ethanol) and are not attracted to monochamol or plant volatiles alone (Teale et al. 2011; Allison et al. 2012). Macias-Samano et al. (2012) further showed that *M. c. clamator* and *M. o. obtusus* Casey also are attracted by mono-chamol released with synergistic plant volatiles and bark beetle kairomones.

Pheromones of cerambycids also may be synergized by volatiles from host plants of the adults. For example, Nakamuta et al. (1997) found that females of the cerambycid *A. subfasciatus* were not attracted by synthetic components of the male-produced pheromone ([*R*]-3-hydroxyhexan-2-one + [*R*]-3-hydroxyoctan-2-one) nor to a floral volatile of the host plant of adults (methyl phenylacetate), but they were attracted by the combination of the pheromone and the floral volatile. This synergism may also occur to a greater or lesser extent with species in which the adult and larval hosts are the same (see earlier examples).

5.8 Chemical Defenses of Cerambycids

Several authors have commented in general terms that many cerambycids produce defensive secretions from metasternal glands or mandibular glands (Linsley 1959; Moore and Brown 1971; Dettner 1987), but there are remarkably few actual cases of defensive compounds being identified from cerambycids. Unfortunately, the first identification of a defensive compound from a cerambycid, the European cer-ambycine species *Aromia moschata* (L.), as 2-hydroxybenzaldehyde (salicylaldehyde) (Hollande 1909), appears to have been incorrect because more than 60 years later, Vidari et al. (1973) identified two rose oxide and two iridodial isomers as the main components of the secretion. The scent of *A. moschata* is perceptible to humans from a distance of many meters and has earned the species the common name of musk beetle (Duffy 1949). Such strong scents are characteristic of the tribe to which *A. moschata* belongs, the Callichromatini, with the glands being associated with the metasternal coxae (Linsley 1959). The scents are believed to be defensive in nature (Švácha and Lawrence 2014). Iridodials have also been identified in the defensive secretions of a Japanese species in that tribe, *Chloridolum loochooanum* Gressitt (Ohmura et al. 2009).

There is anecdotal evidence from cerambycid collectors that the supposed defensive chemicals of callichromatine species are used as alarm pheromones, but to date, there appears to be only a single published report, for the Central American species *Schwarzerion holochlorum* (Bates). Greeney and DeVries (2004) noted that when individuals of that species were handled, resulting in the production of a strong sweet, ammonia-like odor, nearby conspecifics dispersed. The experiment was repeated a number of times, and all indications were that the dispersal signal was indeed a volatile chemical rather than a visual or acoustic signal. Similar behavior, triggered by a similar odor, was also observed in the congener, *S.* nr. *euthalia* Bates (Greeney and DeVries 2004).

Moore and Brown (1972) identified an analog of salicylaldehyde, 2-hydroxy-6-methylbenzaldehyde, from the Australian species *Phoracantha semipunctata*, along with (5-ethylcyclopent-1-enyl)methanol the corresponding aldehyde, and several isomeric aldehydes. The chemicals were exuded when beetles were agitated—such as when handled. Adults of the congener *P. synonyma* Newman also produced 2-hydroxy-6-methylbenzaldehyde as well as several cyclic esters (macrolides) with 10-14 membered rings (Moore and Brown 1976). The "pineapple-like" odor of the secretion was found to be due to a series of short-chain methyl and ethyl esters, such as methyl and ethyl 2-methylbutyrate. As with the callichromatines, the defensive glands of the *Phoracantha* species are in the metasternum.

Moore and Brown (1971) also identified toluene and 2-methylphenol (*o*-cresol), which exude from mandibular glands on the frons of the cerambycids *Stenocentrus ostricilla* (Newman) and *Syllitus grammicus* (Newman), commenting that beetles in these genera are commonly known as "stinking longhorns" because of the distinctive odor of 2-methylphenol. It was thought likely that other members of these genera had similar defensive compounds because congeneric adults have similar pungent odors.

It also had been suggested that cerambycids might sequester toxins from their hosts for use in their own defense. For example, aposematic beetles in the lamiine genus *Tetraopes* are specialists on milkweed plants (*Asclepias* species; Asclepiadaceae), which contain toxic cardenolides. However, analyses of several *Tetraopes* species determined that they contained only microgram amounts of cardenolides, probably not enough to deter predators (Isman et al. 1977). In a second example, cerambycids in the genus *Elytroleptus* mimic and prey upon lycid beetles in the genera *Calopteron* and *Lycus*, which produce the toxin lycidic acid ([5*E*,7*E*]-octadeca-5,7-dien-9-ynoic acid; Eisner et al. 2008). Remarkably, the *Elytroleptus* species do not sequester this toxin but rather may derive indirect protection by mingling within aggregations of their toxic hosts. Similar comingling of cerambycid mimics of lyctid beetles with the lyctids has been reported for a number of South American species (do Nascimento et al. 2010).

There has also been a single report of a cerambycid using a venom. Specifically, the South American species *Onychocerus albitarsis* Pascoe has a sting-like terminal antennal segment with associated glandular tissue (Berkov et al. 2008). When molested, it flicks the antennae at its target to drive the antennal tip home, causing a painful sting.

5.9 Practical Applications of Cerambycid Semiochemicals

Semiochemicals which mediate insect behaviors have four basic practical applications:

1. Detection of a target species to determine presence/absence
2. Sampling or monitoring to follow seasonal phenology and the size, growth, and decline of populations for both invasive and endemic species
3. Sampling insect communities to assess species richness and diversity
4. Management of populations that have become pests

The first three applications are clearly related and so will be discussed together.

5.9.1 Use of Semiochemicals for Detection and Monitoring of Cerambycids

Cerambycid beetles are dichotomous in that they perform essential ecosystem functions by initiating the colonization, decomposition, and recycling of woody biomass, but these characteristics can also render them important pests of living trees, cut timber, and wooden structures (e.g., Solomon 1995). Thus, at one end of the scale, the presence and population sizes of native species reflect the health of ecosystems, whereas at the other end of the scale, they can be some of the most important invasive species, with their introductions into new areas of the world having the potential to cause both enormous economic losses and major disruptions of natural ecosystems. A case in point is the introduction of *Anoplophora glabripennis* into North America and Europe, where it has the potential to devastate natural and urban forests as well as plantation-grown trees and orchards (Haack et al. 2010; Faccoli et al. 2015). In fact, this insect has been listed among the 100 most dangerous invasive insects (Simberloff and Rejmánek 2011). The potential for damage by invasive species is further enhanced if they vector plant pathogens, as is the case with *Monochamus* species that vector pinewood nematode [*Bursaphelenchus xylophilus* (Steiner and Buehrer); Nematoda; Aphelenchoididae], the causative agent of the often lethal pine wilt disease (Mamiya 1983). On the other end of the scale, some cerambycids have been listed as threatened or endangered, such as *Desmocerus c. dimorphus* in central California (Holyoak et al. 2010) or the European species *Rosalia alpina* (L.), which is on the red list of threatened species of the International Union for the Conservation of Nature (World Conservation Monitoring Centre 1996).

For both threatened and invasive species, the basic problem is the same, that is, to detect the presence (and/or monitor population size and phenology) of a target species with the maximum degree of sensitivity and selectivity. As with many other types of insects, semiochemical-baited traps may be the most successful and cost-effective method of accomplishing this for any cerambycid species for which pheromones or related attractants are known. Several specific case studies may serve to emphasize this point. For example, in a recent two-year survey of cerambycid fauna in U.S. forest fragments in the state of Delaware, pheromone-baited traps captured cerambycids of 69 species, including seven species that were new state records (Handley et al. 2015). In another example, 34 male *Desmocerus c. dimorphus* were captured in a small field survey for this threatened species, a number that exceeded the total number of known museum specimens when the species was originally listed as threatened (Ray et al. 2014). A further 63 males were captured in a subsequent pheromone dose-response study (Ray et al. 2014). Other examples of multiple individuals of supposedly rare species being collected in straightforward, short-term field trials include *Curius dentatus* Newman (Lacey et al. 2004), *Anelaphus inflaticollis* Chemsak (Ray et al. 2009b), and *Neoclytus tenuiscriptus* Fall (Ray et al. 2015). Furthermore, the general concept of using pheromone-baited traps as a method of surveying poorly known endemic cerambycid fauna has been tested with success in Russia (Sweeney et al. 2014), China (Wickham et al. 2014), and Australia (Hayes et al. 2016).

Traps baited with combinations of pheromones and host plant volatiles, or other synergists such as bark beetle pheromones, have now been in use for several years for detection and monitoring of range expansions of several invasive cerambycids, including *Anoplophora glabripennis* in North America and Europe (reviewed in Meng et al. 2015), *Tetropium fuscum* in North America (Rhainds et al. 2011), and several *Monochamus* species in Europe (Rassati et al. 2012; Álvarez et al. 2014). To date, these appear to be the only cerambycid pheromones that have been incorporated into operational detection programs, but the great progress in the identification of cerambycid pheromones and attractants in the past decade will undoubtedly provide many more candidate compounds for use in surveillance programs for invasive species. For example, a likely pheromone has been identified for the Asian species *Xylotrechus rufilius* Bates, which recently was intercepted in shipping entering Baltimore, Maryland (Narai et al. 2015), and a pheromone blend was very recently identified for the Asian species *Callidiellum rufipenne*, which invaded the northeastern United States almost two decades ago (Zou et al. 2016). In terms of North American species invading other continents, attractant pheromones are known for North American species such as *Neoclytus a. acuminatus*, which has invaded Europe (Cocquempot and Lindelöw 2010). Pheromone-based attractants have now been identified for at least 12 *Monochamus* species, many of which have the potential to invade or have invaded areas outside their native ranges (Table 5.1). All these *Monochamus* species are attracted to the same, single pheromone component, monochamol, either alone or in combination with host plant volatiles or other synergists.

Ecological studies are also increasingly benefiting from the availability of effective tools for sampling large numbers of cerambycids. Pheromone-baited traps have now been used in several studies of cerambycid species richness and diversity, and these efforts are greatly enhanced by the two complimentary factors that the pheromone structures are often shared by multiple species and that the pheromones can be deployed in blends to attract species from both the same and different subfamilies simultaneously (see the following for further discussion of these points). Thus, a relatively small numbers of traps baited with a few different lure blends can trap beetles of more than 100 species in a given area (e.g., Graham et al. 2012; Hanks and Millar 2013; Wickham et al. 2014; Dodds et al. 2015; Handley et al. 2015).

5.9.1.1 Trap Efficacy

In parallel with the identifications of substantial numbers of pheromones/attractants for cerambycids over the past decade, a number of studies have assessed operational parameters such as trap design, trap placement, and lure design, with the goal of increasing trapping efficiency. In short, attractants are only effective if they are released in appropriate amounts, and the responding insects are indeed trapped. Trap designs have been driven by three main concerns: the efficiency with which an insect that reaches a trap is captured, the retention of the insects once they have been captured, and trap cost. In what follows, we have not attempted an exhaustive recitation of all studies designed to test trap design because essentially all of the earlier papers are cited in the studies described here. Instead, we have focused primarily on the more recent innovations and conclusive studies from the past decade.

From several comparisons of various types and shapes of traps, cross-vane panel traps or multiple funnel Lindgren traps—both of which present a vertical silhouette—have emerged as the most effective general designs (e.g., Graham et al. 2012; Allison et al. 2014; Álvarez et al. 2014; Dodds et al. 2014; Rassati et al. 2014, and references therein), but there are nuances depending upon the specific application. For example, Dodds et al. (2014) found that Malaise traps deployed in the canopy tended to catch larger numbers of species overall, and more rare species, than funnel or panel traps. In addition, Allison et al. (2014) reported that trap efficiency was improved by increasing the diameter of the bottom funnel of traps, which directs insects into the collection bucket, and by using a killing solution in the collection cup to improve retention of captured beetles. However, the single most important factor in improving trap efficacy, demonstrated in a number of trials, was treating the trap surfaces and collection cup with a lubricant such as the Teflon dispersion Fluon® which prevents beetles from clinging to traps when they land and prevents escape from collection buckets (Graham et al. 2010, 2012; Allison et al. 2011, 2014; Álvarez et al. 2014). Various designs of cross-vane intercept traps are available commercially, including traps precoated with Fluon (e.g., Alpha Scents Inc., West Linn, Oregon, USA; Synergy Semiochemical Corp., Burnaby, BC, Canada; ChemTica Internacional SA, San Jose, Costa Rica).

The design of lures is also critically important in order to provide semiochemical release rates that are optimally attractive. Because cerambycid males producing aggregation pheromones can release amounts of many tens of micrograms per hour, or possibly more, pheromone lures should release analogous or greater amounts, which translates to several milligrams per day. Thus, pheromone-impregnated rubber septa that are commonly used as slow release devices for pheromones of Lepidoptera, scale insects, and mealybugs, among other insects, are not used with cerambycid pheromones because the release rates generally are not high enough. Instead, effective lures have been constructed from thin-walled plastic sachets. For example, the authors have used both low-density polyethylene (LDP) resealable plastic bags and heat-sealed plastic tubing as release devices (Hanks et al. 2014). The former have the advantages that they are cheap and readily available anywhere in the world, and they seem to work reasonably well for most cerambycid pheromones, but the disadvantages that they cannot be shipped or stored when loaded, and they tend to leak, so they are obviously not suitable for commercialization. Nevertheless, they have proven very useful and convenient for research-type field trials testing various individual compounds and blends of compounds. In contrast, lures made from heat-sealable plastic tubing avoid most of the disadvantages of the plastic bags, with the added advantages that the tubing comes in a variety of different sizes, wall thicknesses, and chemistries, allowing for easy adjustment of release rates. Most of the cerambycid pheromone lures that pheromone companies are developing are of this type (e.g., Alpha Scents Inc., Synergy Semiochemical Corp., ChemTica Internacional SA). Similar but larger plastic pouches are used to obtain even higher daily release rates (grams per day) of pheromone synergists such as α-pinene and ethanol.

Another factor that may well be neglected in trapping programs is the possible stratification of different cerambycid species from ground level up to the upper canopy. For example, Graham et al. (2012) compared catches in pheromone-baited traps placed at heights of ~1.5 m versus in the upper canopy; from a total of 72 species caught, 21 species were caught exclusively at one of the two heights. Webster et al. (2016) obtained similar results in cerambycid trapping studies carried out in eastern Canada. A number of other studies showed similar trends, but their results may not be as clear-cut because those studies either used passive traps with no attractant (Su and Woods 2001; Vance et al. 2003; Ulyshen and Hanula 2007; Wermelinger et al. 2007; Sugimoto and Togashi 2013) or traps baited with host plant volatiles or kairomones such as bark beetle pheromones (Dodds 2014). Thus, it is unclear to what extent a pheromone lure might alter these results by attracting beetles to canopy strata outside their normal ranges.

5.9.1.2 Generic Lures and Pheromonal Parsimony

One of the bottlenecks of many programs for detecting incursions of invasive insects—and particularly Lepidoptera—is that, as a general rule, lures are formulated to attract only a single species. That is, the pheromones of multiple species are not mixed to form generic lures that might attract multiple species because of the potential for antagonistic interactions among the components (but see Brockerhoff et al. 2013). With cerambycid pheromones, this may be less of a problem for two reasons. First, because of the biosynthetic parsimony that has become evident, it is already common to attract multiple sympatric species to a single compound (e.g., Hanks and Millar 2013; Wickham et al. 2014) in part because the various species use other segregation methods, such as having species-specific, nonoverlapping daily activity cycles of both pheromone production and response (e.g., Mitchell et al. 2015). Second, although the pheromones within a genus, tribe, or even subfamily may be highly conserved, they usually are not conserved across subfamilies. For example, cerambycine species use pheromones (e.g., short-chain alcohols, hydroxyketones, and 2,3-alkanediols) that are in entirely different chemical classes than the pheromones of lamiines (terpenoid derivatives and straight-chain hydroxyethers) or some prionines (prionic acid; Figures 5.1 and 5.2). Two exceptions have been identified to date. First, 2,3-alkanediols are used as female-produced sex pheromones of species in the prionine genera *Tragosoma* and *Megopis* and as male-produced aggregation pheromones by some cerambycine species (see earlier text and Table 5.1). Second, fuscumol and fuscumol acetate are male-produced pheromones for species in both the Spondylidinae and Lamiinae (Table 5.1 and Figure 5.2). In general terms, though, a number of studies on several continents have shown that cerambycid pheromones from different structural classes can be mixed to form generic lures that attract multiple species, thus enabling detection and sampling at far less cost than using multiple traps with a single pheromone per trap (Hanks et al. 2012; Mitchell et al. 2011; Graham et al. 2012; Wong et al. 2012; Hanks and Millar 2013, Sweeney et al. 2014, Wickham et al. 2014; Handley et al. 2015). This concept has now resulted in a commercial version of a multicomponent cerambycid pheromone lure (ChemTica Internacional SA).

However, when the goal is to have the most sensitive trap for a certain species, more caution may be required because there is increasing evidence that there can indeed be inhibition of attraction within cerambycid communities (see Table 5.1), such as inhibition of the (2*S*,3*S*)-2,3-hexanediol pheromone of *Neoclytus a. acuminatus* by (*R*)-3-hydroxyhexan-2-one (Hanks et al. 2012) or inhibition of responses of *Tragosoma* species to their 2,3-hexanediol pheromones by one or more of the "nonnatural" diastereomers for a particular species (Ray et al. 2012a) or inhibition by minor components of the pheromone blends of sympatric species (Mitchell et al. 2015). That being said, generic lures may still be very useful for detecting the presence of a broad spectrum of species because some species are somewhat attracted to blends even though they contain antagonistic components. For example, *N. a. acuminatus* was still attracted in significant numbers to a blend containing its 2,3-hexanediol pheromone despite antagonism from the 3-hydroxyhexan-2-one that was also in the blend (Hanks and Millar 2013). Also, some species apparently will come to relatively weak attractants if there is no stronger attractant in the vicinity. For example, *Prionus californicus* males are attracted to the methyl ester of the female-produced pheromone, prionic acid, if prionic acid itself it not present (Maki et al. 2011b).

5.9.2 Pheromone-Based Tactics for Control of Cerambycids

There are several ways in which pheromones or related semiochemicals could be exploited in programs to control cerambycids. First, it might be possible to disrupt mating using arrays of dispensers releasing pheromones that interfere with mate location, as has been successfully developed for a number of lepidopteran species (e.g., Witzgall et al. 2010). Second, mass trapping with pheromone-baited traps could be used to eliminate a large fraction of the population. This strategy has worked very successfully for groups of large tropical weevils (e.g., Faleiro 2006). Third, pheromone-based attract-and-kill might be possible, using bait stations containing an attractant along with either a toxicant or an entomopathogen such as fungi or nematodes (e.g., Hajek and Bauer 2009; Brabbs et al. 2015).

Pheromone-based mating disruption has been tested twice—both times using the sex pheromone of *Prionus* species. The first trial targeted *P. californicus* in hopyards, where the larvae feed on roots of hop plants. Arrays of lures containing the female sex pheromone resulted in an 84% reduction in males finding a sentinel trap (Maki et al. 2011a). However, because the life cycle of this pest is three to five years, longer-term experiments would be required to determine the efficacy of mating disruption in terms of decreased densities of larvae and damage rates and, ultimately, reduced levels of infestation. The second experiment was done with *P. californicus* in sweet cherries, where a 94% reduction in males locating sentinel traps in an array of pheromone dispensers was achieved (J. D. Barbour, personal communication). Over a two-year period, catches of males in pheromone-baited traps decreased from 234 to 32, suggesting that this strategy could indeed be effective in reducing population densities (Alston et al. 2015).

Pheromone-based mass trapping of cerambycids has been tested several times. For example, Maki et al. (2011a), again working with *P. californicus*, found that surrounding a sentinel trap with a grid of pheromone-baited traps resulted in a very promising 88% reduction in the number of males reaching the sentinel trap. In a longer-term trial with *P. californicus* in sweet cherries, after five years of trapping with pheromone-baited traps, trap catches were reduced from 265 to 2 males, indicating that mass trapping this species has an excellent chance of success if continued for several years (Alston et al. 2015). Similar mass trapping trials targeting *Prionus* species are ongoing in other crops (e.g., pecans; Dutcher 2013) and are likely to be attempted with other prionine species that have powerful female-produced sex phero-mones. For example, the prionine species *Dorysthenes granulosus*, a major pest on more than one mil-lion hectares of sugarcane in southern China, also appears to use prionic acid as its female-produced sex pheromone (Wickham et al. 2016a) and so is another likely target for pheromone-based mass trapping.

Two other tests of pheromone-based mass trapping have been undertaken with species that use male-produced aggregation pheromones. First, Sweeney et al. (2011) tested attractant-baited traps set out in 10 by 10 m grids for the invasive species *Tetropium fuscum* and found lower infestation rates in spruce bait logs and fewer larvae per log in treated plots, indicating that this tactic might be useful for controlling or eliminating beetles from small areas, such as newly established populations of this invasive species. Over larger areas, the initial cost of traps and lures, and the cost of servicing traps, would probably be prohibi-tive. Finally, following up on their development of very effective lures containing a blend of monochamol, bark beetle pheromones, and α-pinene for *Monochamus galloprovincialis*, Pajares and coworkers tested the efficacy of mass trapping this species (Sanchez-Husillos et al. 2015). Using mark-recapture experi-ments to estimate beetle population sizes, these researchers estimated that less than one trap per hectare could reduce beetle populations by 95%, concluding that this method could be highly effective in both controlling the beetle and containing the spread of the beetle-vectored pinewood nematode.

5.10 Summary

It is no understatement to say that our knowledge and general understanding of cerambycid beetle semio-chemistry have changed dramatically, if not diametrically, in the last decade. Specifically, a frequently cited 2004 review of the chemical ecology of cerambycids listed volatile pheromones from only six cer-ambycine and one lamiine species and concluded that most Cerambycidae do not use sex or aggregation pheromones as attractants (Allison et al. 2004). In contrast, volatile sex and aggregation pheromones now have been identified from more than 100 species in the five major cerambycid subfamilies, and likely pheromones have been identified from many more species based on their strong attraction to known pheromone components in field screening studies (Table 5.1). Thus, the large volume of recent studies indicates that lack of volatile pheromones may be the exception rather than the rule in the Cerambycidae. In parallel with the surge in identifications of pheromones, major improvements in lure and trap designs have greatly improved the efficiency with which beetles are attracted and captured. In combination, these advances in pheromone chemistry and trap technology have provided valuable new tools for detec-tion and monitoring of cerambycids—whether it be for detecting invasive species, surveys for rare or endangered species, routine monitoring of endemic species, or surveys to assess species richness. For at least some species, it also may be possible to develop practical control methods using pheromone-based mass trapping or mating disruption. Clearly, we have only scratched the surface of what there is to be learned about the use of semiochemicals by this large insect family, and the next few years should see continuing advancements as the pace of research accelerates.

ACKNOWLEDGMENTS

Much of the work reviewed in this chapter would not have been possible without generous financial support to the authors from the U.S. Department of Agriculture (USDA) Animal and Plant Health Inspection Service, the USDA's National Institute of Food and Agriculture, the USDA's Western Regional Integrated Pest Management Fund, and the Alphawood Foundation of Chicago. We also gratefully acknowledge the help of the numerous students, postdoctoral scholars, visiting scientists, and collaborators who have assisted with so much of the work described here.

REFERENCES

Akino, T., M. Fukaya, H. Yasui, et al. 2001. Sexual dimorphism in cuticular hydrocarbons of the white-spotted longicorn beetle, *Anoplophora malasiaca* (Coleoptera: Cerambycidae). *Entomological Science* 4: 271–277.

Allison, J., J. Borden, R. McIntosh, et al. 2001. Kairomonal response by four *Monochamus* species to bark beetle pheromones. *Journal of Chemical Ecology* 27: 633–646.

Allison, J. D., B. D. Bhandari, J. L. McKenney, et al. 2014. Design factors that influence the performance of flight intercept traps for the capture of longhorned beetles (Coleoptera: Cerambycidae) from the subfamilies Lamiinae and Cerambycinae. *PLoS One* 9: e93203.

Allison, J. D., J. H. Borden, and S. J. Seybold. 2004. A review of the chemical ecology of the Cerambycidae (Coleoptera). *Chemoecology* 14: 123–150.

Allison, J. D., C. W. Johnson, J. R. Meeker, et al. 2011. Effect of aerosol surface lubricants on the abundance and richness of selected forest insects captured in multiple-funnel and panel traps. *Journal of Economic Entomology* 104: 1258–1264.

Allison, J. D., J. L. McKenney, J. G. Millar, et al. 2012. Response of the woodborers *Monochamus carolinensis* and *Monochamus titillator* (Coleoptera: Cerambycidae) to known cerambycid pheromones in the presence and absence of the host plant volatile α-pinene. *Environmental Entomology* 41: 1587–1596.

Allison, J. D., W. D. Morewood, J. H. Borden, et al. 2003. Differential bioactivity of *Ips* and *Dendroctonus* (Coleoptera: Scolytidae) pheromone components for *Monochamus clamator* and *M. scutellatus* (Coleoptera: Cerambycidae). *Environmental Entomology* 32: 23–30.

Alston, D. G., J. D. Barbour, and S. A. Steffen. 2015. California prionus. *Orchard Pest Management Online*. http://jenny.tfrec.wsu.edu/opm/displaySpecies.php?pn=643 (accessed January 28, 2016).

Álvarez, G., B. Ammagarahalli, D. R. Hall, et al. 2015. Smoke, pheromone and kairomone olfactory receptor neurons in males and females of the pine sawyer *Monochamus galloprovincialis* (Olivier) (Coleoptera: Cerambycidae). *Journal of Insect Physiology* 82: 46–55.

Álvarez, G., I. Etxebeste, D. Gallego, et al. 2014. Optimization of traps for live trapping of pine wood nematode vector *Monochamus galloprovincialis*. *Journal of Applied Entomology* 139: 618–626.

Barata, E. N., and J. Araújo. 2001. Olfactory orientation responses of the eucalyptus woodborer, *Phoracantha semipunctata*, to host plant in a wind tunnel. *Physiological Entomology* 26: 26–37.

Barata, E. N., J. A. Pickett, L. J. Wadhams, et al. 2000. Identification of host and nonhost semiochemicals of eucalyptus woodborer *Phoracantha semipunctata* by gas chromatography-electroantennography. *Journal of Chemical Ecology* 26: 1877–1895.

Barbour, J. D., D. E. Cervantes, E. S. Lacey, et al. 2006. Calling behavior in the primitive longhorned beetle *Prionus californicus* Mots. *Journal of Insect Behavior* 19: 623–629.

Barbour, J. D., J. G. Millar, J. Rodstein, et al. 2011. Synthetic 3,5-dimethyldodecanoic acid serves as a general attractant for multiple species of *Prionus* (Coleoptera: Cerambycidae). *Annals of the Entomological Society of America* 104: 588–593.

Bartelt, R., A. Schaner, and L. Jackson. 1985. *cis*-Vaccenyl acetate as an aggregation pheromone in *Drosophila melanogaster*. *Journal of Chemical Ecology* 11: 1747–1756.

Beeson, C. F. C. 1930. Sense of smell of longicorn beetles. *Nature* 126: 12.

Bello, J. E., J. S. McElfresh, and J. G. Millar. 2015. Isolation and determination of absolute configurations of insect-produced methyl-branched hydrocarbons. *Proceedings of the National Academy of Sciences of the United States of America* 112: 1077–1082.

Bello, J. E., and J. G. Millar. 2013. Efficient asymmetric synthesis of long chain methyl-branched hydrocarbons, components of the contact sex pheromone of females of the cerambycid beetle, *Neoclytus acuminatus acuminatus*. *Tetrahedron: Asymmetry* 24: 822–826.

Benham, G. S., and R. J. Farrar. 1976. Notes on the biology of *Prionus laticollis* (Coleoptera: Cerambycidae). *The Canadian Entomologist* 108: 569–576.

Berkov, A., N. Rodriguez, and P. Centeno. 2008. Convergent evolution in the antennae of a cerambycid beetle, *Onychocerus albitarsis*, and the sting of a scorpion. *Naturwissenschaften* 95: 257–261.

Billings, R. F., and R. S. Cameron. 1984. Kairomonal responses of Coleoptera, *Monochamus titillator* (Cerambycidae), *Thanasimus dubius* (Cleridae), and *Temnochila virescens* (Trogositidae), to behavioral chemicals of southern pine bark beetles (Coleoptera: Scolytidae). *Environmental Entomology* 13: 1542–1548.

Boland, D. J., J. J. Brophy, and A. P. N. House. 1991. *Eucalyptus leaf oils: Use, chemistry, distillation and marketing*. Melbourne: Inkata Press.

Bouhelier, R., and E. Hudault. 1936. Un dangereux parasite de la vigne au Maroc *Dorysthenes forficatus* F. (Col. Ceramb.). *Revue de Zoologie Agricole et Appliquée* 35: 145–153.

Boulanger, Y., L. Sirois, and C. Hébert. 2013. Distribution patterns of three long-horned beetles (Coleoptera: Cerambycidae) shortly after fire in boreal forest: Adults colonizing stands versus progeny emerging from trees. *Environmental Entomology* 42: 17–28.

Boyer, F., C. Malosse, P. Zagatti, et al. 1997. Identification and synthesis of vesperal, the female sex pheromone of the longhorn beetle *Vesperus xatarti*. *Bulletin de la Société Chimique de France* 134: 757–764.

Brabbs, T., D. Collins, F. Hérard, et al. 2015. Prospects for the use of biological control agents against *Anoplophora* in Europe. *Pest Management Science* 71: 7–14.

Brockerhoff, E. G., D. M. Suckling, A. Roques, et al. 2013. Improving the efficiency of lepidopteran pest detection and surveillance: Constraints and opportunities for multiple-species trapping. *Journal of Chemical Ecology* 39: 50–58.

Brodie, B. 2008. Investigation of semiochemicals in *Monochamus scutellatus*, the white-spotted pine sawyer beetle (Coleoptera: Cerambycidae). MSc Thesis, State University of New York, Syracuse, NY.

Byers, J. A. 2004. Chemical ecology of bark beetles in a complex olfactory landscape. In *Bark and wood boring insects in living trees in Europe, a synthesis,* eds. F. Lieutier, K. Day, A. Battisti, et al., 89–134. Dordrecht: Kluwer Academic Publishers.

Byers, J. A., and Q. Zhang. 2011. Chemical ecology of bark beetles in regard to search and selection of host trees. In *Recent advances in entomological research,* eds. T. Liu and L. Kang, 150–190. Beijing and Berlin: Higher Education Press, and Springer-Verlag.

Campbell, S. A., and J. H. Borden. 2009. Additive and synergistic integration of multimodal cues of both hosts and non-hosts during host selection by woodboring insects. *Oikos* 118: 553–563.

Cardé, R. T. 2014. Defining attraction and aggregation pheromones: Teleological versus functional perspectives. *Journal of Chemical Ecology* 40: 519–520.

Cervantes, D. E., L. M. Hanks, E. S. Lacey, et al. 2006. First documentation of a volatile sex pheromone in a longhorned beetle (Coleoptera: Cerambycidae) of the primitive subfamily Prioninae. *Annals of the Entomological Society of America* 99: 718–722.

Champlain, A. B., and H. B. Kirk. 1926. Bait pan insects. *Entomological News* 37: 288–291.

Champlain, A. B., and J. N. Knull. 1932. Fermenting bait traps for trapping Elateridae and Cerambycidae (Coleop.). *Entomological News* 43: 253–257.

Chénier, J. V. R., and B. J. R. Philogène. 1989. Field responses of certain forest Coleoptera to conifer monoterpenes and ethanol. *Journal of Chemical Ecology* 15: 1729–1745.

Cocquempot, C., and A. Lindelöw. 2010. Longhorn beetles (Coleoptera, Cerambycidae). In *Alien terrestrial arthropods of Europe,* eds. A. Roques, M. Kenis, D. Lees, et al. *BioRisk* 4: 193–218.

Collignon, R. M., I. P. Swift, Y. Zou, et al. 2016. The influence of host plant volatiles on the attraction of longhorn beetles (Coleoptera: Cerambycidae) to pheromones. *Journal of Chemical Ecology* 42: 215–229.

Costello, S. L., J. F. Negron, and W. R. Jacobi. 2008. Traps and attractants for wood-boring insects in ponderosa pine stands in the Black Hills, South Dakota. *Journal of Economic Entomology* 101: 409–420.

Courchamp, F., T. Clutton-Brock, and B. Grenfell. 1999. Inverse density dependence and the Allee effect. *Trends in Ecology & Evolution* 14: 405–410.

Coyle, D., C. L. Brissey, and K. J. K. Gandhi. 2015. Species characterization and responses of subcortical insects to trap-logs and ethanol in a hardwood biomass plantation. *Agricultural and Forest Entomology* 17: 258–269.

Crook, D. J., J. A. Hopper, S. B. Ramaswamy, et al. 2004. Courtship behavior of the soybean stem borer *Dectes texanus texanus* (Coleoptera: Cerambycidae): Evidence for a female contact sex pheromone. *Annals of the Entomological Society of America* 97: 600–604.

Crook, D. J., A. Khrimian, J. A. Francese, et al. 2008. Development of a host-based semiochemical lure for trapping emerald ash borer, *Agrilus planipennis* (Coleoptera: Buprestidae). *Environmental Entomology* 37: 356–365.

Crook, D. J., D. R. Lance, and V. C. Mastro. 2014. Identification of a potential third component of the male-produced pheromone of *Anoplophora glabripennis* and its effect on behavior. *Journal of Chemical Ecology* 40: 1241–1250.

Dahl, T. M. 2006. Preliminary investigation of semiochemical cues for mate location and recognition in the red oak borer, *Enaphalodes rufulus*. MSc Thesis, University of Arkansas, Fayetteville, AR.

Dettner, K. 1987. Chemosystematics and evolution of beetle chemical defenses. *Annual Review of Entomology* 32: 17–48.

do Nascimento, E. A., K. Del-Claro, and U. R. Martins. 2010. Mimetic assemblages of lycid-like Cerambycidae (Insecta: Coleoptera) from southeastern Brazil. *Revista Brasileira Zoociencias* 12: 187–193.

Dodds, K. J. 2014. Effects of trap height on captures of arboreal insects in pine stands of northeastern United States of America. *The Canadian Entomologist* 146: 80–89.

Dodds, K. J., J. D. Allison, D. R. Miller, et al. 2015. Considering species richness and rarity when selecting optimal survey traps: Comparisons of semiochemical baited flight intercept traps for Cerambycidae in eastern North America. *Agricultural and Forest Entomology* 17: 36–47.

Dodds, K. J., C. Graber, and F. M. Stephen. 2001. Facultative intraguild predation by larval Cerambycidae on bark beetle larvae. *Community and Ecosystem Ecology* 30: 17–22.

Duffy, E. 1949. A contribution towards the biology of *Aromia moschata* L., the "musk" beetle. *Proceedings of the South London Entomological and Natural History Society* 1947–8: 82–110.

Dunn, J. P., and D. A. Potter. 1991. Synergistic effects of oak volatiles with ethanol in the capture of saprophagous wood borers. *Journal of Entomological Science* 26: 425–429.

Dutcher, J. D. 2013. Field trials in commercial pecan orchards in Georgia with California prionus pheromone as an attractant for tilehorned prionus and broad necked root borer. Oral presentation, Annual Meeting of the Entomological Society of America, Nov. 10–13, 2013, Austin, TX, USA.

Eisner, T., F. C. Schroeder, N. Snyder, et al. 2008. Defensive chemistry of lycid beetles and of mimetic cerambycid beetles that feed on them. *Chemoecology* 18: 109–119.

Elliot, W. R., and D. L. Jones. 1990. *Encyclopaedia of Australian plants suitable for cultivation*. Vol. 5: Melbourne: Lothian Publishing Company.

Faccoli, M., R. Favaro, M. T. Smith, et al. 2015. Life history of the Asian longhorn beetle *Anoplophora glabripennis* (Coleoptera: Cerambycidae) in southern Europe. *Agricultural and Forest Entomology* 17: 188–196.

Faleiro, J. 2006. A review of the issues and management of the red palm weevil *Rhynchophorus ferrugineus* (Coleoptera: Rhynchophoridae) in coconut and date palm during the last one hundred years. *International Journal of Tropical Insect Science* 26: 135–154.

Fan, J., L. Kang, and J. Sun. 2007a. Role of host volatiles in mate location by the Japanese pine sawyer, *Monochamus alternatus* Hope (Coleoptera: Cerambycidae). *Environmental Entomology* 36: 58–63.

Fan, J., J. Sun, and J. Shi. 2007b. Attraction of the Japanese pine sawyer, *Monochamus alternatus*, to volatiles from stressed host in China. *Annals of Forest Science* 64: 67–71.

Farrar, R. J. 1967. A study of the life history and control of *Prionus laticollis* (Drury) in Rhode Island. MSc Thesis, University of Rhode Island, Kingston, RI.

Fettköther, R., K. Dettner, F. Schröder, et al. 1995. The male pheromone of the old house borer *Hylotrupes bajulus* (L.) (Coleoptera: Cerambycidae): Identification and female response. *Experientia* 51: 270–277.

Fettköther, R., G. V. P. Reddy, U. Noldt, et al. 2000. Effect of host and larval frass volatiles on behavioural response of the old house borer, *Hylotrupes bajulus* (L.) (Coleoptera: Cerambycidae), in a wind tunnel bioassay. *Chemoecology* 10: 1–10.

Fierke, M. K., D. D. Skabeikis, J. G. Millar, et al. 2012. Identification of a male-produced aggregation pheromone for *Monochamus scutellatus scutellatus* and an attractant for the congener *Monochamus notatus* (Coleoptera: Cerambycidae). *Journal of Economic Entomology* 105: 2029–2034.

Flaherty, L., D. Quiring, D. Pureswaran, et al. 2013. Preference of an exotic wood borer for stressed trees is more attributable to pre-alighting than post-alighting behaviour. *Ecological Entomology* 38: 546–552.

Flaherty, L., J. D. Sweeney, D. Pureswaran, et al. 2011. Influence of host tree condition on the performance of *Tetropium fuscum* (Coleoptera: Cerambycidae). *Environmental Entomology* 40: 1200–1209.

Fonseca, M. G., D. M. Vidal, and P. Zarbin. 2010. Male-produced sex pheromone of the cerambycid beetle *Hedypathes betulinus*: Chemical identification and biological activity. *Journal of Chemical Ecology* 36: 1132–1139.

Francke, W., and K. Dettner. 2005. Chemical signaling in beetles. *Topics in Current Chemistry* 240: 85–166.

Frost, S. W., and H. Dietrich. 1929. Coleoptera taken from bait-traps. *Annals of the Entomological Society of America* 22: 427–437.

Fujiwara-Tsujii, N., H. Yasui, and S. Wakamura. 2013. Population differences in male responses to chemical mating cues in the white-spotted longicorn beetle, *Anoplophora malasiaca*. *Chemoecology* 23: 113–120.

Fujiwara-Tsujii, N., H. Yasui, S. Wakamura, et al. 2012. The white-spotted longicorn beetle, *Anoplophora malasiaca* (Coleoptera: Cerambycidae), with blueberry as host plant, utilizes host chemicals for male orientation. *Applied Entomology and Zoology* 47: 103–110.

Fukaya, M., and H. Honda. 1992. Reproductive biology of the yellow-spotted longicorn beetle, *Psacothea hilaris* (Pascoe) (Coleoptera: Cerambycidae). I. Male mating behaviors and female sex pheromones. *Applied Entomology and Zoology* 27: 89–97.

Fukaya, M., and H. Honda. 1995. Reproductive biology of the yellow-spotted longicorn beetle, *Psacothea hilaris* (Pascoe) (Coleoptera: Cerambycidae). II. Evidence for two female pheromone components with different functions. *Applied Entomology and Zoology* 30: 467–470.

Fukaya, M., and H. Honda. 1996. Reproductive biology of the yellow-spotted longicorn beetle, *Psacothea hilaris* (Pascoe) (Coleoptera: Cerambycidae). IV. Effects of shape and size of female models on male mating behaviors. *Applied Entomology and Zoology* 31: 51–58.

Fukaya, M., T. Akino, T. Yasuda, et al. 1999. Mating sequence and evidence for synergistic component in female contact sex pheromone of the white-spotted longicorn beetle, *Anoplophora malasiaca* (Thomson) (Coleoptera: Cerambycidae). *Entomological Science* 2: 183–187.

Fukaya, M., T. Akino, T. Yasuda, et al. 2000. Hydrocarbon components in contact sex pheromone of the white-spotted longicorn beetle, *Anoplophora malasiaca* (Thomson) (Coleoptera: Cerambycidae) and pheromonal activity of synthetic hydrocarbons. *Entomological Science* 3: 211–218.

Fukaya, M., T. Akino, T. Yasuda, et al. 2004. Visual and olfactory cues for mate orientation behaviour in male white-spotted longicorn beetle, *Anoplophora malasiaca*. *Entomologia Experimentalis et Applicata* 111: 111–115.

Fukaya, M., T. Akino, H. Yasui, et al. 2005. Effects of size and color of female models for male mate orientation in the white-spotted longicorn beetle *Anoplophora malasiaca* (Coleoptera: Cerambycidae). *Applied Entomology and Zoology* 40: 513–519.

Fukaya, M., S. Wakamura, T. Yasuda, et al. 1997. Sex pheromonal activity of geometric and optical isomers of synthetic contact pheromone to males of the yellow-spotted longicorn beetle, *Psacothea hilaris* (Pascoe) (Coleoptera: Cerambycidae). *Applied Entomology and Zoology* 32: 654–656.

Fukaya, M., T. Yasuda, S. Wakamura, et al. 1996. Reproductive biology of the yellow-spotted longicorn beetle, *Psacothea hilaris* (Pascoe) (Coleoptera: Cerambycidae). III. Identification of contact sex pheromone on female body surface. *Journal of Chemical Ecology* 22: 259–270.

Galford, J. R. 1977. Evidence for a pheromone in the locust borer. *United States Department of Agriculture Forest Service Research Note* NE-240: 1–3.

Galford, J. R. 1980a. Use of a pheromone to cause copulation between two species of cerambycids. *United States Department of Agriculture Forest Service Research Note* NE-189: 1–2.

Galford, J. R. 1980b. Bait bucket trapping for red oak borers (Coleoptera: Cerambycidae). *United States Department of Agriculture Forest Service Research Note* NE-293: 1–2.

Gardiner, L. 1957. Collecting wood-boring beetle adults by turpentine and smoke. *Canadian Forest Service, Bi-Monthly Research Notes* 13: 2.

Ginzel, M. D. 2010. Hydrocarbons as contact pheromones of longhorned beetles (Coleoptera: Cerambycidae). In *Insect hydrocarbons: Biology, biochemistry and chemical ecology,* eds. J. G. Blomquist and A. Bagnères, 375–389. Cambridge: Cambridge University Press.

Ginzel, M. D., J. G. Millar, and L. M. Hanks. 2003. (Z)-9-Pentacosene—Contact sex pheromone of the locust borer, *Megacyllene robiniae*. *Chemoecology* 13: 135–141.

Ginzel, M. D., J. A. Moreira, A. M. Ray, et al. 2006. (Z)-9-Nonacosene—Major component of the contact sex pheromone of the beetle *Megacyllene caryae*. *Journal of Chemical Ecology* 32: 435–451.

Godinez-Aguilar, J. L., J. E. Macias-Samano, and A. Moron-Rios. 2009. Notes on biology and sexual behavior of *Tetrasarus plato* Bates (Coleoptera: Cerambycidae), a tropical longhorn beetle in coffee plantations in Chiapas, Mexico. *The Coleopterists Bulletin* 63: 311–318.

Graham, E. E., R. F. Mitchell, P. F. Reagel, et al. 2010. Treating panel traps with a fluoropolymer enhances their efficiency in capturing cerambycid beetles. *Journal of Economic Entomology* 103: 641–647.

Graham, E. E., T. M. Poland, D. G. McCullough, et al. 2012. A comparison of trap type and height for capturing cerambycid beetles (Coleoptera). *Journal of Economic Entomology* 105: 837–846.

Greeney, H. F., and P. J. DeVries. 2004. Experimental evidence for alarm pheromones in the neotropical long-horn beetle, *Schwarzerion holochlorum* Bates (Coleoptera: Cerambycidae). *The Coleopterists Bulletin* 58: 642–643.

Gwynne, D., and B. B. Hostetler. 1978. Mass emergence of *Prionus emarginatus* (Say) (Coleoptera: Cerambycidae). *The Coleopterists Bulletin* 32: 347–348.

Haack, R. A., F. Hérard, J. Sun, et al. 2010. Managing invasive populations of Asian longhorned beetle and citrus longhorned beetle: A worldwide perspective. *Annual Review of Entomology* 55: 521–546.

Hagaman, T. E., and R. T. Cardé. 1984. Effect of pheromone concentration on organization of preflight behaviors of the male gypsy moth, *Lymantria dispar* (L.). *Journal of Chemical Ecology* 10: 17–23.

Hajek, A. E., and L. S. Bauer. 2009. Use of entomopathogens against invasive wood boring beetles in North America. In *Use of microbes for control and eradication of invasive arthropods,* eds. A. E. Hajek, T. R. Glare and M. O. O'Callaghan, 159–179. The Netherlands: Springer.

Hall, D. R., A. Cork, S. J. Phythian, et al. 2006a. Identification of components of male-produced pheromone of coffee white stem borer, *Xylotrechus quadripes*. *Journal of Chemical Ecology* 32: 195–219.

Hall, D. R., D. Kutywayo, C. Chanika, et al. 2006b. Chemistry of the African coffee stemborer, *Monochamus leuconotus*: But where's the ecology? Abstract, *22nd meeting of the International Society of Chemical Ecology,* Barcelona, Spain. p. 306.

Handley, K., J. Hough-Goldstein, L. M. Hanks, et al. 2015. Species richness and phenology of cerambycid beetles in urban forest fragments of northern Delaware. *Annals of the Entomological Society of America* 108: 251–262.

Hanks, L. M. 1999. Influence of the larval host plant on reproductive strategies of cerambycid beetles. *Annual Review of Entomology* 44: 483–505.

Hanks, L. M., and J. G. Millar. 2013. Field bioassays of cerambycid pheromones reveal widespread parsimony of pheromone structures, enhancement by host plant volatiles, and antagonism by components from heterospecifics. *Chemoecology* 23: 21–44.

Hanks, L. M., J. G. Millar, J. A. Mongold-Diers, et al. 2012. Using blends of cerambycid beetle pheromones and host plant volatiles to simultaneously attract a diversity of cerambycid species. *Canadian Journal of Forest Research* 42: 1050–1059.

Hanks, L. M., J. G. Millar, J. A. Moreira, et al. 2007. Using generic pheromone lures to expedite identification of aggregation pheromones for the cerambycid beetles *Xylotrechus nauticus, Phymatodes lecontei,* and *Neoclytus modestus modestus*. *Journal of Chemical Ecology* 33: 889–907.

Hanks, L. M., J. G. Millar, and T. D. Paine. 1995. Biological constraints on host-range expansion by the wood-boring beetle *Phoracantha semipunctata* (Coleoptera: Cerambycidae). *Annals of the Entomological Society of America* 88: 183–188.

Hanks, L. M., J. G. Millar, and T. D. Paine. 1996. Mating behavior of the eucalyptus longhorned borer (Coleoptera: Cerambycidae) and the adaptive significance of long "horns". *Journal of Insect Behavior* 9: 383–393.

Hanks, L. M., P. F. Reagel, R. F. Mitchell, et al. 2014. Seasonal phenology of the cerambycid beetles of east central Illinois. *Annals of the Entomological Society of America* 107: 211–226.

Hansen, L., T. Xu, J. Wickham, et al. 2015. Identification of a male-produced pheromone component of the citrus longhorned beetle, *Anoplophora chinensis*. *PLoS One* 10: e0134358.

Hayes, R. A., M. W. Griffiths, H. F. Nahrung, et al. 2016. Optimizing generic cerambycid pheromone lures for Australian biosecurity and biodiversity monitoring. *Journal of Economic Entomology* 109: 1741–1749.

Heintz, A. 1925. Lepturinernas blombesök och sekundära könskaraktärer. *Entomologisk Tidskrift* 46: 21–34.

Higgs, M. D., and D. A. Evans. 1978. Chemical mediators in the oviposition behaviour of the house longhorn beetle, *Hylotrupes bajulus*. *Experientia* 34: 46–47.

Hill, D. S. 2008. *Pests of crops in warmer climates and their control.* New York: Springer Science & Business Media.

Hollande, A. 1909. Sur la fonction d'excrétion chez les insectes salicicoles et en particulier sur l'existence des dérivés salicylés. *Annales de l'Université de Grenoble* 21: 459–517.

Holyoak, M., T. S. Talley, and S. E. Hogle. 2010. The effectiveness of US mitigation and monitoring practices for the threatened valley elderberry longhorn beetle. *Journal of Insect Conservation* 14: 43–52.

Hoover, K., M. Keena, M. Nehme, et al. 2014. Sex-specific trail pheromone mediates complex mate finding behavior in *Anoplophora glabripennis*. *Journal of Chemical Ecology* 40: 169–180.

Hoshino, K., S. Nakaba, H. Inoue, et al. 2015. Structure and development of male pheromone gland of longi-corn beetles and its phylogenetic relationships within the tribe Clytini. *Journal of Experimental Zoology. Part B: Molecular and Developmental Evolution* 324: 68–76.

Hovore, F. T., and E. F. Giesbert. 1976. Notes on the ecology and distribution of Western Cerambycidae (Coleoptera). *The Coleopterists Bulletin* 30: 349–360.

Hughes, A. L., and M. K. Hughes. 1982. Male size, mating success, and breeding habitat partitioning in the whitespotted sawyer *Monochamus scutellatus* (Say) (Coleoptera: Cerambycidae). *Oecologia* 55: 258–263.

Hughes, G. P., J. E. Bello, J. G. Millar, et al. 2015. Determination of the absolute configuration of female-produced contact sex pheromone components of the longhorned beetle, *Neoclytus acuminatus acuminatus* (F.) (Coleoptera: Cerambycidae). *Journal of Chemical Ecology* 41: 1050–1057.

Hughes, G. P., Y. Zou, J. G. Millar, et al. 2013. (*S*)-Fuscumol and (*S*)-fuscumol acetate produced by male *Astyleiopus variegatus* (Coleoptera: Cerambycidae). *The Canadian Entomologist* 145: 327–332.

Ibeas, F., D. Gallego, J. J. Diez, et al. 2007. An operative kairomonal lure for managing pine sawyer beetle *Monochamus galloprovincialis* (Coleoptera: Cerambycidae). *Journal of Applied Entomology* 131: 13–20.

Ikeda, T., E. Ohya, H. Makihara, et al. 1993. Olfactory responses of *Anaglyptus subfasciatus* Pic and *Demonax transilis* Bates (Coleoptera: Cerambycidae) to flower scents. *Journal of the Japanese Forestry Society* 75: 108–112.

Imrei, Z., J. G. Millar, and M. Tóth. 2013. Field screening of known pheromone components of longhorned beetles in the subfamily Cerambycinae (Coleoptera: Cerambycidae) in Hungary. *Zeitschrift für Naturforschung C* 68: 236–242.

Isman, M., S. Duffey, and G. Scudder. 1977. Cardenolide content of some leaf- and stem-feeding insects on temperate North American milkweeds (*Asclepias* spp.). *Canadian Journal of Zoology* 55: 1024–1028.

Iwabuchi, K. 1986. Mating behavior of *Xylotrechus pyrrhoderus* Bates. III. Pheromone secretion by male. *Applied Entomology and Zoology* 21: 606–612.

Iwabuchi, K. 1988. Mating behavior of *Xylotrechus pyrrhoderus* Bates (Coleoptera: Cerambycidae). VI. Mating system. *Journal of Ethology* 6: 69–76.

Iwabuchi, K. 1999. Sex pheromones of cerambycid beetles. In *Environmental entomology*. eds. T. Hidaka, Y. Matsumoto, K. Honda, et al., 436–451. Tokyo: University of Tokyo Press.

Iwabuchi, K., J. Takahashi, Y. Nakagawa, et al. 1986. Behavioral responses of female grape borer to synthetic male sex pheromone components. *Applied Entomology and Zoology* 21: 21–27.

Iwabuchi, K., J. Takahashi, and T. Sakai. 1987. Occurrence of 2,3-octanediol and 2-hydroxy-3-octanone, possible male sex pheromone in *Xylotrechus chinensis*. *Applied Entomology and Zoology* 22: 110–111.

Jallon, J., C. Antony, and O. Benamar. 1981. Un anti-aphrodisiaque produit par les mâles de *Drosophila melanogaster* est transféré aux femelles lors de la copulation. *Comptes Rendues de l'Academie des Sciences, Paris* 292: 1147–1149.

Jayarama, M. G. V., M. V. Souza, R. Naidu, et al. 1998. Sex pheromone of coffee white stem borer for monitoring and control is on the anvil. *Indian Coffee* 62: 15–16.

Joseph, G., R. G. Kelsey, R. W. Peck, et al. 2001. Response of some scolytids and their predators to ethanol and 4-allylanisole in pine forests of central Oregon. *Journal of Chemical Ecology* 27: 697–715.

Jurc, M., S. Bojovic, M. F. Fernández, et al. 2012. The attraction of cerambycids and other xylophagous beetles, potential vectors of *Bursaphelenchus xylophilus*, to semiochemicals in Slovenia. *Phytoparasitica* 40: 337–349.

Kelsey, R. G., D. Gallego, F. J. Sánchez-García, et al. 2014. Ethanol accumulation during severe drought may signal tree vulnerability to detection and attack by bark beetles. *Canadian Journal of Forest Research* 44: 554–561.

Kelsey, R. G., and G. Joseph. 2003. Ethanol in ponderosa pine as an indicator of physiological injury from fire and its relationship to secondary beetles. *Canadian Journal of Forest Research* 33: 870–884.

Kimmerer, T. W., and T. T. Kozlowski. 1982. Ethylene, ethane, acetaldehyde, and ethanol production by plants under stress. *Plant Physiology* 69: 840–847.

Kobayashi, F., A. Yamane, and T. Ikeda. 1984. The Japanese pine sawyer beetle as the vector of pine wilt disease. *Annual Review of Entomology* 29: 115–135.

Kozlowski, T. T., and S. G. Pallardy. 1997. *Physiology of woody plants (2nd ed.)*. San Diego: Academic Press.

Kuwahara, Y., S. Matsuyama, and T. Suzuki. 1987. Identification of 2,3-octanediol, 2-hydroxy-3-octanone and 3-hydroxy-2-octanone from male *Xylotrechus chinensis* Chevrolat as possible sex pheromones. *Applied Entomology and Zoology* 22: 25–28.

Lacey, E. S., M. D. Ginzel, J. G. Millar, et al. 2004. Male-produced aggregation pheromone of the cerambycid beetle *Neoclytus acuminatus acuminatus*. *Journal of Chemical Ecology* 30: 1493–1507.

Lacey, E. S., M. D. Ginzel, J. G. Millar, et al. 2008a. 7-Methylheptacosane is a major component of the contact sex pheromone of the cerambycid beetle *Neoclytus acuminatus acuminatus*. *Physiological Entomology* 33: 209–216.

Lacey, E. S., J. G. Millar, J. A. Moreira, et al. 2009. Male-produced aggregation pheromones of the cerambycid beetles *Xylotrechus colonus* and *Sarosesthes fulminans*. *Journal of Chemical Ecology* 35: 733–740.

Lacey, E. S., J. A. Moreira, J. G. Millar, et al. 2007a. Male-produced aggregation pheromone of the cerambycid beetle *Neoclytus mucronatus mucronatus*. *Entomologia Experimentalis et Applicata* 122: 171–179.

Lacey, E. S., J. A. Moreira, J. G. Millar, et al. 2008b. A male-produced aggregation pheromone blend consisting of alkanediols, terpenoids, and an aromatic alcohol from the cerambycid beetle *Megacyllene caryae*. *Journal of Chemical Ecology* 34: 408–417.

Lacey, E. S., A. M. Ray, and L. M. Hanks. 2007b. Calling behavior of the cerambycid beetle *Neoclytus acuminatus acuminatus* (F.). *Journal of Insect Behavior* 20: 117–128.

Leal, W. S., J. M. S. Bento, E. F. Vilela, et al. 1994. Female sex pheromone of the longhorn beetle *Migdolus fryanus* Westwood: *N*-(2′*S*)-methylbutanoyl 2-methylbutylamine. *Experientia* 50: 853–856.

Leal, W. S., X. W. Shi, K. Nakamuta, et al. 1995. Structure, stereochemistry, and thermal isomerization of the male sex pheromone of the longhorn beetle *Anaglyptus subfasciatus*. *Proceeding of the National Academy of Sciences of the United States of America* 92: 1038–1042.

Lemay, M. A., P. J. Silk, and J. Sweeney. 2010. Calling behavior of *Tetropium fuscum* (Coleoptera: Cerambycidae: Spondylidinae). *The Canadian Entomologist* 142: 256–260.

Li, S., Y. Fang, and Z. Zhang. 2007. Effects of volatiles of non-host plants and other chemicals on oviposition of *Monochamus alternatus* (Coleoptera: Cerambycidae). *Journal of Pest Science* 80: 119–123.

Li, Y., Q. Meng, J. Sweeney, et al. 2013. Ultrastructure of prothoracic pore structures of longhorn beetles (Coleoptera: Cerambycidae, Cerambycinae) native to Jilin, China. *Annals of the Entomological Society of America* 106: 637–642.

Liendo-Barandiaran, C. V., B. Herrara, F. Morillo, et al. 2010b. Identification of male sexual pheromone in *Steirastoma breve* (Coleoptera: Cerambycidae). Abstract O-7, 1st Latin-American Meeting of Chemical Ecology, Colonia del Sacramento, Uruguay, Oct. 17–20, 2010.

Liendo-Barandiaran, C. V., B. Herrera-Malaver, F. Morillo, et al. 2010a. Behavioral responses of *Steirastoma breve* (Sulzer) (Coleoptera: Cerambycidae) to host plant *Theobroma cacao* L., brushwood piles, under field conditions. *Applied Entomology and Zoology* 45: 489–496.

Lingafelter, S. W. 2007. *Illustrated key to the longhorned woodboring beetles of the eastern United States. Special publication no. 3.* North Potomac, MD: The Coleopterists Society.

Lingafelter, S. W., and E. R. Hoebeke. 2002. *Revision of the genus* Anoplophora *(Coleoptera: Cerambycidae).* Washington DC: The Entomological Society of Washington.

Linsley, E. G. 1959. Ecology of Cerambycidae. *Annual Review of Entomology* 4: 99–138.

Linsley, E. G. 1962. The Cerambycidae of North America: Part II: Taxonomy and classification of the Parandrinae, Prioninae, Spondylinae, and Aseminae. *University of California Publications in Entomology* 19: 1–110.

Linsley, E. G., and J. A. Chemsak. 1997. The Cerambycidae of North America. Part VIII: Bibliography, index, and host plant index. *University of California Publications in Entomology* 117: 1–534.

Lu, W., Q. Wang, M. Y. Tian, et al. 2007. Mate location and recognition in *Glenea cantor* (Fabr.) (Coleoptera: Cerambycidae: Lamiinae): Roles of host plant health, female sex pheromone, and vision. *Environmental Entomology* 36: 864–870.

Luo, S., P. Zhuge, and M. Wang. 2011. Mating behavior and contact pheromones of *Batocera horsfieldi* (Hope) (Coleoptera: Cerambycidae). *Entomological Science* 14: 359–363.

Macias-Samano, J. E., D. Wakarchuk, J. G. Millar, et al. 2012. 2-Undecyloxy-1-ethanol in combination with other semiochemicals attracts three *Monochamus* species (Coleoptera: Cerambycidae) in British Columbia, Canada. *The Canadian Entomologist* 144: 764–768.

Maki, E. C., J. G. Millar, J. Rodstein, et al. 2011a. Evaluation of mass trapping and mating disruption for managing *Prionus californicus* (Coleoptera: Cerambycidae) in hop production yards. *Journal of Economic Entomology* 104: 933–938.

Maki, E. C., J. Rodstein, J. G. Millar, et al. 2011b. Synthesis and field tests of possible minor components of the sex pheromone of *Prionus californicus*. *Journal of Chemical Ecology* 37: 714–716.

Mamiya, Y. 1983. Pathology of the pine wilt disease caused by *Bursaphelenchus xylophilus*. *Annual Review of Phytopathology* 21: 201–220.

Mayo, P. D., P. J. Silk, M. Cusson, et al. 2013. Steps in the biosynthesis of fuscumol in the longhorn beetles *Tetropium fuscum* (F.) and *Tetropium cinnamopterum* Kirby. *Journal of Chemical Ecology* 39: 377–389.

Mazaheri, A., J. Khajehali, and B. Hatami. 2011. Oviposition preference and larval performance of *Aeolesthes sarta* (Coleoptera: Cerambycidae) in six hardwood tree species. *Journal of Pest Science* 84: 355–361.

Meng, P. S., K. Hoover, and M. A. Keena. 2015. Asian longhorned beetle (Coleoptera: Cerambycidae), an introduced pest of maple and other hardwood trees in North America and Europe. *Journal of Integrated Pest Management* 6. doi: 10.1093/jipm/pmv003

Meng, P. S., R. T. Trotter, M. A. Keena, et al. 2014. Effects of pheromone and plant volatile release rates and ratios on trapping *Anoplophora glabripennis* in China. *Environmental Entomology* 43: 1379–1388.

Michelsen, A. 1963. Observations on the sexual behaviour of some longicorn beetles, subfamily Lepturinae (Coleoptera, Cerambycidae). *Behaviour* 22: 152–166.

Millar, J. G., L. M. Hanks, J. A. Moreira, et al. 2009. Pheromone chemistry of cerambycid beetles. In *Chemical ecology of wood-boring insects,* eds. K. Nakamuta, and J. G. Millar, 52–79. Ibaraki: Forestry and Forest Products Research Institute.

Millar, J. G., and K. F. Haynes. 1998. *Methods in chemical ecology. Vol. 2: Chemical methods*. Norwell, MA: Springer Academic Publishers.

Millar, J. G., C. Zhao, G. N. Lanier, et al. 1986. Components of moribund American elm trees as attractants to elm bark beetles, *Hylurgopinus rufipes* and *Scolytus multistriatus*. *Journal of Chemical Ecology* 12: 583–608.

Miller, D. R., and C. Asaro. 2005. Ipsenol and Ipsdienol attract *Monochamus titillator* (Coleoptera: Cerambycidae) and associated large pine woodborers in Southeastern United States. *Journal of Economic Entomology* 98: 2033–2040.

Miller, D. R., C. Asaro, C. M. Crowe, et al. 2011. Bark beetle pheromones and pine volatiles: Attractant kairomone lure blend for longhorn beetles (Cerambycidae) in pine stands of the southeastern United States. *Journal of Economic Entomology* 104: 1245–1257.

Miller, D. R., and J. H. Borden. 1990. Beta-phellandrene: Kairomone for pine engraver, *Ips pini* (Say) (Coleoptera: Scolytidae). *Journal of Chemical Ecology* 16: 2519–2531.

Miller, D. R., C. M. Crowe, K. J. Dodds, et al. 2015a. Ipsenol, ipsdienol, ethanol, and α-pinene: Trap lure blend for Cerambycidae and Buprestidae (Coleoptera) in pine forests of eastern North America. *Journal of Economic Entomology* 108: 1837–1851.

Miller, D. R., C. M. Crowe, P. D. Mayo, et al. 2015b. Responses of Cerambycidae and other insects to traps baited with ethanol, 2,3-hexanediol, and 3,2-hydroxyketone lures in north-central Georgia. *Journal of Economic Entomology* 108: 2354–2365.

Mitchell, R. F., E. E. Graham, J. C. H. Wong, et al. 2011. Fuscumol and fuscumol acetate are general attractants for many species of cerambycid beetles in the subfamily Lamiinae. *Entomologia Experimentalis et Applicata* 141: 71–77.

Mitchell, R. F., J. G. Millar, and L. M. Hanks. 2013. Blends of (*R*)-3-hydroxyhexan-2-one and alkan-2-ones identified as potential pheromones produced by three species of cerambycid beetles. *Chemoecology* 23: 121–127.

Mitchell, R. F., P. F. Reagel, J. C. H. Wong, et al. 2015. Cerambycid beetle species with similar pheromones are segregated by phenology and minor pheromone components. *Journal of Chemical Ecology* 41: 431–440.

Montgomery, M. E., and P. M. Wargo. 1983. Ethanol and other host-derived volatiles as attractants to beetles that bore into hardwoods. *Journal of Chemical Ecology* 9: 181–190.

Moore, B., and W. Brown. 1971. Chemical defence in longhorn beetles of the genera *Stenocentrus* and *Syllitus* (Coleoptera: Cerambycidae). *Australian Journal of Entomology* 10: 230–232.

Moore, B. P., and W. V. Brown. 1972. The chemistry of the metasternal gland secretion of the common eucalypt longicorn, *Phoracantha semipunctata* (Coleoptera: Cerambycidae). *Australian Journal of Chemistry* 25: 591–598.

Moore, B. P., and W. V. Brown. 1976. The chemistry of the metasternal gland secretion of the eucalypt longicorn *Phoracantha synonyma* (Coleoptera: Cerambycidae). *Australian Journal of Chemistry* 29: 1365–1374.

Morewood, W. D., K. E. Simmonds, R. Gries, et al. 2003. Disruption by conophthorin of the kairomonal response of sawyer beetles to bark beetle pheromones. *Journal of Chemical Ecology* 29: 2115–2129.

Morewood, W. D., K. E. Simmonds, I. M. Wilson, et al. 2002. Alpha-pinene and ethanol: Key host volatiles for *Xylotrechus longitarsis* (Coleoptera: Cerambycidae). *Journal of the Entomological Society of British Columbia* 99: 117–122.

Mori, K. 2007. Absolute configuration of gomadalactones A, B and C, the components of the contact sex pheromone of *Anoplophora malasiaca*. *Tetrahedron Letters* 48: 5609–5611.

Muraji, M., S. Wakamura, H. Yasui, et al. 2011. Genetic variation of the white-spotted longicorn beetle *Anoplophora* spp. (Coleoptera: Cerambycidae) in Japan detected by mitochondrial DNA sequence. *Applied Entomology and Zoology* 46: 363–373.

Nakamuta, K., W. S. Leal, T. Nakashima, et al. 1997. Increase of trap catches by a combination of male sex pheromones and floral attractant in longhorn beetle, *Anaglyptus subfasciatus*. *Journal of Chemical Ecology* 23: 1635–1640.

Nakamuta, K., H. Sato, and T. Nakashima. 1994. Behavioral and morphological evidence for a male-produced sex pheromone in the cryptomeria twig borer, *Anaglyptus subfasciatus* Pic (Coleoptera: Cerambycidae). *Japanese Journal of Entomology* 62: 371–376.

Nakashima, T., K. Nakamuta, H. Makihara, et al. 1994. Field response of *Anaglyptus subfasciatus* Pic (Coleoptera: Cerambycidae) to benzyl acetate and structurally related esters. *Applied Entomology and Zoology* 29: 421–425.

Narai, Y., Y. Zou, K. Nakamuta, et al. 2015. Candidate attractant pheromones of two potentially invasive Asian cerambycid species in the genus *Xylotrechus*. *Journal of Economic Entomology* 108: 1444–1446.

Nehme, M., M. Keena, A. Zhang, et al. 2010. Evaluating the use of male-produced pheromone components and plant volatiles in two trap designs to monitor *Anoplophora glabripennis*. *Environmental Entomology* 39: 169–176.

Nehme, M., R. Trotter, M. Keena, et al. 2014. Development and evaluation of a trapping system for *Anoplophora glabripennis* (Coleoptera: Cerambycidae) in the United States. *Environmental Entomology* 43: 1034–1044.

Noldt, U., R. Fettköther, and K. Dettner. 1995. Structure of the sex pheromone-producing prothoracic glands of the male old house borer. *International Journal of Morphology and Embryology* 24: 223–234.

Ohmura, W., S. Hishiyama, T. Nakashima, et al. 2009. Chemical composition of the defensive secretion of the longhorned beetle, *Chloridolum loochooanum*. *Journal of Chemical Ecology* 35: 250–255.

Pajares, J. A., G. Álvarez, D. R. Hall, et al. 2013. 2-(Undecyloxy)-ethanol is a major component of the male-produced aggregation pheromone of *Monochamus sutor*. *Entomologia Experimentalis et Applicata* 149: 118–127.

Pajares, J. A., G. Álvarez, F. Ibeas, et al. 2010. Identification and field activity of a male-produced aggregation pheromone in the pine sawyer beetle, *Monochamus galloprovincialis*. *Journal of Chemical Ecology* 36: 570–583.

Paschen, M. A., N. M. Schiff, and M. D. Ginzel. 2012. Role of volatile semiochemicals in the host and mate location behavior of *Mallodon dasystomus* (Coleoptera: Cerambycidae). *Journal of Insect Behavior* 25: 569–577.

Phillips, T. W., A. J. Wilkening, T. H. Atkinson, et al. 1988. Synergism of turpentine and ethanol as attractants for certain pine-infesting beetles (Coleoptera). *Environmental Entomology* 17: 456–462.

Pilat, T. J. 1972. Copulation in two species of *Tetraopes*: Behavior, mechanics and morphology (Coleoptera, Cerambycidae). MSc Thesis, Loyola University, Chicago, IL.

Rabaglia, R., D. Duerr, R. Acciavatti, et al. 2008. *Early detection and rapid response for non-native bark and ambrosia beetles.* Washington DC: United States Department of Agriculture Forest Service, Forest Health Protection.

Rassati, D., E. P. Toffolo, A. Battisti, et al. 2012. Monitoring of the pine sawyer beetle *Monochamus galloprovincialis* by pheromone traps in Italy. *Phytoparasitica* 40: 329–336.

Rassati, D., E. P. Toffolo, A. Roques, et al. 2014. Trapping wood boring beetles in Italian ports: A pilot study. *Journal of Pest Science* 87: 61–69.

Ray, A. M., R. A. Arnold, I. Swift, et al. 2014. (*R*)-Desmolactone is a sex pheromone or sex attractant for the endangered valley elderberry longhorn beetle *Desmocerus californicus dimorphus* and several congeners (Cerambycidae: Lepturinae). *PLoS One* 9: e115498.

Ray, A. M., J. D. Barbour, J. S. McElfresh, et al. 2012a. 2,3-Hexanediols as sex attractants and a female-produced sex pheromone for cerambycid beetles in the prionine genus *Tragosoma*. *Journal of Chemical Ecology* 38: 1151–1158.

Ray, A. M., E. S. Lacey, and L. M. Hanks. 2006. Predicted taxonomic patterns in pheromone production by longhorned beetles. *Naturwissenschaften* 93: 543–550.

Ray, A. M., J. G. Millar, J. S. McElfresh, et al. 2009a. Male-produced aggregation pheromone of the cerambycid beetle *Rosalia funebris*. *Journal of Chemical Ecology* 35: 96–103.

Ray, A. M., J. G. Millar, J. A. Moreira, et al. 2015. North American species of cerambycid beetles in the genus *Neoclytus* share a common hydroxyhexanone-hexanediol pheromone structural motif. *Journal of Economic Entomology* 108: 1860–1868.

Ray, A. M., I. P. Swift, J. S. McElfresh, et al. 2012b. (*R*)-Desmolactone, a female-produced sex pheromone component of the cerambycid beetle *Desmocerus californicus californicus* (subfamily Lepturinae). *Journal of Chemical Ecology* 38: 157–167.

Ray, A. M., I. P. Swift, J. A. Moreira, et al. 2009b. (*R*)-3-hydroxyhexan-2-one is a major pheromone component of *Anelaphus inflaticollis* (Coleoptera: Cerambycidae). *Environmental Entomology* 38: 1462–1466.

Ray, A. M., A. Žunič, R. L. Alten, et al. 2011. *cis*-Vaccenyl acetate, a female-produced sex pheromone component of *Ortholeptura valida*, a longhorned beetle in the subfamily Lepturinae. *Journal of Chemical Ecology* 37: 173–178.

Reagel, P. F., M. D. Ginzel, and L. M. Hanks. 2002. Aggregation and mate location in the red milkweed beetle (Coleoptera: Cerambycidae). *Journal of Insect Behavior* 15: 811–830.

Reddy, G. V. P., R. Fettköther, U. Noldt, et al. 2005. Enhancement of attraction and trap catches of the old-house borer, *Hylotrupes bajulus* (Coleoptera: Cerambycidae), by combination of male sex pheromone and monoterpenes. *Pest Management Science* 61: 699–704.

Rhainds, M., W. E. Mackinnon, K. B. Porter, et al. 2011. Evidence for limited spatial spread in an exotic longhorn beetle, *Tetropium fuscum* (Coleoptera: Cerambycidae). *Journal of Economic Entomology* 104: 1928–1933.

Rodstein, J., J. S. McElfresh, J. D. Barbour, et al. 2009. Identification and synthesis of a female-produced sex pheromone for the cerambycid beetle *Prionus californicus*. *Journal of Chemical Ecology* 35: 590–600.

Rodstein, J., J. G. Millar, J. D. Barbour, et al. 2011. Determination of the relative and absolute configurations of the female-produced sex pheromone of the cerambycid beetle *Prionus californicus*. *Journal of Chemical Ecology* 37: 114–124.

Rotrou, P. 1936. Le *Cyrtognathus forficatus* F. et ses moeurs. *Bulletin de la Societé des Sciences Naturelles du Maroc* 16: 246–251.

Rutledge, C. E., J. G. Millar, C. M. Romero, et al. 2009. Identification of an important component of the contact sex pheromone of *Callidiellum rufipenne* (Coleoptera: Cerambycidae). *Environmental Entomology* 38: 1267–1275.

Ryall, K., P. Silk, R. P. Webster, et al. 2015. Further evidence that monochamol is attractive to *Monochamus* (Coleoptera: Cerambycidae) species, with attraction synergised by host plant volatiles and bark beetle (Coleoptera: Curculionidae) pheromones. *The Canadian Entomologist* 147: 564–579.

Sakai, T., Y. Nakagawa, J. Takahashi, et al. 1984. Isolation and identification of the male sex pheromone of the grape borer *Xylotrechus pyrrhoderus* Bates (Coleoptera: Cerambycidae). *Chemistry Letters* 1984: 263–264.

Sakakibara, Y., A. Kikuma, R. Iwata, et al. 1998. Performances of four chemicals with floral scents as attractants for longicorn beetles (Coleoptera: Cerambycidae) in a broadleaved forest. *Journal of Forest Resources* 3: 221–224.

Sanchez-Husillos, E., I. Etxebeste, and J. Pajares. 2015. Effectiveness of mass trapping in the reduction of *Monochamus galloprovincialis* Olivier (Col.: Cerambycidae) populations. *Journal of Applied Entomology* 139: 747–758.

Schröder, F., R. Fettköther, U. Noldt, et al. 1994. Synthesis of (3*R*)-3-hydroxy-2-hexanone, (2*R*,3*R*)-2,3-hexanediol and (2*S*,3*R*)-2,3-hexanediol, the male sex pheromone of *Hylotrupes bajulus* and *Pyrrhidium sanguineum* (Cerambycidae). *Liebigs Annalen der Chemie* 1994: 1211–1218.

Schröder, F. C. 1996. Identizifierung und synthese neuer alkaloide, hydroxyketone und bicyclischer acetale aus insekten. Ph.D. Thesis, University of Hamburg, Hamburg, Germany.

Schroeder, L., and Å. Lindelöw. 1989. Attraction of scolytids and associated beetles by different absolute amounts and proportions of α-pinene and ethanol. *Journal of Chemical Ecology* 15: 807–817.

Schroeder, L. M., and J. Weslien. 1994. Interactions between the phloem-feeding species *Tomicus piniperda* (Col.: Scolytidae) and *Acanthocinus aedilis* (Col.: Cerambycidae) and the predator *Thanasimus formicarius* (Col.: Cleridae) with special reference to brood production. *Entomophaga* 39: 149–157.

Seybold, S. J., T. Ohtsuka, D. L. Wood, et al. 1995. Enantiomeric composition of ipsdienol: A chemotaxonomic character for North American populations of *Ips* spp. in the *pini* subgeneric group (Coleoptera: Scolytidae). *Journal of Chemical Ecology* 21: 995–1016.

Silk, P. J., J. Sweeney, J. Wu, et al. 2007. Evidence for a male-produced pheromone in *Tetropium fuscum* (F.) and *Tetropium cinnamopterum* (Kirby) (Coleoptera: Cerambycidae). *Naturwissenschaften* 94: 697–701.

Silk, P. J., J. Sweeney, J. Wu, et al. 2011. Contact sex pheromones identified for two species of longhorned beetles (Coleoptera: Cerambycidae) *Tetropium fuscu*m and *T. cinnamopterum* in the subfamily Spondylidinae. *Environmental Entomology* 40: 714–726.

Simberloff, D., and M. Rejmánek. 2011. *Encyclopedia of biological invasions*. Berkeley, CA: University of California Press.

Solomon, J. D. 1995. *Guide to insect borers in North American broadleaf trees and shrubs*. Washington DC: United States Department of Agriculture Forest Service Handbook 706.

Sousa, E., M. A. Bravo, J. Pires, et al. 2001. *Bursaphelenchus xylophilus* (Nematoda; Aphelenchoididae) associated with *Monochamus galloprovincialis* (Coleoptera; Cerambycidae) in Portugal. *Nematology* 3: 89–91.

Su, J., and S. Woods. 2001. Importance of sampling along a vertical gradient to compare the insect fauna in managed forests. *Environmental Entomology* 30: 400–408.

Suckling, D. M., A. R. Gibb, J. M. Daly, et al. 2001. Behavioral and electrophysiological responses of *Arhopalus tristis* to burnt pine and other stimuli. *Journal of Chemical Ecology* 27: 1091–1104.

Sugimoto, H., and K. Togashi. 2013. Canopy-related adult density and sex-related flight activity of *Monochamus alternatus* (Coleoptera: Cerambycidae) in pine stands. *Applied Entomology and Zoology* 48: 213–221.

Švácha, P., and J. F. Lawrence. 2014. Cerambycidae Latreille, 1802. In *Handbook of zoology: Arthropoda: Insecta: Coleoptera, beetles. Vol. 3: Morphology and systematics (Phytophaga)*, eds. R. A. B. Leschen, and R. G. Beutel, 77–177. Berlin/Boston: Walter de Gruyter.

Sweeney, J., P. de Groot, L. Macdonald, et al. 2004. Host volatile attractants and traps for detection of *Tetropium fuscum* (F.), *Tetropium castaneum* (L.) and other longhorned beetles (Coleoptera: Cerambycidae). *Environmental Entomology* 33: 844–854.

Sweeney, J. D., P. J. Silk, and V. Grebennikov. 2014. Efficacy of semiochemical-baited traps for detection of longhorn beetles (Coleoptera: Cerambycidae) in the Russian Far East. *European Journal of Entomology* 111: 397–406.

Sweeney, J. D., P. J. Silk, J. M. Gutowski, et al. 2010. Effect of chirality, release rate, and host volatiles on response of *Tetropium fuscum* (F.), *Tetropium cinnamopterum* Kirby, and *Tetropium castaneum* (L.) to the aggregation pheromone, fuscumol. *Journal of Chemical Ecology* 36: 1309–1321.

Sweeney, J. D., P. Silk, M. Rhainds, et al. 2011. Mass trapping for population suppression of an invasive longhorned beetle, *Tetropium fuscum* (F.) (Coleoptera: Cerambycidae). Proc. 22nd U. S. Dept. Ag. interagency research forum on invasive species. USDA Gen. Tech. Rep. NRS-P-52, USDA-Forest Service, Newton Square, PA, p. 92.

Swift, I., and A. M. Ray. 2008. A review of the genus *Tragidion* Audinet-Serville, 1834 (Coleoptera: Cerambycidae: Cerambycinae: Trachyderini). *Zootaxa* 1892: 1–25.

Teale, S. A., J. D. Wickham, F. Zhang, et al. 2011. A male-produced aggregation pheromone of *Monochamus alternatus* (Coleoptera: Cerambycidae), a major vector of pine wood nematode. *Journal of Economic Entomology* 104: 1592–1598.

Tsubaki, R., N. Hosoda, H. Kitajima, et al. 2014. Substrate-borne vibrations induce behavioral responses in the leaf-dwelling cerambycid, *Paraglenea fortunei*. *Zoological Science* 31: 789–794.

Ulyshen, M. D., and J. L. Hanula. 2007. A comparison of the beetle (Coleoptera) fauna captured at two heights above the ground in a North American temperate deciduous forest. *The American Midland Naturalist* 158: 260–278.

Vance, C. C., K. R. Kirby, J. R. Malcolm, et al. 2003. Community composition of longhorned beetles (Coleoptera: Cerambycidae) in the canopy and understory of sugar maple and white pine stands in south-central Ontario. *Environmental Entomology* 32: 1066–1074.

Vidal, D. M., M. G. Fonseca, and P. Zarbin. 2010. Enantioselective synthesis and absolute configuration of the sex pheromone of *Hedypathes betulinus* (Coleoptera: Cerambycidae). *Tetrahedron Letters* 51: 6704–6706.

Vidari, G., M. De Bernardi, M. Pavan, et al. 1973. Rose oxide and iridodial from *Aromia moschata* L. (Coleoptera: Cerambycidae). *Tetrahedron Letters* 14: 4065–4068.

Wang, Q., and L. Chen. 2005. Mating behavior of a flower-visiting longhorn beetle *Zorion guttigerum* (Westwood) (Coleoptera: Cerambycidae: Cerambycinae). *Naturwissenschaften* 92: 237–241.

Wang, Q., L. Chen, J. Li, et al. 1996. Mating behavior of *Phytoecia rufiventris* Gautier (Coleoptera: Cerambycidae). *Journal of Insect Behavior* 9: 47–60.

Wang, Q., and R. A. B. Leschen. 2003. Identification and distribution of *Arhopalus* species (Coleoptera: Cerambycidae: Aseminae) in Australia and New Zealand. *New Zealand Entomologist* 26: 53–59.

Wang, Q., W. Zeng, L. Chen, et al. 2002. Circadian reproductive rhythms, pair-bonding, and evidence for sex-specific pheromones in *Nadezhdiella cantori* (Coleoptera: Cerambycidae). *Journal of Insect Behavior* 15: 527–539.

Wang, Q., W. Zeng, and J. Li. 1990. Reproductive behavior of *Paraglenea fortunei* (Coleoptera: Cerambycidae). *Annals of the Entomological Society of America* 83: 860–866.

Webster, R. P., C. A. Alderson, V. L. Webster, et al. 2016. Further contributions to the longhorn beetle (Coleoptera: Cerambycidae) fauna of New Brunswick and Nova Scotia, Canada. *Zookeys* 552: 109–122.

Wermelinger, B., P. F. Flückiger, M. K. Obrist, et al. 2007. Horizontal and vertical distribution of saproxylic beetles (Col., Buprestidae, Cerambycidae, Scolytinae) across sections of forest edges. *Journal of Applied Entomology* 131: 104–114.

White, M. G. 1954. The house longhorn beetle *Hylotrupes bajulus* L. (Col. Cerambycidae) in Great Britain. *Forestry* 27: 31–40.

Wickham, J. D., R. D. Harrison, W. Lu, et al. 2014. Generic lures attract cerambycid beetles in a tropical montane rain forest in southern China. *Journal of Economic Entomology* 107: 259–267.

Wickham, J. D., W. Lu, T. Jin, et al. 2016a. Prionic acid: An effective sex attractant for an important pest of sugarcane, *Dorysthenes granulosus* (Coleoptera: Cerambycidae: Prioninae). *Journal of Economic Entomology* 109: 484–486.

Wickham, J. D., J. G. Millar, L. M. Hanks, et al. 2016b. (2*R*,3*S*)-2,3-Octanediol, a female-produced sex pheromone of *Megopis costipennis* (Coleoptera: Cerambycidae: Prioninae). *Environmental Entomology* 45: 223–228.

Witzgall, P., P. Kirsch, and A. Cork. 2010. Sex pheromones and their impact on pest management. *Journal of Chemical Ecology* 36: 80–100.

Wong, J. C. H., R. F. Mitchell, B. L. Striman, et al. 2012. Blending synthetic pheromones of cerambycid beetles to develop trap lures that simultaneously attract multiple species. *Journal of Economic Entomology* 105: 906–915.

Wood, D. L. 1982. The role of pheromones, kairomones, and allomones in the host selection and colonization behavior of bark beetles. *Annual Review of Entomology* 27: 411–446.

World Conservation Monitoring Centre. *Rosalia alpina*. 1996. The IUCN Red List of Threatened Species 1996. http://dx.doi.org/10.2305/IUCN.UK.1996.RLTS.T19743A9009447.en (accessed January 28, 2016).

Yamasaki, T., M. Sato, and H. Sakoguchi. 1997. (-)-Germacrene D: Masking substance of attractants for the cerambycid beetle, *Monochamus alternatus*. *Applied Entomology and Zoology* 32: 423–429.

Yasui, H. 2009. Chemical communication in mate location and recognition in the white-spotted longicorn beetle, *Anoplophora malasiaca* (Coleoptera: Cerambycidae). *Applied Entomology and Zoology* 44: 183–194.

Yasui, H., T. Akino, M. Fukaya, et al. 2008. Sesquiterpene hydrocarbons: Kairomones with a releaser effect in the sexual communication of the white-spotted longicorn beetle, *Anoplophora malasiaca* (Thomson) (Coleoptera: Cerambycidae). *Chemoecology* 18: 233–242.

Yasui, H., T. Akino, T. Yasuda, et al. 2003. Ketone components in the contact sex pheromone of the white-spotted longicorn beetle, *Anoplophora malasiaca*, and pheromonal activity of synthetic ketones. *Entomologia Experimentalis et Applicata* 107: 167–176.

Yasui, H., T. Akino, T. Yasuda, et al. 2007a. Gomadalactones A, B, and C: Novel 3-oxabicyclo[3.3.0]octane compounds in the contact sex pheromone of the white-spotted longicorn beetle, *Anoplophora malasiaca*. *Tetrahedron Letters* 48: 2395–2400.

Yasui, H., N. Fujiwara-Tsujii, and S. Wakamura. 2011. Volatile attractant phytochemicals for a population of white-spotted longicorn beetles *Anoplophora malasiaca* (Thomson) (Coleoptera: Cerambycidae) fed on willow differ from attractants for a population fed on citrus. *Chemoecology* 21: 51–58.

Yasui, H., T. Yasuda, M. Fukaya, et al. 2007b. Host plant chemicals serve intraspecific communication in the white-spotted longicorn beetle, *Anoplophora malasiaca* (Thomson) (Coleoptera: Cerambycidae). *Applied Entomology and Zoology* 42: 255–268.

Yatagai, M., H. Makihara, and K. Oba. 2002. Volatile components of Japanese cedar cultivars as repellents related to resistance to *Cryptomeria* bark borer. *Journal of Wood Science* 48: 51–55.

Zarbin, P. H., M. G. Fonseca, D. Szczerbowski, et al. 2013. Biosynthesis and site of production of sex pheromone components of the cerambycid beetle, *Hedypathes betulinus. Journal of Chemical Ecology* 39: 358–363.

Zeh, D. W., J. A. Zeh, and G. Tavakilian. 1992. Sexual selection and sexual dimorphism in the harlequin beetle *Acrocinus longimanus. Biotropica* 24: 86–96.

Zhang, A., J. E. Oliver, J. R. Aldrich, et al. 2002. Stimulatory beetle volatiles for the Asian longhorned beetle, *Anoplophora glabripennis* (Motschulsky). *Zeitschrift für Naturforschung* 57c: 553–558.

Zhang, A., J. E. Oliver, K. Chauhan, et al. 2003. Evidence for contact sex recognition pheromone of the Asian longhorned beetle, *Anoplophora glabripennis* (Coleoptera: Cerambycidae). *Naturwissenschaften* 90: 410–413.

Zou, Y., J. G. Millar, J. S. Blackwood, et al. 2015. (2*S*,4*E*)-2-hydroxy-4-octen-3-one, a male-produced attractant pheromone of the cerambycid beetle *Tylonotus bimaculatus. Journal of Chemical Ecology* 41: 670–677.

Zou, Y., C. F. Rutledge, K. Nakamuta, et al. 2016. Identification of a pheromone component and a critical synergist for the invasive beetle *Callidiellum rufipenne* (Coleoptera: Cerambycidae). *Environmental Entomology* 45: 216–222.

6

Cerambycids as Plant Disease Vectors with Special Reference to Pine Wilt

Süleyman Akbulut
Düzce University
Düzce, Turkey

Katsumi Togashi
The University of Tokyo
Tokyo, Japan

Marc J. Linit
University of Missouri
Columbia, Missouri

CONTENTS

6.1 Introduction

The genus *Monochamus* Dejean belongs to the subfamily Lamiinae of the family Cerambycidae, which is one of the largest and most important among wood-boring families (Linsley 1961). Species from this genus are called "pine sawyers" because the larvae make a loud noise as they tunnel within the wood of host trees. They feed on conifers and broad-leaved trees. In Japan, some species feed on both (Kojima and Nakamura 2011). Many species attack dying, newly dead, or freshly cut trees of the genera *Pinus* L., *Picea* Mill., and *Abies* Mill., and are considered secondary pests.

Several *Monochamus* species, however, are primary vectors of the pinewood nematode, *Bursaphelenchus xylophilus* (Steiner & Buhrer) Nickle (Mamiya 1983; Linit 1988), the causal pathogen of pine wilt disease. The nematode has caused the destruction of a large number of pine trees in East Asia since the early 1900s and in Europe since 1999 (Mamiya 1976; Robertson et al. 2011). Several genera of the Cerambycidae, other than *Monochamus*, are known to carry *B. xylophilus*. So far, none of these genera has been shown to be a vector of *B. xylophilus*. In addition, other beetle families such as Curculionidae, Buprestidae, and Elateridae can also carry the pinewood nematode, but no evidence for their transmission of this nematode disease is documented (Linit et al. 1983; Kobayashi et al. 1984; Linit 1988). Known vector species in *Monochamus* are *M. carolinensis* (Olivier), *M. titillator* (Fabricius), *M. mutator* LeConte, *M. notatus* Drury, *M. scutellatus* (Say), *M. obtusus* Casey, *M. marmorator* Kirby, *M. alternatus* Hope, *M. saltuarius* Gebler, *M. nitens* Dates, and *M. galloprovincialis* (Olivier) (Akbulut and Stamps 2012). Detailed information on vector *Monochamus* species is provided in Table 6.1. The most important and effective vectors are *M. alternatus* in East Asia, *M. carolinensis* in the United States, and *M. galloprovincialis* in Europe.

Monochamus alternatus was first identified as an important vector of the pathogenic nematode in Japan in 1971 (Mamiya and Enda 1972; Morimoto and Iwasaki 1972). Since then, numerous studies have addressed the life histories, population dynamics, chemical ecology, and population genetics of the vectors, and the relationships among the vectors, nematodes, and host trees. Mathematical models have been developed to describe the beetle's biology, to assist in the control of pine wilt disease epidemics, and to evaluate the risks of the disease spread. This chapter reviews the current knowledge about the vector–nematode–pine system and its management measures in East Asia, North America, and Europe.

6.2 Biology of *Monochamus* species

This section deals with (1) life cycles, (2) larval diapause involved in the regulation mechanism of life history, (3) survival of immatures within trees, (4) adult life span, (5) fecundity and host selection, (6) flight and host search, and (7) mating behavior of *Monochamus* species important as vectors of the pinewood nematode.

6.2.1 Life Cycles of *Monochamus* Species

It takes several months to two years to complete development in *Monochamus* species. The developmental time may differ within a species depending on environmental conditions and time when eggs

TABLE 6.1

Monochamus Species Known to Carry *B. xylophilus*

Monochamus Species	Distribution	Host Trees
M. carolinensis	Canada, Mexico, USA	*Pinus* spp.
M. scutellatus	Canada, Mexico, USA	*Abies* spp., *Larix* spp., *Pinus* spp., *Picea* spp.
M. titillator	Canada, USA	*Abies* spp., *Pinus* spp., *Picea* spp.
M. mutator	Canada, USA	*Pinus* spp.
M. obtusus	Canada, USA	*Abies* spp., *Pinus* spp., *Pseudotsuga menziesii*
M. notatus	Canada, USA	*Pinus strobus*
M. marmorator	Canada, USA	*Abies* spp., *Picea* spp.
M. alternatus	China, Japan, Korea, Laos, Taiwan, Vietnam	*Abies* spp., *Cedrus* spp., *Larix* spp., *Picea* spp., *Pinus* spp.
M nitens	Japan	*Pinus* spp , *Picea* spp , *Abies* spp., *Larix* spp.
M. saltuarius	Austria, China, Croatia, Czech Republic, Germany, Italy, Japan, Korea, Lithuania, Mongolia, Poland, Romania, Russia, Slovakia, Switzerland, Ukraine	*Pinus* spp., *Picea* spp., *Larix* spp., *Tsuga* spp.
M. galloprovincialis	Albania, Algeria, Armenia, Austria, Azerbaijan, Belarus, Bosnia-Herzegovina, Bulgaria, China, Croatia, Czech Republic, Estonia, Finland, France, Georgia, Germany, Greece, Hungary, Italy, Kazakhstan, Latvia, Lithuania, Macedonia, Moldova, Mongolia, Montenegro, Morocco, The Netherlands, Poland, Portugal, Romania, Russia, Serbia, Slovakia, Slovenia, Spain, Switzerland, Tunisia, Turkey, UK, Ukraine	*Pinus* spp., *Abies* spp., *Picea* spp.

Source: Data are gathered from European and Mediterranean Plant Protection Organization (EPPO) plant quarantine information retrieval system and references cited in the text.

are deposited, which affect the induction of larval diapause. For example, *M. carolinensis* completes up to two generations per year in Missouri, and a two-year generation is common in the northern part of the range (Pershing and Linit 1986a). In the laboratory, the total developmental time from oviposition to adult emergence ranges from 8 to 12 weeks in logs when eggs are deposited under long-day conditions before mid-July, allowing more than one generation per year for *M. carolinensis* if the climate is favorable (Pershing and Linit 1986b). In Japan, *M. alternatus* usually has one generation per year (Mamiya 1984; Kobayashi 1988; Togashi 1989a, 1989b). However, when the eggs are deposited near the end of flight season, they take two years to complete their development (Togashi 1989a, 1989b). In cooler areas, 20–30% of the beetles complete their life cycle in two years (Kobayashi 1988). By contrast, *M. alternatus* is bivoltine in southern China (Song et al. 1991). *Monochamus galloprovincialis* has one generation per year in Italy (Francardi and Pennachio 1996), but its larval development takes two years in Finland (Tomminen 1993). Other studies have also reported the variations in life cycle durations of *M. galloprovincialis*. For example, most beetles complete their life cycle in one year with only a small portion of them (5%) developing through a two-year life cycle (Naves et al. 2008). About 8.1% of individuals need two years to complete development (Koutroumpa et al. 2008).

In the laboratory, the within-wood development period differs among tree species provided to larvae for feeding. The females of *M. galloprovincialis* develop more rapidly in *P. sylvestris* L. logs (199 days) than in *P. nigra* J. F. Arnold logs (223 days) at 24–25°C, 70–80% RH, and a photoperiod of 14:10 h (light:dark, L:D). The shortest developmental time is 74 and the longest 340 days (Akbulut 2009). When reared on *P. sylvestris* logs, the within-wood developmental period of *M. carolinensis* ranges from 83 to 108 days under constant conditions (30°C, 70–80% RH, and a photoperiod of 14:10h [L:D]) (Akbulut and Linit 1999a, 1999b).

Monochamus eggs are white, elongate, cylindrical, and slightly flattened, with rounded ends (Figure 6.1). The average egg size is 2.5–3.0 mm long and 0.9 mm wide for *M. carolinensis* (Walsh and Linit 1985), and 4.03 ± 0.12 mm long and 1.22 ± 0.04 mm wide for *M. galloprovincialis*

FIGURE 6.1   An egg of *Monochamus galloprovincialis*. (Courtesy of S. Akbulut.)

FIGURE 6.2   Oviposition slits of *Monochamus galloprovincialis*. (Courtesy of S. Akbulut.)

(Koutroumpa et al. 2008). After maturation feeding, female beetles make oviposition slits in the bark using their mandibles and lay eggs in the slits (Figure 6.2). Egg incubation period ranges from six to nine days under natural conditions but can vary from 10 to 12 days according to temperature (Kobayashi et al. 1984).

Young larvae are soft-bodied, elongate, and dirty white in color, with a light yellow thorax and an amber brown head. The final instar larvae have 10 abdominal segments, and the length of mature larvae is between 25 and 50 mm (Figure 6.3). The number of larval instars is variable among *Monochamus* species,

FIGURE 6.3   A larva of *Monochamus* sp. (Courtesy of S. Akbulut.)

FIGURE 6.4   A larval entrance hole of *Monochamus galloprovincialis*. (Courtesy of S. Akbulut.)

with three to eight instars for *M. carolinensis* reared on a meridic diet (Pershing and Linit 1986a) and four instars for both *M. alternatus* (Yamane 1975) and *M. galloprovincialis* (Koutroumpa et al. 2008).

The early instars of larvae develop entirely in the subcortical zone of the host tree (Pershing and Linit 1986b). Late instars construct galleries in the sapwood. The entrance hole into the sapwood is oval-shaped (Figure 6.4). They pupate at the upper end of the gallery after plugging the gallery with wood shavings from gallery construction (Kobayashi et al. 1984; Togashi 1989a, 1989b) and creating a pupal chamber that is closed to the surface of the wood. Pupal chambers are usually 10–60 mm long for *M. carolinensis* and 20–150 mm long for *M. alternatus* (Togashi et al. 2005, 2008).

The average length of *M. galloprovincialis* pupae is about 20 mm (Koutroumpa et al. 2008). The adult ecloses within the pupal chamber and remains there until cuticular tanning is complete (Pershing and Linit 1986b). In *M. alternatus*, the length of the pupal period is 12–13 days at 25°C (Yamane 1974) and

FIGURE 6.5   An adult exit hole of *Monochamus galloprovincialis*. (Courtesy of S. Akbulut.)

may vary further depending on the temperature (Mamiya 1984). Eclosed *M. alternatus* adults remain in the pupal chambers for four to five days before emergence (Yamane 1975).

Monochamus carolinensis usually overwinters at the larval stage but sometimes in the egg or pupal stages (Pershing and Linit 1986b). *Monochamus alternatus* and *M. galloprovincialis* overwinter exclusively as larvae (Togashi 1989a, 1989b; Koutroumpa et al. 2008). Most late instar larvae of *M. alternatus* overwinter in pupal chambers within the xylem, and young larvae overwinter under the bark (Togashi 1989a, 1989b).

Adults emerge by chewing a round exit hole in the wood and bark (Figure 6.5). The first emergence of adult beetles usually starts in the spring. Adult emergence continues until late summer in some *Monochamus* species such as *M. carolinensis*. *M. galloprovincialis* adults start to emerge in mid-May and continue until early September with a peak emergence in July (Naves et al. 2008). By contrast, the adult emergence from dead trees ceases in July in the Japanese *M. alternatus*, although its flight season lasts until the end of September (Togashi and Magira 1981). The variation in the emergence period is related to the number of generations per year.

6.2.2 Larval Diapause in *Monochamus* Species

Diapause, physiologically arrested development, is commonly involved in the regulatory mechanism of insect life cycles under seasonal environments with cold winter and hot summer (Tauber et al. 1986; Danks 1987, 2006, 2007). Larval development is arrested at the final instar when the larvae of the Japanese *M. alternatus* and *M. urussovi* Fischer von Waldheim are reared on host tree bolts at 25°C and a photoperiod of 16:8 h (L:D) (Togashi et al. 2008, 2010). The larval diapause of both species is terminated by chilling at 5 or 10°C in the dark (Togashi 1995a; Togashi et al. 2010). In the field, *M. alternatus* larval diapause is terminated by mid-February (Togashi 1991a). Interestingly, some larvae of a Taiwanese *M. alternatus* population reared on pine bolts at 25°C and 16:8 h (L:D) do not enter diapause (Togashi 2014). In addition, food shortage has an inhibitory effect on diapause induction under these conditions (Togashi 2014). Unlike Japanese *M. alternatus*, North American *M. carolinensis* develops continuously when reared on pine bolts at 30°C, 14:10 h (L:D), and 75% RH (Linit 1985).

Some larvae of Taiwanese *M. alternatus* reared on an artificial diet do not enter diapause under combined conditions of two air temperatures (26°C and 29°C) and three fixed photoperiodic regimes of 20:4,

12:12, and 0:24 h (L:D), but all larvae enter diapause at 23°C and 12:12 h (L:D) (Enda and Kitajima 1990). Some larvae of *M. galloprovincialis* reared on an artificial diet do not enter diapause at 23°C under four photoperiodic regimes of 15:9, 12:12, 9:15, and 0:24 h (L:D) (Naves et al. 2007a). In a recent study, Togashi (2014) suggested that a facultative diapause may be involved in the life history regulation of *M. galloprovincialis* and Taiwanese *M. alternatus* even when the larvae are under the bark.

The life history regulation mechanism of Japanese *M. alternatus* has been well studied (Togashi 2013). In central Japan, *M. alternatus* adults deposit eggs between June and September. Consequently, the larvae overwinter at the first to fourth (final) instars within trees depending on when the eggs are deposited in a season (Togashi 1989a, 1989b). For example, before overwintering, fully grown fourth instar larvae enter diapause, whereas the first to early fourth instar larvae do not (Togashi 1991c).

Diapaused larvae of Japanese *M. alternatus* resume development in the spring without feeding and then pupate (Togashi 1991c, 1995a). By contrast, nondiapaused fourth instar and some third instar larvae resume feeding after overwintering and pupate without entering diapause the year following oviposition. Other third and second instar larvae enter diapause after feeding, leading to a two-year life history because of diapause termination by cold temperatures in the second winter (Togashi 1991c, 1995b, 2013).

6.2.3 Survival of *Monochamus* Immatures within Trees

In *M. alternatus*, when two larvae encounter one another, one kills the other by biting (Anbutsu and Togashi 1997), resulting in the so-called contest-type competition regardless of the presence or absence of natural enemies (Togashi 1986). Akbulut et al. (2004) reported that the number of adults produced per log is not related to the number of eggs laid. Late instar larval and adult numbers are significantly correlated with log size possibly because of the contest-type competition. Only 12% of eggs develop into adults. The high within-wood mortality may be attributed to intraspecific competition and cannibalism among larvae (Akbulut et al. 2004).

In North America, the within-wood survival rate of *M. carolinensis* ranges from 6% to 15% in the laboratory (Linit 1985; Akbulut and Linit 1999a; Akbulut et al. 2004), and 15% of *M. carolinensis* and *M. titillator* in logs maintained in outdoor cages survived to adults in North Carolina (Alya and Hain 1985). The within-wood survival rates are 14% and 13% for *M. galloprovincialis* that develop in *P. sylvestris* and *P. nigra* logs, respectively (Akbulut 2009). Within-wood survivorship from egg to adult is 53%, 36.5%, and 14% for *M. galloprovincialis* that develop in *P. pinaster* Aiton (Naves et al. 2008), *P. sylvestris*, and *P. pinea* L. logs (Husillos-Sanchez et al. 2013), respectively. This difference may be related to the study conditions, host trees, or to the origins of *M. galloprovincialis* populations used in these studies.

In a field population of *M. alternatus* in Japan, within-wood survival rates are between 13% and 30%, with a mean of 25% among four consecutive generations (Togashi 1990a). The survival rate of immature stages also differs depending on when pine trees are diseased. Key factor analysis indicates that the third and fourth instar larval mortality within pupal chambers determines the between-generation difference in the within-wood survival rate to adult emergence. There is evidence for predation by insects during this key stage. The mean stage-specific mortalities (q_x) are 24.6%, 11.9%, 23.9%, 21.6%, 6.2%, and 33.1% for eggs, first to second instar larvae, third to fourth instar larvae before completing construction of pupal chambers, mature larvae in pupal chambers, pupae in pupal chambers, and adults in pupal chambers, respectively.

Numerous natural enemies of *M. alternatus* have been identified (Nakamura-Matori 2008): bacteria *Serratia marcescens* Bizio and *Serratia* spp.; fungus *Beauveria bassiana* (Bals.-Criv.) Vuill; parasitoid wasps *Atanycolus initiator* Fabricius, *Spathius* Nees., *Doryctus* Haliday, *Ecphylus hattorii* Kono & Watanabe, and *Iphiaulax impostor* Scopoli (Hymenoptera: Braconidae), *Sclerodermus nipponicus* Yuasa and *S. guani* Xiao & Wu (Hymenoptera: Bethylidae), *Cleonymus* Lat. (Hymenoptera: Pteromalidae), *Dolichomitus* Smith and *Megarhyssa* Ashmead (Hymenoptera: Ichneumonidae), *Dastarcus helophoroides* Fairmaire and *D. kurosawai* (Coleoptera: Bothrideridae), and *Billaea* Robineau-Desvoidy (Diptera: Tachinidae); predators *Monomorium intrudens* Smith (Hymenoptera: Formicidae), *Stenagostus umbratilis* Lewis, *Paracalais berus* Candèze, and *P. larvatus* Candèze (Coleoptera: Elateridae), *Platysoma lineicollis* Marseul (Coleoptera: Histeridae), *Mimemodes emmerichi* Mader and *Rhizophagus* Herbst. (Coleoptera: Rhizophagidae), *Trogossita japonica* Reitter (Coleoptera: Trogossitidae), *Thanasimus lewisi* Jacobson (Coleoptera: Cleridae), *Velinus nodipes* Uhler (Hemiptera: Reduviidae), *Inocellia japonica* Okamoto

(Neuroptera: Inocellidae), and *Anisolabella marginalis* Dohrn (Dermaptera: Anisolabididae); and the woodpecker *Dendrocopos major* L. (Piciformes: Picidae). Furthermore, Naves et al. (2005) have discovered three braconid (Hymenoptera) parasitoid species in larval galleries of *M. galloprovincialis* in Portugal: *Cyanopterus flavator* Fabricius, *Iphiaulax impostor* Scopoli, and *Coeloides sordidator* Ratzeburg, among which the most common species is *C. flavator*, consisting of 58% of the parasitoid individuals counted. Exclusion of insect natural enemies by meshed cages shows that insect natural enemies are density-independent mortality factors of *M. alternatus* immature stages, reducing their survival rate (Togashi 1986).

6.2.4 Life Span of *Monochamus* Adults

After emergence from dead trees, *Monochamus* adults feed on fresh bark of host tree twigs during the day (Figure 6.6). This feeding is obligatory for maturation of the female reproductive system (Yamane 1981; Wingfield and Blanchette 1983). The preovipositional period ranges from 5 to 30 days for *M. alternatus* females at 25°C (Togashi 1997). Generally, maturation feeding occurs for 10–14 days in *M. alternatus, M. carolinensis*, and *M. galloprovincialis* (Kobayashi 1988; Togashi 1997; Naves et al. 2006a; Akbulut and Stamps 2012). Katsuyama et al. (1989) reported that the ovaries of *M. alternatus* develop more quickly when females are fed with the current-year twigs of *P. densiflora* Siebold & Zucc. than when fed with one- or two-year-old twigs. That is also the case for *M. saltuarius* females when fed with *Pinus koraiensis* Siebold & Zucc. twigs (Yoon et al. 2011). *M. alternatus* males require at least five days of feeding after emergence for sexual maturation (Nobuchi 1976). Both sexes of *Monochamus* species continue to feed following sexual maturity.

There is a great variation in the longevity of beetles among species and within species. For example, the mean longevity of *M. alternatus* females is 82.8 days, with the longest being 145 days at 25°C and 16:8 h (L:D) (Togashi 1997). In field cages, early-emerged adults of *M. alternatus* live for an average of 68.0 days compared to late-emerged adults of 43.1 days (Togashi and Magira 1981). *M. carolinensis* adults that emerge from *P. sylvestris* L. logs cut in autumn and spring live for 38 and 59 days, respectively, under laboratory conditions (30°C, 70–80% RH, and a photoperiod of 14:10 h [L:D]) (Akbulut and Linit 1999b). The mean longevity of males and females in *M. carolinensis* is 51 and

FIGURE 6.6 Maturation feeding of *Monochamus galloprovincialis*. (Courtesy of S. Akbulut.)

56 days, respectively, in field cages (Togashi et al. 2009). Males and females of *M. galloprovincialis* have mean longevity of 61.2 and 64.0 days, respectively, when developing in *P. pinaster* logs and feeding on *P. pinaster* twigs under laboratory conditions in Portugal (Naves et al. 2006a). When reared on *P. sylvestris* at larval and adult stages, the mean adult longevity of *M. galloprovincialis* is 47.8 days, whereas it is 33.0 days when reared on *P. nigra* (Akbulut 2009). Variations in adult longevity can be attributed to the nutritional quality of the phloem and cambium of pine logs consumed at the larval stage and also the quality of the twig subcortical tissues consumed at the adult stage (Akbulut and Linit 1999b).

6.2.5 Fecundity and Host Selection of *Monochamus* Adults

The oviposition behavior of the family Cerambycidae have been classified into two types: (1) exclusively with the aid of ovipositor and (2) with the aid of both ovipositor and mandibles (Butovitsch 1939). *Monochamus* species fall into the latter category. Edwards and Linit (1991) divided the oviposition behavior of *M. carolinensis* into four stages: the excavation of the oviposition niche or slit with mandibles, a 180-degree twist to position the ovipositor over the niche, insertion of the ovipositor, and egg deposition. Okamoto (1984) and Anbutsu and Togashi (2000) showed that *M. alternatus* exhibits oviposition behavior similar to *M. carolinensis*. Immediately after oviposition, *M. alternatus* females deposit deterrents in the entrance of the oviposition wound to inhibit oviposition by other females, resulting in a uniform dispersion of eggs over a unit area of bark (Shibata 1984; Anbutsu and Togashi 2001). Larval frass of *M. alternatus* and its methanol extracts also suppresses oviposition (Anbutsu and Togashi 2002). The deterrent has been identified as a mixture of monoterpenes and butylated hydroxytoluene (Li and Zhang 2006).

Monochamus females leave two types of oviposition marks as a result of egg deposition: a slit-like mark on thin bark and a pit-like mark on thick bark. Oviposition occurs on tree trunks, branches, and twigs as small as 2 cm in diameter. *M. alternatus* females prefer to deposit their eggs on thin barked areas; however, they will also utilize less favorable, thicker bark sites (Kobayashi et al. 1984). Oviposition occurs at night on dead or dying pine trees and on recently cut logs.

The number of eggs laid in an oviposition wound differs both within and among species: one egg in most cases, eggless and more than one egg in others (Togashi and Magira 1981; Yamane 1981; Kobayashi et al. 1984; Mamiya 1984; Walsh and Linit 1985; Linit 1988; Akbulut 1998). In the United States, dissection of 566 oviposition sites created by a laboratory colony of *M. carolinensis* females gave an average of 1.014 eggs per oviposition wound (Akbulut and Linit 1999a). Togashi et al. (2009) dissected 1,758 oviposition sites made by *M. carolinensis* in field cages, demonstrating 1,173 slits with one egg, 51 with two, 3 with three, 1 with four, and 530 with no egg. The mean number of eggs per oviposition slit in the field cages was lower than that in the laboratory study. The highest reported number of eggs per oviposition slit for a *Monochamus* species is nine in *M. titillator* (Luzzi et al. 1984). In Portugal, 99.2% of 1,928 oviposition slits made by *M. galloprovincialis* on *P. pinaster* logs contained one egg per slit (Naves et al. 2006a). In Turkey, Akbulut et al. (2008) showed that, of 2,128 oviposition slits made by a laboratory colony of *M. galloprovincialis*, 175 had no eggs, 1,780 bore one egg each, 152 contained two eggs each, 17 possessed three eggs each, and 4 held four eggs each. In another study, however, all oviposition slits made by *M. galloprovincialis* on different *Pinus* species contained either one or no egg (Husillos-Sanchez et al. 2013).

Fecundity, the total number of eggs laid by a female, varies among females within a population of a given species. For example, it ranges from 0 to 343 for *M. alternatus* (Kobayashi et al. 1984; Kobayashi 1988; Togashi 1997) and 0 to 384 for *M. carolinensis* (Walsh and Linit 1985; Akbulut 1998). Lifetime fecundity is correlated positively to oviposition period, oviposition rate, and female body size in *M. alternatus* (Togashi 1997); to female body size (Naves et al. 2006a) or adult longevity (Koutroumpa et al. 2008) in *M. galloprovincialis*; and to female life span, oviposition period, and oviposition rate in *M. saltuarius* (Jikumaru et al. 1994). In relations among female body size, oviposition period, oviposition rate, and lifetime fecundity, path analysis reveals that female body size contributes greatly to fecundity through the oviposition rate but little through the oviposition period in a *M. alternatus* laboratory population (Togashi 1997). The host plant quality of larval and adult diets may also impact fecundity through its influence on adult longevity; for example, *M. galloprovincialis* females have significantly greater lifetime fecundity when reared on logs and twigs of *P. sylvestris* than when reared on those of *P. nigra* (Akbulut 2009).

Fecundity also varies among and within species of the genus *Monochamus*. For example, the mean fecundity ranges from 33 to 581 eggs in five *M. alternatus* populations, from 117 to 451 eggs in six *M. carolinensis* populations, and from 15 to 78 eggs in four *M. saltuarius* populations (Togashi 2007; Togashi et al. 2009). In *M. galloprovincialis*, the mean fecundity is 67 eggs on *P. pinaster* logs in Portugal (Naves et al. 2006a), 126 eggs on *P. sylvestris* and 57 eggs on *P. nigra* logs with the highest value of 359 eggs in Turkey (Akbulut 2009) and 138.2 eggs in *P. sylvestris* logs in France (Koutroumpa et al. 2008). Togashi et al. (2009) demonstrated that the mean female body size, longevity, and fecundity are positively correlated for 16 populations of *M. alternatus*, *M. carolinensis*, *M. galloprovincialis*, and *M. saltuarius*.

Monochamus species associated with the pinewood nematode mainly colonize *Pinus* species, although they may sometimes attack other conifers including *Abies* Mill. and *Picea* Link (Table 6.1). For example, *M. alternatus* attacks 18 *Pinus* species, 3 *Picea* species, and 1 species of *Abies, Cedrus* Trew, and *Larix* Philip Miller (Kobayashi 1988). The most common host trees of *M. galloprovincialis* in Europe are *P. sylvestris*, *P. pinaster*, and *P. nigra*.

Host preference by *Monochamus* species has been studied in different regions. For example, *P. sylvestris* is the most preferred host among five pine species for both adult feeding and oviposition of *M. galloprovincialis* (Naves et al. 2006b). Although both *P. sylvestris* and *P. nigra* are suitable hosts of *M. galloprovincialis*, *P. sylvestris* has a significantly positive effect on the reproductive potential of female beetles (Akbulut 2009). However, Kobayashi (1988) suggested that trees preferred for adult maturation feeding may not be the best for larval development.

6.2.6 Flight and Host Search of *Monochamus* Adults

Flight activity of *Monochamus* species varies with initial body weight, reproductive maturity, mating, pine density in a stand, and nematode load (Ito 1982; Humphry and Linit 1989a, 1989b; Togashi 1990b, Togashi 1990c; Akbulut and Linit 1999c). Premature *M. alternatus* adults have greater flight activity than sexually mature adults both in the laboratory (Ito 1982) and in field studies (Togashi 1990b). In *M. carolinensis*, virgin females are better fliers than mated females (Akbulut and Linit 1999c). In *M. alternatus* (Ito 1982) and *M. carolinensis* (Humphry and Linit 1989a), there is no difference in flight activity between sexes, except for five-day-old adults, the males of which are stronger fliers than the females. In newly emerged adults of *M. alternatus*, as the healthy pine tree density decreases in a stand due to the pine wilt disease epidemic, the activity of beetle movement increases and the duration of their stay on a healthy tree decreases (Togashi 1990c). In a *P. densiflora* stand with an epidemic of pine wilt disease, sticky screen trap captures indicate that the activity of *M. alternatus* adults is six times higher in the canopy compared to the sub-canopy (Sugimoto and Togashi 2013).

Monochamus beetles are considered to be poor fliers, although there is evidence that they can perform both short- and long-distance dispersals. It has been estimated that one- to five-day-old adults can travel 7.1–37.8 m per week (Togashi 1990c). Male and female adults are estimated to traverse 10.6 and 12.3 m, respectively, among healthy trees between first and final captures (Shibata 1986). On the other hand, most marked beetles are captured within 100 m of the release site and a few beetles captured 2.4 km away from the site (Ido et al. 1975). Distance between newly diseased trees and the site where infested logs are introduced suggests that *M. alternatus* adults can disperse for up to 3.3 km (Kobayashi et al. 1984). Mas et al. (2013) carried out mark–release–recapture trials with *M. galloprovincialis*, and marked adults flew a maximum distance of 22,100 m and about 2% of released beetles traveled more than 3,000 m. Another mark–recapture study on *M. galloprovincialis* revealed that the lifetime adult dispersal distance was 107–122 m on average with a maximum of 464 m (Torres-Villa et al. 2015).

In flight mill experiments, five-day-old *M. alternatus* adults continue to fly for a mean of 16.5 and 20.4 min for females and males, respectively (Ito 1982). The longest duration of flight is 32.8 and 39.0 min for females and males, respectively. Unlike *M. alternatus, M. carolinensis* females fly up to 10 km in 115 min (Akbulut and Linit 1999c), and *M. galloprovincialis* females are able to fly cumulatively over several tens of kilometers during their life span (David et al. 2013). Flight mill experiments also show that the flight performance of *M. galloprovincialis* is highly variable within a population, and that the beetles can fly a mean distance of 16 km during their entire adult life span (David et al. 2014).

In cerambycid beetles, host-plant recognition is mediated by chemical cues originating from plants (Allison et al. 2004). Primary attractants for some cerambycids, including some *Monochamus* species, are smoke volatiles (smoke) (*M. galloprovincialis, M. sutor* L., *M. scutellatus* Say) and trunk and leaf volatiles (*M. alternatus, M. carolinensis, M. titillator, M. scutellatus, M. clamator* LeConte, *M. notatus* Drury) comprising α-pinene, β-pinene, β-phellandrene, ethanol, turpentine, (+)-juniperol, (+)-primaral, and (+)-*cis*-3-pinene-2-ol (Allison et al. 2004). *Monochamus alternatus* adults are attracted to a mixture of monoterpene hydrocarbons such as α-pinene, β-pinene, and β-myrcene and ethanol released from dying or recently felled host trees (Ikeda et al. 1980, 1981).

The blend of host volatiles (turpentine, α-pinene, ethanol) and bark beetle pheromones (ipsenol, ipsdienol, *cis*-verbenol, methyl-butenol) are more attractive to *M. galloprovincialis* adults than host volatiles alone (Pajares et al. 2004). Ipsenol appears to be the strongest kairomonal signal to *M. galloprovincialis*, particularly when it is released together with the host monoterpene α-pinene or with methyl butenol (Ibeas et al. 2007). Several *Monochamus* species (*M. clamator, M. scutellatus, M. mutator, M. notatus, M. obtusus, M. titillator*) from North America also have similar strong responses to ipsenol (Allison et al. 2003; De Groot and Nott 2004; Miller and Asaro 2005). It has been demonstrated that the blend composed of α-pinene, ethanol, ipsenol, ipsdienol, and 2-methyl-3-buten-2-ol is very effective for trapping some cerambycid species, including *M. galloprovincialis* (Francardi et al. 2009).

Intriguingly, bark beetle pheromones as well as inner bark extracts have been identified as oviposition stimulants of some *Monochamus* species (*M. alternatus, M. titillator, M. clamator, M. notatus, M. obtusus, M. scutellatus*) (Islam et al. 1997; Sato et al. 1999a, 1999b; Allison et al. 2004). Some compounds are isolated from inner bark extracts of *P. densiflora* and identified as oviposition stimulants for *M. alternatus*. These include: D-catechin, the flavanonol glucoside of (−)-2,3-trans-dihydroquercetin-3′-*O*-β-D-glucopyranoside, polymeric proanthocyanidins, and two dimeric proanthocyanidins of epicatechin-(4β→8′)-catechin and epicatechin-(4α→8′)-catechin, the phenylpropanoid glucoside of dihydroconiferyl alcohol-9-*O*-β-D-glucopyranoside, and four neolignan glucosides of cedrusin-4′-*O*-β-D-glucopyranoside, cedrusin-4′-*O*-α-L-rhamnopyranoside, 7-*O*-methyl cedrusin-4′-*O*-α-L-rhamnopyranoside, and 1-(4′-hydroxy-3′-methoxyphenyl)-2-[4″-(3-hydroxypropyl)-2″-hydroxyphenoxy]-1,3-propanediol-4′-*O*-β-D-xylopyranoside (Islam et al. 1997; Sato et al. 1999a, 1999b).

6.2.7 Mating Behavior and Male Competition for Mates in *Monochamus*

Both genders of sexually mature *Monochamus* beetles are attracted to stressed trees and/or recently cut logs for mating and oviposition during the night. In *M. alternatus*, the female plays an active role by approaching a male, and upon antennal contact, the male attempts to mate (Fauziah et al. 1987). According to Fauziah et al. (1987) and Kim et al. (1992), the mating behavior of *M. alternatus* is divided into two steps: (1) males attract females and (2) males mount females to copulate upon making contact with the female's body surface. However, *M. carolinensis* males search for females (Edwards and Linit 1991). No sex pheromone is known to exist, and antennal contact between sexes is sufficient for mate recognition in *M. carolinensis*, although sex or aggregation pheromones elicited by males may be related to mating in other *Monochamus* species (see following text). *M. galloprovincialis* mating behavior shares aspects of Asian and North American beetle behavior. Most males of this species encounter females while both sexes are walking, and antennal contact occurs prior to the onset of copulation—although the behavioral patterns vary (Ibeas et al. 2008). Antennal contact is made by both sexes prior to copulation. In some cases, copulation occurs following an initial antennal contact; whereas in other cases, repeated antennal contact occurs before copulation begins (Ibeas et al. 2008).

Sex pheromones to attract the opposite sex have not been thought to exist in *Monochamus* species because adult beetles of both sexes are brought together by attraction to volatiles released by stressed trees and recently cut logs. Several studies, however, have suggested the existence of sex pheromones. Two types of pheromones have been reported for *M. alternatus*: (1) a male sex pheromone that attracts the female and (2) a contact pheromone on the body surface of the female that elicits mate recognition and initiation of courtship (Fauziah et al. 1987; Kim et al. 1992). Kim et al. (1992) also provided weak evidence for a long-range sex pheromone in *M. alternatus*. However, the combination of volatile compounds from injured pine trees and the presence of a contact sex pheromone for mate recognition may obviate the need for long-range sex pheromones. Recently, Ibeas et al. (2008) reported that mature female *M. galloprovincialis* under laboratory conditions were attracted to volatile compounds produced by mature male beetles; however, the males did not respond to female-produced volatiles. A volatile of 2-undecyloxy-1-ethanol produced specifically by mature males attracts female and male *M. galloprovincialis* in the presence of α-pinene and two bark beetle pheromone components (ipsdienol and 2-methyl-3-buten-2-ol) in the field (a male produced aggregation pheromone) (Pajares et al. 2010). A chemical substance of 2-(undecyloxy)-ethanol has been proven to be a male-produced aggregation pheromone for *M. galloprovincialis* (Pajares et al. 2010), *M. alternatus* (Teale et al. 2011), *M. carolinensis* and *M. titillator* (Allison et al. 2012; Rasatti et al. 2012), *M. scutellatus scutellatus* (Say) (Fierke et al. 2012), and *M. sutor* (L.) (Pajares et al. 2013). Macias-Samano et al. (2012) suggested that 2-(undecyloxy)-ethanol (monochamol) is a likely pheromone component for *M. clamator* (LeConte) and *M. obtusus* Casey and an attractant for *M. notatus* (Drury) (Fierke et al. 2012).

Female *M. scutellatus* prefer to oviposit on large-diameter pine logs (Hughes and Hughes 1982). Males of this species compete intensely for access to females and for positions on the portion of an oviposition substrate preferred by females. When mounting a female, the male uses his antennae to ward off other males (Hughes 1979) and, while maintaining the pair-bond, the male copulates with the female repeatedly and the female continues to oviposit. The number of eggs deposited during a pair-bond increases as the duration of pair-bond increases (Hughes 1981). Large males keep the pair-bond for a longer period than smaller males, resulting in greater reproductive fitness gain. Females of coexistent *M. scutellatus* and *M. notatus* compete for oviposition sites on large-diameter portions of felled trunks (Hughes and Hughes 1987). Where this occurs, the larger species, *M. notatus*, outcompetes the smaller *M. scutellatus*.

6.3 Biology of *Bursaphelenchus xylophilus*

The genus *Bursaphelenchus* was described in 1937 by Fuchs. The nematodes in the genus are mycophagous or phytophagous or both. Currently, there are more than 100 species of *Bursaphelenchus* described (Table 6.2). Almost 70% of known species are associated with conifers, primarily *Pinus* species (Braasch et al. 2004). The most commonly known virulent plant pathogens are *B. xylophilus* (Figure 6.7) and *B. cocophilus* (Cobb) Baujard. This section deals with the life cycles of *B. xylophilus* and the disease development process by the pathogenic nematode.

6.3.1 Life Cycle of *B. xylophilus*

Adult females of *B. xylophilus* copulate with males as with all *Bursaphelenchus* species except *B. okinawaensis* Kanzaki, Maehara, Aikawa, and Togashi (Kanzaki 2008; Kanzaki et al. 2008). Futai (2013) has documented attraction between males and females of *B. xylophilus*. Each inseminated female lays between 80 and 150 eggs over a 28-day oviposition period (Mamiya 1975; Kishi 1995; Vicente et al. 2012).

Bursaphelenchus xylophilus develops through four juvenile stages. Development is temperature dependent and lasts 3–11 days (Ishibashi and Kondo 1977; Mamiya 1984; Futai 2013). The first stage juvenile, designated J_1, develops within the egg; thus, the nematode hatches as a second-stage juvenile (J_2) (Figure 6.8). Further juvenile development follows one of two distinct pathways: propagative or dispersal (Figure 6.8). When within-wood conditions are favorable, development follows the propagative pathway. The nematode develops through two additional juvenile stages: the third (J_3) and fourth (J_4) propagative stages (Figure 6.8). When the within-wood conditions for juvenile development deteriorate due to low or high moisture content or low nutritional reserves (Vicente et al. 2012), nematode development switches to the dispersal pathway passing through the third (J_{III}) and fourth (J_{IV}) dispersal juvenile stages (Figure 6.8). The J_{III} and J_{IV} stages are characterized by a thickened cuticle and accumulated lipid droplets to enhance survival within the deteriorating environment of the dying tree or log (Futai 2013).

TABLE 6.2

Bursaphelenchus Species, Host Trees, and Distribution

Bursaphelenchus Species	Original Description	Host Tree	Distribution
B. aberrans	Fang et al. (2002a)	*Pinus massoniana, Pinus merkusii*	China, Thailand
B. abietinus	Braasch and Schmutzenhofer (2000)	*Abies alba*	Austria
B. abruptus	Giblin-Davis et al. (1993)	No plant host (isolated from soil-dwelling bee, *Anthophora abrupta*)	USA
B. africanus	Braasch et al. (2006)	Packaging wood (*Pinus radiata*)	South Africa
B. anamurius	Akbulut et al. (2007)	*P. brutia, P. nigra*	Turkey
B. anatolius	Giblin-Davis et al. (2005)	No plant host (isolated from soil-dwelling bee, *Halictus* sp.)	Turkey
B. andrassyi	Dayi et al. (2014)	*Picea* spp., *Abies cilicica, Pinus brutia*	Romania, Turkey
B. antoniae	Penas et al. (2006)	*Pinus pinaster*	Portugal
B. arthuri	Burgermeister et al. (2005)	Conifer wood	Taiwan, South Korea
B. arthuroides	Gu et al. (2012)	*Pinus* sp.	Brazil
B. baujardi	Walie et al. (2003)	*Bombax ceiba*	India
B. bestiolus	Massey (1974)	*Pinus ponderosa*	USA
B. borealis	Korenchenko (1980)	*Larix dahurica, Picea abies, Pinus sylvestris, P. brutia*	Russia, Germany, Cyprus
B. braaschae	Gu and Wang (2010)	Dunnage	Thailand
B. burgermeisteri	Braasch et al. (2007)	Packaging wood (*Pinus radiata*)	Japan
B. chengi	Li et al. (2008)	Packaging material	Taiwan
B. chitwoodi	Rühm (1956)	*Pinus sylvestris*	Germany, Georgia
B. clavicauda	Kanzaki et al. (2007)	*Castanopsis cuspidata*	Japan
B. cocophilus	Cobb (1919)	*Cocos nucifera, Elaeis guineensis, Phoenix dactylifera, Roystonea oleracea, Acrocomia aculeata, Mauritia flexuosa, Maximiliana maripa, Oenocarpus distichus*	Grenada, West Indies, Belize, Brazil, Costa Rica, Ecuador, El Salvador, Granada, Guayana, Honduras, Mexico, Panama, St. Vincent, Tobago, Trinidad, Venezuela, Colombia, Surinam, Peru
B. conicaudatus	Kanzaki et al. (2000)	*Ficus carica, Morus* spp.	Japan
B. corneolus	Massey (1966)	*Pinus ponderosa*	USA
B. crenati	Rühm (1956)	*Fraxinus excelsior*	Germany, Georgia
B. cryphali	Fuchs (1930)	*Abies alba*	Germany, Slovakia
B. curvicaudatus	Wang et al. (2005)	Packaging materials (Conifer species)	?

(Continued)

TABLE 6.2 (Continued)
Bursaphelenchus Species, Host Trees, and Distribution

Bursaphelenchus Species	Original Description	Host Tree	Distribution
B. debrae	Hazir et al. (2007)	No plant host (isolated from soil-dwelling bee, Halictus brunnescens)	Turkey
B. dongguanensis	Fang et al. (2002b)	Pinus massoniana	China
B. digitulus	Loof (1964)	Cocos nucifera	Venezuela
B. doui	Braasch et al. (2005)	Pinus densiflora	Taiwan, South Korea, China, Japan
B. eggersi	Rühm (1956)	Larix leptolepis, Picea excelsa, Picea orientalis, Pinus sylvestris, P. strobus Abies sp., Larix sp., Pinus cedrus, Picea abies, P. pinaster, Abies alba, P. radiata, P. brutia	Germany, Switzerland, Georgia, Greece, Austria, Spain
B. eidmanni	Rühm (1956)	Picea abies, P. excelsa, P. sitchensis, Abies sp., Larix sp., P. orientalis, Pinus cedrus, P. sosnowskyi, Beta vulgaris	Germany, Georgia, Slovakia, Uzbekistan
B. elytrus	Massey (1971)	P. resinosa	USA
B. eremus	Rühm (1956)	Populus gracilis, Salix sp., Castanea vulgaris, Quercus iberica, Q. pedunculata, Q. sessiliflora, Ulmus folicea	Germany, Georgia, Czech Republic
B. eroshenkii	Kolossova (1997)	Pinus sibirica	Russia
B. erosus	Kurashvili et al. (1980)	Abies sp., Picea orientalis, Pinus sosnowskyi	Georgia
B. eucarpus	Rühm (1956)	Malus silvestris, Pyrus communis, M. domestica, Prunus sp., Sorbus sp.	Germany, Georgia
B. fagi	Tomalak and Filipiak (2014)	Fagus silvatica	Poland
B. firmae	Kanzaki et al. (2012)	Abies firma	Japan
B. fraudulentus	Rühm (1956)	Populus nigra, P. tremula, Prunus avium, Fagus silvatica, Quercus robur, Alnus glutinosa, Betula pendula, B. pubescens, Quercus petraea, Prunus cerasus, Pinus monticola, Thuja plicata, Picea sp.	Germany, Georgia, Austria, Hungary, USA
B. fuchsi	Kruglik and Eroshenko (2004)	Pinus koraiensis	Russia
B. fungivorus	Franklin and Hooper (1962)	Gardenia sp., Pinus pinaster	UK, Germany, Czech Republic, Spain
B. gerberae	Giblin-Davis et al. (2006a)	Cocos nucifera	Trinidad
B. gillanii	Schönfeld et al. (2014)	Conifer wood	China
B. glochis	Brzeski and Baujard (1997)	Pinus sylvestris	Poland
B. gonzalezi	Loof (1964)	Allium sativum var. vulgare, Solanum tuberosum	Venezuela
B. hellenicus	Skarmoutsos et al. (1998)	Pinus brutia, P. sylvestris, P. pinaster, P. yunnanensis, P. halepensis	Greece, Germany, China, Portugal, Turkey

(Continued)

TABLE 6.2 (Continued)

Bursaphelenchus Species, Host Trees, and Distribution

Bursaphelenchus Species	Original Description	Host Tree	Distribution
B. hildegardae	Braasch et al. (2006)	*Pinus sylvestris*	Germany
B. hofmanni	Braasch (1998)	*Picea abies, Abies alba, Pinus pinaster, Pinus armandii*	Germany, Austria, Portugal, Slovenia, China, Romania
B. hunti	Steiner (1935)	*Lilium tigrinum*	Japan
B. hylobianum	Korenchenko (1980)	*Larix dahurica, L. sibirica, Pinus sylvestris, Pinus merkusi, P. pinaster, P. radiata, P. massoniana*	Russia, Thailand, Portugal, Spain, Portugal, China
B. idius	Rühm (1958)	*Picea excelsa, Carpinus caucasica, Juglans sp., Populus tremula, Quercus iberica, Picea abies, Pinus brutia*	Germany, Georgia, Slovakia, Cyprus
B. incurvus	Rühm (1956)	*Abies alba, Picea excelsa, P. sitchensis, P. breweriana, P. orientalis, Pinus pungens, Abies sp., P. sylvestris*	Germany, Georgia
B. kesiyaen	Kanzaki et al. (2015)	*Pinus kesiya*	Vietnam
B. kevini	Giblin et al. (1984)	No plant host (isolated from soil-dwelling bee, *Halictus* spp.)	USA
B. kiyoharai	Kanzaki et al. (2011)	Isolated from *Xyleborus seriatus* emerged from *Fagus crenata*	Japan
B. kolymensis	Korentchenko (1980)	*Larix dahurica*	Russia
B. koreanus	Gu et al. (2013)	*Pinus* sp.	South Korea
B. leoni	Baujard (1980)	*Pinus pinaster; P. halepensis, P. pinea, P. brutia, Pinus nigra, P. sylvestris, P. radiata*	France, Italy, Cyprus, South Africa, Germany, Greece, Austria, Portugal, Spain
B. luxuriosae	Kanzaki and Futai (2003)	*Aralia elata*	Japan
B. macromucronatus	Gu et al. (2008)	Packaging materials	Taiwan, India
B. masseyi	Tomalak et al. (2013)	*Populus tremuloides*	USA
B. maxbassiensis	Massey (1971)	*Fraxinus pennsylvanica*	USA
B. mazandaranense	Pedram et al. (2011)	*Fagus orientalis*	Iran
B. minutes	Walia et al. (2003)	*Pinus wallichiana*	India
B. mucronatus	Mamiya and Enda (1979)	*Pinus densiflora, P. thunbergii, P. pentaphylla, P. pinaster, P. elliotti, P. massoniana, P. serotina, P. rigida, P. taeda, P. nigra, P. brutia, P. sylvestris, P. halepensis, P. armandii, P. yunnanensis, P. koraiensis, P. taiwanensis, P. strobus, Cedrus deodara, Abies balsamea, A. alba, A. bornmülleriana, Larix decidua, L. sibirica, Picea abies*	Japan, France, Norway, China, Korea, Sweden, Finland, Germany, Russia, Italy, Poland, Taiwan, Greece, Bulgaria, Czech Republic, Spain, Thailand, Portugal, Turkey
B. newmexicanus	Massey (1974)	*Pinus ponderosa*	USA
B. niphades	Tanaka et al. (2014)	*Pinus* sp.	Japan
B. nuesslini	Rühm (1956)	*Abies alba*	Germany, Georgia, Slovakia

(Continued)

TABLE 6.2 (Continued)

Bursaphelenchus Species, Host Trees, and Distribution

Bursaphelenchus Species	Original Description	Host Tree	Distribution
B. obeche	Braasch et al. (2007)	*Triplochiton scleroxylon*	Indonesia
B. okinawaensis	Kanzaki et al. (2008)	Isolated from *Monochamus maruokai*, feed on broad-leaved trees	Japan
B. paracorneolus	Braasch (2000)	*Picea abies, P. sylvestris, Larix sibirica*	Germany, Russia?
B. paraburgeri	Wang and Gu (2012)	Deciduous trees	Malaysia
B. paraluxuriosae	Gu et al. (2012)	*Alphitonia* sp.	Indonesia
B. parantoniae	Maria et al. (2015)	*Pinus* sp. (*packaging material*)	Belgium
B. paraparvispicularis	Gu et al. (2010)	*Pinus* sp. (*packaging material*)	China
B. parvispicularis	Kanzaki and Futai (2005)	*Quercus mongolica var grosseserrata*	Japan
B. piceae	Tomalak and Pomorski (2015)	*Picea abies*	Poland
B. pinasteri	Baujard (1980)	*Pinus pinaster, P. sylvestris, P. pinea*	France, Germany, Spain
B. piniperdae	Fuchs (1937)	*Pinus sylvestris, P. montana, Picea excelsa*	Germany, The Netherlands, Georgia, Slovakia
B. pinophilus	Brzeski and Baujard (1997)	*P. sylvestris, P. pinaster, P. nigra*	Poland, Germany, Portugal, Turkey
B. pityogeni	Massey (1974)	*P. ponderosa*	USA
B. platzeri	Giblin-Davis et al. (2006)	(Isolated from insect in rotting orange fruits [Nitidulid beetle])	USA
B. poligraphi	Fuchs (1937)	*Picea abies, P. excelsa*	Germany, Slovakia, Romania
B. populi	Tomalak and Filipiak (2010)	*Populus tremula*	Poland
B. posterovulvus	Gu et al. (2013)	Packaging wood (deciduous trees)	Singapore
B. rainulfi	Braasch and Burgermeister (2002)	*Pinus caribaea*	Malaysia, China
B. ratzeburgii	Rühm (1956)	*Betula verrucosa*	Germany, Georgia
B. rufipennis	Kanzaki et al. (2008)	*Pinus* sp. (isolated from *Dendroctonus rufipennis* emerged from pine tree)	Alaska, Canada
B. sachsi	Rühm (1956)	*Picea excelsa, P. abies*	Germany, Slovakia
B. sakishimanus	Kanzaki et al. (2015)	No plant host (isolated from a stag beetle, *Dorcus titanus sakishimanus*)	Japan
B. scolyti	Massey (1974)	*Ulmus americana*	USA
B. seani	Giblin and Kaya (1983)	No plant host (isolated from soil-dwelling bee, *Anthophora bomboides stanfordiana*)	USA

(Continued)

TABLE 6.2 (Continued)

Bursaphelenchus Species, Host Trees, and Distribution

Bursaphelenchus Species	Original Description	Host Tree	Distribution
B. sexdentati	Rühm (1960)	*Picea orientalis, Pinus sosnowskyi, P. pinaster, P. halepensis, P. pinea, P. sylvestris, P. brutia, P. nigra, P. radiata, Abies alba*	Germany, Georgia, Lithuania, Russia, Italy, Greece, Austria, Bulgaria, Spain, Cyprus, Portugal, Turkey
B. silvestris	Lieutier and Lamond (1978)	*P. sylvestris*	France
B. sinensis	Palmisano et al. (2004)	*Pinus* sp.	China
B. singaporensis	Gu et al. (2005)	Packaging wood	Singapore
B. steineri	Rühm (1956)	*Alnus glutinosa*	Germany
B. sutoricus	Devdariani (1974)	*Pinus* sp.	Georgia
B. sychnus	Rühm (1956)	*Quercus pedunculata*	Germany
B. tadamiensis	Kanzaki et al. (2012)	Isolated from *Dorcus striatipennis* (associated with dead deciduous trees)	Japan
B. talonus	Thorne (1935) and Kaisa (2003)	*Pinus contorta*	USA
B. thailandae	Braasch and Braasch-Bidasak (2002)	*Pinus merkusi*	Thailand, China
B. tiliae	Tomalak and Malewski (2014)	*Tilia cordata*	Poland
B. tokyoensis	Kanzaki et al. (2009)	*Pinus densiflora*	Japan
B. tritrunculus	Massey (1974)	*Pinus taeda*	USA
B. trypophloei	Tomalak and Filipiak (2011)	*Populus tremula*	Poland
B. tusciae	Ambrogioni and Palmisano (1998)	*Pinus pinea, P. sylvestris, P. pinaster*	Italy, Germany, Portugal
B. typographi	Kakuliya (1967)	*Picea orientalis*	Georgia
B. ulmophilus	Ryss et al. (2015)	*Ulmus glabra*	Russia
B. uncispicularis	Zhou et al. (2007)	*Pinus yunnanensis*	China
B. vallesianus	Braasch et al. (2004)	*P. sylvestris, P. nigra*	Switzerland, Turkey, Romania
B. varicauda	Thong and Webster (1983)		Canada
B. wilfordi	Massey (1964)	*Abies concolor*	USA
B. willi	Massey (1974)	*Pinus ponderosa*	USA
B. willibaldi	Schönfeld et al. (2006)	*Abies* sp.	Romania

(Continued)

TABLE 6.2 (Continued)

Bursaphelenchus Species, Host Trees, and Distribution

Bursaphelenchus Species	Original Description	Host Tree	Distribution
B. xerokarterus	Rühm (1956)	*Ulmus campestris, U. pedunculata, U. foliacea, Zelkova* sp., *Carpinus caucasica, Juglans* sp., *Populus nigra*	Germany, Georgia
B. xylophilus Syn. *B. lignicolus*	Steiner and Buhrer (1934) Mamiya and Kiyohora (1972)	*Pinus echinata, P. palustris, P. nigra, P. sylvestris, P. banksiana, P. cembra, P. clausa, P. contorta, P. densiflora, P. elliotti, P. mugo, P. ponderosa, P. radiata, P. resinosa, P. strobus, P. taeda, P. thunbergii, P. virginiana, P. rigida, P. echinata, P. estevesii, P. halepensis, P. pseudostrobus, P. bungeana, P. massoniana, P. engelmannii, P. greggii, P. leiophylla, P. luchuensis, P. muricata, P. oocarpa, P. palustris, P. parviflora, P. armandii, P. taiwanensis, P. pinaster, P. caribaea, P. kesiya, P. merkusi, Cedrus deodora, C. atlantica, Larix laricina, L. decidua, L. europaea, L. americana, Picea glauca, P. pungens, P. mariana, P. rubens, P. excelsa, P. canadensis, Abies balsamea, Pseudotsuga menziesii, P. douglasii*	USA, Canada, Mexico, Japan, China, Korea, Taiwan, Portugal, Spain
B. yongensis	Gu et al. (2006)	*Pinus massoniana, P. thunbergii*	China, Japan
B. yuyaoensis	Gu et al. (2013)	*Pinus massoniana*	China

Source: Data are gathered from Ryss et al. 2005; Kanzaki, N., In *Pine Wilt Disease*, Springer, Dordrecht, 2008; and references cited in the text and table.

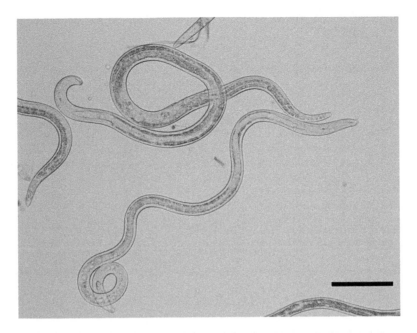

FIGURE 6.7 Adult female (upper) and male (lower) of the pinewood nematode *Bursaphelenchus xylophilus*. Bar = 0.1 mm. (Courtesy of K. Togashi.)

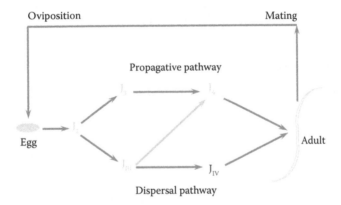

FIGURE 6.8 Life cycles of *Bursaphelenchus xylophilus*. (Courtesy of S. Akbulut.)

The J_{III}s aggregate around *Monochamus* pupal chambers (Figure 6.9) (Mamiya 1983). They then molt to the J_{IV} stage. Several compounds including linoleic acid, oleic acid, 1-monoolein, and toluene have been reported to attract J_{III}s (Maehara and Tokoro 2010; Futai 2013); however, their effectiveness remains uncertain. Zhao et al. (2007) reported that a specific ratio of three terpenes (α-pinene, β-pinene, longifolene) released by the larvae of *M. alternatus* strongly attracts J_{III}s. However, a different ratio of these terpenes, found in the xylem of *P. massoniana* Lamb., is not attractive to the dispersal juveniles of *B. xylophilus* (Zhao et al. 2007). The fungal microflora present around the beetle's pupal chamber has also been implicated in the aggregation of dispersal juveniles (Maehara and Futai 1997; Maehara and Futai 2002; Maehara 2008; Futai 2013).

The J_{IV}s, also known as the dauer juveniles (Mamiya 1984), are a nonfeeding dispersal stage adapted to survive under unfavorable conditions. Within the pupal chamber of a newly eclosed adult beetle, the J_{IV}s follow a carbon dioxide (CO_2) gradient and enter the respiratory system of the beetle, primarily through the first abdominal spiracles (Futai 2013). The J_{IV}s occupy the tracheae of the beetle and remain

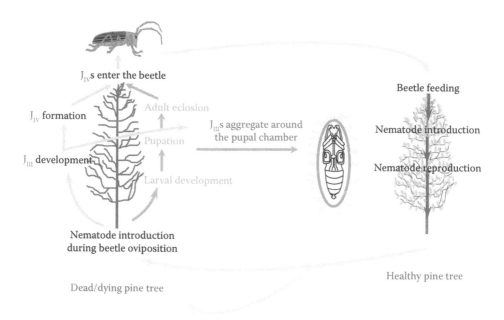

J$_{IV}$s enter the beetle

J$_{IV}$ formation

J$_{III}$ development

Adult eclosion

Pupation

Larval development

J$_{III}$s aggregate around
the pupal chamber

Beetle feeding

Nematode introduction

Nematode reproduction

Nematode introduction
during beetle oviposition

Dead/dying pine tree

Healthy pine tree

FIGURE 6.9   Interrelationships between *B. xylophilus* and its beetle vector *Monochamus* species. (Courtesy of S. Akbulut.)

there until the beetle starts feeding and oviposition. The J$_{IV}$s are transmitted to healthy trees through feeding wounds of their beetle vectors and to stressed or dying trees through the oviposition wounds (Figure 6.9). Upon entering a new host tree, J$_{IV}$s molt to the adult stage, the propagative cycle is resumed, and the within-tree population expands rapidly (Futai 2013).

6.3.2 *Bursaphelenchus xylophilus* and Pine Wilt Disease

Three to four weeks after infection of the nematodes, susceptible host trees start to exhibit wilting symptoms—such as decreased oleoresin exudation and increased emission of volatiles (Futai 2013). The pine wilt symptom develops in four stages (Malek and Appleby 1984): reduction and then cessation of oleoresin flow from fresh wounds; change in foliage coloration from green to yellowish-green with some brown needles; yellowish-brown foliage and abundance of bluestain fungi in the xylem; and all foliage becoming brown and the fungal contamination of xylem. During this process, the propagative pathway nematodes switch from phytophagy to mycophagy. Trees may die within 30–40 days of artificial infection at which time they harbor millions of nematodes. In most of Japan, naturally infected trees die within one year of nematode infection, but in cooler, northern parts of the country, some infected trees show external symptoms of death in the spring of the following year (Kobayashi 1988). In the United States, infected trees usually die three to six weeks after the infection (Malek and Appleby 1984).

Several physical and chemical traits have been implicated in the pathogenicity of the pinewood nematode, including cell wall–degrading enzymes, excreted proteins, phytotoxins–bacterial toxins, and surface coat proteins (Futai 2013). Cell wall–degrading enzymes play an important role in the interaction between *B. xylophilus* and its host, but they are also present in nonpathogenic *B. mucronatus* Mamiya and Enda (Kikuchi et al. 2006; Futai 2013), suggesting that they are not likely the sole source of pathogenicity. Some studies suggest that the nematode produces phytotoxins that cause cell death, although others propose that these toxins originate from the bacteria carried by the pinewood nematode (Oku et al. 1980; Zhao et al. 2003). The functional role of bacteria is still not clear and needs further investigation (Vicente et al. 2012; Futai 2013). The carbohydrate and protein patterns of surface coat proteins of *B. xylophilus* continue to change during the development of the nematode (Shinya et al. 2009), suggesting that these proteins are probably involved in modulating or evading the host immune-like response (Shinya et al. 2010). Futai (2013) summarized the mechanism of disease development as: "Infection of the host tree by the pinewood nematode, cell death, leakage of cell content, inflow into the tracheids, adhesion to the pit membrane to impair valve function, choke off water conduction and wilt to death," implicating that water blockage is the ultimate cause of tree mortality.

6.4 The Association between *Monochamus* Species and *B. xylophilus*

There is a remarkable synchrony in the life cycles and interactions between the pinewood nematodes and their vector beetles. This section details aspects of that association, including documented impacts of *B. xylophilus* on the fitness of *Monochamus* adults.

6.4.1 Nematode Phoresy

The phoretic relationship begins when J_{III}s of *B. xylophilus* aggregate around the pupal chamber of developing *Monochamus* vectors. The aggregated dispersal juveniles molt to nonfeeding J_{IV}s (Mamiya 1984) and then enter the tracheal system of the newly formed adult beetle (Pershing and Linit 1986b). The majority of nematodes enter through the first abdominal spiracle of the beetle (Togashi 2008), whereas a small number remain on the abdomen, appendages, and the head immediately after beetles exit dead trees (Linit 1988).

The number of nematodes carried by a beetle upon emergence from a host tree (the initial nematode load) varies greatly among different studies. For example, in *M. carolinensis*, 94% of 234 dissected adults carry J_{IV}s, ranging from 0 to 79,000 nematodes per beetle (Linit et al. 1983), and male and female adults average 42,000 and 28,000 nematodes per beetle, respectively (Akbulut 1998). In the Japanese *M. alternatus*, mean nematode loads range from 63 to 35,000, and 30.2% of 63 beetle populations examined have a mean load of more than 10,000 *B. xylophilus* (Kishi 1995). Kobayashi et al. (1984) and Kobayashi (1988) reported that 90% of the nematodes are recovered from 20% of beetles in vector populations, suggesting an aggregated dispersion across a population of beetles in which most beetles carry few or no nematodes. The highest nematode load reported so far is 289,000 for an individual *M. alternatus* (Linit 1988). High variation in nematode load among beetles likewise suggests an aggregated distribution of the nematodes among *M. carolinensis* adults (Linit 1988).

Initial nematode load is influenced by abiotic and biotic factors (Aikawa 2008) and is positively correlated with nematode density within the xylem of the host tree (Togashi 1989e). Several investigators reported that nematode load is not gender-related (Hosoda et al. 1974; Linit et al. 1983; Wingfield and Blanchette 1983) but is positively correlated with the body size of the vector beetle (Linit et al. 1983; Aikawa and Togashi 1998). However, Hosoda (1974) found no relationship between beetle body size and nematode load. Japanese scientists have found that *M. alternatus* that emerge early in the season carry more nematodes than those that emerge later in the same year (Kobayashi et al. 1984).

Environmental conditions in the wood surrounding the beetle pupal chamber can impact the initial nematode load (Mamiya 1984). Beetles that emerge from relatively dry wood bear fewer nematodes than those that emerge from moist wood. Fungi in wood are a food source that can enhance the growth of *B. xylophilus* populations and accelerate the development from the propagative to the dispersal forms. If the fungi *Verticillium* Nees and *Trichoderma* Persoon, which are unfavorable for the development of the nematode, infect host trees, development from propagative to dispersal form is very low, and consequently, the initial nematode load of *M. alternatus* is low (Maehara and Futai 1996, 1997). By contrast, the presence of the blue stain fungus *Ophiostoma minus* (Hedgc.) Syd. & P. Syd. in the xylem enhances the initial nematode load on *M. alternatus* (Maehara and Futai 1996, 1997). Warren et al. (1995) documented that *M. carolinensis* adults that emerge from pine bolts inoculated with *O. minus* carry significantly more nematodes than those from noninoculated bolts. Initial nematode load has also been reported to be positively related to the virulence of the nematode population (Aikawa et al. 2003; Aikawa 2008).

6.4.2 Exit of *B. xylophilus* from *Monochamus* Adults

The number of J_{IV}s carried by the beetle decreases with time following its emergence from the tree (Togashi and Sekizuka 1982; Linit 1988). J_{IV}s emerge from the beetle by exiting the spiracles and moving to the posterior tip of the abdomen where they travel down the setae on the terminal abdominal sclerites and drop off (Kobayashi et al. 1984; Mamiya 1984). The temporal pattern of nematode exit from an adult beetle has been well documented. The exit rate of *B. xylophilus* from *M. alternatus* is low during the first week after beetle emergence, reaches a maximum during the second and third weeks, and then declines during the rest of the vector's adult life (Enda 1972; Togashi and Sekizuka 1982; Togashi 1985; Aikawa and Togashi 1998). For example, 25% of J_{IV}s exit their beetle vectors within 10 days, 87% by 20 days, and 94% by 30 days, after beetle emergence (Enda 1972). Only 4.5% of the total nematode load exit from *M. carolinensis* during the first week following the adult emergence, 20.5% during the second week, and 13.1% during the third week (Linit 1989). The life span of the adult beetles also influences the departure of J_{IV}s (Aikawa 2008). For example, the proportion of J_{IV}s that successfully departs from beetles is higher in long-lived than short-lived *M. alternatus* (Togashi 1985).

Internal and external factors appear to mediate J_{IV} exit from beetle vectors. When *P. densiflora* Siebold & Zucc. twig sections are autoclaved, they do not emit any volatiles such as α-pinene and β-myrcene; when *M. alternatus* is provided with such twig sections, unimodal nematode transmission curves are observed (Aikawa and Togashi 1998). This suggests that internal factors of *B. xylophilus* may be involved in the exit behavior. J_{IV}s within *M. carolinensis* tracheae have greater content of neutral lipids than those that have left the beetles (Stamps and Linit 2001). The J_{IV}s attracted to β-myrcene (a pine volatile) have lower lipid content, whereas those attracted to a toluene (a beetle-associated hydrocarbon) possess higher lipid content (Stamps and Linit 2001). It is thus likely that the response of J_{IV}s to exogenous chemical cues is mediated by endogenous lipid storage, so that newly phoretic J_{IV}s that have a high lipid content are stimulated to remain in the newly emerged beetle (Stamps 1995). This behavior switches as their stored lipids decrease and the nonfeeding J_{IV}s gradually become more attracted to plant volatiles that initiate departure from the vector. Aikawa (2008) has summarized this phenomenon as follows: J_{IV}s in the tracheal system of newly emerged beetles are filled with neutral lipid, are attracted to toluene, and remain in the trachea for a while; as the quantity of neutral lipid in J_{IV}s decreases over time, they exit the tracheal system in response to this intrinsic cue; J_{IV}s with a very low lipid content also respond strongly to pine volatiles, especially β-myrcene, and their departure rate is accelerated.

6.4.3 Transmission of *B. xylophilus* from *Monochamus* Adults to Host Trees

Transmission occurs when J_{IV}s successfully exit their host beetle and enter a host tree through either feeding wounds or oviposition wounds made by *Monochamus* beetles (Linit 1988). The temporal pattern of J_{IV} transmission into host trees is correlated with the departure pattern of the nematodes from the beetle (Aikawa 2008).

Two patterns of nematode transmission through feeding wounds have been documented: L-shaped and unimodal (Kishi 1978; Togashi 1985). The unimodal transmission pattern is more common in Japan (Togashi 1985) and is discussed here. In the unimodal pattern, the highest rate of nematode transmission occurs during the second week following *M. alternatus* emergence (Shibata 1985), or between days 10 and 35 postemergence (Togashi 1985; Jikumaru and Togashi 2000). Similar results have been reported for *M. carolinensis* and *M. galloprovincialis* (Linit 1990; Naves et al. 2007b). Transmission rates through feeding wounds are affected by the initial nematode load of a beetle and by the ambient temperature (Togashi 1985; Jikumaru and Togashi 2000; Aikawa 2008). The transmission rates in both *M. carolinensis* and *M. alternatus* are correlated with the initial nematode load (Togashi 1985; Linit 1990; Jikumaru and Togashi 2000). A decrease in ambient temperature usually results in a decline in the transmission efficiency (Jikumaru and Togashi 2000).

The wounds made by female *Monochamus* beetles during oviposition also enable the pinewood nematodes to enter host trees. This is likely the pathway through which the nematode–beetle–tree association has evolved in North America (Dwinell and Nickle 1989). Transmission through oviposition wounds has received less attention than transmission through feeding wounds because the former does not lead to disease development. Transmission of nematodes through oviposition wounds may be influenced by three factors: age of female beetles, nematode load per beetle, and egg deposition in the oviposition wound (Linit 1990; Edwards and Linit 1992). Two other pathways have been reported. One is the transmission of J_{IV}s by *M. alternatus* males through oviposition wounds across which they crawl. Male nematode-infested beetles have transmitted nematodes into fresh pine bolts containing one to six oviposition wounds. J_{IV}s can also successfully colonize pine bolts when they are placed on the bark 5 or 10 cm away from a small, artificial wound (Arakawa and Togashi 2002). The other is the transmission of nematodes from nematode-infested beetles to nematode-free beetles (between sexes), which may increase the number of vectors (Arakawa and Togashi 2004; Aikawa 2008).

Within a pine stand where pine wilt is expressed, the interaction between vector beetles and *B. xylophilus* primarily is mutualistic, although sublethal impacts of heavy nematode loads have been documented (see following text). The beetles help the nematodes in dispersal by transporting them from infested to uninfested hosts and, in return, the nematodes create new substrates for beetles' oviposition and progeny development as a result of pine wilt disease. Unlike East Asia and Europe, where *B. xylophilus* is the invasive pest, in the indigenous *B. xylophilus* range of North America in the absence of pine wilt, *Monochamus* species and the pinewood nematode have a commensal relationship. The nematode is dependent upon the beetle for transport from an infested to a new, dying, or stressed host tree. Thus, the persistence of the nematode population within a tree is dependent on the presence of developing beetles on which to be phoretic. The beetle populations, in contrast, can persist without the presence of the nematode.

6.4.4 Effects of *B. xylophilus* on the Fitness Components of *Monochamus* Adults

The impact of high numbers of nematodes in the tracheae of *Monochamus* beetles has been studied by several researchers. For example, the longevity of *M. alternatus* beetles is positively correlated with their fecundity but negatively related to the initial number of J_{IV}s carried by them (Togashi and Sekizuka 1982; Togashi 1985; Kishi 1995). Adult *M. alternatus* with heavy initial loads of more than 10,000 J_{IV}s live for a shorter time and lay fewer eggs than those that carry fewer nematodes. *M. carolinensis* beetles with a high nematode load tend to have lower reproductive potential than those with a medium or low nematode load, but the differences are not significant (Akbulut and Linit 1999a).

An adult vector population could be divided into three subpopulations based on nematode load: (1) resource producers—adults that carry a high number of J_{IV}s and effectively transmit them to healthy trees creating a habitat for their progeny; (2) progeny producers—adults that carry a low number of J_{IV}s and maximize their reproductive potential, and (3) adults of intermediate characters that carry a moderate number of J_{IV}s (Togashi 1985). In Japan, 30% of *M. alternatus* field populations bear a mean load of more than 10,000 J_{IV}s per beetle (Kishi 1995), greatly facilitating pine wilt disease epidemics. In the United States, however, most newly emerged adults carry fewer than 10,000 J_{IV}s per beetle (Linit et al. 1983; Wingfield and Blanchette 1983; Malek and Appleby 1984; Akbulut and Linit 1999c). Therefore, the reproductive potential of the majority of beetles in the United States may not be negatively affected by J_{IV}s (Akbulut and Linit 1999a).

Beetle flight performance is also affected by the nematode load. For example, in a laboratory study, *M. carolinensis* adults carrying more than 10,000 J_{IV}s have significantly shorter flight distance and duration than those carrying fewer than 10,000 J_{IV}s (Akbulut and Linit 1999c). In eastern Asia, where pine wilt disease is epidemic and susceptible host trees are abundant, the nematode-infested beetles do not need to fly long distances to locate feeding and oviposition sites. Therefore, the negative impact of nematode load on the beetle flight performance may not be significant in reducing epidemics of pine wilt disease in eastern Asian forest stands but may be important in North America, where susceptible pine trees, planted largely as ornamentals, are sparsely distributed (Akbulut and Linit 1999c).

6.5 Dispersal of *M. alternatus* on the Three Spatial Scales with Reference to Pine Wilt Disease Incidence

The dispersal patterns of *M. alternatus* have been intensively studied using analytical and modeling approaches due to the severity of pine wilt disease in Japan, China, and the other adjacent areas of Asia. This section reviews aspects of vector dispersal and the pattern of pine wilt disease incidence.

6.5.1 Population Dynamics of *M. alternatus* in a Pine Stand

Monochamus alternatus has one- and two-year life cycles, and the adults emerge from dead trees in June and July in central Japan (Togashi and Magira 1981; Togashi 1989a, 1989b). In a mark–recapture study, the density of *M. alternatus* adults in flight per tree increases from early June, peaks in early July, and then remains almost constant for a month in a *B. xylophilus*-infested *P. thunbergii* Parl. stand (Togashi 1988). It begins to decrease in mid- or late August, and no adult activity is found by October. A similar pattern has also been recorded in the other *P. thunbergii* stands (Shibata 1981).

Monochamus alternatus adults exhibit a clumped distribution in a pine stand of high tree density (Shibata 1981; Togashi 1989c). Adult beetles and diseased pine trees do not overlap spatially within a pine stand in June—whereas overlaps occur in July and after (Togashi 1989c), suggesting that the beetles are not attracted to diseased trees before reproductive maturation and then are attracted to diseased trees once mature. Diseased trees show a reduced capacity of oleoresin exudation and emit monoterpenes and ethanol that are attractive to mature adults (Ikeda and Oda 1980; Ikeda et al. 1980). There is a positive correlation between mature adult density per tree during June–July and pine wilt disease incidence at quadrat size of 25 m² within a pine stand (Togashi 1989c).

In one field study (Kishi 1995), the number of emerging adult vectors increased over three years following the initiation of the disease and then decreased as the epidemics progressed. Togashi (1988, 1989d) also showed that the population of *M. alternatus* adults in flight peaks three years after the first disease incidence in a *P. thunbergii* stand. Furthermore, a significantly positive correlation is found between pine wilt disease incidence and mean beetle density per tree per day during June–August (Togashi 1988).

Yoshimura et al. (1999) have developed a deterministic model for the dynamics of the vector beetles, nematodes, and host trees, where H_t and P_t are the healthy tree density and vector density before nematode infestation (nematode transmission) at year t, respectively. Assuming a constant transmission rate for each vector, a proportion of healthy trees, $\exp(-\alpha P_t)$, escapes nematode infestation, where α is the product of the average rate at which a pine tree gets infested by a unit density of beetles per unit time and the period of maturation feeding. Consequently, the density of healthy trees in the following year H_{t+1} and density of diseased trees $\tilde{H}_t$ are

$$H_{t+1} = \exp(-\alpha P_t)H_t, \tag{6.1}$$

$$\tilde{H}_t = \left\{1 - \exp(-\alpha P_t)\right\}H_t. \tag{6.2}$$

Because vector density at year $t+1$ is determined by the densities of vector and diseased trees at year t,

$$P_{t+1} = (1-\theta)F\left(P_t, \tilde{H}_t\right)\tilde{H}_t, \tag{6.3}$$

where θ is the eradication rate of beetles and $F\left(P_t, \tilde{H}_t\right)$ is the density of adults that for the immature stages:

$$F\left(P_t, \tilde{H}_t\right) = \frac{(0.98\sigma SKP_t)}{S\{a + \tilde{H}_t\} + 0.065\sigma KP_t}, \tag{6.4}$$

where σ is the sex ratio of the beetle, K is the mean number of eggs that a female beetle can deposit maximally, S is the mean surface area of a pine tree, and $1/a$ is the efficiency of oviposition. The parameters are estimated as $\alpha = 7.7/m^2$, $\sigma = 0.48$, $K = 80$, $S = 2.4$ m², and $a = 0.022/m^2$, respectively, based on data from a pine stand on the northwest coast of Japan (Togashi and Magira 1981; Togashi 1988, 1989c).

The results using the model presented by Yoshimura et al. (1999) are suggestive. There is a minimum pine density below which the disease always fails to become established (Figure 6.10); however, at pine densities that exceed the threshold, disease establishment can fail due to the Allee effect when the density of beetles is extremely low. The Allee effect is caused by the low production of wilt-diseased trees and poor reproduction of vectors at low vector density. When the beetle density is reduced by control, the

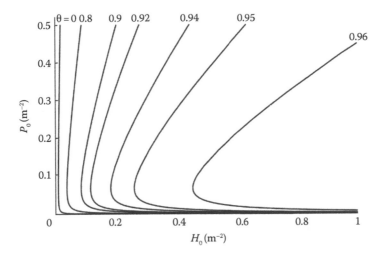

FIGURE 6.10 Boundary curves for the successful establishment of *Monochamus alternatus* in the (H_0, P_0) plane for various values of eradication rate, θ; H_0 and P_0 are the initial densities of healthy pine trees and adult beetles, respectively. On the right side of each curve, the beetle density increases the following year, while on the left side, it decreases. (After Fig. 6 from Yoshimura, A., et al., *Ecology*, 80, 1691–1702, 1999. With permission.)

minimum pine density increases disproportionately with an increase in the eradication rate. The probability that a healthy tree escapes infestation through the course of the epidemic decreases sharply with an increase in initial pine density or initial beetle density.

6.5.2 Seasonal Change in the Spatial Pattern of Pine Wilt-Diseased Trees in a Pine Stand

Seasonal changes in the number of newly diseased trees can differ among geographical areas. For example, in Chiba Prefecture, bordering the Pacific Ocean, 87% of newly diseased trees occur between June and August in a *P. thunbergii* stand (Mamiya 1976), whereas in Ishikawa Prefecture, bordering the Sea of Japan, the proportion of newly diseased trees increases from June to August and then remains constant until October (Togashi 1989d). After the invasion of a pine stand by *B. xylophilus*, the annual number of killed trees increases in an accelerated manner, reaches a peak, and then decreases in the absence of any control measures (Kishi 1995).

The spatial distribution pattern of wilt-diseased trees within a pine stand varies depending on the season and number of years after the initial nematode invasion. During the initial invasion, diseased trees occur at limited sites within a pine stand in the early half of the beetle flight season, and then the disease spreads around these affected trees in the late half of the season (Togashi 1991b). If *M. alternatus* adults carry more than 10,000 nematodes initially, they transmit an average of 15,000 nematodes per five-day interval between 15 and 25 days after emergence from dead trees (Togashi 1985). Because adults stay on a pine tree for an average of two to three days (Togashi 1990c) and 300 or more nematodes are required to establish a disease incidence (Hashimoto and Sanui 1974), reproductively immature, single beetles with a high nematode load are more likely to cause the early occurrence of diseased trees. With initial loads of between 1,000 and 9,999 nematodes, beetles transmit an average of 350 nematodes per five days between 25 and 30 days after the emergence. After reproductive maturation, these beetles tend to concentrate on diseased trees and neighboring healthy trees and probably are responsible for the subsequent occurrence of diseased trees. With initial loads of fewer than 1,000 nematodes, beetles can only transmit extremely small numbers of nematodes. One year after the diseased trees die in a pine stand, many *M. alternatus* adults may emerge from the dead trees and transmit a sufficient number of nematodes to neighboring healthy trees.

Based on empirical data on beetle emergence, survival, short-distance dispersal, fecundity, and nematode transmission rates (Togashi and Magira 1981; Togashi and Sekizuka 1982; Togashi 1985, 1986,

1988, 1990c, 1991b), Togashi (1989c) has developed a simulation model for expressing the population dynamics of *M. alternatus* and trees in a pine stand. Assuming a defined frequency distribution of initial nematode load on beetles, a small number of beetles with low dispersal ability can cause an epidemic of pine wilt in a stand, whereas those with high dispersal ability do not—even in the absence of control measures. When the epidemic occurs, diseased trees show a clumped distribution regardless of the level of susceptibility of trees to pine wilt. Among highly susceptible trees, the degree of aggregation increases over the season. However, it decreases over the season among trees with some resistance. The simulation also shows that the time taken to decimate pine stands without control extends from four to six years even when the initial number of trees in a stand increases from 400 to 2,500 at 100 beetles initially introduced, and it reduces from six to three years when the initial beetle number increases from 25 to 1,000 at a pine stand of 900 trees.

6.5.3 Range Expansion of the Pinewood Nematode with Reference to the Long-Distance Dispersal of *M. alternatus*

Long-distance dispersal of *B. xylophilus*-infested beetles accelerates the spread of pine wilt. Local spread of the disease from infested pine stands to surrounding, uninfested stands is probably caused by long-distance dispersal of infested beetles. The nematode range expansion across pine stands has been determined in several areas in Japan and China (Table 6.3) through mapping the front of pine wilt disease incidence over a course of nine years. Results show that the disease expansion speed is between 2 and 15 km per year with an average of 6 km per year.

Mathematical models incorporating the reproduction and long-distance dispersal of vector beetles within pine stands estimate that the mean disease expansion speed is several kilometers per year (Takasu et al. 2000; Yamamoto et al. 2000). Let $H_t(x)$, $P_t(x)$, and $P'_t(x)$ be the densities of healthy trees, predispersal beetles, and postdispersal beetles at year t at site x in a one-dimensional space, respectively. When $f(|x-y|)$ is the possibility that a beetle that emerges at site y reaches site x, the relation between $P'_t(x)$ and $P_t(x)$ can be expressed as follows:

$$P'_t(x) = \int_{-\infty}^{\infty} f(|x-y|)P_t(y)\,dy = \int_{-\infty}^{\infty} f(z)P_t(y)\,dy, \tag{6.5}$$

where travel distance, z, is used as $z = |x-y|$ for simplicity. By substituting P_t in Equations 6.1 through 6.4 with $P'_t(x)$ in Equation 6.5, the disease incidence and beetle reproduction at site x can be described after dispersal. Some beetles disperse over a long distance and others over a short distance. Thus,

$$f(z) = (1-\sigma_L)f_s(z) + \sigma_L f_L(z), \tag{6.6}$$

TABLE 6.3

Spread Rate of *Bursaphelenchus xylophilus* in Japan and China

Speed (km/year)	Study Period (years)	Location, Country	References
3–15	4	Chiba, Japan	Matsubara 1976
4–5	5	Shizuoka (east), Japan	Fujishita 1978
9–10	5	Shizuoka (west), Japan	Fujishita 1978
4	3	Aichi, Japan	Kato and Okudaira 1977
2–3	1	Fukuoka, Japan	Hagiwara et al. 1975
4.2	9	Ibaraki, Japan	Takasu et al. 2000
7.5	6	Nanjing, China	Robinet et al. 2009

Source: Modified from Togashi, K. 2008. In *Pine wilt disease,* Springer, Tokyo, 2008 with added data.

where $\sigma_L, f_S,$ and f_L are the proportion of long-distance dispersers, the distribution of traveling distance of short-distance dispersers, and the distribution of traveling distance of long-distance dispersers, respectively. The following functions are used as f_S and f_L:

$$f_s(z) = \sigma_s \frac{u}{2\upsilon\Gamma(1/u)} \exp\left\{-\left(\frac{z}{\upsilon}\right)^u\right\} + (1-\sigma_s)\delta(x), \qquad (6.7)$$

where σ_S, Γ, and δ represent the proportion of short-distance dispersers, gamma function, and delta function, respectively. The parameter values are estimated as $u = 2.554$ and $\upsilon = 35.69$ (m) using data obtained by Shibata (1986). For long-distance dispersal,

$$f_L(z) = \frac{\mu}{2} exp(-\mu z), \qquad (6.8)$$

where the μ value is estimated to be 5.5×10^{-4}/m from the data of Fujioka (1992), resulting in a mean dispersal distance of 1,820 m.

Takasu et al. (2000) simulate the effect of short-distance dispersal on the range expansion of *B. xylophilus* using Equations 6.1 through 6.7. They indicate that, when the initial tree density (H_0) is low, no range expansion occurs because the beetle cannot establish itself and becomes extinct. As H_0 increases beyond a threshold, the beetle can establish itself and the rate of range expansion is transformed into a traveling wave. The range expansion of *B. xylophilus* is sensitive to changes in initial tree density around the threshold value. After an initial sharp rise, this rate gradually reaches a point where it cannot increase further because the beetle's reproduction has an upper limit due to the effect of density, as included in Equation 6.4. At an equal, initial tree density, the range expansion decreases with an increasing eradication rate and reaches zero at a certain eradication rate above which the beetle cannot establish itself. In every case, the range expansion rate is, at most, 50 m/year.

With long-distance dispersal, the range expansion rate can be several kilometers per year (Takasu et al. 2000; Figure 6.11). The expansion rate increases from zero with the increasing proportion of long-distance dispersers and suddenly drops to zero beyond a certain proportion where the beetle population density is too low to establish (Figure 6.11a). The threshold proportion decreases with increasing eradication rate (Figure 6.11a) and with decreasing initial tree density (Figure 6.11b). A mechanistic individual-based model also shows the Allee effect (Takasu 2009).

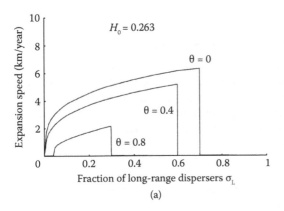 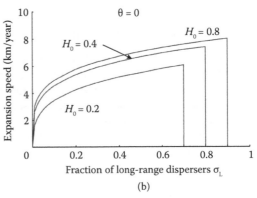

(a) (b)

FIGURE 6.11 Dependence of the range expansion rate of *Bursaphelenchus xylophilus* on the fraction of long-distance dispersers σ_L. (a) The initial pine density is fixed as $H_0 = 0.263$/m², while θ is varied as indicated; (b) the eradication rate is fixed as $\theta = 0$, while H_0 is varied as indicated. (After Fig. 7 from Takasu, F., et al., *Biol. Invasions*, 2, 141–150, 2000. With permission.)

6.5.4 Description of Regional Dispersal of *M. alternatus* and *B. xylophilus* Using an Analytic Approach

The long-range dispersal of *B. xylophilus* from infested areas to distant noninfested areas is thought to be the result of human-mediated, long-distance transportation of infested pine logs containing *B. xylophilus* and its immature vector beetles. This has been observed in Japan and South Korea (Togashi et al. 2004). For example, dispersal of *B. xylophilus* over a distance of more than 200 km occurred four times between 1925 and 1982 in Japan (Togashi et al. 2004).

Japan is divided into 47 prefectures (administrative districts), the area of which ranges from 1,876 to 83,453 km^2 with a mean area being 8,038 km^2. Pine wilt disease has spread in western prefectures and then to northeastern prefectures based on observations at 20-year intervals (Togashi et al. 2004). Assuming that *B. xylophilus* invades each prefecture from the nearest already-invaded prefecture and that pine wilt is found in the year when the infestation (effective invasion) occurs, the time required for invasion, the distance transported, and the year of invasion can be determined for each newly invaded prefecture. The time required for *B. xylophilus* to invade a prefecture from the nearest already-invaded prefecture varies substantially, ranging from 1 to 34 years with a mean of 9.6 years. In 44% of invaded prefectures, the time required for *B. xylophilus* invasion is one to four years, suggesting that the human transportation of infested logs plays an important role in the disease spread (Togashi and Shigesada 2006). The distance between a newly invaded prefecture and the nearest already-invaded prefecture ranges from 33 to 924 km with a mean of 125.4 km.

Assuming that *B. xylophilus* is spread from an already-invaded prefecture to an uninvaded prefecture at a rate of p per year and that the spread from a newly invaded prefecture begins τ years after the invasion, a simple model can be formulated to express the cumulative number of prefectures invaded by *B. xylophilus* from 1920 to 2000 in Japan. For example, fitting the model to observations results in $p = 0.0035$ and $\tau = 2$ (Togashi et al. 2004; Togashi and Shigesada 2006).

Redistribution of *B. xylophilus* two or more years after the introduction into a new prefecture may be due to the rapid initial increase in tree mortality and the subsequent increase in local beetle density. Tree mortality from *B. xylophilus* infestation increased from 6 to 67 during the first two consecutive years in a *P. densiflora* stand of 858 healthy trees without any control measures in Ibaraki Prefecture (Kishi 1995). This suggests that the removal and transport of a substantial number of infested logs could begin a few years after *B. xylophilus* invasion of a pine stand. For example, a portion of wilt-killed trees have been harvested and hauled, together with healthy trees, to lumber mills and then distributed for use in construction (e.g., Kuniyoshi 1974). Wood chips of *B. xylophilus*-infested pine trees have also been used in the pulp mills. Therefore, the abundance of wilt-killed pine trees and the ease of distribution of infested wood have shortened the time required for a newly invaded prefecture to export the nematode to noninvaded prefectures.

6.5.5 Description of Regional Dispersal of *M. alternatus* Using DNA Markers

Molecular biology has been used to determine the dispersal routes of *M. alternatus*. The Ōu mountain range runs north–south through the Tohoku district, the northeastern part of Honshu Island in Japan. Between 1980 and 2004, pine wilt–infested forests expanded from south to north into an area located 100 km on the east and 200 km on the west sides of the long mountain range. Using five microsatellite (SSR) loci, Shoda-Kagaya (2007) showed that six *M. alternatus* frontier populations on the west side differ in genetic structures from the three populations on the east side. F_{ST} index, the standardized variance of allele frequencies between populations, reveals a measure of genetic differentiation. She indicates that the genetic differences ($F_{ST}/[1 - F_{ST}]$) and geographic distance in the six west populations are significantly positively correlated, suggesting that *M. alternatus* populations spread northward independently on each side of the mountain range, which forms a geographical barrier.

Population genetic analyses using other DNA markers also contribute to our understanding of the geographical population structures of *M. alternatus* in northeast Asia (Kawai et al. 2006). Mitochondrial DNA fragments of 565 bp, including parts of cytochrome *c* oxidase II (CO II) and tRNA[Lcu-UUR], demonstrate that there are seven haplotypes in 33 local populations (31 in Japan, 1 in China, and

1 in Taiwan). Phylogenetic analyses and haplotypic network divide the haplotypes into two clades, each representing a monophyletic group of taxa sharing a close, common ancestry. One clade consisting of one haplotype is distributed in central Japan, whereas the other clade of six haplotypes is present across northeastern Asia. Mean sequence divergence between the two clades of haplotypes is 2.18 ± 0.21%, suggesting that the two clades began to diverse from each other some 1.45 million years ago, based on an estimate of 1.5% per million years in the genus *Tetraopes* Schoenherr (Coleoptera: Cerambycidae) (Farrell 2001). This study offers two possible explanations for the spatial distribution of the genetic structure of *M. alternatus* populations in northeastern Asia: (1) two or more independent colonization events may have occurred in Japan and (2) a random loss of genetic variants may have taken place in different lineages that have derived from a polymorphic common ancestry.

Monochamus alternatus is also native to Mainland China, where it occurs primarily in the southern, eastern, and central provinces, and only a few isolated populations are present in the western provinces such as Shaanxi and Tibet (Chen et al. 1959). Pine wilt disease epidemics have increased *M. alternatus* population densities since the introduction of *B. xylophilus* to Nanjing in 1982. Hu et al. (2013) have determined DNA fragment sequences of two mitochondrial genes, CO I and CO II, for 14 *M. alternatus* populations across 14 *B. xylophilus*-infested provinces in China in 2010 and 2011. They suggest that long-distance dispersal of *M. alternatus* over hundreds of kilometers has occurred through transportation of insect-containing wood and wood-packaging material.

6.6 Management of *Monochamus* Species with Special Reference to *M. alternatus*

Based on the survival rate from eggs to adults, sex ratio, and female fecundity, *M. alternatus* has a high net reproduction rate of 10.7 (Togashi and Magira 1981; Togashi 1990a). The beetle population size is determined by the population size of the previous generation, by food resources of immature stages, and by the impact of mortality factors. The latter two factors create a contest-type of intraspecific competition so that the number of emerging adults per unit area of bark surface of dead trees increases to a maximum and then plateaus as the egg density increases (Morimoto and Iwasaki 1974; Togashi 1986). Because the epidemic level of pine wilt disease is usually positively correlated with *M. alternatus* population size (Togashi 1988; Yoshimura et al. 1999), effective control of this beetle is a key to the suppression of the nematode disease.

6.6.1 Control Measures for *Monochamus* Immatures

Various control techniques have been developed to suppress immature stages because *Monochamus* beetles spend most of their lives as immatures within dead host trees and branches. These include cutting diseased trees; followed by crushing, burning, or burying the wood in the soil; or treating the wood with insecticides (Kamata 2008). A simulation model for the integrated management of pine wilt disease shows that destruction of 80% or more within-wood immature vector stages can achieve successful control of this disease (Togashi 1989c).

The overall mortality of immature stages by any control measure in a pine stand varies depending on the successful location of diseased trees within the stand. It is often difficult to find all *B. xylophilus*-infested trees due to the closed structure of a pine forest canopy or weather conditions. However, a survey by helicopter can locate three times as many wilt-diseased trees with brown foliage as a ground survey (Togashi 1989c). In addition, branches harboring beetle immature stages may fall from the tree and be left in the forest after the dead trees are felled and treated or removed. As a result, it may be a challenge to effectively suppress populations of immature stages using the aforementioned control measures.

The role of natural enemies has been studied to evaluate their potential as biocontrol agents. The coleopteran ectoparasitoid *Dastarcus helophoroides* (Fairmaire) (Coleoptera: Bothrideridae) has been reared on an artificial diet developed by Ogura et al. (1999) and released at 40–120 adults per standing *P. densiflora* tree infested by *M. alternatus* larvae in late April and early May for three consecutive years (Urano 2006; Shimazu 2008). This has achieved 30.1–47.3% *M. alternatus* larval mortalities between early June and mid-July. In China, more than 1.6 billion adults and eggs of *D. helophoroides* have been produced and released so far (Yang et al. 2014). The mean parasitism rates are 88.6% and 92.6% in two experimental sites in Anhui and Hunan Provinces, respectively, resulting in successful control of the spread of pine wilt disease (Yang et al. 2014).

Three *Sclerodermus* (Hymenoptera: Bethylidae) species, parasitoids of *M. alternatus* larvae, have been tested for efficacy as biological control agents in China. Results show that the release of *Sclerodermus* sp., *S. guani* Xiao and Wu, and *S. sichuanensis* Xiao has achieved 38.6%, 32.6%, and 25.4% *M. alternatus* larval mortalities, respectively (Yang et al. 2014).

Steinernematidae and Heterorhabditidae (Nematoda) are entomopathogenic nematodes vectoring the bacteria *Xenorhabdus* Thomas and Poinar and *Photorhabdus* (Boemare et al.) emend. Fischer-Le Saux et al., respectively, when they enter the host insect bodies (Phan 2008). The bacteria multiply rapidly in the hemolymph within the insect hemocoel and kill the insects within 24–48 h. The nematodes feed on bacteria after insect death and transmit them to new host insects. Laboratory tests show that 4,000 infective juveniles of *Steinernema kushidai* Mamiya, *S. carpocapsae* Weiser, and *S. intermedia* Poinar can cause 83–100% mortalities of *M. alternatus* larvae placed in Petri dishes (Mamiya 1989). Injection of a water suspension (20,000 infective juveniles per 0.1 mL water) of *S. kushidai* and *S. carpocapsae* into every entrance hole of *M. alternatus* pupal chambers on pine log surfaces in late May results in an average of 15.5% and 66.1% *M. alternatus* mortalities, respectively (Mamiya 1989). Application of 600 mL of a *S. carpocapsae* (strain Mexican) suspension of 20×10^6 infective juveniles per m^2 to the upper surface of eight pine logs resulted in a larval mortality of 71.4% (Yamanaka 1993). Other application rates provided less effective control. Entomopathogenic nematodes, however, have not been used in the field control of *M. alternatus* in Japan.

Beauveria bassiana (Bals.-Criv.) Vuill. (Hypocreales: Clavicipitaceae) is an important fungal pathogen of *M. alternatus*. Shimazu (2008) has evaluated the following methods to introduce the pathogen under the bark of dead pine trees: spraying conidial suspension of *B. bassiana*, putting wheat bran pellets with *B. bassiana* conidia into holes drilled on dead trees, releasing the pine bark beetle *Cryphalus fulvus* Niisima (Coleoptera: Scolytidae) loaded with *B. bassiana* conidia, and placing nonwoven fabric strips with *B. bassiana* conidia on dead trees and logs. Results reveal that wheat bran pellets with *B. bassiana* conidia is the most effective. Interestingly, *B. bassiana* conidia attached to fabric strips caused high mortalities of *M. alternatus* adults (Shimazu 2008). In China, an attempt has been made to load *B. bassiana* and another entomopathogenic fungus *Metarhizium anisopliae* (Metchnikoff) Sorokin on the hymenopteran parasitoid *Scleroderma guani* Xiao and Wu (Xu et al. 2015).

6.6.2 Control Measures for *Monochamus* Adults

Monochamus alternatus females have a mean pre-oviposition period of about three weeks, and the transmission rate of *B. xylophilus* by *M. alternatus* adults peaks during a period between 15 and 25 days after beetle emergence (Togashi 1985). Therefore, effective application of preventive insecticidal sprays that target *M. alternatus* adults in flight can reduce the incidence of pine wilt disease (Kishi 1995) through the reduction of pinewood nematode transmission and beetle oviposition (Togashi 1980).

Reproductively mature adults of *M. alternatus* are attracted to dying, recently dead, and wilt-diseased *P. thunbergii* trees (Ikeda and Oda 1980; Ikeda et al. 1980). Alpha-pinene and ethanol are volatiles emitted from such trees and are attractants for *M. alternatus* adults (Ikeda et al. 1986). Commercially available traps containing α-pinene and ethanol effectively capture *M. alternatus* in flight; however, pine trees surrounding the traps are likely to succumb to pine wilt disease in areas infested with *B. xylophilus* (Shimazu 2008).

Adult males of Chinese *M. alternatus* emit the aggregation pheromone 2-undecyloxy-1-ethanol (Teale et al. 2011). The efficacy of the pheromone, if used alone, is low, but this increases markedly when augmented by α-pinene and ethanol emitted from pine branches triggered by beetle feeding (Teale et al. 2011). Mature males of *M. galloprovincialis* also release the aggregation pheromone 2-undecyloxy-1-ethanol (Pajares et al. 2010). The number of *M. galloprovincialis* adults captured in traps containing the aggregation pheromone is greatly enhanced by the addition of kairomones such as α-pinene and pheromone components of *Ips* bark beetles, ipsenol, ipsdienol, and 2-methyl-3-buten-2-ol (Pajares et al. 2010).

Although effective in trapping beetles, the use of the traps may increase the incidence of pine wilt disease in trees in the trapping area.

Oviposition deterrence has potential as a control measure but has not yet been evaluated. Adult females of *M. alternatus* use a jelly-like secretion to seal the oviposition scars immediately after the end of egg deposition (Anbutsu and Togashi 2001). Females are deterred from using preexistent oviposition scars following the maxillary and labial palpation of sites sealed with secretion. In contrast, females utilize the secretion-free scars for oviposition. Methanol extracts of the spermathecal gland or other reproductive organs and methanol extracts from larval frass can also deter females from ovipositing (Anbutsu and Togashi 2001, 2002). The oviposition deterrent in larval frass has been identified as a mixture of monoterpenes and butylated hydroxytoluene (Li and Zhang 2006).

Members of the genus *Wolbachia* Hertig and Wolbach are endosymbiotic bacteria (Rickettsiales: Rickettsiaceae) and infect more than 60% of insect species (Hilgenboecker et al. 2008). They are vertically transmitted via the host maternal germline and manipulate host reproduction through cytoplasmic incompatibility, parthenogenesis, male killing, or feminization (Bourtzis and Miller 2003). Reduced vector adult life span caused by *Wolbachia* infection may be exploited in insect pest management practices (e.g., McMeniman et al. 2009). In Japan, the *Wolbachia ftsZ* gene is present in a *M. alternatus* population from Honshu Island but absent in a population from Miyako Island (Aikawa et al. 2009). An extensive survey shows that *Wolbachia* genes are carried by 11 *M. alternatus* populations across Japan and one Taiwanese *M. alternatus* population (Aikawa et al. 2014). Intriguingly, two populations of the congeneric species *M. saltuarius* do not carry any of 214 *Wolbachia* genes examined (Aikawa et al. 2014). Further studies on interactions between *M. alternatus* and *Wolbachia* might produce a novel technique for controlling the insect pest population.

6.6.3 Suppression of the Food Resources of *Monochamus* Immatures

Pine trees succumbing to pine wilt disease become an oviposition substrate for many *Monochamus* species. Various nematicides can be injected into healthy pine trees in winter as a prophylactic treatment to protect the trees during the infective season (e.g., Matsuura 1984). Nematicide-injected trees have extremely high survival rates for four to seven years. The treatments, however, are expensive and are used only for the protection of high-value trees. It is not a cost-effective treatment for the protection of forest stands.

Induced resistance against virulent isolates of *B. xylophilus* occurs in susceptible species of *P. densiflora* and *P. thunbergii* through inoculation with avirulent isolates (Kosaka et al. 2001). This strategy, however, has not been used in the management of pine wilt disease.

The susceptibility to pine wilt disease can be highly variable among individual trees of susceptible pine species (Toda and Kurinobu 2002). In Japan, a national project to select resistant individuals of *P. densiflora* and *P. thunbergii* was initiated in 1978 in western Japan. The project has identified 108 resistant trees from 26,066 candidate trees that have survived the pine wilt epidemics (Toda and Kurinobu 2002). Open pollinated seedlings of resistant *P. thunbergii* and *P. densiflora* trees have been planted for reforestation efforts following destruction of existing stands by pine wilt. Unfortunately, pine wilt now occurs in the forests selected for disease resistance. Tree resistance, however, has proven more successful in China. For example, three resistant provenances of *P massoniana* were selected from 40 provenance candidates by inoculation of *B. xylophilus*, and then one of the resistant provenances was planted over an area of 1,340 ha in south Nanjing in 1984 (Xu 2008). This forest retains the resistance 24 years later. In addition, because *Pinus* species differ in susceptibility to pine wilt disease (Futai and Furuno 1979), cross-breeding native susceptible *Pinus* species with exotic resistant species is underway in Japan (Nose and Shiraishi 2008).

6.6.4 Integrated Management of *M. alternatus*

Togashi (1989c) has developed a simulation model to evaluate various pine wilt control strategies. Simulated aerial application of insecticides to suppress adults is more effective than control of immature life stages within the trees. The model predicts that pine wilt disease is under complete control within four years after two preventive insecticide applications against *M. alternatus* adults or one preventive insecticide application and one measure that causes more than 70% of mortality rate of immature insects. However, it will be difficult to manage pine wilt disease successfully through the destruction of immatures alone according to the model. For example, a simulated 80% annual mortality of immature stages will require 14–20 years for complete control to occur. Reduction in the susceptibility level or an increase in the resistance level of trees can enhance control efficacy of various strategies as long as the virulence of *B. xylophilus* does not increase.

It is well documented that *B. xylophilus* can be transmitted among *M. alternatus* adults (both male to female and female to male) (Togashi and Arakawa 2003, Arakawa and Togashi 2004). The male-to-female transmission of *B. xylophilus* has been found in *M. carolinensis* (Edwards and Linit 1992). In addition, the two-way horizontal transmission of *B. mucronatus* is recorded in *M. saltuarius* adults (Togashi and Jikumaru 1996). Horizontally transmitted *B. xylophilus* is known to successfully invade small pine branch bolts through oviposition scars (Arakawa and Togashi 2002; Togashi and Arakawa 2003). Lee and Lashari (2014) have developed a mathematical model to evaluate the effectiveness of various control measures including nematicide injection, removal of infested trees, preventive insecticide spraying, and restraint of beetle mating. The model utilizes five differential equations that express the dynamics of tree and vector beetle populations, with horizontal nematode transmission incorporated in the model. Pontryagin's maximum principle indicates that preventive insecticide spraying and control of beetle mating (= horizontal nematode transmission) are highly effective in reducing the incidence of infested trees and vector beetles.

In recent years, an integrated management program has achieved successful control of pine wilt disease at experimental sites in Anhui and Guizhou Provinces, China (Yang et al. 2014). This program combines measures including the release of the coleopteran parasitoid *D. helophoroides*, trapping of adult beetles using light traps with a wavelength below 365 nm, and removal of infested trees over a two-year period.

There are some examples of successful management of pine wilt disease in Japan (Kamata 2008). *Monochamus alternatus* and *B. xylophilus* have been eradicated by aerial application of insecticides and destruction of immature beetles in Okinoerabu Island, Kagoshima Prefecture. Also, pine stands in Wakayama and Saga Prefectures have been protected through aggressive elimination of *M. alternatus* populations surrounding the target stand, eradication of immature vectors within trees in the target stands, and aerial spraying of insecticides for control of adult vectors.

6.7 Concluding Remarks: *Monochamus* Species with Reference to Pine Wilt Disease in the Future

Cerambycidae have received increasing attention from researchers and forest managers over the past 30–40 years due to (1) the increasing economic importance of cerambycids in forests, crops, log yards, and urban locations; (2) their role as vectors of the pinewood nematode; (3) a serious increase in the number of interceptions and introduction of exotic cerambycids, particularly in North America and Europe; and (4) the role of cerambycids in forest ecosystems (biodiversity, disturbance, and degradation) (Allison et al. 2004). Several *Monochamus* species transmit the pinewood nematode (Mamiya 1983), and other species are implicated in the transmission of fungal tree pathogens including Dutch elm disease, chestnut blight, dieback of balsam fir, oak wilt, and hypoxylon canker (Donley, 1959; Linsley 1961; Nord and Night 1972; Ostry and Anderson 1995; Alisson et al. 2004).

Monochamus species have long been known for their destructive larval galleries made in coniferous trees killed by natural forces such as wind, fire, and outbreak of herbivorous insects in North America (Richmond and Lejeune 1945; Rose 1957; Ross 1960). Since *B. xylophilus* invaded Asia and Europe and epidemics of pine wilt disease occurred, many workers in North America, East Asia, and Europe have studied the ecology of *Monochamus* species as the vectors of this important nematode. For example, investigations into *M. alternatus* population dynamics in the field have enhanced our understanding of the spatiotemporal dynamics of adult beetles (Shibata 1981, 1986; Togashi 1988, 1989c, 1990c, 1991b) and immature stages (Togashi 1990a). Based on these studies, mathematical and simulation models have been constructed for description of pine wilt disease systems, forecast of disease spread, and development of an integrated disease management system (Togashi 1989c; Yoshimura et al. 1999; Takasu et al. 2000; Yamamoto et al. 2000; Togashi and Shigesada 2006; Takasu 2009; Robinet et al. 2009, 2011; Økland et al. 2010; Park et al. 2013; Lee and Lashari 2014). In Europe, studies on the performance of *M. galloprovincialis* on different pine species (Akbulut et al. 2008) and the flight distance of its adults (Mas et al. 2013; David et al. 2014; Torres-Villa et al. 2015) may help to predict the spread of pine wilt disease in the future. Our knowledge on chemical ecology of *Monochamus* species has grown rapidly (e.g., Ikeda et al. 1980; Pajeras et al. 2010; Teale et al. 2011; Alisson et al. 2012; Fierke et al. 2012; Macias-Samano et al. 2012; Pajares et al. 2013; Ryall et al. 2015). In addition, molecular biology techniques have been utilized to determine the dispersion routes of *Monochamus* species on a large spatial scale (Shoda-Kagaya 2007; Hu et al. 2013).

The interspecific association of *Monochamus* species and *B. xylophilus* has been studied in Japan and North America (Togashi and Sekizuka 1982; Togashi 1986; Akbulut and Linit 1999a, 1999c), demonstrating a negative correlation between *B. xylophilus* load and *M. alternatus* adult life span. Virulent isolates of *B. xylophilus* have a higher transfer rate from trees to vectors compared to avirulent isolates (Aikawa et al. 2003). To better understand the dynamics of pine wilt disease within a pine stand, it is necessary to analyze the temporal change in virulence in the course of an epidemic in a pine stand (Togashi and Jikumaru 2007).

Since the introduction of *B. xylophilus* into Japan in the early 1900s (Mamiya 1983), foresters there have attempted to screen and select resistant trees from susceptible indigenous *P. densiflora* and *P. thunbergii* and have planted open-pollinated seedlings for reforestation following the epidemics of pine wilt disease (Toda and Kurinobu 2002). However, outcomes from these efforts have been disappointing because the mortality rate of the seedlings caused by the nematode infestation still ranges from 17% to 63%. Future management of pine wilt disease has to take account of the arms race between pine trees and *B. xylophilus* and its impact on a vector beetle population.

In Europe, the range of *B. xylophilus* has expanded from Portugal to Spain (Robertson et al. 2011) and may spread as far north as Scandinavia in the future (Robinet et al. 2011). Pine wilt disease, however, is not expressed under cool summer climatic conditions (Rutherford et al. 1990). Although inoculation of *B. xylophilus* does not cause susceptible *P. sylvestris* trees to succumb to pine wilt disease in Vermont, the within-tree nematode populations may persist for more than 10 years in this region of the United States (Halik and Bergdahl 1994; Bergdahl and Halik 1999). A simulation model suggests that it is difficult to eradicate *B. xylophilus* in Scandinavia through removal of all logging residues where the pathogenic nematode is detected (Økland et al. 2010; Bergseng et al. 2012). Thus, global warming may trigger outbreaks of *Monochamus* beetles and cause future pine wilt disease in areas that are currently cooler (Robinet et al. 2011).

Extensive plantations of pine species have been established in Southern Hemisphere countries such as South Africa, New Zealand, Australia, Chile, and Brazil since the early twentieth century (Le Maitre 1998). So far, *Monochamus* species and *B. xylophilus* have not occurred in these pine plantations, which include susceptible species such as *P. radiata* and *P. pinaster* (e.g., Estay et al. 2014). Because *Monochamus* larvae and *B. xylophilus* can survive in wood during transportation among continents (e.g., Tomminen 1991; Gu et al. 2006; Sousa et al. 2011), and climate in these southern countries permits *Monochamus* species to persist (Estay et al. 2014), strict biosecurity measures should be in place to prevent invasion of the vectors and nematodes.

To date, research on *Monochamus* species has been conducted primarily to control pine wilt disease. The causative agent of the disease, *B. xylophilus*, is an invasive pest in Asia and Europe, whereas the vectors are native there. This is interesting from the view of biological invasion because the pathogenic nematode established itself but American *Monochamus* species did not—even though both organisms are believed to have arrived in Japan at the same time. This is also the case in Europe. What mechanism works to exclude the original vector species in the newly invaded areas? It is significant to clarify the mechanism involved in the Allee effect. A mathematical model of Yoshimura et al. (1999) points out the lack of disease incidence because of low host tree density and low vector beetle density (Allee effect). In Japan, the absence of disease incidence is observed in some pine stands even when vector adults are captured. Thus, it is necessary to determine the mechanism of the Allee effect and the Allee threshold of beetle density, below which the basic reproduction rate of infected trees is lower than unity, with relation to the host tree density and other factors.

There are many vector candidates of *B. xylophilus* among *Monochamus* species. The developmental time and the adult emergence season vary depending on *Monochamus* species and climate. The epidemic or incidence pattern of pine wilt disease, including the spatiotemporal distribution of diseased trees, is expected to change according to the life cycle of vector populations. However, there has been insufficient information about a life cycle regulating mechanism for vector candidates and the distribution of nematode load in relation to the time of adult emergence. It is important to characterize the epidemic pattern of pine wilt according to climate and vector life cycles. In addition, there is a large variation in the virulence against host trees among the pathogenic nematodes. The short life cycle of the pine wood nematode suggests that the rapid evolution of virulence has occurred and one of the virulence determinants is transmission efficiency (e.g., Anderson and May 1992; Dieckman et al. 2002). Thus, the epidemiological and ecological study of pine wilt may help reduce the mean virulence.

ACKNOWLEDGMENTS

Katsumi Togashi was supported in part by JSPS KAKENHI grants (nos. 22380081 and 26292080) when writing this chapter.

REFERENCES

Aikawa, T. 2008. Transmission biology of *Bursaphelenchus xylophilus* in relation to its insect vector. In *Pine wilt disease*, eds. B. G. Zhao, K. Futai, J. R. Sutherland, and Y. Takeuchi, 123–138. Tokyo: Springer.

Aikawa, T., H. Anbutsu, N. Nikoh, T. Kikuchi, F. Shibata, and T. Fukatsu. 2009. Longicorn beetle that vectors pinewood nematode carries many *Wolbachia* genes on an auto some. *Proceedings of the Royal Society B* 276: 3791–3798.

Aikawa, T., N. Nikoh, H. Anbutsu, and K. Togashi. 2014. Prevalence of laterally transferred *Wolbachia* genes in Japanese pine sawyer, *Monochamus alternatus* (Coleoptera: Cerambycidae). *Applied Entomology and Zoology* 49: 337–346.

Aikawa, T., and K. Togashi. 1998. An effect of pine volatiles on departure of *Bursaphelenchus xylophilus* (Nematoda: Aphelenchoididae) from *Monochamus alternatus* (Coleoptera: Cerambycidae). *Applied Entomology and Zoology* 33: 231–237.

Aikawa, T., K. Togashi, and H. Kosaka. 2003. Different developmental responses of virulent and avirulent isolates of the pinewood nematode, *Bursaphelenchus xylophilus* (Nematoda: Aphelenchoididae), to the insect vector, *Monochamus alternatus* (Coleoptera: Cerambycidae). *Environmental Entomology* 32: 96–102.

Akbulut, S. 1998. Effect of *Bursaphelenchus xylophilus* (Nematoda: Aphelenchoididae) Fourth stage dispersal juveniles and log seasonality on life processes of *Monochamus carolinensis* (Coleoptera: Cerambycidae). PhD Dissertation, Department of Entomology, University of Missouri, Columbia, MO, USA.

Akbulut, S. 2009. Comparison of the reproductive potential of *Monochamus galloprovincialis* Olivier (Coleoptera: Cerambycidae) on two pine species under laboratory conditions. *Phytoparasitica* 37: 125–135.

Akbulut, S., A. Keten, and W. T. Stamps. 2008. Population dynamics of *Monochamus galloprovincialis* Olivier (Coleoptera: Cerambycidae) in two pine species under laboratory conditions. *Journal of Pest Science* 81: 115–121.

Akbulut, S., and M. J. Linit. 1999a. Reproductive potential of *Monochamus carolinensis* (Coleoptera: Cerambycidae) with respect to nematode phoresis. *Environmental Entomology* 28: 407–411.

Akbulut, S., and M. J. Linit. 1999b. Seasonal effect on reproductive performance of *Monochamus carolinensis* (Coleoptera: Cerambycidae) reared in pine logs. *Journal of Economic Entomology* 92: 631–637.

Akbulut, S., and M. J. Linit. 1999c. Flight performance of *Monochamus carolinensis* (Coleoptera: Cerambycidae) with respect to nematode phoresis and beetle characteristics. *Environmental Entomology* 28: 1014–1020.

Akbulut, S., and W. T. Stamps. 2012. Insect vectors of the pinewood nematode: A review of the biology and ecology of *Monochamus* species. *Forest Pathology* 42: 89–99.

Akbulut, S., W. T. Stamps, and M. J. Linit. 2004. Population dynamics of *Monochamus carolinensis* (Coleoptera: Cerambycidae) under laboratory conditions. *Journal of Applied Entomology* 128: 17–21.

Allison, J. D., J. H. Borden, and S. J. Seybold. 2004. A review of the chemical ecology of the Cerambycidae (Coleoptera). *Chemoecology* 14: 123–150.

Allison, J. D., J. L. McKenney, J. G. Millar, J. S. McElfresh, R. F. Mitchell and L. M. Hanks. 2012. Response of the woodborers Monochamus carolinensis and Monochamus titillator (Coleoptera: Cerambycidae) to known cerambycid pheromones in the presence and absence of the host volatile α-pinene. *Environmental Entomology* 41: 1587–1596.

Allison, J. D., W. D. Morewood, J. H. Borden, K. E. Hein and I. M. Wilson. 2003. Differential bio-activity of *Ips* and *Dendroctonus* (Coleoptera: Scolytidae) pheromone components for *Monochamus clamator* and *M. scutellatus* (Coleoptera: Cerambycidae). *Environmental Entomology* 32: 23–30.

Alya, A. B., and F. P. Hain. 1985. Life histories of *Monochamus carolinensis* and *Monochamus titillator* (Coleoptera: Cerambycidae) in the Piedmont of North Carolina. *Journal of Entomological Science* 20: 390–397.

Anbutsu, H., and K. Togashi. 1997. Effects of spatio-temporal intervals between newly-hatched larvae on larval survival and development in *Monochamus alternatus* (Coleoptera: Cerambycidae). *Researches on Population Ecology* 39: 181–189.

Anbutsu, H., and K. Togashi. 2000. Deterred oviposition response of *Monochamus alternatus* (Coleoptera: Cerambycidae) to oviposition scars occupied by eggs. *Agricultural and Forest Entomology* 2: 217–223.

Anbutsu, H., and K. Togashi. 2001. Oviposition deterrent by female reproductive gland secretion in Japanese pine sawyer, *Monochamus alternatus*. *Journal of Chemical Ecology* 27: 1151–1161.

Anbutsu, H., and K. Togashi. 2002. Oviposition deterrence associated with larval frass of the Japanese pine sawyer, *Monochamus alternatus* (Coleoptera: Cerambycidae). *Journal of Insect Physiology* 48: 459–465.

Anderson, R. M., and R. M. May. 1992. *Infectious diseases of humans: Dynamics and control.* Oxford: Oxford University Press.

Arakawa, Y., and K. Togashi. 2002. Newly discovered transmission pathway of *Bursaphelenchus xylophilus* from males of the beetle *Monochamus alternatus* to *Pinus densiflora* trees via oviposition wounds. *Journal of Nematology* 34: 396–404.

Arakawa, Y., and K. Togashi. 2004. Presence of the pinewood nematode, *Bursaphelenchus xylophilus*, in the spermatheca of female *Monochamus alternatus*. *Nematology* 6: 157–159.

Bergdahl, D. R., and S. Halik. 1999. Inoculated *Pinus sylvestris* serve as long-term hosts for *Bursaphelenchus xylophilus*. In *Sustainability of pine forests in relation to pine wilt and decline*, eds. K. Futai, K. Togashi, T. Ikeda, 73–78. Tokyo: Shokado.

Bergseng E., B. Økland, T. Gobakken, C. Magnusson, T. Rafoss, and B. Solberg. 2012. Combining ecological and economic modelling in analysing a pest invasion contingency plan – The case of pine wood nematode in Norway. *Scandinavian Journal of Forest Research* 27: 337–349.

Braasch, H., W. Burgermeister, U. Schönfeld, K. Metge, and M. Brandstetter. 2004. *Bursaphelenchus vallesianus* sp.n.—A new species of the *Bursaphelenchus sexdentati* group (Nematoda: Parasitaphelenchoididae). *Nematologia Mediterranea* 32: 71–79.

Bourtzis, K., and T. A. Miller. 2003. *Insect symbiosis*. Boca Raton, FL: CRC Press.

Butovitsch, V. 1939. Zur kenntnis der paarung, eiablage und ernahrung der Cerambyciden. *Entomologisk Tidskrift* 60: 206–258.

Chen, S., Y. Xie, and G. Deng. 1959. *Economic Insect Fauna of China. Fasc. 1. Coleoptera: Cerambycidae I.* Beijing: Science Press [in Chinese].

Danks, H. V. 1987. *Insect dormancy: An ecological perspective.* Ottawa: Biological Survey of Canada (Terrestrial Arthropods).

Danks, H. V. 2006. Key themes in the study of seasonal adaptations in insects, Part II. Life-cycle patterns. *Applied Entomology and Zoology* 41: 1–13.

Danks, H. V. 2007. The elements of seasonal adaptations in insects. *The Canadian Entomologist* 139: 1–44.

David, G., J. Hervé, D. Piou, P. Naves, and E. Sousa. 2013. Flight performances of *Monochamus galloprovincialis*, insect vector of the pinewood nematode. In *Pine wilt disease conference 2013*, ed. T. Schröder, p. 20. Braunschweig, Germany.

David, G., B. Giffard, D. Piou, and H. Jactel. 2014. Dispersal capacity of *Monochamus galloprovincialis*, the European vector of the pinewood nematode, on flight mills. *Journal of Applied Entomology* 138: 566–576.

De Groot, P., and R. Nott. 2004. Response of the whitespotted sawyer beetle, *Monochamus scutellatus*, and associated woodborers to pheromones of some *Ips* and *Dendroctonus* bark beetles. *Journal of Applied Entomology* 128: 483–487.

Dieckman, U., J. A. J. Metz, M. W. Sabelis, and K. Sigmund (ed.). 2002. *Adaptive dynamics of infection disease: In pursuit of virulence management*. Cambridge: Cambridge University Press.

Donley, D. E. 1959. Studies of wood boring insects as vectors of the oak wilt fungus. PhD Dissertation. Ohio State University, Columbus, OH, USA.

Dwinell, L. D., and W. R. Nickle. 1989. An Overview of the pinewood nematode ban in North America. General Technical Report SE-55. North American Forestry Commission Publication No: 2. USDA, Forest Service, Southeastern Forest Experiment Station. 13 pp.

Edwards, O. R., and M. J. Linit. 1991. Oviposition behavior of *Monochamus carolinensis* (Coleoptera: Cerambycidae) infested with the pinewood nematode. *Annals of Entomological Society of America* 84: 319–323.

Edwards, O. R., and M. J. Linit. 1992. Transmission of *B. xylophilus* through oviposition wounds of *M. carolinensis*. *Journal of Nematology* 24: 133–139.

Enda, N., 1972. Removing dauerlarvae of *Bursaphelenchus lignicolus* from the body of *Monochamus alternatus*. *Transactions of Annual Meeting of Kanto Branch Japanese Forestry Society* 24: 32.

Enda, N., and H. Kitajima. 1990. Rearing of adults and larvae of the Taiwanese pine sawyer (*Monochamus alternatus* Hope, Coleoptera, Cerambycidae) on artificial diets. *Transactions of the 101st Meeting of Japanese Forestry Society*, pp. 503–504 [in Japanese].

Estay, S. A., F. A. Labra, R. D. Sepulveda, and L. D. Bacigalupe. 2014. Evaluating habitat suitability for the establishment of *Monochamus* spp. through climate-based niche modeling. *PLoS One* 9(7): e102592. doi:10.1371/journal.pone.0102592

Farrell, B. D. 2001. Evolutionary assembly of the milkweed fauna: Cytochrome oxidase I and the age of *Tetraopes* beetles. *Molecular Phylogenetics and Evolution*. 18: 467–478.

Fauziah, B. A., T. Hidaka, and K. Tabata. 1987. The reproductive behavior of *Monochamus alternatus* Hope (Coleoptera: Cerambycidae). *Applied Entomology and Zoology* 22: 272–285.

Fierke, M. K., D. D. Skabeikis, J. G. Millar, et al. 2012. Identification of a male-produced aggregation pheromone for *Monochamus scutellatus scutellatus* and an attractant for the congener *Monochamus notatus* (Coleoptera: Cerambycidae). *Journal of Economic Entomology* 105: 2029–2034.

Francardi, V., J., de Silva, F. Pennacchio, and P. F. Roversi. 2009. Pine volatiles and terpenoid compounds attractive to European xylophagous species, vectors of *Bursaphelenchus spp.* nematodes. *Phytoparasitica*. 37: 295–302.

Francardi, V., and F. Pennacchio. 1996. Note sulla bioecologia di *Monochamus galloprovincialis galloprovincialis* (Olivier) in Toscana e in Liguria (Coleoptera: Cerambycidae). *Redia* 79: 153–169.

Fujioka, H. 1992. A report on the habitat of *Monochamus alternatus* Hope in Akita prefecture. *Bulletin of the Akita Prefecture Forest Technical Center* 2: 40–56 [in Japanese with English summary].

Fujishita, A. 1978. Infestation of pine wilt disease in Shizuoka Prefecture. *Transactions of the 26th Annual Meeting of Chubu Branch Japanese Forestry Society*, pp. 193–198 [in Japanese].

Futai, K. 2013. Pine Wood Nematode, *Bursaphelenchus xylophilus*. *Annual Review of Phytopathology* 51: 61–83.

Futai, K., and T. Furuno. 1979. The variety of resistances among pine species to pine wood nematode, *Bursaphelenchus lignicolus*. *Bulletin of Kyoto University Forests* 51: 23–36.

Gu, J., H. Braasch, W. Burgermeister, and J. Zhang. 2006. Records of *Bursaphelenchus* spp. intercepted in imported packaging wood at Ningbo, China. *Forest Pathology* 36: 323–333.

Hagiwara, Y., S. Ogawa, and H. Takeshita. 1975. Range expansion of pine wilt disease. *Transactions of the 28th Annual Meeting of Kyushu Branch Japanese Forestry Society*, pp. 153–154 [in Japanese].

Halik, S., and D. R. Bergdahl. 1994. Long-term survival of *Bursaphelenchus xylophilus* in living *Pinus sylvestris*. *European Journal of Forest Pathology* 24: 357–363.

Hashimoto, H., and T. Sanui. 1974. Influence of inoculum quantity of *Bursaphelenchus xylophilus* on wilting disease development in *Pinus thunbergii* trees. *Transactions of the 85th Annual Meeting of Japanese Forestry Society*, pp. 251–253 [in Japanese].

Hilgenboecker, K., P. Hammerstein, P. Schlattmann, A. Telschow, and J. H. Werren. 2008. How many species are infected with *Wolbachia*? A statistical analysis of current data. FEMS *Microbiology Letters* 281: 215–220.

Hosoda, R. 1974. Relationship between the body size of adult *Monochamus alternatus* and the number of pine wood nematodes carried by the beetle. *Transactions of the Annual Meeting of Kansai Branch Japanese Forestry Society*. 25: 306–309 [in Japanese].

Hosoda, R., M. Okuda, A. Taketani, and K. Kobayashi. 1974. Number of pinewood nematode held in the pine sawyer adult emerging from dead pine trees in the late stage of heavy infestation stands. *Transactions of the Annual Meeting of Japanese Forestry Society* 85: 231–233 [in Japanese].

Hu, S. J., T. Ning, D. Y. Fu, et al. 2013. Dispersal of the Japanese pine sawyer, *Monochamus alternatus* (Coleoptera: Cerambycidae), in mainland China as inferred from molecular data and associations to indices of human activity. *PLoS One* 8: e57568. doi: 10.1371/journal.pone.0057568.

Hughes, A. L. 1979. Reproductive behavior and sexual dimorphism in the white-spotted sawyer *Monochamus scutellatus* (Say). *The Coleopterists Bulletin* 33: 45–47.

Hughes, A. L. 1981. Differential male mating success in the white-spotted sawyer *Monochamus scutellatus* (Coleoptera: Cerambycidae). *Annals of Entomological Society of America* 74: 180–184.

Hughes, A. L., and M. K. Hughes. 1982. Male size, mating success, and breeding habitat partitioning in the whitespotted sawyer *Monochamus scutellatus* (Say) (Coleoptera: Cerambycidae). *Oecologia* 55: 258–263.

Hughes, A. L., and M. K. Hughes. 1987. Asymmetric contests among sawyer beetles (Cerambycidae: *Monochamus notatus* and *M. scutellatus*). *Canadian Journal of Zoology* 65: 823–827.

Humphry, S. J., and M. J. Linit. 1989a. Tethered flight of *Monochamus carolinensis* (Coleoptera: Cerambycidae) with respect to beetle age and sex. *Environmental Entomology* 18: 124–126.

Humphry, S. J., and M. J. Linit. 1989b. Effect of pinewood nematode density on tethered flight of *Monochamus carolinensis* (Coleoptera: Cerambycidae). *Environmental Entomology* 18: 670–673.

Husillos-Sanchez, E., G. Alvarez-Baz, I. Etxebeste, and J. A., Pajares. 2013. Shoot feeding, oviposition, and development of *Monochamus galloprovincialis* on *Pinus pinea* relative to other pine species. *Entomologia Experimentalis et Applicata* 149: 1–10.

Ibeas, F., D. Gallego, J. J. Diez, and J. A. Pajares. 2007. An operative kairomonal lure for managing pine sawyer beetle *Monochamus galloprovincialis* (Coleoptera: Cerambycidae). *Journal of Applied Entomology* 131: 13–20.

Ibeas, F., J. J., Diez, and J. A. Pajares. 2008. Olfactory sex attraction and mating behaviour in the pine sawyer *Monochamus galloprovincialis* (Coleoptera: Cerambycidae). *Journal of Insect Behavior* 21: 101–110.

Ido, N., J. Takeda, K. Kobayashi, A. Taketani, M. Okuda, and R. Hosoda. 1975. Investigation of the dispersal of pine sawyer adults. *Transactions of the 86th. Annual Meeting of Japanese Forestry Society* 335–36.

Ikeda T, N. Enda, A. Yamane, K. Oda, and T. Toyoda. 1980. Attractants for the Japanese pine sawyer, *Monochamus alternatus* Hope (Coleoptera: Cerambycidae). *Applied Entomology and Zoology* 15: 358–361.

Ikeda, T., and K. Oda. 1980. The occurrence of attractiveness for *Monochamus alternatus* Hope (Coleoptera: Cerambycidae) in nematode-infected pine trees. *Journal of Japanese Forestry Society* 62: 432–434.

Ikeda, T., A. Yamane, N. Enda, K. Matsuura, and K. Oda. 1981. Attractiveness of chemical treated pine trees for *Monochamus alternatus* Hope (Coleoptera: Cerambycidae). *Journal of Japanese Forestry Society* 63: 201–207.

Ikeda, T., A. Yamane, N. Enda, K. Oda, H. Makihara, K. Ito, and I. Okochi. 1986. Attractiveness of volatile components of felled pine trees for *Monochamus alternatus* (Coleoptera: Cerambycidae). *Journal of Japanese Forestry Society* 68: 15–19.

Ishibashi, N., and E. Kondo. 1977. Occurrence and survival of the dispersal forms of pine wood nematode, *Bursaphelenchus lignicolus* Mamiya and Kiyohara. *Applied Entomology and Zoology* 12: 293–302.

Islam, S. Q., J. Ichiryu, M. Sato, and T. Yamasaki. 1997. D-Catechin: An oviposition stimulant for the cerambycid beetle, *Monochamus alternatus*, from *Pinus densiflora*. *Journal of Pesticide Science* 22: 338–341.

Ito, K. 1982. The tethered flight of the Japanese pine sawyer, *Monochamus alternatus* Hope (Coleoptera: Cerambycidae). *Journal of Japanese Forestry Society* 64: 395–397.

Jikumaru, S., and K. Togashi. 2000. Temperature effects on the transmission of *Bursaphelenchus xylophilus* (Nematoda: Aphelenchoididae) by *Monochamus alternatus* (Coleoptera: Cerambycidae). *Journal of Nematology* 3: 325–333.

Jikumaru, S., K. Togashi, A. Taketsune, and F. Takahashi. 1994. Oviposition biology of *Monochamus saltuarius* (Coleoptera: Cerambycidae) at a constant temperature. *Applied Entomology and Zoology* 29: 555–561.

Kamata, N. 2008. Integrated pest management of pine wilt disease in Japan: Tactics and strategies. In *Pine wilt disease*, eds. B. G. Zhao, K. Futai, J. R. Sutherland, Y. Takeuchi, 304–322. Tokyo: Springer.

Kanzaki, N. 2008. Taxonomy and systematics of the nematode genus *Bursaphelenchus* (Nematoda: Parasitaphelenchidae). In *Pine Wilt Disease*, eds. B. Zhao, K. Futai, J. R. Sutherland & Y. Takeuchi, 44–66. Dordrecht: Springer.

Kanzaki, N., N. Maehara, T. Aikawa and K. Togashi. 2008. First report of parthenogenesis in the genus *Bursaphelenchus* Fuchs, 1937: A description of *Bursaphelenchus okinawaensis* sp. nov. isolated from *Monochamus maruokai* (Coleoptera: Cerambycidae). *Zoological Science* 25: 861–873.

Kato, R., and T. Okudaira. 1977. Range expansion of the pinewood nematode in Aichi Prefecture. *Transactions of the 26th Annual Meeting of Chubu Branch Japanese Forestry Society*, pp. 159–164 [in Japanese].

Katsuyama, N., H. Sakurai, K. Tabata, and S. Takeda. 1989. Effect of age of post-feeding twig on the ovarian development of Japanese pine sawyer, *Monochamus alternatus*. *Research Bulletin Faculty of Agriculture Gifu University* 54: 81–89 [in Japanese with English summary].

Kawai, M., E. Shoda-Kagaya, T. Maehara, et al. 2006. Genetic structure of pine sawyer *Monochamus alternatus* (Coleoptera: Cerambycidae) population in Northeast Asia: Consequences of the spread of pine wilt disease. *Environmental Entomology* 35: 569–579.

Kikuchi, T., H. Shibuya, T. Aikawa, and J. T. Jones. 2006. Cloning and characterization of pectate lyases expressed in the esophageal gland of the pine wood nematode *Bursaphelenchus xylophilus*. *Molecular Plant-Microbe Interactions*. 19: 280–287.

Kim, G., J. Takabayashi, and K. Tabata. 1992. Function of pheromones in mating behavior of the Japanese pine sawyer beetle, *Monochamus alternatus* Hope. *Applied Entomology and Zoology* 27: 489–497.

Kishi, Y. 1978. Invasion of pine trees by *Bursaphelenchus lignicolus* M & K. (Nematoda: Aphelenchoidae) from *Monochamus alternatus* Hope (Coleoptera: Cerambycidae) *Journal of the Japanese Forestry Society* 60: 179–82.

Kishi, Y. 1995. *The Pinewood Nematode and the Japanese Pine Sawyer. Forest Pest in Japan.* No. 1. Thomas Company Limited, Tokyo, Japan.

Kobayashi, F. 1988. The Japanese pine sawyer. In *Dynamics of Forest Insect Populations: Patterns, Causes, Implications*, ed. A. A. Beryman, 431–454. New York: Plenum Press.

Kobayashi, F., A. Yamane, and T. Ikeda. 1984. The Japanese pine sawyer beetle as the vector of pine wilt disease. *Annual Review of Entomology* 29: 115–135.

Kojima, K., and S. Nakamura. 2011. *Food plants of cerambycid beetles (Cerambycidae, Coleoptera) in Japan. Revised and enlarged edition.* Shobara: Hiba Society of Natural History.

Kosaka, H., T. Aikawa, N. Ogura, K. Tabata, and T. Kiyohara, 2001. Pine wilt disease caused by the pine wood nematode: The induced resistance of pine trees by the avirulent isolates of nematode. *European Journal of Plant Pathology* 107: 667–675.

Koutroumpa, F. A., B. Vincent, G. Roux-Morabito, C. Martin, and F. Lieutier. 2008. Fecundity and larval development of *Monochamus galloprovincialis* (Coleoptera: Cerambycidae) in experimental breeding. *Annals of Forest Science* 65: 707.

Kuniyoshi, S. 1974. Occurrence of the pinewood nematode in Okinawa Prefecture. *Forest Pests* 23: 40–42 [in Japanese].

Lee, K. S., and A. A. Lashari. 2014. Stability analysis and optimal control of pine wilt disease with horizontal transmission in vector population. *Applied Mathematics and Computation* 226: 793–804.

Le Maitre, D. C. 1998. Pines in cultivation: A global view. In: *Ecology and biogeography of Pinus*, ed. D. M. Richardson. 407–431. Cambridge: Cambridge University Press.

Li, S. Q., and Z. N. Zhang. 2006. Influence of larval frass extracts on the oviposition behaviour of *Monochamus alternatus* (Col., Cerambycidae). *Journal of Applied Entomology* 130: 177–182.

Linit, M. J. 1985. Continuous laboratory culture of *Monochamus carolinensis* with notes on larval development. *Annals of Entomological Society of America* 78: 212–213.

Linit, M. J. 1988. Nematode-vector relationships in the pine wilt system. *Journal of Nematology* 20: 227–235.

Linit, M. J. 1989. Temporal pattern of pinewood nematode exit from the insect vector *Monochamus carolinensis*. *Journal of Nematology* 21: 105–107.

Linit, M. J. 1990. Transmission of pinewood nematode through feeding wounds of *Monochamus carolinensis* (Coleoptera: Cerambycidae). *Journal of Nematology* 22: 231–236.

Linit, M. J., E. Kondo, and M. T. Smith. 1983. Insect Associated with the Pinewood Nematode, *Bursaphelenchus xylophilus* (Nematoda: Aphelenchoididae), in Missouri. *Environmental Entomology* 12: 467–470.

Linsley, E. G. 1961. The Cerambycidae of North America. Part I: Introduction. *University of California Publications in Entomology* 18: 1–135.

Luzzi, M. A., R. C. Wilkinson, and A. C. Tarjan. 1984. Transmission of the pinewood nematode, *Bursaphelenchus xylophilus*, to Slash pine trees and bolts by a cerambycid beetle, *Monochamus titillator*, in Florida. *Journal of Nematology* 16: 37–40.

Macias–Samano, J. E., D. Wakarchuk, J. G. Millar, et al. 2012. 2-Undecyloxy-1-ethanol in combination with other semiochemicals attracts three *Monochamus* species (Coleoptera: Cerambycidae) in British Columbia. *The Canadian Entomologist* 144: 764–768.

Maehara, N. 2008. Reduction of *Bursaphelenchus xylophilus* (Nematoda: Parasitaphelenchidae), population by inoculating *Trichoderma spp.* into pine wilt killed trees. *Biological Control* 44: 61–66.

Maehara, N., and K. Futai. 1996. Factors affecting both the number of the pinewood nematode, *Bursaphelenchus xylophilus* (Nematoda: Aphelenchoididae), carried by the Japanese pine sawyer, *Monochamus alternatus* (Coleoptera: Crembycidae), and the nematode's life history. *Applied Entomology and Zoology* 31: 443–452.

Maehara, N., and K. Futai. 1997. Effect of fungal interactions on the numbers of the pinewood nematode, *Bursaphelenchus xylophilus* (Nematoda: Aphelenchoididae), carried by the Japanese sawyer, *Monochamus alternatus* (Coleoptera: Cerambycidae). *Fundamental and Applied Nematology* 20: 611–617.

Maehara, N., and K. Futai. 2002. Factor affecting the number of *Bursaphelenchus xylophilus* (Nematoda: Aphelenchoididae), carried by the several species of beetles. *Nematology* 4: 653–658.

Maehara, N., and M. Tokora. 2010. Effect of unsaturated fatty acid around pupal chambers of *Monochamus alternatus* (Coleoptera: Cerambycidae) on the number of *Bursaphelenchus xylophilus* (Nematoda: Parasitaphelenchidae) carried by the beetles. *Nematology* 12: 721–729.

Malek, R. B., and J. E. Appleby. 1984. Epidemiology of pine wilt in Illinois. *Plant Disease* 68: 180–186.

Mamiya, Y. 1975. The life history of the pinewood nematode, *Bursaphelenchus lignicolus*. *Japanese Journal of Nematology*. 5: 16–25.

Mamiya, Y. 1976. Pine wilting disease caused by the pine wood nematode, *Bursaphelenchus lignicolus*, in Japan. *Japan Agricultural Research Quarterly* 10: 206–212.

Mamiya, Y. 1983. Pathology of the pine wilt disease caused by *Bursaphelenchus xylophilus*. *Annual Review Phytopathology* 21: 201–220.

Mamiya, Y. 1984. The pine wood nematode. In *Plant and insect nematodes*, ed. W. R. Nickle, 589–627. New York: Marcel Dekker Inc.

Mamiya, Y. 1989. Comparison of the infectivity of Steinernema kushidai (Nematoda: Steinernematidae) and other steinernematid and heterohabditid nematodes for three different insects. *Applied Entomology and Zoology* 24: 302–308.

Mamiya, Y., and N. Enda. 1972. Transmission of *Bursaphelenchus lignicolus* (Nematoda: Aphelenchoididae) by *Monochamus alternatus* (Coleoptera: Cerambycidae). *Nematologica* 18: 159–162.

Mas, H., R. Hernández, M. Villarova, et al. 2013. Dispersal behavior and long distance flight capacity of *Monochamus galloprovincialis* (Olivier 1795). In *Pine wilt disease conference 2013*, ed. T. Schröder, p. 22. Braunschweig, Germany.

Matsubara, I. 1976. Observations of the epidemic mortality of pine trees in Chiba prefecture. *Transactions of the 87th Annual Meeting of Japanese Forestry Society*, pp. 307–308 [in Japanese].

Matsuura, K. 1984. Preventive and therapeutic effects of certain systemic nematicides upon the pine wilt disease caused by pine-wood nematodes, *Bursaphelenchus xylophilus* (I) Chemotherapeutic effect of six systemic nematicides. *Journal of Japanese Forestry Society* 66: 1–9.

McMeniman, C. J., R. V. Lane, B. N. Cass, et al. 2009. Stable introduction of a life-shortening *Wolbachia* infection into the mosquito *Aedes aegypti*. *Science* 323: 141–144.

Miller, D. R., and C. Asaro. 2005. Ipsenol and Ipsdienol attract *Monochamus titillator* (Coleoptera: Cerambycidae) and associated large pine woodborers in Southeastern United States. *Journal of Economic Entomology* 98: 2033–2040.

Morimoto, K., and A. Iwasaki. 1972. Rôle of *Monochamus alternatus* (Coleoptera: Cerambycidae) as a vector of *Bursaphelenchus lignicolus* (Nematoda: Aphelenchoididae). *Journal of Japanese Forest Society* 54: 177–183 [in Japanese with English summary].

Morimoto, K., and A. Iwasaki. 1974. Studies on the pine sawyer (XI). Density effects on the emergence rate. *Transaction of the 85th Annual Meeting of the Japanese Forestry Society*, pp 299–300 [in Japanese].

Nakamura-Matori, K. 2008. Vector-host tree relationships and the abiotic environment. In *Pine wilt disease*, eds. B. G. Zhao, K. Futai, J. R. Sutherland, and Y. Takeuchi, 144–161. Tokyo: Springer.

Naves, P., E. De Sousa, and J. A. Quartau. 2006a. Reproductive traits of *Monochamus galloprovincialis* (Coleoptera: Cerambycidae) under laboratory conditions. *Bulletin of Entomological Research* 96: 289–294.

Naves, P., M. Kenis, and E. Sousa. 2005. Parasitoids associated with *Monochamus galloprovincialis* (Oliv.) (Coleoptera: Cerambycidae) within the pine wilt nematode-affected zone in Portugal. *Journal of Pest Science* 78: 57–62.

Naves, P. M., S. Camacho, E. M. Sousa, and J. A. Quartau. 2007b. Transmission of the pinewood nematode *Bursaphelenchus xylophilus* through feeding activity of *Monochamus galloprovincialis* (Coleoptera: Cerambycidae). *Journal of Applied Entomology* 131: 21–25.

Naves, P. M., E. M. de Sousa, and J. A. Quartau. 2006b. Feeding and oviposition preferences of *Monochamus galloprovincialis* for certain conifers under laboratory conditions. *Entomologia Experimentalis et Applicata* 120: 99–104.

Naves, P. M., E. M. de Sousa, and J. A. Quartau. 2007a. Winter dormancy of the pine sawyer *Monochamus galloprovincialis* (Col., Cerambycidae) in Portugal. *Journal of Applied Entomology* 131: 669–673.

Naves, P. M., E. Sousa, and J. M. Rodrigues. 2008. Biology of *Monochamus galloprovincialis* (Coleoptera: Cerambycidae) in the pine wilt disease affected zone, southern Portugal. *Silva Lusitana* 16: 133–148.

Nobuchi, A. 1976. Fertilization and oviposition of *Monochamus alternatus* Hope. *Transaction of the 87th Meeting Japanese Forest Society*, pp. 247–248 [in Japanese].

Nord, J. C., and F. B. Knight. 1972. The importance of *Saperda inornata* and *Oberea schaumii* (Coleoptera: Cerambycidae) galleries as infection courts of *Hypoxylon pruinatum* in trembling aspen, *Populus tremuloides*. *Great Lakes Entomologist* 5: 87–92.

Nose, M., and S. Shiraishi. 2008. Breeding for resistance to pine wilt disease. In *Pine wilt disease*, eds. B. G. Zhao. K. Futai. J. R. Sutherland, and Y. Takeuchi, 334–350. Tokyo: Springer.

Ogura, N., K. Tabata, and W. Wang. 1999. Rearing of the colydiid beetle predator, *Dastarcus helophoroides*, on artificial diet. *Biocontrol* 44: 291–299.

Okamoto, H. 1984. Behavior of the adult of Japanese pine sawyer, *Monochamus alternatus*, Hope. In *Proceedings of the United States-Japan seminar: Resistance mechanisms of pines against pine wilt disease*, ed. V. H. Dropkin, 82–90. Columbia, MO: University of Missouri.

Økland, B., O. Skarpaas, M. Schroeder, C. Magnusson, Å. Lindelöw, and K. Thunes. 2010. Is eradication of the pinewood nematode (*Bursaphelenchus xylophilus*) likely? an evaluation of current contingency plans. *Risk Analysis* 30: 1424–1439.

Oku, H., T. Shiraishi, S. Outchi, S. Kurozumi, and H. Ohta. 1980. Pine wilt toxin, the metabolite of a bacterium associated with a nematode. *Naturwissenschaften* 67: 198–199.

Ostry, M. E., and N. A. Anderson. 1995. Infection of *Populus tremuloides* by *Hypoxilon mammatum* ascospores through *Saperda inornata* galls. *Canadian Journal of Forest Research* 25: 813–816.

Pajares, J. A., G. Álvarez, F. Ibeas, D. Gallego, D. R. Hall, and D. I. Farman. 2010. Identification and field activity of a male-produced aggregation pheromone in the pine sawyer beetle, *Monochamus galloprovincialis*. *Journal of Chemical Ecology* 36: 570–583.

Pajares, J. A., G. Álvarez, D. R. Hall, et al. 2013. 2-(Undecyloxy)-ethanol is a major component of the male-produced aggregation pheromone of *Monochamus sutor*. *Entomologia Experimentalis et Applicata* 149: 118–127.

Pajares, J. A., F. Ibeas, J. J. Diez, and D. Gallego. 2004. Attractive responses by *Monochamus galloprovincialis* (Coleoptera: Cerambycidae) to host and bark beetle semiochemicals. *Journal of Applied Entomology* 128: 633–638.

Park, Y. S., Y. J. Chung, and Y. S. Moon. 2013. Hazard ratings of pine forests to a pine wilt disease at two spatial scales (individual trees and stands) using self-organizing map and random forest. *Ecological Informatics* 13: 40–46.

Pershing, J. C., and M. J. Linit. 1986a. Development and seasonal occurrence of *Monochamus carolinensis* (Coleoptera: Cerambycidae) in Missouri. *Environmental Entomology* 15: 251–253.

Pershing, J. C., and M. J. Linit. 1986b. Biology of *Monochamus carolinensis* (Coleoptera: Cerambycidae) on Scotch pine in Missouri. *Journal of Kansas Entomological Society* 59: 706–711.

Phan, K. L. 2008. Potential of entomopathogenic nematodes for controlling the Japanese pine sawyer, *Monochamus alternatus*. In *Pine wilt disease*, eds. B. G. Zhao, K. Futai. J. R. Sutherland, and Y. Takeuchi, 371–379. Tokyo: Springer.

Rasatti, D., E. P. Toffolo, A. Battisti, M. Faccoli. 2012. Monitoring of the pine sawyer beetle *Monochamus galloprovincialis* by pheromone traps in Italy. *Phytoparasitica* 40: 329–336.

Richmond, H. A., and R. R. Lejeune. 1945. The deterioration of fire-killed white spruce by wood boring insects in northern Saskatchewan. *Forestry Chronicle* 21: 168–192.

Robertson, L., S. C. Arcos, M. Escuer, et al. 2011. Incidence of the pinewood nematode *Bursaphelenchus xylophilus* (Steiner & Buhrer, 1934) Nickle, 1970 in Spain. *Nematology* 13: 755–757.

Robinet, C., A. Roques, H. Pan, et al. 2009. Role of human-mediated dispersal in the spread of the pinewood nematode in China. *PLoS One* 4: e4646. doi: 10.1371/journal.pone.0004646.

Robinet, C., N. van Opstal, R. Baker, and A. Roques. 2011. Applying a spread model to identify the entry points from which the pine wood nematode, the vector of pine wilt disease, would spread most rapidly across Europe. *Biological Invasion* 13: 2981–2995.

Rose, A. H. 1957. Some notes on the biology of *Monochamus scutellatus* (Say) (Coleoptera: Cerambycidae). *The Canadian Entomologist* 89: 547–553.

Ross, D. A. 1960. Damage by long-horned wood borers in fire-killed white spruce, central British Columbia. *Forestry Chronicle* 36: 355–361.

Rutherford, T. A., Y. Mamiya, and J. M. Webster. 1990. Nematode-induced pine wilt disease: Factors influencing its occurrence and distribution. *Forest Science* 36: 145–155.

Ryall, K., P. Silk, R. P. Webster, et al. 2015. Further evidence that monochamol is attractive to *Monochamus* (Coleoptera: Cerambycidae) species, with attraction synergised by host plant volatiles and bark beetle (Coleoptera: Curculionidae) pheromones. *The Canadian Entomologist* 147: 564–579.

Ryss, A., P. Vieira, M. Mota, and O. Kulinich. 2005. A synopsis of the genus *Bursaphelenchus* Fuchs, 1937 (Aphelenchida: Parasitaphelenchidae) with keys to species. *Nematology* 7(3): 393–458.

Sato, M., S. Q. Islam, S. Awata, and T. Yamasaki. 1999a. Flavanonol glucoside and proanthocyanidins: Oviposition stimulants for cerambycid beetle, *Monochamus alternatus*. *Journal of Pesticide Science* 24: 123–129.

Sato, M., S. Q. Islam, and T. Yamasaki. 1999b. Glycosides of a phenylpropanoid and neolignans: Oviposition stimulants in pine inner bark for the cerambycid beetle, *Monochamus alternatus*. *Journal of Pesticide Science* 24: 397–400.

Shibata, E. 1981. Seasonal fluctuation and spatial pattern of the adult population of the Japanese pine sawyer, *Monochamus alternatus* (Coleoptera: Cerambycidae), in young pine forests. *Applied Entomology and Zoology* 16: 306–309.

Shibata, E. 1984. Spatial distribution pattern of the Japanese pine sawyer, *Monochamus alternatus* Hope (Coleoptera: Cerambycidae), on dead pine trees. *Applied Entomology and Zoology* 19: 361–366.

Shibata, E. 1985. Seasonal fluctuation of the pine wood nematode, *Bursaphelenchus xylophilus* (Steiner and Buhrer) Nickle (Nematoda: Aphelenchoididae), transmitted to pine by the Japanese pine sawyer, *Monochamus alternatus* Hope (Coleoptera: Cerambycidae). *Applied Entomology and Zoology* 20: 241–245.

Shibata, E. 1986. Dispersal movement of the adult Japanese pine sawyer, *Monochamus alternatus* (Coleoptera: Cerambycidae), in a young pine forest. *Applied Entomology and Zoology* 21: 184–186.

Shimazu, M. 2008. Biological control of the Japanese pine sawyer beetle, *Monochamus alternatus*. In *Pine wilt disease*, eds. B. G. Zhao, K. Futai, J. R. Sutherland, Y. Takeuchi, 351–370. Tokyo: Springer.

Shinya, R., H. Morisaka, Y. Takeuchi, M. Ueda, and K. Futai. 2010. Comparison of the surface coat proteins of the pine wood nematode appeared during host pine infection and in vitro culture by a proteomic approach. *Phytopathology* 100: 1289–1297.

Shinya, R., Y. Takeuchi, N. Miura, K. Kuroda, M. Ueda, and K. Futai. 2009. Surface coat proteins of the pine wood nematode, *Bursaphelenchus xylophilus*: Profiles of stage- and isolate-specific characters. *Nematology* 11: 429–438.

Shoda-Kagaya, E. 2007. Genetic differentiation of the pine wilt disease vector *Monochamus alternatus* (Coleoptera: Cerambycidae) over a mountain range—Revealed from microsatellite DNA markers. *Bulletin of Entomological Research* 97: 167–174.

Song, S. H., L. Q. Zhang, H. H. Huang, and X. M. Cui. 1991. Preliminary study of biology of *Monochamus alternatus* Hope. *Forest Science and Technology* 6: 9–13.

Sousa, E., P. Naves, L. Bonifácio, L. Inácio, J. Henriques, and H. Evans. 2011. Survival of *Bursaphelenchus xylophilus* and *Monochamus galloprovincialis* in pine branches and wood packaging material. *EPPO Bulletin* 41: 203–207.

Stamps, W. T. 1995. Factors regulating exit of *Bursaphelenchus xylophilus* (Nematoda: Aphelenchoididae) fourth-stage dispersal juveniles from their beetle vector *Monochamus carolinensis* (Coleoptera: Cerambycidae). PhD Dissertation, Department of Entomology, University of Missouri, Columbia, MO, 124 pp.

Stamps, W. T., and M. J. Linit. 2001. Interaction of intrinsic and extrinsic chemical cues in the behavior of *Bursaphelenchus xylophilus* (Aphelenchida: Aphelenchoididae) in relation to its beetle vectors. *Nematology* 3: 295–301.

Sugimoto, H., and K. Togashi. 2013. Canopy-related adult density and sex-related flight activity of *Monochamus alternatus* (Coleoptera: Cerambycidae) in pine stands. *Applied Entomology and Zoology* 48: 213–221.

Takasu, F. 2009. Individual-based modeling of the spread of pine wilt disease: Vector beetle dispersal and the allee effect. *Population Ecology* 51: 399–409.

Takasu, F., N. Yamamoto, K. Kawasaki, K. Togashi, Y. Kishi, and N. Shigesada. 2000. Modeling the expansion of an introduced tree disease. *Biological Invasions* 2: 141–150.

Tauber, M. J., C. A. Tauber, and S. Masaki. 1986. *Seasonal adaptations of insects*. Oxford: Oxford University Press.

Teale, S. A., J. D. Wickham, F. Zhang, et al. 2011. A male-produced aggregation pheromone of *Monochamus alternatus* (Coleoptera: Cerambycidae), a major vector of pine wood nematode. *Journal of Economic Entomology* 104: 1592–1598.

Toda, T., and S. Kurinobu. 2002. Realized genetic gains observed in progeny tolerance of selected red pine (*Pinus densiflora*) and black pine (*P. thunbergii*) to pine wilt disease. *Silvae Genetica* 51: 42–44.

Togashi, K. 1980. A simulation model for the optimal time for utilizing insecticide spray against the Japanese pine sawyer, *Monochamus alternatus* Hope (Coleoptera: Cerambycidae). *Journal of Japanese Forest Society* 62: 381–387 [in Japanese with English summary].

Togashi, K. 1985. Transmission curves of *Bursaphelenchus xylophilus* (Nematoda: Aphelenchoididae) from its vector, *Monochamus alternatus* (Coleoptera: Cerambycidae), to pine trees with reference to population performance. *Applied Entomology and Zoology* 25: 246–251.

Togashi, K. 1986. Effects of the initial density and natural enemies on the survival rate of the Japanese pine sawyer, *Monochamus alternatus* Hope (Coleoptera: Cerambycidae), in pine logs. *Applied Entomology and Zoology* 21: 244–251.

Togashi, K. 1988. Population density of *Monochamus alternatus* adult (Coleoptera: Cerambycidae) and incidence of pine wilt disease caused by *Bursaphelenchus xylophilus* (Nematoda: Aphelenchoididae). *Researches on Population Ecology* 30: 177–192.

Togashi, K. 1989a. Development of *Monochamus alternatus* Hope (Coleoptera: Cerambycidae) in relation to oviposition time. *Japanese Journal of Applied Entomology and Zoology* 33: 1–8 [in Japanese with English abstract].

Togashi, K. 1989b. Development of *Monochamus alternatus* Hope (Coleoptera: Cerambycidae) in *Pinus thunbergii* trees weakened at different times. *Journal of Japanese Forest Society* 71: 383–386 [in Japanese with English abstract].

Togashi, K. 1989c. Studies on population dynamics of *Monochamus alternatus* Hope (Coleoptera: Cerambycidae) and spread of pine wilt disease caused by *Bursaphelenchus xylophilus* (Nematoda: Aphelenchoididae). *Bulletin of Ishikawa Forest Experiment station* 20: 1–142 [in Japanese with English summary].

Togashi, K. 1989d. Temporal pattern of the occurrence of weakened *Pinus thunbergii* trees and causes for mortality. *Journal of Japanese Forest Society* 71: 323–328.

Togashi, K. 1989e. Factors affecting the number of *Bursaphelenchus xylophilus* (Nematoda: Aphelenchoididae) carried by newly emerged adults of *Monochamus alternatus* (Coleoptera: Cerambycidae). *Applied Entomology and Zoology* 24: 379–386.

Togashi, K. 1990a. Life table for *Monochamus alternatus* (Coleoptera: Cerambycidae) within dead trees of *Pinus thunbergii*. *Japanese Journal of Entomology* 58: 217–230.

Togashi, K. 1990b. Change in the activity of adult *Monochamus alternatus* Hope (Coleoptera: Cerambycidae) in relation to age. *Applied Entomology and Zoology* 25: 153–159.

Togashi, K. 1990c. A field experiment on dispersal of newly emerged adults of *Monochamus alternatus* (Coleoptera: Cerambycidae). *Researches on Population Ecology* 32: 1–13.

Togashi, K. 1991a. Larval diapause termination of *Monochamus alternatus* Hope (Coleoptera: Cerambycidae) under natural conditions. *Applied Entomology and Zoology* 26: 381–386.

Togashi, K. 1991b. Spatial pattern of pine wilt disease caused by *Bursaphelenchus xylophilus* (Nematoda: Aphelenchoididae) within a *Pinus thunbergii* stand. *Researches on Population Ecology* 33: 245–256.

Togashi, K. 1991c. Different developments of overwintered larvae of *Monochamus alternatus* (Coleoptera: Cerambycidae) under a constant temperature. *Japanese Journal of Entomology* 59: 149–154.

Togashi, K. 1995a. Interacting effects of temperature and photoperiod on diapause in larvae of *Monochamus alternatus* (Coleoptera: Cerambycidae). *Japanese Journal of Entomology* 63: 243–252.

Togashi, K. 1995b. Diapause avoidance for life cycle regulation of *Monochamus alternatus* (Coleoptera: Cerambycidae). In *International symposium on pine wilt disease caused by pine wood nematode*, 119–127. Beijing: Chinese Society of Forestry.

Togashi, K. 1997. Lifetime fecundity and body size of *Monochamus alternatus* (Coleoptera: Cerambycidae) at a constant temperature. *Japanese Journal of Entomology* 65: 458–470.

Togashi, K. 2007. Lifetime fecundity and female body size in *Paraglenea fortunei* (Coleoptera: Cerambycidae). *Applied Entomology and Zoology* 42: 549–556.

Togashi, K. 2008. Vector-nematode relationships and epidemiology in pine wilt disease. In *Pine wilt disease*, eds. B. G. Zhao, K. Futai, J. R. Sutherland, and Y. Takeuchi, 162–183. Tokyo: Springer.

Togashi, K. 2013. Nematode-vector beetle relation and the regulatory mechanism of vector's life history in pine wilt disease. *Formosan Entomologist* 33: 189–205.

Togashi, K. 2014. Effects of larval food shortage on diapause induction and adult traits in Taiwanese *Monochamus alternatus alternatus*. *Entomologia Experimentalis et Applicata* 151: 34–42.

Togashi, K., J. E. Appleby, and R. B. Malek. 2005. Host tree effect on the pupal chamber size of *Monochamus carolinensis* (Coleoptera: Cerambycidae). *Applied Entomology and Zoology* 40: 467–474.

Togashi, K., J. E. Appleby, H. O. Sadeghi, and R. B. Malek. 2009. Age specific survival rate and fecundity of *Monochamus carolinensis* (Coleoptera: Cerambycidae) under field conditions. *Applied Entomology and Zoology* 44: 249–256.

Togashi, K., and Y. Arakawa. 2003. Horizontal transmission of *Bursaphelenchus xylophilus* between sexes of *Monochamus alternatus*. *Journal of Nematology* 35: 7–16.

Togashi, K., Y. J. Chung, and E. Shibata, 2004. Spread of an introduced tree pest organism—The pinewood nematode. In *Ecological issues in a changing world: Status, response and strategy*, eds. S. K. Hong, J. A. Lee, B. S. Ihm, et al., 173–188, Dordrecht: Kluwer Academic Publishers.

Togashi, K., and S. Jikumaru. 1996. Horizontal transmission of *Bursaphelenchus mucronatus* (Nematoda: Aphelenchoididae) between insect vectors of *Monochamus saltuarius* (Coleoptera: Cerambycidae). *Applied Entomology and Zoology* 31: 317–320.

Togashi, K., and S. Jikumaru. 2007. Evolutionary change in a pine wilt system following the invasion of Japan by the pinewood nematode, *Bursaphelenchus xylophilus*. *Ecological Research* 22: 862–868.

Togashi, K., H. Kasuga, H. Yamashita, and K. Iguchi. 2008. Effect of host tree species on larval body size and pupal-chamber tunnel of *Monochamus alternatus* (Coleoptera: Cerambycidae). *Applied Entomology and Zoology* 43: 235–240.

Togashi, K., H. Kasuga, H. Yamashita, and K. Iguchi. 2010. Larval diapause of *Monochamus urussovi* and photoperiodic effects on larval development in tree bolts. *Journal of Applied Entomology* 134: 672–674.

Togashi, K., and H. Magira. 1981. Age-specific survival rate and fecundity of the adult Japanese pine sawyer, *Monochamus alternatus* Hope (Coleoptera: Cerambycidae), at different emergence times. *Applied Entomology and Zoology* 16: 351–361.

Togashi, K., and H. Sekizuka. 1982. Influence of the pinewood nematode, *Bursaphelenchus lignicolus* (Nematoda: Aphelenchoididae) on longevity of its vector, *Monochamus alternatus* (Coleoptera: Cerambycidae). *Applied Entomology and Zoology* 17: 160–165.

Togashi, K., and N. Shigesada. 2006. Spread of the pinewood nematode vectored by the Japanese pine sawyer: Modeling and analytical approaches. *Population Ecology* 48: 271–283.

Tomminen, J. 1991. Pinewood nematode, *Bursaphelenchus xylophilus*, found in packing case wood. *Silva Fennica* 25: 109–111.

Tomminen, J. 1993. Development of *Monochamus galloprovincialis* Olivier (Coleoptera: Cerambycidae) in cut trees of young pines (*Pinus sylvestris* L.) and log bolts in southern Finland. *Entomologica Fennica* 4: 137–142.

Torres-Villa, L. M., C. Zugasti, J. M. De-Juan, et al. 2015. Mark-recapture of *Monochamus galloprovincialis* with semiochemical-baited traps: Population density, attraction distance, flight behaviour and mass trapping efficiency. *Forestry: An International Journal of Forest Research* 88: 224–236.

Urano, T. 2006. Experimental release of an adult *Dastarcus helophoroides* (Coleoptera: Bothrideridae) in a pine stand damaged by pine wilt disease: Effects on *Monochamus alternatus* (Coleoptera: Cerambycidae). *Bulletin of Forestry and Forest Product Research Institute* 5: 257–263.

Vicente, C., M. Espada, P. Vieira, and M. Mota. 2012. Pine wilt disease: A threat to European forestry. *European Journal of Plant Pathology* 133: 89–99.

Walsh, K. D., and M. J. Linit, 1985. Oviposition biology of the pine sawyer *Monochamus carolinensis* (Coleoptera: Cerambycidae). *Annals of Entomological Society of America* 78: 81–85.

Warren, J. E., O. R. Edwards, and M. J. Linit. 1995. Influence of bluestain fungi on laboratory rearing of pine-wood nematode infested beetles. *Fundamental and Applied Nematology* 18: 95–98.

Wingfield, M. J., and R. A. Blanchette. 1983. The pinewood nematode, *Bursaphelenchus xylophilus*, in Minnesota and Wisconsin: Insect associates and transmission studies. *Canadian Journal of Forest Research* 13: 1068–1076.

Xu, F. 2008. Recent advances in the integrated management of the pine wood nematode in China. In: *Pine wilt disease*, eds. B. G. Zhao, K. Futai, J. R. Sutherland, Y. Takeuchi, 323–333. Tokyo: Springer.

Xu, M., F. Y. Xu, Y. P. Liu, Y. S. Pan, and X. Q. Wu. 2015. Assessment of *Metarhizium anisopliae* (Clavicipitaceae) and its vector, *Scleroderma guani* (Hymenoptera: Bethylidae), for the control of *Monochamus alternatus* (Coleoptera: Cerambycidae). *The Canadian Entomologist* 147: 628–634.

Yamamoto, N., F. Takasu, K. Kawasaki, K. Togashi, Y. Kishi, and N. Shigesada. 2000. Local dynamics and global spread of pine wilt disease. *Japanese Journal of Ecology* 50: 269–276 [in Japanese].

Yamanaka, S. 1993. Field control of the Japanese pine sawyer, *Monochamus alternatus* (Coleoptera: Cerambycidae) Larvae by *Steinernema carpocapsae* (Nematoda: Rhabditida). *Japanese Journal of Nematology* 23: 71–78.

Yamane, A. 1974. Temporal changes in pupal morphology and body mass in *Monochamus alternatus* Hope. *Transaction of the 85th Annual Meeting Japanese Forest Society* 239–240 [in Japanese].

Yamane, A. 1975. Behaviors of the pine sawyers, *Monochamus alternatus* Hope, the main vector of the pine-killing wood nematode. Volunteer paper at the second FAO/IUFRO *World Technical Consultation on Forest Disease and Insects*, New Delhi, India, April 7–12.

Yamane, A. 1981. The Japanese pine sawyer, *Monochamus alternatus* Hope (Coleoptera: Cerambycidae): *Bionomics and control*. *Review of Plant Protection Research* 14: 1–25.

Yang, Z. Q., X. Y. Wang, and Y. N. Zhang. 2014. Recent advances in biological control of important native and invasive forest pests in China. *Biological Control* 68: 117–128.

Yoon, C., Y. H. Shin, J. O. Yang, J. H. Han, and G. H. Kim. 2011. *Pinus koraiensis* twigs affect *Monochamus saltuarius* (Coleoptera: Cerambycidae) longevity and reproduction. *Journal of Asia-Pacific Entomology* 14: 327–333.

Yoshimura, A., K. Kawasaki, F. Takasu, K. Togashi, K. Futai, and N. Shigesada. 1999. Modeling the spread of pine wilt disease caused by nematodes with pine sawyers as vector. *Ecology* 80: 1691–1702.

Zhao, B. G., H. L. Wang, S. F. Han, and Z. M. Han. 2003. Distribution and pathogenicity of bacteria species carried by *Bursaphelenchus xylophilus* in China. *Nematology* 5: 899–906.

Zhao, L. L., W. Wei, L. Kang, and J. H. Sun. 2007. Chemotaxis of the pinewood nematode, *Bursaphelenchus xylophilus*, to volatiles associated with host pine, *Pinus massoniana*, and its vector *Monochamus alternatus*. *Journal of Chemical Ecology* 33: 1207–1216.

7

Laboratory Rearing and Handling of Cerambycids*

Melody A. Keena
USDA Forest Service
Hamden, Connecticut

CONTENTS

7.1 Introduction

Lack of suitable rearing and handling techniques has hampered research on the biology and control of many species of cerambycids that feed on host species of economic importance. Furthermore, because cerambycids spend most or all of their pre-adult life cycle inside the host plant, the biology of many is not well-known and would be difficult to study in nature. This is especially true for species with extended life cycles where adults only appear briefly in the field either annually, biannually, or less often. The development of laboratory rearing methods, either partially or completely independent of host material, can and has been used to rapidly advance our knowledge of cerambycids.

Several diets and rearing protocols have been published for cerambycid species. These larval diets include artificial diets with no host material incorporated into them, artificial diets with pulverized host material incorporated, and natural diets of cut host material. Diets and handling methods have been formulated for species with adverse economic impacts to help understand their biology and to develop management options. There is one case where methods have been developed as a way to preserve a threatened cerambycid species (Dojnov et al. 2012). Methods have also been developed to produce individuals for biological control of weeds and to allow larvae found in host material to complete development so an

---

* The mention of trade names or commercial products is solely for the purpose of providing specific information about what was used and does not imply recommendation or endorsement by the U.S. Department of Agriculture.

adult specimen can be procured for identification purposes. Species that can easily be reared on cut host material or on an artificial diet developed for another species have also been reared in a laboratory setting. This chapter will document and summarize the diets and rearing protocols developed for Cerambycidae and provide guidance on how these can be used as models for developing new methods for related species.

7.2 Larval Rearing and Handling Methods

Methods for rearing 140 species of cerambycid larvae, for all or part of their developmental period, have been published. These include species from six of the eight subfamilies of Cerambycidae; 67 lamiines, 41 cerambycines, 20 lepturines, 8 spondylidines, 4 prionines, and 1 parandrine. Table 7.1 provides information on the species reared, their feeding preferences, generation time, geographic distribution, and diets used. A total of 30 species have been reared on artificial diets that contain no host material; another 57 species have been reared on artificial diets that contain dried-pulverized host material; and 8 species have been reared on both types of artificial diet. An additional 33 species have been reared on cut host material, and 12 species have been reared on both cut host material and one or more artificial diets.

In nature, larvae of species reared in the laboratory inhabit stems, roots, inner bark, or wood of living, dying, or dead hosts. The majority of these species feed on multiple hosts (95%), 64 species feed exclusively on deciduous trees, 35 on coniferous trees, 16 on herbaceous hosts, and 21 species feed on plants from two of these groups. Most of the documented species that have been reared are from the Northern Hemisphere, but this could be biased by the accessible literature or resources for conducting this type of work. The species that have been reared inhabit all four host conditions presented in Hanks (1999): healthy host (HH), stressed host (SH), weakened host (WH), and dead host (DH).

7.2.1 Larval Rearing Using Artificial Diets

7.2.1.1 Diets Containing No Host Material

Rearing larvae on an artificial diet that does not contain host material can have its advantages: eliminating labor-intensive and time-consuming efforts needed to harvest and prepare host material for use, producing large numbers of even aged/sized larvae for bioassays, potentially reducing diet contamination and changes needed, rearing year round, and providing a substrate for testing host-derived compounds that may confer resistance. However, a successful artificial diet can take a long time to develop. Therefore, it is important to utilize information, such as is summarized here, from previous artificial diets created for closely related species or species inhabiting the same host (in the same host condition).

Harley and Wilson (1968) were the first to develop an artificial diet for rearing cerambycid larvae that contained no host material. Twelve additional diets have been developed, often based on previously published diets. Six of these diets (Table 7.2: 3, 4, 7, 9, 10, and 13) have been used to rear only one species, and only one (6) has been evaluated for rearing larvae from five different cerambycid subfamilies. These artificial diets have been used to rear larvae that have multiple generations per year, a single generation per year, and those that take multiple years to complete their development. The larvae successfully reared on these diets come from all host types and conditions.

The ingredients and amounts required to make 1-L batches of each of the 13 artificial diets that contain no host material are given in Table 7.2. All but one of these diets is made by first bringing an agar and water solution to a boil (diet 12 is baked), mixing in the remaining ingredients either while boiling or at different points in the cooling process, pouring or spooning the diet into containers, and then letting the diet cool and dry for varying amounts of time before use. Several of the diets can then be held at 5–10°C for several weeks until needed without affecting larval performance on the diet.

All of these diets contain the following components: a nitrogen source, lipids (except 6 and 8), carbohydrates, vitamins and minerals, protective agents, and a bulking agent. All but two of these diets include cellulose as a bulking agent. Most of the diets contain large amounts of crude ground or powdered plant products (from corn, wheat, or soybeans), which provide both bulk and other essential components. These plant-based products provide protein, lipids, carbohydrates, and vitamins and minerals, but they

TABLE 7.1

Species of Cerambycid Larvae Reared in the Laboratory with Notes on Hosts Utilized, Generation Time, Occurrence, and Artificial Diets Used[*]

Subfamily	Species	Host Type	Host Condition Code	Generation	Occurrence	Diet	Diet/Rearing References
Cerambycinae	*Anelaphus villosus* (Fabricius, 1792)	Deciduous	HH	1–2 years	Eastern North America	11	Keena, unpublished data
Cerambycinae	*Aromia moschata* (Linnaeus, 1758)	Deciduous	WH	N/A	Europe, North Africa, Asia	C	Georgiev et al. 2004
Cerambycinae	*Callidium frigidum* Casey, 1912	Coniferous	DH	1 year	North America	1H	Gardiner 1970
Cerambycinae	*Cerambyx cerdo* Linnaeus, 1758	Deciduous	SH	2–5 years	Europe except north, Turkey	12	Pavlovic et al. 2012; Dojnov et al. 2010
Cerambycinae	*Cerambyx welensii* (Küster, 1846)	Deciduous	WH	N/A	Europe	6	De Viedma et al. 1985
Cerambycinae	*Ceresium guttaticolle* (Fairmaire 1850)	Deciduous	WH	1 year	Oceania (Fiji)	C	Waqa-Sakiti et al. 2014
Cerambycinae	*Ceresium scutellaris* Dillon & Dillon 1952	Deciduous	WH	1 year	Oceania (Fiji)	C	Waqa-Sakiti et al. 2014
Cerambycinae	*Ceresium vacillans* Dillon & Dillon 1952	Deciduous	WH	1 year	Oceania (Fiji)	C	Waqa-Sakiti et al. 2014
Cerambycinae	*Clytus ruricola* (Oliver, 1795)	Deciduous	DH	N/A	North America	1H	Gardiner 1970
Cerambycinae	*Enaphalodes rufulus* (Haldeman, 1847)	Deciduous	WH	2 years	North America	2, 7	Galford 1985; Galford 1969
Cerambycinae	*Eurphagus lundii* (Fabricius, 1793)	Deciduous	N/A	1 year	Indonesia, Malaysia, Thailand, Puerto Rico, Borneo, Myanmar, Laos, Sumatra, Burma, India, China, Vietnam, Cambodia	10H	Higashiyama et al. 1984
Cerambycinae	*Hylotrupes bajulus* (Linnaeus, 1758)	Coniferous	DH	3–6 years, 1–32 range	Originating in Europe, world distribution	6, 8, 8H	Cannon and Robinson 1982; Notario et al. 1993; Iglesias et al. 1989
Cerambycinae	*Megacyllene caryae* (Gahan, 1908)	Coniferous	SH	1 year	North America	2	Galford 1969
Cerambycinae	*Megacyllene melyi* (Chevrolat, 1862)	Herbaceous	HH	2–3 per year	South America, imported to Australia	11H	Boldt 1987
Cerambycinae	*Megacyllene robiniae* (Forster, 1771)	Deciduous	WH	1 year	North America	2, 2H	Galford 1969; Wollerman et al. 1969

(Continued)

TABLE 7.1 (Continued)

Species of Cerambycid Larvae Reared in the Laboratory with Notes on Hosts Utilized, Generation Time, Occurrence, and Artificial Diets Used[*]

Subfamily	Species	Host Type	Host Condition Code	Generation	Occurrence	Diet	Diet/Rearing References
Cerambycinae	*Meriellum proteus* (Kirby in Richardson, 1837)	Coniferous	DH	2 year	North America	1H	Gardiner 1970
Cerambycinae	*Neoclytus acuminatus* (Fabricius, 1775)	Deciduous	WH	1–3 per year	North America, introduced to Europe	2	Galford 1969
Cerambycinae	*Neoclytus caprea* (Say, 1824)	Deciduous	WH	1 year	North America	2	Galford 1969
Cerambycinae	*Neoclytus leucozonus* (Laporte and Gory, 1835)	Coniferous	DH	N/A	North America	1H	Gardiner 1970
Cerambycinae	*Oemona hirta* (Fabricius, 1775)	Deciduous	WH	2 years	New Zealand	21H	Wang et al. 2002
Cerambycinae	*Phoracantha recurva* Newman, 1840	Deciduous	SH	2–3 per year	Australia, introduced pest in many countries	C	Bybee et al. 2004
Cerambycinae	*Phoracantha semipunctata* (Fabricius, 1775)	Deciduous	SH	2–3 per year	Australia, introduced pest in many countries	6, C	De Viedma et al. 1985; Hanks et al. 1993; Bybee et al. 2004
Cerambycinae	*Phymatodes testaceus* (Linnaeus, 1758)	Deciduous and coniferous	DH	1–2 years	Europe, Japan, North Africa, North America	6	De Viedma et al. 1985
Cerambycinae	*Plagionotus arcuatus* (Linnaeus, 1758)	Deciduous	DH	2 years	Europe, North Africa, Caucasus, Iran	6	De Viedma et al. 1985
Cerambycinae	*Plagionotus detritus* (Linnaeus, 1758)	Deciduous	DH	1–2 years	Europe, Russia, Caucasus, North Kazakhstan, Transcaucasia, Iran, Near East	6	De Viedma et al. 1985
Cerambycinae	*Plagithmysus bilineatus* Sharp, 1896	Herbaceous	HH	1 year	Oceania (Hawaii)	9H	Stein and Haraguchi 1984
Cerambycinae	*Plagithmysus funebris* Sharp, 1896	Deciduous	HH	1 year	Oceania (Hawaii)	9H	Stein and Haraguchi 1984
Cerambycinae	*Plagithmysus varians* Sharp, 1896	Deciduous	HH	1 year	Oceania (Hawaii)	9H	Stein and Haraguchi 1984
Cerambycinae	*Purpuricenus ferrugineus* (Fairmaire, 1851)	Herbaceous	N/A	N/A	Southern Europe	6	De Viedma et al. 1985
Cerambycinae	*Sarosesthes fulminans* (Fabricius, 1775)	Deciduous	SH	N/A	North America	1H	Gardiner 1970
Cerambycinae	*Scituloglaucytes muriri* (Gressitt, 1940)	Deciduous	WH	1 year	Oceania (Fiji)	C	Waqa-Sakiti et al. 2014

(Continued)

TABLE 7.1 (Continued)

Species of Cerambycid Larvae Reared in the Laboratory with Notes on Hosts Utilized, Generation Time, Occurrence, and Artificial Diets Used[*]

Subfamily	Species	Host Type	Host Condition Code	Generation	Occurrence	Diet	Diet/Rearing References
Cerambycinae	*Semanotus japonicus* Lacordaire, 1869	Coniferous	WH	1 year	East Asia, Japan, Taiwan, China, introduced into Canada	20H, C	Kitajima 1999; Shibata 1995
Cerambycinae	*Semanotus ligneus* (Fabricius, 1787)	Coniferous	WH	1 year	North America	1H	Gardiner 1970
Cerambycinae	*Semanotus litigiosus* (Casey, 1891)	Coniferous	SH	N/A	North America	1H	Gardiner 1970
Cerambycinae	*Trichoferus holosericeus* (Rossi, 1790)	Deciduous	DH	2–3 years	Mediterranean region	21H	Palanti et al. 2010
Cerambycinae	*Xylotrechus arvicola* (Olivier, 1795)	Deciduous and herbaceous	WH	2 years	Europe, Russia, Caucasus, North Kazakhstan, Transcaucasia, Turkey, Iran, Near East, North Africa	12H,6	García-Ruiz et al. 2012; De Viedma et al. 1985
Cerambycinae	*Xylotrechus colonus* (Fabricius, 1775)	Deciduous and coniferous	WH	1 year	Eastern North America, west to central Texas	1H, 3, C	Galford 1969; Haack and Petrice 2009; Gardner 1970
Cerambycinae	*Xylotrechus pyrrhoderus* Bates, 1873	Herbaceous	HH	1 year	Korea, China, Japan	C	Ashihara 1982
Cerambycinae	*Xylotrechus sagittatus* (Germar, 1821)	Coniferous	WH	1 year	North America	1H	Gardiner 1970
Cerambycinae	*Xylotrechus undulatus* (Say, 1824)	Coniferous	WH	1 year	Canada, plus Alaska and northern United States	1H	Gardiner 1970
Lamiinae	*Acalolepta australis* (Boisduval, 1835)	Deciduous and coniferous	N/A	1 year	Australia, Papua New Guinea, Moluccas, Ireland, New Britain	10H	Higashiyama et al. 1984
Lamiinae	*Acalolepta holotephra* (Boisduval, 1835)	Deciduous	N/A	1 year	Papua New Guinea, Samoa, Vanuatu, Solomon Islands	10H	Higashiyama et al. 1984
Lamiinae	*Acalolepta luxuriosa* (Bates, 1873)	Deciduous	N/A	1 year	Russia, Korea, China, Taiwan, Japan	5H	Akutsu et al. 1980
Lamiinae	*Acalolepta tincturata* (Pascoe, 1866)	Deciduous	N/A	1 year	Papua New Guinea, Indonesia	10H	Higashiyama et al. 1984
Lamiinae	*Acalolepta vastator* (Newman, 1847)	Deciduous	WH	1 year	Australia	1	Goodwin and Pettit 1994

(Continued)

TABLE 7.1 (Continued)

Species of Cerambycid Larvae Reared in the Laboratory with Notes on Hosts Utilized, Generation Time, Occurrence, and Artificial Diets Used[*]

Subfamily	Species	Host Type	Host Condition Code	Generation	Occurrence	Diet	Diet/Rearing References
Lamiinae	*Acanthocinus aedilis* (Linnaeus, 1758)	Coniferous	WH	1–2 years	Europe, Russia, Caucasus, Kazakhstan, Korea, Mongolia, northern China	6	Notario et al. 1993
Lamiinae	*Acanthocinus obsoletus* (Olivier, 1795)	Coniferous	N/A	1 year	North America, Caribbean	1H, C	Haack and Petrice 2009; Gardiner 1970
Lamiinae	*Acanthocinus pusillus* Kirby, 1837	Coniferous	N/A	1 year	North America	1H	Gardiner 1970
Lamiinae	*Aegomorphus modestus* (Gyllenhal in Schoenherr, 1817)	Deciduous	N/A	N/A	North America	11, 1H	M. A. Keena, unpublished data; Gardiner 1970
Lamiinae	*Aerenicopsis championi* Bates, 1885	Herbaceous	HH	1 year	Middle America, Oceania, introduced to Hawaii and Australia	C	Palmer et al. 2000
Lamiinae	*Agapanthia asphodeli* (Latreille, 1804)	Herbaceous	HH	2–3 per year	Europe, Algeria, Tunisia, Turkey	6	De Viedma et al. 1985; Baragaño Galan et al. 1981; Notario and Baragaño 1987
Lamiinae	*Agapanthia cardui* (Linnaeus, 1767)	Herbaceous	HH	1 year	Europe and near East	6	De Viedma et al. 1985
Lamiinae	*Agapanthia irrorata* (Fabricius, 1787)	Herbaceous	HH	1 year	Europe, North Africa	6	De Viedma et al. 1985
Lamiinae	*Agapanthia villosoviridescens* (De Geer, 1775)	Herbaceous	HH	1 year	Europe, Russia, Caucasus, Kazakhstan, Turkey, Near East	6	De Viedma et al. 1985
Lamiinae	*Anoplophora glabripennis* (Motschulsky, 1854)	Deciduous	WH	1–2 years	China and Korea	15H, 19H, 11, 10, C	Zhang and Hu 1993; Zhao et al. 1999; Keena 2005; Dubois et al. 2002; Ludwig et al. 2002; Morewood et al. 2005
Lamiinae	*Anoplophora malasiaca* (Thomson, 1865)	Deciduous and coniferous	HH	1–2 years	Japan, Korea, Taiwan, China, Hong Kong, Malaysia, Myanmar, Viet Nam	7H, 8H, 18H	Murakoshi and Aono 1981; Lee and Lo 1998; Adachi 1994

(Continued)

TABLE 7.1 (Continued)

Species of Cerambycid Larvae Reared in the Laboratory with Notes on Hosts Utilized, Generation Time, Occurrence, and Artificial Diets Used[*]

Subfamily	Species	Host Type	Host Condition Code	Generation	Occurrence	Diet	Diet/Rearing References
Lamiinae	*Apriona germari* (Hope 1831)	Deciduous	HH	1–3 years	China, Korea, Japan, India, Laos, Malaysia, Myanmar, Nepal, Pakistan, Taiwan, Thailand, Vietnam, Far east Russia	17H	Yoon et al. 1997
Lamiinae	*Batocera laena* Thomson 1858	Deciduous	N/A	N/A	Australia, Papua New Guinea, Moluccas, New Britain	10H	Higashiyama et al. 1984
Lamiinae	*Dectes texanus* LeConte, 1852	Herbaceous	HH	1 year	North America	4, C	Hatchett et al. 1973; Niide et al. 2006
Lamiinae	*Dorcadion pseudopreissi* Breuning, 1962	Herbaceous	HH	2 years	Turkey	C	Kumral et al. 2012
Lamiinae	*Eupogonius pauper* LeConte, 1852	Deciduous and herbaceous	DH	2 years	North America	C	Purrington and Horn 1993
Lamiinae	*Glenea cantor* (Fabricius, 1787)	Deciduous	HH	5 per year	Southern Asia	C	Lu et al. 2011
Lamiinae	*Gracisybra flava* (Dillion & Dillon)	Deciduous	WH	1 year	Oceania (Fiji)	C	Waqa-Sakiti et al. 2014
Lamiinae	*Graphisurus fasciatus* (DeGeer, 1775)	Deciduous and coniferous	DH	N/A	North America	1H, C	Gardiner 1970; Haack and Petrice 2009
Lamiinae	*Hestimidus humeralis* Breuning 1939	Deciduous	WH	1 year	Oceania (Fiji)	C	Waqa-Sakiti et al. 2014
Lamiinae	*Hyperplatys aspersa* (Say, 1824)	Deciduous and herbaceous	DH	1.5 per year	North America	1H, C	Gardiner 1970; Purrington and Horn 1993
Lamiinae	*Leiopus nebulosus* (Linnaeus, 1758)	Deciduous	N/A	1–2 years	Europe, Russia	6	De Viedma et al. 1985
Lamiinae	*Microgoes oculatus* (LeConte, 1862)	Deciduous	DH	2–4 years	North America	1H	Gardiner 1970
Lamiinae	*Moechotypa diphysis* (Pascoe, 1871)	Deciduous and herbaceous	DH	1 year	Japan	13H	Kosaka 2011

(Continued)

TABLE 7.1 (Continued)

Species of Cerambycid Larvae Reared in the Laboratory with Notes on Hosts Utilized, Generation Time, Occurrence, and Artificial Diets Used*

Subfamily	Species	Host Type	Host Condition Code	Generation	Occurrence	Diet	Diet/Rearing References
Lamiinae	*Monochamus alternatus* Hope, 1842	Coniferous	SH	1–2 years	China, Japan and Korea	10H, 13H, C	Higashiyama et al. 1984; Kosaka and Ogura 1990; Togashi 1989
Lamiinae	*Monochamus carolinensis* (Oliver, 1795)	Coniferous	SH	1–2 years	North America	1, 16H, C	Alya and Hain 1987; Necibi and Linit 1997; Linit 1985
Lamiinae	*Monochamus clamator latus* Casey, 1924	Coniferous	SH	1–2 years	North America	1H	Gardiner 1970
Lamiinae	*Monochamus galloprovincialis* (Oliver, 1795)	Coniferous	SH	1–2 years	Europe	3H, C, 13	Carle 1969; Koutroumpa et al. 2008; Akbulut et al. 2007; Petersen-Silva et al. 2014
Lamiinae	*Monochamus notatus* (Drury, 1773)	Coniferous	SH	2 years	North America	1H, C	Gardiner 1970; Haack and Petrice 2009
Lamiinae	*Monochamus scutellatus* (Say, 1824)	Coniferous	SH	1–2 years	North America	1H, C	Gardiner 1970; Breton et al. 2013
Lamiinae	*Monochamus titillator* (Fabricius, 1775)	Coniferous	SH	up to 3 gen a year	North America	1	Alya and Hain 1987
Lamiinae	*Morimus funereus* (Mulsant, 1862)	Deciduous and coniferous	DH	3–4 years	Europe (IUCN Threatened Species)	12, 9	Dojnov et al. 2012; Ivanovic et al. 2002
Lamiinae	*Musaria affinis affinis* (Harrer, 1784)	Herbaceous	HH	1 year	Europe except north, Russia, Caucasus, Transcaucasia	6	De Viedma et al. 1985
Lamiinae	*Neosciadella brunnipes* Dillon & Dillon 1952	Deciduous	WH	1 year	Oceania (Fiji)	C	Waqa-Sakiti et al. 2014
Lamiinae	*Neosciadella inflexa* Dillon & Dillon 1952	Deciduous	WH	1 year	Oceania (Fiji)	C	Waqa-Sakiti et al. 2014
Lamiinae	*Neosciadella multivittata* Dillon & Dillon 1952	Deciduous	WH	1 year	Oceania (Fiji)	C	Waqa-Sakiti et al. 2014
Lamiinae	*Neosciadella spixi* Dillon & Dillon 1952	Deciduous	WH	1 year	Oceania (Fiji)	C	Waqa-Sakiti et al. 2014

(Continued)

TABLE 7.1 (Continued)

Species of Cerambycid Larvae Reared in the Laboratory with Notes on Hosts Utilized, Generation Time, Occurrence, and Artificial Diets Used[*]

Subfamily	Species	Host Type	Host Condition Code	Generation	Occurrence	Diet	Diet/Rearing References
Lamiinae	*Oberea oculata* (Linnaeus, 1758)	Deciduous	WH	1–3 years	West Palearctic	C	Georgiev et al. 2004
Lamiinae	*Oberea quadricallosa* LeConte, 1874	Deciduous	HH	2–3 years	North America	1H	Gardiner 1970
Lamiinae	*Oberea tripunctata* (Swederus, 1787)	Deciduous and herbaceous	HH	1–3 years	North America	1H	Gardiner 1970
Lamiinae	*Oncideres cingulata* (Say, 1826)	Deciduous	HH	1 year	North America	C	Rice 1995
Lamiinae	*Oopsis velata* Dillon & Dillon 1952	Deciduous	WH	1 year	Oceania (Fiji)	C	Waqa-Sakiti et al. 2014
Lamiinae	*Oopsis zitja* Dillon & Dillon 1952	Deciduous	WH	1 year	Oceania (Fiji)	C	Waqa-Sakiti et al. 2014
Lamiinae	*Oplosia nubila* (LeConte, 1862)	Deciduous	DH	N/A	North America	1H	Gardiner 1970
Lamiinae	*Palame anceps* (Bates, 1864)	Deciduous	WH	N/A	French Guyana, Brazil, Equator	C	Berkov and Tavakilian 1999
Lamiinae	*Palame crassimana* Bates, 1864	Deciduous	WH	N/A	Venezuela, French Guyana, Brazil, Peru, Guyana, Equator, Surinam	C	Berkov and Tavakilian 1999
Lamiinae	*Palame mimetica* Monné, 1985	Deciduous	WH	N/A	Venezuela, French Guyana, Brazil, Surinam	C	Berkov and Tavakilian 1999
Lamiinae	*Paraglenea fortunei* (Saunders, 1853)	Deciduous	HH	N/A	Indochina, continental China, and Taiwan, invasive in Japan	20H	Kitajima and Makihara 2011
Lamiinae	*Phytoecia icterica* (Schaller, 1783)	Herbaceous	HH	1 year	Europe, Turkey, Caucasus, Russia	6	De Viedma et al. 1985
Lamiinae	*Phytoecia rufiventris* Gautier des Cottes, 1870	Herbaceous	HH	1 year	Russia, Mongolia, Korea, China, Taiwan, Japan, Vietnam	22H	Shintani 2011
Lamiinae	*Plagiohammus spinipennis* (Thomson, 1860)	Herbaceous	HH	N/A	Central America, Oceania, South America, introduced into Hawaii	1, C	Harley and Willson 1968; Hadlington and Johnson 1973
Lamiinae	*Plectrodera scalator* (Fabricius, 1792)	Deciduous	WH	2 years	North America	11	M. A. Keena, unpublished data
Lamiinae	*Pogonocherus mixtus* Halderman, 1847	Coniferous and deciduous	DH	1 year	North America	1H	Gardiner 1970

(Continued)

TABLE 7.1 (Continued)

Species of Cerambycid Larvae Reared in the Laboratory with Notes on Hosts Utilized, Generation Time, Occurrence, and Artificial Diets Used[*]

Subfamily	Species	Host Type	Host Condition Code	Generation	Occurrence	Diet	Diet/Rearing References
Lamiinae	*Pogonocherus penicillatus* LeConte in Agassiz, 1850	Coniferous	DH	1 year	North America	1H	Gardiner 1970
Lamiinae	*Psacothea hilaris* Pascoe, 1857	Deciduous	WH	1–2 years	Asia, invasive in Europe	14H	Fukaya and Honda 1992
Lamiinae	*Psenocerus supernotatus* (Say, 1823)	Herbaceous and deciduous	DH	N/A	North America	C	Purrington and Horn 1993
Lamiinae	*Saperda calcarata* Say, 1824	Deciduous	WH	2–5 years	North America	1H	Gardiner 1970
Lamiinae	*Saperda discoidea* Fabricius, 1798	Deciduous	WH	1 year	North America	1H	Gardiner 1970
Lamiinae	*Saperda populnea* (Linnaeus, 1758)	Deciduous	WH	2 years	Europe	C	Georgiev et al. 2004
Lamiinae	*Saperda similis* Laicharting, 1784	Deciduous	WH	2–3 years	Europe, Near East	C	Georgiev et al. 2004
Lamiinae	*Saperda vestita* Say, 1824	Deciduous	WH	2–3 years	North America	1H	Gardiner 1970
Lamiinae	*Sormida cinerea* Dillon & Dillon 1952	Deciduous	WH	1 year	Oceania (Fiji)	C	Waqa-Sakiti et al. 2014
Lamiinae	*Sormida maculicollis* (Thomson 1865)	Deciduous	WH	1 year	Oceania (Fiji)	C	Waqa-Sakiti et al. 2014
Lamiinae	*Sybra alternans* (Wiedemann, 1825)	Deciduous	N/A	N/A	Southeast Asia, Introduced to United States (Hawaii, California and Florida)	9H	Stein and Haraguchi 1984
Lamiinae	*Sybromimus obliquatus* Breuning 1940	Deciduous	WH	1 year	Oceania (Fiji)	C	Waqa-Sakiti et al. 2014
Lamiinae	*Tetraopes tetrophthalmus* (Forster, 1771)	Herbaceous	HH	1 year	North America	1H	Gardiner 1970
Lamiinae	*Urgleptes querci* (Fitch, 1858)	Deciduous and coniferous	DH	N/A	North America	C	Purrington and Horn 1993
Lepturinae	*Acmaeops proteus* (Kirby in Richardson, 1837)	Coniferous	DH	2–3 years	North America	1H	Gardiner 1970
Lepturinae	*Anastrangalia sanguinolenta* (Linnaeus, 1761)	Coniferous	WH/DH	2–3 years	Europe, Turkey, Caucasus, Transcaucasia	6	De Viedma et al. 1985; Notario et al. 1993
Lepturinae	*Anastrangalia sanguinea* (LeConte, 1859)	Coniferous	DH	1+ years	North America	1H	Gardiner 1970

(Continued)

TABLE 7.1 (Continued)

Species of Cerambycid Larvae Reared in the Laboratory with Notes on Hosts Utilized, Generation Time, Occurrence, and Artificial Diets Used[a]

Subfamily	Species	Host Type	Host Condition Code	Generation	Occurrence	Diet	Diet/Rearing References
Lepturinae	*Anthophylax attenuatus* (Haldeman, 1847)	Deciduous	DH	N/A	North America	1H	Gardiner 1970
Lepturinae	*Anthophylax cyaneus* (Haldeman, 1847)	Deciduous	N/A	N/A	North America	1H	Gardiner 1970
Lepturinae	*Centrodera decolorata* (Harris, 1838)	Deciduous	DH	N/A	North America	1H	Gardiner 1970
Lepturinae	*Evodinus monticola* (Randall, 1838)	Coniferous	N/A	N/A	North America	1H	Gardiner 1970
Lepturinae	*Grammoptera subargentata* (Kirby in Richardson, 1837)	Deciduous	N/A	N/A	North America	1H	Gardiner 1970
Lepturinae	*Judolia montivagans montivagans* (Couper, 1864)	Coniferous and deciduous	DH	2 years	North America	1H	Gardiner 1970
Lepturinae	*Paracorymbia stragulata* (Germar, 1824)	Coniferous	DH	2 years	Europe	12H, 6, 8	Iglesias et al. 1989; De Viedma et al. 1985; Notario et al. 1993
Lepturinae	*Pidonia ruficollis* (Say, 1824)	Deciduous	N/A	N/A	North America	1H	Gardiner 1970
Lepturinae	*Rhagium bifasciatum* (Linnaeus, 1758)	Coniferous and deciduous	DH	2–3 years	Europe except NE, Caucasus, Transcaucasia, Turkey	6	De Viedma et al. 1985; Notario et al. 1993
Lepturinae	*Rhagium inquisitor* (Linnaeus, 1758)	Coniferous and deciduous	DH	2 years	Europe and Northern Asia (excluding China), North America	12H, 6, 8	Iglesias et al. 1989; De Viedma et al. 1985; Notario et al. 1993
Lepturinae	*Stictoleptura scutellata* (Fabricius, 1781)	Deciduous	DH	2–3 years	Europe except north, North Africa, Turkey, Caucasus, Transcaucasia, North Iran	6	De Viedma et al. 1985
Lepturinae	*Stictoleptura canadensis* (Oliver, 1795)	Coniferous	DH	N/A	North America	1H	Gardiner 1970
Lepturinae	*Stictoleptura fontenayi* (Mulsant, 1839)	Coniferous and deciduous	DH	2–3 years	Africa, Europe	6	De Viedma et al. 1985
Lepturinae	*Stictoleptura rubra* (Linnaeus, 1758)	Coniferous	DH	2–3 years	Europe, Russia, North Africa	6	De Viedma et al. 1985; Notario et al. 1993

(Continued)

TABLE 7.1 (Continued)

Species of Cerambycid Larvae Reared in the Laboratory with Notes on Hosts Utilized, Generation Time, Occurrence, and Artificial Diets Used[*]

Subfamily	Species	Host Type	Host Condition Code	Generation	Occurrence	Diet	Diet/Rearing References
Lepturinae	*Trachysida mutabilis* (Newman, 1841)	Coniferous and deciduous	DH	N/A	North America	1H	Gardiner 1970
Lepturinae	*Trigonarthris minnesotana* (Casey, 1913)	Ceciduous	DH	N/A	North America	1H	Gardiner 1970
Lepturinae	*Typocerus velutinus* (Olivier, 1795)	Deciduous	DH	1+ years	North America	1H	Gardiner 1970
Parandrinae	*Neandra marginicollis marginicollis* Schaeffer, 1929	Deciduous and coniferous	DH	2–3 years	North America, introduced into Europe	1H	Gardiner 1970
Prioninae	*Ergates faber* (Linnaeus, 1761)	Coniferous	DH	3+ years	Central and South Europe, North Africa	12H, 6, 8	Iglesias et al. 1989; De Viedma et al. 1985; Notario and Baragaño 1978; Notario et al. 1993
Prioninae	*Orthosoma brunneum* (Forster, 1771)	Deciduous and coniferous	DH	2–3 years	North America	1H	Gardiner 1970
Prioninae	*Prionoplus reticularis* White, 1843	Coniferous and deciduous	DH	2+ years	New Zealand	21H	Rogers et al. 2002
Prioninae	*Prionus imbricornis* (Linnaeus, 1767)	Deciduous	WH	3–5 years	North America	4H	Payne et al. 1975
Spondylidinae	*Arhopalus ferus* (Mulsant, 1839)	Coniferous	DH	2–4 years	Europe, North Africa, Northern Asia, New Zealand	6	Notario et al. 1993
Spondylidinae	*Arhopalus foveicollis* Halderman, 1847	Coniferous	DH	N/A	Eastern North America	1H	Gardiner 1970
Spondylidinae	*Arhopalus rusticus* Linnaeus, 1758	Coniferous	WH	2 years	North America, Northern and Central Europe, Siberia, Korea, Mongolia, Japan and the Northern China	12H,21H, 6, 8	Iglesias et al. 1989; Rogers et al. 2002; De Viedma et al. 1985

(Continued)

TABLE 7.1 (Continued)

Species of Cerambycid Larvae Reared in the Laboratory with Notes on Hosts Utilized, Generation Time, Occurrence, and Artificial Diets Used*

Subfamily	Species	Host Type	Host Condition Code	Generation	Occurrence	Diet	Diet/Rearing References
Spondylidinae	*Arhopalus syriacus* Reitter, 1895	Coniferous	DH	2–3 years	Mediterranean region, Canary Islands	3H, 12H, 6, 8	Carle 1969; Iglesias et al. 1989; De Viedma et al. 1985
Spondylidinae	*Asemum striatum* Linnaeus, 1758	Coniferous	WH	1–3 years	Europe, Siberia, Korea, Mongolia, Japan and Sakhalin Island	1H	Gardiner 1970
Spondylidinae	*Atimia confusa* (Say, 1826)	Coniferous	DH	1–2 years	North America	1H, 12H	Gardiner 1970, Iglesias et al. 1989
Spondylidinae	*Spondylis buprestoides* Linnaeus, 1758	Coniferous	DH	2–3 years	Europe	6	De Viedma et al. 1985; Notario et al. 1993
Spondylidinae	*Tetropium cinnamopterum* Kirby, 1837	Coniferous	SH	N/A	North America	1H	Gardiner 1970

* Specie names were verified using the Integrated Taxonomic Information System (http://www.itis.gov) or Cerambycidae of the World (http://cerambycidae.org/search). Other information on the species was obtained from Craighead (1923), Jeniš (2001), Linsley (1961, 1962a, 1962b, 1963, 1964), Linsley and Chemsak (1972, 1976, 1984, 1995, 1997), http://wiki.bugwood.org, http://www.cerambyx.uochb.cz/, http://www.cerambycoidea.com, or links to online references given in the previously mentioned Web sites. Host condition codes were assigned based on the system presented in Hanks (1999) where HH = healthy host, SH = stressed host, WH = weakened host, and DH = dead host. Diet numbers are assigned based on the diets in Tables 7.2 (numbers only) and 7.3 (numbers followed by an H), and C = reared in host material. N/A = information not available.

TABLE 7.2

Diet Ingredients (in grams, unless specified) Required to Make One Liter of Each of 12 Artificial Diets for Cerambycid Larvae that Contain No Host Material

Ingredient	Harley and Willson 1968	Galford 1969	Galford 1969	Hatchett et al. 1973	Payne et al. 1975	Notario and Baragaño 1978	Galford 1985	Iglesias et al. 1989	Ivanovic et al. 2002	Dubois et al. 2002	Keena 2005	Dojnov et al. 2012	Petersen-Silva et al. 2014
	1	2	3	4	5	6	7	8	9	10	11	12	13
Agar	24.3	24.6	25.9	26.8	23.5	40.9	21.1	37.0	8.6	35.4	21.8	62.0	30.0
Lecithin	1.2	0.6	–	–	–	–	–	–	–	–	–	–	–
Alphacel (cellulose)	51.3	246.0	226.7	22.3	144.1	20.5	198.0	52.9	–	176.4	121.5	–	160.0
corn meal	–	–	–	–	–	–	–	–	107.5	–	–	–	–
Palenta	–	–	–	–	–	–	–	–	–	–	–	620.3	–
Wheat germ	–	15.4	97.2	52.1	28.2	–	79.2	–	–	–	52.9	–	–
Soybean powder	–	–	–	–	–	–	–	–	–	60.0	–	–	50.0
Soybean protein	–	12.3	–	–	–	–	–	–	–	–	–	–	–
Sucrose	11.7	12.3	–	40.2	32.8	25.6	6.6	66.1	21.5	25.3	15.6	155.1	20.0
Dextrose	11.7	–	–	–	–	–	–	–	–	–	–	–	–
Fructose	–	6.2	–	–	–	–	–	–	–	–	–	–	–
Glucose	–	6.2	–	–	–	15.3	–	39.7	–	–	–	–	–
Potato starch	11.7	–	–	–	–	–	–	–	–	25.3	–	–	20.0
Cholesterol	0.5	0.6	–	–	–	–	0.1	–	0.1	1.5	0.9	–	–
Linoleic acid	0.5	–	–	–	–	–	–	–	–	–	–	–	–
Soybean oil (mL)	–	–	–	–	–	–	–	–	–	5.9	–	–	4.0
Wheat germ oil (mL)	–	3.1	–	7.7	–	–	2.6	–	–	–	4.6	–	–
10% potassium hydroxide (mL)	9.7	–	–	2.7	–	–	–	–	–	–	–	–	–
Aureomycin	–	–	–	0.3	–	–	–	–	–	–	–	–	–
Benzoic acid	–	–	–	–	–	1.0	–	2.6	–	–	–	–	–
Chloramphenicol	–	–	–	–	–	–	0.4	–	–	–	–	–	–
Ethyl alcohol 95%	12.1	–	–	9.1	–	–	4.0	13.2	–	–	–	6.2	3.0
Formaldehyde 10% (mL)	–	–	–	–	7.5	–	–	–	–	–	–	–	–

(Continued)

TABLE 7.2 (Continued)

Diet Ingredients (in grams, unless specified) Required to Make One Liter of Each of 12 Artificial Diets for Cerambycid Larvae that Contain No Host Material

Ingredient	Harley and Willson 1968	Galford 1969	Galford 1969	Hatchett et al. 1973	Payne et al. 1975	Notario and Baragaño 1978	Galford 1985	Iglesias et al. 1989	Ivanovic et al. 2002	Dubois et al. 2002	Keena 2005	Dojnov et al. 2012	Petersen-Silva et al. 2014
	1	2	3	4	5	6	7	8	9	10	11	12	13
Formaldehyde 40% (mL)	–	–	–	0.9	–	–	–	–	–	–	–	–	–
Methyl paraben	1.1	0.8	0.8	1.6	1.9	–	1.3	2.6	2.2	2.5	1.6	1.2	–
Potassium sorbate	–	–	–	2.4	–	–	–	–	–	–	–	–	–
Propionic acid (mL)	–	–	–	–	–	–	–	–	–	1.7	–	–	1.0
Sodium propionate	–	–	–	–	–	–	–	–	–	–	1.2	–	–
Sorbic acid	1.5	1.5	1.6	–	1.9	0.5	1.3	–	–	2.5	1.6	–	–
Brewer's yeast	–	30.8	–	–	–	30.7	–	79.4	–	45.6	–	155.1	40.0
Torula yeast	–	–	–	–	–	–	13.2	–	–	–	28.1	–	–
Casein	23.3	–	–	37.5	32.8	12.3	6.6	31.7	–	15.2	9.4	–	–
L-Methionine	–	–	–	–	–	–	–	2.6	–	–	–	–	–
Ascorbic acid	–	–	–	–	3.8	4.1	–	10.6	–	2.5	–	–	–
Choline chloride	–	–	–	–	0.9	–	–	–	–	–	–	–	–
Citric acid	–	–	–	–	–	–	–	–	–	2.0	–	–	1.0
USDA Vitamin pre mix 26862	–	–	–	–	–	–	–	–	–	–	4.4	–	–
Vanderzant's vitamin mix (mL)	19.4	–	–	–	9.4	20.5	–	52.9	–	–	–	–	–
Vitamin B	–	0.1	–	–	–	–	–	–	–	–	–	–	–
Vitamin diet fortification mix	–	9.2	–	10.7	–	–	3.3	–	–	–	–	–	–
Wesson's salt mixture	3.5	15.4	–	10.7	9.4	10.2	2.0	26.5	–	4.6	2.8	–	–
Water (mL)	816.7	615.1	647.8	774.7	703.9	818.9	660.2	582.0	860.1	653.6	685.3	–	671.0

Note: Numbers 1–13 indicate diet number. Dashes indicate the ingredient was not part of that diet.

need to be processed or heated to denature enzymes and reduce unpalatable components (Cohen 2003). These ingredients can also be of variable quality among vendors or from the same vendor, so monitoring quality of the product and tracking quality control measures for insect colonies can help identify new problems before they have a negative effect. For example, *Anoplophora glabripennis* larvae will develop a gut obstruction if the cellulose fibers are too short (M. A. Keena, unpublished data), so purchasing from a company that has tested their product for use with insects is best.

All but five of the diets contain casein, which provides protein containing all but three of the essential amino acids in high quantities and well-balanced proportions (Reinecke 1985; Cohen 2003). Sucrose or various monosaccharaides are added to all but one of the diets because they are important as building materials and fuels, and they may also be phagostimulants. Yeast is added to eight of the diets and provides a good source of protein as well as vitamins and minerals. Only four diets (3, 9, 12, 13) do not include both Wesson salt mixture (a mineral mixture) and vitamins. There are some indications that more work on the dietary requirements of cerambycids with regard to particular vitamins and minerals is needed. For example, unlike *Lymantria dispar* (a caterpillar that feeds on leaves), *A. glabripennis* (wood feeder) larvae do not require higher levels of available iron in the diet. In fact, not adding any iron in the salt mix improves development of these larvae (Keena 2005). Additionally, ensuring that the vitamin A in the diet is either released slowly (incorporated in starch beadlets) or is maintained in the diet over time improves *A. glabripennis* larval development and reduces the number of diet changes needed (M. A. Keena, unpublished data).

Protective agents are added to all the diets to prevent microbial contamination, oxidation, or other means of destruction of essential nutrients. The function of these agents is antibacterial (e.g., Aureomycin or Chloramphenicol), antifungal (e.g., sorbic acid, methyl paraben, propionic acid, and formaldehyde) or antioxidant (e.g., ascorbic acid). Although cerambycids seem to be fairly tolerant of many of these compounds, it is always a fine balance between incorporating enough protective ingredients to preserve the diet but not so much that the insect is adversely affected. Some of these ingredients are also considered dangerous for human health (e.g., formaldehyde) and safer alternatives may need to be investigated. If problems with microbial contamination arise and modifications to the diet are needed, Cohen (2003) provided information on the protective agents. For more information on how to deal with microbial contaminants and how to identify which ones are present, see the control of pathogens and microbial contaminants in insect rearing section in *Advances and Challenges in Insect Rearing* (King and Leppla 1984).

7.2.1.2 Diets Containing Pulverized Host Material

The earliest artificial cerambycid diets containing host material were two developed in 1969 (Carle 1969; Wollerman et al. 1969) and a diet originally developed in 1965 for rearing spruce budworm that was modified a year later (Gardiner 1970). Twenty additional diets have been developed, often based on previously published diets (Table 7.3). Five of the listed diets use commercially available formulations, with minor modifications, originally developed for *Bombyx mori* (10H, 13H, and 14H), phytophagus insects (22H), and *Vanessa cardui* (16H, Table 7.3). Nine of these diets (2H, 4H, 5H, 7H, 14H, 16H, 17H, 19H, and 21H) have been used to rear only one species, and only one (1H) has been evaluated for rearing larvae from six different cerambycid subfamilies. These artificial diets have been used to rear larvae that have multiple generations per year, a single generation per year, and those that take multiple years to complete their development. The larvae successfully reared on these diets come from all host types and conditions.

The ingredients and amounts required to make one liter batches of each of the 23 artificial diets containing host material are given in Table 7.3. The host material to be incorporated into the diet is first dried and then pulverized before being used. The dried pulverized host material added to these diets averages 14% by weight (range 1–40%). All of these diets (except 13H, 14H, 22H, and 23H) are made by first bringing an agar and water solution to a boil, mixing in the remaining ingredients either while boiling or at different points in the cooling process, pouring or spooning the diet into containers, and then letting the diet cool and dry for varying amounts of time before use.

About half of these diets have components added that specifically provide a nitrogen source (casein or peptone), carbohydrates (sucrose or potato starch), vitamins, and minerals (via mixtures), and four have lipids added (cholesterol or oils). Thirteen of these diets have cellulose added as a bulking agent,

TABLE 7.3

Diet Ingredients (in grams, unless specified) Required to Make One Liter of Each of the 23 Artificial Diets for Cerambycid Larvae that Contain Dried, Pulverized Host Material

Ingredient	Gardiner 1970	Wollerman et al. 1969	Carle 1969	Payne et al. 1975	Akutsu et al. 1980	Baragaño Galán et al. 1981	Murakoshi and Aono 1981	Cannon and Robinson 1982	Stein and Haraguchi 1984	Higashiyama et al. 1984	Boldt 1987	Iglesias et al. 1989
	1H	2H	3H	4H	5H	6H	7H	8H	9H	10H	11H	12H
Water (mL)	797.4	666.1	779.2	684.0	745.7	687.5	710.7	516.4	851.4	727.3	816.7	572.1
Agar	24.3	22.2	37.1	22.8	33.6	27.5	17.8	34.4	49.9	31.1	24.3	28.6
Lecithin	–	–	–	–	–	–	–	–	0.7	–	1.2	–
Alphacel (cellulose)	4.9	88.8	33.4	91.2	67.1	–	40.9	129.1	–	–	–	–
Powdered host material	48.9	106.6	111.3	77.1	134.2	60.5	63.5	129.1	10.1	103.9	51.3	207.7
Soybean powder	–	–	–	–	–	–	102.8	44.8	–	–	–	–
Corn meal	–	–	–	–	–	60.5	–	–	–	–	–	–
Wheat germ	29.3	26.6	–	27.4	–	121.0	–	39.6	4.4	–	–	116.3
Sucrose	34.2	31.1	–	31.9	–	–	–	–	5.1	20.8	11.7	0.0
Dextrose	–	–	–	–	–	–	–	–	–	–	11.7	–
Glucose	–	–	–	–	–	–	–	–	–	–	–	–
Potato starch	–	–	–	–	–	–	–	–	–	10.4	11.7	–
Casein	34.2	31.1	–	31.9	–	–	–	–	5.1	–	23.3	–
Ascorbic acid	3.9	3.6	–	3.6	–	1.7	–	44.8	–	–	–	1.7
Choline chloride	1.0	0.9	–	0.9	–	–	–	5.2	0.1	–	–	–
Citric acid	–	–	–	–	–	–	–	3.4	–	–	–	–
Vitamin mix (see reference for mix)	9.8	8.9	14.8	9.1	–	–	7.4	–	51.1	–	19.4	–
Wesson's salt mixture	9.8	8.9	–	9.1	–	–	8.9	–	1.5	–	3.5	–
Bacto-peptone	–	–	–	–	–	–	–	10.3	–	–	–	–
Peptone	–	–	–	–	–	–	–	–	2.0	–	–	–
Cholesterol	–	–	–	–	–	–	–	13.8	–	–	0.5	–
Linoleic acid	–	–	–	–	–	–	–	–	–	–	0.5	–
Soybean oil (mL)	–	–	–	–	–	–	–	–	–	–	–	–

(Continued)

TABLE 7.3 (Continued)

Diet Ingredients (in grams, unless specified) Required to Make One One Liter of Each of the 23 Artificial Diets for Cerambycid Larvae that Contain Dried, Pulverized Host Material

Ingredient	Gardiner 1970	Wollerman et al. 1969	Carle 1969	Payne et al. 1975	Akutsu et al. 1980	Baragaño Galán et al. 1981	Murakoshi and Aono 1981	Cannon and Robinson 1982	Stein and Haraguchi 1984	Higashiyama et al. 1984	Boldt 1987	Iglesias et al. 1989
	1H	2H	3H	4H	5H	6H	7H	8H	9H	10H	11H	12H
10% potassium hydroxide (mL)	—	—	—	—	—	—	—	—	—	—	9.7	—
Acetic acid (25%)	0.3	—	—	—	—	—	—	—	1.5	—	—	—
Aureomycin	—	—	—	—	—	—	—	—	0.1	—	—	—
Chloramphenicol	—	—	—	—	—	—	—	—	—	0.01	—	—
Pyruvic acid	—	—	—	—	—	—	—	—	—	0.5	—	—
Benzoic acid	—	—	0.7	—	—	2.8	—	—	—	—	—	2.9
Butyl paraben	—	1.8	—	—	—	—	—	—	—	—	—	—
Ethyl alcohol 95% (mL)	—	—	—	7.3	—	5.5	—	—	11.2	—	12.1	35.8
Formaldehyde 10% (mL)	—	—	—	—	—	—	—	—	0.6	—	—	—
Formaldehyde 37% (mL)	0.5	—	—	—	—	—	—	13.8	—	—	—	—
Methyl paraben	1.5	1.8	1.1	1.8	2.2	2.8	—	1.7	1.2	2.1	1.1	3.6
Potassium hydroxide	—	—	—	—	—	—	—	—	2.2	—	—	—
Potassium sorbate	—	—	—	—	—	—	—	—	—	—	—	—
Propionic acid (mL)	—	—	—	—	—	—	—	3.4	—	—	—	—
Propylene glycol (mL)	—	—	—	—	—	—	—	—	—	—	—	—
Sorbic acid	—	1.8	—	1.8	2.2	—	48.2	—	1.6	—	1.5	—
Bacto yeast extract	—	—	—	—	—	—	—	10.3	—	—	—	—
Dry yeast (see reference for type)	—	—	22.3	—	15.0	30.2	—	—	—	—	—	31.5
Dry mix Insecta F-II, Nihon Nosan Co.	—	—	—	—	—	—	—	—	—	—	—	—
Dry commercial diet for Bombyx mori	—	—	—	—	—	—	—	—	—	103.9	—	—
Dry commercial diet for Vanessa cardui	—	—	—	—	—	—	—	—	—	—	—	—

(Continued)

TABLE 7.3 (Continued)

Diet Ingredients (in grams, unless specified) Required to Make One Liter of Each of the 23 Artificial Diets for Cerambycid Larvae that Contain Dried, Pulverized Host Material

Ingredient	Kosaka and Ogura 1990	Fukaya and Honda 1992	Zhang and Hu 1993	Necibi and Limit 1997	Yoon et al. 1997	Lee and Lo 1998	Zhao et al. 1999	Kitajima 1999	Rogers et al. 2002	Palanti et al. 2010	Shintani 2011
	13H	14H	15H	16H	17H	18H	19H	20H	21H	22H	23H
Water (mL)	375.0	–	604.7	374.6	842.1	541.5	499.9	480.8	697.9	518.2	375.0
Agar	–	–	10.1	–	–	36.1	35.0	25.7	32.5	–	–
Lecithin	–	–	–	–	–	–	–	–	–	–	–
Alphacel (cellulose)	–	–	100.8	–	–	–	70.0	109.0	73.0	45.5	–
Powdered host material	250.0	–	141.1	249.8	140.4	415.2	200.0	96.2	48.7	172.7	156.3
Soybean powder	–	–	40.3	–	–	–	60.0	–	–	–	–
Corn meal	–	–	–	–	–	–	–	–	–	–	–
Wheat germ	–	–	–	–	–	–	–	–	24.3	–	–
Sucrose	–	–	25.2	–	–	–	25.1	88.1	20.3	–	–
Dextrose	–	–	–	–	–	–	–	–	–	–	–
Glucose	–	–	–	–	–	–	–	–	12.2	–	–
Potato starch	–	–	–	–	–	–	25.1	32.8	–	145.5	–
Casein	–	–	45.4	–	–	–	15.0	98.4	24.3	–	–
Ascorbic acid	–	–	–	–	–	–	3.0	4.4	3.5	–	–
Choline chloride	–	–	1.0	–	–	–	–	–	–	–	–
Citric acid	–	–	4.0	–	–	–	2.0	4.3	–	–	–
Vitamin mix (see reference for mix)	–	–	10.1	–	–	–	–	–	16.2	–	–
Wesson's salt mixture	–	–	10.1	–	–	–	4.5	6.5	8.1	–	–
Bacto-peptone	–	–	–	–	–	–	–	–	–	–	–
Peptone	–	–	–	–	–	–	–	–	–	–	–
Cholesterol	–	–	–	–	–	–	1.5	1.0	–	–	–

(Continued)

TABLE 7.3 (*Continued*)

Diet Ingredients (in grams, unless specified) Required to Make One Liter of Each of the 23 Artificial Diets for Cerambycid Larvae that Contain Dried, Pulverized Host Material

Ingredient	Kosaka and Ogura 1990	Fukaya and Honda 1992	Zhang and Hu 1993	Necibi and Linit 1997	Yoon et al. 1997	Lee and Lo 1998	Zhao et al. 1999	Kitajima 1999	Rogers et al. 2002	Palanti et al. 2010	Shintani 2011
	13H	14H	15H	16H	17H	18H	19H	20H	21H	22H	23H
Linoleic acid	–	–	–	–	–	–	–	–	–	–	–
Soybean oil (mL)	–	–	–	–	–	–	7.0	6.7	–	–	–
10% potassium hydroxide (mL)	–	–	2.0	–	–	–	–	–	–	–	–
Acetic acid (25%)	–	–	–	–	–	–	–	–	–	–	–
Aureomycin	–	–	–	–	–	–	–	–	–	–	–
Chloramphenicol	–	–	–	–	–	–	–	–	–	–	–
Pyruvic acid	–	–	–	–	–	–	–	–	–	–	–
Benzoic acid	–	–	1.2	–	–	–	–	–	–	–	–
Butyl paraben	–	–	–	–	–	–	–	–	–	–	–
Ethyl alcohol 95% (mL)	–	–	–	–	–	–	–	–	12.1	–	–
Formaldehyde 10% (mL)	–	–	–	–	–	–	–	–	–	–	–
Formaldehyde 37% (mL)	–	–	–	–	7.0	–	–	–	–	–	–
Methyl paraben	–	–	2.0	–	2.1	3.6	2.5	1.0	1.1	–	–
Potassium hydroxide	–	–	–	–	–	–	–	–	–	–	–
Potassium sorbate	–	–	–	1.0	–	–	–	–	–	–	–
Propionic acid (mL)	–	–	–	–	–	–	2.0	–	–	–	–
Propylene glycol (mL)	–	–	–	–	7.0	–	–	–	–	–	–
Sorbic acid	–	–	2.0	–	1.4	3.6	2.5	1.0	1.4	–	–
Bacto yeast extract	40.0	–	–	–	–	–	–	–	–	–	–
Dry yeast (see reference for type)	–	–	–	40.0	–	–	45.0	44.2	24.3	118.2	–
Dry mix Insecta F-II, Nihon Nosan Co	–	–	–	–	–	–	–	–	–	–	468.8
Dry commercial diet for *Bombyx mori*	335.0	1000.0	–	–	–	–	–	–	–	–	–
Dry commercial diet for *Vanessa cardui*	–	–	–	334.7	–	–	–	–	–	–	–

Numbers 1–13 indicate diet number. Dashes indicate the ingredient was not part of that diet.

and almost all have protective agents added—with methyl paraben being the most common. Twelve of the diets contain some crude ground or powdered plant products (from wheat, corn, or soybeans), which provide additional bulk and other essential components; 11 have yeast added, which provides protein, vitamins, and minerals. Thus, most do not rely solely on the host material for vital nutrients. Instead, the host material may provide some unidentified critical elements or more likely phagostimulants needed to get the larvae to establish and develop normally.

7.2.1.3 Larval Handling Methods and Holding Conditions

The artificial diets are poured into, scooped and pressed into, or cut into chunks and then placed into a variety of containers for rearing the larvae. Container size is often changed as larvae grow. Some larvae can be reared in plastic petri dishes, cups, or tubes for their whole developmental period, while others must be transferred to glass containers when they are larger so that they will not chew their way out. Most cerambycid larvae must be grown individually or they will mutilate or cannibalize each other. There are only two published cases where larvae are reared in groups for all (Cannon and Robinson 1982) or part (Rogers et al. 2002) of their development but, in each case, some mortality due to the grouping occurred. Each time the larvae are placed in the diet (at initial setup or after a diet change), they are either inserted into a hole made in the diet that is just large enough to accommodate them (Figures 7.1 and 7.2) or dropped into a tight space between a diet cube and the side of the container. There is only one case in which surface-sterilized eggs are used to initially infest the diet (Kosaka and Ogura 1990).

The artificial diet generally has to be changed one or more times while larvae are growing. About half the diets used for cerambycids have to be changed monthly, but the range of intervals between changes goes from three days to eight months. If larvae become contaminated during rearing or before establishment in the diet, they can be surface sterilized. Two methods for larval surface sterilization have been published: a quick dip (10s) in a 5% bleach solution (Dubois et al. 2002, M. A. Keena, unpublished data) or dipping in 3% formalin (Higashiyama et al. 1984), both being followed by multiple rinses in distilled water and then allowing the larvae to dry before inserting them in the diet.

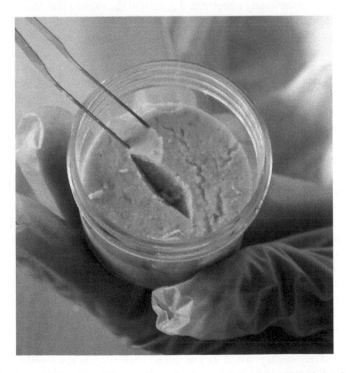

FIGURE 7.1 *Anoplophora glabripennis* rearing container showing a wedge cut out and newly hatched larva in the grove. The wedge will be placed back over the larva before the lid is secured. (Courtesy of M. Keena.)

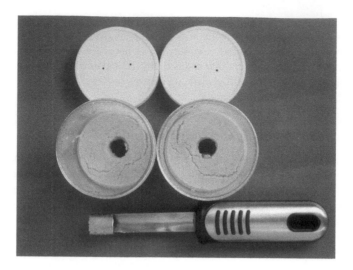

FIGURE 7.2 *Anoplophora glabripennis* rearing container for larger larvae with a hole cut in the center using the apple corer shown. The large larva is placed head down in the hole, which is just larger than its body width. (Courtesy of M. Keena.)

The majority of larvae reared in an artificial diet are held at temperatures between 23°C and 27°C; the larvae of two species were held at 20°C (Payne et al. 1975; Rogers et al. 2002), and the species reared on three other diets are held at 28°C (Palanti et al. 2010), 29°C (Alya and Hain 1987), and 30°C (Galford 1969), respectively. The optimal temperature (maximum growth rate without increased mortality) for species where the effects of temperature on development have been studied is between 25°C and 30°C. For example, the developmental rate of *A. glabripennis* larvae increases linearly between 10°C and 30°C; at higher temperatures, the larvae are adversely affected and most will die (Keena and Moore 2010). Tolerance to higher and lower temperatures can also be greatly affected by relative humidity (Linsley 1959). Generally, there is better survival at higher humidities. When humidity of the environmental chamber where cerambycid larvae are held is controlled, relative humidities between 50% and 70% are commonly used for rearing them. However, the moisture content of cerambycid diets for those where it has been evaluated varies between 50% and 65% (Galford 1969; Wollerman et al. 1969), and most diets are dried for a period of time before use. Diets with a high moisture content may reduce first instar larval establishment and survival in some species. Variability in moisture needs of larvae is to be expected due to the wide range of host plant conditions they inhabit, from living hosts to processed lumber to decaying wood. Further evaluation of differences in larval resistance to desiccation could improve our understanding of the role moisture plays in larval survival and development and improve laboratory rearing.

Although cerambycid larvae spend their time inside host plant tissues, some have been shown to respond to photoperiod changes. For example, larvae of *Xylotrechus pyrrhoderus* (Ashihara 1982) and *Psacothea hilaris* (Shintani et al. 1996) held at 25°C will proceed to pupation under long-day conditions, but under short-day conditions, they will undergo a few supernumerary molts and stop feeding; they will not pupate until undergoing a chill followed by increased temperatures and long-day conditions. Photoperiod has also been shown to affect larval development and pupation timing in three other species of cerambycids (Honda et al. 1981; Ueda and Enda 1995; Shintani 2011). About half the larvae reared in an artificial diet are held in the dark, and the other half are held in containers that let in some light (Figure 7.3), and these are exposed to various light cycles (14L:10D and 16L:8D being the most common). When developing a rearing system for cerambycids, some consideration should be given to the photoperiod that larvae are exposed to because it can impact metamorphosis timing.

Although not all cerambycid larvae require a chill period to proceed to pupation, chilling the larvae once they have reached the weight at which they would normally pupate can speed pupation and synchronize adult emergence (Gardiner 1970; Hanks et al. 1991; Keena 2005; Keena and Moore 2010). This also can be beneficial for stockpiling larvae and building colonies. Methods for rearing about half the species

FIGURE 7.3 Opaque containers with holes in the side that are used to hold *Anoplophora glabripennis* larvae in diet containers during rearing. (Courtesy of M. Keena.)

on an artificial diet call for a larval chill period of one to six months in length. The temperatures used for chill range from 1°C–10°C and, in most cases, the larvae are moved straight from the rearing temperature to the chill temperature. In three cases, the holding temperatures for the larvae are ramped down and then back up after the chill period. For example, *Paraglenea fortunei* larvae are held at 15°C for 10 days both before and after a 96-day chill period at 10°C (Kitajima and Makihara 2011). The proportion of larvae that require chilling is often high in the first generation in the laboratory but can quickly decline as they adapt to the rearing conditions (Hatchett et al. 1973, M. A. Keena, unpublished data). However, chilling larvae before they reach the right size can pose problems. For example, if *A. glabripennis* larvae are chilled when they are too small (after only 6–9 weeks at 25°C), they will resume feeding and growth after the chill and then require a second chill before pupation (Keena 2005). Chilling prepupae at temperatures close to 0°C can delay development after chill in one species (Hanks et al. 1991). In *A. glabripennis*, providing the larvae with drier than normal diet can reduce or eliminate the need for a larval chill period (M. A. Keena, unpublished data).

7.2.1.4 Preventing Mite Infestation of Artificial Diets

There are many grain and cheese mites that commonly infest stored food products that can get into artificial diets and cause problems. They proliferate under high moisture conditions and are often found in conjunction with fungal growth. When the infestation is severe, the surface of the diet will appear to have dust on it and, on closer inspection, the dust will be moving. The mites are 0.3–0.6 mm in length, pale or whitish in color, with spines on both body and legs. They have eight legs (except in the first larval stage when they have only six legs).

Prevention is the best way to avoid dealing with the mites. Good sanitation of the facilities and rearing containers is important because any diet crumbs or residue will attract mites and provide them with a place to become established. Work surfaces should be regularly disinfected with antimicrobial cleaners, floors should be mopped (not swept because that sends potential contaminants into the air), and clean protective clothing and gloves used. Work done with host plant material should be kept separate from work with artificial diets because it can be a source of fungal and mite contamination. Keeping diet ingredients in tight, dry containers and/or in a freezer is also important to prevent contamination. Mites do not survive below ~50% relative humidity or in a freezer. To prevent artificial diets with larvae in them from becoming contaminated, use either rearing or holding containers with tight fitting lids (no holes) or filter paper inside of screw-on lids if there are air holes. Once the diet becomes contaminated, there are no good ways to get rid of the mites that will not harm the larvae, so it is best to just dispose of the containers unopened.

7.2.2 Larval Rearing Using Cut Host Material

Rearing larvae exclusively on host material may be advantageous in some cases by eliminating challenges associated with artificial diets, such as requiring a long time to develop a diet on which larvae will develop normally, labor-intensive making of the diet and transferring larvae to fresh diets periodically, possibly changing the biology or behavior of the insects as they adapt to the laboratory, and requiring relatively clean conditions to keep the diet from becoming contaminated. Although using cut host material to rear larvae can overcome many of these problems, it has some of its own, including the selection of host material in optimal condition for the normal development of larvae (some cerambycids cannot complete development in cut host material), fungal and mite contamination of host material, desiccation of host material, predators and parasitoids that may be present if host material is naturally infested, and the need to obtain fresh host material under less than ideal weather conditions if rearing year round.

Larvae from 45 species, 10 cerambycines and 35 lamiines, have been reared on cut host material (Table 7.1). Those reared on naturally infested wood brought into the laboratory include species that are twig girdlers or pruners (Rice 1995; Clarke and Zamalloa 2009), ones that infest turf grass roots (Kumral et al. 2012) or herbaceous plants (Niide et al. 2006), and those that prefer recently cut or dead wood (Tyson 1966; Purrington and Horn 1993; Berkov and Tavakilian 1999; Georgiev et al. 2004; Haack and Petrice 2009; Waqa-Sakiti et al. 2014). Generally, the infested material is held in a cage, tube, or other type of enclosure, either under controlled laboratory conditions or outdoors in a sheltered area (Figures 7.4 through 7.6). Several *Monochamus* and one *Aerenicopsis* species have been reared by exposing recently cut (usually partially dried) bolts/steams of the preferred host to ovipositing females and then letting the eggs hatch and larvae complete development in the bolts (Linit 1985; Anbutsu and Togashi 1997;

FIGURE 7.4   Sheltered outdoor adult emergence cage containing *Eucalyptus* bolts infested with *Phoracantha*. (Courtesy of L. Hanks.)

FIGURE 7.5 Cardboard tubes often used to hold infested wood until adults emerge. The lids have clear cups that let in light that attracts the adults and makes removing them easier. (Courtesy of M. Keena.)

FIGURE 7.6 Garbage barrels with wire screens in the lids used to either hold infested wood and contain adults that emerge until the can be removed or to hold large oviposition bolts and mating pairs when natural infestation of host material is desired.

Zhang and Linit 1998; Palmer et al. 2000; Akbulut et al. 2007; Koutroumpa et al. 2008; Breton et al. 2013). Species that prefer stressed or minimally weakened hosts have been reared by inserting larvae either into cut bolts/twigs (Figure 7.7) (Hanks et al. 1993; Shibata 1995; Bancroft et al. 2002; Bybee et al. 2004; Lu et al. 2011) or into potted trees in a greenhouse (Figure 7.8) (Ludwig et al. 2002; Morewood et al. 2005). Larvae are inserted into notches or under bark flaps (a hole or depression may be cut into the wood under it), and then the bark is held in place with plastic wrap and tape until the larvae establish. The bolts used

FIGURE 7.7 Bolt showing site of larval insertion after the larva has established and begun pushing out wood shavings and frass. (Courtesy of D. Long.)

FIGURE 7.8 Potted tree in a quarantine greenhouse that has had *Anoplophora glabripennis* larvae inserted. (Courtesy of D. Long.)

for insertion or oviposition have either one or both ends sealed with hot paraffin wax to slow desiccation (Figure 7.9). If only one end is sealed, the other end is put in either damp sand or a damp sand–peat mixture to help maintain the wood moisture content. Both these bolts and naturally infested host material are often misted with water one or more times a week to help maintain moisture levels. Two species have also been reared by providing larvae with fresh twigs of a preferred host in a glass tube and changing the twigs weekly (Adachi 1994; Bancroft et al. 2002). The cut host material for larval rearing is held at temperatures

FIGURE 7.9 Bolts that have had larvae inserted. Both ends have been dipped in hot paraffin to seal them and prevent moisture loss. (Courtesy of M. Keena.)

between 20°C and 30°C, most often at 25°C. The humidity, when controlled, is generally 70–80% RH and the light:dark (L:D) cycles used include 12L:12D, 14L:10D, and 16L:8D.

7.2.2.1 Preventing Fungal Infestation of Cut Host Material

Fungal spores present on cut host material can germinate and compromise rearing operations by degrading the host material and potentially harming the insects themselves. Sealing the ends of woody material with hot paraffin (which is already used to prevent moisture loss) can help prevent fungal growth on the exposed tissue. Cut twigs without leaves used for *Anoplophora glabripennis* food are treated with ultraviolet light for 30 minutes, and the bolts used for oviposition are stood on end and treated with ultraviolet light for 60 minutes to kill bacteria and fungi that are on the surface. This has eliminated entomophagus fungal infections of the adults and greatly reduced fungal infestation of the oviposition bolts (M. A. Keena, unpublished data). Alternative surface sterilization methods may be possible (e.g., wiping them with ethanol or dusting them with sulfur), but their efficacy and effects on the beetle have not been documented.

7.3 Prepupal and Pupal Handling Methods

Prepupae and pupae are delicate and can be easily damaged, so handling usually is kept to a minimum. Many rearing protocols for cerambycids indicate that larvae are allowed to pupate in the diet or host material; some are removed after a few days, but most remain in the diet until emergence as an adult. There are at least three reasons why removing these stages from artificial diets may be advantageous: exact timing of pupation or adult emergence is needed, diet changes disrupt pupal chamber formation, or pupation or adult emergence does not proceed normally in the diet. When prepupae or pupae are removed from the diet, they usually are placed in a container with moistened paper (e.g., filter paper, paper towel) or vermiculite to help maintain moisture and to provide a rough surface on which to pupate (Hatchett et al. 1973; Galford 1985; Kosaka and Ogura 1990; Keena 2005; Shintani 2011). Determining when larvae reach the right stage to remove them from the diet is not difficult. For example, once *A. glabripennis* larvae have reached full size and cease feeding, they form a pupal chamber (wider area than the normal tunnels) in the diet or a trough open to the top and then begin to show signs they are becoming prepupae. As larvae become prepupae, the body segments shorten, a section near the head becomes translucent and, just before pupation, they become flaccid (Figure 7.10).

FIGURE 7.10 *Anoplophora glabripennis* prepupa on top of the diet showing the translucent section just behind the head and the shrinkage as evidenced by the narrower body rings. (Courtesy of M. Keena.)

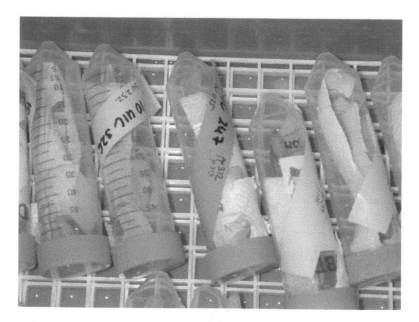

FIGURE 7.11 *Anoplophora glabripennis* pupae in 50-mL containers with two pinholes in the lid and a piece of paper towel to wick moisture. The tubes are placed on grates over water to maintain high humidity. (Courtesy of M. Keena.)

Prepupae can be carefully moved to the holding container as shown in Figure 7.11. This holding container is a 50-mL centrifuge tube with two pin holes in the lid and a piece of paper towel along one side—the end of which sticks out under the lid to act as a wick. The tubes are then placed in an opaque box with a thin layer of water and a grate on the bottom to hold the tubes above the water (keeping the humidity near 90% RH). This keeps the prepupae and pupae moist but not wet, which is critical to the successful shedding of the larval and pupal skins. For species that are held in the diet, the diet provides this critical moisture.

7.4 Adult Handling Methods and Oviposition Substrates

7.4.1 Teneral Adults and Maturation Feeding

Teneral adults generally require several days to sclerotize before they are ready to feed. Rearing methods where the pupae are taken off the diet indicate that new adults are held in the dark without food during this time (e.g., Keena 2005). This period of time is when the newly enclosed adult would be in the pupal chamber within the host before it starts chewing out. A recent study on *A. glabripennis* has indicated that the period of time before they start chewing will vary with holding temperature (Sánchez and Keena 2013), and it likely will also vary with species. Males usually emerge before females and can be distinguished from the females by their longer antennae.

Some feeding is a prerequisite to egg maturation and oviposition for most cerambycids (Linsley 1959). The length of the maturation feeding ranges from a few days to more than a month and varies among species (e.g., Hadlington and Johnson 1973; Koutroumpa et al. 2008; Lu et al. 2011) and with temperature (Keena 2006). During this period, various protocols indicate the adults are either held singly or in groups (usually by sex) (Figure 7.12). Possible food sources for adults include honeybee pollen, 10% honey solution, 30% sucrose solution, thin bark on twigs or branches, leaves, petioles, needles, cones, sap, fruit chunks, blocks of artificial diet (Gardiner 1970), and roots. A source of water is also often added when the moisture content of the host material is too low, such as moist filter paper on the container bottom, moist cotton or sand, or misting the containers periodically. If twigs or branches without leaves are used, they generally are changed weekly, and foliage and other food sources that can mold or dry out quickly are changed more frequently. Choice of food source is based on the individual species' natural food preferences or from information on closely related species if their biology is not known.

FIGURE 7.12 Wide mouth jars used for holding single adults during maturation feeding on top and larger mating jars below. (Courtesy of M. Keena.)

7.4.2 Mating and Oviposition Substrates

Cerambycid species belonging to the subfamilies Prioninae, Spondylidinae, and primitive Cerambycinae mate at night or late in the day, while those from other subfamilies mate during bright sunlight (Linsley 1959). Ensuring that adults have mated may be more difficult in the night mating species, but males may still be in the mate guarding mounted position, which makes the pairing obvious. Because male competition for females is often violent, resulting in mutilation, most species are mated in single pairs. In a few species, group cages of 5 pairs (Fukaya and Honda 1992), 10 pairs (Hanks et al. 1993), 20 pairs (Akbulut et al. 2007; Lu et al. 2011), 25 pairs (Linit 1985), or 200 individuals (Hadlington and Johnston 1973) are used. The pairs are housed in glass jars, screen cages, organdy bags, or plastic cages of varying sizes. The mating containers are held outdoors, in greenhouses, at room temperature, or in environmental chambers set at 18–28°C and 40–80% RH. Choice of temperature and humidity for holding pairs can be critical. At lower temperatures within the 10–30°C range, adults generally live longer but fecundity will be reduced. In *A. glabripennis*, for example, the optimum temperature for survival is 18°C, but maximum fecundity occurs at about 24°C and no oviposition occurs at <10°C and >35°C (Keena 2006). If humidity is too low, at either higher or lower temperatures in particular, and adequate sources of moisture are not available, adult survival will be reduced.

In addition to food and water sources, an oviposition substrate is added. Oviposition in cerambycids occurs on host surfaces, in bark/wood cracks, in the soil, or under the bark in specially prepared egg niches (Linsley 1959). Species that oviposit on surfaces or in host cracks are either supplied with host material (or a host surrogate, Hatchett et al. 1973) to oviposit on or some other cracks to use. Cracks or tight places provided for oviposition include rolled paper (García-Ruiz and Pérez-Moreno 2012); 1–2 cm wide cotton fabric wrapped in a spiral (with gaps between wraps) around a host bolt/stem (Galford 1969; Wollerman et al. 1969; Boldt 1987); tightly rolled leaves (Fukaya and Honda 1992); polyethylene sheeting cut in strips, stacked, folded, and stapled together and placed so that the folds are available to the female (Hanks et al. 1993); or tight places, such as between a damp filter paper and the cage bottom (Gardiner 1970; Shibata 1995). A freshly cut bolt of host material with both ends dipped in paraffin wax (to retain moisture) generally is provided as an oviposition substrate to species that oviposit in specially prepared egg niches. The bolts are changed weekly. An example of a mating container setup used for *A. glabripennis* is shown in Figure 7.13. These bolts subsequently will have to have the bark carefully peeled off to retrieve eggs or newly hatched larvae depending on how long the bolt is held (Figure 7.14).

FIGURE 7.13 *Anoplophora glabripennis* mating pair in a jar with maple twigs for food and a bolt for oviposition. (Courtesy of M. Keena.)

FIGURE 7.14 Bolt with some of the bark peeled back to show the *Anoplophora glabripennis* eggs laid underneath. (Courtesy of M. Keena.)

7.5 Egg Handling and Stockpiling Methods

Once eggs are extracted from the oviposition substrate, they can be held for hatch either with or without surface disinfection. Four surface disinfecting methods are listed in the published literature: dipped in 70% ethanol for 10 seconds, in 0.05% benzalkonium chloride for 5 minutes, and washed three times with distilled water (Kosaka and Ogura 1990); dipped in 70% ethanol, then three to five times in 1% methanol, and washed four times with distilled water (Zhao et al. 1999); dipped in 1% sodium hydroxide solution for 20 seconds, soaked in 10% formalin for 10 minutes, and then washed three times with distilled water (Kitajima 2003); and dipped in 4% bleach solution for 20 seconds, soaked in 70% ethanol for 12 seconds, and then washed with distilled water (Akutsu et al. 1980). Generally, eggs are held in groups in petri dishes or other plastic or glass containers with or without damp filter paper in the bottom. In one case, the filter paper had to be wet with Ringer's solution to prevent egg rupturing (Wollerman et al. 1969). *A. glabripennis* eggs also can be held singly with no disinfection, as in Figure 7.15, to prevent newly hatched larvae from mutilating each other (Keena and Moore 2010). In some cases, the eggs are not removed from the oviposition substrate. Instead, they are held until the larvae hatch and then they are retrieved.

Eggs can be held at similar temperatures to those used for rearing larvae or holding adults if hatch is desired immediately. If hatch needs to be delayed, synchronized, or egg stockpiling is desired, cerambycid eggs can be held at lower temperatures. Egg hatch can be delayed by holding eggs or the oviposition substrate with eggs at 15–20°C (Hanks et al. 1993; Keena 2006; Shintani 2011), or egg hatch can be prevented by holding them at 5–10°C for up to 80 days without substantial mortality (Keena 2005, 2006). The humidity in the holding location must be maintained at >80% RH to ensure egg survival when holding them for longer periods of time.

7.6 Developing Rearing Methods for Additional Species

When no published method for rearing a species of cerambycid exists, begin by assembling all available knowledge of the biology and hosts used by the species of interest. Next search for methods or diets that exist for other species (Table 7.1) that are either most closely related (e.g., same genus) or that inhabit

FIGURE 7.15 Section of a 24-well plate showing *Anoplophora glabripennis* eggs and newly hatched larvae. Well plates are held on grates over water in boxes to maintain near 100% RH. (Courtesy of M. Keena.)

similar hosts that are in the same condition (e.g., weakened or dead hosts). If there are no likely candidates based on these two criteria or there is insufficient knowledge about the species you wish to rear, the best diets to try initially are those that have been evaluated for use with a number of species from several different subfamilies.

Start by using the most commonly utilized rearing methods and environmental conditions as documented in this chapter. Another good source of information and ideas are the researchers that have already successfully developed methods and diets for cerambycids. The two most common problem areas for rearing cerambycids are diet moisture content and environmental conditions not being favorable for pupation. Using a drier diet (too much moisture can have adverse effects on establishment and development) and mimicking a natural overwintering environment (temperature and light cycle) are potential ways to avoid these problems. Start with a diet that has 40–50% moisture content if working with a larva that develops in a live, stressed, or moist decaying host and with a drier diet (e.g., 30%) if the larva develops in a weakened or dry, dead host. To determine the best environmental conditions to use, obtain the climatic data for the native habitat and take the modulating effect of the wood environment into account (add at least 2°C). Average weekly temperatures in the summer will work for many species as the larval rearing temperature. Larvae generally survive chill well and progress to pupation if chilled at 5–10°C for one to three months. The correct photoperiods to use will depend on the timing of pupation and the light cycle in the native habitat, both during larval rearing and chill. Be aware that in all rearing method and diet development programs, it may take several years to fully evaluate and modify existing methods to adequately rear a new species.

ACKNOWLEDGMENTS

I wish to thank J. Hansen, P. Moore, and V. Sánchez for their reviews of earlier versions of the chapter and A. Ray for assistance in reviewing and filling in some missing data in the species table. I also thank K. Hoover and L. Hanks for supplying photos.

REFERENCES

Adachi, I. 1994. Development and life-cycle of *Anoplophora malasiaca* (Thomson) (Coleoptera, Cerambycidae) on citrus trees under fluctuating and constant-temperature regimes. *Applied Entomology and Zoology* 29: 485–497.

Akbulut, S., A. Keten, I. Baysal, and B. Yuksel. 2007. The effect of log seasonality on the reproductive potential of *Monochamus galloprovincialis* Olivier (Coleoptera: Cerambycidae) reared in black pine logs under laboratory conditions. *Turkish Journal of Agriculture and Forestry* 31: 413–422.

Akutsu, K., K. Honda, and S. Arai. 1980. Mass rearing method of the Udo longicorn beetle, *Acalolepta luxuriosa* Bates (Coleoptera: Cerambycidae) on artificial diet. *Japanese Journal of Applied Entomology and Zoology* 24: 119–121.

Alya, A. B., and F. P. Hain. 1987. Rearing *Monochamus* species larvae on artificial diet (Coleoptera, Cerambycidae). *Journal of Entomological Science* 22: 73–76.

Anbutsu, H., and K. Togashi. 1997. Effects of spatio-temporal intervals between newly-hatched larvae on larval survival and development in *Monochamus alternatus* (Coleoptera: Cerambycidae). *Researches on Population Ecology* 39: 181–189.

Ashihara, W. 1982. Effects of temperature and photoperiod on the development of the grape borer, *Xylotrechus pyrrhoderus* Bates (Coleoptera, Cerambycidae). *Japanese Journal of Applied Entomology and Zoology* 26: 15–22 [in Japanese].

Bancroft, J. S., M. T. Smith, E. K. Chaput, and J. Tropp. 2002. Rapid test of the suitability of host-trees and the effects of larval history on *Anoplophora glabripennis* (Coleoptera: Cerambycidae). *Journal of Kansas Entomological Society* 75: 308–316.

Baragaño Galán, J. R., A. Notario Gómez, and C. Sa Montero. 1981. *Agapanthia asphodeli* Latereille (Col.: Cerambycidae): Cria artificial y estudio cariologico. *Boletín Servicios Plagas* 7: 161–167.

Berkov, A., and G. Tavakilian. 1999. Host utilization of the brazil nut family (Lecythidaceae) by sympatric wood-boring species of palame (Coleoptera, Cerambycidae, Lamiinae, Acanthocinini). *Biological Journal of the Linnean Society* 67: 181–198.

Boldt, P. E. 1987. Host specificity and laboratory rearing studies of *Megacyllene mellyi* (Coleoptera, Cerambycidae), a potential biological-control agent of Baccharis-Neglecta Britt (Asteraceae). *Proceedings of the Entomological Society of Washington* 89: 665–672.

Breton, Y., C. Hebert, J. Ibarzabal, R. Berthiaume, and E. Bauce. 2013. Host tree species and burn treatment as determinants of preference and suitability for *Monochamus scutellatus scutellatus* (Coleoptera: Cerambycidae). *Environmental Entomology* 42: 270–276.

Bybee, L. F., J. G. Millar, T. D. Paine, K. Campbell, and C. C. Hanlon. 2004. Seasonal development of *Phoracantha recurva* and *P. semipunctata* (Coleoptera: Cerambycidae) in Southern California. *Environmental Entomology* 33: 1232–1241.

Cannon, K. F., and W. H. Robinson. 1982. An artificial diet for the laboratory rearing of the old house borer, *Hylotrupes bajulus* (Coleoptera, Cerambycidae). *The Canadian Entomologist* 114: 739–742.

Carle, P. 1969. Milieux artificiels pour l'élevage des larves de *Pissodes notatus* L. (Col. Curculionidae) et autres xylophages du Pin maritime. *Annales des Sciences Forestières* 26: 397–406.

Clarke, R. O. S., and S. Zamalloa. 2009. Life cycle of *Phoebemima ensifera* Tippmann (Coleoptera, Cerambycidae). *Revista Brasileira de Entomologia* 53: 287–290.

Cohen, A. C. 2003. Function of insect diet components. In *Insect diets: Science and technology*, ed. A. C. Cohen, 21–46. Boca Raton, FL: CRC Press.

Craighead, F. C. 1923. North American cerambycid larvae. *Dominion of Canada Department of Agriculture Bulletin* 27: 1–239.

De Viedma, M. G., A. Notario, and J. R. Baragano. 1985. Laboratory rearing of lignicolous Coleoptera (Cerambycidae). *Journal of Economic Entomology* 78: 1149–1150.

Dojnov, B., N. Lončar, N. Božić, V. Nenadović, J. Ivanović, and Z. Vujčić. 2010. Comparison of alpha-amylase isoforms from the midgut of *Cerambyx cerdo* L. (Coleoptera: Cerambycidae) larvae developed in the wild and on an artificial diet. *Archives of Biological Sciences, Belgrade* 62: 575–583.

Dojnov, B., Z. Vujčić, N. Božić, et al. 2012. Adaptations to captive breeding of the longhorn beetle *Morimus funereus* (Coleoptera: Cerambycidae); application on amylase study. *Journal of Insect Conservation* 16: 239–247.

Dubois, T., A. E. Hajek, and S. Smith. 2002. Methods for rearing the Asian longhorned beetle (Coleoptera: Cerambycidae) on artificial diet. *Annals of the Entomological Society of America* 95: 223–230.

Fukaya, M., and H. Honda. 1992. Reproductive-biology of the yellow-spotted longicorn beetle, *Psacothea hilaris* (Pascoe) (Coleoptera: Cerambycidae). I: Male mating behaviors and female sex-pheromones. *Applied Entomology and Zoology* 27: 89–97.

Galford, J. R. 1969. *Artificial rearing of 10 species of wood-boring insects*. Research Note NE-102. Upper Darby, PA: U.S. Department of Agriculture, Forest Service, Northeastern Forest Experiment Station.

Galford, J. R. 1985. *Enaphalodes Rufulus*. In *Handbook of insect rearing Vol. 1*, eds. P. Singh and R. F. Moore, 255–64. New York: Elsevier.

García-Ruiz, E., V. Marco, and I. Perez-Moreno. 2012. Laboratory rearing and life history of an emerging grape pest, *Xylotrechus arvicola* (Coleoptera: Cerambycidae). *Bulletin of Entomological Research* 102: 89–96.

Gardiner, L. M. 1970. Rearing wood-boring beetles (Cerambycidae) on artificial diet. *The. Canadian Entomologist* 102: 113–117.

Georgiev, G., V. Sakalian, K. Ivanov, and P. Boyadzhiev. 2004. Insects reared from stems and branches of goat willow (*Salix caprea* L.) in Bulgaria. *Journal of Pest Science* 77: 151–153.

Goodwin, S., and M. A. Pettit. 1994. *Acalolepta vastator* (Newman) (Coleoptera, Cerambycidae) infesting grapevines in the Hunter Valley, New-South-Wales. 2. biology and ecology. *Journal of the Australian Entomological Society* 33: 391–397.

Haack, R. A., and T. R. Petrice. 2009. Bark- and wood-borer colonization of logs and lumber after heat treatment to ISPM 15 specifications: The role of residual bark. *Journal of Economic Entomology* 102: 1075–1084.

Hadlington, P. W., and J. A. Johnson. 1973. Mass rearing of *Plagiohammus spinipennis* (Thoms.) (Col: Cerambycidae) for the biological control of *Lantan camara* L. (Verbenaceae). *Journal of the Australian Entomological Society* 8: 32–36.

Hanks, L. M. 1999. Influence of the larval host plant on reproductive strategies of cerambycid beetles. *Annual Review of Entomology* 44:483-505.

Hanks, L. M., J. S. McElfresh, J. G. Millar, and T. D. Paine. 1991. Evaluation of cold temperatures and density as mortality factors of the Eucalyptus longhorned borer (Coleoptera, Cerambycidae) in California. *Environmental Entomology* 20: 1653–1658.

Hanks, L. M., J. S. McElfresh, J. G. Millar, and T. D. Paine. 1993. *Phoracantha semipunctata* (Coleoptera, Cerambycidae), a serious pest of *Eucalyptus* in California—Biology and laboratory-rearing procedures. *Annals of the Entomological Society of America* 86: 96–102.

Harley, K. L. S., and B. W. Willson. 1968. Propagation of a cerambycid borer on a meridic diet. *Canadian Journal of Zoology* 46: 1265–1266.

Hatchett, J. H., R. D. Jackson, and R. M. Barry. 1973. Rearing a weed cerambycid, *Dectes texanus*, (Coleoptera: Cerambycidae) on an artificial medium, with notes on biology. *Annals of the Entomological Society of America* 66: 519–522.

Higashiyama, N., M. Ishida, and Y. Abe. 1984. Artificial medium for larval rearing of longhorn beetles (Coleoptera: Cerambycidae). *Research Bulletin of Plant Protection of Japan* 20: 39–45.

Honda, K., K. Akutsu, and S. Arai. 1981. Photoperiodic response of *Acalolepta luxuriosa* Bates (Coleoptera: Cerambycidae): Effect of photoperiodic change for induction of larval diapause. *Japanese Journal of Applied Entomology and Zoology* 25: 108–112 [in Japanese].

Iglesias, C., A. Notario, and J. R. Baragano. 1989. Evaluación de las condiciones de cría Y datos bionómicos de coleópteros lignícolas de tocón de pino. *Boletin de Sanidad Vegetal—Plagas* 15: 9–16.

Ivanovic, J., S. Dordevic, L. Ilijin, M. Jankovic-Tomanic, and V. Nenadovic. 2002. Metabolic response of cerambycid beetle (*Morimus funereus*) larvae to starvation and food quality. *Comparative Biochemistry and Physiology Part A* 132: 555–566.

Jeniš, I. 2001. *Long-horned beetles. Distenidae, Oxypeltidae, Vesperidae, Anoplodermatidae & Cerambycidae I. Vesperidae & Cerambycidae of Europe I.* Zlín: Ateliér Regulus.

Keena, M. A. 2005. Pourable artificial diet for rearing *Anoplophora glabripennis* (Coleoptera: Cerambycidae) and methods to optimize larval survival and synchronize development. *Annals of the Entomological Society of America* 98: 536–547.

Keena, M. A. 2006. Effects of temperature on *Anoplophora glabripennis* (Coleoptera: Cerambycidae) adult survival, reproduction, and egg hatch. *Environmental Entomology* 35: 912–921.

Keena, M. A., and P. M. Moore. 2010. Effects of temperature on *Anoplophora glabripennis* (Coleoptera: Cerambycidae) larvae and pupae. *Environmental Entomology* 39: 1323–1335.

King, E. G., and N. C. Leppla. 1984. *Advances and challenges in insect rearing.* New Orleans, LA: USDA Agricultural Research Service.

Kitajima, H. 1999. Effect of sucrose levels in artificial diet on larval growth of Cryptomeria bark borer, *Semanotus japonicus* (Lacordaire) (Coleoptera: Cerambycidae). *Japanese Journal of Applied Entomology and Zoology* 43: 203–205 [in Japanese].

Kitajima, H. 2003. Aseptic rearing of larvae of the cryptomeria bark borer, *Semanotus japonicus* (Lacordaire) (Coleoptera: Cerambycidae), on an artificial diet. *Japanese Journal of Applied Entomology and Zoology* 47: 36–38 [in Japanese].

Kitajima, H., and H. Makihara. 2011. Rearing of *Paraglenea fortunei* (Coleoptera: Cerambycidae) larvae on artificial diets. *Japanese Journal of Applied Entomology and Zoology* 55: 67–70 [in Japanese].

Kosaka, H. 2011. Artificial diets for the larval oak longicorn beetle, *Moechotypa diphysis* (Coleoptera: Cerambycidae). *Applied Entomology and Zoology* 46: 581–584.

Kosaka, H., and N. Ogura. 1990. Rearing of the Japanese pine sawyer, *Monochamus alternatus* (Coleoptera, Cerambycidae) on artificial diets. *Applied Entomology and Zoology* 25: 532–534.

Koutroumpa, F. A., B. Vincent, G. Roux-Morabito, C. Martin, and F. Lieutier. 2008. Fecundity and larval development of *Monochamus galloprovincialis* (Coleoptera: Cerambycidae) in experimental breeding. *Annals of Forest Science* 65: 707–717.

Kumral, N. A., U. Bilgili, and E. Acikgoz. 2012. *Dorcadion pseudopreissi* (Coleoptera: Cerambycidae), a new turf pest in Turkey, the bio-ecology, population fluctuation and damage on different turf species. *Türkiye Entomoloji Dergisi (Turkish Journal of Entomology)* 36: 123–133 [in Turkish with English abstract].

Lee, C. Y., and K. C. Lo. 1998. Rearing of *Anoplophora macularia* (Thomson) (Coleoptera: Cerambycidae) on artificial diets. *Applied Entomology and Zoology* 33: 105–109.

Linit, M. J. 1985. Continuous laboratory culture of *Monochamus carolinensis* (Coleoptera, Cerambycidae) with notes on larval development. *Annals of the Entomological Society of America* 78: 212–213.

Linsley, E. G. 1959. Ecology of Cerambycidae. *Annual Review Entomology* 4: 99–138.

Linsley, E. G. 1961. *The Cerambycidae of North America. Part I: Introduction*. Berkeley, CA: University of California Press.

Linsley, E. G. 1962a. *The Cerambycidae of North America. Part II: Taxonomy and classification of the Parandrinae, Prioninae, Spondylinae, and Aseminae*. Berkeley, CA: University of California Press.

Linsley, E. G. 1962b. *The Cerambycidae of North America. Part III: Taxonomy and classification of the subfamily Cerambycinae, tribes Opsimini through Megaderini*. Berkeley, CA: University of California Press.

Linsley, E. G. 1963. *The Cerambycidae of North America. Part IV: Taxonomy and classification of the subfamily Cerambycinae, tribes Elaphidionini through Rhinotragini*. Berkeley, CA: University of California Press.

Linsley, E. G. 1964. *The Cerambycidae of North America. Part V: Taxonomy and classification of the subfamily Cerambycinae, tribes Callichromatini through Ancylocerini*. Berkeley, CA: University of California Press.

Linsley, E. G., and J. A. Chemsak. 1972. *The Cerambycidae of North America. Part VI, No. 1: Taxonomy and classification of the subfamily Lepturinae*. Berkeley, CA: University of California Press.

Linsley, E. G., and J. A. Chemsak. 1976. *The Cerambycidae of North America. Part VI. No. 2: Taxonomy and classification of the subfamily Lepturinae*. Berkeley, CA: University of California Press.

Linsley, E. G., and J. A. Chemsak. 1984. *The Cerambycidae of North America. Part VII, No. 1: Taxonomy and classification of the subfamily Lamiinae, tribes Parmenini through Acanthoderini*. Berkeley, CA: University of California Press.

Linsley, E. G., and J. A. Chemsak. 1995. *The Cerambycidae of North America, Part VII, No. 2: Taxonomy and classification of the subfamily Lamiinae, tribes Acanthocinini through Hemilophini*. Berkeley, CA: University of California Press.

Linsley, E. G., and J. A. Chemsak. 1997. *The Cerambycidae of North America. Part VIII: Bibliography, index, and host plant index*. Berkeley: University of California Press.

Lu, W., Q. Wang, M. Y. Tian, J. Xu, and A. Z. Qin. 2011. Phenology and laboratory rearing procedures of an Asian longicorn beetle, *Glenea cantor* (Coleoptera: Cerambycidae: Lamiinae). *Journal of Economic Entomology* 104: 509–516.

Ludwig, S. W., L. Lazarus, D. G. McCullough, K. Hoover, S. Montero, and J. C. Sellmer. 2002. Methods to evaluate host tree suitability to the Asian longhorned beetle, *Anoplophora glabripennis*. *Journal of Environmental Horticulture* 20: 175–180.

Monné, M. A. 1995. *Catalogue of the Cerambycidae (Coleoptera) of the western hemisphere. Part XVIII: Subfamily Lamiinae: Tribe Acanthocinini*. São Paulo: Sociedade Brasileira de Entomologia.

Morewood, W. D., K. Hoover, P. R. Neiner, and J. C. Sellmer. 2005. Complete development of *Anoplophora glabripennis* (Coleoptera: Cerambycidae) in Northern red oak trees. *The Canadian Entomologist* 137: 376–379.

Murakoshi, S., and N. Aono. 1981. Rearing of the white-spotted longicorn beetle, *Anoplophora malasiaca* Thomson (Coleoptera: Cerambycidae) on an artificial diet. *Japanese Journal of Applied Entomology and Zoology* 25: 55–56.

Necibi, S., and M. J. Linit. 1997. A new artificial diet for rearing *Monochamus carolinensis* (Coleoptera: Cerambycidae). *Journal of the Kansas Entomological Society* 70: 145–146.

Niide, T., R. D. Bowling, and B. B. Pendleton. 2006. Morphometric and mating compatibility of *Dectes texanus texanus* (Coleoptera: Cerambycidae) from soybean and sunflower. *Journal of Economic Entomology* 99: 48–53.

Notario, A., and R. Baragaño. 1978. *Ergates faber* Linnaeus (Col. Cerambycidae): Descripcion, cria artificial y estudio cariologico. *Anales INIA, Service de la Protection de Végétaux, Madrid* 8: 45–57.

Notario, A., R. Baragaño and L. Castresana. 1993. Estudio de cerambicidos xilofagos de *Pinus sylvestris* L. utilizando dietas artificiales. *Ecologia* 7: 499–502.

Palanti, S., B. Pizzo, E. Feci, L. Fiorentino, and A. M. Torniai. 2010. Nutritional requirements for larval development of the dry wood borer *Trichoferus holosericeus* (Rossi) in laboratory cultures. *Journal of Pest Science* 83: 157–164.

Palmer, W. A., B. W. Willson, and K. R. Pullen. 2000. Introduction, rearing, and host range of *Aerenicopsis championi* Bates (Coleoptera: Cerambycidae) for the biological control of *Lantana camara* L. in Australia. *Biological Control* 17: 227–233.

Pavlovic, R., M. Grujic, B. Dojnov, et al. 2012. Influence of nutrient substrates on the expression of cellulases in *Cerambyx cerdo* L. (Coleoptera: Cerambycidae) larvae. *Archives of Biological Sciences, Belgrade* 64: 757–765.

Payne, J. A., H. Lowman, and R. R. Pate. 1975. Artificial diets for rearing tilehorned Prionus. *Annals of the Entomological Society of America* 68: 680–682.

Petersen-Silva, R., P. Naves, E. Sousa, and J. Pujade-Villar. 2014. Rearing the pine sawyer *Monochamus galloprovincialis* (Oliver, 1795) (Coleoptera: Cerambycidae) on artificial diets. *Journal of the Entomological Research Society* 16: 61-70.

Purrington, F. F., and D. J. Horn. 1993. Canada moonseed vine (Menispermaceae)—Host of 4 roundheaded wood borers in central Ohio (Coleoptera, Cerambycidae). *Proceedings of the Entomological Society of Washington* 95: 313–320.

Reinecke, J. P. 1985. Nutrition: Artificial diets. In *Comprehensive Insect Physiology Biochemistry and Pharmacology* Vol. 4, eds. G. A. Kerkut and L. I. Gilbert, 391–419. New York: Pergamon Press.

Rice, M. E. 1995. Branch girdling by *Oncideres cingulata* (Coleoptera: Cerambycidae) and relative host quality of persimmon, hickory, and elm. *Annals of the Entomological Society of America* 88: 451–455.

Rogers, D. J., S. E. Lewthwaite, and P. R. Dentener. 2002. Rearing Huhu beetle larvae, *Prionoplus reticularis* (Coleoptera: Cerambycidae) on artificial diet. *New Zealand Journal of Zoology* 29: 303–310.

Sánchez, V., and M. A. Keena. 2013. Development of the teneral adult *Anoplophora glabripennis* (Coleoptera: Cerambycidae): Time to initiate and completely bore out of maple wood. *Environmental Entomology* 42: 1–6.

Shibata, E. 1995. Reproductive strategy of the Sugi bark borer, *Semanotus japonicus* (Coleoptera: Cerambycidae) on Japanese cedar, *Cryptomeria japonica*. *Researches on Population Ecology* 37: 229–237.

Shintani, Y. 2011. Quantitative short-day photoperiodic response in larval development and its adaptive significance in an adult-overwintering cerambycid beetle, *Phytoecia rufiventris*. *Journal of Insect Physiology* 57: 1053–1059.

Shintani, Y., Y. Ishikawa, and S. Tatsuki. 1996. Larval diapause in the yellow-spotted longicorn beetle, *Psacothea hilaris* (Pascoe) (Coleoptera: Cerambycidae). *Applied Entomology and Zoology* 31: 489–494.

Stein, J. D., and J. E. Haraguchi. 1984. Meridic diet for rearing of the host specific tropical wood-borer *Plagithmysus bilineatus* (Coleoptera, Cerambycidae). *The Pan-Pacific Entomologist* 60: 94–96.

Togashi, K. 1989. Development of *Monochamus alternatus* Hope (Coleoptera, Cerambycidae) in relation to oviposition time. *Japanese Journal of Applied Entomology and Zoology* 33: 1–8 [in Japanese].

Tyson, W. H. 1966. Notes on reared Cerambycidae. *The Pan-Pacific Entomologist* 42: 201–207.

Ueda, A., and N. Enda. 1995. Photoperiodic and thermal control on diapause of the Japanese pine sawyer, *Monochamus alternates* Hope (Coleoptera: Cerambycidae). *Transactions of Kansai Branch of the Japanese Forestry Society* 4: 163–166 [in Japanese].

Wang, Q., G. L. Shi, D. Song, D. J. Rogers, L. K. Davis, and X. Chen. 2002. Development, survival, body weight, longevity, and reproductive potential of *Oemena hirta* (Coleoptera: Cerambycidae) under different rearing conditions. *Journal of Economic Entomology* 95: 563–569.

Waqa-Sakiti, H., A. Stewart, L. Cizek, and S. Hodge. 2014. Patterns of tree species usage by long-horned beetles (Coleoptera: Cerambycidae) in Fiji. *Pacific Science* 86: 57–64.

Wollerman, H., C. Adams, and G. C. Heaton. 1969. Continuous laboratory culture of locust borer *Megacyllene robiniae*. *Annals of the Entomological Society of America* 62: 647–649.

Yoon, H. J., I. G. Park, Y. I. Mah, and K. Y. Seol. 1997. Larval development of mulberry longicorn beetle, *Apriona germari* Hope, on the artificial diet. *Korean Journal of Applied Entomology* 36: 317–322.

Zhang, K., and M. Hu. 1993. Study on the artificial diets of *Anoplophora nobilis*. *Scientia Silvae Sinicae* 29: 227–233 [in Chinese with English abstract].

Zhang, X., and M. J. Linit. 1998. Comparison of oviposition and longevity of *Monochamus alternatus* and *M. carolinensis* (Coleoptera: Cerambycidae) under laboratory conditions. *Environmental Entomology* 27: 885–891.

Zhao, J., N. Ogura, and M. Isono. 1999. Artificial rearing of *Anoplophora glabripennis* (Motsch.) (II). *Journal of Beijing Forestry University* 21: 62–66 [in Chinese with English abstract].

8

Natural Enemies and Biological Control of Cerambycid Pests

Timothy D. Paine
University of California
Riverside, California

CONTENTS

8.1 Introduction

Cerambycid wood borers are integral components of both hardwood and conifer forest insect communities around the globe. They are critical contributors to recycling of nutrients in these communities and forest succession in natural forest stands. Typically, they are colonizers of weakened, dead, and dying trees, or they are secondary invaders of trees colonized by other insects and diseases. With some important exceptions, they are not considered to be primary pests in native environments. However, the pest status of the insects is changing.

Two critical changes have occurred over the last half century that contributed to the recognition of longhorned borers as significant economic pests. The first is associated with the tremendous increase in plantation forestry. This can reflect reforestation or afforestation of large areas with native trees as part of efforts for both environmental restoration and economic resources. Alternatively, plantations of fast-growing introduced timber species (e.g., pines and eucalypts) have been planted throughout the world to satisfy increasing demand for paper pulp and wood product. The plantations of exotic trees growing in new environments represent a substantial increase in resources available to wood borers, particularly if invasive species capable of utilizing those trees are introduced without their natural enemies.

The second critical change is the dramatic growth in global trade over the same time interval. This increase in movement of goods and people around the world has resulted in an intensification in the introductions of forest insect pests from one continent to another (Aukema et al. 2010). The introductions of herbivores into new areas has, in some cases, reestablished contact between an insect and its native host tree or, in other cases, brought a herbivore into contact with a new community of hosts. In many of these examples, the herbivore has been introduced into the new environment without the communities of predators, parasitoids, and disease organisms that are important mortality factors holding the populations below pest status. Consequently, populations of the invasive species can reach damaging levels in the novel environment.

It is critical to understand the roles of these natural enemies in the population dynamics of insect herbivores. Because many of the cerambycid species are not economic concerns in native forests, there often is limited information available on the community of predators and parasitoids. In addition, a large proportion of a lengthy larval life cycle of the beetles is typically concealed within the host. As a result, generating meaningful information on the impact of natural enemies on cerambycid population dynamics is difficult.

Successful introductory biological control programs require identification, collection, and rearing of effective natural enemies from the native range of the target species, environmental risk assessment and testing for potential nontarget effects, mass rearing, release, and evaluation. All of these steps must be undertaken in compliance with a complex regulatory environment. It is not surprising that with the ecology and natural history characteristics of the beetles and the limited information available on the natural enemy communities, there are very few examples of successful biological control of cerambycids. The eucalyptus longhorned borer, *Phoracantha semipunctata* (F.), is under excellent biological control in a number of parts of the world following the introduction of an encyrtid egg parasitoid, *Avetianella longoi* Siscaro (Hanks et al. 1995a). However, that beetle had been moved around the world from Australia beginning in the early part of the twentieth century and was not brought under biological control until the last decade of that century. With the increasing interest in invasive longhorned borers, the damage caused in plantation forests and tree crops, and the natural enemies associated with the beetles in their native ranges, hopefully it will not take another 70 years to achieve the next biological control success.

8.2 Natural Enemies in Relation to Cerambycid Life History

The natural enemies of longhorned beetles are diverse and may be highly specialized for a particular life stage of the host insects. Ovipositing female beetles search out suitable host plants, typically woody plants. These hosts may be dead or dying, stressed, or vigorously growing. The natural enemies of the beetles must be able to also locate the suitable host plants at a time when susceptible life stages of their insects are present and available.

8.2.1 Oviposition Sites

Female beetles will lay eggs singly or in groups on the plant surface, or they will cut a slit or niche in the outer bark with their mandibles and lay their eggs in the slit. If eggs are laid on the bark surface, they may be in crevices or under loose, exfoliated bark. Although still on the surface, these locations do provide some level of protection from searching natural enemies. However, specialized egg parasitoids may be small enough to also get into the cryptic sites (Hanks et al. 1995a). Similarly, searching ants may also remove exposed eggs (Way et al. 1992).

8.2.2 Larval and Pupal Habitats

The larval and pupal habitats are at the interface of the wood and cambial tissues or in the wood. Typically, the larvae construct long winding galleries that are protected under the bark. The larval galleries are packed with frass, limiting access to the feeding immature beetle life stages. When larval feeding is complete, many species will construct pupation chambers deep in the wood that are often plugged with wood shavings and frass. Access of natural enemies to these life stages is limited. However, many specialized arthropod predators and parasitoids have evolved morphological structures and searching behaviors that allow them to exploit them as hosts. In addition, there are large vertebrate predators—for example, woodpeckers and other bird species—that have the capability of removing bark and wood with their specialized bills to reach developing larvae and pupae.

8.2.3 Host Location Cues

Many cerambycids are restricted to colonizing a narrow range of host species, and within that host range, they may be restricted to a particular vigor classification of trees (Saint-Germain et al. 2004a, 2004b, 2006). The first challenge is location of a suitable host habitat. In some cases, beetles use visual or odor cues to locate suitable or susceptible hosts. For example, the white spotted sawyer, *Monochamus scutellatus* Say, may use a combination of ethanol and alpha-pinene to locate suitable habitats containing stressed host trees (Chénier and Philogène. 1989; Peddle et al. 2002), particularly fire damaged hosts. Location of an individual tree may be as a result of random landing within the suitable habitat or directed landings in response to volatile cues. *Phoracantha semipunctata* is attracted to a complex mixture of host volatiles from individual stressed *Eucalyptus* trees (Hanks et al. 1998; Barata et al. 2000; Barata and Araujo 2001).

8.3 Taxonomic Range of Natural Enemies

8.3.1 Classes

There is a wide range of natural enemies attacking longhorned borers. Arthropod natural enemies include parasites (e.g., mites [Hanks et al. 1992]), hymenopteran and dipteran parasitoids, and a range of coleopteran, raphidiid, diperan, and dermapteran predators (Kenis and Hilszczanski 2007 and references therein). Cerambycid larvae can be important components of the diets of some vertebrate predators including woodpeckers (Hall 1954; Soloman 1968; Soloman and Morris 1971; Soloman 1977; Costello et al. 2011). Entomopathogenic fungi, including species of *Metarhizium* Sorokīn (Shanley and Hajek 2008; Ugine et al. 2013) and *Beauvaria* Vuillemin (Shimazu et al. 1995; Dubois et al. 2004), have been explored as a means to manage longhorned beetles, as have entomopathogenic nematodes (Liu et al. 1992).

8.3.2 Enemies of Life Stages

All life stages may be exploited by natural enemies. The egg laid on the bark surface may be exposed to both predation (Way et al. 1992) and parasitization (Hanks et al. 1995a; Luhring et al. 2000, 2004; Reed et al. 2007; Wang et al. 2008). Although cryptic, the larval stages serve as hosts to a wide range of parasitoids and predators. Kenis and Hilszczanski (2007) provided long lists of larval parasitoids that have been associated with both conifer (66 parasitoid species) and broad-leaved (54 parasitoid species) feeding cerambycids in Europe. Austin et al. (1994) listed 17 species of parasitoids attacking *P. semipunctata* larvae. Records for natural enemies using pupae for hosts are not as readily available; this probably is a consequence of the inaccessibility of the pupal chambers constructed deep in the wood compared to the larvae feeding under the bark at the interface of the cambium and xylem tissues. There may be predation pressure on the adult stages, although again this is difficult to document. However, the adults may be subject to predation by birds (Chittenden 1910, also see Soloman 1995) and spiders (Soloman 1995).

8.3.3 Facultative Predators

Although subject to significant pressure from natural enemies, cerambycids may also play a role as facultative predators. Red oak borer *Enaphalodes rufulus* (Haldeman) larvae can be cannibalistic with predatory larvae showing significant weight gains relative to solely phytophagous larvae (Ware and Stephen 2006). Consequently, there may be a fitness advantage accruing to facultative cannibalism by an individual larva. However, if the larvae are cannibalistic on siblings, the fitness of that generation may be reduced. Mechanisms to avoid or limit consumption of siblings could be the subject of further investigation. *Monochamus carolinensis* (Olivier) is a very effective intraguild predator of bark beetle larvae in conifers (Dodds et al. 2001). Intraspecific competition and intraspecific larval predation are critical causes of mortality for *P. semipunctata* (Paine et al. 2001; Hanks et al. 2005).

8.3.4 Location of Host

Natural enemies that arrive on the host tree at the same time as their beetle host may use similar cues to find the suitable habitat. The encyrtid egg parasitoid, *A. longoi*, arrives at stressed *Eucalyptus* at the same time as its *P. semipunctata* host in response to the same suite of host volatiles as the beetle. Use of the same set of cues by the beetle and the natural enemy ensures that the parasitoid will arrive at the same time that the beetle is laying eggs. Natural enemies that use other life stages are likely to utilize different long-range cues to arrive when the appropriate life stage of their host is available.

Once on the host plant, searching natural enemies may use a variety of cues to locate the beetle life stages. Typically, the larvae are concealed, but volatile cues may be present. More importantly, there may be vibration cues that can be detected on the plant surface. The feeding larvae create a vibration signature as the mandibles cut at the plant fibers and woody tissue. These may be detected through chordotonal organs. For example, the braconid larval parasitoid, *Syngaster lepidus* Brullé, detects the exact location of feeding *Phoracantha* larvae under the bark of the host tree using vibrational cues and then oviposits onto the larvae through the bark (Joyce et al. 2002, 2011; Paine et al. 2004).

8.4 Impact of Natural Enemies on the Population Dynamics of Cerambycids

There are very few studies that can document the impact of the community of natural enemies on population dynamics of cerambycid species. Many studies simply record the species of natural enemies associated with the host beetle. However, where data are available, it appears that the influence of components of the community can be significant. Woodpeckers removed 30% of white oak borer *Goes tigrinus* (DeGeer) larvae during winter (Soloman and Donley 1983), 39% of beech borer *G. pulverulentus* (Haldeman) larvae (Soloman and Morris 1971), and 65% of oak branch borer *G. debilis* LeConte larvae (Soloman 1977). In North America, at the larval stage, 50–80% of dogwood twig borer *Oberea tripunctata* (Swederus) (Ruggles 1915; Driggers 1929), 22% of red headed ash borer *Neoclytus acuminatus* (F.) (Waters 1981), 14–22% of the population of poplar-gall saperda *Saperda inornata* Say (McLeod and Wong 1967; Grimble and Knight 1970), and 50% of banded alder borer *Rosalia funebris* (Motschulsky) (Craighead 1923) have been reported to be parasitized. *Tetropium* sp. larvae were parasitized at rates ranging from 20% to 75% in Central Europe (Schimitschek 1929 cited in Kenis and Hilszczanski 2007) and 41% to 68% in Italy (Hellrigl 1985 cited in Kenis and Hilszczanski 2007). In Asia, about 50% of brown mulberry borer *Apriona germari* (Hope) eggs were parasitized by *Aprostocetus fukutai* Miwa & Sonan (Yan et al. 1996) and 23% of pistachio longicorn *Calchaenesthes pistacivora* Holzschuh larvae parasitized by *Megalommum pistacivorae* Achterberg & Mehrnejad (Achterberg and Mehrnejad 2011). In Australasia, three parasitoid wasps attacked lemon tree borer *Oemona hirta* (F.) larvae, but the parasitism rates were low (<5%) (Wang and Shi 1999, 2001).

8.5 Biological Control as a Management Strategy

Natural enemies have proven to be very effective at reducing populations of many different arthropods, including a long list of pest species. The naturally occurring mortality attributed to natural enemies of cerambycids infesting hardwood species in North America can range from very small levels to levels that seem to hold the populations in check (Soloman 1995). Similar ranges in natural enemy-caused mortality have been reported from European cerambycids (Kenis and Hilszczanski 2007). Consequently, because the impacts of predators and parasitoids can be significant, it is important to protect and enhance the populations of natural enemies.

8.5.1 Conservation

Conservation of existing populations of natural enemies may be significantly enhanced through habitat protection and conservation (Schimitschek 1929 cited in Kenis and Hilszczanski 2007). Many adult parasitoids require sources of food, often pollen and nectar, to enhance their reproductive fitness. These sources may come from a range of plant species that normally occur in the community. In that case, conservation of the plant diversity and structure may be critical to maintaining adult parasitoid populations. In addition, providing supplemental or alternate sources of food through augmentation or supplementation or sheltering sites through vegetation efforts may also function to help conserve natural enemy populations.

8.5.2 Introduction of Natural Enemies

Many of the critical problems associated with cerambycid pest species are the result of accidental introductions of wood borers into new environments. Frequently, there are no natural enemies in the adventive environment that can contribute to population regulation. Introduction of natural enemies from the native environment to establish biological control may be a critical option. However, classical biological control must be approached in a careful and deliberate manner. The candidate natural enemies must be able to survive in the new environment and must be selected for efficacy and selectivity. Efficacy assumes both a capacity to search and find the target species and a reproductive capacity that enables these natural enemies to reproduce at a rate that limits the capability of the target to respond. Selectivity assumes that the candidate natural enemy is restricted to the target species as a host and the risk of nontarget impacts is minimized. The candidate species must be evaluated for efficacy and selectivity either in the country of origin or through importation into a secure quarantine facility that prevents accidental escapes.

The screening process allows evaluation of candidate natural enemies and elimination of any unwanted diseases or hyperparasitoids. However, a significant practical problem still remains in the selection of a suitable candidate for release to establish biological control. It must be possible to rear the natural enemy species in colony culture. Colonies are required for the specificity testing, and they are needed to build up sufficient numbers of individuals that it will be possible to establish populations in the field releases. If populations cannot be reared in large numbers, then there is little hope for successful evaluation, mass release, and subsequent establishment. This limitation raises a very large practical barrier to potential biological control efforts. If, however, that barrier can be surmounted, then the final step is evaluation of the impact on populations of the target species. With specific regard to biological control of cerambycid species, evaluation can be a long and difficult process because the beetles may be cryptic and the host resources may be scattered across a wide area.

8.6 Example Case Studies

8.6.1 The Asian Longhorned Beetle

The Asian longhorned beetle, *Anoplophora glabripennis* (Motschulsky), is native to China and to South Korea (Lingafelter and Hoebeke 2002) where it colonizes *Acer, Populus, Salix,* and *Ulmus* trees (Williams et al. 2004; Haack 2006; Hu et al. 2009; Haack et al. 2010). Repeated attacks with extensive boring in the cambial and wood tissues by larvae can cause death of the stressed and healthy host trees (Gao et al. 1993). Outbreaks of the beetle in the last 30 years have been associated with extensive plantings of windbreaks and reforestation efforts across many areas of northwestern China with susceptible host trees (Haack et al. 2010). Management in China is focused on the use of foliar insecticides (Hu et al. 2009), physical control, and internal quarantines (Haack et al. 2010). The role of natural enemies, including coleopteran parasitoids (Wei and Jiang 2011), entomopathogenic fungi (Dubois et al. 2008), nematodes (Liu et al. 1992), and woodpeckers (Hu et al. 2009), in the control of this pest has been explored.

The beetle was first detected in North America in 1996 in New York (Hu et al. 2009). The movement from China to North America probably was associated with solid wood packing material (Haack et al. 2010). It has subsequently been detected in the urban areas of Chicago and Massachusetts in the United States and in Toronto, Canada. In Europe, the beetle was discovered in Austria in 2001 (Tomiczek et al. 2002) and in France in 2003 (Herard et al. 2007). Infestations were later discovered in Germany and in Italy (Hu et al. 2009). Management in both North America and in Europe has focused on detection, delimitation, and eradication of established infestations using systemic insecticide applications, removal, chipping, and burning (Haack et al. 2010).

Biological control options are being investigated in the invaded regions to support the eradication efforts. Entomopathogenic nematodes (Fallon et al. 2004) and fungi, including *Beauveria and Metarhizium* species, applied in bands around the trunks of trees (Ugine et al. 2013) have been tested for efficacy. The generalist parasitoid *Dastarcus helophoroides* (Coleoptera: Bothrideridae) is recognized as a critical natural enemy in China (Hu et al. 2009; Yang et al. 2014) but would be an unlikely candidate for importation and release in invaded environments. This parasitoid has been used for the control of *Batocera horsfieldi* Hope on walnut trees in China, achieving between 40% and 80% control (Li et al. 2009). The arthropod natural enemies of the beetle in Europe are being evaluated for future introduction to establish classical biological control (Pan 2005). Surveys conducted in Italy suggest that a community of native parasitoids, including an egg parasitoid and eight species of larval parasitoids, appears to be capable of making a host shift onto *A. glabripennis*.

8.6.2 Sawyer Beetles

Sawyer beetles refer to those in the cerambycid genus *Monochamus* Dejean. They infest a range of conifer species. For example, the white spotted sawyer, *M. scutellatus*, is transcontinental in North America with a host range that includes weakened pine, fir, larch, and spruce trees and cut logs (Furniss and Carolin 1977; Anonymous 1985). *Monochamus galloprovincialis* (Oliver) infests a range of pine species across Europe (Kenis and Hilszcanski 2007; Naves and Sousa 2009). The Japanese pine sawyer, *M. alternatus* Hope, colonizes pines across a broad geographic area including China, Japan, and Korea (Togashi 1986; Yang et al. 2014). The natural enemy community associated with the beetles has been, until recently, relatively unstudied (Kenis and Hilszcanski 2007). In some cases, there were no reported natural enemies. This may be, in part, because the insects were considered secondary, colonizing weakened trees or trees that had been previously colonized by tree-killing species (e.g., *Dendroctonus* bark beetles).

The nematode, *Bursaphelenchus xylophilus* (Steiner and Buhrer) Nickle (Nematoda; Aphelenchoididae), is the causal organism of pine wilt disease. The nematode is vectored among host trees by *Monochamus* species (Dropkin et al. 1981). Although the origins of the nematode were obscured for some time, it is now recognized as native to North America. Because of the long association between pathogen and host tree, pines in the native range appear to be less susceptible to pine wilt disease than pines in the invaded parts of the world. Although there has been a significant effort to understand the vector pathogen relationships in North America (e.g., Akbulut and Linit, 1999a, 1999b; Stamps and Linit 2001), there has been little impetus to document the natural enemy communities that might contribute to regulating the populations of beetle vectors there. In stark contrast, introduction of the nematode into Asia and Europe has stimulated great interest in the natural enemy communities of the *Monochamus* vector species in those areas.

There had been reports of three ichneumonid, two braconid, and one tachinid parasitoid species associated with *M. galloprovincialis* prior to the introduction of the pine wilt nematode (Tomminen 1993; Martikainen and Kopenen 2001; Kenis and Hilszcanski 2007). Studies in Portugal by Naves et al. (2005) identified three additional braconid larval parasitoids of the beetle. However, parasitism rates did not exceed 10.5% of the larvae sampled. No egg parasitoids were discovered in their survey.

Extensive efforts to characterize the natural enemy community and to determine the impact of the parasitoids on the population dynamics of *M. alternatus* have been undertaken in China and Japan

(Urano 2010; Yang et al. 2014). Studies of parasitoid biology and host finding have added important understanding about the relationships of the interactions between beetles and natural enemies (Li and Sun 2011; Liu et al. 2011; Wei and Jiang 2011; Wei et al. 2013). These discoveries are leading to recommendations for the use by natural enemies for beetle management to reduce the incidence disease transmission (Yang et al. 2014).

8.6.3 Eucalyptus Longhorned Borers

Trees in the genus *Eucalyptus* are native to Australia and New Guinea. They have been widely planted in many parts of the world for their rapid growth, tolerance of a wide range of growing conditions, and high-quality cellulose fiber that is in great demand for superior paper production (Doughty 2000). *Phoracantha semipunctata* and *P. recurva* Newman are cerambycids native to Australia but have been introduced throughout many parts of the world where eucalyptus is grown (Paine et al. 2011a). The beetles normally colonize broken branches and damaged or stressed trees. Mass colonization can rapidly kill trees. If trees are not maintained in a vigorous state, they can be at risk of infestation and death (Hanks et al. 1993a; Paine et al. 2002).

Management of the beetles in new environments initially focused on cultural practices that included enhancing host resistance of the trees through appropriate irrigation to reduce moisture stress (Hanks et al. 1991, 1999) and planting less preferred species (Hanks et al. 1993b, 1995b, 1995c; Paine et al. 2000a). These approaches provided tools that could be used to reduce risk of infestation, but it was not clear what impact they would have on the population dynamics of the cerambycid species. A classical biological control approach was undertaken in an effort to permanently reduce population size.

There have been limited reports of natural enemies shifting onto *Phoracantha* species in the colonized environments. Way et al. (1992) reported foraging by ants on *P. semipunctata* eggs. In Italy, the encyrtid egg parasitoid, *Avetianella longoi* Siscaro, was first described on *P. semipunctata* eggs (Siscaro 1992). Although the initial recoveries of the egg parasitoid were made in Europe, it was subsequently determined that the wasp had been accidentally introduced from its native range in Australia (Austin et al. 1994).

The parasitoid community associated with the beetles has been reasonably well characterized in Australia. Austin et al. (1994) recognized 18 species of primary or hyperparasitoid species associated with *P. semipunctata* but added an additional larval parasitoid in 1997 (Austin and Dangerfield 1997). The parasitoids partition the larval hosts by host size and bark thickness (Paine et al. 2000b; Hanks et al. 2001). The base of information available regarding the parasitoid community in the native range made it possible to efficiently collect parasitoids for evaluation as candidates for release in California.

Three species of parasitoid initially were collected in Australia as candidates for a classical biological control program (Hanks et al. 1995a; Millar et al. 2002). *Jarra phoracantha* Austin and Marsh (Figure 8.1) (Hymenoptera: Braconidae) is a gregarious idiobiont larval ectoparasitoid. *Syngaster lepidus* Brulle (Figure 8.2) (Hymenoptera: Braconidae) is also an idiobiont larval ectoparasitoid but is solitary. The females of these species assess the size of the larval host under the bark, probably using a combination of surface vibrational cues and chemical cues perceived when the host is stung, and lay male eggs on small larvae and female eggs on large larvae they encounter (Joyce et al. 2002, 2011). *Avetianella longoi* (Figure 8.3) is an encyrtid egg parasitoid that may lay one to three eggs in the eggs of the host. All three species were evaluated, mass-reared, and released in Southern California. Unfortunately, despite repeated intensive efforts, the larval parasitoids were never recovered, and it is unlikely that they established. The egg parasitoid was released in excess of 200,000 individual per year for a number of years. These parasitoids established broadly across the state and are responsible for complete biological control of *P. semipunctata* (Hanks 1995a).

Although *A. longoi* was a success against *P. semipunctata*, it was ineffective for control of *P. recurva*. The parasitoid was capable of finding and parasitizing egg masses of *P. recurva*, but the rate of parasitization was much lower than in *P. semipunctata* (Luhring et al. 2000). The eggs of *P. recurva* are capable of mounting an encapsulation response that kills the *A. longoi* eggs (Luhring et al. 2004; Reed et al. 2007). Wang et al. (2008) collected a second race of *A. longoi* from successfully parasitized *P. recurva* eggs in Australia. Large numbers of individuals of this race of parasitoids have been released

FIGURE 8.1   *Jarra phoracantha* is a gregarious larval ectoparasitoid. The shorter ovipositor confines the wasp to larvae that are feeding under thin host bark.

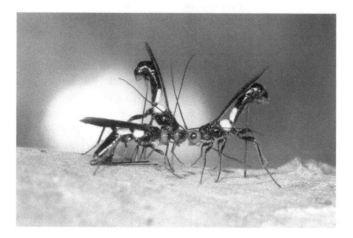

FIGURE 8.2   *Syngaster lepidus* is a solitary larval ectoparasitoid. A female can detect chewing larvae under the bark, determine host size, and allocate the sex of her offspring based on that assessment. The longer ovipositor makes it possible for females to exploit larvae feeding under thicker bark.

FIGURE 8.3   *Avetianella longoi* is an egg parasitoid that arrives on the host tree at the same time as ovipositing female *Phoracantha*. Although they normally lay a single egg within a beetle egg, as many as four parasitoid eggs have been observed. This species is responsible for excellent biological control of the beetle host in many parts of the world.

in Southern California. Although recoveries have been limited, the number of trees killed by the beetle appears to have declined.

The classical biological control program directed against the two *Phoracantha* species has been successful. The parasitoids of *P. semipunctata* have held the populations of the beetle to low levels for almost two decades with a very significant economic benefit from tree protection (Paine et al. 2015). The effect on *P. recurva* is much more recent. However, the stability of the biological control effort may be under threat. In addition to the introduction of longhorned borers, members of at least three other guilds of eucalyptus-feeding insects have been introduced into the state (Paine et al. 2011a). Several of these are also under excellent biological control, but the principal control strategy used against others is application of systemic pesticides. Unfortunately, these insecticides can accumulate in the *Eucalyptus* floral nectar, an important food source for adult parasitoids (Paine et al. 2011b). It will be critical to develop alternatives to the use of the systemic insecticides to protect the community of natural enemies that are effectively controlling the damaging pest species.

REFERENCES

Achterberg, K. van, and M, Mehrnejad. 2011. A new species of *Megalommum* Szepligeti (Hymenoptera: Braconidae: Braconinae), a parasitoid of the pistachio longhorn beetle (*Calchaenesthes pistacivora* Holzschuh; Coleoptera, Cerambycidae) in Iran. *Zookeys* 112: 21–38.

Akbulut, S., and M. J. Linit. 1999a. Reproductive potential of *Monochamus caroninensis* (Coleoptera: Cerambycidae) with respect to pinewood nematode phoresis. *Environmental Entomology* 28: 407–411.

Akbulut, S., and M. J. Linit. 1999b. Flight performance of *Monochamus carolinensis* (Coleoptera: Cerambycidae) with respect to nematode phoresis and beetle characteristics. *Environmental Entomology* 28: 1014–1020.

Anonymous. 1985. *Insects of eastern forests*. Misc. Publication 1426. US Department of Agriculture, Forest Service, Washington, DC, 608 pp.

Aukema, J. E., D. G. McCullough B. Von Holle, A. M. Liebhold, K. Britton, and J. Frankel. 2010. Historical accumulation of nonindigenous forest pests in the continental United States. *Bioscience* 60: 886–897.

Austin, A. D., and P. C. Dangerfield. 1997. A new species of *Jarra* Marsh and Austin (Hymenoptera: Braconidae) with comments on other parasitoids associated with the eucalypt longicorn *Phoracantha semipunctata* (F.) (Coleoptera: Cerambycidae). *Australian Journal of Entomology* 36: 327–331.

Austin, A. D., D. L. J. Quicke, and P. M. Marsh. 1994. The hymenopterous parasitoids of eucalypt longicorn beetles, *Phoracantha* spp. (Coleoptera: Cerambycidae) in Australia. *Bulletin of Entomological Research* 84: 145–174.

Barata, E. N., and J. Araujo. 2001. Olfactory orientation responses of the eucalyptus woodborer, *Phoracantha semipunctata*, to host plant in a wind tunnel. *Physiological Entomology* 26: 26037.

Barata, E. N., J. A. Pickett, L. J. Wadhams, C. M. Woodcock, and H. Mustaparta. 2000. Identification of host and nonhost semiochemicals of eucalyptus woodborer *Phoracantha semipunctata* by gas chromatography-electroantennography. *Journal of Chemical Ecology* 26: 1877–1895.

Chénier, J. V. R., and B. J. R. Philogène. 1989. Field responses of certain forest Coleoptera to conifer monoterpenes and ethanol. *Journal of Chemical Ecology* 15: 1729–1745.

Chittenden, F. H. 1910. The oak pruner. Circ. 130. U.S. Department of Agriculture, Bureau of Entomology. Washington, DC, 7 p.

Costello, S. L., J. F. Negron, and W. R. Jacobi. 2011. Wood-boring insect abundance in fire-injured ponderosa pine. *Agriculture and Forest Entomology* 13: 373–393.

Craighead, F. C. 1923. North American cerambycid larvae. In *A classification and the biology of the North American cerambycid larvae*. Bull. 27. Canadian Department of Agriculture, Entomology Branch, Ottawa, ON, 238 p.

Dodds, K. J., C. Graber, and F. M. Stephen. 2001. Facultative intraguild predation by larval Cerambycidae (Coleoptera) on bark beetle larvae (Coleoptera: Scolytidae). *Environmental Entomology* 30: 17–22.

Doughty, R. W. 2000. *The Eucalyptus: A Natural and Commercial History of the Gum Tree*. The Johns Hopkins University Press, Baltimore, MD, 237 pp.

Driggers, B. F. 1929. Notes on the life history and habits of the blueberry stem borer, *Oberea myops* Hald., on cultivated blueberries. *Journal of the New York Entomological Society* 37: 67–75.

Dropkin, V. H., A. Foudin, E. Kondo, M. Linit, M. Smith, and K. Robbins, K. 1981. Pinewood nematode: A threat to United States forests. *Plant Disease* 65: 1022–1027.

Dubois, T., A. E. Hajek, J. F. Hu, and Z. Z. Li. 2004. Efficacy of fiber bands impregnated with *Beauveria brongniartii* cultures against the Asian longhorned beetle, *Anoplophora glabripennis* (Coleoptera: Cerambycidae). *Biological Control* 31: 320–328.

Dubois, T., J. Lund, L. S. Bauer, and A. E. Hajek. 2008. Virulence of entomopathogenic hypocrealean fungi infecting *Anoplophora glabripennis*. *Biocontrol* 53: 517–528.

Fallon, D. J., L. F. Solter, M. Keena, M. McManus, J. R. Cate, and L. M. Hanks. 2004. Susceptibility of Asian longhorned beetle, *Anoplophora glabripennis* (Motchulsky) (Coleoptera: Cerambycidae) to entomopathogenic nematodes. *Biological Control* 30: 430–438.

Furniss, R. L., and V. M. Carolin. 1977. *Western forest insects*. Misc. Publication 1399. US Department of Agriculture, Forest Service, Washington, DC, 654 pp.

Gao, R., X. Qin, D. Chen, and W. Chen. 1993. A study on the damage to poplar caused by *Anoplophora glabripennis*. *Forest Research* 6: 189–193.

Grimble, D. G., and F. B. Knight. 1970. Life tables and mortality factors for *Saperda inornata* (Coleoptera: Cerambycidae). *Annals of the Entomological Society of America* 63: 1309–1319.

Haack, R. A. 2006. Exotic bark- and wood-boring Coleoptera in the United States: Recent establishments and interceptions. *Canadian Journal of Forest Research* 36: 269–288.

Haack, R. A., F. Herard, J. Sun, and J. J. Turgeon. 2010. Managing invasive populations of Asian Longhorned Beetle and Citrus Longhorned Beetle: A worldwide perspective. *Annual Review of Entomology* 55: 521–546.

Hall, R. C. 1954. *Control of the locust borer*. Circ. 626. U.S. Department of Agriculture, Forest Service, Washington, DC, 19 p.

Hanks, L. M., J. R. Gould, T. D. Paine, J. G. Millar, and Q. Wang. 1995a. Biology and host relations of *Avetianella longoi* (Hymenoptera: Encyrtidae), an egg parasitoid of the eucalyptus longhorned borer (Coleoptera: Cerambycidae). *Annals of the Entomological Society of America* 88: 666–671.

Hanks, L. M., J. S. McElfresh, J. G. Millar, and T. D. Paine. 1992. Control of the straw itch mite (Acari: Pyemotidae) with sulfur in an insect rearing facility. *Journal of Economic Entomology* 85: 683–686.

Hanks, L. M., J. S. McElfresh, J. G. Millar, and T. D. Paine. 1993a. *Phoracantha semipunctata* F. (Coleoptera: Cerambycidae), a serious pest of *Eucalyptus* in California: Biology and laboratory rearing procedures. *Annals of the Entomological Society of America* 86: 96–102.

Hanks, L. M., J. G. Millar, and T. D. Paine. 1995c. Biological constraints on host range expansion by the wood-boring beetle *Phoracantha semipunctata* F. (Coleoptera: Cerambycidae). *Annals of the Entomological Society of America* 88: 183–188.

Hanks, L. M., J. G. Millar, and T. D. Paine. 1998. Aerial dispersal of the eucalyptus longhorned borer (Coleoptera: Cerambycidae) in an urban landscape. *Environmental Entomology* 27: 1418–1424.

Hanks, L. M., J. G. Millar, T. D. Paine, Q. Wang, and E. O. Paine. 2001. Patterns of host utilization by two parasitoids (Hymenoptera: Braconidae) of the eucalyptus longhorned borer (Coleoptera: Cerambycidae). *Biological Control* 21: 152–159.

Hanks, L. M., T. D. Paine, and J. G. Millar. 1991. Mechanisms of resistance in *Eucalyptus* against larvae of the eucalyptus longhorned borer (Coleoptera: Cerambycidae). *Environmental Entomology* 20: 1583–1588.

Hanks, L. M., T. D. Paine, and J. G. Millar. 1993b. Host species preference and larval performance in the wood-boring beetle *Phoracantha semipunctata* F. *Oecologia* (Berl.) 95: 22–29.

Hanks, L. M., T. D. Paine, and J. G. Millar. 2005. Influence of the larval environment on performance and adult body size of the wood-boring beetle *Phoracantha semipunctata* (F.) (Coleoptera: Cerambycidae). *Entomologia Experimentalis et Applicata* 114: 25–34.

Hanks, L. M., T. D. Paine, J. G. Millar, C. D. Campbell, and U. K. Schuch. 1999. Water relations of host trees and resistance to the phloem-boring beetle *Phoracantha semipunctata* F. (Coleoptera: Cerambycidae). *Oecologia* 119: 400–407.

Hanks, L. M., T. D. Paine, J. G. Millar, and J. L. Hom. 1995b. Variation among *Eucalyptus* species in resistance to eucalyptus longhorned borer in Southern California. *Entomologia Experimentalis et Applicata* 74: 185–194.

Hellrigl, K. 1985. Uber Parasitierung und Farbformen des Larchenbockes *Tetropium gabrieli* Weise (Col, Cerambycidae) in Sudtirol. *Anzeiger furScchadlingskunde, Pflanzenschutz, Umweltschutz* 58: 88–90.

Herard, F., M. Ciampitti, M. Maspero, et al. 2007. New associations between the Asian pests *Anoplophora* spp. and local parasitoids, in Italy (2005). In *Proc. 17th U.S. Dep. Agric. Interagency Res. Forum on Gypsy Moth and Other Invasive Species 2006*. Gen. Tech. Rep. NRS-P-10, eds. KW Gottschalk, U. S. Department of Agriculture, Forest Service Northeastern Research Station, Newtown Square, PA, 50 p.

Hu, J., S. Angeli, S. Schuetz, Y. Luo, and A. E. Hajek. 2009. Ecology and management of exotic and endemic Asian longhorned beetle *Anoplophora glabripennis*. *Agriculture and Forest Entomology* 11: 359–375.

Joyce, A. L., J. G. Millar, J. S. Gill, M. Singh, D. Tanner, and T. D. Paine. 2011. Do acoustic cues mediate host finding by *Syngaster lepidus* Brullé (Hymenoptera: Braconidae)? *Biocontrol* 56: 145–153.

Joyce, A. L., J. G. Millar, and T. D. Paine. 2002. The effect of host size on the sex ratio of *Syngaster lepidus*, a parasitoid of eucalyptus longhorned borers (*Phoracantha* spp.) *Biological Control* 24: 207–213.

Kenis, M., and J. Hilszczanski. 2007. Natural enemies of Cerambycidae and Buprestidae infesting living trees. In *Bark and Wood Boring Insects in Living Trees in Europe, a Synthesis*, eds. F. Lieutier, F. R. Day, A. Battisti, J.-C. Gregoire, and H. F. Evans. Dordrecht, The Netherlands: Springer, 475–798 p.

Li, L., and J. Sun. 2011. Host suitability of a gregarious parasitoid on beetle hosts: Flexibility between fitness of adult and offspring. *PLoS One* 6(4): e18563. doi:10.1371/journal.pone.0018563

Li, J. Q., Z. Q. Yang, Y. L. Zhang, Z. X. Mei, Y. R. Zhang, and X. Y. Wang. 2009. Biological control of *Batocera horsfieldi* (Coleoptera: Cerambycidae) by releasing its parasitoid *Dastarcus helophoroides* (Coleoptera: Bothrideridae). *Scientia Silvae Sinicae* 45: 94–100 [in Chinese with English abstract].

Lingafelter, S. W., and E. R. Hoebeke. 2002. *Revision of Anoplophora (Coleoptera: Cerambycidae)*. Entomological Society of Washington, Washington, DC, 236 p.

Liu, S. R., C. X. Zhu, and X. P. Lu. 1992. Field trials of controlling several cerambycid larvae with entomopathogenic nematodes. *Chinese Journal of Biological Control* 8: 176.

Liu, Z., B. Xu, L. Li, and J. Sun. 2011. Host-size mediated trade-off in a parasitoid *Sclerodermus harmandi*. *PLoS One* 6(8): e23260. doi:10.1371/ journal.pone.0023260

Luhring, K. A., J. G. Millar, T. D. Paine, D. Reed, L. M. Hanks, and H. Christiansen. 2004. Oviposition preferences and progeny development of the egg parasitoid *Avetianella longoi*: Factors mediating replacement of one species by a congener in a shared habitat. *Biological Control* 30: 382–391.

Luhring, K. A., T. D. Paine, J. G. Millar, and L. M. Hanks. 2000. Suitability of the eggs of two species of eucalyptus longhorned borers (*Phoracantha recurva* and *P. semipunctata*) as hosts for the parasitoid *Avetianella longoi* Siscaro. *Biological Control* 19: 95–104.

Martikainen, P., and M. Kopenen. 2001. *Meteorus corax* Marshall, 1898 (Hymenoptera: Braconidae) a new species to Finland and Russian Karelia, with an overview of Northern species of *Meteorus* parasitizing beetles. *Entomologica Fennica* 12: 169–172.

McLeod, B. B., and H. R. Wong. 1967. Biological notes on *Saperda concolor* LeC. in Manitoba and Saskatchewan (Coleoptera: Cerambycidae). *Manitoba Entomologist* 1: 27–33.

Millar, J. G., T. D. Paine, C. D. Campbell, and L. M. Hanks. 2002. Methods for rearing *Syngaster lepidus* and *Jarra phoracantha* (Hymenoptera: Braconidae), larval parasitoids of the phloem-colonizing longhorned beetles *Phoracantha semipunctata* and *P. recurva* (Coleoptera: Cerambycidae). *Bulletin of Entomological Research* 92: 141–146.

Naves, P., and E. De Sousa. 2009. Threshold temperatures and degree-day estimates for development of postdormancy larvae of *Monochamus galloprovincialis* (Coleoptera: Cerambycidae). *Journal Pest Science* 82: 1–6.

Naves, P., M. Kenis, and E. De Sousa. 2005. Parasitoids associated with *Monochamus galloprovincialis* (Oliv.) (Coleoptera: Cerambycidae) within the pine wilt nematode-affected zone in Portugal. *Journal Pest Science* 78: 57–62.

Paine, T. D., L. M. Hanks, J. G. Millar, and E. O. Paine. 2000a. Attractiveness and suitability of host tree species for colonization and survival of *Phoracantha semipunctata* F. (Coleoptera: Cerambycidae). *Canadian Entomologist* 132: 907–914.

Paine, T. D., C. C. Hanlon, and F. J. Byrne. 2011b. Potential risks of systemic imidacloprid to parasitoid natural enemies of a cerambycid attacking *Eucalyptus*. *Biological Control*. 56: 175–178.

Paine, T. D., A. L. Joyce, J. G. Millar, and L. M. Hanks. 2004. The effect of variation in host size on the sex ratio, size, and survival of *Syngaster lepidus*, a parasitoid of eucalyptus longhorned borers (*Phoracantha* spp.). II. *Biological Control* 30: 374–381.

Paine, T. D., and J. G. Millar. 2002. Insect pests of eucalypts in California: Implications of managing invasive species. *Bulletin of Entomological Research* 92: 147–151.

Paine, T. D., J. G. Millar, L. M. Hanks, et al. 2015. Cost–benefit analysis for biological control programs that targeted insect pests of eucalypts in urban landscapes of California. *Journal of Economic Entomology* 108: 2497–2504.

Paine, T. D., J. G. Millar, E. O. Paine, and L. M. Hanks. 2001. Influence of host log age and refuge from natural enemies on colonization and survival of *Phoracantha semipunctata*. *Entomologia Experimentalis et. Applicata* 98: 157–163.

Paine, T. D., E. O. Paine, L. M. Hanks, and J. G. Millar. 2000b. Resource partitioning among parasitoids (Hymenoptera: Braconidae) of *Phoracantha semipunctata* in their native range. *Biological Control* 19: 223–231.

Paine, T. D., M. J. Steinbauer, and S. A. Lawson. 2011a. Native and exotic pests of *Eucalyptus*: A worldwide perspective. *Annual Review of Entomology* 56: 181–201.

Pan, H. Y. 2005. Review of the Asian longhorned beetle: Research, biology, distribution and management in China. For. Dep. Work. Pap., FBS/6E, Food Agric. Org., Rome, Italy.

Peddle, S., P. de Groot, and S. Smith. 2002. Oviposition behaviour and response of *Monochamus scutellatus* (Coleoptera: Cerambycidae) to conspecific eggs and larvae. *Agriculture and Forest Entomology* 4: 217–222.

Reed, D. A., K. A. Luhring, C. A. Stafford, et al. 2007. Host defensive response against an egg parasitoid involves cellular encapsulation and melanization. *Biological Control* 41: 214–222.

Ruggles, A. G. 1915. Life history of *Oberea tripunctata* Swed. *Journal of Economic Entomology* 8: 79–85.

Saint-Germain, M., C. M. Buddie, and P. Drapeau. 2006. Sampling saproxylic Coleoptera: Scale issues and the importance of behavior. *Environmental Entomology* 35: 478–487.

Saint-Germain, M., P. Drapeau, and C. Hébert. 2004a. Xylophagous insects of fire-killed black spruce in central Quebec: Species composition and substrate use. *Canadian Journal of Forest Research* 34: 677–685.

Saint-Germain, M., P. Drapeau, and C. Hébert. 2004b. Landscape-scale habitat selection patterns of *Monochamus scutellatus* (Coleoptera: Cerambycidae) in a recently burned black spruce forest. *Environmental Entomology* 33: 1703–1710.

Schimitschek, E. 1929. *Tetropium gabrieli* Weise and *Tetropium fuscum* L. Ein Beitrag zu ihrer Lebensgeschichte und Lebensgemeinschaft. *Zeitschrift fur Angewandte Entomologie* 22: 558–564.

Shanley, R. P., and A. E. Hajek. 2008. Environmental contamination with *Metarhizium anisopliae* from fungal bands for control of the Asian longhorned beetle, *Anoplophora glabripennis* (Coleoptera: Cerambycidae). *Biocontrol Science and Technology* 18: 109–120.

Shimazu, M., D. Tsuchiya, H. Sato, and T. Kushida. 1995. Microbial control of *Monochamus alternatus* Hope (Coleoptera: Cerambycidae) by application of nonwoven fabric strips with *Beauvaria bassiana* (Deuteromycotina: Hyphomycetes) on infested tree trunks. *Applied Entomology and Zoology* 30: 207–213.

Siscaro, G. 1992. *Avetianella longoi*, new species (Hymenoptera Encyrtidae) egg parasitoid of *Phoracantha semipunctata* F. (Coleoptera Cerambycidae) *Bollettino di Zoologia Agraria e di Bachicoltura* 24: 205–212.

Soloman, J. D. 1968. Cerambycid borer in mulberry. *Journal of Economic Entomology*. 61: 1023–1025.

Soloman, J. D. 1977. Biology and habits of the oak branch borer (*Goes debilis*). *Annals of the Entomological Society of America*. 70: 57–59.

Solomon, J. D. 1995. Guide to insect borers on North American broadleaf trees and shrubs. Agric. Handbook. 706. US Department of Agriculture, Forest Service. Washington, DC, 735 p.

Soloman, J. D., and D. E. Donley. 1983. *Bionomics and control of the white oak borer*. Res. Pap. SO-198. U. S. Department of Agriculture, Forest Service, Southern Forest Experiment Station. New Orleans, LA, 5 p.

Soloman, J. D., and R. C. Morris. 1971. Woodpeckers in the ecology of southern hardwood borers. In *Proceedings, Tall, Timbers Conference on Ecological Control by Habitat Management*. Tall Timbers Research Station, Tallahassee, FL. Vol 2: 109–115.

Stamps, W. T., and M. J. Linit. 2001. Interaction of intrinsic and extrinsic chemical cues in the behaviour of *Bursaphelenchus xylophilus* (Aphelenchida: Aphelenchoididae) in relation to its beetle vectors. *Nematology* 3: 295–301.

Togashi, K. 1986. Effects of the initial density and natural enemies on the survival rate of the Japanese pine sawyer, *Monochamus alternatus* Hope (Coleoptera, Cerambycidae), in pine logs. *Applied Entomology and Zoology* 21: 244–251.

Tomiczek, C., H. Krehan, and P. Menschhorn. 2002. Dangerous Asiatic longicorn beetle found in Austria: New danger for our trees? *Allgemeine Forst Zeitschrift fur Waldwirtschaft und Umweltvorsorge* 57: 52–54.

Tomminen, J. 1993. Development of *Monochamus galloprovincialis* Oliver (Coleoptera: Cerambycidae) in cut trees of young pines (*Pinus sylvestris* L.) and log bolts in southern Finland. *Entomologica Fennica* 4: 137–142.

Ugine, T. A., N. E. Jenkins, S. Gardescu, and A. E. Hajek. 2013. Comparing fungal band formulations for Asian longhorned beetle biological control. *Journal of Invertebrate Pathology* 113: 240–246.

Urano, T. 2010. Effects of host stage and number of feeding larvae on parasitism success and fitness in the coleopteran parasitoid, *Dastarcus longulus* Sharp (Coleoptera: Bothrideridae). *Applied Entomology and Zoology* 45: 215–223.

Wang, Q., J. G. Millar, D. A. Reed, et al. 2008. Development of a strategy for selective collection of a parasitoid attacking one member of a large herbivore guild. *Journal of Economic Entomology* 101: 1771–1778.

Wang, Q., and G. L. Shi. 1999. Parasitic natural enemies of lemon tree borer. *Proceedings of New Zealand Plant Protection Conference* 52: 60–64.

Wang, Q., and G. L. Shi. 2001. Host preference and sex allocation of three hymenopteran parasitoids of a longicorn pest, *Oemona hirta* (Fabricius) (Coleoptera: Cerambycidae). *Journal of Applied Entomology* 125: 463–468.

Ware, V. L., and F. M. Stephen. 2006. Facultative intraguild predation of red oak borer larvae (Coleoptera: Cerambycidae). *Environmental Entomology* 35: 443–447.

Waters, D. J. 1981. Life history of *Neoclytus acuminatus* with notes on other cerambycids associated with dead or dying trees. MS Thesis. Auburn University, Auburn, AL. 103 p.

Way, M. J., M. E. Cammell and M. R. Paiva. 1992. Studies on egg predation by ants (Hymenoptera: Formicidae) on the eucalyptus borer *Phoracantha semipunctata* (Coleoptera: Cerambycidae) in Portugal. *Bulletin of Entomological Research* 82: 425–432.

Wei, J. R., and L. Jiang. 2011. Olfactory response of *Dastarcus helophoroides* (Coleoptera: Bothrideridae) to larval frass of *Anoplophora glabripennis* (Coleoptera: Cerambycidae) on different host tree species. *Biocontrol Science and Technology* 21: 1263–1272.

Wei, J. R., X. P. Lu, and L. Jiang. 2013. Monoterpenes from larval frass of two cerambycids as chemical cues for a parasitoid, *Dastarcus helophoroides*. *Journal of Insect Science* 13: 59.

Williams, D. W., H. P. Lee, and I. K. Kim. 2004. Distribution and abundance of *Anoplophora glabripennis* (Coleoptera: Cerambycidae) in natural Acer stands in South Korea. *Environmental Entomology* 33: 540–45.

Yan, Y. H., D. Z. Huang, Z. G. Wang, D. R. Ji, and J. J. Yan. 1996. Preliminary study on the biological characteristics of *Aprostocetus futukai* Miwa et Sonan. *Journal of Hebei Agricultural University* 19(2): 41–46 [in Chinese with English abstract].

Yang, Z. Q., X. Y. Wang, and Y. N. Zhang. 2014. Recent advances in biological control of important native and invasive forest pests in China. *Biological Control* 68: 117–128.

9

Cultural Control of Cerambycid Pests

Qiao Wang
Massey University
Palmerston North, New Zealand

CONTENTS

9.1 Introduction

The Cerambycidae is a very large family of beetles with more than 36,000 species in the world (see Chapter 1). Nevertheless, only a small proportion of cerambycid species are plant pests of economic importance. They may damage plants by direct feeding or by transmission of plant diseases. With the growth of international trade, an increasing number of cerambycid species have become established outside their natural distribution range (Haack et al. 2010). Examples include the establishment of the Asian species *Anoplophora glabripennis* (Motschulsky) in North America (Haack et al. 2010) and Europe (Straw 2015a, 2015b) and *Anoplophora chinensis* (Forster) in Europe (Haack et al. 2010; Peverieri et al. 2012), as well as the Australian species *Phoracantha semipunctata* (Fabricius) and *Phoracantha recurva* Newman in California and many parts of the world (CABI 2015a, 2015b).

Cultural controls are probably the oldest management practices aimed at reducing pest establishment, reproduction, dispersal, and survival. These involve making the habitat more suitable for natural enemies of the pests (see Chapter 8) or less favorable for the pests. The focus of this chapter is on measures available for achieving the latter goal. Cultural controls, although often simple to undertake, are complex in action, relying on our knowledge of behavior, life cycle, and habits of the target pests and our ability to predict consequences from actions in a system with many interactive variables. Therefore, cultural controls require a long-term management approach. There are a number of cultural control techniques that can be used to render the environment unfavorable to the pests, such as pruning, sanitation, irrigation, rotation, plant density, intercropping, traps, planting and harvest times, and plant resistance.

In addition, climate change has been recognized as a major cause of frequent extreme weather events worldwide, such as drought and high temperatures. Insect migration due to such environmental stressors can lead to novel biotic interactions and trophic changes, causing problems such as pest invasions (Blois et al. 2013) and less effective biological control (Thomson et al. 2010). Evidence shows that the range of some insect pests in agriculture and forestry has expanded in recent decades due to climate change (Battisti and Larsson 2015). Moreover, poleward range shifts have been faster for pest species than for nonpest species, probably reflecting both global warming and human-assisted movement (Lindstrom and Lehmann 2015). The species that remain in their local ecosystems may alter their life history strategies to adapt to these changes, such as increasing their rate of food consumption (Karuppaiah and Sujayanad 2012) and population size (Ozgul et al. 2010; Youngsteadt et al. 2015), which can create more pest problems. The effects of climate change are likely to result in exotic pests such as *A. glabripennis* becoming more prevalent outside their native range (Evans et al. 2002; Straw et al. 2015a) and native species that were minor pests historically becoming serious pests, such as *P. semipunctata* in Australia (Seaton et al. 2015). Therefore, utilizing cultural controls as environmental measures is particularly important for the management of cerambycid pests at the present and in the future.

Cerambycids spend most of their lifetime inside host plants or in the soil and thus often escape the impact of biological control agents (see Chapter 8) and chemical control (see Chapter 10), making cultural controls more relevant and probably more effective and practical in many cases. Furthermore, cerambycid pests may damage forest and urban trees (see Chapter 11) and woody and herbaceous crops (see Chapter 12) in very different ecosystems. Therefore, some cultural control techniques may be effective in one ecosystem but totally impractical in others. For example, to control cerambycid pests of plants that are under water stress due to drought, irrigation may be practical and effective for crops and street trees, although for forest trees selection of plant species or varieties that are tolerant to drought may be more practical. This chapter outlines general principles and methods of commonly used cultural control techniques with special reference to cerambycid pest control. Table 9.1 summarizes techniques used for cerambycid pest control with examples from around the world.

9.2 Mechanical and Sanitary Techniques

These techniques aim to reduce the pest population size and to prevent pest spread and reinfestation using mechanical and sanitary treatments. Commonly used measures for the control of cerambycid pests in agriculture, forestry, and horticulture are illustrated and discussed here.

9.2.1 Pruning

Pruning is a practice of removing infested plant parts to reduce pest populations and damage and is most commonly used in tree and vine crops for the control of cerambycid pests. This technique requires frequent monitoring of infestation signs and symptoms such as frass ejection, girdle wounds, and wilting. Most pruning takes place during summer, fall, and winter. To avoid disease infection and further longicorn infestation, pruning wounds should be sealed with paint or similar protective material. Pruned infested parts are treated using sanitary measures (see Section 9.2.5).

For longicorn pests that damage twigs and small branches, pruning infested plant parts is probably the most effective control method and can achieve almost complete control if practiced properly. For example, the females of *Linda nigroscutata* (Fairmaire), an apple twig borer in Asia, girdle twigs and tender shoots before laying their eggs in the wound and the larvae begin boring in the twigs and shoots (Duffy 1968; Cao 1981). Careful examination of twigs and shoots should be carried out between June

TABLE 9.1

Examples of Cultural Control Techniques for Cerambycid Pest Control

Cultural Control Techniques	Suitable for	Examples and References
Pruning of twigs and branches	Tree twig and branch borers	*Linda nigroscutata* (Fairmaire) (Duffy 1968, Cao 1981)
	Tree trunk borers with early instars in twigs or branches	*Apriona germari* (Hope) (Hussain and Chishti 2012), *Oemona hirta* (Fabricius) (Clearwater and Muggleston 1985), *Acalolepta mixta* (Hope) (Dunn and Zurbo 2014)
Selective plant destruction	Eradication of invaded pests	*Anoplophora glabripennis* (Motschulsky) (Smith and Wu 2008, Anonymous 2013)
	Defense against dispersal or reinfestation	*Phoracantha semipunctata* (Fabricius) (El-Yousfi 1989, Paine 2009), *Enaphalodes rufulus* Haldeman (Donley 1981, 1983; Riggins et al. 2009), *Monochamus alternatus* Hope (Wu et al. 2000), *Semanotus japonicus* Lacordaire (Sato 2007)
	Heavily infested crop trees or vines that cannot be saved	*Prionus imbricornis* (Linnaeus) (Agnello 2013), *Aeolesthes sarta* Solsky (Gupta and Tara 2014), *Acalolepta mixta* (Dunn and Zurbo 2014)
Plowing	Field crops	*Dorysthenes granulosus* (Thomson) (Yang 1992, Gong et al. 2008), *Dectes texanus* LeConte (Campbell and Vanduyn 1977, Rogers 1985), *Obereopsis brevis* (Gahan) (Sharma et al. 2014)
	Root feeders of removed tree crops	*Prionus californicus* Motschulsky (Alston et al. 2007)
Removal of loose bark	Borers that prefer to oviposit under loose bark of crop tree trunks or main branches	*Monochamus leuconotus* (Pascoe) (Rutherford and Phiri 2006), *Xylotrechus javanicus* (Castelnau & Gory) (Venkatesha and Dinesh 2012)
Destruction of eggs and neonate larvae	Lamiine borers that make visible oviposition slits on tree trunks	*Anoplophora chinensis* (Forster) (Lieu 1945), *Pseudonemorphus versteegi* (Ritsema) (Banerjee and Nath 1971, Shylesha et al. 1996)
Extraction and destruction of older larvae	Borers that have entered tree trunks	*Aromia bungii* (Faldermann) (Gressitt 1942), *Saperda candida* Fabricius (Agnello 2013), *Nitakeris nigricornis* (Olivier) (Crowe 1963, 2012), *Pseudonemorphus versteegi* (Shylesha et al. 1996)
Sanitary treatment	Borers in plant residues	*Psyrassa unicolor* (Randall) (Solomon 1985), *Oncideres cingulata* (Say) (Baker and Bambara 2001), *Anoplophora glabripennis* (Anonymous 2013), *Phoracantha semipunctata* (Paine 2009), *Saperda vestita* Say (Johnson and Williamson 2007), *Xylotrechus javanicus* (Venkatesha and Dinesh 2012)
Periodic irrigation	Field crops and trees in streets, parks, and orchards	*Dectes texanus* (Charlet et al. 2007a, Knodel et al. 2010), *Phoracantha acanthocera* (Macleay) (Wang 1995a), *Phoracantha semipunctata* (Paine 2009), *Calchaenesthes pistacivora* Holzschuh (Rad 2006, Mehrnejad 2014)
Flooding irrigation	Root feeders	*Dorysthenes granulosus* (Yu et al. 2006)
Plant density management	Field crops, tree crops and forest plantations	*Dectes texanus* (Qureshi et al. 2007, Michaud et al. 2009), *Oncideres cingulata* (Forcella 1984), *Semanotus japonicus* (Yoshino 2004), *Phoracantha acanthocera* (Abbott et al. 1991, Farr et al. 2000)
Planting time	Field crops	*Dectes texanus* (Rogers 1985, Knodel et al. 2010), *Obereopsis brevis* (Parsai and Shrivastava 1993, Meena and Sharma 2006), *Rhytiphora diva* (Brier 2007)
Harvest time	Field crops and forest plantations	*Dectes texanus* (Michaud and Grant 2005, Michaud et al. 2009, Niide et al. 2006), *Monochamus scutellatus* (Say) (Raske 1973)

(Continued)

TABLE 9.1 *(Continued)*

Examples of Cultural Control Techniques for Cerambycid Pest Control

Cultural Control Techniques	Suitable for	Examples and References
Physical barriers (whitewashing, tree wrapping, netting, and root covering)	Pests that lay eggs on crop tree trunks and exposed roots	*Anoplophora chinensis* (Lieu 1945, Adachi 1990, Ho et al. 1995), *Nadezhdiella cantori* (Lieu 1947), *Aphrodisium gibbicolle* (White) (Zhang et al. 2010), *Cerambyx dux* (Faldermann) (Lozovoi 1954, Sharaf 2010), *Saperda candida* (Agnello 2013), *Aeolesthes induta* (Newman) (Yang and Liu 2010, Lv et al. 2011)
Light traps	Prionine *Dorysthenes*	*Dorysthenes granulosus* (Yu et al. 2007, 2009), *Dorysthenes hydropicus* (Pascoe) (Qin et al. 2008), *Dorysthenes hugelii* (Redtenbacher) (Sharma and Khajuria 2005, Singh et al. 2010)
Trap logs	Tree crops and forest plantations	*Steirastoma breve* (Sulzer) (Guppy 1911, Duffy 1960, Entwistle 1972), *Phoracantha semipunctata* (Gonzalez Tirado 1986, 1990; El-Yousfi 1989)
Trap crops	Field crops	*Dectes texanus* (Michaud et al. 2007b), *Obereopsis brevis* (Sharma et al. 2014)
Pheromone traps	Pests that produce long-range pheromones	Many species (see Chapter 5 of this book)
Crop rotation	Field crops	*Dorysthenes granulosus* (Yu et al. 2007), *Rhytiphora diva* (Duffy 1963)
Intercropping	Tree crops and forest plantations	*Acalolepta cervina* (Hope), *Xylotrechus javanicus* (Waller et al. 2007, Lan and Wintgens 2012), *Anoplophora glabripennis* (Sun et al. 1990)
Plant resistance	Field crops, tree crops, and forest plantations	*Dectes texanus* (Michaud and Grant 2009, Charlet et al. 2009, Niide et al. 2012), *Dorysthenes granulosus* (Pliansinchai et al. 2007), *Obereopsis brevis* (Patil et al. 2006), *Apriona germari* (Hussain and Chishti 2012), *Neoplocaederus ferrugineus* (L.) (Sahu et al. (2012), *Anoplophora glabripennis* (Li et al. 2003, Morewood et al. 2004, Li et al. 2008), *Anoplophora chinensis* (Zeng et al. 2014), *Semanotus japonicus* (Kato and Taniguchi 2003)
Climate change preparedness	Potentially many crops and forest plantations but particularly forest plantations	*Phoracantha semipunctata* (Paine et al. 2011, Seaton et al. 2015), *Phoracantha solida* (Blackburn) (Nahrung et al. 2014), *Enaphalodes rufulus* (Haavik et al. 2012, 2015; White 2015), *Monochamus* spp. (Roques et al. 2015)

and August for signs of girdle wounds made by female adults and frass ejected from small holes made by larvae in twigs. Infested twigs and shoots should then be pruned. Pruning is probably the only effective measure for the control of this pest (Cao 1981).

In species where females prefer to lay eggs in or on twigs and branches of trees and larvae bore their way downward toward the trunk, removal of infested plant parts before the larvae reach the trunk can prevent them from weakening and even killing the trees and also can substantially reduce the pest population for reinfestation in the next generation. A number of cerambycid pests fall into this category. For instance, the females of *Apriona germari* (Hope), a serious pest of mulberry trees used for silkworm production in China (Zhao et al. 1994) and northern India (Hussain and Chishti 2012), prefer to lay their eggs in primary branches (Yoon et al. 1997). Most larvae bore downward through the branches to the trunk (Yoon et al. 1997), damaging the trunk and even the roots. The larvae eject fine, wet, and sawdust-like frass at about 10-cm intervals as they tunnel downward (Duffy 1968). Growers should look for these signs and symptoms of infestation during the summer and prune infested branches. Hussain and Chishti (2012) demonstrated that this practice can reduce the pest infestation in mulberry farms by up to 60%.

Oemona hirta (Fabricius) is an important pest of citrus and many other fruit trees in New Zealand (Clearwater and Muggleston 1985; Wang et al. 1998; Lu and Wang 2005). The females prefer to lay their eggs on twigs and branches (Lu and Wang 2005). In the first year of infestation, the major signs and symptoms are ejected frass and dieback of the infested twigs and small branches. In the second year, the larvae move downward and damage the main branches and trunk, weakening the tree. Infestation is monitored, and infested twigs and branches are pruned during the summer to prevent the larvae from entering the main branches and trunk. Pruning during fall and winter can further reduce the borer population size. So far, appropriate pruning is probably the most effective control method for this pest (Clearwater and Muggleston 1985).

Pruning is also effective for the control of cerambycid pests in vineyards. For example, *Acalolepta mixta* (Hope) infests up to 70% of grapevines in some vineyards in southeast Australia (Goodwin et al. 1994; Dunn and Zurbo 2014). The females prefer to lay their eggs at the base of young vine canes on the main or secondary arms. The larvae bore into the vine wood and tunnel throughout the trunk and into the roots. Sawdust-like frass is often visible in tunnels and around the infested vine trunk. Pruning and destroying the infested vines may remove many larvae (Dunn and Zurbo 2014). Similar measures can be used to control *Xylotrechus arvicola* (Olivier), a relatively new but serious pest in Spanish vineyards (Ocete et al. 2002, 2008).

9.2.2 Selective Plant Destruction

Selective removal of infested plants usually is carried out for eradication of newly introduced pests (see Chapter 13) and defense against dispersal and reinfestation of established (either native or exotic) pests. This technique is applied to orchards, vineyards, and urban and forest trees. Its success in forest pest management depends on careful surveillance and early detection of infestations. Massive tree removal programs in large-scale forests usually are not practical but are being attempted in some eradication programs for *A. glabripennis* in the United States (see Chapter 11). Removed trees or vines are subject to proper sanitary treatment (see Section 9.2.5).

In eradication programs for newly invaded cerambycid pests, removal and destruction of infested trees are most commonly used as the first line of defense against establishment and spread. In Europe (Anonymous 2013), for example, trees infested by *A. glabripennis* are felled immediately upon discovery and destroyed with suggested sanitary measures (see Section 9.2.5). Root material should be removed if larval galleries are found on the cut surface of the stump. Immediately after tree felling, an intensive survey of at least a 2-km radius around the felled trees should be carried out. In the United States (Smith and Wu 2008), infested trees and those considered to be at high risk of attack within a radius of 400 m from the edge of the known infested area are felled and destroyed. High-risk trees within a radius of a second 400 m may also be removed and destroyed or treated with insecticide.

For established exotic pests such as *P. semipunctata*, selective removal of infested trees can suppress the pest population and slow down spread. In eucalypt forests (El-Yousfi 1989) and urban plantations (Paine 2009), trees that are infested or killed by *P. semipunctata* are felled immediately. All cut tree materials are subject to appropriate sanitary treatment to kill borers inside the trunks because they can complete development in felled logs.

The red oak borer, *Enaphalodes rufulus* Haldeman, is native to North America and is thought to be responsible at least in part for the widespread oak decline in Arkansas, Missouri, Oklahoma, and Ohio (Donley 1981, 1983; Riggins et al. 2009). Selective tree felling appears to be a very effective cultural control measure for this pest. For example, removal of infested oaks in Ohio has reduced the borer population by 50% within the following generation and about 90% during the second generation (Donley 1981). According to Donley (1983), felling infested trees results in 63–68% borer reduction. Such treatment only sacrifices <1% of the potential crop trees in each management unit, with costs ranging from $16 to $18/ha and benefits from $528 to $1,232/ha, assuming an 80-year timber management regimen. These practices may help lessen future widespread red oak borer activity (Riggins et al. 2009). In an attempt to control *Monochamus alternatus* Hope in a pine forest in China, Wu et al. (2000) demonstrate that felling infested trees and removing all logging residue from the forest reduce the pest population by 91%. Furthermore, felling and removal of trees severely damaged by *Semanotus japonicus* Lacordaire in a Japanese red cedar forest proved to be effective for the control of this pest (Sato 2007).

Removal of heavily infested or killed trees or vines is a common practice in horticulture. This usually applies to cerambycid pests that attack the trunks and roots because recovery of individual plants is seldom successful. For example, *Prionus imbricornis* (Linnaeus) is an important pest of apple and various crop trees in the United States (Agnello 2013). The neonate larvae feed on the root bark, and older larvae enter, bore, girdle, or even kill larger roots. Heavily infested trees cannot be saved and should be removed and destroyed before the following spring to prevent developing borers from completing their life cycle. The larvae of a serious apple pest from India, *Aeolesthes sarta* Solsky, bore into the large branches and trunks, significantly weakening and even killing the trees (Tara et al. 2009). When trunks are heavily infested, the trees are felled and removed before November (Gupta and Tara 2014). In southeast Australian vineyards, grape vines heavily infested by *A. mixta* are cut and replaced with new vines (Dunn and Zurbo 2014).

9.2.3 Plowing

Plowing or tillage practices are adopted in the control of some cerambycid pests in herbaceous field crops such as soybean and sugarcane. This mechanical approach can remarkably reduce the population size by (1) bringing soil-dwelling larvae and pupae onto the soil surface, thereby exposing them to desiccation and predation, freezing, and thawing; (2) damaging the pests in their soil-inhabiting stages; (3) destroying crop residues that harbor pests that could invade new crops, and (4) burying residues so deep that emergence from pupae is greatly reduced.

In the past decade, there have been frequent outbreaks of *Dorysthenes granulosus* (Thomson) in major sugarcane-growing regions of Southeast Asia. In China's Guangxi Province, infestation rates can reach up to 60% of sugarcane plants, causing up to 50% yield loss (Liao et al. 2006). Because the pest spends most of its life cycle below the ground surface with larvae feeding on roots and lower stems (Liao et al. 2006; Long and Wei 2007) and pupation taking place in soil (Yu et al. 2012), plowing can significantly suppress its population and infestation rate in the next season. For example, after harvest or before planting, deep plowing using a rotary cultivator can destroy infested sugarcane roots and stumps and kill more than 56% of larvae and pupae (Yang 1992; Gong et al. 2008).

In the management of two soybean and sunflower pests, *Obereopsis brevis* (Gahan) and *Dectes texanus* LeConte, plowing is recommended as an important measure. *O. brevis* is a South Asian species, infesting up to 90% of soybean (Sharma et al. 2014) and sunflower crops (Veda and Shaw 1994) in some regions in India and causing substantial losses. Although the main economic losses result from girdle formation by adult females (Jayanti et al. 2014), the larval tunneling in the stem weakens or even kills the whole plant. Therefore, deep plowing and destroying the plant residues during the summer can kill some larvae and pupae, reducing the population in the next generation.

D. texanus is a North American pest, particularly in the U.S. Midwest (Michaud and Grant 2010; Sloderbeck and Buschman 2011), infesting up to 90% of sunflower plants (Michaud and Grant 2009) and 100% of soybean plants (Tindall et al. 2010). For the control of the pest on soybeans (Campbell and Vanduyn 1977), winter burial of stems, at a depth of about 5 cm, with deep plowing or row bedding methods is effective in reducing larval survival and adult emergence. In sunflowers, a single disking and sweep tillage of the sunflower stubble in October and January can kill up to 73.5% and 39.7% of overwintering larvae, respectively (Rogers 1985).

Plowing may also be used for the control of root-feeding cerambycids in orchards. For example, *P. californicus* Motschulsky is a serious pest of stone fruit trees in North America (Alston et al. 2014). The larvae stay in the soil for three to five years, feeding on roots and root-crowns. Heavily infested trees cannot be saved and should be removed. The infested field should then be plowed for at least two years before planting new orchard crops, during which period annual (nonhost) crops should be planted as ground cover and tilled under the crops each year to stimulate microbial activity; *Prionus* larval population declines as a result (Alston et al. 2007).

9.2.4 Hand Removal of Loose Bark and Destruction of Eggs and Larvae

These methods can be labor-intensive but very effective for the control of certain cerambycid pests in tree crops. They usually are not practical for forest trees. Removing loose bark before adult emergence can reduce future oviposition sites. Destruction of eggs and young larvae by hand and of older larvae that have entered trunks using metal wires can lower the infestation rate and sometimes save infested trees. Knowledge of the biology of target pests, particularly their oviposition behavior, is the key to the success of these measures. Monitoring adult phenology and examination of infestation signs and symptoms are also critical.

Two coffee pests, *Xylotrechus javanicus* (Castelnau & Gory) in Asia (Lan and Wintgens 2012; Venkatesha and Dinesh 2012) (as *X. quadripes* Chevrolat in Venkatesha and Dinesh 2012) and *Monochamus leuconotus* (Pascoe) in Africa (Waller et al. 2007; Crowe 2012), are used to illustrate the practice of loose bark removal. Both species are serious pests of Arabica coffee trees, causing substantial losses to growers each year; *X. javanicus* lays eggs on the trunk and main branches (Venkatesha and Dinesh 2012), while *M. leuconotus* oviposits on the lower trunk (usually the lower 50 cm of trunk) (Duffy 1957). Before the beginning of adult emergence, the loose or scaly bark on the trunk and main branches is rubbed off, using a hard or metal brush or hands in garden gloves, to prevent oviposition by *X. javanicus*. If this practice is carried out during the oviposition period, it can also destroy eggs and young larvae when present. To control *M. leuconotus*, the same method is only applied to the lower section of the trunk before the beginning of adult emergence to reduce oviposition (Rutherford and Phiri 2006). However, Egonyu et al. (2015) claim that the stem-smoothing treatment has little effect on infestation rate by *M. leuconotus* and on coffee yield. Similar methods also can be used to control cerambycid pests on other tree crops such as *Nadezhdiella cantori* (Hope) on citrus (Lieu 1947) and even vine crops such as *A. mixta* on grapevines (Dunn and Zurbo 2014).

Inspection of trees during the peak adult activity period can help detect eggs in crevices, under loose barks and in oviposition slits (in the case of pests from the Lamiinae), and discover frass ejected by early instar larvae. Eggs and young larvae can be destroyed using hand pressure or knives. For example, *A. chinensis* females make oviposition slits with their mandibles in the living bark of lower trunks and in exposed roots of citrus trees (Lieu 1945). The exposed roots and lower trunks up to 50 cm (particularly the lower 25 cm) from the ground are examined for egg slits and exuding frass at intervals of 10–14 days from early July to early October. The egg slits are pressed with the thumb to crush the eggs, and the bark near where frass is extruded is cut with a knife to kill young larvae. A similar method has been used to control *Pseudonemorphus versteegi* (Ritsema), a serious pest of citrus trees in South Asia (Banerjee and Nath 1971; Shylesha et al. 1996).

Batocera horsfieldi Hope is an important pest of walnut trees and several other tree crops in Asia (Duffy 1968; Li et al. 1997). In some Chinese walnut orchards, the infestation rate can be as high as 70% (Wang et al. 2004). The females lay eggs in living bark of the main branches and the trunk. Careful examination of trunks and main branches during the spring and summer helps locate early infestations, including oviposition slits and young larvae. Eggs and young larvae are then destroyed by cutting into or probing the bark with these symptoms using a knife (Wang et al. 2004).

When the larvae have entered deep within the trunks, the aforementioned measures are no longer effective. Extraction and destruction of larvae in wood using hooked wires can reduce damage and reinfestation. This approach is applied to tree crops worldwide. Examples include a stone fruit pest, *Aromia bungii* (Faldermann), in East Asia (Gressitt 1942; Li et al. 1997); an apple pest, *Saperda candida* Fabricius, in North America (Agnello 2013); a coffee pest, *Nitakeris nigricornis* (Olivier) (= *Dirphya nigricornis* Olivier), in Africa (Crowe 1963, 2012); and the citrus pest, *P. versteegi*, in South Asia (Shylesha et al. 1996). However, when trees are beyond recovery, they should be removed and subject to sanitary treatment described as follows.

9.2.5 Sanitary Treatment of Plant Residues

Felled trees and human- or cerambycid-pruned twigs and branches should be treated properly for two reasons: (1) many cerambycid borers in these materials can complete their development and emerge as adults and (2) some cerambycid adults are highly attracted to newly felled trees for oviposition, and their offspring can develop into adults in these recently cut logs.

Burning of all removed infested plant residues, when permitted, is recommended for the control of all cerambycid pests. This can be done immediately after pruning and felling (for tree crops) or after harvest (for field crops). In some cerambycid pests, larvae bore in small twigs until intersecting a larger branch, which they girdle. For example, *Psyrassa unicolor* (Randall) is a pest of pecan and oak trees in the United States (Solomon 1985), and its larvae girdle twigs. Girdled twigs fall to the ground from January to May. Severed twigs on the ground under trees in orchards and ornamental plantings should be picked up in early spring and burned or buried before the adults emerge in late spring and early summer. To be most effective, the removal and destruction should be done for the entire orchard, woodlot, or neighborhood. Another pecan tree pest in the United States, *Oncideres cingulata* (Say) (Forcella 1984; Baker and Bambara 2001), damages the trees by adult girdling. The females girdle the twigs and small branches using their mandibles before oviposition to create proper conditions for the development of their larvae. The girdled twigs and branches break off or hang loosely on the tree. Collection, removal, and burning of all severed twigs and branches on the ground as well as those lodged in the trees in orchards and nurseries during fall, winter, and spring can substantially reduce the pest populations in one or two seasons (Baker and Bambara 2001).

As a general postharvest measure for the control of cerambycid pests, the wastes left after logging activities in Australian eucalypt forests are burned within six months of logging (Wang 1995a). This can reduce the load of dead/dying tree consumers and minimize the possibility that they return to attack adjacent forests. Infested plant residues, particularly twigs and branches, from orchards also can be buried at a depth that can prevent adults from emerging.

Chipping also is an effective method of destroying cerambycid larvae and pupae in plant residues of all sizes from small twigs to large trees. However, it often is used for treating larger branches and trunks. For example, chipping felled trees infested by *A. glabripennis* (Anonymous 2013) and *P. semipunctata* (Paine 2009) has been a common practice in sanitary treatment of felled logs for eradication and control programs. In the past decade, *Saperda vestita* Say has become an increasingly important pest of linden trees (*Tilia* spp.) in streets and nurseries throughout Wisconsin (Johnson and Williamson 2007). According to Johnson and Williamson (2007), chipping infested linden trees can effectively destroy the larvae, pupae, and adults of these pests. Arborists and landscape managers should consider chipping felled trees infested with *S. vestita* to prevent adults from potentially attacking nearby susceptible trees. Chip size should be taken into consideration depending on cerambycid species; for instance, the chipped fragments should be less than 2.5 cm in any dimension for *A. glabripennis* (Anonymous 2013). Chipped materials can make excellent mulch.

Firewood should be treated appropriately before storage or sale. For example, if the uprooted or felled coffee trees infested by *X. javanicus* are used for firewood, they should be submerged in water for 7–10 days to kill the borers that can otherwise complete development in the cut trunks and branches (Venkatesha and Dinesh 2012). To reduce *P. semipunctata*'s primary breeding sites, eucalypt logs should be debarked immediately, cut into sections, and split (Paine 2009). In commercial logging operations, the eucalypt logs should be debarked within two months of being felled to prevent cerambycid pests such as *Phoracantha acanthocera* (Macleay) from boring into hardwood—thus minimizing degrading of timber (Wang 1995a). *P. acanthocera* is an important pest of eucalypt forests in southern Australia (Wang 1995b).

9.3 Irrigation

Irrigation is a good management approach for maintaining the general vigor of plants. It can significantly reduce cerambycid infestation and damage because (1) many cerambycid pests are attracted to specific semiochemicals released by water-stressed plants and (2) plants under water stress are less resistant to damage after infestation (Hanks 1999; Hanks et al. 1999, 2005; Caldeira et al. 2002; Wotherspoon et al. 2014). This technique usually is practiced in the control of cerambycid pests in orchard, field, and vine crops as well as in trees planted in streets and parks.

In Western Australia, *P. acanthocera* activity is positively correlated with water stress level in stands of eucalypts, and dry sites are prone to borer attack (Abbott et al. 1991; Farr et al. 2000). Thus, regular irrigation is recommended for trees planted in streets and parks (Wang 1995a). Irrigation for the control of *P. semipunctata* has been widely practiced in regions where eucalyptus trees have been introduced. In Portugal, Caldeira et al. (2002) show that *P. semipunctata* larvae cannot survive in regularly irrigated eucalypt trees. Paine (2009) has prepared a good summary of the application of irrigation for the control of this pest in California. Irrigation on a regular schedule during prolonged dry periods should be carried out for urban eucalypt trees. If a tree has received regular irrigation, prolonged interruptions to watering should be avoided, particularly during the summer when insect pests are most active. Suddenly cutting off irrigation to trees that have been receiving water regularly will cause trees to become water-stressed and susceptible to insect attack. During irrigation, water should be applied to the ground below the edge of the outer canopy and not close to the trunk. Paine's (2009) general recommendation is to irrigate eucalyptus trees infrequently (possibly once a month during drought periods) but with sufficient amounts of water so that the water penetrates into the soil at least 30 cm below the surface. This can be achieved by applying water slowly through drip emitters that run continuously for several days. However, Paine and Millar (2005) suggest that, although irrigation of eucalyptus trees to maintain bark moisture content greater than 60% is effective in making trees resistant to *P. semipunctata*, it may increase susceptibility of the trees to psyllids (Hemiptera).

In orchards and vineyards, scheduled irrigation for crop production can also help control cerambycid pests. Crops will become water-stressed during periods of extended drought; thus, during such periods, regular irrigation is important to avoid heavy borer attack. In the Middle East, for example, *Calchaenesthes pistacivora* Holzschuh has become an increasingly important pest of pistachio trees because of lengthy drought periods in that region (Rad 2006; Mehrnejad 2014). The pest prefers to attack trees weakened by water stress, and its larvae kill twigs and branches if not controlled. Rad (2006) and Mehrnejad (2014) suggest that maintenance of tree vigor through appropriate and regular irrigation can effectively control the pest in pistachio orchards.

For annual and perennial field crops, irrigation can be used as a measure for the control of various pests, including cerambycids. For example, adequate irrigation of sunflower fields during the growing season can substantially reduce *D. texanus* populations (Charlet et al. 2007a). Irrigated sunflowers have thick stalks that may reduce lodging and yield losses from the stem borer (Knodel et al. 2010). The larvae of *D. granulosus* cause more severe damage to sugarcane planted in sandy soil because the larvae survive better in well-drained sandy soils (Zeng and Huang 1981; Long and Wei 2007). Based on this knowledge, Yu et al. (2006) recommend the practice of irrigation by flooding to drown the larvae. Flooding may be effective for the control of soil-dwelling cerambycid pests in other crops as well, such as fruit trees and pastures.

9.4 Plant Density Management

Plant density is one of the most important factors that influence soil moisture and nutrition levels, plant vigor and size, and plant maturation time. Therefore, plant spacing may affect the relative rate of growth of the plant and the behavior of the insect pest in searching for food or oviposition site. Establishing appropriate plant density has long been used as a cultural control technique for various insect pests. The primary objective of this cultural method is to maximize yield per unit area without reducing crop quality. In many agroecosystems, an increase in plant density (i.e., reduction in plant spacing) generally lowers pest numbers (Dent 1991). However, depending on the cerambycid pest species and ecosystems under investigation, an increase in plant density may reduce or increase infestations, and higher infestation may not necessarily increase yield loss.

Reduction in plant spacing may lower the infestation rate by some cerambycid pests but may increase yield losses caused by these pests. For example, *D. texanus* on sunflowers has a lower infestation rate in the high-density fields (40,340 plants/ha) compared with the low-density fields (33,140 plants/ha) (Qureshi et al. 2007). The plants in the high-density fields have smaller stalk circumferences than in the low-density fields—probably due to lower soil moisture. The lower *D. texanus* infestation in the high plant population is probably because stalks of smaller circumference are less favored for oviposition by the beetle (Michaud et al. 2007a; Qureshi et al. 2007). However, the lower infestation rate does not translate into higher yields because infestation itself does not reduce yield loss (Michaud et al. 2007a). These authors show that smaller stalks are more prone to lodging following larval girdling than are larger stalks that are often not fully girdled by virtue of their girth. More importantly, the higher rate of larval girdling in the high plant population plots may reflect the faster desiccation of slender stalks. Larvae of *D. texanus* do not girdle plants until they cease feeding and prepare for overwintering, a sequence of events that appears cued by the stalk tissues drying beyond some critical point where they are no longer edible (Michaud et al. 2007a). Therefore, larger stalks in low-density plots can delay stalk desiccation and, thus, pest girdling. Michaud et al. (2009) recommend that farmers consider increasing plant spacing to reduce pest damage to stalks by girdling before harvest.

In a study on the damage to the Japanese red cedar by *S. japonicus*, Yoshino (2004) examined trees in plots with low, medium, and high planting density (1,700, 3,200, and 7,300 trees/ha, respectively). Results show that radial growth at breast height in the low-density plot is greater than that in the high-density plot but that the pest infestation is higher and the damage more severe in the low-density plot than in the high-density plot. The author thus suggests that restriction of radial growth of trees in the juvenile period by high-density planting can reduce infestation by *S. japonicus*.

High crop tree density may increase infestation and damage by certain cerambycids. For example, in the United States, the pecan pest *O. cingulata* girdles and destroys more twigs in higher host tree density orchards (Forcella 1984). The outbreak of *P. acanthocera* in Western Australian eucalypt forests could be caused by dense regrowth of trees after logging (Abbott et al. 1991; Farr et al. 2000). This suggests that increasing plant spacing and thinning regrowth stands may be good management options for reducing infestation by these pests.

9.5 Planting and Harvest Times

These techniques aim to desynchronize the plants with their insect pests so that major yield loss can be avoided. Appropriate management of sowing or harvest times for many crops may avoid invasion by migrating insects or the oviposition period of particular pests or may synchronize the pest attack with weather conditions that are adverse for the pest. Timing can be used to allow young plants to establish a tolerant stage before attack occurs, to reduce the susceptible period of attack, or to mature the crop before a pest becomes abundant.

Evidence indicates that sowing time has a significant effect on damage to sunflowers by *D. texanus*. For example, sunflowers sown before mid-April or after late May in Texas suffer significantly lower stalk infestation (Rogers 1985). However, delayed sowing may be a more reliable and effective management tool for growers in the central (Charlet et al. 2007b) and southern U.S. Great Plains (Knodel et al. 2010). So far, there is little information on the effect of sowing time on soybean yield in relation to *D. texanus*, and more research on soybeans is warranted.

The girdle formation by the adult females of the soybean pest *O. brevis* results in significant reduction in the number of pods and seeds and in seed weight (Jayanti et al. 2014). Between 37 and 44 days after germination, the soybean crop appears to be most vulnerable to infestation. The severe yield loss occurs in the plants infested in the last week of August (40.0%) followed by the first and second weeks of September (37.6% and 32.3%, respectively). No more yield reduction takes place in the plants infested during the fourth week of September and the first week of October. Therefore, soybeans should not be sown before July because sowing in July results in the lowest infestation rate and highest yield (Parsai and Shrivastava 1993; Meena and Sharma 2006). In another soybean pest, *Rhytiphora diva* (Thomson) in Australia, later soybean plantings are also recommended to shorten the crop development period and reduce losses due to the pest (Brier 2007).

Harvest time for both sunflowers and soybeans can affect the damage level caused by *D. texanus*. For example, severe girdling (thus yield loss) of sunflower stalks by the pest can be reduced via early harvest; but early harvest can significantly reduce seed weight and oil content (Michaud et al. 2009), a balance growers have to consider in their sowing and harvest plans. In soybeans, economic injury occurs when the plants become susceptible to late-season lodging due to the larval girdling. Therefore, any delay in harvest can result in significant crop losses caused by *D. texanus* (Michaud and Grant 2005; Niide et al. 2006).

Although management of harvest times primarily is used for pest control in annual crops, it may also be effective for reduction of cerambycid damage in forests. For example, the conifer pest *Monochamus scutellatus* (Say) in North America may be managed to some extent by alteration of logging dates (Raske 1973). Raske (1973) shows that, in spruce and pine plantations, losses caused by the pest are lower when felling operations are carried out in autumn and winter rather than in spring and summer. It is also recommended that logs felled in spring and summer should be processed without delay.

9.6 Physical Barriers

Application of physical barriers to prevent infestations can be an excellent measure for the control of agricultural and horticultural pests. The objective of this technique for the management of cerambycid pests is to cover the sites where the pests prefer to lay eggs to reduce infestation.

Whitewashing with water-based lime (lime:water ≈ 1:6) is one of the most commonly used methods for the control of cerambycids that lay eggs on the trunk and main branches of crop trees and vines. If sulfur is added (lime:sulfur:water ≈ 1:1.5:8), the suspension can function as both a physical and a chemical barrier. For example, females of two common citrus pests, *A. chinensis* (Forster) and *A. macularia* (Thomson) in Asia (see Chapter 12), make oviposition slits with their mandibles in the living bark of lower trunks and in exposed roots. Painting the exposed roots and lower trunks up to 50 cm from the ground with the lime suspension before adult emergence in the spring can substantially reduce egg laying (Lieu 1945; Ho et al. 1995). The coating will remain for about two months. If the paste is washed away by heavy rain during the oviposition period, the treated trees should be retreated. Whitewashing is also recommended for the control of other citrus pests such as *N. cantori* (Lieu 1947) and *Aphrodisium gibbicolle* (White) (Zhang et al. 2010) in Asia, the cashew tree pest *Neoplocaederus obesus* (Gahan) (Liu et al. 1998) in Asia, and apple and stone fruit tree pest *Cerambyx dux* (Faldermann) in the Mediterranean region (Lozovoi 1954; Sharaf 2010). In addition, for the control of an apple pest in North America, *S. candida*, painting the bottom 60 cm of the trunks with white latex paint can effectively prevent oviposition (Agnello 2013). The author suggests that this approach works better on the smooth trunk surfaces of younger trees and should be repeated each year because the paint layer cracks with normal tree growth.

Other protective coverings also have been applied to reduce oviposition by cerambycid pests. For example, the management program for *N. cantori* in China involves covering the trunks with flat straw ropes during the spring and summer, which probably is a better approach than whitewashing for the control of this pest (Chen 1942). In the control of *S. candida*, Agnello (2013) recommends that in early May protective coverings of various materials be wrapped around the trunks up to 60 cm above the ground and then removed in September after all egg-laying activity is finished. The protective coverings can be mosquito netting, fine mesh hardware cloth, tree wrap, tarpaper, cotton batting, or even layers of newspaper that are tied at the top with twine and covered at the bottom with soil (Agnello 2013). In Japan, Adachi (1990) has tested the application of wire netting to prevent oviposition by *A. chinensis* in a mature citrus grove, achieving effective control. The net was wrapped loosely around the lower trunk, tied at the top with wire, and covered at the bottom with soil. The netting is removed during September–October.

Exposed roots are preferred oviposition sites of some cerambycid pests, such as the tea tree pest *Aeolesthes induta* (Newman) and the citrus pest *A. chinensis* in China. Yang and Liu (2010) show that in Zhejiang Province, eastern China, about 51% of tea trees planted in shallow soil with roots partially exposed to the air are infested by *A. induta* compared with 17% of trees planted in deep soil with roots completely covered with soil. Covering exposed roots with soil is recommended as an effective approach to preventing oviposition on roots and reducing infestation rates (Yang and Liu 2010; Lv et al. 2011). Similar measures also should be effective for the control of other cerambycid pests that lay eggs on exposed roots.

9.7 Traps

Measures to lead unwanted pests away from desirable plants are useful in insect pest management programs. Based on the behaviors of individual pest species, traps of various types can be designed and applied to attract adults and destroy them and their offspring. Commonly used traps for cerambycid pest management include light traps, trap logs, trap crops, and pheromone traps.

9.7.1 Light Traps

Because adults of many cerambycid species are nocturnal or crepuscular and attracted to light, the use of light traps as a pest control measure for these species has been proposed by many authors. For example, the use of light traps has been suggested for controlling *N. cantori* (Gressitt 1942) and an African coffee pest, *Bixadus sierricola* (White) (Duffy 1957; Waller et al. 2007). So far, there is little evidence for practicality and effectiveness of light traps for management of these and many other cerambycid pests whose adults are attracted to light.

However, light traps appear to be very useful for the control of pests in the prionine genus *Dorysthenes*. For instance, the application of frequency vibration-killing lamps together with pitfall traps underneath the lamps during April–June achieves very promising results in control programs for the sugarcane pest *D. granulosus* in China. Briefly, the light traps are set up about 100 m apart and 1.5 m high, and under each light trap a pitfall trap (40 cm × 40 cm × 60 cm) is dug in the soil and filled with water (Yu et al. 2007) or no water (Yu et al. 2009). One trap can attract up to 59 adults per day. Trapped adults are collected and destroyed every morning.

D. hydropicus (Pascoe) is another sugarcane pest in China (Li et al. 1997), and it recently has become a serious pest of common reed [*Phragmites australis* (Cav.) Trin. ex Steud.] and pomelo [*Citrus maxima* (Burm. f.) Merr.] trees (Qin et al. 2008). In the pomelo orchards in Guangdong, the adults are attracted to light with one lamp being able to trap more than 1,000 adults per night (Qin et al. 2008), suggesting a great potential for the use of light traps for the control of this pest. A similar approach has been taken for the control of *D. hugelii* (Redtenbacher), an apple pest in India (Sharma and Khajuria 2005; Singh et al. 2010).

9.7.2 Trap Logs

Adults of many cerambycid species are attracted to plants of certain species or conditions for feeding (see Chapter 3) and reproduction (see Chapter 4). Trees or branches of preferred hosts can be cut and used to attract and kill cerambycid pests.

Steirastoma breve (Sulzer) has been one of the most serious cerambycid pests of cocoa trees in the New World in the past 100 years (Liendo-Barandiaran et al. 2010). The preventative cultural control using trap logs probably still is the most effective control measure for this pest (Guppy 1911; Duffy 1960; Entwistle 1972). Briefly, fresh branches (3–12 cm in diameter and 90–120 cm long) of *Pachira aquatica* Aublet, one of the borer's preferred hosts, are cut and hung on the cocoa tree branches 150–250 cm above the ground. Adults are attracted to the trap logs for oviposition. The traps are replaced with newly cut branches every two to three weeks in the rainy season and every 10–12 days in the dry season. Used trap logs should be burned to kill eggs and larvae. The attracted adults also can be collected and destroyed.

In control programs for *P. semipunctata* outside its native range, the application of trap logs can substantially reduce the pest population size. For example, in Spain, eucalypt trees are felled and used as trap logs (Gonzalez Tirado 1986, 1990). Results show that the duration of the attractive effect depends on the time of year when trees are felled. The logs felled and installed in the spring and at the beginning of summer are attractive for the longest time, drawing in 50% of total beetles within 12 days of felling, compared within five and eight days for summer and autumn trap logs, respectively. Used logs should be destroyed to kill eggs and larvae. These findings agree with this author's personal observations in Australia that trees felled in January (summer) are the most attractive to *P. semipunctata* in the first few days after felling. Similar measures also have been used in Morocco, achieving promising outcomes (El-Yousfi 1989).

9.7.3 Trap Crops

Trap crops are preferred host plants of an insect pest that are grown near the primary crop to protect it from oviposition. The pest infests the preferred or trap crop instead of the primary crop. The pest can then be killed in the trap crop using sanitary measures (described earlier) or insecticides.

In North America, *D. texanus* uses both soybeans and sunflowers as hosts but significantly prefers sunflowers to soybeans for oviposition if given a choice (Michaud and Grant 2005). On the basis of this knowledge, sunflowers can be planted as the trap crops in a soybean field to control the pest on soybeans. Michaud et al. (2007b) demonstrate that planting sunflowers in the nonirrigated corners of a center-pivot irrigated soybean field reduced infestation of soybeans by 65% compared with a control field without adjacent sunflowers. Surrounding a 0.33-ha soybean field with six rows of sunflowers also reduced soybean infestation to <5% of plants, compared with a 96% infestation of the sunflower plants.

For the control of another soybean pest, *O. brevis*, Sharma et al. (2014) suggest planting *Dhaincha* (which the pest prefers to attack) adjacent to soybean fields and then the trap crop will be destroyed. However, the effectiveness of this measure for the control of this pest needs further evaluation. The use of trap crops should also be evaluated in other crop systems such as vineyards and orchards.

9.7.4 Pheromone Traps

Discoveries in the past decade demonstrate that the use of volatile pheromones is widespread in cerambycids. In Chapter 5, Millar and Hanks provide a detailed treatment of cerambycid chemical ecology including the potential application of semiochemical-baited traps for pest monitoring and control. Several recent studies also focus on designing traps to maximize their effectiveness (e.g., Graham and Poland 2012; Allison et al. 2014; Nehme et al. 2014; Torres-Vila et al. 2015). Major developments in this field are expected to occur in the near future, when pheromone traps are widely used for cerambycid pest management programs.

9.8 Crop Rotation and Intercropping

Crop rotation refers to an approach where botanically unrelated plants are grown in successive years or growing seasons; intercropping refers to growing two or more plant species in association with one another in the same or nearby fields. Both techniques are widely used for annual field crops to reduce pest infestation, and their success relies heavily on growers' knowledge of pest behavior and host range.

Crop rotation usually is effective for the control of pests that are monophagous or oligophagous and either univoltine or multivoltine. So far, the application of this technique for management of cerambycid pests is relatively rare—probably because many cerambycid species are polyphagous and semivoltine. However, there is some evidence that crop rotation with rice or sweet potato can reduce sugarcane infestation by *D. granulosus* in southern China (Yu et al. 2007). For the control of the soybean pest *R. diva* in Australia (Romano and Kerr 1977; Strickland 1981; Heap 1991), it is recommended that an infested legume field be ploughed and sown with botanically unrelated crops the next season (Duffy 1963). Furthermore, planting susceptible soybean crops close to alfalfa should be avoided (Bailey and Goodyer 2007; Brier 2007).

The intercropping approach has been used for the control of several cerambycid pests that infest trees. For example, to control *A. glabripennis* in poplar plantations in northern China, Sun et al. (1990) suggest that planting both *Melia azedarach* L. (which is highly resistant to the beetle) and *Acer negundo* L. (which is highly susceptible to the beetle) trees among poplar plantations achieves a 60–70% reduction in damage to the poplars. Females of two coffee tree pests in Asia, *Acalolepta cervina* (Hope) and *X. javanicus*, prefer to lay eggs on sun-exposed coffee plants (Waller et al. 2007; Lan and Wintgens 2012). Therefore, planting appropriate shade trees alongside coffee trees can substantially reduce infestation. Species selected as shade trees should be nonhost plants of these pests.

9.9 Plant Resistance

Most authorities consider true plant resistance as being primarily under genetic control (i.e., the mechanisms of resistance are derived from preadapted inherited characters). These mechanisms include antixenosis and antibiosis. Antixenosis refers to plant characteristics that reduce the likelihood of insect feeding and oviposition on the host through either allelochemic or morphological features. Antibiosis is a mechanism by which a colonized plant adversely affects insect development, reproduction, and survival.

Evidence of plant resistance against cerambycid pests has been widely reported. However, in many cases, if not most, resistance mechanisms are not clearly defined. In field crops, for example, sugarcane cultivars Uthong 3 and K 88-92 are resistant to *D. granulosus* attack (Pliansinchai et al. 2007) (*D. buqueti* Guerin mentioned by these authors should be *D. granulosus*; also see Chapter 12), and soybean cultivars JS-335 and Pusa-20 have lower infestation by *O. brevis* and better yield potential as compared with other cultivars tested in India (Patil et al. 2006). In tree crops, the coffee variety KP423 has significantly lower *M. leuconotus* incidence and higher yield than two other varieties, KP162 and SL28, in Africa (Egonyu et al. 2015), and there is evidence where mulberry cultivars vary in resistance against *A. germari* in India (Hussain and Chishti 2012). In screening programs for potentially resistant genotypes of poplar trees against *A. glabripennis* in China, *Populus euramericana* cv. 'Jiqin-1', *P. euramericana* cv. 'Jiqin-2' (Li et al. 2003), and *P. deltoids* cv. 'Zhonghuai2' (Li et al. 2008) are found to be highly resistant.

Antixenosis against *D. texanus* has been identified in both soybeans and sunflowers. In a study on why the infestation of wild sunflowers by *D. texanus* is remarkably rare whereas 80–90% of plants are infested in the cultivated varieties, Michaud and Grant (2009) have tested adult feeding and oviposition preference. They show that females feed less and make significantly fewer ovipositional punctures on wild plants than they do on cultivated ones. Furthermore, only 18.4% of ovipunctures in wild plants result in oviposition, compared with 44.9% in cultivated plants, suggesting that additional resistance factors reduce the probability of oviposition during ovipositor insertion. These authors demonstrate that petioles of wild sunflowers exude more than four times the weight of resinous material when severed as do cultivated plants, are significantly less succulent as determined by water content, and require significantly more force to penetrate the epidermis in a standardized puncture test. On the basis of these findings, Michaud and Grant (2009) conclude that a combination of physical and chemical properties of wild sunflowers confers substantial resistance to oviposition by *D. texanus* and that this natural antixenosis has been inadvertently diminished in the course of breeding cultivated varieties. Charlet et al. (2009) have evaluated resistance of 61 oilseed sunflower accessions and 31 interspecific crosses against *D. texanus* and other pests in a seven-year study. They show that four accessions (PI 386230, PI 431542, PI 650497, and PI 650558) reduce the infestation rate by *D. texanus*. In soybeans, two genotypes (PI171451 and PI165676) appear to have oviposition antixenosis (Niide et al. 2012). In the field, the preferred oviposition site is localized to leaf petioles in the upper four or five nodes of the plant canopy (Niide et al. 2012). Histomorphological analyses of petiole cross-sections of these genotypes indicate that leaf petiole morphology may be related to reduced *D. texanus* oviposition.

Evidence for antibiosis against cerambycid pests is increasing. For example, Kato and Taniguchi (2003) have screened 200 *Cryptomeria japonica* (Thunb. ex L.f.) D. Don clones for resistance to the bark borer *S. japonicus* in Japan, indicating that 100% larval mortality occurs in 46 clones. However, it is not yet clear whether the antibiosis operates through biochemical or physical mechanisms. In the United States, Morewood et al. (2004) evaluated the resistance potential and mechanisms of four tree species against *A. glabripennis*. They show that adult *A. glabripennis* feed more on golden-rain trees (*Koelreuteria paniculata* Laxmann) and river birches (*Betula nigra* L.) than on London planetrees (*Platanus x acerifolia* (Aiton) Willdenow) or Callery pears (*Pyrus calleryana* Decaisne). Although females prefer to lay eggs in a golden-rain tree, larval mortality is relatively high and larval growth is slow in this species. The oviposition rate is lowest in Callery pears, and larvae fail to survive in this tree species. Adult beetles feeding on Callery pears have reduced longevity, and females feeding only on Callery pears fail to develop any eggs. The resistance of golden-rain trees to the larvae operates primarily through the physical mechanism of abundant sap flow. The resistance of Callery pears to both larvae and adults operates through the chemical composition of the tree, which may include compounds that are toxic or that otherwise interfere with normal growth and development of the beetle. Studies suggest that both golden-rain trees and Callery pears are present in the native range of *A. glabripennis* and may therefore have developed resistance to the beetle by virtue of exposure to attack during their evolutionary history. In soybeans, one antibiosis factor has been identified in the genotype PI165673 for resistance against *D. texanus* (Niide et al. 2012).

Like soybean resistance to *D. texanus* (Niide et al. 2012), tree resistance against cerambycid pests can also operate through both antixenosis and antibiosis. For example, *A. chinensis* causes significant damage to *Casuarina* trees that are planted as a main coastal shelter forest in southeast China. Zeng et al. (2014) evaluated the resistance to *A. chinensis* of 48 *C. equisetifolia* L. clones in the germplasm banks at Huian in Fujian Province. They show that three clones, Hui13, Hui76, and Hui83, are highly resistant to *A. chinensis*, significantly reducing oviposition and larval survival. In a study on infestation of cashew trees by *Neoplocaederus ferrugineus* (L.) (= *Plocaederus ferrugineus*) in India, Sahu et al. (2012) have tested relative resistance levels of 17 cultivars. They conclude that one cultivar (Sel-2) suffers no infestation, and fewer than 3% of two cultivars (Ullal-2 and VRI-2) are infested. Although these authors use the term tolerance, the lack of infestation or the low infestation levels may well be conferred by antixenosis and/or antibiosis.

9.10 Climate Change and Pest Management

The effect of climate change on plant resistance and insect damage is often difficult to predict because long-term data are largely lacking, and the impact of climate change involves multiple biotic and abiotic factors. However, if extreme drought and high temperature become more widespread or prolonged, plant stress and pest damage are expected to increase, and traditional planting sites may no longer be suitable for certain plant species (Wylie and Speight 2012). For example, extreme climatic events, including droughts and heat waves, can trigger outbreaks of woodboring beetles by compromising host defenses and creating habitats conducive for beetle development (Seaton et al. 2015). Under the scenario of increasing temperature, it is expected that insect abundance and diversity will increase at higher elevations and latitudes, resulting in novel interactions among previously nonoverlapping plant and insect species (Rasmann and Pellissier 2015). Therefore, the ability of the plant to adjust its defense and tolerance strategy in response to such new interactions is expected to determine whether outbreaks of new pests will take place.

Using *A. glabripennis* infestation as one of the factors, Foran et al. (2015) developed an approach for assessing the impacts of multiple extreme weather scenarios likely to become more frequent under climate change and subsequently to influence the composition of urban trees and pest infestation in Cambridge, Massachusetts. Their results are a reasonable indication of the more tolerant tree species in Cambridge and their locations, which can be used for proactive management of the urban forest in the future. A number of reports show that outbreaks of the Australian longicorn *P. semipunctata* have occurred in water-stressed eucalypt plantations overseas that have either been planted in unsuitable sites or have experienced prolonged drought (Paine et al. 2011).

In their native Australia, *Phoracantha* species normally attack less vigorous trees growing on poor sites or at the extreme of a species' distribution or recently killed trees. However, the outbreaks of *P. solida* (Blackburn) in *Corymbia* forests of subtropical Australia could be attributed to prolonged drought (Nahrung et al. 2014). The recent outbreaks of *P. semipunctata* in eucalypt forests of southwest Australia are another example of drought-related pest problems (Seaton et al. 2015). The authors predict that, as a result of climate change, drought-prone areas are expected to become more common and could result in more frequent outbreaks. A similar situation has also taken place in North America. For example, U.S. oak-hickory forests recently have experienced episodes of oak mortality in concert with an outbreak of the native *E. rufulus* (Haavik et al. 2012, 2015; White 2015). Studies suggest that drought or heat and drought are events that trigger outbreaks of this oak pest.

Climate change also may be responsible for recent outbreaks of plant diseases vectored by cerambycids. For example, the pinewood nematode, *Bursaphelenchus xylophilus* (Steiner & Buhrer) Nickle, which causes pine wilt disease, is a major threat to pine forests across the world, particularly in Asia and Europe (Roques et al. 2015; see Chapter 6). The nematode is carried by local *Monochamus* species, which vector it to pine trees during maturation feeding and oviposition. Roques et al. (2015) indicate that global warming has enlarged the vectors' distribution range and accelerated their development. Warmer temperatures also may increase the development rate and reproduction of the nematode. Furthermore, the expression of pine wilt disease is also modulated by warm temperatures and drought stress of pine trees. Recent evidence shows that pine wilt disease has expanded its range in areas of East Asia that previously were considered as unsuitable (Roques et al. 2015).

There is an urgent need to investigate the effects of climate change on new pest problems and strategies to handle them. Fundamental knowledge is badly needed for selection of plant species or varieties that are drought and/or heat resistant and of new planting sites that are more suitable for growing certain plant species or varieties. It also is important to establish models that can predict the potential distribution expansion of cerambycids and their host plants and novel cerambycid-plant interactions in response to climate change.

9.11 Concluding Remarks

There are a number of cultural control techniques that can be used for management of cerambycid pests. Their success often depends heavily on a good understanding of these pests' biology and host plants, and their interactions with biotic and abiotic environments. However, our knowledge of cerambycid biology is surprisingly poor at present for many species. Much of the information we use today to formulate control strategies is based on careful observations made decades ago. Future studies should focus on the general biology of these beetles and their interactions with abiotic and biotic environments and on the potential impact of climate change on cerambycid pests and their damage to forests and crops. Furthermore, the integration of cultural control with biological and chemical controls is critical to successful suppression of many cerambycid pests.

ACKNOWLEDGMENTS

I would like to thank Robert A. Haack for constructive comments on an earlier draft of this chapter.

REFERENCES

Abbott, I., R. Smith, M. Williams, and R. Voutier. 1991. Infestation of regenerated stands of karri *(Eucalyptus diversicolor)* by bullseye borer *(Tryphocaria acanthocera;* Cerambycidae) in Western Australia. *Australian Forestry* 54: 66–74.

Adachi, I. 1990. Control methods for *Anoplophora malasiaca* (Thomson) (Coleoptera: Cerambycidae) in citrus groves. 2: Application of wire netting for preventing oviposition in a mature grove. *Applied Entomology and Zoology* 25: 79–83.

Agnello, A. M. 2013. Apple-boring beetles. New York State Integrated Pest Management Program, Cornell University. http://nysipm.cornell.edu/factsheets/treefruit/pests/ab/ab.asp (accessed January 10, 2016).

Allison, J. D., B. D. Bhandari, J. L. McKenney, and J. G. Millar. 2014. Design factors that influence the performance of flight intercept traps for the capture of longhorned beetles (Coleoptera: Cerambycidae) from the subfamilies Lamiinae and Cerambycinae. *PLos One* 9: e93203.

Alston, D. G., J. D. Barbour, and S. A. Steffan. 2007. California prionus *Prionus californicus* Motschulsky (Coleoptera: Cerambycidae). Washington State University Tree Fruit Research and Extension Center, Orchard Pest Management Online. http://jenny.tfrec.wsu.edu/opm/displayspecies.php?pn=643 (accessed January 13, 2016).

Alston, D. G., S. A. Steffan, and M. Pace. 2014. *Prionus* root borer *(Prionus californicus)*. Utah Pests Fact Sheet. http://utahpests.usu.edu/ipm/htm/fruits/fruit-insect-disease/prionus-borers10 (accessed January 13, 2016).

Anonymous. 2013. *Anoplophora glabripennis*: Procedures for official control. *EPPO Bulletin* 43: 510–517.

Bailey, P. T., and G. Goodyer. 2007. Lucerne. In *Pests of field crops and pastures: Identification and control*, ed. P. T. Bailey, pp. 425–440. Collingwood: CSIRO Publishing.

Baker, J. R., and S. Bambara. 2001. Twig girdler *Oncideres cingulata* (Say) (Cerambycidae: Coleoptera). Publication No. Ent/ort-96, North Carolina Cooperative Extension Service, North Carolina State University. http://www.ces.ncsu.edu/depts/ent/notes/O&T/trees/note96/note96.html (accessed January 10, 2016).

Banerjee, S. N., and Nath, D. K. 1971. Life-history, habits and control of the trunk borer of orange, *Anoplophora versteegi* (Ritsema) (Cerambycidae: Coleoptera). *Indian Journal of Agricultural Sciences* 41: 765–771.

Battisti, A., and S. Larsson. 2015. Climate change and insect pest distribution range. In *Climate change and insect pests*, eds. C. Bjorkman, and P. Niemela, 1–15. Wallingford, CT: CABI.

Blois, J. L., P. L. Zarnetske, M. C. Fitzpatrick, and S. Finnegan. 2013. Climate change and the past, present, and future of biotic interactions. *Science* 341: 499–504.

Brier, H. 2007. Pulses—Summer (including peanuts). In *Pests of field crops and pastures: Identification and control*, ed. P. T. Bailey, 169–258. Collingwood: CSIRO Publishing.

CABI. 2015a. *Phoracantha recurva* (eucalyptus longhorned borer). Invasive Species Compendium. http://www.cabi.org/isc/datasheet/40371 (accessed February 13, 2016).

CABI. 2015b. *Phoracantha semipunctata* (eucalyptus longhorned borer). Invasive Species Compendium, online at: http://www.cabi.org/isc/datasheet/40372 (accessed February 13, 2016).

Caldeira, M. D., V. Fernandez, J. Tome, and J. S. Pereira. 2002. Positive effect of drought on longicorn borer larval survival and growth on eucalyptus trunks. *Annals of Forest Science* 59: 99–106.

Campbell, W. V., and J. W. Vanduyn. 1977. Cultural and chemical control of *Dectes texanus texanus* (Coleoptera: Cerambycidae) on soybeans. *Journal of Economic Entomology* 70: 256–258.

Cao, C. Y. 1981. Biology and control of *Linda nigroscutata*. *Acta Phytophylacica Sinica* 8: 193–196 [in Chinese with English abstract].

Charlet, L. D., R. M. Aiken, R. F. Meyer, and A. Gebre-Amlak. 2007a. Impact of irrigation on larval density of stem-infesting pests of cultivated sunflower in Kansas. *Journal of Economic Entomology* 100: 1555–1559.

Charlet, L. D., R. M. Aiken, R. F. Meyer, and A. Gebre-Amlak. 2007b. Impact of combining planting date and chemical control to reduce larval densities of stem-infesting pests of sunflower in the central plains. *Journal of Economic Entomology* 100: 1248–1257.

Charlet, L. D., R. M. Aiken, J. F. Miller, and G. J. Seiler. 2009. Resistance among cultivated sunflower germplasm to stem-infesting pests in the Central Great Plains. *Journal of Economic Entomology* 102: 1281–1290.

Chen, F. G. 1942. *A preliminary study on citrus cerambycid, Nadezhdiella cantori (Hope)*. Chengtu: Szechwan Agricultural Research Station.

Clearwater, J. R., and S. J. Muggleston. 1985. Protection of pruning wounds: Favoured oviposition sites for the lemon tree borer. *Proceedings of New Zealand Weed and Pest Control Conference* 38: 199–202.

Crowe, T. J. 1963. The biology and control of *Dirphya nigricornis* (Olivier), a pest of coffee in Kenya (Coleoptera: Cerambycidae). *Journal of the Entomological Society of Southern Africa* 25: 304–312.

Crowe, T. J. 2012. Coffee pests in Africa. In *Coffee: Growing, processing, sustainable production—A guidebook for growers, processors, traders and researchers*, ed. J. N Wintgens, 425–462 (2nd rev. ed.). Hoboken, NJ: Wiley, John & Sons Incorporated.

Dent, D. 1991. *Insect pest management*. CABI: UK.

Donley, D. E. 1981. Control of the red oak borer by removal of infested trees. *Journal of Forestry* 79: 731–733.

Donley, D. E. 1983. Cultural control of the red oak borer (Coleoptera, Cerambycidae) in forest management units. *Journal of Economic Entomology* 76: 927–929.

Duffy, E. A. J. 1957. *A monograph of the immature stages of African timber beetles (Cerambycidae)*. London: British Museum (Natural History).

Duffy, E. A. J. 1960. *A monograph of the immature stages of Neotropical timber beetles (Cerambycidae)*. London: British Museum (Natural History).

Duffy, E. A. J. 1963. *A monograph of the immature stages of Australasian timber beetles (Cerambycidae)*. London: British Museum (Natural History).

Duffy, E. A. J. 1968. *A monograph of the immature stages of Oriental timber beetles (Cerambycidae)*. London: British Museum (Natural History).

Dunn, G., and B. Zurbo. 2014. Grapevine pests and their management. PRIMEFACT 511 (2nd ed.), NSW Department of Primary Industries. http://www.dpi.nsw.gov.au/__data/assets/pdf_file/0010/110998/Grapevine-pests-and-their-management.pdf (accessed January 12, 2016).

Egonyu, J. P., P. Kucel, G. Kagezi, et al. 2015. Coffea arabica variety KP423 may be resistant to the cerambycid coffee stemborer *Monochamus leuconotus*, but common stem treatments seem ineffective against the pest. *African Entomology* 23: 68–74.

El-Yousfi, M. 1989. Las bases de la lucha servicola contra *Phoracantha semipunctata* Fabr. *Boletin de Sanidad Vegetal, Plagas* 15: 129–137.

Entwistle, P. 1972. *Pests of Cocoa*. London: Longman.

Evans, H., N. Straw, and A. Watt. 2002. Climate change: Implications for insect pests. *Forestry Commission Bulletin* 125: 99–118.

Farr, J. D., S. G. Dick, M. R. Williams, and I. B. Wheeler. 2000. Incidence of bullseye borer (*Phoracantha acanthocera* (Macleay) Cerambycidae) in 20–35 year old regrowth karri in the south west of Western Australia. *Australian Forestry* 63: 107–123.

Foran, C. M. K. M. Baker, M. J. Narcisi, and I. Linkov. 2015. Susceptibility assessment of urban tree species in Cambridge, MA, from future climatic extremes. *Environment Systems and Decisions* 35: 389–400.

Forcella, F. 1984. Tree size and density affect twig girdling intensity of *Oncideres cingulata* (Coleoptera: Cerambycidae). *The Coleopterists Bulletin* 38: 37–42.

Gong, H. L., Y. X. An, C. X. Guan, et al. 2008. Pest status and control strategies of *Dorysthenes granulosus* (Thomson) on sugarcane in China. *Sugarcane and Canesugar* (5): 1–5, 38 [in Chinese].

Gonzalez Tirado, L. 1986. *Phoracantha semipunctata* dans le sud-ouest espagnol: Lutte et dégâts. *EPPO Bulletin* 16: 289–292.

Gonzalez Tirado, L. 1990. Algunos aspectos practicos sobre la utilizacion de arboles cebo en la lucha contra el perforado del eucalipto *Phoracantha semipunctata* Fab. (Coleoptera: Cerambycidae). *Boletin de Sanidad Vegetal, Plagas* 16: 529–542.

Goodwin, S., M. A. Pettit, and L. J. Spohr 1994. *Acalolepta vastator* (Newman) (Coleoptera, Cerambycidae) infesting grapevines in the Hunter Valley, New South Wales. 1. Distribution and dispersion. *Journal of the Australian Entomological Society* 33: 385–390.

Graham, E. E., and T. M. Poland. 2012. Efficacy of fluon conditioning for capturing cerambycid beetles in different trap designs and persistence on panel traps over time. *Journal of Economic Entomology* 105: 395–401.

Gressitt, J. L. 1942. Destructive long-horned beetle borers at Canton, China. *Special Publications of lingnan Natural History Survey Museum* 1: 1–60.

Guppy, P. L. 1911. The life-history and control of the cacao beetle. *Circular of the Board of Agriculture in Trinidad* 1: 1–35.

Gupta, R., and J. S. Tara. 2014. Management of apple tree borer, *Aeolesthes holosericea* Fabricius on apple trees (*Malus Domestica* Borkh.) in Jammu Province, Jammu and Kashmir State, India. *Journal of Entomology and Zoology Studies* 2: 96–98.

Haack, R. A., F. Hérard, J. H. Sun, and J. J. Turgeon. 2010. Managing invasive populations of Asian long-horned beetle and citrus longhorned beetle: A worldwide perspective. *Annual Review of Entomology* 55: 521–546.

Haavik, L. J., S. A. Billings, J. M. Guldin, and F. M. Stephen. 2015. Emergent insects, pathogens and drought shape changing patterns in oak decline in North America and Europe. *Forest Ecology and Management* 354: 190–205.

Haavik, L. J., J. S. Jones, L. D. Galligan, J. M. Guldin, and F. M. Stephen. 2012. Oak decline and red oak borer outbreak: Impact in upland oak-hickory forests of Arkansas, USA. *Forestry* 85: 341–351.

Hanks, L. M. 1999. Influence of the larval host plant on reproductive strategies of cerambycid beetles. *Annual Review of Entomology* 44: 483–505.

Hanks, L. M., T. D. Paine, and J. G. Millar. 2005. Influence of the larval environment on performance and adult body size of the woodboring beetle *Phoracantha semipunctata*. *Entomologia Experimentalis et Applicata* 114: 25–34.

Hanks, L. M., T. D. Paine, J. G. Millar, C. D. Campbell, and U. K. Schuch. 1999. Water relations of host trees and resistance to the phloem-boring beetle *Phoracantha semipunctata* F. (Coleoptera: Cerambycidae). *Oecologia* 119: 400–407.

Heap, M. A. 1991. *The Lucerne crown borer, Zygrita diva Thomson (Coleoptera: Cerambycidae), a pest of soybean.* Perth: University of Western Australia.

Ho, K. Y., K. C. Lo., C. Y. Lee., A. S. Hwang. 1995. Ecology and control of the white-spotted longicorn beetle on citrus. *Taiwan Agricultural Research Institute Special Publication* 51: 263–278.

Hussain, A., and M. Z. Chishti. 2012. Cultural control of *Apriona germari* Hope (Coleoptera: Cerambycidae: Lamiinae) infesting mulberry plants in Jammu and Kashmir (India). *Journal of Entomological Research* (New Delhi) 36: 207–210.

Jayanti, L., R. G. Shali, and K. Nanda. 2014. Studies on seasonal incidence, nature of damage and assessment of losses caused by girdle beetle on *Glycine max*. *Annals of Plant Protection Sciences* 22: 320–323.

Johnson, T. A., and R. C. Williamson 2007. Potential management strategies for the linden borer (Coleoptera: Cerambycidae) in urban landscapes and nurseries. *Journal of Economic Entomology* 100: 1328–1334.

Kato, K., and T. Taniguchi. 2003. Ten years examination in the primary screening test in a project for selecting Japanese cedar resistant to *Semanotus japonicus* (Coleoptera: Cerambycidae) conducted in Kanto Breeding Region. *Bulletin of the Forest Tree Breeding Institute* 19: 13–24.

Karuppaiah, V., and G. K. Sujayanad. 2012. Impact of climate change on population dynamics of insect pests. *World Journal of Agricultural Sciences* 8: 240–246.

Knodel, J. J., L. D. Charlet, and J. Gavloski. 2010. Integrated pest management of sunflower insect pests in the Northern Great Plains. *NDSU Extension Service, Fargo, ND, E-1457*: 1–20.

Lan, C. C., and J. N. Wintgens. 2012. Major pests of coffee in the Asia-Pacific region. In *Coffee: Growing, processing, sustainable production—A guidebook for growers, processors, traders and researchers* (2nd rev. ed.), ed. J. N. Wintgens, 463–477. Hoboken, NJ: Wiley, John & Sons Incorporated.

Li, H. P., D. Z. Huang, M. S. Yang, and S. H. Zhang. 2003. Selection of super poplar clones with high resistance to *Anoplophora glabripennis*. *Journal of Northeast Forestry University* 31: 30–32 [in Chinese with English abstract].

Li, L., J. J. Hu, S. M. Li, Z. C. Zhao, and Y. F. Han. 2008. A new insect-resistant poplar variety 'Zhonghuai2'. *Scientia Silvae Sinicae* 44: 174–174.

Li, L. Y., R. Wang, and D. F. Waterhouse. 1997. *The distribution and importance of arthropod pests and weeds of agriculture and forestry plantations in southern China.* Canberra: ACIAR.

Liao, H. H., M. Lu, L. Y. Bu, H. H. Lin, and Z. X. Mo. 2006. *Dorysthenes granulosus* (Thomson) causes serious damage to sugarcanes in Beihai City, Guangxi. *Guangxi Plant Protection* 19: 22–24 [in Chinese].

Liendo-Barandiaran, C. V., B. Herrera-Malaver, F. Morillo, P. Sánchez, and J. V. Hernández. 2010. Behavioral responses of *Steirastoma breve* (Sulzer) (Coleoptera: Cerambycidae) to host plant *Theobroma cacao* L., brushwood piles, under field conditions. *Applied Entomology and Zoology* 45: 489–496.

Lieu, K. O. V. 1945. The study of wood borers in China. I: Biology and control of the citrus-root-cerambycids, *Melanauster chinensis*, Forster (Coleoptera). *Florida Entomologist* 27: 62–101.

Lieu, K. O. V. 1947. The study of wood borers in China.II: Biology and control of the citrus-trunk-cerambycid, *Nadezhdiella cantori* (Hope) (Coleoptera). *Notes d'Entomologie Chinoise* 11: 69–119.

Lindstrom, L., and P. Lehmann. 2015. Climate change effects on agricultural insect pests in Europe. In *Climate change and insect pests*, eds. C. Bjorkman, and P. Niemela, 136–153. Wallingford, CT: CABI.

Liu, K. D., S. B. Liang, and S. S. Deng. 1998. Integrated production practices of cashew in China. In *Integrated production practices of cashew in Asia*, eds. M. K. Papademetriou, and E. M. Herath Bangkok: Food and Agriculture Organization of the United Nations Regional Office for Asia and Pacific. http://www.fao.org/docrep/005/ac451e/ac451e03.htm#fn3 (accessed January 10, 2016).

Long, H. N., and Q. X. Wei. 2007. Occurrence and control of *Dorysthenes granulosus* in Liuzhou. *Guangxi Plant Protection* 20: 41–42 [in Chinese].

Lozovoi, D. J. 1954. Control of *Cerambyx dux* Fald on stone fruits. *Sad i Ogorod* 6: 79–79.

Lu, W., and Q. Wang. 2005. Systematics of the New Zealand longicorn beetle genus *Oemona* Newman with discussion of the taxonomic position of the Australian species, *O. simplex* White (Coleoptera: Cerambycidae: Cerambycinae). *Zootaxa* 971: 1–31.

Lv, K. S., X. M. Lu, Z. P. Huang, Y. L. Liu, and Z. H. Pang. 2011. The preliminary survey on the pest and diseases of *Camellia oleifera* branches in Liuzhou City. *Guangdong Forestry Science and Technology* 27: 58–62 [in Chinese with English abstract].

Meena, N. L., and U. S. Sharma. 2006. Effect of sowing date and row spacing on incidence of major insect-pests of soybean, *Glycine max* (L.) Merrill. *Soybean Research* 4: 73–76.

Mehrnejad, M. R. 2014. Pest problems in pistachio producing areas of the world and their current means of control. *Acta Horticulturae (ISHS)* 1028: 163–169.

Michaud, J. P., and A. K. Grant. 2005. The biology and behaviour of the longhorned beetle, *Dectes texanus* on sunflower and soybean. *Journal of Insect Science* 5: 25.

Michaud, J. P., and A. K. Grant. 2009. The nature of resistance to *Dectes texanus* (Col., Cerambycidae) in wild sunflower, *Helianthus annuus*. *Journal of Applied Entomology* 133: 518–523.

Michaud, J. P., and A. K. Grant. 2010. Variation in fitness of the longhorned beetle, *Dectes texanus*, as a function of host plant. *Journal of Insect Science* 10: 206.

Michaud, J. P., A. K. Grant, and J. L. Jyoti. 2007a. Impact of the stem borer, *Dectes texanus*, on yield of the cultivated sunflower, *Helianthus annuus*. *Journal of Insect Science* 7: 21.

Michaud, J. P., J. A. Qureshi, and A. K. Grant. 2007b. Sunflowers as a trap crop for reducing soybean losses to the stalk borer *Dectes texanus* (Coleoptera: Cerambycidae). *Pest Management Science* 63: 903–909.

Michaud, J. P., P. W. Stahlman, J. L. Jyoti, and A. K. Grant. 2009. Plant spacing and weed control affect sunflower stalk insects and the girdling behavior of *Dectes texanus* (Coleoptera: Cerambycidae). *Journal of Economic Entomology* 102: 1044–1053.

Morewood, W. D., K. Hoover, P. R. Neiner, J. R. McNeil, and J. C. Sellmer. 2004. Host tree resistance against the polyphagous wood-boring beetle *Anoplophora glabripennis*. *Entomologia Experimentalis et Applicata* 110: 79–86.

Nahrung, H. F., T. E. Smith, A. N. Wiegand, S. A. Lawson, and V. J. Debuse. 2014. Host tree influences on longicorn beetle (Coleoptera: Cerambycidae) attack in subtropical *Corymbia* (Myrtales: Myrtaceae). *Environmental Entomology* 43: 37–46.

Nehme, M. E. R., T. Trotter, M. A. Keena, et al. 2014. Development and evaluation of a trapping system for *Anoplophora glabripennis* (Coleoptera: Cerambycidae) in the United States. *Environmental Entomology* 43: 1034–1044.

Niide, T., R. D. Bowling, and B. B. Pendleton. 2006. Morphometric and mating compatibility of *Dectes texanus texanus* (Coleoptera: Cerambycidae) from soybean and sunflower. *Journal of Economic Entomology* 99: 48–53.

Niide, T., R. A. Higgins, R. J. Whitworth, W. T. Schapaugh, C. M. Smith, and L. L. Buschman. 2012. Antibiosis resistance in soybean plant introductions to *Dectes texanus* (Coleoptera: Cerambycidae). *Journal of Economic Entomology* 105: 598–607.

Ocete, R., M. Lara, L. Maistrello, A. Gallardo, and M. A. Lopez. 2008. Effect of *Xylotrechus arvicola* (Olivier) (Coleoptera: Cerambycidae) infestations on flowering and harvest in Spanish vineyards. *American Journal of Enology and Viticulture* 59: 88–91.

Ocete, R., M. A. Lopez, C. Prendes, C. D. Lorenzo, J. L. Gonzalez-Andujar, and M. Lara. 2002. *Xylotrechus arvicola* (Olivier) (Coleoptera, Cerambycidae), a new impacting pest on Spanish vineyards. *Vitis* 41: 211–212.

Ozgul, A., D. Z. Childs, M. K. Oli, et al. 2010. Coupled dynamics of body mass and population growth in response to environmental change. *Nature* 466: 482–485.

Paine, T. D. 2009. *Eucalyptus longhorned borers*. Davis: UC Statewide IPM Program, University of California. http://www.ipm.ucdavis.edu/PMG/PESTNOTES/pn7425.html (accessed January 10, 2016).

Paine, T. D., and J. G. Millar. 2005. Invasion biology and management of insects feeding on eucalyptus in California. In *Proceedings of the Entomological Research in Mediterranean Forest Ecosystems MAY 06–11, 2002, International Symposium on Entomological Research in Mediterranean Forest Ecosystems*, eds. F. Lieutier, and D. Ghaioule, 221–228. Rabat: Morocco.

Paine, T. D., M. J. Steinbauer, and S. A. Lawson. 2011. Native and exotic pests of Eucalyptus: A worldwide perspective. *Annual Review of Entomology* 56: 181–201.

Parsai, S. K., and S. K. Shrivastava. 1993. Effect of dates of sowing and different soyabean varieties on infestation by the girdle beetle, *Oberiopsis brevis* Swed. *Bhartiya Krishi Anusandhan Patrika* 8: 1–4.

Patil, A. K., S. R. Kulkarni, R. V. Kulkarni, and G. L. Chunale. 2006. Field screening of soybean genotypes for resistance to girdle beetle. *Research on Crops* 7: 321–323.

Peverieri, G. S., G. Bertini, P. Furlan, G. Cortini, and P. F. Roversi. 2012. *Anoplophora chinensis* (Forster) (Coleoptera: Cerambycidae) in the outbreak site in Rome (Italy): Experiences in dating exit holes. *Redia-Giornale di Zoologia* 95: 89–92.

Pliansinchai, U., V. Jarnkoon, S. Siengsri, C. Kaenkong, S. Pangma, and P. Weerathaworn. 2007. Ecology and destructive behaviour of cane boring grub (*Dorysthenes buqueti* Guerin) in north eastern Thailand. *XXVI: Congress, International Society of Sugar Cane Technologists, ICC, Durban, South Africa, 29 July – 2 August,* 2007: 863–867.

Qin, R. M., J. Chen, C. Q. Xu, et al. 2008. Preliminary studies on biological characteristics of *Dorysthenes hydropicus. Zhongguo Zhongyao Zazhi* (China Journal of Chinese Materia Medica) 33: 2887–2891 [in Chinese with English abstract].

Qureshi, J. A., P. W. Stahlman, J. P. Michaud. 2007. Plant population and weeds influence stalk insects, soil moisture, and yield in rainfed sunflowers. *Insect Science* 14: 425–435.

Rad, H. H. 2006. Study on the biology and distribution of long-horned beetles *Calchaenesthes pistacivora* n. sp (Col.: Cerambycidae): A new pistachio and wild pistachio pest in Kerman Province. *Acta Horticulturae (ISHS)* 726: 425–430.

Raske, A. G. 1973. Relationship between felling date and larval density of *Monochamus scutellatus. Bi-monthly Research Notes* 29: 23–24.

Rasmann, S., and L. Pellissier. 2015. Adaptive responses of plants to insect herbivores under climate change. In *Climate change and insect pests*, eds. C. Bjorkman, and P. Niemela, 38–53. Wallingford, CT: CABI.

Riggins, J. J., L. D. Galligan, and F. M. Stephen. 2009. Rise and fall of red oak borer (Coleoptera: Cerambycidae) in the Ozark Mountains of Arkansas, USA. *Florida Entomologist* 92: 426–433.

Rogers, C. E. 1985. Cultural management of *Dectes texanus* (Coleoptera, Cerambycidae) in sunflower. *Journal of Economic Entomology* 78: 1145–1148.

Romano, I., and J. C. Kerr. 1977. Soybeans in the central and upper Burnett. *Queensland Agricultural Journal* 103: 64–67.

Roques, A., L. L. Zhao, J. H. Sun, and C. Robinet. 2015. Pine wood nematode, pine wilt disease, vector beetle and pine tree: How a multiplayer system could reply to climate change. In *Climate change and insect pests*, eds. C. Bjorkman, and P. Niemela, 220–234. Wallingford, CT: CABI.

Rutherford, M. A., and N. Phiri. 2006. *Pests and diseases of coffee in Eastern Africa: A technical and advisory manual.* Wallingford, UK: CAB International.

Sahu, K. R., B. C. Shukla, B. S. Thakur, S. K. Patil, and R. R. Saxena. 2012. Relative tolerance of cashew cultivars against cashew stems and root borer (CSRB), *Plocaederus ferrugineus* L. in Chhattisgarh. *Uttar Pradesh Journal of Zoology* 32: 311–316.

Sato, S. 2007. Relationships between insect and disease pest damages and thinning practice in plantations of Japanese red cedar and Hinoki cypress in Japan. *Bulletin of the Forestry and Forest Products Research Institute, Ibaraki* 404: 135–143.

Seaton, S., G. Matusick, K. X. Ruthrof, and G. E. S. J. Hardy. 2015. Outbreak of *Phoracantha semipunctata* in response to severe drought in a Mediterranean Eucalyptus forest. *Forests* 6: 3868–3881.

Sharaf, N. S. 2010. Colonization of *Cerambyx dux* Faldermann (Coleoptera: Cerambycidae) in stone-fruit tree orchards in Fohais Directorate, Jordan. *Jordan Journal of Agricultural Sciences* 6: 560–578.

Sharma, A. N., G. K. Gupta, R. K. Verma, et al. 2014. *Integrated pest management for soybean.* New Delhi: National Centre for Integrated Pest Management.

Sharma, J. P., and D. R. Khajuria. 2005. Distribution and activity of grubs and adults of apple root borer *Dorysthenes hugelii* Redt. *Acta Horticulturae* 696: 387–393.

Shylesha, A. N., N. S. A. Thakur, and Mahesh Kumar. 1996. Bioecology and management of citrus trunk borer *Anoplophora (Monohammus) versteegi* Ritsema (Coleoptera: Cerambycidae): A major pest of citrus in North Eastern India. *Pest Management in Horticultural Ecosystems* 2: 65–70.

Singh, M., J. P. Sharma, and D. R. Khajuria. 2010. Impact of meteorological factors on the population dynamics of the apple root borer, *Dorysthenes hugelii* (Redt.), adults in Kullu valley of Himachal Pradesh. *Pest Management and Economic Zoology* 18: 134–139.

Sloderbeck, P. E., and L. L. Buschman. 2011. Aerial insecticide treatments for management of Dectes stem borer, *Dectes texanus*, in soybean. *Journal of Insect Science* 11: 49.

Smith, M. T., and Wu, J. 2008. Asian longhorned beetle: Renewed threat to northeastern USA and implications worldwide. *International Pest Control* 50: 311–316.

Solomon, J. D. 1985. Observations on a branch pruner, *Psyrassa unicolor* Randall Coleoptera Cerambycidae, in pecan trees. *Journal of Entomological Science* 202: 258–262.

Straw, N. A., N. J. Fielding, C. Tilbury, D. T. Williams, and D. Inward 2015a. Host plant selection and resource utilisation by Asian longhorn beetle *Anoplophora glabripennis* (Coleoptera: Cerambycidae) in southern England. *Forestry* 88: 84–95.

Straw, N. A., C. Tilbury, N. J. Fielding, D. T. Williams, and T. Cull. 2015b. Timing and duration of the life cycle of Asian longhorn beetle *Anoplophora glabripennis* (Coleoptera: Cerambycidae) in southern England. *Agricultural and Forest Entomology* 17: 400–411.

Strickland, G. R. 1981. Integrating insect control for Ord soybean production. *Journal of Agriculture, Western Australia* 22: 81–82.

Sun, J. Z., Z. Y. Zhao, T. Q. Ru, Z. G. Qian, and X. J. Song. 1990. Control of *Anoplophora glabripennis* by using cultural methods. *Forest Pest and Disease* 2: 10–12 [in Chinese with English abstract].

Tara, J. S., R. Gupta, and M. Chhetry. 2009. A study on *Aeolesthes holosericea* Fabricius, an important pest of apple plantations (*Malus domestica* Borkh.) in Jammu region. *Asian Journal of Animal Science* 3: 222–224.

Thomson, L. J., S. Macfadyen, and A. A. Hoffmann. 2010. Predicting the effects of climate change on natural enemies of agricultural pests. *Biological Control* 52: 296–306.

Tindall, K. V., S. Stewart, F. Musser, et al. 2010. Distribution of the long-horned beetle, *Dectes texanus*, in soybeans of Missouri, Western Tennessee, Mississippi, and Arkansas. *Journal of Insect Science* 10: 178.

Torres-Vila, L. M., C. Zugasti, J. M. De-Juan, et al. 2015. Mark-recapture of *Monochamus galloprovincialis* with semiochemical-baited traps: Population density, attraction distance, flight behaviour and mass trapping efficiency. *Forestry* 88: 224–236.

Veda, O. P., and S. S. Shaw. 1994. Record of stem boring beetle on sunflower in Jhabua Hills of M. P. *Current Research—University of Agricultural Sciences (Bangalore)* 23: 118–119.

Venkatesha, M. G., and A. S. Dinesh. 2012. The coffee white stemborer *Xylotrechus quadripes* (Coleoptera: Cerambycidae): Bioecology, status and management. *International Journal of Tropical Insect Science* 32: 177–188.

Waller, J. M., M. Bigger, and R. J. Hillocks. 2007. Stem- and branch-borers. In *Coffee pests, diseases and their management*, eds. J. M. Waller, M. Bigger, and R. J. Hillocks, 41–67. Oxfordshire (UK): CABI Publishing.

Wang, Q. 1995a. Australian longicorn beetles plague our eucalypts. *Australian Horticulture* 93: 34–37.

Wang, Q. 1995b. A taxonomic revision of the Australian genus *Phoracantha* (Coleoptera: Cerambycidae). *Invertebrate Taxonomy* 9: 865–958.

Wang, Q., G. Shi, and L. K. Davis. 1998. Reproductive potential and daily reproductive rhythms of *Oemona hirta* (Coleoptera: Cerambycidae). *Journal of Economic Entomology* 91: 1360–1365.

Wang, S. L., A. Q. Wang, M. H. Xia, and S. H. Dong. 2004. Study on the occurrence of *Batocera horsfieldi* on walnut and its control. *China Fruits* 2: 11–13 [in Chinese].

White, T. C. R. 2015. Are outbreaks of cambium-feeding beetles generated by nutritionally enhanced phloem of drought-stressed trees? *Journal of Applied Entomology* 139: 567–578.

Wotherspoon, K., T. Wardlaw, R. Bashford, and S. A. Lawson 2014. Relationships between annual rainfall, damage symptoms and insect borer populations in mid-rotation *Eucalyptus nitens* and *Eucalyptus globulus* plantations in Tasmania: Can static traps be used as an early warning device? *Australian Forestry* 77: 15–24.

Wu, Z. L., G. M. Xu, G. X. Zhou, W. B. Jiang, and R. L. Li. 2000. Effects of forest management measures on the population of *Monochamus alternatus*. *Journal of Zhejiang Forestry Science and Technology* 20: 36–37, 49 [in Chinese with English abstract].

Wylie, F. R., and M. R. Speight. 2012. *Insect pests in tropical forestry* (2nd ed.). Oxfordshire: CABI.

Yang, C. 1992. Integrated management of *Dorysthenes granulosus* Thomson. *Sugarcane and Canesugar* 1: 11–15 [in Chinese].

Yang, D. W., and Q. Z. Liu. 2010. Tea longhorn beetle damage and its control emphasis in organic tea gardens. *Journal of Anhui Agricultural Sciences* 38: 1305, 1312 [in Chinese with English abstract].

Yoon, H. J., I. G. Park, Y. I. Mah, S. B. Lee, and S. Y. Yang. 1997. Ecological characteristics of mulberry longicorn beetle, *Apriona germari* Hope, at the hibernation stage in mulberry fields. *Korean Journal of Applied Entomology* 36: 67–72 [in Korean with English abstract].

Yoshino, Y. 2004. Difference in damage caused by the sugi bark borer (*Semanotus japonicus* Lacordaire) with planting density in a Japanese cedar (*Cryptomeria japonica*) plantation. *Journal of the Japanese Forestry Society* 86: 1–4.

Youngsteadt, E., A. G. Dale, A. J. Terando, R. R. Dunn, and S. D. Frank. 2015. Do cities simulate climate change? A comparison of herbivore response to urban and global warming. *Global Change Biology* 21: 97–105.

Yu, Y. H., G. J. Chen, D. W. Wei, X. R. Zeng, and T. Zeng. 2012. Division of larval instars of *Dorysthenes granulosus* based on Crosby growth rule. *Journal of Southern Agriculture* 43: 1485–1489 [in Chinese with English abstract].

Yu, Y. H., R. Z. Chen, D. W. Wei, X. F. Chen, T. Zeng, and Z. Y. Wang. 2009. Trapping and killing of *Dorysthenes granulosus* adults using frequency vibration-killing lamps. *Guangxi Agricultural Sciences* 40: 1552–1554 [in Chinese with English abstract].

Yu, Y. H., T. Zeng, D. W. Wei, X. R. Zeng, L. F. Li, and Z. Y. Wang. 2006. Damage status and control strategies of *Dorysthenes granulosus* (Thomson) in Guangxi sugarcane region. *Guangxi Agricultural Sciences* 37: 545–547 [in Chinese with English abstract].

Yu, Y. H., T. Zeng, D. W. Wei, X. R. Zeng, L. F. Li, and Z. Y. Wang. 2007. Sugarcanes in Guangxi are seriously damaged by *Dorysthenes granulosus* in recent years. *Plant Protection* 33: 136–137 [in Chinese].

Zeng, C. F., and Q. S. Huang. 1981. Preliminary study on occurrence and control of *Dorysthenes granulosus*. *Entomological Knowledge (Kunchong Zhishi)* 1: 18–20 [in Chinese].

Zeng, L. Q., J. S. Huang, S. P. Cai, et al. 2014. Resistance identification of 48 *Casuarina* clones to *Anoplophora chinensis*. *Journal of Nanjing Forestry University (Natural Sciences Edition)* 38: 51–56 [in Chinese with English abstract].

Zhang, Q. D., B. Z. Ji, T. Xu, et al. 2010. Progress and perspectives on the genus *Aphrodisium*. *Scientia Silvae Sinicae* 46: 144–151 [in Chinese with English abstract].

Zhao, K. J., Y. M. Zhou, H. B. Zhu, K. Z. Xu, and H. X. Sun. 1994. Study on the control of the mulberry borer (*Apriona germari*) using a fumigating rod. *Plant Protection* 20: 10–11 [in Chinese].

10

Chemical Control of Cerambycid Pests

Qiao Wang
Massey University
Palmerston North, New Zealand

CONTENTS

10.1 Introduction

Insecticides are relatively less commonly used for cerambycid pest control as compared to many other insect pests because cerambycids spend most of their lifetime inside host plants or in soil, making it difficult to directly apply insecticides to them. As a result, cultural control probably is the most widely used tactic for cerambycid pest management (see Chapter 9). Nevertheless, chemical control using insecticides still is an important and sometimes an essential tactic used in integrated pest management of cerambycids.

Insecticides have been used for control of a number of cerambycid pests that infest high-value urban trees (see Chapter 11) and tree and annual field crops (see Chapter 12). Table 10.1 exemplifies chemical control of cerambycid pests from three main subfamilies, Cerambycinae, Lamiinae, and Prioninae, in Africa, Asia, Australasia, Europe, North America, and South America. Chemical control also may play an important role in biosecurity treatments of postharvest products and programs for eradication of non-native invasive pests (Meng et al. 2015; see Chapter 13). The application of insecticides appears to be impractical for control of cerambycid pests in large-scale forest situations.

Although insecticides that are botanical, microbial, nematodal, and hormonal in nature have been tested for controlling cerambycid pests, there is little evidence that these insecticides currently are used commercially on a large scale. Therefore, this chapter focuses on synthetic organic insecticides and, to a lesser extent, on some inorganic chemicals that are or have been used commercially for cerambycid pest control. Matthews et al. (2014) provided excellent information on treatments, formulations, and application methods of various insecticides. This chapter begins with the main classes of insecticides commonly used for cerambycid pest control and their modes of action and then provides examples to discuss the application methods, including field spray, bark treatment, trunk injection, and soil and root treatment.

Because doses of insecticides vary enormously in the control of different pests in different countries or regions, only insecticide classes, formulations, and application methods for various cerambycid pests are discussed in this chapter. Growers can decide on the appropriate insecticide dosage to use based on local pest species, insecticide label instructions, and regulations. When more than one application method is used for control of a given species, then the species is discussed under each application technique.

TABLE 10.1

Examples of Cerambycid Pests from Different Subfamilies that Have Been Controlled Using Insecticides in Various World Regions

Subfamily	Species	Main Host Plant where Insecticides Applied	Geographic Location
Lamiinae	*Acalolepta cervina*	Coffee trees	Asia
	Acalolepta mixta	Grape vines	Australasia
	Anoplophora chinensis	Many tree species	Asia, Europe
	Anoplophora glabripennis	Many tree species	Asia, North America
	Apriona rugicollis	Mulberry and apple trees	Asia
	Bacchisa atritarsis	Tea and oil tea trees	Asia
	Batocera horsfieldi	Walnut trees	Asia
	Batocera rufomaculata	Tropical fruit trees	Asia
	Bixadus sierricola	Coffee trees	Africa
	Celosterna scabrator	Grape vines	Asia
	Dectes texanus	Soybeans	North America
	Monochamus leuconotus	Coffee trees	Africa
	Obereopsis brevis	Soybeans	Asia
	Nitakeris nigricornis	Coffee trees	Africa
	Plagiohammus maculosus	Coffee trees	South America
	Saperda vestita	Linden trees	North America
Cerambycinae	*Aeolesthes induta*	Tea and oil tea trees	Asia
	Aeolesthes sarta	Apple trees	Asia
	Calchaenesthes pistacivora	Pistachio trees	Asia
	Neoplocaederus ferrugineus	Cashew trees	Asia
	Rhytidodera bowringii	Mango trees	Asia
	Xylotrechus javanicus	Coffee trees	Asia
Prioninae	*Dorysthenes granulosus*	Sugarcanes; longan, citrus, and cucalypt trees	Asia
	Dorysthenes paradoxus	Turf grasses	Asia

10.2 Main Classes of Chemical Insecticides

There are a number of references on chemical insecticides used for control of various insect pests. This section describes classes of chemical insecticides that are or have been used for cerambycid pest control in the field. Fumigation treatment of post-harvest products such as wood is discussed in Chapter 13.

Chemical insecticides usually are divided into five major classes: organochlorines, organophosphates, carbamates, pyrethroids, and neonicotinoids. In addition to these main classes, there are a number of organic and inorganic compounds representing smaller classes such as avermectins, phenylpyrazoles, and phosphines. Much of the information on insecticides described in this section was extracted from Dent (2000), Devine and Furlong (2007), BASF (2013), Brown and Ingianni (2013), and Matthews et al. (2014). Table 10.2 summarizes the main classes of insecticides that have been used for cerambycid pest control in the world.

10.2.1 Organochlorines

Organochlorine insecticides are synthetic organic compounds that affect the sodium or chloride channel by inhibiting the gamma-aminobutyric acid (GABA) receptors in insect nerve cells, thereby affecting neurotransmission. When an organochlorine insecticide binds to the GABA molecule, the neurotransmitter can no longer close the channel, resulting in a continuous electrical charge from the neurons, overstimulation of the nervous system, and death. Common organochlorine insecticides include aldrin, chlordane, chlordecone, DDT, dieldrin, endosulfan, endrin, heptachlor, hexachlorobenzene, lindane (gamma-hexachlorocyclohexane), methoxychlor, mirex, pentachlorophenol, and TDE.

TABLE 10.2

Main Classes of Insecticides that Have Been Used for Cerambycid Pest Control in Various Areas of the World

Class	Poison and Property	Time Introduced	Pest Example
Organochlorines	Contact and ingestion	1940s	*Apriona rugicollis, Calchaenesthes pistacivora, Neoplocaederus ferrugineus*
Organophosphates	Contact and ingestion with some having systemic and fumigant properties	1950s	*Acalolepta mixta, Anoplophora glabripennis, Obereopsis brevis, Xylotrechus javanicus*
Carbamates	Contact and ingestion with some having systemic and fumigant properties	1950s	*Apriona rugicollis, Dorysthenes granulosus*
Pyrethroids	Contact and ingestion	1970s	*Acalolepta mixta, Anoplophora glabripennis, Obereopsis brevis, Dectes texanus*
Neonicotinoids	Contact and ingestion with strong systemic properties	1990s	*Acalolepta mixta, Aeolesthes sarta, Anoplophora chinensis, Anoplophora glabripennis*
Phenylpyrazoles	Contact and ingestion with strong systemic properties	1990s	*Acalolepta mixta, Dorysthenes granulosus*
Avermectins	Contact and ingestion with some having systemic and fumigant properties	1980s	*Monochamus leuconotus, Anoplophora glabripennis*
Phosphine	Fumigation	1940s	*Anoplophora glabripennis, Apriona rugicollis, Bixadus sierricola, Celosterna scabrator*

Beginning in the mid-1940s, organochlorines were widely used for the control of insect pests including cerambycids for about three decades. They act by contact, ingestion, and sometimes vapor action but none is systemic within the plant. These insecticides are broad spectrum and have prolonged persistence. Because of the nature of their stability and low solubility in water, they are highly persistent and lead to long-term environmental contamination and gradual accumulation in animals at the higher end of food chains. They can accumulate in body fat and gradually kill mammals, including humans. As a result, this class of insecticides has been banned by most developed countries and is in the process of being phased out in many developing countries for use in agricultural and forest pest control.

10.2.2 Organophosphates

Organophosphate insecticides are synthetic organic chemicals known as cholinesterase inhibitors. They bind to the enzyme cholinesterase, which is responsible for breaking down acetylcholine (ACh) after it has carried its message across the synapse. As a result, the insect's cholinesterase is unavailable to break down the ACh, and the electrical charge from neurons is not able to stop. This causes overstimulation of the nervous system, and the insect dies. The cholinesterase inhibition by organophosphates is not reversible, and poisoned insects usually cannot recover. Commonly used organophosphate insecticides include acephate, azinphos-methyl, bensulide, chlorethoxyfos, chlorpyrifos, chlorpyriphos-methyl, diazinon, dichlorvos (DDVP), dicrotophos, dimethoate, disulfoton, ethoprop, fenamiphos, fenitrothion, fenthion, fosthiazate, malathion, methamidophos, methidathion, mevinphos, monocrotophos, naled, omethoate, oxydemeton-methyl, parathion, parathion-methyl, phorate, phosalone, phosmet, phostebupirim, phoxim, pirimiphos-methyl, profenofos, terbufos, tetrachlorvinphos, triazophos, tribufos, and trichlorfon.

This class of insecticides was introduced in early 1950s and still is widely used for the control of many insect pests, including cerambycids. There currently are many organophosphate insecticides on the market—each differing in persistence, effectiveness against different insects, and toxicity to humans. However, they are much less persistent and thus less of a threat to the environment than are organochlorines. The fast breakdown of organophosphate insecticides fits well with the principles of integrated pest management, but it also means that the timing of application is critical to ensure good control. Therefore, an effective monitoring and action threshold strategy is required to ensure timely application and maximum economic return. Organophosphates usually act by contact and ingestion. Some have systemic properties, such as acephate and dimethoate, and a few, such as dichlorvos, possess a considerable fumigant effect. Some organophosphates, such as parathion, are extremely toxic to nontarget animals, including humans, and are thus banned for use in many countries.

10.2.3 Carbamates

Like organophosphates, carbamate insecticides are synthetic cholinesterase inhibitors. The cholinesterase inhibition by carbamates is somewhat reversible, and thus the poisoned insects can sometimes recover. Carbamates have a much shorter duration of action than organophosphates. Some of the main carbamate products are aldicarb, bendiocarb, carbofuran, carbaryl, dioxacarb, fenobucarb, fenoxycarb, isoprocarb, methomyl, and 2-(1-methylpropyl)phenyl methylcarbamate. Some have very low toxicity, such as carbaryl, which is recommended for household use. However, aldicarb and carbofuran are highly toxic, usually available as granules, and their use is strictly controlled.

Carbamate insecticides were introduced and developed in late 1950s–1960s. Most have good contact action, and some also are fumigant and slightly systemic (such as methomyl). The most commonly used carbamate is carbaryl, which is a broad-spectrum insecticide with both contact and systemic activity.

10.2.4 Pyrethroids

Pyrethroids are synthetic versions of natural insecticide pyrethrins. Pyrethroids are more stable in the environment than pyrethrins. They can provide longer-lasting control—from a few days to a few weeks—depending on products and environmental conditions. They prevent the sodium channels from closing, resulting in continual nerve impulse transmission, tremors, and death. These compounds generally are less toxic to mammals than organophosphates and carbamates, but they may relate to short-term toxicity and particularly toxicity to fish and nontarget invertebrates. Commercially available pyrethroid products are allethrin, bifenthrin, cyhalothrin (lambda-cyhalothrin), cypermethrin, cyfluthrin, deltamethrin, etofenprox, fenvalerate, permethrin, phenothrin, prallethrin, resmethrin, tetramethrin, tralomethrin, and transfluthrin.

Pyrethroids were introduced in the mid-1970s. They have proven to be powerful insecticide control tools. These chemicals act by contact and ingestion and normally have no systemic activity. Furthermore, pyrethroids usually are effective for the control of various insect pests at a very low dose, such as deltamethrin and lambda-cyhalothrin.

10.2.5 Neonicotinoids

Neonicotinoids are synthetic analogs of the natural insecticide nicotine but with much lower acute mammalian toxicity and greater field persistence. Neonicotinoid insecticides act as agonists of the acetylcholine receptor by mimicking the action of the neurotransmitter acetylcholine (ACh). Although cholinesterase is not affected by these insecticides, the nerve is continually stimulated by the neonicotinoid itself, and the end result is similar to that caused by cholinesterase inhibitors—overstimulation of the nervous system leading to poisoning and death. However, the neonicotinoids are a closer mimic for the insect's ACh than for human ACh, giving this class of insecticides more specificity for insects and less ability to poison humans. Common neonicotinoid products available on the market are acetamiprid, clothianidin, dinotefuran, imidacloprid, nithiazine, thiacloprid, and thiamethoxam.

Neonicotinoids were introduced to insect pest control in the early 1990s. They are broad-spectrum systemic insecticides with rapid contact and ingestion action. Due to their excellent systemic properties, they can be applied as sprays, drenches, injections, and seed and soil treatments. Neonicotinoids recently have come under scrutiny for possible negative effects on honeybees because various studies have suggested that the widespread use of neonicotinoid insecticides threatens the survival of bees (Kessler et al. 2015; Lundin et al. 2015; Rundlof et al. 2015). Furthermore, Williams et al. (2015) demonstrated for the first time that exposure to field-realistic concentrations of neonicotinoid pesticides during development can severely affect queens of western honeybees. As a result of these concerns, some neonicotinoid insecticides recently have been banned in European Union countries and in the United States (Pearce 2015).

10.2.6 Other Classes of Insecticides

There are many other classes of insecticides available on the market for control of various insect pests. This section describes a few, both organic and inorganic, that have been used for cerambycid pest control.

Phenylpyrazole is a group of synthetic organic compounds that disrupt the insect's central nervous system by blocking GABA-gated chloride and glutamate-gated chloride (GluCl) channels, resulting in hyperexcitation of nerves and muscles and eventually the death of insects. This unique multitarget mode of action may reduce the potential for resistance development. Fipronil, a member of this chemical family, was introduced in early 1990s as a commercial insecticide. It offers low-dose, highly effective insect control against a broad range of pests through both contact and ingestion actions with strong systemic properties. Fipronil is a slow-acting poison. Its half-life in the soil is from four months to one year, but it is much shorter on the soil surface due to its sensitivity to sunlight. This insecticide has low toxicity for mammals and waterfowl but is highly toxic to fish and other birds. It also is highly toxic for some aquatic invertebrates and honeybees. Low application rates combined with formulation innovations and improved delivery methods are intended to reduce the risks associated with this chemical's use.

The chemical avermectin activates glutamate-gated chloride channels (GluCls) on insect muscle and nerve cells, leading to flaccid paralysis and death. The most commonly used products of this chemical family are abamectin and emamectin benzoate, which were introduced to the insecticide market in the mid-1980s. These insecticides are broad spectrum with contact, ingestion, and systemic properties.

These compounds are rapidly absorbed by plants, and the remaining surface residues are quickly degraded by ultraviolet light, which reduces their effect on bees. There is no accumulation potential because chloride channel activators bind tightly to the soil and do not leach. The doses for insect pest control are not toxic for mammals because they do not possess glutamate-gated chloride channels. However, these chemicals are highly toxic to bees when they are exposed to direct treatment or to residues.

There are also a number of inorganic chemical insecticides on the market. Two commonly used inorganics are aluminum phosphide and zinc phosphide, usually available in the formulation of pellets or sticks. Upon contact with atmospheric moisture, they release phosphine gas, which is colorless, odorless, flammable, and highly toxic. The pellets also contain agents to reduce the potential for ignition or explosion of the released phosphine. Phosphines often are fumigants used to control a broad spectrum of insect pests found in stored products in sealed containers or structures. They are also applied via insertion into trees to kill cerambycid larvae inside trees.

10.3 Field Spray

Field spray (Table 10.3), with an attempt to cover the entire affected area, is one of the most commonly used insecticide application methods for the control of various insect pests. However, because most stages of the cerambycid life cycle pass inside the host plant or in the soil, this method is rather less common in cerambycid pest control. Following close monitoring, field spray to target adults and sometimes neonate larvae can be effective for some cerambycid species. If the adult emergence period expands over a long period, such as many months in a season, which occurs in most cerambycid pest species, the field spray approach may not be effective. Furthermore, field spray for cerambycid control in large-scale forests may not be practical.

This section discusses two pests, *Acalolepta mixta* (Hope) and *Obereopsis brevis* (Gahan), to illustrate how field spray of insecticides can control cerambycids. Chemical control has been adopted as a routine tactic for these two cerambycids. The section concludes with a summary on the application of field spray for other cerambycid pests.

10.3.1 *Acalolepta mixta* (Hope)

Acalolepta mixta [= *A. vastator* (Newman)] is a serious pest of grapevines in southeastern Australia (Dunn and Zurbo 2014) and in some vineyards of New South Wales it infests up to 70% of vines (on average, 36–46%) (Goodwin 2005a). The larvae bore into the vine wood and tunnel throughout the trunk and into the roots, which makes the larvae inaccessible for contact insecticides. However, the annual life cycle of the pest provides two windows of opportunity for contact insecticides to reach these insects (Goodwin and Pettit 1994): (1) during peak adult emergence, in January and February, when the adults chew emergence holes to exit the host plant and later as they walk on the bark surface; and (2) during peak oviposition, between late January and March, with egg hatching in two to four days and young larvae of the first three instars feeding on the surface of oviposition sites for the first four to five weeks before penetration into the nonconductive heartwood tissue.

In the late 1980s, initial field trials on the control of *A. mixta* evaluated chemicals that were registered for use in grapevines in New South Wales to provide immediate access to a treatment for this pest. The chemicals for these trials included three organophosphate insecticides: chlorpyrifos, methidathion, and azinphos-methyl. The introduction of these chemical treatments for *A. mixta* had an immediate impact on the pest population with mean annual infestations falling from 39.7% in 1990 to 7.7% in 1991 and to 5.8% in 1992 (Goodwin et al. 1994). Growers spray the insecticide thoroughly to grapevine trunks and arms to ensure wetting of the barked area where adults emerge and oviposit and where the feeding of young larvae occurs. Seasonal spraying of organophosphate insecticides at two- to four-week intervals during the aforementioned windows in vineyards to control adults and young larvae raises some concerns about insecticide residues detected in grapes and wine (Goodwin and Ahmad 1998). It was recommended that monthly applications commence in November up to the withholding period prior to vintage

TABLE 10.3

Insecticide Classes, Application Methods, and Target Life Stages in the Control of Cerambycid Pests

Application Method	Class of Insecticide	Main Life Stage Targeted	Example
Field spray	Organophosphates, pyrethroids, phenylpyrazoles, or neonicotinoids	Adults and young larvae	*Acalolepta mixta*
	Organophosphates or organochlorines	Adults	*Calchaenesthes pistacivora*
	Organophosphates or pyrethroids or both	Adults	*Obereopsis brevis*
	Pyrethroids	Adults	*Dectes texanus*
Bark treatment	Organophosphates	Adults and neonate larvae	*Acalolepta cervina*
	Organophosphates, neonicotinoids, or pyrethroids	Adults	*Anoplophora chinensis*
	Organophosphates or pyrethroids or both	Adults	*Anoplophora glabripennis*
	Organochlorines, organophosphates, carbomates, pyrethroids, or organophosphates+pyrethroids	Adults	*Apriona rugicollis*
	Organophosphates	Young larvae	*Bacchisa atritarsis*
	Organophosphates, pyrethroids, or phenylpyrazoles	Adults and neonate larvae	*Monochamus leuconotus*
	Organophosphates	Adults and neonate larvae	*Neoplocaederus ferrugineus*
	Organophosphates	Adults and neonate larvae	*Xylotrechus javanicus*
Injection into predrilled holes	Organophosphates and/or neonicotinoids	Adults and young larvae	*Aeolesthes sarta*
	Organophosphates, neonicotinoids, or emamectin benzoate	Adults and young larvae	*Anoplophora glabripennis*
Insertion/injection into frass holes	Organophosphates or phosphines	Larvae	*Aeolesthes sarta*
	Organophosphates	Larvae	*Aeolesthes induta*
	Phosphine	Larvae	*Anoplophora glabripennis*
	Organophosphates	Larvae	*Anoplophora chinensis*
	Phosphine, organophosphates, or pyrethroids	Larvae	*Apriona rugicollis*
	Organophosphates	Larvae	*Batocera horsfieldi*
	Phosphine	Larvae	*Batocera rufomaculata*
	Phosphine	Larvae	*Bixadus sierricola*
	Organophosphates or phosphines	Larvae	*Celosterna scabrator*
	Organophosphates	Larvae	*Neoplocaederus ferrugineus*
	Organophosphates	Larvae	*Nitakeris nigricornis*
	Organophosphates	Larvae	*Plagiohammus maculosus*
	Organophosphates	Larvae	*Rhytidodera bowringii*
Soil and root treatment	Neonicotinoids	Adults and larvae	*Anoplophora chinensis*
	Organophosphates	Adults and larvae	*Celosterna scabrator*
	Carbamates, organophosphates, or phenylpyrazoles	Larvae	*Dorysthenes granulosus*
	Organophosphates	Neonate larvae	*Dorysthenes paradoxus*
	Organophosphates	Adults and larvae	*Obereopsis brevis*
	Neonicotinoids	Adults and larvae	*Saperda vestita*

and resume after vintage. However, the immediate reduction in the pest population from the seasonal spraying scheme can be followed by pest resurgence (Goodwin 2005a).

In a trial of some new chemicals, the neonicotinoid, imidacloprid, and pyrethroid, λ-cyhalothrin, give excellent control (Goodwin 2005a). Two sprays of concentrated λ-cyhalothrin, the first early in the season and the second post-vintage, are the most successful control strategy against this pest. In a separate trial, Goodwin (2005b) demonstrated that a single application of insecticide (either bifenthrin, fipronil, or imidacloprid), at a higher rate than that normally registered for crop use, during dormancy after pruning and before budburst, is highly effective in killing emerging adults and young larvae. This strategy overcomes a number of problems experienced with previous chemical application methods for this pest.

10.3.2 *Obereopsis brevis* (Gahan)

Obereopsis brevis (= *Oberea brevis* Gahan) is a serious pest of soybean (Kumar et al. 2012; Netam et al. 2013; Jayanti et al. 2014; Sharma et al. 2014) and sunflower (Veda and Shaw 1994; Pachori and Sharma 1997) crops in India. The damage mainly is caused by female adults that girdle the stems, petioles, petiolets, and branches before laying eggs. The plant parts above the girdle wilt and die. The girdle formation by the adult females results in significant reduction in the number of pods and seeds and in the seed weight (Jayanti et al. 2014).

Although many control measures have been explored (see Chapter 12), so far, the use of insecticides appears to be the most effective control method for managing this pest (Chaudhary and Tripathi 2011). Since the 1960s, almost all classes of insecticides mentioned in Section 10.2 have been used for *O. brevis* control but, presently, the organophosphate (such as dimethoate and triazophos) and pyrethroid (such as deltamethrin and lambda-cyhalothrin) insecticides are the most often applied (Singh and Singh 1966; Singh 1986; Pande et al. 2000; Upadhyay and Sharma 2000; Yadav et al. 2001; Purwar and Yadav 2004; Gupta 2008; Chaudhary and Tripathi 2011; Jain and Sharma 2011; Sharma et al. 2014; Lonagre et al. 2015). Organochlorine insecticides have now been banned for use in agriculture in most regions of the world and are no longer commonly utilized in India.

In general, two sprays are recommended for the control of this pest (Singh and Dhamdhere 1983; Parsai et al. 1990; Pande et al. 2000; Yadav et al. 2001; Purwar and Yadav 2004; Gupta 2008; Chaudhary and Tripathi 2011; Jain and Sharma 2011; Sharma et al. 2014). The first spray is applied 30–45 days after sowing or within 7–10 days after the girdling symptoms are present. The crops are sprayed the second time about two weeks after the first spray. However, if the pest is under control after the first spray, there is no need to apply the second spray. If the infestation is observed 75–80 days after sowing, spraying will not improve the yield (Sharma et al. 2014). Most recently, Lonagre et al. (2015) trialed the effectiveness of pyrethroids alone and combined with organophosphate insecticides for the control of this pest. Their results show that three sprays of mixed deltamethrin and dimethoate beginning 30 days after sowing at an internal of 15 days are the most effective approach.

10.3.3 Other Cerambycid Pests

Insecticide spray has proven to be effective in the control of several cerambycid pests in the world. For example, *Calchaenesthes pistacivora* Holzschuh is an increasingly serious pest of pistachio trees in Iran, the largest pistachio producer in the world (Rad et al. 2005; Rad 2006; Achterberg and Mehrnejad 2011; Mehrnejad 2014). The damage is caused by the larval boring inside the twigs and branches. The infested twigs and branches eventually die. Rad et al. (2005) tested the effectiveness of spraying insecticides to kill adults in the field during the adult activity period and suggest that, if used during April, the adult beetles could be effectively controlled in the field. However, these findings are still in the experimental stage rather than used on a large commercial scale, and one of the insecticides used for this experiment— endosulfan (organochlorine)—has been banned.

Dectes texanus LeConte is a serious pest of soybean and sunflower crops in the United States (Michaud and Grant 2010; Sloderbeck and Buschman 2011). It damages these crops by larval girdling of the stem prior to seed maturation (Campbell and Vanduyn 1977; Rogers 1985). Its lengthy larval stage (about

10 months) inside the stems or in stubbles makes chemical control difficult in general (Campbell and Vanduyn 1977; Rystrom 2015). An attempt to suppress adults by spraying an insect growth regulator and a pyrethroid at two-week intervals also failed (Andrews and Williams 1988), probably due to a prolonged window of adult activity (Rystrom 2015). However, trials in central Kansas with aerial applications of a pyrethroid (lambda-cyhalothrin) three times at a weekly interval in July have successfully suppressed the adults (Sloderbeck and Buschman 2011). Further trials are needed before large-scale commercial applications of this tactic can be put in practice.

10.4 Bark Treatment

The method of spraying, swabbing, or painting insecticides on tree trunks or branches has been used successfully to control a number of cerambycid pests (Table 10.3). The insecticides applied usually possess contact and ingestion properties, and some are systemic. The main rationale for these methods is to kill adults and neonate larvae through contact and ingestion during the breeding season. In the following sections, *Anoplophora glabripennis* (Motschulsky) and *Apriona rugicollis* Chevrolat are two examples used to illustrate details of this method; a review of the application of this method with several other cerambycid pests from around the world is also presented.

10.4.1 *Anoplophora glabripennis* (Motschulsky)

Anoplophora glabripennis of Asian origin is a polyphagous borer of many tree species and now occurs in Europe and North America (Haack et al. 2010; Meng et al. 2015). It can cause massive tree mortality because it attacks healthy trees and causes tree death after several years of successive attack (Haack 2016). Damage is made by larval tunneling in the cambial region and then in the sapwood and heartwood. In its native China, this pest costs the country about US$1.5 billion annually (Hu et al. 2009), and in the United States, about 12–61% of all urban trees are at risk, with a total estimated replacement value of US$669 billion (Nowak et al. 2001). After emergence, adults need to feed on tender twigs and leaves of host trees before sexual maturation, and then they search for mates and begin oviposition. Females make oviposition slits using mandibles before laying eggs.

Adults often walk along tree trunks and branches after emergence, moving to and from feeding sites, searching for mates and oviposition sites. This behavior provides an opportunity for development of control measures such as coating tree trunks with contact insecticides (Hu et al. 2009; Smith et al. 2009; Haack et al. 2010). In China, Zhang et al. (1999a) trialed the application of a mixture of encapsulated pyrethroid and organophosphate insecticides (cypermethrin + fenthion or fenvalerate + fenthion) to suppress populations within high-value areas and to protect high-value trees in urban landscapes from attack. The authors indicate that spraying the lower 3-m of trunks with these insecticides provides 100% control for about 38 days. Similar trials have also achieved promising results (Pan et al. 2001; Sun et al. 2003). Xu et al. (2003) developed a control method based on coating tree trunks with cypermethrin, where each tree trunk is painted with a complete insecticide band and adults that move across the band are killed within 24 hours. The treatment remains effective for 30–60 days. This method is particularly effective for the control of *A. glabripennis* on smaller trees and can be applied to trap trees.

In the United States, Wu and Smith (2015) tested the potential of pyrethroid-treated Denier (polyvinyl chloride–polyester) bands for controlling *A. glabripennis* in specialized areas where current control strategies are not desirable or feasible. Briefly, Denier is cut into 12 by 22.5 cm bands and dipped into different concentrations of a commercial formulation of lambda-cyhalothrin. Adult beetles are then allowed to walk across the surface of the treated Denier bands that have been exposed to the natural environment for 10, 20, 45, 69, and 90 days. The authors show that the formulation is highly effective against adult *A. glabripennis*, immediately disabling adult beetles on contact that then die within minutes and doing so for approximately 60–90 days. Therefore, placement of insecticide-treated bands on infested and uninfested trees may offer a method for suppression of adult populations during emergence and initial colonization, respectively.

10.4.2 *Apriona rugicollis* Chevrolat

Based on the most recent taxonomic revision on *Apriona* (Jiroux 2011), *A. rugicollis* (= *A. japonica* Thomson) is widely distributed in China, Korea, and Japan, while *A. germari* (Hope) is found in southern China, India, and Southeast Asian countries. Therefore, most published information on the biology and control of *A. germari* in China (EPPO 2014) should refer to *A. rugicollis*. This species is a serious pest of mulberry trees that are used for silkworm production in China (Gao et al. 1988; Zhao et al. 1994; Zhu et al. 1995) and Japan (Kikuchi 1976; Esaki 1995), causing significant economic losses. The damage is made by larval tunneling in branches and trunks. It is also an important pest of apple (Yang et al. 2005), fig (Qin et al. 1997), paper mulberry (Jin et al. 2011), walnut (Zhang et al. 1999b), poplar (Li 1996; Shui et al. 2009), and hackberry and locust (Esaki 2007) trees. Infested trees may die, and trunks may be broken by wind and other forces.

Like many other lamiine species, adults feed on tender barks and leaves of host trees before sexual maturation and search for mates and oviposition sites; females make oviposition slits using mandibles before laying eggs. These biological features make coating trunks and main branches with contact and ingestion insecticides highly effective. In China, Gao et al. (1988) sprayed or brushed trunks of mulberry trees with organochlorine insecticides, achieving between 80% and 90% control. In a field trial of trunk spray effectiveness using different insecticide formulations for the control of this borer on mulberry trees, Zhu et al. (1995) found that the controlled release microcapsules with pyrethroid+organophosphate insecticides are more effective and persistent than the emulsifiable concentrates. In a separate field trial of the effectiveness of spraying sugar maple tree branches with organophosphates, carbamates, and pyrethroids, respectively, Lian et al. (1997) demonstrated that pyrethroids and carbamates perform better in killing adults. On apple trees, Yang et al. (2005) reported that spraying cypermethrin in a microcapsule solution on the trunk and large branches during peak adult activity and then again in 20 days provides successful control. One spray remains effective for about 30 days.

In a field trial in Japan, Esaki (2007) tested the effectiveness of branch sprays. Results show that spraying the branches of Chinese hackberry with fenitrothion (organophosphate) twice at an interval of three weeks during the early adult activity period can achieve 100% adult mortality over a period of nine weeks. However, this method has not been applied to mulberry trees. Further investigations into the effectiveness of insecticide sprays on trunks and branches of mulberry trees and the potential impact of such sprays on silkworm are warranted.

10.4.3 Other Cerambycid Pests

Neoplocaederus ferrugineus (Linnaeus) [= *Plocaederus ferrugineus* (Linnaeus)] is a serious pest of cashew trees in India and several Southeast Asian countries (Maruthadurai et al. 2012; Sahu et al. 2012; Vasanthi and Raviprasad 2013) as well as in Nigeria (Asogwa et al. 2009). Portions of the trunk below 50 cm are most frequently attacked (Mohapatra and Mohapatra 2004). Field trials indicate that swabbing, painting, or spraying the exposed roots and the trunk up to a height of about 1 m with organophosphate insecticides such as chlorpyriphos, monocrotophos, and dimethoate one to three times a year during the peak adult activity period can successfully suppress adults and neonate larvae (Mohapatra and Jena 2007, 2008; Chakraborti 2011). Organochlorine insecticides such as lindane are still used occasionally in India (Mohapatra and Jena 2008).

Bacchisa atritarsis (Pic) is an important pest of two tea tree species (one for drinking tea leaves and another for tea oil seeds) commercially grown in southern and southeastern China (Chen 1958; Qian 1984; Chen et al. 2002; Tan et al. 2003). The damage is caused by the larval tunneling in the branches and trunks, which show characteristic symptoms (spiral-shaped swollen marks). Field trials indicate that swabbing the trunk as a band just below the damage marks with a systemic organophosphate insecticide omethoate during August kills almost all larvae 20–30 days after application (Chen et al. 2002).

Swabbing or brushing the trunks with synthetic insecticides also proves to be effective for the control of several important coffee tree borers in the major coffee growing regions of the world: *Acalolepta cervina* (Hope) and *Xylotrechus javanicus* (Castelnau & Gory) in Asia (Waller et al. 2007; Venkatesha and Dinesh 2012) and *Monochamus leuconotus* (Pascoe) in Africa (Rutherford and Phiri 2006; Waller et al. 2007). For control of *A. cervina* and *X. javanicus*, Lan and Wintgens (2012) suggested that swabbing the bark of the trunk with a solution of the organophosphate dichlorvos from the base to 1 m above the ground during the oviposition period can reduce egg laying and during the egg hatch period may kill neonate larvae and prevent them from boring into the trunk. The most effective control method for *M. leuconotus* is the application of insecticides to the trunk with a brush or sprayer as a band around the lower 50-cm section of the trunk and the root collar region. Common insecticides used include organophosphates (Schoeman and Pasques 1993; Waller et al. 2007), pyrethroids (Schoeman 1995), and phenylpyrazoles (Rutherford and Phiri 2006), and applications usually are carried out just before the start of the rains (Rutherford and Phiri 2006).

Anoplophora chinensis (Forster) of Asian origin is a highly polyphagous borer of many tree species and now is found in Europe (Eyre et al. 2010; Haack et al. 2010). Spraying trunks and crowns of maple trees planted in parks using imidacloprid microcapsule solution appears to be effective for adult control in China (Li 2008). To treat bonsai maples for export to Europe, Zhong et al. (2012) tested the effectiveness of spraying entire plants, covering crowns, trunks, and exposed roots with solutions of several insecticides, demonstrating that cypermethrin, omethoate, and bisultap can cause 100% adult mortality within 48 hours after application. In China, Wang et al. (2015) showed that coating trunks from the base to 1.5 m above the ground with neonicotinoid (dinotefuran) solutions can reduce adult emergence. In Italy, spraying or brushing the basal trunks of trees using pyrethroid solutions in late May–early June and then treating the trunks again 20 days later with a mixture of pyrethroid and neonicotinoid solutions have proven to be effective in the control of this pest (Cavalieri 2013).

10.5 Injection and Insertion

Injection and insertion of insecticides into tree trunks are two commonly used methods for cerambycid pests (Table 10.3). Various organic insecticides with systemic properties, such as organophosphates and neonicotinoids, are injected into predrilled holes or borer-created "frass ejection holes" to kill adults and/or larvae. After treatment, adults feeding on twigs and leaves and early instar larvae feeding in cambium or cambium–phloem interface are killed. Inorganic insecticides such as aluminum phosphide and zinc phosphide are inserted into borer holes to kill larvae via the fumigation property of phosphine released from these chemicals.

Three holes usually are drilled into the lower trunk or around the root flare of the tree for insecticide injection. If borer frass holes are used for insecticide injection or insertion, the holes selected are those closest to a root flare. After treatment, holes are sealed immediately with mud or Bordeaux paste. Here, *A. glabripennis* and *A. rugicollis* are presented as two examples to exemplify details of these methods, and then their applications with a number of other cerambycid pests from around the world are reviewed.

10.5.1 *Anoplophora glabripennis*

Injection and insertion of insecticides into tree trunks to kill adults and early instar larvae have been widely used for the control of this pest (Hu et al. 2009; Haack et al. 2010; Meng et al. 2015; Wang 2016). In Asia, Zhang et al. (1994) demonstrate that, during breeding season, injection of systemic organophosphate solutions into poplar trees planted for windbreaks and on urban streets achieves 90% mortality of early instar larvae and 99% mortality of adults feeding on twigs and leaves. One treatment remains effective for 38 days. In a separate trial in China, approximately 90% of the first- to fourth-instar larvae and 65% of feeding adults are killed by injecting methamidophos into the trunks of poplar trees (Zhu et al. 1998). Injection of dichlorvos into willow trunks also shows promising control (Liu et al. 1999). Injection of omethoate solutions into poplar trunks kills about 88% of early instar larvae and up to 78% of adults, with one treatment being effective for two months (Wang 2016). Another commonly used method is to insert bamboo or wooden sticks containing aluminum phosphide into larval frass holes, which are then sealed with mud or other materials (Zhao et al. 1995; Liu et al. 1999). Upon contact with moisture, aluminum phosphide produces phosphine that kills the larvae. Generally, this method is safe for the public, and it is widely used in China where the treated sticks are inserted deeply into larval frass holes.

Xu et al. (1999) tested the impact of organophosphate and neonicotinoid injections on adults of congeneric pests *A. glabripennis* and *A. nobilis* Ganglbauer feeding on the bark of twigs, leaves, and on leaf petioles of treated poplar trees. They conclude that imidacloprid is the most effective and persistent, killing 70–80% of adults for about three months. In a large field trial in China, Poland et al. (2006) injected trunks with various systemic insecticides, azadirachtin, emamectin benzoate, imidacloprid, and thiacloprid, for the control of *A. glabripennis* in naturally infested elms, poplars, and willows. These authors demonstrate that imidacloprid is translocated rapidly in infested trees and is persistent at lethal levels for several months. Application of imidacloprid may also affect early instar larvae feeding in the cambium–phloem interface as well as ovipositing adults.

In North America, the neonicotinoid-based insecticides (clothianidin, dinotefuran, and thiamethoxam) are tested for the control of *A. glabripennis* via ingestion, contact, and antifeedant effects (Wang et al. 2005). Field evaluation of the effectiveness of injecting the neonicotinoid product, imidacloprid, in the United States shows very promising outcome: *A. glabripennis* escaped or survived the treatment in only 11 of almost 250,000 at-risk, treated trees in New York, Illinois, and New Jersey (Sawyer, 2007). Neonicotinoids have been selected for use in the U.S. eradication programs for *A. glabripennis* (USDA APHIS. 2008). Ugine et al. (2012) evaluated the effect of imidacloprid on survival and reproduction of adults that feed on twigs and leaves from trunk-injected trees treated once in the spring of 2010 in Worcester, Massachusetts. They demonstrate that adult reproductive output and survival are significantly reduced when beetles feed on twig bark or leaves from treated trees. However, their results vary widely, with many twig samples having no detectable imidacloprid and little effect on the beetles. When twigs with >1ppm imidacloprid in the bark are fed to mated beetles, the number of larvae produced is reduced by 94%, and median adult survival is reduced to 14 days. For twigs with <1ppm imidacloprid, 68% of reproductively mature mated beetles survive 21 days and 56% of unmated recently eclosed beetles survive 42 days.

Systemic insecticides have been widely used in the United States, and the treatment of trees with imidacloprid injection is considered a key component in the eradication of *A. glabripennis* in Chicago (Meng et al. 2015). However, the potential nontarget effects of neonicotinoid-based insecticides, especially on pollinators, raise concerns (Meng et al. 2015; Kessler et al. 2015), and further evaluation is warranted.

10.5.2 *Apriona rugicollis*

So far, the injection and insertion method has only been used to control *A. rugicollis* larvae in apple (Zhao et al. 1994) and poplar (Pan 1999; Shui et al. 2009) trees in China. This method has not been used in mulberry trees. This probably is due to concerns about its potential harm to silkworms, which are fed with mulberry foliage.

In a large trial in apple orchards in several provinces of China, Zhao et al. (1994) compared the effectiveness of organophosphate injection and aluminum phosphide insertion into larval frass holes for the control of *A. rugicollis* larvae. Briefly, dichlorvos solutions are injected directly into the frass holes using a syringe, while bamboo sticks containing aluminum phosphide are inserted into larval frass holes, which are then sealed with mud. In the latter case, a metal wire is used to clean away frass before insertion is made. Seven days after treatment, dichlorvos injection achieves 82.5% larval mortality and aluminum phosphide insertion results in 97.3% larval mortality. These authors suggest that frass holes with fresh frass should be selected for injection or insertion. These authors have also examined phosphine residue in apple fruit, suggesting that the residue in fruit at harvest is likely lower than what is permitted in China.

In a poplar forest planted for windbreak in eastern China, Pan (1999) selected a number of trees infested by *A. rugicollis* and tested the effectiveness of three organophosphates (triazophos, omethoate, and dichlorvo) and one pyrethroid (deltamethrin) for the control of larvae. Chemicals are injected into fresh larval frass holes using a syringe, with 15 trees treated for each chemical. Results show that,

12 days after injection, triazophos, omethoate, and deltamethrin achieve more than 90% larval mortality. In a separate trial for the control of *A. rugicollis* in poplar trees, Shui et al. (2009) inserted sticks containing zinc phosphide into larval frass holes on trunks with three treatments: (1) insertion of one stick into the lowest hole, (2) insertion of one stick into each of the two lowest holes, and (3) insertion of one stick into each of the three lowest holes. They removed the frass around the holes before inserting the sticks. Holes were sealed with mud immediately. All treatments achieved excellent larvae control. As a result, the authors recommend implementation of the first treatment as the most cost-effective control option in the field.

10.5.3 Other Cerambycid Pests

Methods similar to those described earlier for *A. glabripennis* and *A. rugicollis* have been used throughout the world for a number of other cerambycid pests. Examples include *Aeolesthes sarta* Solskys, a pest of apple and many other tree crops (Bhat et al. 2010a, 2010b; Mazaheri et al. 2015); *Batocera horsfieldi* Hope, a pest of walnut and several forest and fruit tree species (Jia et al. 2001; Wang et al. 2004; Li 2009); *Batocera rufomaculata* (DeGeer), a pest of a number of tropical and subtropical fruit tree crops (Ahmed et al. 2013, 2014; Upadhyay et al. 2013); *Neoplocaederus ferrugineus,* a pest of cashew trees (Vasanthi and Raviprasad 2013; Jalgaonkar et al. 2015); *Rhytidodera bowringii* White, a pest of mango trees (Ding et al. 2014); *Aeolesthes induta* (Newman), a pest of tea and oil tea trees (Yang and Liu 2010; Lv et al. 2011), and *Celosterna scabrator* (Fabricius), a pest of various forest tree species (Duffy 1968), pomegranate trees (Naik et al. 2011), and grapevines (Kumari and Vijaya 2015) in Asia; *Anoplophora chinensis*, a pest of many tree species in Asia and Europe (Eyre et al. 2010; Haack et al. 2010); *Bixadus sierricola* (White), a pest of coffee trees (Padi and Adu-Acheampong 2001; Crowe 2012) and *Nitakeris nigricornis* (Olivier), a pest of coffee trees (Waller et al. 2007; Crowe 2012) in Africa, and *Plagiohammus maculosus* (Bates), a pest of coffee trees (Espinosa 2000; Barrera 2008) in South America.

Injection of systemic insecticides into tree trunks via either predrilled or larval frass holes proves to be effective for control of several cerambycid pests. For example, in separate field trials for the control of *C. scabrator,* Jagginavar et al. (2008), Naik et al. (2011), and Kumari and Vijaya (2015) injected dichlorvos into borer frass holes and achieved 90–100% larval mortality. The same method is used for *P. maculosus,* resulting in good control (Espinosa 2000; Barrera 2008).

Mazaheri et al. (2015) reported that the application of both imidacloprid and oxydemeton-methyl through trunk injection is highly effective for the control of *A. sarta* but recommended the use of imidacloprid. Injection of dichlorvos into frass holes provides good control of *A. sarta* larvae for more than 45 days after the treatment (Bhat et al. 2010b). In a field trial on *N. ferrugineus* control, trunk injection of chlorpyrifos or DDVP results in almost 100% larval mortality (Jalgaonkar et al. 2015). Trunk injections of omethoate into frass holes result in 96–100% larval mortality in *A. chinensis* (Wu et al. 2000). Waller et al. (2007) and Crowe (2012) recommended that injection of persistent insecticides into borer frass holes should be used for *N. nigricornis* control. Furthermore, insertion of cotton wool saturated with organophosphates into larval frass holes once or twice a year is also highly successful for controlling several cerambycid pests, such as *A. chinensis* (Chen 2002), *R. bowringii* (Luo et al. 1990; Ding et al. 2014), *B. horsfieldi* (Wang et al. 2004), *A. induta* (Yang and Liu 2010; Lv et al. 2011), and *P. maculosus* (Espinosa 2000; Barrera 2008).

Trunk insertions of aluminum phosphide, using various different formulations, are also recommended for the control of cerambycid pests. For example, a paste containing aluminum phosphide is squeezed into fresh borer frass holes of *B. sierricola,* achieving 100% larval mortality within 15 days after treatment and. no reinfestation within nine months after the treatment (Padi and Adu-Acheampong 2001). Tablets containing the chemical are more widely used for cerambycid pest control due to their ease of application. For example, insertion of one tablet containing 1 g of aluminum phosphide per hole gives larval mortality of 79% for *A. sarta* (Bhat et al. 2010b), 83% for *B. rufomaculata* (Ahmed et al. 2013), and 100% for *C. scabrator* (Kumari and Vijaya 2015). One treatment usually remains effective for about two months. A second treatment may be necessary if larval activity continues two months after the first treatment.

10.6 Soil and Root Treatment

Synthetic organic insecticides with systemic, contact, and ingestion actions are used to control several cerambycid pests that attack trees (particularly young trees) and field crops (Table 10.3). Chemicals in granular or liquid formulations may be applied for soil or root treatment. Here, *Dorysthenes granulosus* (Thomson) serves as an example of how soil and root treatments can successfully control cerambycid pests, which is followed by a review of these methods and their use for controlling several other cerambycid pests throughout the world.

10.6.1 *Dorysthenes granulosus* (Thomson)

This is an increasingly more serious pest of sugarcane in southern China (Yang et al. 1992; Liao et al. 2006; Yu et al. 2006, 2007, 2009, 2010, Long and Wei 2007; Gong et al. 2008; Xiong et al. 2010) and northern Thailand (Sommartya and Suasa-ard 2006; Pliansinchai et al. 2007; Sommartya et al. 2007; Thongphak and Hanlaoedrit 2012). [The Thai species mentioned in these papers should be *D. granulosus* rather than *D. buqueti*—see Chapter 12.] For example, its damage causes 10–50% yield loss in Guangxi Province (Liao et al. 2006). Young larvae feed on tender roots and then bore into main roots, tunnel upward in the stems to as high as 30–60 cm above the ground, and hollow out the stems (Zeng and Huang 1981; Liao et al. 2006; Long and Wei. 2007).

Several kinds of insecticides in granular formulations are used to control this pest at the time of planting or immediately after plowing along the ratoons of cane stumps. Prior to and during the 1990s, carbofuran and phorate were widely used individually or in combination; in the latter case, the ratio was about 1:2 (Zeng and Huang 1981; Yang et al. 1992). The insecticide granules are scattered in the soil, then thoroughly mixed in and covered with additional soil. One application can reduce dead seedlings by 55% and bored stems by 43.7% and can increase yield by 14.8%. However, these two insecticides are highly toxic to mammals and humans and are no longer widely used for the control of this pest. Alternative insecticides such as ethoprop, isofenphos-methyl, and isazofos can be used in the same manner, achieving good control (Yu et al. 2006, 2007; Xiao 2012).

Gong et al. (2008) suggested that there are two windows for insecticide applications. First, at the time of planting or after plowing, insecticide granules are applied around the cane sections/ratoons and covered with soil so that larvae are killed through contact with the chemicals when they move to feed on roots. Second, they are used around the seedlings and covered with soil during June–July when egg hatch peak occurs so that neonate larvae are killed via contact with the chemicals when moving in surface soil toward the roots. Depending on the severity status of the pest, the application can be undertaken once in one window (Gong et al. 2008) or twice in both windows (Xiong et al. 2010). Any of the following insecticides can result in good control: chlorpyrifos, terbufos, and isofenphos-methyl. In the sugarcane fields where *D. granulosus* causes severe damage, Gong et al. (2008) and Xiao (2012) recommended that, before planting, seed cane sections be soaked in liquid insecticides for 30 minutes and allowed to dry. Persistent chlorpyrifos EC and fipronil SC with systemic properties are most commonly used for this purpose. Larvae are killed through ingestion of treated canes.

The pest also damages roots and stumps of various woody plants in southern China, particularly if they have been planted in previous sugarcane fields. For example, about 10% of less than one-year-old mango seedlings are infested, becoming wilted and dying (Zhang 1992); longan seedlings and young citrus trees are heavily damaged with up to 12% mortality in longan seedlings and about 30% of citrus trees losing economic value (Zhu and Xu 1996). More recently, this pest has caused economic damage to cassava (Chen et al. 2012).

When longan and citrus seedlings are transplanted to previous sugarcane fields, the planting holes should be treated with insecticides. For example, carbofuran granules can be applied to planting holes and mixed with the soil before seedlings are planted (Zhu and Xu 1996). For trees that are already infested by *D. granulosus*, dichlorvos or methamidophos solutions can be poured into the soil under the tree crown to kill larvae underneath (Zhang 1992). To reduce the damage to young eucalypt trees, Zhu (1995) trialed the application of carbofuran granules under the tree crown in early June, achieving reasonable control of neonate larvae. Also in early June, the application of trichlorfon granules under the citrus tree crown is effective in killing neonate larvae (Zhu 2012).

10.6.2 Other Cerambycid Pests

Soil and/or root treatments using insecticides also prove effective in controlling other cerambycid pests. Examples include *Dorysthenes paradoxus* (Faldermann), a serious pest of turfgrass in northern China that feeds on grass roots (Yan et al. 1997; Hao et al. 1999); *Celosterna scabrator,* a pest of pomegranate trees (Naik et al. 2011) and grapevines (Kumari and Vijaya 2015) in Asia; *Anoplophora chinensis*, an important pest of many tree species in Asia and Europe (Eyre et al. 2010; Haack et al. 2010); *Obereopsis brevis*, a serious pest of soybean (Kumar et al. 2012; Netam et al. 2013; Jayanti et al. 2014; Sharma et al. 2014) and sunflower (Veda and Shaw 1994; Pachori and Sharma 1997) crops in India; and *Saperda vestita* Say, an important pest of linden trees in north-central and northeastern North America (Johnson and Williamson 2007).

Granular organophosphates with systemic properties are widely used to control *O. brevis* in Indian soybean fields. For example, scattering phorate in the field and thoroughly mixing the insecticide with the soil at the time of sowing achieve a 32–75% reduction of infestation later in the season (Sharma 1994) and result in the highest yield (Dahiphale et al. 2007). In a separate field trial, Singh and Singh (1989) demonstrated that, if the same chemical and application method are used 25 days after sowing, maximum grain yield and the best cost/benefit ratio are obtained. In the management program for *D. paradoxus* damaging lawns, Yan et al. (1997) recommended that organophosphate (phoxim or isofenphos-methyl) solutions be mixed with soil and fertilizers and scattered on the turf surface, which is then watered. This application should be made during late July–late August to kill neonate larvae.

Systemic insecticide solutions can be injected or poured into the soil within the periphery of tree crowns to control cerambycid pests in nurseries. In a Wisconsin linden nursery, Johnson and Williamson (2007) demonstrated that spring and autumn injections of imidacloprid solutions into the soil at a depth of about 20 cm at evenly spaced injection points around the periphery of the dripline of each tree (<11.4 cm in diameter at breast height) achieve more than 90% control of *S. vestita*. However, similar treatment is not effective for controlling the pest in larger and established linden trees (>22 cm in diameter at breast height). Wang et al. (2015) tested the effectiveness of several neonicotinoid insecticides for the control of *A. chinensis* via soil treatment in a red maple nursery in China, demonstrating that imidacloprid is the most effective and persistent insecticide. During late April–early May, about one month before the adult emergence peak, a ring groove at the periphery of the tree crown is dug and the insecticide solution is poured into the groove, which is then covered with soil. One application remains effective for the control of both adults and larvae for one year. If the insecticide is applied in two successive years, the pest is under complete control for three years.

A systemic organophosphate, monocrotophos, has been used to treat roots of woody crops for cerambycid pest control. In a field trial for the control of *C. scabrator* in an Indian vineyard, Chandrasekaran et al. (1990) demonstrate that immersing one or two roots of each vine into the insecticide EC achieves 76–96% larval mortality 15 days after the treatment. However, due to its acute toxicity to birds and humans and its long persistence, this insecticide has been banned in many countries.

10.7 Concluding Remarks

Although chemical control of cerambycid pests generally is more difficult than that of insect pests that feed externally, this tactic is used successfully for cerambycid pest management in postharvest treatments, invasive pest eradication programs (see Chapter 13), and in plant production programs. This chapter describes the main classes of insecticides and illustrates their application methods and formulations in cerambycid pest management. For example, field sprays and trunk coatings during the early breeding season of cerambycids can control adults and sometimes neonate larvae; injection or insertion of organic or inorganic insecticides with systemic or fumigant properties into trunks can kill adults and/ or larvae, and soil and root treatments with contact, ingestion, and systemic insecticides can kill both larvae and adults. When insecticides are applied to food crops, such as fruit trees and field crops, insecticide residues in food should be monitored closely. Future studies also should address issues about pest resistance and resurgence as well as nontarget effects (such as on bees) in relation to insecticide applications for cerambycid pest control.

ACKNOWLEDGMENTS

I thank W. Lu for providing numerous references in Chinese and R.A. Haack for constructive comments on an earlier manuscript.

REFERENCES

Achterberg, K. van, and M. R. Mehrnejad. 2011. A new species of *Megalommum* Szepligeti (Hymenoptera: Braconidae: Braconinae), a parasitoid of the pistachio longhorn beetle (*Calchaenesthes pistacivora* Holzschuh; Coleoptera, Cerambycidae) in Iran. *Zookeys* 112: 21–38.

Ahmed, K. U., M. M. Rahman, M. Z. Alam, M. M. Hossain, and M. G. Miah. 2013. Evaluation of some control methods against the jackfruit trunk borer, *Batocera rufomaculata* De Geer (Cerambycidae: Coleoptera). *Bangladesh Journal of Zoology* 41: 181–187.

Ahmed, K. U., M. M. Rahman, M. Z. Alam, M. M. Hossain, and M. G. Miah. 2014. Evaluation of relative host preference of *Batocera rufomaculata* De Geer on different age stages of jackfruit trees. *Asian Journal of Scientific Research* 7: 232–237.

Andrews, G. L., and R. L. Williams. 1988. An estimate of the effect of soybean stem borer on yields. *Technical Bulletin, Mississippi Agricultural and Forestry Experiment Station* 153: 1–5.

Asogwa, E. U., J. C. Anikwe, T. C. N. Ndubuaku, and F. A. Okelana. 2009. Distribution and damage characteristics of an emerging insect pest of cashew, *Plocaederus ferrugineus* L. (Coleoptera: Cerambycidae) in Nigeria: A preliminary report. *African Journal of Biotechnology* 8: 53–58.

Barrera, J. F. 2008. Coffee pests and their management. In *Encyclopedia of entomology*, ed. J. L. Capinera, 961–998 (2nd ed.). Dordrecht: Springer.

BASF. 2013. *Insecticide mode of action—Technical training manual*. Research Triangle Park, NC: BASF Corporation.

Bhat, J. A., N. A. Wani, G. M. Lone, and M. S. Pukhta. 2010a. Influence of apple cultivars, age, and edaphic factors on infestation by the stem borer, *Aeolesthes sarta* Solsky. *Indian Journal of Applied Entomology* 24: 89–92.

Bhat, M. A., M. A. Mantoo, and F. A. Zaki. 2010b. Management of tree trunk borer, *Aeolesthes sarta* (Solsky) infesting apple in Kashmir. *Journal of Insect Science (Ludhiana)* 23: 124–128.

Brown, A. E., and E. Ingianni. 2013. Mode of action of insecticides and related pest control chemicals for production agriculture, ornamentals, and turf. *University of Maryland Extension Pesticide Information Leaflet* 41: 1–13.

Campbell, W. V., and J. W. Vanduyn. 1977. Cultural and chemical control of *Dectes texanus texanus* (Coleoptera: Cerambycidae) on soybeans. *Journal of Economic Entomology* 70: 256–258.

Cavalieri, G. 2013. Summary of 2008–2011 trials on the possibility of control of *Anoplophora chinensis* with pesticides. *Journal of Entomological and Acarological Research* 45: 19–20.

Chakraborti, S. 2011. Management of the early stages of cashew stem and root borer, *Plocaederus ferrugineus* L. (Cerambycidae, Coleoptera). *Journal of Entomological Research* (New Delhi) 35: 203–207.

Chandrasekaran, J., R. S. Azhakiamanavalan, and I. H. Louis. 1990. Control of grapevine borer, *Coelosterna scabrator* F. (Cerambycidae: Coleoptera). *South Indian Horticulture* 38: 108–108.

Chaudhary, H. R., and N. N. Tripathi. 2011. Evaluation of some insecticides against major insect pests infesting soybean. *Indian Journal of Applied Entomology* 25: 53–55.

Chen, J. T. 1958. Morphology and habit of six main borers of the tea bush. *Acta Entomologica Sinica* 8: 272–280 [in Chinese].

Chen, W.F. 2002. The occurrence characters of citrus star borer and its integrated control method. *South China Fruits* 31(2): 14–15 [in Chinese].

Chen, H. L., L. Y. Dong, C. L. Zhou, X. L. Liu, and F. Lian. 2002. Studies on the biology and control of *Bacchisa atritarsis* Pic. *Jiangxi Plant Protection* (3): 65–67 [in Chinese].

Chen, Q., F. P. Lu, H. Lu, et al. 2012. Phenology and control of *Dorysthenes granulosus* and *Anomala corpulenta* on cassava. *Chinese Journal of Tropical Crops* 33: 332–337 [in Chinese].

Crowe, T. J. 2012. Coffee pests in Africa. In *Coffee: Growing, processing, sustainable production—A guidebook for growers, processors, traders and researchers*, ed. J. N Wintgens, 425–462 (2nd rev. ed.). Hoboken, NJ: Wiley.

Dahiphale, K. D., D. S. Suryawanshi, S. K. Kamble, S. P. Pole, and I. A. Madrap. 2007. Effect of new insecticides against the control of major insect pests and yield of soybean [*Glycine max* (L.) Merrill]. *Soybean Research* 5: 87–90.

Dent, D. 2000. *Insect pest management*, 2nd ed. New York: CABI.

Devine, G. J., and M. J. Furlong. 2007. Insecticide use: Context and ecological consequences. *Agriculture and Human Values* 24: 281–306.

Ding, L. F., J. Y. Ma, H. B. Du, Y. N. Chen, and H. M. Chen. 2014. Comparison of several pesticides in the effectiveness for the control of mango stem borer (*Rhytidodera bowringi* White). *China Tropical Agriculture* 2014(1): 49–50 [in Chinese].

Duffy, E. A. J. 1968. *A monograph of the immature stages of Oriental timber beetles (Cerambycidae)*. London: British Museum (Natural History).

Dunn, G., and B. Zurbo. 2014. Grapevine pests and their management. PRIMEFACT 511 (2nd ed.), NSW Department of Primary Industries. http://www.dpi.nsw.gov.au/__data/assets/pdf_file/0010/110998/Grapevine-pests-and-their-management.pdf (accessed April 30, 2016).

EPPO. 2014. *Apriona germari*. *EPPO Bulletin* 44: 155–158.

Esaki, K. 1995. Ovipositional characteristics of *Apriona japonica* Thomson (Coleoptera: Cerambycidae) in a *Zelkova serrata* Makino plantation. *Journal of the Japanese Forestry Society* 77: 596–598 [in Japanese].

Esaki, K. 2007. Management of *Apriona japonica* Thomson (Coleoptera: Cerambycidae) adults by spraying feeding trees with fenitrothion. *Journal of the Japanese Forest Society* 89: 61–65 [in Japanese].

Espinosa, A. E. 2000. Control of the coffee shoot borer *Plagiohammus maculosus* in the region of Santa Clara. II: *Simposio Latinoamericano sobre caficultura*, 81–83. Costa Rica: Instituto Interamericano de Ciencias Agricolas.

Eyre, D., R. Cannon, D. McCann, and R. Whittaker. 2010. Citrus longhorn beetle, *Anoplophora chinensis*: An invasive pest in Europe. *Outlooks on Pest Management* 21: 195–198.

Gao, Z. S., J. Chen, C. S. Zhu, D. Y. Zhou, and P. C. Lou. 1988. Control of cerambycids on mulberry trees. *Acta Sericologica Sinica* 14: 188–190 [in Chinese].

Gong, H. L., Y. X. An, C. X. Guan, et al. 2008. Pest status and control strategies of *Dorysthenes granulosus* (Thomson) on sugarcane in China. *Sugarcane and Canesugar* (5): 1–5, 38 [in Chinese].

Goodwin, S. 2005a. Chemical control of fig longicorn, *Acalolepta vastator* (Newman) (Coleoptera: Cerambycidae), infesting grapevines. *Australian Journal of Entomology* 44: 71–76.

Goodwin, S. 2005b. A new strategy for the chemical control of fig longicorn, *Acalolepta vastator* (Newman) (Coleoptera: Cerambycidae), infesting grapevines. *Australian Journal of Entomology* 44: 170–174.

Goodwin, S., and N. Ahmad. 1998. Relationship between azinphos-methyl usage and residues on grapes and in wine in Australia. *Pesticide Science* 53: 96–100.

Goodwin, S., and M. A. Pettit. 1994. *Acalolepta vastator* (Newman) (Coleoptera, Cerambycidae) infesting grapevines in the Hunter Valley, New South Wales. 2. Biology and ecology. *Journal of the Australian Entomological Society* 33: 391–397.

Goodwin, S., M. A. Pettit, and L. J. Spohr. 1994. *Acalolepta vastator* (Newman) (Coleoptera: Cerambycidae) infesting grapevines in the Hunter Valley, New South Wales. 1. Distribution and dispersion. *Journal of the Australian Entomological Society* 33: 385–390.

Gupta, M. P. 2008. Efficacy and economics of biorationals and their admixtures against incidence of major insect pests of soybean. *Annals of Plant Protection Sciences* 16: 282–288.

Haack, R. A., F. Hérard, J. H. Sun, and J. J. Turgeon. 2010. Managing invasive populations of Asian longhorned beetle and citrus longhorned beetle: A worldwide perspective. *Annual Review of Entomology* 55: 521–546.

Hao, C., C. X. Yan, G. H. Zhang, and G. F. Yu. 1999. Morphological anatomy of alimentary canal and reproductive system of *Dorysthenes paradoxus* Faldermann. *Journal of Shanxi Agricultural University* 19: 7–13 [in Chinese with English abstract].

Hu, J., S. Angeli, S. Schuetz, Y. Luo, and A. E. Hajek. 2009. Ecology and management of exotic and endemic Asian longhorned beetle *Anoplophora glabripennis*. *Agricultural and Forest Entomology* 11: 359–375.

Jagginavar, S. B., N. D. Sunitha, and D. R. Paitl. 2008. Management strategies for grape stem borer *Celosterna scabrator* Fabr. (Coleoptera: Cerambycidae). *Indian Journal of Agricultural Research* 42: 307–309.

Jain, N., and D. Sharma. 2011. Evaluation of yield losses by girdle beetle and its management in soybean crop in Kota region, Rajasthan, India. *Life Science Bulletin* 8: 123–125.

Jalgaonkar, V. N., S. A. Chavan, P. D. Patil, and K. V. Naik. 2015. Comparative efficacy of insecticides for control of cashew stem and root borer (*Plocaederus ferruginous* Feb.). *Acta Horticulturae* 1080: 445–448.

Jayanti, L., R. G. Shali, and K. Nanda. 2014. Studies on seasonal incidence, nature of damage and assessment of losses caused by girdle beetle on *Glycine max*. *Annals of Plant Protection Sciences* 22: 320–323.

Jia, X. Y., G. Y. Ma, L. G. Wang, W. Liang, and H. Wen. 2001. Integrated control of walnut pests. *China Fruits* (1): 39–40 [in Chinese].

Jin, F., B. Z. Ji, S. W. Liu, L. Tian, and J. Gao. 2011. Influence of oviposition secretion of *Apriona germari* Hope (Coleoptera: Cerambycidae) on water content, pH level and microbial quantity in its incisions on paper mulberry (*Broussonetia papyrifera*). *Acta Entomologica Sinica* 54: 477–482 [in Chinese].

Jiroux, E. 2011. Révision du genre *Apriona* Chevrolat, 1852 (Coleoptera: Cerambycidae: Lamiinae: Batocerini). *Les Cahiers Magellanes* 5: 1–103.

Johnson, T. A., and R. C. Williamson. 2007. Potential management strategies for the linden borer (Coleoptera: Cerambycidae) in urban landscapes and nurseries. *Journal of Economic Entomology* 100: 1328–1334.

Kessler, S. C., E.J. Tiedeken, K.L. Simcock, et al. 2015. Bees prefer foods containing neonicotinoid pesticides. *Nature* 521: 74–76.

Kikuchi, M. 1976. Control of insect pests of mulberry in Japan. *Japan Pesticide Information* 29: 9–11.

Kumar, U., S. P. Kumar, and S. Surabhi. 2012. Spectrum of insect pest complex of soybean (*Glycine max* (L) Merrill) at Lambapeepal village in Kota region, India. *International Research Journal of Biological Sciences* 1: 80–82.

Kumari, D. A., and D. Vijaya. 2015. Management of stem borer, *Coelosterna scrabrator* Fabr. in grapevine. *Plant Archives* 15: 1089–1091.

Lan, C. C., and J. N. Wintgens. 2012. Major pests of coffee in the Asia-Pacific region. In *Coffee: Growing, processing, sustainable production*, ed. J.N. Wintgens, 459–473. Weinheim: WILEY-VCH Verlag GmbH.

Li, J. Q. 2009. Biocontrol of *Batocera horsfieldi* (Coleoptera: Cerambycidae) with parasitoid *Dastarcus helophoroides* (Coleoptera: Bothrideridae). Ph.D. Thesis, Northwest A&F University of Science and Technology, Yanglin, China [in Chinese].

Li, K. Z. 1996. Poplar stem-boring pests and their control. *Journal of Northeast Forestry University* 7: 51–61 [in Chinese].

Li, Y. F. 2008. Comparative experiment on controlling *Anoplophora chinensis* on the trees such as Buerger maple with several insecticides. *Journal of Anhui Agricultural Science* 36: 4166–4169 [in Chinese].

Lian, C. J., G. H. Li, G. W. Li, R. T. Gao, Z. Y. Zhao, and J. Z. Sun. 1997. Study on the use of systemic and pyrethroid insecticides to control *Anoplophora glabripennis* and *Apriona germari*. *Forest Research* 10: 189–193 [in Chinese].

Liao, H. H., M. Lu, L.Y. Bu, H. H. Lin, and Z. X. Mo. 2006. *Dorysthenes granulosus* (Thomson) causes serious damage to sugarcanes in Beihai City, Guangxi. *Guangxi Plant Protection* 19: 22–24 [in Chinese].

Liu, Z., J. Fan, and K. Zhang. 1999. The experiments of prevention and control of *Anoplophora glabripennis* in *Salix matsudana* with insecticide. *Journal of Hubei Agricultural University* 19: 313–314 [in Chinese].

Lonagre, S. G., A. Y. Thakare, and P. B. Salunke. 2015. Efficacy of different pyrethroid alone and in combination with dimethoate against girdle beetle of soybean. *Journal of Food Legumes* 28: 66–68.

Long, H. N., and Q. X. Wei. 2007. Occurrence and control of *Dorysthenes granulosus* in Liuzhou. *Guangxi Plant Protection* 20: 41–42 [in Chinese].

Lundin, O., M. Rundlof, H. G. Smith, I. Fries, and R. Bommarco. 2015. Neonicotinoid insecticides and their Impacts on bees: A systematic review of research approaches and identification of knowledge gaps. *PLoS One* 10: e0136928.

Luo, Y. M., S. M. Cai, and Q. A. Jin. 1990. *Rhytidodera bowringii* White in Hainan Island. *Chinese Journal of Tropical Crops* 11: 107–112 [in Chinese].

Lv, K. S., X. M. Lu, Z. P. Huang, Y. L. Liu, and Z. H. Pang. 2011. The preliminary survey on the pests and diseases of *Camellia oleifera* branches in Liuzhou City. *Guangdong Forestry Science and Technology* 27: 58–62 [in Chinese].

Maruthadurai, R., A. R. Desai, H. R. C. Prabhu, and N. P. Singh. 2012. Insect pests of cashew and their management. *Technical Bulletin of ICAR (Research Complex for Goa, India)* 28: 1–16.

Matthews, G. A., R. Bateman, and P. Miller. 2014. *Pesticide application methods* (4th ed.). Hoboken, NJ: Wiley.

Mazaheri, A., J. Khajehali, M. Kashkouli, and B. Hatami. 2015. Laboratory and field evaluation of insecticides for the control of *Aeolesthes sarta* Solsky (Col.: Cerambycidae). *Journal of Crop Protection* 4: 257–266.

Mehrnejad, M. R. 2014. Pest problems in pistachio producing areas of the world and their current means of control. *Acta Horticulturae (ISHS)* 1028: 163–169.

Meng, P. S., K. Hoover, and M. A. Keena. 2015. Asian longhorned beetle (Coleoptera: Cerambycidae), an introduced pest of maple and other hardwood trees in North America and Europe. *Journal of Integrated Pest Management* 6(1): article 4.

Michaud, J. P., and A. K. Grant. 2010. Variation in fitness of the longhorned beetle, *Dectes texanus*, as a function of host plant. *Journal of Insect Science* 10: article 206.

Mohapatra, L. N., and R. N. Mohapatra. 2004. Distribution, intensity and damage of cashew stem and root borer, *Plocaederus ferrugineus* in Orissa. *Indian Journal of Entomology* 66: 4–7.

Mohapatra, R. N., and B. C. Jena. 2007. Evaluation of granular and spray formulations of insecticides against cashew stem and root borer *Plocaederus ferrugineus* L. *Journal of Entomological Research* 31: 205–207.

Mohapatra, R. N., and B. C. Jena. 2008. Effect of prophylactic measures in management of cashew stem and root borer (*Plocaederus ferrugineus* L.). *Journal of Plantation Crops* 36: 140–141.

Naik, L. K., S. B. Jagginavar, and A. P. Biradar. 2011. Beetle enemies of pomegranate and their management. *Acta Horticulturae* 890: 565–567.

Netam, H. K., R. Gupta, and S. Soni. 2013. Seasonal incidence of insect pests and their biocontrol agents on soybean. *IOSR Journal of Agriculture and Veterinary Science* 2: 7–11.

Nowak, D. J., J. E. Pasek, R. A. Sequeira, D. E. Crane, and V. C. Mastro. 2001. Potential effect of *Anoplophora glabripennis* (Coleoptera: Cerambycidae) on urban trees in the United States. *Journal of Economic Entomology* 94: 116–122.

Pachori, R., and D. Sharma. 1997. Changing host preference of *Obereopsis brevis* (Swed.) as a new pest of sunflower in India. *Journal of Entomological Research* 21: 195–196.

Padi, B., and R. Adu-Acheampong. 2001. An integrated pest management strategy for the control of the coffee stem borer, *Bixadus sierricola* White (Coleoptera: Lamiidae). *19eme Colloque Scientifique International sur le Cafe*, Trieste, Italy, 14–18 Mai 2001: 1–5.

Pan, C. R. 1999. Experiment on prevention and cure of *Apriona germarii* in poplar by injection method. *Journal of Zhejiang Forestry Science and Technology* 19: 56–57 [in Chinese].

Pan, H., Y. Lian, L. Yang, and C. Wang. 2001. Experiment on the application of microcapsules against *Anoplophora glabripennis* in poplars. *Forest Pest and Disease* 20: 17–18 [in Chinese].

Pande, A. K., S. V. Mahajan, A. T. Munde, and S. M. Telang. 2000. Evaluation of insecticides against major pests of soybean (*Glycine max* L.). *Journal of Soils and Crops* 10: 244–247.

Parsai, S. K., S. K. Shrivastava, and S. S. Shaw. 1990. Comparative efficacy and economics of synthetic pyrethroids in the management of soybean girdle beetle. *Current Research—University of Agricultural Sciences (Bangalore)* 19: 134–136.

Pearce, F. 2015. Bees win as US court rules against neonicotinoid pesticide. *New Scientist.* https://www.newscientist.com/article/dn28167-bees-win-as-us-court-rules-against-neonicotinoid-pesticide/ (accessed April 28, 2016).

Pliansinchai, U., V. Jarnkoon, S. Siengsri, C. Kaenkong, S. Pangma, and P. Weerathaworn. 2007. Ecology and destructive behaviour of cane boring grub (*Dorysthenes buqueti* Guerin) in north eastern Thailand. *XXVI: Congress, International Society of Sugar Cane Technologists, ICC, Durban, South Africa, 29 July–2 August,* 2007: 863–867.

Poland, T. M., R. A. Haack, T. R. Petrice, D. L. Miller, L. S. Bauer, and R. Gao. 2006. Field evaluations of systemic insecticides for control of *Anoplophora glabripennis* (Coleoptera: Cerambycidae) in China. *Journal of Economic Entomology* 99: 383–392.

Purwar, J. P, and S. R. Yadav. 2004. Effect of bio-rational and chemical insecticides on stem borers and yield of soybean (*Glycine max* (L.) Merrill). *Soybean Research* 2: 54–60.

Qian, T. Y. 1984. Descriptions of the larvae of four species of tea tree borers (Coleoptera: Cerambycidae). *Wuyi Science Journal* 4: 189–193 [in Chinese].

Qin, H. Z., Y. M. Ju, and W. Z. Wang. 1997. A study on control of damage of *Apriona germari* (Hope) on *Ficus carica*. *Journal of Shanghai Agricultural College* 15: 239–242 [in Chinese].

Rad, H. H. 2006. Study on the biology and distribution of long-horned beetles *Calchaenesthes pistacivora* n. sp (Col.: Cerambycidae): A new pistachio and wild pistachio pest in Kerman Province. *Acta Horticulturae (ISHS)* 726: 425–430.

Rad, H. H., R. Zeydabadi, A. Rajabi, H. Bahreyney, and M. Morovati. 2005. Study of some insecticide effects on the long-horned beetle *Calchaenesthes pistacivora*. AGRIS, FAO, http://agris.fao.org/agris-search/search.do?recordID=IR2008000725 (accessed April 25, 2016).

Rogers, C. E. 1985. Cultural management of *Dectes texanus* (Coleoptera, Cerambycidae) in sunflower. *Journal of Economic Entomology* 78: 1145–1148.

Rundlof, M., G. K. S. Andersson, R. Bommarco, et al. 2015. Seed coating with a neonicotinoid insecticide negatively affects wild bees. *Nature* 521: 77–80.

Rutherford, M. A., and N. Phiri. 2006. *Pests and diseases of coffee in Eastern Africa: A technical and advisory manual*. Wallingford: CAB International.

Rystrom, Z. D. 2015. Seasonal activity and sampling methods for the Dectes stem-borer, *Dectes texanus* LeConte, in Nebraska soybeans. MSc. Thesis, University of Nebraska, Lincoln, Nebraska.

Sahu, K. R., B. C. Shukla, B. S. Thakur, S. K. Patil, and R. R. Saxena. 2012. Relative tolerance of cashew cultivars against cashew stems and root borer (CSRB), *Plocaederus ferrugineus* L. in Chhattisgarh. *Uttar Pradesh Journal of Zoology* 32: 311–316.

Sawyer, A. 2007. Incidence of Asian longhorned beetle infestation among treated trees in New York. In *Proceedings of the 2006 Emerald Ash Borer and Asian Longhorned Beetle Research and Technology Development Review Meeting*. USDA Forest Service, Forest Health Enterprise Team, Morgantown, WV, 106–107. FHTET-2007-04.

Schoeman, P. S. 1995. Integrated control of arabica coffee pests, Part II: An overview of measures to control the white coffee stemborer, *Monochamus leuconotus* (Pascoe) (Coleoptera: Cerambycidae). *Inligtingsbulletin—Instituut vir Tropiese en Subtropiese Gewasse* 280: 11–20.

Schoeman, P. S., and B. Pasques. 1993. Chemicals tested against the white coffee stemborer. *Inligtingsbulletin—Instituut vir Tropiese en Subtropiese Gewasse* 252: 16–18.

Sharma, A. N. 1994. Quantifying girdle beetle resistance in soybean. *Soybean Genetics Newsletter* 21: 124–127.

Sharma, A. N., G. K. Gupta, R. K. Verma, et al. 2014. *Integrated pest management for soybean*. New Delhi: National Centre for Integrated Pest Management.

Shui, S. Y., J. B. Wen, M. Chen, X. I. Hu, F. Liu, and J. Li. 2009. Chemical control of *Apriona germari* (Hope) larvae with zinc phosphide sticks. *Forestry Studies in China* 11: 9–13.

Singh, D. P., and K. M. Singh. 1966. Certain aspects of bionomics and control of *Oberea brevis* S., a new pest of beans and cow peas. *Labdev Journal of Science and Technology* 4: 174–177.

Singh, O. P. 1986. Toxicity of insecticides on the eggs and grubs of girdle beetle *Oberea brevis* Swed. a pest of soybean in Madhya Pradesh. *Legume Research* 9: 11–15.

Singh, O. P., and S. V. Dhamdhere. 1983. Girdle beetle a major pest of soybean. *Indian Farming* 32(10): 35–36.

Singh, O. P., and K. J. Singh. 1989. Efficacy and economics of granular insecticides in the management of the stem borers of soybean in Madhya Pradesh, India. *Pestology* 13: 8–12.

Sloderbeck, P. E., and L. L. Buschman. 2011. Aerial insecticide treatments for management of Dectes stem borer, *Dectes texanus*, in soybean. *Journal of Insect Science* 11: article 49.

Smith, M. T., J. J. Turgeon, P. de Groot, and B. Gasman. 2009. Asian longhorned beetle *Anoplophora glabripennis* (Motschulsky): Lessons learned and opportunities to improve the process of eradication and management. *American Entomologist* 55: 21–25.

Sommartya, P., and W. Suasa-ard. 2006. Sugarcane longhorn stem borer *Dorysthenes buqueti* (Coleoptera: Cerambycidae) and its natural enemies in Thailand (abstract). *ISSCT Entomology Workshop*, 14–24 May 2006, Cairns, Australia.

Sommartya, P., W. Suasa-ard, and A. Puntongcum. 2007. Natural enemies of sugarcane longhorn stem borer, *Dorysthenes buqueti* Guerin (Coleoptera: Cerambycidae), in Thailand. *XXVI: Congress, International Society of Sugar Cane Technologists, ICC*, Durban, South Africa, 29 July–2 August, 2007: 858–862.

Sun, X., B. Wang, H. Cong, J. Li, and R. Yan. 2003. The experiment of contact-broken microcapsule to control longhorned beetle. *Henan Forestry Science and Technology* 23: 9–10 [in Chinese].

Tan, J. C., J. W. Zhang, N. W. Xiao, T. Z. Yuan, and X. Deng. 2003. A list of tea pest insects and mites in Hunan Province. *Journal of Hunan Agricultural University (NaturalSciences)* 29: 296–307 [in Chinese].

Thongphak, D., and C. Hanlaoedrit. 2012. Taxonomy of sugarcane longhorn beetle *Dorysthenes* spp. derived from North and Northeast of Thailand. *KKU Research Journal* 17: 1–8 [in Thai].

Ugine, T. A., S. Gardescu, P. A. Lewis, and A. E. Hajek. 2012. Efficacy of imidacloprid, trunk-injected into *Acer platanoides*, for control of adult Asian longhorned beetles (Coleoptera: Cerambycidae). *Journal of Economic Entomology* 105: 2015–2028.

Upadhyay, S., and S. Sharma. 2000. Efficacy of some insecticides against eggs and grubs of soybean girdler (*Oberea brevis*) in the field. *Indian Journal of Agricultural Sciences* 70: 57–58.

Upadhyay, S. K., B. Chaudhary, and B. Sapkota. 2013. Integrated management of mango stem borer *(Batocera rufomaculata* Dejan) in Nepal. *Global Journal of Biology, Agriculture and Health Sciences* 2: 132–135.

USDA APHIS. 2008. New pest response guidelines—Asian longhorned beetle *Anoplophora glabripennis*. USDA APHIS Plant Protection and Quarantine, Riverdale, MD.

Vasanthi, P., and T. N. Raviprasad. 2013. Biology and morphometrics of cashew stem and root borers (CSRB) *Plocaederus ferrugenius* and *Plocaederus obesus* (Coleoptera: Cerambycidae) reared on cashew bark. *International Journal of Scientific and Research Publications* 3: 1–7.

Veda, O. P., and S. S. Shaw. 1994. Record of stem boring beetle on sunflower in Jhabua Hills of M. P. *Current Research—University of Agricultural Sciences* 23: 118–119.

Venkatesha, M. G., and A. S. Dinesh. 2012. The coffee white stemborer *Xylotrechus quadripes* (Coleoptera: Cerambycidae): Bioecology, status and management. *International Journal of Tropical Insect Science* 32: 177–188.

Waller, J. M., M. Bigger, and R. J. Hillocks. 2007. Stem- and branch-borers. In *Coffee pests, diseases and their management*, eds. J. M. Waller, M. Bigger, and R. J. Hillocks, 41–67. Oxfordshire (UK): CABI Publishing.

Wang, B., R. Gao, V. C. Mastro, and R. C. Reardon. 2005. Toxicity of four systemic neonicotinoids to adults of *Anoplophora glabripennis* (Coleoptera: Cerambycidae). *Journal of Economic Entomology* 98: 2292–300.

Wang, F. X., J. B. Hong, and X. S. Cheng. 2015. Control effects of several neonicotinoid insecticides on *Anoplophora chinensis*. *Journal of Henan Agricultural Sciences* 44(10): 100–103 [in Chinese].

Wang, H. Z. 2016. Experiment on application of omethoate injection for the control of *Anoplophora glabripennis* adults in poplars. *Protection Forest Science and Technology* (1): 33–34 [in Chinese].

Wang, S. L., A. Q. Wang, M. H. Xia, and S. H. Dong. 2004. Study on the occurrence of *Batocera horsfieldi* on walnut and its control. *China Fruits* 211–13 [in Chinese].

Williams, G. R., A. Troxler, G. Retschnig, et al. 2015. Neonicotinoid pesticides severely affect honey bee queens. *Scientific Reports* 5: 14621.

Wu, J. Q., and M. T. Smith. 2015. Lethal effects of lambda-cyhalothrin and its commercial formulation on Asian longhorned beetle (Coleoptera: Cerambycidae): Implications for population suppression, tree protection, eradication, and containment. *Journal of Economic Entomology* 108: 150–156.

Wu, Z. D., X. X. He, and L. L. Fang. 2000. Chemical control of *Populus nigra* trunk borer (*Anoplophora chinensis*). *Journal of Zhejiang Forestry College* 17: 106–108 [in Chinese].

Xiao, Y. Y. 2012. How to control *Dorysthenes granulosus*? *Information on Pesticide Market* (4): 45–45 [in Chinese].

Xiong, G. R., Z. P. Li, G. L. Feng, et al. 2010. Occurrence and control of sugarcane pests in Hainan Province. *Chinese Journal of Tropical Crops* 31: 2243–2249 [in Chinese].

Xu, Z. C., X. Chen, and H. Tian. 1999. Experiment on the control of *Anoplophora glabripennis* (Motsch.) and *A. nobilis* Ganglbauer with confidor. *Forest Pest and Disease* 18: 11–13 [in Chinese].

Xu, Z. C., H. Tian, X. Y. Chen, J. B. Wen, and S. C. Yan. 2003. Efficacy of trunk coating insecticides in the control of poplar longhorned beetles. *Forest Pest and Disease* 22: 17–20 [in Chinese].

Yadav, M. K., S. M. Matkar, A. N. Sharma, M. Billore, K. N. Kapoor, and G. L. Patidar. 2001. Efficacy and economics of some new insecticides against defoliators and stem borers of soybean (*Glycine max* (L.) Merrill). *Crop Research* (Hisar) 21: 88–92.

Yan, C. X., C. Hao, and M. Q. Wang. 1997. A preliminary study on *Dorysthenes paradoxus*. *Journal of Shanxi Agricultural University* 17: 342–345 [in Chinese].

Yang, C., H. Y. Chen, and Z. Q. Huang. 1992. Integrated management of *Dorysthenes granulosus* Thomson. *Sugarcane and Canesugar* (1): 11–15 [in Chinese].

Yang, D. W., and Q. Z. Liu. 2010. Tea longhorn beetle damage and its control emphasis in organic tea gardens. *Journal of Anhui Agricultural Sciences* 38(3): 1305, 1312 [in Chinese].

Yang, Q. L., M. C. Zhang, G. M. Wang, and H. P. Zhang. 2005. Effectiveness of using 8% cypermethrin for control of *Apriona germarii*. *China Fruits* (1): 54–54 [in Chinese].

Yu, Y. H., R. Z. Chen, D. W. Wei, X. F. Chen, T. Zeng, and Z. Y. Wang. 2009. Trapping and killing of *Dorysthenes granulosus* adults using frequency vibration-killing lamps. *Guangxi Agricultural Sciences* 40: 1552–1554 [in Chinese].

Yu, Y. H., T. Zeng, D. W. Wei, X. R. Zeng, L. F. Li, and Z. Y. Wang. 2006. Damage status and control strategies of *Dorysthenes granulosus* (Thomson) in Guangxi sugarcane region. *Guangxi Agricultural Sciences* 37: 545–547 [in Chinese].

Yu, Y. H., T. Zeng, D. W. Wei, X. R. Zeng, L. F. Li, and Z. Y. Wang. 2007. Sugarcanes in Guangxi are seriously damaged by *Dorysthenes granulosus* in recent years. *Plant Protection* 33: 136–137 [in Chinese].

Yu, Y. H., T. Zeng, X. R. Zeng, D. W. Wei, Z. Y. Wang, and S. X. Ren. 2010. Virulence of *Metarhizium* strains to the larvae of *Dorysthenes granulosus* Thomson. *Chinese Journal of Biological Control* 26(Suppl.): 8–13 [in Chinese].

Zeng, C. F., and Q. S. Huang. 1981. Preliminary study on occurrence and control of *Dorysthenes granulosus*. *Entomological Knowledge (Kunchong Zhishi)* (1): 18–20 [in Chinese].

Zhang, C. 1992. Damage to mango seedlings by *Dorysthenes granulosus*. *Guangxi Agricultural Sciences* (6): 259–259 [in Chinese].

Zhang, C., A. Yan, C. Xia, J. Tang, and Z. Li. 1999a. A study on controlling tests of contacted-breaking micro-capsules to *Anoplophora nobilis* Ganglbauer and *A. glabripennis* (Motch). *Journal of Nanjing Forestry University* 23: 73–75 [in Chinese].

Zhang, S. H., X. Xia, and H. Shu. 1994. An experimental study on the control of *Anoplophora glabripennis* (Motschulsky) with stem injection of insecticides. *Journal of Inner Mongolia Institute of Agriculture and Animal Husbandry* 15: 15–20 [in Chinese].

Zhang, S. P., D. Z. Huang, and J. J. Yan. 1999b. The sequence choice of the main tree species to *Apriona germarii* in northern part of China. *Journal of Northeast Forestry University* 27: 17–21 [in Chinese].

Zhao, K. J., Y. M. Zhou, H. B. Zhu, K. Z. Xu, and H. X. Sun. 1994. Study on the control of the mulberry borer (*Apriona germari*) using a fumigating rod. *Plant Protection* 20: 10–11 [in Chinese].

Zhao, X. R., G. Li, J. Zhang, Z. Z. Li, J. Z. Zhao, and X. Q. Guan. 1995. Study on harm behavior of *Anoplophora glabripennis* Motsth and control effect by new type of insecticide–Mothcide II, IV. *Journal of Ningxia Agricultural College* 16: 24–26 [in Chinese].

Zhong, G. X., Y. Wu, L. Chen, J. W. Wang, R. T. Qian, and J. Zhu. 2012. A chemical method for controlling *Anoplophora chinensis* in bonsai plants. *Plant Quarantine* (Shanghai) 26(2): 28–31.

Zhu, J. H., and M. Y. Xu. 1996. Damage to longan and citrus by Dorysthenes granulosus. *Plant Protection* 22: 38–38 [in Chinese].

Zhu, J. W. 1995. Damage to *Eucalyptus citriodora* by *Dorysthenes granulosus*. *Guangxi Forestry* (1): 19 [in Chinese].

Zhu, J. W. 2012. Preliminary study on damage to *Citrus sinensis* (L.) cv. Hong Jiang Cheng by *Dorysthenes granulosus*. *Agriculture and Technology* (10): 109 [in Chinese].

Zhu, M. Y., L. Q. Duan, S. J. Feng, L. Cong, and D. G. Liu. 1998. The chemical control of *Anoplophora glabripennis* (Coleoptera: Cerambycidae). *Journal of Inner Mongolia Forestry College* 20: 64–67 [in Chinese].

Zhu, Z. C., X. B. Wang, J. G. Tang, and S. F. Min. 1995. Studies on the relationship between insecticide formulation and its toxicity and persistence in controlling forest insect pests. *Scientia Silvae Sinicae* 31: 330–337 [in Chinese].

11

Cerambycid Pests in Forests and Urban Trees

Robert A. Haack
USDA Forest Service
Lansing, Michigan

CONTENTS

11.1 Introduction

There are more than 36,000 species of Cerambycidae (Coleoptera) recognized worldwide (see Chapter 1), and they are found on all continents except Antarctica (Linsley 1959, 1961). Nearly all cerambycids are phytophagous, feeding primarily on woody plants, although some species do feed on herbaceous plants (see Chapter 3). Cerambycids develop in nearly all parts of woody plants, especially in roots, trunks, and branches, but occasionally also in seeds, pods, cones, and leaves. In addition, cerambycid larvae develop in nearly all major tissues in woody plants, including outer bark, inner bark, cambium, sapwood, heartwood, and pith (see Chapter 3).

Cerambycids utilize a wide diversity of woody plants as larval hosts, but certain plant families serve as hosts to many cerambycid species, while others are rarely used. For example, the number of cerambycid species that utilize various plant families as larval hosts is listed in Table 11.1 for four distinct world regions in the Northern Hemisphere where there is good knowledge of the larval hosts for most cerambycids: Montana, Fennoscandia, Israel, and Korea. (Fennoscandia refers to the countries of Norway, Sweden, Finland, and a small part of neighboring Russia.) The sources used to obtain the host data are listed in the footnotes for Table 11.1. Overall, 44 plant families were identified as larval hosts for the cerambycids of Montana, 23 for Fennoscandia and Denmark, 45 for Israel, and 44 for Korea (Table 11.1). Among the top 10 plant families in each of these world regions were five plant families that all regions had in common (Fagaceae, Pinaceae, Rosaceae, Salicaceae, and Ulmaceae), and an additional three families that were common to at least three of the four world regions (Betulaceae, Fabaceae, and Juglandaceae). Many of the species in these plant families are common trees that dominate the temperate forests in the Northern Hemisphere (Daubenmire 1978).

Cerambycids infest trees in a wide variety of host conditions (Haack and Slansky 1987; Mattson and Haack 1987; Hanks 1999). Some cerambycids infest living trees that vary in condition from healthy to stressed, including many species of *Anoplophora, Enaphalodes, Goes, Lamia, Megacyllene, Oberea, Oncideres, Plectrodera*, and *Saperda*. By contrast, many species of *Arhopalus, Ergates, Parandra*, and *Rhagium* commonly infest dead trees (Craighead 1923; Linsley 1959; Bílý and Mehl 1989; Solomon 1995). In addition, some dead-wood infesting species prefer moist wood (*Mallodon* and *Rhagium*), while others prefer dry wood (*Chlorophorus, Hylotrupes*, and *Stromatium*) (Craighead 1923; Duffy 1953; Linsley 1959; Bense 1995). Because of their requirements for specific host conditions, there is a succession of cerambycids and other wood borers that occur as a living tree first declines, then dies, and later decays (Blackman and Stage 1924; Graham 1925; Savely 1939; Haack et al. 1983; Khan 1985; Harmon et al. 1986; Hanula 1996; Saint-Germain et al. 2007; Lee et al. 2014). With respect to forestry, such host requirements reveal why some cerambycids are pests primarily of living trees, while

TABLE 11.1

Number of Cerambycid Species that Utilize Various Plant Families as Larval Hosts in Four World Regions, Ranked from Highest Number of Cerambycid Species to Lowest and Arranged Alphabetically for Families with the Same Number of Species[*]

Montana		Finland		Israel		Korea	
Family	No.	Family	No.	Family	No.	Family	No.
Pinaceae	69	Fagaceae	63	Fagaceae	34	Ulmaceae	49
Salicaceae	28	Pinaceae	57	Fabaceae	26	Fagaceae	46
Fagaceae	25	Betulaceae	52	Rosaceae	18	Pinaceae	45
Rosaceae	23	Salicaceae	42	Asteraceae	14	Betulaceae	42
Juglandaceae	15	Rosaceae	25	Moraceae	14	Rosaceae	37
Aceraceae	14	Tiliaceae	22	Anacardiaceae	13	Salicaceae	36
Ulmaceae	14	Ulmaceae	20	Ulmaceae	11	Juglandaceae	26
Anacardiaceae	11	Aceraceae	10	Pinaceae	10	Moraceae	23
Fabaceae	11	Juglandaceae	10	Rhamnaceae	10	Leguminosae	22
Betulaceae	10	Oleaceae	10	Salicaceae	10	Vitaceae	14
Vitaceae	10	Celastraceae	5	Boraginaceae	5	Rutaceae	13
Asteraceae	9	Caprifoliaceae	4	Aceraceae	4	Aceraceae	11
Cornaceae	9	Apiaceae	2	Juglandaceae	4	Euphorbiaceae	9
Magnoliaceae	7	Asteraceae	2	Myrtaceae	4	Taxodiaceae	8
Moraceae	7	Araliaceae	2	Rutaceae	4	Cornaceae	7
Caprifoliaceae	6	Rhamnaceae	2	Apiaceae	3	Ebenaceae	7
Celastraceae	5	Aquifoliaceae	1	Oleaceae	3	Oleaceae	7
Grossulariaceae	5	Cornaceae	1	Poaceae	3	Tiliaceae	7
Oleaceae	5	Cupressaceae	1	Arecaceae	2	Araliaceae	6
Tiliaceae	5	Fabaceae	1	Betulaceae	2	Cupressaceae	6
Hamamelidaceae	4	Geraniaceae	1	Brassicaceae	2	Lauraceae	6
Rutaceae	4	Hippocastanaceae	1	Cupressaceae	2	Scrophulariaceae	5
Cupressaceae	3	Moraceae	1	Dipsacaceae	2	Compositae	3
Ericaceae	3	–	–	Euphorbiaceae	2	Gramineae	3
Lauraceae	3	–	–	Labiatae	2	Actinidiaceae	2
Poaceae	3	–	–	Platanaceae	2	Anacardiaceae	2
Asclepiadaceae	2	–	–	Apocynaceae	1	Caprifoliaceae	2
Ebenaceae	2	–	–	Asphodeloideae	1	Elaeagnaceae	2
Hippocastanaceae	2	–	–	Campanulaceae	1	Lythraceae	2
Myrtaceae	2	–	–	Casuarinaceae	1	Meliaceae	2
Scrophulariaceae	2	–	–	Celastraceae	1	Platanaceae	2
Aquifoliaceae	1	–	–	Cornaceae	1	Punicaceae	2
Cactaceae	1	–	–	Cucurbitaceae	1	Rhamnaceae	2
Caryophyllaceae	1	–	–	Elaeagnaceae	1	Styracaceae	2
Elaeagnaceae	1	–	–	Lamiaceae	1	Cannabaceae	1
Euphorbiaceae	1	–	–	Lauraceae	1	Casuarinaceae	1
Malvaceae	1	–	–	Malvaceae	1	Cucurbitaceae	1
Menispermaceae	1	–	–	Meliaceae	1	Daphniphyllaceae	1
Platanaceae	1	–	–	Ranunculaceae	1	Ginkgoaceae	1
Polygonaceae	1	–	–	Sapindaceae	1	Malvaceae	1
Ranunculaceae	1	–	–	Scrophulariaceae	1	Myrtaceae	1
Rhamnaceae	1	–	–	Styracaceae	1	Simaroubaceae	1
Smilacaceae	1	–	–	Tamaricaceae	1	Sterculiaceae	1
Solanaceae	1	–	–	Valerianaceae	1	Urticaceae	1
Unknown	17	–	–	Vitaceae	1	–	–
Total cerambycid species	152		123		104		181

[*] The information for the cerambycids of Montana was based on Hart et al. (2013) and the accompanying online database http://www.mtent.org/Cerambycidae.html (accessed December 30, 2015), and supplemented as needed with host data from Linsley (1962a, 1962b, 1963, 1964) and Linsley and Chemsak (1972, 1976, 1984, 1995). Similarly, Bílý and Mehl (1989) was the source of the host data for Fennoscandia and Denmark, Sama et al. (2010) for Israel, and Lim et al. (2014) for Korea. These references listed host plants for 89% of the cerambycid species from Montana, 100% for Fennoscandia, 91% for Israel, and 57% for Korea.

others are mostly pests of stressed or recently felled trees or logs, and still others are pests of lumber and wood products.

Although only a small percentage of the world's cerambycids are considered economic pests, there are nevertheless several species that are well-recognized as pests of forest and urban trees as well as wood products. The cerambycids selected for discussion in this chapter represent just a few of the many species reported as economic tree pests worldwide. Their selection was based primarily on a review of forest entomology textbooks, regional cerambycid guides, and major reviews from several world regions. For example, Duffy (1957), Roberts (1969), Wagner et al. (1991), Akanbi and Ashiru (2002), and Schabel (2006) were reviewed for Africa; Zhuravlev and Osmolovskii (1964), Duffy (1968), Gressitt et al. (1970), Rozhkov (1970), Cherepanov (1988a, 1988b, 1990), Xiao (1992), and Shin et al. (2008) for Asia; Froggatt (1923), Duffy (1963), and Elliot et al. (1998) for Australasia; Duffy (1953), Novák (1976), Bílý and Mehl (1989), and Bense (1995) for Europe; Duffy (1960) and Rivas (1992) for Central and South America; Craighead (1923, 1950), Linsley (1962a, 1962b, 1963, 1964), Linsley and Chemsak (1972, 1976, 1984, 1995), Furniss and Carolin (1977), Drooz (1985), Cibrián Tovar et al. (1995), and Solomon (1995) for North America including Mexico; and Browne (1968), Gray (1972), Nair (2007), and Wylie and Speight (2012) for the tropics in general.

For each of the species listed here, details are presented on the insect's native and introduced geographic range, larval hosts, adult size and general appearance, life history, and economic impact. The 43 selected species illustrate the wide range of life-history strategies found among tree-infesting cerambycids throughout the world. The species are listed alphabetically by subfamily, including 28 species of Cerambycinae, 10 Lamiinae, 1 Parandrinae, 2 Prioninae, and 2 Spondylidinae. Geographically, of the 43 treated species, 6 are native primarily to Africa, 11 to Asia, 7 to Australia and nearby areas, 5 to Eurasia, 1 to Europe, 11 to North America, and 2 to South and Central America. In addition, there are several cerambycids that are forest pests but that will be discussed in Chapter 12 as pests of agricultural and horticultural crops, including 2 Cerambycinae: *Aromia bungii* (Faldermann) and *Strongylurus thoracicus* (Pascoe), and 11 Lamiinae: *Analeptes trifasciata* (Fabricius), *Anoplophora chinensis* (Forster), *Apriona germari* (Hope), *Bacchisa atritarsis* (Pic), *Batocera horsfieldi* (Hope), *Batocera lineolata* Chevrolat, *Batocera rufomaculata* (DeGeer), *Celosterna scabrator* (Fabricius), *Oncideres cingulata* (Say), *Plagiohammus spinipennis* (Thomson), and *Saperda candida* Fabricius. Relatively few details are provided on control options in this chapter given that they can vary widely from country to country and over time. Readers with an interest in this topic are directed to the references listed in this chapter as well as to Chapter 8 for a discussion of biological control options, Chapter 9 for cultural control options, Chapter 10 for chemical control options, and Chapter 13 for phytosanitary options.

11.2 Cerambycinae

11.2.1 *Anelaphus parallelus* (Newman)

Anelaphus parallelus is native to eastern North America, including both Canada and the United States. This species often has been confused with *Anelaphus villosus* (Fabricius), another North American twig pruner, and therefore caution must be used when reading the literature (Gosling 1981). *Quercus* is the principal host genus, but occasionally *Carya* and *Juglans* serve as larval hosts (Gosling 1981; Haack 2012). Adults are 10–15 mm long and generally brown in color with lighter-colored patches (Linsley 1963; Figure 11.1).

The typical life cycle is completed in two years and is highly synchronous with adults emerging primarily in odd-numbered years (Gosling 1978; Haack 2012; Figure 2.11 in this book). Eggs usually are laid singly on small twigs. After eclosion, the larva enters the twig, feeds on the wood, and tunnels to the node of the adjoining larger branch during the first summer (Gosling 1981). In the second summer, the larva extends the gallery in the direction of the trunk, feeding in the center of the branch. Eventually, the larva consumes a disc of wood, leaving only the bark intact, and then plugs its gallery with wood fibers. The "pruned" branch eventually breaks where the disc of wood was eaten and falls to

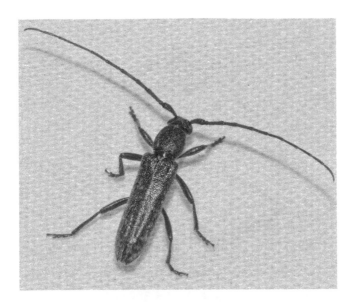

FIGURE 11.1 *Anelaphus parallelus* adult from Missouri (United States). (Courtesy of Jack Foreman at bugguide.net.)

the ground with the larva inside. The larva pupates inside the fallen branch in late summer of the second season, overwinters as a pupa, and emerges as an adult the following spring (Gosling 1978).

Larval damage becomes noticeable as the fallen branches accumulate under infested trees (Gosling 1978; Haack 2012). Although the pruned branches often are a concern to homeowners and park managers, this level of damage is seldom detrimental to trees (Solomon 1995). Control efforts usually are aimed at gathering infested branches from the ground and destroying them by burning or chipping (Solomon 1995; see Chapter 9).

11.2.2 *Callichroma velutinum* (Fabricius)

Callichroma velutinum is native to northern South America (Bolivia, Brazil, French Guiana, Guyana, Peru, and Suriname) and the West Indies (Trinidad) (Duffy 1960; Monné et al. 2012; Bezark 2015a, 2015b). The principal host trees include species of *Achras, Manilkara*, and *Pouteria* (Duffy 1960; Tavakilian et al. 1997). Adults measure 24–42 mm in length and have elytra that are metallic bluish green or violet in color, with two darker longitudinal bands (Duffy 1960; Figure 11.2). The antennae are about twice the body length in males but about the same length as the body in females.

The time needed to complete one generation appears to be less than a year. According to Duffy (1960), adults are active from mid-morning to mid-afternoon. Eggs are laid singly on the bark surface within crevices. After eclosion, larvae tunnel through the bark and feed initially in the cambial region, with late instars tunneling in the sapwood and eventually the heartwood.

In Trinidad, Duffy (1960) reported that stumps and recently cut logs of *Manilkara* are quickly colonized by *C. velutinum*. At times, infestations can be so high that the wood becomes unsuitable for use as lumber or railway ties (Duffy 1960; Gray 1972). Control efforts should be aimed at quickly utilizing newly felled trees to minimize oviposition and early larval development (Duffy 1960).

11.2.3 *Callidiellum rufipenne* (Motschulsky)

Callidiellum rufipenne is native to Asia, occurring in China, Japan, Korea, the Russian Far East, and Taiwan (Duffy 1968; Danilevsky 2015; EPPO 2015). In addition, by the end of 2015, established populations of *C. rufipenne* had been reported in Europe (Belgium, Croatia, France, Italy, and Spain; Campadelli and Sama 1989; Bahillo and Iturrondobeitia 1995; Cocquempot and Lindelöw 2010; Loś and Plewa 2011; Van Meer and Cocquempot 2013; Drumont et al. 2014), North America

FIGURE 11.2   *Callichroma velutinum* adult from French Guiana. (Courtesy of Daniel Prunier at insecterra.forumactif.com.)

FIGURE 11.3   *Callidiellum rufipenne* adult male from Caucasus, Russia. (Courtesy of Maxim Smirnov at www.zin.ru\ Animalia\Coleoptera.)

(United States; Maier and Lemmon 2000; Maier 2007), and South America (Argentina; Turienzo 2006). The principal host genera are conifers in the family Cupressaceae (which now contains genera of the former family Taxodiaceae), including *Chamaecyparis, Cryptomeria, Cupressus, Juniperus*, and *Thuja*. In Asia, the Pinaceae genera *Abies* and *Pinus* are also listed as occasional hosts (Duffy 1968); however, in North America, no Pinaceae have yet been documented as hosts (Maier 2007). Adults measure 6–13 mm in length, with males usually being blackish-blue in color and females being reddish-brown (Hoebeke 1999; Humphreys and Allen 2000; Maier and Lemmon 2000; Figure 11.3).

Callidiellum rufipenne is univoltine. Adults emerge in spring, mate on the bark surface of host trees, and soon begin to oviposit in bark cracks and crevices. Adults apparently do not feed, and typically live for two to three weeks. Eggs hatch in about two weeks and larvae immediately tunnel through the bark and feed in the cambial region, packing their galleries with frass. Mature larvae enter the wood in late summer and construct a cell at the end of their galleries in which they pupate. Pupation occurs in autumn, with adults overwintering within the hosts and emerging through oval-shaped exit holes the following spring (Shibata 1994; Hoebeke 1999; Humphreys and Allen 2000; Maier and Lemmon 2000).

Callidiellum rufipenne generally is considered a secondary pest, primarily infesting weakened or recently dead trees (Shibata 1994). However, in the eastern United States, *C. rufipenne* occasionally has infested living *Chamaecyparis, Juniperus*, and *Thuja* trees and shrubs (Maier and Lemmon 2000; Maier 2007). Heavy infestations of live hosts can result in tree or branch death, but more typically economic impact results from lowering the quality of the wood due to larval feeding. Given its life history traits, *C. rufipenne* can easily move in barked logs, wood packaging material, cut branches, and live plants. Therefore, depending on the product being considered, control options should include insecticidal treatment, inspection, and certification programs for nursery stock, as well as rapid utilization, debarking, and proper heat treatment or fumigation of logs and wood products.

11.2.4 *Chlorophorus carinatus* Aurivillius

Chlorophorus carinatus is native to East Africa, including Kenya, Tanzania, and Uganda (Browne 1968). Both hardwoods (*Acacia, Allophylus, Coffea, Dombeya, Eucalyptus, Fagaropsis, Hagenia, Laguniaria, Premna*, and *Theobroma*) and conifers (*Cupressus, Juniperus, Pinus*, and *Podocarpus*) have been reported as larval hosts (Duffy 1957; Schabel 2006). Tavakilian and Chevillotte (2013) give the length of one adult specimen as 15 mm. Adults are blackish in color with a pattern of gray transverse stripes.

Generation time can be as short as six months (Gardner 1957; Schabel 2006). Adults typically lay eggs on stressed and dying trees as well as on recently cut logs and stumps. Eggs are laid in bark crevices and, after eclosion, the larvae develop primarily in the cambial region. Frass is packed tightly in the larval galleries. Mature larvae enter the sapwood to pupate.

Most damage caused by *C. carinatus* larvae is restricted to the sapwood surface; however, the pupal cells are constructed deeper in the sapwood. In addition, *C. carinatus* will oviposit near bark wounds on apparently healthy *Cupressus* trees, with the resulting larvae tunneling deeper in the sapwood (Browne 1968; Schabel 2006). Given that early larval development occurs primarily in the cambial region, control efforts should focus on rapid utilization and debarking of recently felled trees.

11.2.5 *Citriphaga mixta* Lea

Citriphaga mixta is native to Australia, being first reported in New South Wales where it infests *Citrus* (= *Eremocitrus*) *glauca* (Lindl.) Burkill (desert lime), which is a thorny shrub or small tree native to semiarid areas of Australia (Froggatt 1923; Hawkeswood 1993). Adults are 2–3 cm long, dark brown in color, with white spots on the elytra (Froggatt 1919, 1923).

The life cycle is completed in about a year, with larval development spanning about 10 months and the pupal period taking about four to six weeks (Froggatt 1919, 1923). Eggs are laid on the bark of the trunk near the ground. Larvae tunnel into the wood and upward near the center of the stem for 1–1.5 m. When multiple larvae develop within the same stem, the wood will be riddled with galleries.

Larval tunneling can lead to breakage of branches and stems. In addition, infested trees respond to the larval feeding by secreting gum-like compounds into the wood surrounding the galleries (Froggatt 1923). Given the beetle's life history traits, individual trees, especially ornamentals, would likely require protection of the lower trunk with insecticides or some device to exclude ovipositing adults.

11.2.6 *Cordylomera spinicornis* (Fabricius)

Cordylomera spinicornis is native to Africa, being reported from many countries such as Angola, Benin, Cameroon, Central African Republic, Democratic Republic of Congo, Equatorial Guinea, Gabon,

Gambia, Ghana, Guinea, Ivory Coast, Liberia, Malawi, Mozambique, Niger, Nigeria, Republic of Congo, Senegal, Sierra Leone, Sudan, Tanzania, Togo, Uganda, and Zaire (Duffy 1957, 1980; GBIF 2014). Although *C. spinicornis* apparently has not yet become established outside of Africa, adults have commonly been intercepted in foreign countries, especially on imported logs (O'Connor and Nash 1984; Cocquempot and Mifsud 2013; Rassati et al. 2015). Throughout its native range, *C. spinicornis* primarily infests species in the family Meliaceae such as *Entandrophragma, Guarea, Khaya, Lovoa, Trichilia*, and *Turraeanthus*. In addition, trees in the genera *Acacia, Baphia, Celtis, Funtumia, Guarea, Lasiodiscus, Teclea*, and *Theobroma* have been infested in Africa (Duffy 1957, 1980; Roberts 1969; Wagner et al. 1991). Adults measure 13–25 mm in length and have a metallic coloration that can be various shades of green, blue, and bronze (Duffy 1957; Figure 11.4).

The generation time of *C. spinicornis* has not been reported. Adults typically emerge during the dry season, November to February, and lay eggs on the bark of living host trees and recently cut logs (Roberts 1969). Multiple eggs are laid together in bark crevices. Larvae tunnel primarily in the cambial region, entering the sapwood usually no more than 5 cm to pupate (Duffy 1957).

Cordylomera spinicornis has been reported as a pest of street trees, forest trees, and recently cut logs (Duffy 1957; Roberts 1969; Wagner et al. 1991). Populations can reach high levels in timber yards where logs are sorted and stored (Duffy 1957; Roberts 1969). Although infestation levels can be high, most damage is restricted to the outer 5 cm of the logs (Duffy 1957). Control efforts should focus on rapid debarking and utilization of logs as well as on destruction of logging residue, which also can harbor larvae (Duffy 1957).

11.2.7 *Diotimana undulata* (Pascoe)

Diotimana undulata is native to the Oceania area, including Australia and Papua New Guinea (Duffy 1963; Gray 1968). Note that in Duffy (1963), the species name was spelled as *Diotimana undata*, and a few other authors have followed this spelling. The principal hosts are coniferous species in the genus *Araucaria* (Froggatt 1925; Duffy 1963; Gray 1968); but occasionally species of *Cryptomeria* and *Pinus*, both of which were introduced to Australia, have also been infested by *D. undulata*

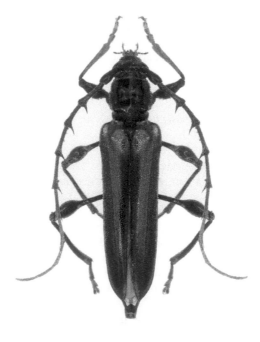

FIGURE 11.4 *Cordylomera spinicornis* adult collected at the port of Ravenna, Italy. (Courtesy of Paolo Paolucci and reported in Rassati et al. (2015).)

(Hawkeswood 1993). Adults average about 20 mm long and have a brown–gray vertical pattern on the elytra (Froggatt 1925, 1927).

There appears to be one generation per year. Eggs are laid on the bark surface in crevices. Early instar larvae tunnel primarily in the cambial region, while late instars tunnel into the wood for several centimeters and construct individual pupal cells.

Adults typically oviposit on storm-damaged trees as well as on recently dead trees or cut trunks, branches, and stumps (Froggatt 1925, 1927; Hawkeswood 1993). In Papua New Guinea, however, *D. undulata* was reported to infest living *Araucaria* trees and occasionally cause tree mortality (Gray 1968), but a later report by Gray and Wylie (1974) noted that damage by this beetle to living trees had never become widespread. Control options may require chemical treatment when live trees are at risk, but most efforts should focus on rapid utilization of felled trees.

11.2.8 *Eburia quadrigeminata* (Say)

Eburia quadrigeminata is native to eastern North America and may also occur on some Caribbean islands (Bezark 2015a, 2015b). A large number of hardwood trees serve as larval hosts, including species of *Acer, Carya, Castanea, Fagus, Fraxinus, Gleditsia, Prunus, Robinia, Quercus,* and *Ulmus* (Linsley 1962b). There are a few reports of *E. quadrigeminata* infesting the conifer bald cypress [*Taxodium distichum* (L.) Rich.]; however, this could have resulted from confusion with *Eburia distincta* Haldeman, a species that regularly uses *T. distichum* as a larval host in the southeastern United States (Thomas 1999). Adults are 14–24 mm long and light brown in color, and on each elytron, there are two pairs of longitudinal, ivory-colored spots with one pair located at the base of each elytron and the other near the center (Baker 1972; Figure 11.5).

The typical life cycle is considered to be two years (Baker 1972). Adults emerge during summer and are mostly nocturnal. Eggs often are laid near wounds on trees where the wood is exposed (Craighead 1923). Larvae tunnel primarily in the heartwood, showing a preference for dry, solid wood. The galleries are tightly packed with frass.

The larval galleries of *E. quadrigeminata* result in great economic loss because they are relatively large and penetrate to the heartwood. In addition, this species has the ability to complete development in finished wood products, often many years after manufacture. For example, *E. quadrigeminata* adults were reported to have emerged from flooring at least 14 years after manufacture (Webster 1889) and

FIGURE 11.5 *Eburia quadrigeminata* adult from Illinois (United States). (Courtesy of Tony Gerard at www.inaturalist.org.)

similarly at least 19 years after construction of a doorsill (McNeil 1886) and a wardrobe (Hickin 1951), at least 20 years after construction of a bedstead (Troop 1915), and at least 40 years after construction of a bookcase (Jaques 1918). Given this ability for protracted development, it is not surprising that *E. quadrigeminata* has emerged from furniture constructed in the United States and shipped to other countries such as Argentina (Di Iorio 2004b) and the United Kingdom (Hickin 1951). With respect to control options, damage to the lower trunk of residual trees should be minimized to reduce future infestations, while infested lumber can be fumigated or simply replaced, if possible (Craighead 1950).

11.2.9 *Enaphalodes rufulus* (Haldeman)

Enaphalodes rufulus is native to eastern North America. Oaks (*Quercus*) in both the white oak group and red oak group are infested by *E. rufulus* (Donley and Acciavatti 1980). Adults measure 23–33 mm in length and are brown in color with a mottled appearance (Solomon 1995; Figure 11.6).

The life cycle of *E. rufulus* is completed in two years and is highly synchronous with adults emerging primarily in odd-numbered years (Donley and Acciavatti 1980). Adults are active during the summer months and are mostly nocturnal. Females lay eggs in bark cracks as well as under lichen and vines that are attached to the host trees. On average, females lay 119 eggs during their lifetime (Donley 1978a). Larvae tunnel primarily in the cambial region during their first year of development; during the second year, they eventually tunnel into the sapwood and heartwood and construct individual pupal cells. Larvae pupate the following spring, with the adults emerging through oval-shaped exit holes (Donley and Acciavatti 1980).

Given that *E. rufulus* oviposits primarily along the lower trunk, which is where the highest quality wood occurs, considerable economic damage can result from the larval galleries (Donley and Rast 1984). Fortunately, populations of *E. rufulus* tend to be low, but in recent decades a major outbreak of *E. rufulus* occurred in the central United States (Stephen et al. 2001). For example, prior to the recent outbreak, Hay (1974) considered even one exit hole per tree to be high. By contrast, during the recent outbreak in the central United States, Riggins et al. (2009) estimated that populations peaked at 174 adults emerging

FIGURE 11.6 *Enaphalodes rufulus* adult from Illinois (United States). (Courtesy of Natasha Wright [Bugwood image 5205017].)

per tree for the cohort that emerged in 2001. In forest stands, there often are a few heavily infested trees referred to as "brood" trees. Forest managers should remove brood trees and thereby lower the local borer population (Solomon 1995).

11.2.10 *Glycobius speciosus* (Say)

Glycobius speciosus is native to eastern North America, being found throughout the range of its only known host *Acer saccharum* Marshall (Solomon 1995). Adults measure 22–27 mm in length and are black and yellow in color, having a wasp-like appearance. Some of the key characteristics used to identify the adults include a yellow head and yellow legs, a yellow-banding pattern in the design of a "W" across the elytra near the pronotum, and a black dot near the tip of each elytron (MacAloney 1968; Hoffard and Marshall 1978; Figure 11.7).

The typical life cycle of *G. speciosus* is completed in two years (MacAloney 1968; Hoffard and Marshall 1978; Solomon 1995). Adults are active in summer, laying eggs in bark crevices, under bark scales, and near wounds. Oviposition occurs primarily along the main trunk, usually below 10 m. During the first summer, larvae tunnel primarily in the cambial region and then enter the sapwood a short distance to overwinter. During the second summer, the larvae continue to tunnel in the cambial region at first and then enter the sapwood, tunneling upward for several centimeters. Larvae typically expel frass from their galleries. The larva overwinters at the end of the gallery, which is enlarged into a pupal cell. In spring, before pupating, the larva extends its gallery to near the bark surface to create a tunnel that it will use to exit once an adult. Once the exit tunnel is complete, the larva moves back to the pupal cell, pupates, and then emerges as an adult through an oval-shaped exit hole.

Tree mortality seldom occurs as a result of *G. speciosus* infestation, although individual branch death and crown thinning can occur (Solomon 1995). Most economic losses result from the defects caused by the larval tunnels in the wood of the lower trunk. Besides the actual galleries, the nearby wood often has a twisted grain pattern and can show signs of decay and discoloration (MacAloney 1968; Shigo et al. 1973; Hesterberg et al. 1976; Hoffard and Marshall 1978). Bark ridges often form over the larval galleries that were created in the cambial region, and at times the bark sloughs off, exposing the underlying sapwood and old larval galleries (Hoffard and Marshall 1978). Silvicultural methods are the main control option used for *G. speciosus* in forest stands, including removal of large, low-vigor trees—especially those currently infested (Hoffard and Marshall 1978; Solomon 1995). Insecticidal control is an option for high-value *Acer* trees, such as ornamentals or veneer-quality trees (Solomon 1995).

FIGURE 11.7 *Glycobius speciosus* adult from Canada. (Courtesy of David Cheung at www.dkbdigitaldesigns.com.)

11.2.11 *Hesthesis cingulata* (Kirby)

Hesthesis cingulata is native to Australia, including Tasmania, where it infests various species of *Eucalyptus* (Moore 1966; Elliott and de Little 1984; Elliott et al. 1998). Adults average about 25 mm in length, are wasp-like in appearance with reduced elytra and black coloration, and have one to three white to yellow transverse bands on the abdomen (Elliott et al. 1998).

Hesthesis cingulata is univoltine. Adults are active during summer, with eggs deposited on the lower stem of young eucalypts. Typically, one egg is laid per stem, usually within 5–20 cm of the soil, near the junction of a shoot and the main stem (Moore 1966). Early larval instars tunnel downward in a spiral fashion primarily within the cambial region. Older instars tunnel deeper into the sapwood near the ground but leave a column of woody tissue in the center that supports the stem upright. In response to this feeding, the lower stem becomes swollen and gall-like. The larva ejects frass from holes it makes in the stem (Moore 1966). The larva continues to tunnel downward into the taproot where it later reverses its position and overwinters. Pupation takes place the following spring with the new adults emerging through the stem wall above ground level (Moore 1966).

Hesthesis cingulata usually infests young trees that are 0.5–2.0 m tall (Elliott et al. 1998). Infestation levels vary from plantation to plantation. Infested trees often die as a result of the larval tunneling, although some trees survive, developing new shoots from the base (Moore 1966). Considering that females oviposit on the lower trunk of young trees, protection of the lower trunk by means of pesticides or exclusion techniques may be warranted at times.

11.2.12 *Hoplocerambyx spinicornis* Newman

Hoplocerambyx spinicornis is a widespread Asian species, being found from Afghanistan, Pakistan, India, and Nepal eastward to Borneo, Indonesia, New Guinea, and the Philippine Islands (Duffy 1968). It is a major pest of several tree genera in the family Dipterocarpaceae, including *Anisoptera, Dipterocarpus, Hopea, Parashorea, Pentacme,* and *Shorea,* as well as genera in other families such as *Duabanga* and *Hevea* (Duffy 1968). Adult *H. spinicornis* vary greatly in size from 20 to 65 mm in length and vary in color from reddish brown to brownish black with some gray pubescence (Duffy 1968; Figure 11.8).

FIGURE 11.8 *Hoplocerambyx spinicornis* adult from Vietnam. (Courtesy of Udo Schmidt at www.kaefer-der-welt.com.)

Hoplocerambyx spinicornis is univoltine, with adult emergence occurring during the early monsoon season in June and July (Beeson and Bhatia 1939). Adults are active during the day, feeding on sap and inner bark of their host trees. Adults are strongly attracted to the odors associated with newly felled *Shorea* trees and will fly distances of 1–2 km (Duffy 1968; Singh and Misra 1981). Adults live for about a month, with oviposition beginning about a week after mating. Eggs are laid individually along the trunk and major branches, usually in bark cracks and crevices or in holes or cuts along the bark. Females often lay 100–300 eggs in their lifetime, with the maximum number recorded being 468 eggs (Duffy 1968; Nair 2007). Larvae initially feed in the cambial region, while older instars tunnel in the sapwood and heartwood. Frass is pushed out of the galleries and accumulates around the tree. Mortality of early larval instars can be high, especially when tunneling in vigorous hosts that often exude large amounts of sap in response to the larval feeding. By contrast, survival is relatively high in drought-stressed trees, apparently because of reduced sap flow (Negi and Joshi 2009). In November, in India, the mature larva will construct an exit hole that it will use as an adult the following year; then, it returns to the end of its gallery in the heartwood and seals itself inside by secreting a calcareous solution to plug the gallery. Pupation occurs the next year, usually in April and May, with the adult remaining quiescent until the onset of the monsoon season (Duffy 1968; Nair 2007).

Typically, *H. spinicornis* infests trees that are stressed by drought, defoliation, or disease, or trees that have recently died or were cut. Adults oviposit preferentially on large, mature trees, but during outbreaks smaller trees down to 30 cm in diameter are infested. Both the trunk and major branches are infested. Larval galleries can greatly reduce wood quality because they can be numerous and can extend deep into the heartwood. Tree mortality can occur during outbreaks as a result of high larval densities (over 300 larvae per tree) and the extensive tunneling that effectively girdles the infested trees (Nair 2007). Major outbreaks of *H. spinicornis* have been recorded in India and Pakistan for more than a century, with *Shorea robusta* Roth being the principal host (Beeson and Bhatia 1939; Nair 2007). In India, some *H. spinicornis* outbreaks during the 1900s covered thousands of hectares and resulted in millions of trees killed (Nair 2007). In India and Pakistan, where major outbreaks regularly occur, several integrated control measures have been developed, including monitoring infestation levels in individual stands, selective cutting and removal of infested trees, debarking of trees that cannot be removed promptly, protecting logs in mill yards with cover sprays or tarps, and using trap logs to attract and kill adult beetles (Beeson 1941; Nair 2007).

11.2.13 *Hylotrupes bajulus* (L.)

Hylotrupes bajulus is native to Europe and North Africa but has been introduced to many other countries around the world, including Argentina, Australia, Brazil, China, Madagascar, South Africa, Uruguay, and the United States (Duffy 1968; Di Iorio 2004b). In Australia, efforts were undertaken to eradicate *H. bajulus* (French 1969; Eldridgea and Simpson 1987). Reports that *H. bajulus* occurs in New Zealand are incorrect (Bain 2009). *H. bajulus* primarily infests seasoned coniferous wood, especially *Pinus*, but also *Abies, Picea,* and *Pseudotsuga* (Duffy 1953, 1968; Bense 1995). In addition, there have been occasional reports of *H. bajulus* infesting hardwoods in the genera *Acacia, Alnus, Corylus, Juglans, Populus, Quercus,* and *Tamarix* (Duffy 1968). Adults measure 7–21 mm in length and vary in color from reddish brown to brownish black; they have two conspicuous tubercles on the pronotum and have one or two whitish bands or spots on each elytron (Duffy 1968; Baker 1972; Bense 1995; Figure 11.9).

The life cycle of *H. bajulus* can be completed in as short as two to three years, but in seasoned wood that is low in moisture, the larval developmental period can be greatly protracted (Bense 1995). For example, Baker (1972) suggested generation times of three to eight years for *H. bajulus* developing inside structures in the United States, and in one case in the United Kingdom, Bayford (1938) reported the occurrence of *H. bajulus* in *Pinus* furniture that had been manufactured (and apparently infested) at least 17 years earlier.

Adults emerge primarily in the summer months under natural conditions (Bense 1995). Eggs are laid in groups in cracks and crevices in the wood or between pieces of wood that are stacked (Baker 1972). Females typically lay 30–100 eggs in their lifetime, but some females have laid as many as

FIGURE 11.9   *Hylotrupes bajulus* adult female from Spain. (Courtesy of Udo Schmidt at www.kaefer-der-welt.com.)

200–400 eggs (Duffy 1953, 1968). Larvae feed primarily in the sapwood, packing their galleries with fine powdery frass. The larvae seldom break through the surface of the wood when tunneling; thus the frass is concealed within the infested timbers. The sounds produced as the larvae feed usually are audible to the homeowner. Mature larvae tunnel near to the surface of wood and then prepare to pupate deeper in the wood. The pupal period can be as short as two to three weeks. New adults emerge through an oval-shaped exit hole.

Although originally a forest insect, infesting dead branches tree stumps, *H. bajulus* is now primarily a domestic pest of structural timbers (Duffy 1968; Baker 1972; Robinson and Cannon 1979). Heavy infestations of homes and buildings can result in serious damage to roofing, framing, and flooring timbers, as well as to furniture. To remedy such infestations, homeowners often incur high costs for repair and, at times, fumigation. Because of their ability to infest dry wood and their long developmental time, *H. bajulus* has been spread to many countries around the world in infested timber, wood products, and solid wood packaging material.

11.2.14 *Megacyllene robiniae* (Forster)

Megacyllene robiniae is native to the eastern and central United States but has spread to Canada and the southern and western United States with the widespread planting of its only host *Robinia pseudoacacia* L. (Galford 1984). Adults measure 12–19 mm in length and are black in color with transverse yellow bands across the pronotum and elytra (Solomon 1995; Figure 11.10).

Megacyllene robiniae is univoltine, with adults being most active during late summer, when they are observed on host trees or when feeding on pollen of goldenrod (*Solidago*) flowers (Galford 1984). Eggs are laid singly in bark cracks and crevices and around pruning wounds and callus tissue. Oviposition occurs primarily along the trunk and the lower portion of major branches (Harman and Harman 1990). In about a week, the eggs hatch and the new larvae enter the inner bark, construct a small cell, and prepare to overwinter. In spring, the larvae first tunnel into the sapwood; then in summer, they move deeper into the heartwood. The larvae push frass out of their galleries, which tends to be white in color when the

FIGURE 11.10 *Megacyllene robiniae* adult from the United States. (Courtesy of Clemson University–USDA Cooperative Extension Slide Series [Bugwood image 1235175].)

larvae are tunneling in sapwood and yellow when tunneling in heartwood. Prior to pupation, the larva constructs a small tunnel that it will use to exit when an adult. Pupation occurs in late summer and lasts about two weeks. Adults emerge through oval-shaped exit holes (Solomon 1995).

Heavy infestations can weaken trees and lead to trunk and branch breakage. Individual trees can be infested for a number of years before stem breakage occurs; therefore, such trees can serve as hosts to several generations of the borer. Swelling often occurs along the infested portions of the trunk. Adults initially oviposit preferentially on stressed trees—but when populations are high, even dominant, apparently healthy trees can be infested (Galford 1984). Trees weakened by drought, fire, and soil compaction are especially prone to *M. robiniae* infestation (Galford 1984; Solomon 1995). Silvicultural control options include removal of heavily infested trees or, when several infested trees are removed, stump sprouting should be encouraged followed by selection of the most vigorous sprouts. Some *Robinia* selections have shown resistance to *M. robiniae*. Insecticidal sprays are at times warranted and are most effective when applied during the oviposition period in late summer (Galford 1984; Solomon 1995).

11.2.15 *Neoclosterus boppei* Quentin & Villiers

Neoclosterus boppei is native to countries in both East and West Africa, including Guinea, Nigeria, the Republic of the Congo, Sierra Leone, and Tanzania (Roberts 1969; Duffy 1980; Schabel 2006). The common host trees include species of *Brachystegia, Ficus*, and *Isoberlinia* (Schabel 2006). Adults vary in length from 34 to 58 mm, are dark brown to reddish in color, have pectinate (comb-like) antennae, and have stout lateral spines on the pronotum (Quentin and Villiers 1969; Duffy 1980; Figure 11.11).

Details on voltinism have not been reported. Eggs are laid singly on the bark of branches and trunks of living host trees (Roberts 1969). At first, larvae tunnel vertically within the sapwood and eventually enter the heartwood. The larvae expel frass from several sites along their galleries, which accumulates on the forest floor. In addition to frass, large amounts of tree-produced sap also flow from the same holes where the frass is ejected (Roberts 1969; Schabel 2006). Before pupating, larvae construct a tunnel from which to exit once transformed to adults. Pupation occurs in the heartwood at the end of the gallery in a cell lined with wood fibers (Roberts 1969).

FIGURE 11.11   *Neoclosterus boppei* adult from the Democratic Republic of the Congo. (Courtesy of Thierry Bouyer.)

Infestation by *N. boppei* often leads to death of individual branches and stem sections. Recently cut logs can also be infested (Roberts 1969). The larval galleries, which penetrate into the heartwood, lower wood quality. Rapid utilization of logs should help minimize oviposition and larval tunneling.

11.2.16 *Neoclytus acuminatus* (Fabricius)

Neoclytus acuminatus is native to North America, including both Canada and the United States. In addition, *N. acuminatus* has been established in Europe since the early 1900s, being first reported in Italy in 1908 and later spreading to several European countries including the Czech Republic, Croatia, France, Germany, Hungary, Montenegro, Portugal, Serbia, Slovenia, and Switzerland (Cocquempot and Lindelöw 2010). In South America, *N. acuminatus* has been reported as likely established in Argentina after two adults were collected in the field on branches of *Prosopis* in 1998 (Di Iorio 2004b). *N. acuminatus* is highly polyphagous on hardwood trees, shrubs, and vines. In North America, species of *Fraxinus* are preferred followed by *Carya, Celtis, Diospyros*, and *Quercus* (Solomon 1995). Other occasional hosts in North America include species of *Acer, Betula, Carpinus, Castanea, Cercis, Cercocarpus, Cornus, Fagus, Gleditsia, Ilex, Juglans, Liquidambar, Liriodendron, Maclura, Malus, Prunus, Pyrus, Robinia, Sassafras, Syringa, Ulmus*, and *Tilia* (Solomon 1995). In Europe, species of *Fraxinus, Juglans*, and *Ulmus* are common hosts (Cocquempot and Lindelöw 2010), but other European dicot hosts include *Acer, Betula, Carpinus, Castanea, Cercis, Corylus, Euonymus, Fagus, Ficus, Hibiscus, Lonicera, Morus, Ostrya, Populus, Prunus, Pyrus, Quercus, Robinia*, Rosa, *Salix, Tilia, Ulmus*, and *Vitis*, as well as occasionally conifers such as *Abies* (Bense 1995). Adults vary greatly in size from 4 to 18 mm in length (Bense 1995; Solomon 1995). They have a wasp-like appearance, being reddish brown in color with four transverse yellow bands across the elytra (Solomon 1995; Figure 11.12).

In the southern United States, *N. acuminatus* completes two to three generations per year, whereas in the northern United States, this species is mostly univoltine (Solomon 1995). Adults begin emerging in spring and, after mating, they lay eggs in bark cracks and crevices and under lichen and bark scales along the trunk and branches of host plants (Solomon 1995). Adults are active during the day. After egg hatch, larvae penetrate the bark and at first tunnel in the cambial region and then enter the sapwood where they complete their larval development (Solomon 1995). Larval galleries usually do not extend deeper than

FIGURE 11.12 *Neoclytus acuminatus* adult. (Courtesy of Gyorgy Csoka [Bugwood image 1231138].)

the outer heartwood. Most individuals overwinter as larvae. The larvae pack their galleries with granular frass and later, when preparing to pupate, they construct a pupal cell near the outer sapwood surface. New adults chew a circular exit hole through the bark when emerging.

Neoclytus acuminatus often infests stressed, dying, and recently dead trees, as well as recently cut logs with bark. When infestations are high, the sapwood can be riddled with galleries, which greatly reduces the quality of the wood, often making it unmarketable (Solomon 1995). This species has also been reported to infest apparently healthy *Robinia* trees that were planted as windbreaks and in woodlots and, at times, nursery stock and recently planted trees (Barr and Manis 1954; Solomon 1995). During harvest operations, logs should be moved quickly to sawmills for processing to minimize infestation (Solomon 1995).

11.2.17 *Neoclytus rufus* (Olivier)

Neoclytus rufus is known to occur in parts of the Caribbean (Grenada, Trinidad), Central America (Panama), and South America (Argentina, Bolivia, Colombia, French Guiana, Guyana, Paraguay, and Venezuela) (Duffy 1960; Giesbert 1989; Di Iorio 1995; Bezark 2015a, 2015b). Duffy (1960) reporte that *N. rufus* is a pest of *Mora* in Trinidad and *Inga* in Venezuela and has also been reported from *Peltophorum*. In Argentina, Gonzalez and Di Iorio (1996) reported that *N. rufus* infests species of *Celtis, Paullinia, Schinus*, and *Ziziphus*. Adults are 6–13 mm in length, reddish-brown in color, without transverse bands on the pronotum, but with white transverse bands or spots across the elytra (Di Iorio 1995; Figure 11.13).

Neoclytus rufus likely completes one or more generations per year throughout its range given that Duffy (1960) reports that a single generation can be completed in as little as three months. Adults are active during the day, often being observed on the branches, trunks, and recently cut logs of host trees. Eggs are laid in bark cracks and crevices on trees as well as on logs and lumber where some bark is still present (Duffy 1960). Larvae tunnel primarily in the cambial region when bark is present but will penetrate the outer sapwood when bark is lacking. The larvae pack frass within their galleries. After pupation, the new adults emerge through circular exit holes.

Neoclytus rufus commonly infests freshly cut logs and recently sawn lumber, especially when some bark is retained. Adults are commonly observed around sawmills in Trinidad. Larval galleries in the sapwood can lower the quality of affected lumber (Duffy 1960). Logs should be utilized quickly and debarking should be as complete as possible to minimize infestation (Duffy 1960).

FIGURE 11.13   *Neoclytus rufus* adult from Paraguay. (Courtesy of Karina Diarte.)

11.2.18 *Neoplocaederus viridipennis* (Hope)

Neoplocaederus viridipennis is native to several countries in Africa, including Angola, Benin, Cameroon, Central Africa, Guinea-Bissau, Ivory Coast, Nigeria, Sierra Leone, Togo, Uganda, and Zaire (Duffy 1957; Roberts 1969; Vitali 2015). Larval hosts include species of *Canarium, Daniellia, Hevea, Khaya, Milicia* (= *Chlorophora*), *Mitragyna, Phyllanthus, Pseudospondias, Ricenodendron, Spondianthus, Terminalia*, and *Trilepisium* (= *Bosqueia*), and possibly *Theobroma* (Duffy 1957, 1980; Roberts 1969). Adults are 18–22 mm in length, having a black head and prothorax, and elytra that are metallic blue or green (Duffy 1957; Figure 11.14).

Generation time of *N. viridipennis* has not been reported specifically but appears to be univoltine based on data from Duffy (1957) and Roberts (1969). Eggs are laid in bark cracks and crevices along the trunks of mature trees. Larvae feed primarily in the outer sapwood but near to the bark surface. As they tunnel, the larvae construct a number of holes in the bark through which they eject frass. Mature larvae tunnel deeper into the sapwood or heartwood and there construct a pupal chamber that they line with a calcareous secretion. In addition, prior to pupation, larvae plug their galleries with wood fibers (Duffy 1957).

The economic impact of *N. viridipennis* can be high given that living, mature trees are infested and the sapwood is often heavily mined. However, because the heartwood is less frequently tunneled, losses are kept to a minimum (Duffy 1957; Roberts 1969). Because both live trees and recently cut logs can be infested, rapid utilization of logs should help minimize infestation and tunneling.

11.2.19 *Oemida gahani* (Distant)

Oemida gahani is native to East Africa, being reported from the countries of Kenya, Tanzania, Uganda, Zimbabwe, and possibly South Africa (Duffy 1957; Schabel 2006). *O. gahani* is highly polyphagous, developing in both hardwoods and conifers, including both native and exotic species. For example, more than 173 species of woody plants in Africa, representing at least 56 plant families, have been reported as larval hosts of *O. gahani* (Jones and Curry 1964; Schabel 2006; Duffy 1980). Some of the common larval hosts include species of *Acacia, Acrocarpus, Artocarpus, Celtis, Croton, Cupressus, Erythrina, Eucalyptus, Grevillea, Juniperus, Olea, Podocarpus*, and *Trichocladus* (Duffy 1957; Rathore 1995; Schabel 2006). Adults are 8–23 mm in length, brownish in color, with two longitudinal ridges on each elytron.

FIGURE 11.14 *Neoplocaederus viridipennis* adult female from Cameroon. (Courtesy of Francesco Vitali at www. cerambycoidea.com.)

The life cycle of *O. gahani* usually spans one to three years but can extend to 10 years in dry, structural timbers (Curry 1965; Schabel 2006). Adults emerge and are active at night—usually peaking in April and November during the two rainy seasons in Kenya. Adults live for about two to four weeks and apparently do not feed (Curry 1965). With regard to native plants, *O. gahani* typically infests stressed or recently dead or cut trees and logs. By contrast, when infesting exotic trees, *O. gahani* commonly oviposits on living trees (Schabel 2006). Females lay eggs in batches of 50–100, often preferring bark wounds and pruning scars as oviposition sites, as well as old emergence holes in structural timbers (Curry 1965; Schabel 2006). Larvae tunnel in both the sapwood and heartwood and pack their galleries with frass (Curry 1965). The pupal chamber is constructed near the wood surface, and the new adult emerges through an oval-shaped exit hole (Duffy 1957; Curry 1965).

Before the widespread planting of exotic tree plantations in East Africa, *O. gahani* was mostly a pest of stressed and recently cut native trees as well as stumps used for coppicing (Schabel 2006). At times, *O. gahani* would infest apparently healthy native trees, primarily *Juniperus procera* Hochstetter ex Endlicher. As for the exotic plantation trees planted widely in East Africa, species of *Cupressus* were the most heavily infested by *O. gahani*, and to a lesser degree plantations of various *Acacia, Acrocarpus, Eucalyptus*, and *Grevillea* species (Curry 1965; Schabel 2006). Upon felling mature *Cupressus* trees in Kenya during the mid-1900s, evidence of *O. gahani* infestation was found in as many as 34–80% of the trees (Gardner and Evans 1957; Schabel 2006). More recently, the impact of this beetle has been greatly lowered as a result of several changes, such as reduced frequency of pruning, treating pruning scars when pruning does occur, rapid stand sanitation after logging operations, stand conversion to less susceptible tree species such as pines, and kiln-drying wood to reduce possible active infestations in structural timbers (Gardner and Evans 1957; Jones and Curry 1964; Schabel 2006).

11.2.20 *Phoracantha recurva* Newman

Phoracantha recurva is native to Australia and Papua New Guinea (Duffy 1963) and has been considered to be the most common *Phoracantha* species in Australia (Froggatt 1923). This species has been introduced to several countries throughout the world, including parts of Africa (Libya, Malawi, Morocco, South Africa, Tunisia, and Zambia), Asia (Israel and Turkey), Europe (Cyprus, France, Greece, Italy, Malta, Portugal, and Spain), North America (United States), and South America (Argentina, Brazil, Chile, and Uruguay) (Di Iorio 2004a; CABI 2015b). Although *P. recurva* has been reported to occur in New Zealand in various publications (Wang 1995; CABI 2015b), this information is considered incorrect based on misidentification (Bain 2012). Several species of *Eucalyptus* serve as the primary larval hosts for *P. recurva* (Froggatt 1923; Duffy 1963), although a few other tree genera (*Angophora, Cupressus,* and *Syncarpia*) have been reported as occasional hosts in South Africa (Kliejunas et al. 2001). Adults measure 15–30 mm in length, with yellowish-brown antennae and legs, reddish-brown head and pronotum, and yellowish-brown elytra with a dark blackish-brown spot on the basal portion and a transverse band on the apical portion (CABI 2015b; Figure 11.15).

Phoracantha recurva generally completes one generation a year, but multiple generations can occur in tropical and subtropical countries (Löyttyniemi 1983). In Australia, *P. recurva* generally oviposits on stressed or recently cut trees. Similarly, throughout its introduced range, *P. recurva* oviposits on living trees, including both stressed and apparently healthy trees (Ivory 1977; Paine and Millar 2002). Adults are nocturnal. Eggs are laid in batches of up to 40 under bark scales along the trunk and main branches (Duffy 1963; Ivory 1977). Larvae tunnel primarily in the cambial region and pack their galleries with frass. Individual galleries can exceed 1 m in length (Froggatt 1923). Mature larvae tunnel into the sapwood and heartwood to form a pupal cell as well as construct a short tunnel toward the bark that is opposite the entrance into the wood. Larvae then plug the gallery with frass and wood fibers and prepare to pupate. New adults reach the wood surface by tunneling through the gallery they had constructed as mature larvae and emerge through an oval-shaped exit hole. In areas where *P. recurva* coexists with *Phoracantha semipunctata* (Fabricius), with time *P. recurva* tends to largely replace *P. semipunctata* (Paine and Millar 2002; Paine et al. 2011)

Although considered primarily a secondary pest in Australia, *P. recurva* has caused significant mortality of eucalypts in many other parts of the world where these trees have been introduced (Froggatt 1923; Drinkwater 1975; Ivory 1977; Paine et al. 2011; CABI 2015b). Trees infested by *P. recurva* initially display wilted foliage and crown dieback, but extensive larval feeding eventually girdles the tree and leads to tree death. Extensive mortality of eucalypts has occurred in both ornamental and plantation settings (Paine et al. 2011). International movement of *P. recurva* has likely resulted from trade in eucalypt logs, wood products and associated wood packaging materials, and possibly live plants (Haack 2006; CABI 2015b). For example, *P. recurva* and *P. semipunctata* were likely introduced to South Africa in the early 1900s on *Eucalyptus* timbers used as railroad crossties that were imported from Australia (Drinkwater 1975; Cillie and Tribe 1991). Control options often focus on irrigation to reduce tree stress, avoiding trunk injuries, planting *Eucalyptus* species that are more resistant to borer attack, rapid debarking or utilization of logs, and release of biological control agents (FAO 2009; also see Chapter 8).

FIGURE 11.15 *Phoracantha recurva* adult from Gibraltar. (Courtesy of Charles Perez at The Gibraltar Ornithological & Natural History Society [www.gonhs.org].)

11.2.21 *Phoracantha semipunctata* (Fabricius)

Phoracantha semipunctata is native to Australia and Papua New Guinea (Duffy 1963) and has become established in several countries throughout the world, including parts of Africa (Algeria, Egypt, Lesotho, Libya, Malawi, Mauritius, Morocco, Mozambique, South Africa, Swaziland, Tunisia, Zambia, and Zimbabwe), Asia (Israel, Lebanon, and Turkey), Europe (Cyprus, France, Italy, Malta, Portugal, and Spain), North America (Mexico, United States), Oceania (New Zealand), and South America (Argentina, Bolivia, Brazil, Chile, Peru, and Uruguay) (CABI 2015b). Several species of *Eucalyptus* have been recorded as larval hosts for *P. semipunctata* (Duffy 1963; CABI 2015c), although a few other tree genera in the family Myrtaceae have also been recorded as occasional hosts in Australia (*Angophora, Corymbia,* and *Syncarpia*) (Duffy 1963; Kliejunas et al. 2003). Adults measure 16–35 mm in length, being primarily dark reddish-brown in color with each elytron having a yellowish zigzag band near the center and a yellowish apical spot (Duffy 1963; Solomon 1995; CABI 2015c; Figure 11.16).

Phoracantha semipunctata generally completes one to two generations per year, but three annual generations can occur under tropical conditions (Löyttyniemi 1983; Bense 1995; Solomon 1995). Adults are nocturnal. Oviposition typically occurs on stressed or recently cut trees, but in some countries where *P. semipunctata* has been introduced, both stressed and apparently healthy trees can be infested (Drinkwater 1975; Ivory 1977). Eggs usually are laid in bark cracks and under bark scales in batches of 3–30 eggs, and at times batches of 110 eggs have been observed (Duffy 1963; Scriven et al. 1986). Oviposition occurs mostly along the main trunk, but eggs are also laid on large branches (Schabel 2006). Larvae tunnel mostly in the cambial region often constructing galleries over a meter in length and packed with frass (Solomon 1995; Elliott et al. 1998). Mature larvae tunnel 6–10 cm into the wood and construct a pupal cell as well as a short tunnel into the outer bark that they use when exiting the tree as adults (Duffy 1963).

In Australia, *P. semipunctata* typically is a minor pest, infesting stressed, dying, and recently dead or cut trees, but it does not successfully infest dry eucalypt logs and lumber (Duffy 1963). In many countries where *P. semipunctata* has been introduced, this beetle has become a serious pest, resulting in widespread mortality of ornamental and plantation eucalypts (Drinkwater 1975; Löyttyniemi 1983;

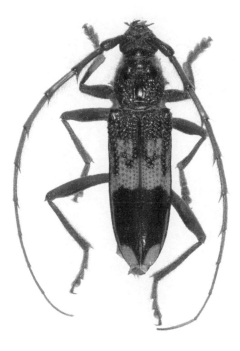

FIGURE 11.16 *Phoracantha semipunctata* adult collected in Gibraltar from timber imported from Spain. (Courtesy of Charles Perez at The Gibraltar Ornithological & Natural History Society [www.gonhs.org].)

Scriven et al. 1986; Elliott et al. 1998; Schabel 2006). As mentioned earlier, in many areas where both *P. recurva* and *P. semipunctata* have been introduced, *P. recurva* tends to displace *P. semipunctata* (Paine and Millar 2002; Paine et al. 2011). Control efforts for *P. semipunctata* are similar to those listed earlier for *P. recurva*.

11.2.22 *Phymatodes testaceus* (L.)

Phymatodes testaceus is native to Eurasia and North Africa and has been reported as introduced to North America (Canada and United States) since the mid-1800s—and apparently to Japan as well (LeConte 1850; Cherepanov 1988a; Bense 1995; LaBonte et al. 2005; Swift and Ray 2010). Several hardwood species serve as larval hosts, including *Carpinus, Carya, Castanea, Corylus, Fagus, Fraxinus, Malus, Prunus, Quercus, Salix,* and *Ulmus* (Linsley 1964; Bense 1995). *Quercus* appears to be the preferred host in Europe and North America (Craighead 1923; Duffy 1953; Cherepanov 1988a). In addition, Craighead (1923) stated that others had recorded the conifers *Picea* and *Tsuga* as larval hosts of *P. testaceus* in North America. Adults measure 7–17 mm in length and are variable in color including solid shades of yellow, brown, red, blue, black, violet, and green, or are of two colors where the elytra and pronotum differ (Linsley 1964; Novák 1976; Cherepanov 1988a; Figure 11.17).

Generally, one to three years is required for *P. testaceus* to complete one generation (Duffy 1953; Linsley 1964). Eggs are laid in bark cracks or crevices on recently dead trees or cut logs (Duffy 1953). Larvae tunnel primarily within the bark or in the cambial region and retain frass within their galleries (Craighead 1950). Mature larvae tunnel a few centimeters into the outer sapwood to construct pupal cells when the outer bark is relatively thin, or if the bark is sufficiently thick, they will construct an oval cell within the cambial region in which to pupate (Duffy 1953). Pupation and adult emergence usually occur in spring to early summer (Duffy 1953).

This cerambycid has been a major pest in the tan bark industry in both Europe and North America (Craighead 1923; Duffy 1953). Traditionally, tannins are extracted from tree bark and used to treat animal skins in the process of making leather. *Quercus* bark is a common source of tannins in both Europe and North America. High populations of *P. testaceus* are able to form in the tan bark industry

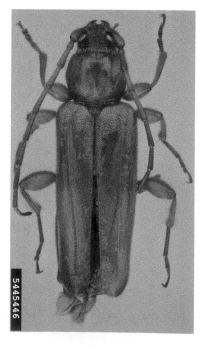

FIGURE 11.17 *Phymatodes testaceus* adult from the United States. (Courtesy of Steven Valley [Bugwood image 5445446].)

because bark is stored either on cut logs or separated from the logs, and *P. testaceus* larvae can complete development in both situations (Craighead 1950). Given that most galleries do not penetrate deeply into the sapwood, *P. testaceus* generally is not considered a major wood pest (Duffy 1953). In the tanbark industry, long-term storage of bark should be avoided to reduce population buildups of *P. testaceus* (Craighead 1950).

11.2.23 *Semanotus bifasciatus* (Motschulsky)

Semanotus bifasciatus is native to China, Japan, Korea, and the Russian Far East (Duffy 1968; Cherepanov 1988a). The primary larval hosts are species of *Juniperus*, but species of *Chamaecyparis, Platycladus, Thuja,* and *Thujopsis* are also infested (Duffy 1968; Cherepanov 1988a; Ma et al. 2010). Adults vary in length from 8 to 14 mm (Cherepanov 1988a), and their head and elytra are shades of brown to black with white transverse bands across the elytra (Cherepanov 1988a; Figure 11.18).

Semanotus bifasciatus generally completes its life cycle in one year in the southern part of its range, although two years is common in its northern range (Kim and Park 1984; Cherepanov 1988a). Adults are most active during spring and summer, emerging earlier in the southern portion of their range. Adults are active at dusk when they mate and lay eggs on the trunks of recently dead and dying trees (Cherepanov 1988a; Iwata et al. 2007). Adults apparently do not feed (Cherepanov 1988a; Yan 2003). Eggs are laid in bark crevices. Average fecundity has been reported by Cherepanov (1988a) as 48 eggs per female and by Yan (2003) as 69 eggs. Kim and Park (1984) reported that in Korea, the egg stage lasted 16–20 days, the larval stage 112–126 days, the pupal stage 15–21 days, and adult lifespan 19 days for females and 16 days for males. After hatching from the eggs, larvae penetrate the bark and feed primarily in the cambial region and later enter the sapwood to pupate (Cherepanov 1988a; Yan 2003). Iwata et al. (2007) reported that, in fallen logs, larvae develop best on the lower portion of the log that is in contact with the soil where the wood moisture content is higher. Pupation occurs in the sapwood in frass-plugged cells constructed at a depth of about 2 cm below the cambial region (Cherepanov 1988a). Pupation occurs in late summer, with new adults remaining within their pupal cells to overwinter until the next year when they chew an oval-shaped exit hole and emerge (Cherepanov 1988a; Yan 2003).

FIGURE 11.18 *Semanotus bifasciatus* adult. (Courtesy of the Pest and Diseases Image Library [Bugwood image 5488621].)

Semanotus bifasciatus typically is considered a secondary pest of stressed and recently dead trees (Iwata et al. 2007). However, occasionally it has been reported as a major pest of living trees, especially in urban and ornamental settings (Kim and Park 1984; Yan 2003; Gao et al. 2007). For example, Gao et al. (2007) reported that *S. bifasciatus* was infesting ancient cypress trees in several areas of China. To minimize losses, logs should be transported from the forest and utilized quickly. For ornamental plantings, various biological control options have been investigated, including mites, nematodes, and parasitoids (Qiu 1999; Sun 2000; Ma et al. 2010).

11.2.24 *Stromatium barbatum* F.

Stromatium barbatum is native to Asia, including Bangladesh, India, Myanmar, Nepal, Pakistan, Sri Lanka, and Thailand as well as the Andaman and Nicobar Islands (Duffy 1968; Vitali 2015). In addition, *S. barbatum* has been introduced to the African countries of Somalia and Tanzania as well as to various nearby islands in the Indian Ocean, including Madagascar, Mauritius, Réunion, Rodrigues, and the Seychelles (Duffy 1968; Duffy 1980; Vitali 2015). Schabel (2006) stated that *S. barbatum* was likely introduced to Africa in the 1950s. Given that *S. barbatum* is a drywood-infesting species, it is capable of developing in a wide array of hosts, with more than 350 species recorded as hosts, including both conifer and hardwood trees as well as bamboo and some woody vines such as grape (Beeson and Bhatia 1939; Duffy 1968; Salini and Yadav 2011). Adults vary in length from 12 to 29 mm, and their general color varies from reddish brown to brownish black (Duffy 1968; Figure 11.19).

Stromatium barbatum completes its life cycle in one to several years. For example, in India, Beeson and Bhatia (1939) reported that adults of the same cohort emerged in one to seven years from the same host material, *Albizia stipulata* (DC.) Boivin, with most (93%) requiring two to four years. The longest development period reported was 10 years (Duffy 1968). Based largely on Beeson and Bhatia (1939) and Duffy (1953), it is known that eggs of *S. barbatum* are laid on rough wood surfaces or in cracks and holes in wood either with or without bark. Eggs are laid singly or in small groups. On average, each female lays about 100 eggs, with a maximum of 246 eggs recorded. Larvae create powdery frass that is packed tightly in their galleries, which occur in both sapwood and heartwood. At times, the galleries are so numerous in individual pieces of wood that only the exterior wood surfaces are left intact. Larvae pupate at the end of their galleries without constructing any discrete pupal cell.

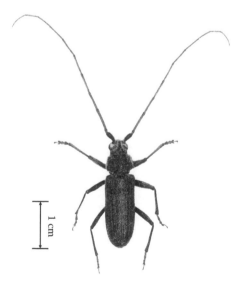

1 cm

UGA2154055

FIGURE 11.19 *Stromatium barbatum* adult. (Courtesy of Christopher Pierce [Bugwood image 2154055].)

The economic impact of *S. barbatum* can be significant given this insect's wide host range and its ability to infest dry, seasoned wood. This species can infest a wide array of wood products, including packaging materials, lumber, furniture, bamboo stakes, and even museum specimens (Duffy 1953; Schabel 2006), and therefore is frequently transported in international trade (Cocquempot et al. 2014). Control efforts often focus on impregnation, fumigation, and the application of various insecticides and protectants (Duffy 1968).

11.2.25 *Strongylurus decoratus* (McKeown)

Strongylurus decoratus is native to Australia (Duffy 1963). The principal host is *Araucaria cunninghamii* Aiton ex D.Don, with *A. bidwillii* Hook. listed as a minor host (Duffy 1963; Elliott et al. 1998). Adults measure 18–28 mm in length and are brown in color, with five white markings on the thorax and black forward-pointing chevron markings on each elytron (Elliott et al. 1998).

Strongylurus decoratus is univoltine. Females oviposit on branches and the terminal shoots of their host plants, typically laying one egg per branch (Wylie 1982; Elliott et al. 1998; Wylie and Speight 2012). The larvae first tunnel within the cambial region and later move into the wood and tunnel in the center of the branch. Prior to pupation, the larva creates a spiral tunnel where the branch usually breaks, with the larva remaining in the tree. The larva then prepares a pupal chamber but first creates a set of exit holes close to where the spiral gallery was originally made as well as another set about 3–11 cm below the first set. It plugs both ends of the gallery between these two sets of potential exit holes with wood fibers and then pupates. The new adult departs the branch using one of the premade exit holes.

The oviposition behavior of *S. decoratus* results in branch pruning. Damage usually is minor, especially when older trees are infested, but on occasion severe damage can occur—especially in young *Araucaria* plantations when the main leader is infested (Wylie 1982; Elliott et al. 1998; Wylie and Speight 2012). This insect has been found infesting trees that varied in age from four to more than 30 years, with heights ranging from 3 to 31 m (Wylie 1982; Wylie and Speight 2012). Wylie (1982) stated that proper site and choice of tree provenance can reduce infestation levels.

11.2.26 *Trichoferus campestris* (Faldermann)

Trichoferus campestris is native to Asia, including Armenia, China, Japan, Kazakhstan, Korea, Kyrgyzstan, Mongolia, Russia, Tajikistan, and Uzbekistan (Cherepanov 1988a; Smith 2009). In addition, *T. campestris* has become established in several countries in Europe (Dascalu et al. 2013) and in both Canada (Grebennikov et al. 2010; Bullas-Appleton et al. 2014) and the United States (Burfitt et al. 2015). Much of the early literature for this species was published under the synonym *Hesperophanes campestris* (Faldermann). Several genera of mostly hardwoods and a few conifers are reported as hosts of *T. campestris* in Asia (Cherepanov 1988a; Iwata and Yamada 1990), but mostly hardwoods have been recorded as hosts in Europe and North America (Dascalu et al. 2013; Bullas-Appleton et al. 2014; Burfitt et al. 2015). Some of the common genera infested include *Acer, Betula, Broussonetia, Gleditsia, Malus, Morus, Prunus, Salix,* and *Sorbus*. Larvae often are associated with stressed and dying hosts but are also able to complete development in dry wood (Smith 2009). Adults measure 11–20 mm in length and are various shades of brown (Smith 2009; Figure 11.20).

Most reports suggest that *T. campestris* requires one to two years to complete a single generation (Cherepanov 1988a; Švácha and Danilevsky 1988; Smith 2009). Adults are active during the summer months and are mostly nocturnal (Cherepanov 1988a; Smith 2009). Eggs usually are laid singly under bark flakes (Cherepanov 1988a). Early instar larvae tunnel primarily in the cambial region, with late instars entering the outer sapwood to pupate. Frass is often extruded from the galleries and can collect at the base of the tree (Bullas-Appleton et al. 2014). Bark appears to be necessary for oviposition and early instar development (Iwata and Yamada 1990). Individuals overwinter as larvae and pupate in spring (Cherepanov 1988a).

Although tree death has not been attributed to *T. campestris*, heavy infestation can reduce tree vigor (Smith 2009). In addition, given that larvae can develop in dry wood (Iwata and Yamada 1990), this species has been found in association with wood rafters and lumber (Dascalu et al. 2013) as well as

UGA2154045

FIGURE 11.20   *Trichoferus campestris* adult. (Courtesy of Christopher Pierce [Bugwood image 2154045].)

wood packaging material used in international trade (Allen and Humble 2002; Cocquempot 2006; Haack 2006). Given that *T. campestris* can infest living, stressed, and recently cut trees, as well as complete development in dry wood, control requires an integrated approach that includes inspection and possible treatment of live trees, rapid utilization of cut trees, and proper heat treatment or fumigation of wood packaging material.

11.2.27 *Xylotrechus altaicus* Gebler

Xylotrechus altaicus is native to Russia (primarily east of the Urals in Siberia) and northern Mongolia (Rozhkov 1970; Cherepanov 1988b) and apparently has not become established elsewhere in the world (as of 2015). The only reported larval host is larch (*Larix*). Adults are 11–24 mm in length, with reddish legs and antennae, blackish head and thorax, and elytra that are brownish in color with two or three lighter-colored transverse bands (Rozhkov 1970; Cherepanov 1988b; EPPO 2005; Figure 11.21).

Xylotrechus altaicus typically requires two years to complete a single generation in Siberia. Adults are active during the summer months, living two to three weeks, and not feeding as adults (Rozhkov 1970; Cherepanov 1988b). Females oviposit in bark crevices, usually laying 50–70 eggs in their lifetime, with a maximum of 145 eggs recorded, and preferring to oviposit on the sunlit side of the tree trunk (Rozhkov 1970; Cherepanov 1988b). Oviposition occurs throughout the trunk. Larvae eclose from the eggs in one to two weeks and at first tunnel back and forth between the cambial region and the outer bark, a behavior that seems to allow the young larvae to escape from host resin (Cherepanov 1988b). During their first winter, larvae reside within the cambial region or outer bark. During the second summer, the larvae first create galleries in the cambial region that are perpendicular to the trunk, which collectively can girdle the host tree. Later that summer, the larvae enter the sapwood and continue to tunnel perpendicular to the trunk and therein spend the second winter. During the second spring, the larvae tunnel close to the bark and construct a pupal cell at the end of their gallery in the outer bark or sapwood. Pupation occurs in early summer and lasts two to three weeks (Cherepanov 1988b).

Xylotrechus altaicus usually is a secondary pest, infesting trees that have been weakened by fire, defoliation, and other stressors. However, during outbreaks of this cerambycid, apparently healthy larch trees can be infested and killed (USDA Forest Service 1991). Rozhkov (1970) reported that some outbreaks can last for decades and extend over large areas. Given that this borer can cause widespread tree

FIGURE 11.21 *Xylotrechus altaicus* adult from Mongolia. (Courtesy of Vladimir Petko [Bugwood image 5174039].)

mortality, it could easily be transported in cut logs and wood packaging (Rozhkov 1970; USDA Forest Service 1991; EPPO 2005). For example, in one study in Russia, an average of 36 larval galleries was recorded within the sapwood per meter-length of trunk (Cherepanov 1988b). Given that *X. altaicus* commonly infests stressed trees, sanitation efforts should be initiated promptly following major stress events such as fire and defoliation (EPPO 2005). In addition, rapid transport and debarking of logs would help reduce borer populations, especially during the first year of infestation.

11.2.28 *Xystrocera festiva* Thomson

Xystrocera festiva is native to Indonesia, Malaysia, and Myanmar (Duffy 1968; Nair 2007). Several broadleaf woody plants serve as hosts, such as species of *Acacia, Albizia, Coffea, Paraserianthes, Pithecolobium*, and *Theobroma* (Duffy 1968). *Paraserianthes falcataria* (L.) Nielsen, for example, is a fast-growing leguminous tree native to Southeast Asia where it is commonly used in plantation forestry and often infested by *X. festiva* (Endang and Haneda 2010). Adults are 30–35 mm in length and overall are brownish in color with blue–green lateral margins along the prothorax and elytra (Nair 2007; Figure 11.22).

Xystrocera festiva is reported to complete a generation in about 190 days (Duffy 1968) and thus potentially could complete two generations per year. However, in nature, all life stages are present at any one time given that there are overlapping generations (Nair 2007). Adults are nocturnal and live for about four to 10 days (Duffy 1968; Nair 2007). Females lay clusters of eggs in bark crevices on live trees, usually 3–8 m above groundline (Nair 2007; Endang and Haneda 2010). The egg clusters usually contain 20–40 eggs each (Johki and Hidaka 1987) but can contain more than 100 eggs at times (Nair 2007). Individual females can lay as many as 170–400 eggs in their lifetime (Duffy 1968; Nair 2007). After hatching, the larvae enter the cambial region and feed gregariously, which is unusual for cerambycids (Johki and Hidaka 1987). While feeding within the cambial region, larvae tunnel mostly in a downward direction, but later, after the larvae enter the sapwood, they construct individual galleries in an upward direction and therein pupate (Nair 2007). The pupal period averages 18 days (Duffy 1968).

Xystrocera festiva can infest *P. falcataria* trees when they are only two to three years old. As a result of the gregarious feeding behavior by larvae in the cambial region, young trees are easily girdled and killed (Nair 2007). Infestations can lead to growth reduction, stem breakage, and tree death (Nair 2007). Trees of all ages can be infested (Endang and Haneda 2010). Pest populations can be lowered by removing heavily infested trees during thinning operations (Nair 2007). In young plantations, annual inspections should be conducted during the first six years with all infested trees removed. Adults can be collected during the tree removal process, with any eggs produced being placed in the plantations to encourage parasitism (Nair 2007).

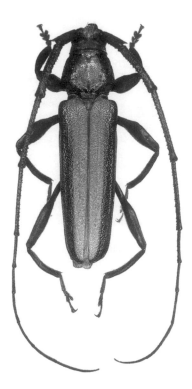

FIGURE 11.22   *Xystrocera festiva* adult from Sumatra. (Courtesy of Udo Schmidt at www.kaefer-der-welt.com.)

11.3 Lamiinae

11.3.1 *Anoplophora glabripennis* (Motschulsky)

Anoplophora glabripennis is native to Asia, primarily in China and Korea; however, during the past 20 years, *A. glabripennis* has become established in several countries in both North America and Europe (Hérard et al. 2006; Haack et al. 2010a; see Chapter 13). *A. glabripennis* is highly polyphagous, infesting broadleaf tree species in at least 15 plant families (Haack et al. 2010a). In Asia, the most commonly infested tree genera include *Acer, Populus, Salix*, and *Ulmus*; and similarly, in North America and Europe, the most commonly infested host genera include *Acer, Aesculus, Betula, Populus, Salix*, and *Ulmus* (Haack et al. 2006; Haack et al. 2010a; Sjöman et al. 2014; Straw et al. 2014; Faccoli et al. 2015; Table 11.2). Adults measure 17–40 mm in length and are glossy black in color, with usually 10–20 distinct irregular-shaped patches on the elytra that range in color from white to shades of yellow and orange (Haack et al. 2010a; Figure 11.23).

The life cycle of *A. glabripennis* usually is completed in one year, but two years may be needed in colder parts of its range (Haack et al. 2010a). Adults are most active in the summer months. After emergence, adults conduct maturation feeding for one to two weeks on twigs and foliage of host trees. Adults usually mate on the trunks and branches of host trees and then females chew funnel-shaped oviposition pits through the bark into which a single egg is inserted. Oviposition usually starts in the upper trunk and branches and then occurs lower along the trunk in succeeding years (Haack et al. 2006). Eggs typically hatch in one to two weeks, and the emerging larvae first tunnel in the cambial region and later move into the sapwood and heartwood. Overwintering usually occurs in the larval stage, with pupation occurring in late spring to early summer. Pupation occurs at the end of the larval gallery. The new adults emerge through circular exit holes that are 10–15 mm in diameter (Haack et al. 2010a).

Anoplophora glabripennis can cause widespread tree mortality because it successfully infests apparently healthy trees and causes tree death after several years of successive attack. Extensive tree mortality

TABLE 11.2

Summary Data (as of January 2016) for the Status of Infestations of *Anoplophora glabripennis* in Five U.S. States.

State	Year of Discovery	Status in December 2015	No. Infested Trees Cut[a]	No. High Risk Trees Cut[b]	Top Five Infested Tree Genera as of April 2015
Illinois	1998	Eradicated	1,551	220	*Acer, Ulmus, Fraxinus, Salix, Aesculus*
Massachusetts	2008	Ongoing	24,404	10,679	*Acer, Betula, Ulmus, Populus, Fraxinus*
New Jersey	2002	Eradicated	730	21,251	*Acer, Ulmus, Salix, Betula*
New York	1996	Ongoing	7,062	16,658	*Acer, Ulmus, Salix, Aesculus, Betula*
Ohio	2011	Ongoing	16,446	61,557	*Acer, Aesculus, Ulmus, Salix, Populus*

[a] Total number of trees cut during the eradication program that were considered infested with *A. glabripennis* by state.
[b] Total number of trees cut during the eradication program that were considered high risk, which meant that they were potential host trees growing close to infested trees but not known to be infested themselves.

FIGURE 11.23 *Anoplophora glabripennis* adult from United States. (Courtesy of Steven Valley [Bugwood image 5449526].)

has occurred in China, involving millions of trees, especially in northern China where large-scale tree planting efforts occurred (Pan 2005; Zhao et al. 2007). Given this threat, active eradication programs have been undertaken in all European and North American countries where *A. glabripennis* has been introduced. For example, in the United States, established populations of *A. glabripennis* had been found in five states as of January 2016 (Illinois, Massachusetts, New Jersey, New York, and Ohio), with the first populations being found in New York in 1996 and the most recent in Ohio in 2011. Of these five states, all *A. glabripennis* infestations have been eradicated in only Illinois and New Jersey, while efforts are still ongoing in the other three states (Table 11.2).

Details on the control efforts for *A. glabripennis* in China, Europe, and North America are provided in Hérard et al. (2006) and Haack et al. (2010a). Eradication is the goal for all outbreak areas in Europe and North America and generally include tree surveys of varying intensities in concentric zones around known infested areas, removal of all infested trees, and at times removal of nearby host trees that are not obviously infested but could be (i.e., so-called high-risk trees; see Table 11.2). In the United States, observers at first only surveyed trees for signs of infestation from the ground using binoculars, but later, tree climbers and bucket trucks were used to improve efficiency in finding *A. glabripennis* exit holes and oviposition pits. After infested trees are cut, they are chipped and sometimes burned. In China, *A. glabripennis* greatly expanded its natural range in the late 1900s as a result of large-scale tree planting efforts that used many species that were highly susceptible to *A. glabripennis* (Haack et al. 2010a). As a result, millions of infested trees were cut in China as well as an internal quarantine was established to reduce the likelihood of human-assisted spread (Haack et al. 2010a).

11.3.2 *Apriona swainsoni* (Hope)

Apriona swainsoni is native to China, India, Korea, Laos, Myanmar, Thailand, and Vietnam (Duffy 1968; Liu et al. 2006). The principal hosts include woody tree and vine species of *Butea, Caesalpinia, Dalbergia, Ligustrum, Paulownia, Salix, Sophora*, and *Tectona* (Duffy 1968; EPPO 2013). In China, Liu and Tang (2002) reported that *Sophora japonica* L. was one of the most highly susceptible tree species. Adults measure 25–40 mm in length and are brown in color with white specks (Beeson 1941; Figure 11.24).

Apriona swainsoni tends to have a two-year life cycle (Duan 2001). Adults are active during the summer months, mating and ovipositing primarily at night (Duan 2001). Adults conduct maturation feeding on host bark (Duan 2001; Wang 2011). Females lay 27–62 eggs each, placing them in bark crevices along the trunk and main branches (Duan 2001; Wang 2011). Larval galleries are constructed initially in the cambial region and later in the sapwood with larvae first tunneling toward the center of the stem or branch and then upward and finally outward to nearly the wood surface (Duan 2001). Individuals overwinter as larvae in both years of development. Pupation occurs in late spring to early summer.

FIGURE 11.24 *Apriona swainsoni* adult from Vietnam. (Courtesy of Ben Sale at commons.wikimedia.org.)

Beeson (1941) and Duffy (1968) report that when *Butea* vines are present along tree trunks, females often oviposit in the vines, with the resulting larvae first developing in the vine and then tunneling into the supporting tree to complete development.

Apriona swainsoni can cause severe damage to trees growing in natural forests, plantations, and in urban settings (Tang and Liu 2000; Liu et al. 2006; EPPO 2013). Infested trees usually are at least 7–8 cm in diameter (Duan 2001; Liu and Tang 2002); however, Beeson (1941) reported that some teak trees of more than 1 m in diameter were infested. In China, urban plantings of *S. japonica* have been heavily infested, with many trees exhibiting 60–70 exit holes each (Liu et al. 2006). In addition, given that *A. swainsoni* is not found throughout all of China, an internal quarantine has been in place since 1996 (Liu et al. 2006). In teak plantations, it is recommended that vines be removed from the trunks to reduce oviposition on the vines (Beeson 1941; Duffy 1968). Various biological control agents have been tested for their ability to control *A. swainsoni* in China (Lu et al. 2011; Wang et al. 2014).

11.3.3 *Aristobia horridula* Hope

Aristobia horridula occurs in China, India, Myanmar, Nepal, Thailand, and Vietnam (Duffy 1968; Dhakal et al. 2005; Nair 2007; Wylie and Speight 2012). The principal hosts include species of *Dalbergia, Pterocarpus*, and *Xylia* (Beeson 1941; FAO 2007; Nair 2007; Wylie and Speight 2012). Adults measure 27–32 mm in length and are brown in color, with elytra studded with bristle-like long bluish hairs and dense tufts of hairs on the distal portion of the first and second antennal segments (Nair 2007).

Aristobia horridula generally is univoltine, with adults most active during the summer months in India but apparently year-round in Thailand (Hutacharern and Panya 1996; Nair 2007). Adults are active during the day, feeding on bark of young branches and laying eggs singly in transverse cuts made in the bark of living trees (Nair 2007). Adult movement usually occurs in short bursts of flight of just 10–20 m (Hutacharern and Panya 1996). Most oviposition occurs along the lower trunk (Nair 2007). The larvae construct galleries that measure 50–75 cm in total length and penetrate both sapwood and heartwood (Hutacharern and Panya 1996; Nair 2007). As larvae feed, infested trees often become swollen and exude resin near the oviposition site (Nair 2007). Pupation occurs at the end of the gallery in the heartwood. Adults produce circular exit holes.

Aristobia horridula has caused extensive damage in young plantations of *Dalbergia cochinchinensis* Pierre ex Laness, *D. sissoo* Roxb., and *Pterocarpus macrocarpus* Kurz in India and Nepal, and similarly to plantations of *D. cochinchinensis, P. macrocarpus*, and *Pterocarpus indicus* Willd. in Thailand (Mishra et al. 1985; Hutacharern and Panya 1996; Dhakal et al. 2005; Nair 2007; Wylie and Speight 2012). When small-diameter trees are infested, they are prone to wind throw (Nair 2007). Infestation levels in plantations often reach 30–90% and, in some stands, all trees can be infested (Mishra et al. 1985; Hutacharern and Panya 1996; Dhakal et al. 2005). To protect trees, the lower 2 m of trunk often are treated with insecticides or other solutions (Hutacharern and Panya 1996; Nair 2007).

11.3.4 *Goes tigrinus* (DeGeer)

Goes tigrinus is found throughout the eastern United States where it infests species of *Quercus*, especially members of the white oak group (Solomon 1995). Adults measure 22–44 mm in length and have an overall brownish-grayish mottled appearance that forms an irregular dark band across the lower half of the elytra (Linsley and Chemsak 1984; Figure 11.25).

Goes tigrinus usually requires three to four years to complete a single generation (Solomon 1995). Adults become active in spring to early summer in the southern United States but not until mid-summer in the northern United States (Solomon 1995). Adults feed on twigs and leaf petioles for one to two weeks after emergence before mating and initiating oviposition. Females are reported to lay only 9–15 eggs in their lifetime (Solomon 1995). Oviposition involves first chewing an oval pit through the bark before inserting a single egg through the base of the pit into the cambial region. Larvae tend to tunnel upward for 11–23 cm, first in the sapwood and then in the heartwood (Solomon 1995). Larvae expel frass and wood shavings from near the origin of their gallery that accumulate at the base of the tree (Solomon 1977, 1995). Individuals overwinter as larvae in all years of development. Pupation occurs at the end of the

FIGURE 11.25   *Goes tigrinus* adult from Texas (United States). (Courtesy of Mike Quinn at bugguide.net.)

gallery in spring or early summer, after which the new adult extends the gallery from the pupal chamber to the bark, chewing a circular exit hole (Solomon 1995).

Goes tigrinus infests *Quercus* trees in both urban and natural forest settings. Although infestation by *G. tigrinus* seldom causes tree death, the value of lumber from heavily infested trees is greatly reduced given that larval galleries penetrate both sapwood and heartwood (Solomon 1995). In Ohio, Donley (1978b) demonstrated that young oak trees, 10–25 cm in diameter at breast height (DBH), were most often infested. Moreover, it was shown that within most forest stands, there usually are just a few "brood" trees (often less than 5% of all host trees present) that produce most of the borers, and that by removing these brood trees, the borer population could be greatly reduced (Donley 1978b; Solomon 1995).

11.3.5 *Monochamus alternatus* Hope

Monochamus alternatus is native to China, Japan, Korea, Laos, Taiwan, and Vietnam (Duffy 1968; Akbulut and Stamps 2012). The larval hosts of *M. alternatus* are conifers in the genera *Abies, Cedrus, Larix, Picea*, and *Pinus* (Akbulut and Stamps 2012). Adults measure 18–31 mm in length and have an overall reddish-brown color along with two pinkish longitudinal stripes on the pronotum and several small white and black patches on the elytra (Duffy 1968; Figure 11.26).

Monochamus alternatus typically is univoltine; however, in southern China, it may have two generations per year, and occasionally, two years are required to complete a single generation, especially when eggs are laid late in the year (Song et al. 1991; Akbulut and Stamps 2012; Togashi 2013). Adults are active during the summer months throughout much of their range and nearly year-round in their southern range (Song et al. 1991). Adults live for several weeks, feeding at first on twigs of healthy host trees to become sexually mature, and then ovipositing on trunks and branches of stressed, dying, or recently cut host trees (Togashi and Magira 1981). Females chew an oviposition pit in the bark and then usually lay a single egg within the cambial region (Togashi and Magira 1981). Depending on the time of year of adult emergence, females lay (on average) 23–86 eggs in their lifetime (Togashi and Magira 1981). Eggs hatch in about a week. The early larval instars feed and tunnel in the cambial region, with late instars entering the sapwood for a few centimeters and then turning to follow the wood grain and constructing

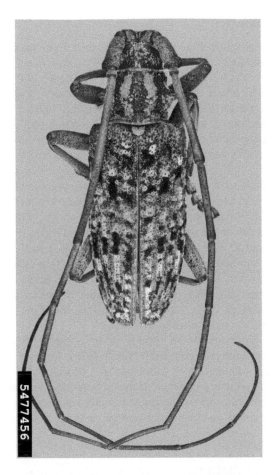

FIGURE 11.26 *Monochamus alternatus* adult. (Courtesy of Steven Valley [Bugwood image 5477456].)

a pupal chamber. Individuals overwinter in the larval stage. Adults create a circular exit hole when they emerge (Kobayashi et al. 1984).

The galleries of *M. alternatus* often are numerous, and because they penetrate deep within the wood, they lower the value of logs used for lumber (Duffy 1968). In addition, *M. alternatus* is the principal vector in Asia of the pinewood nematode, *Bursaphelenchus xylophilus* (Steiner & Buhrer) Nickle, which is the causal agent of pine wilt disease (Kobayashi et al. 1984; Togashi 2013). When adults emerge from infected trees, they can carry thousands of nematodes in their tracheal system, which are later released into healthy and dying pine trees when the adult beetles feed and oviposit (Kobayashi et al. 1984; Akbulut and Stamps 2012).

A thorough review of the control options available for *M. alternatus* has been prepared by the Centre for Agriculture and Biosciences International (CABI 2015a). Infested logs can be chipped, burned, or buried under at least 15 cm of soil. In outbreak areas of pine wilt disease, localized clear-cutting of infested trees followed by log treatment has reduced disease spread. Several pesticides have been tested and include insecticides applied to tree trunks and foliage as well as to target adults because they conduct maturation feeding and oviposit. Nematicides can also be used to inhibit development of pine wilt disease. Various biological control strategies have been developed using entomophagous fungi, nematodes, and larval parasitoids. Trap trees can be used to attract ovipositing adults followed by removal and destruction of the trees or logs. Traps baited with pheromones or with alpha-pinene and ethanol can be used to capture local adults and to serve as a tool for monitoring and control. In addition, breeding programs for resistance to pine wilt disease in Asian pines have been developed in China and Japan (Nose and Shiraishi 2008).

11.3.6 *Monochamus sutor* (L.)

Monochamus sutor is native to much of the Eurasian conifer belt from Spain through China and Russia to Korea and Japan (Bílý and Mehl 1989; Cherepanov 1990). The primary larval hosts of *M. sutor* in Europe are *Picea* and *Pinus* (Duffy 1953; Bílý and Mehl 1989; Bense 1995), with *Larix* being a major host in Siberia (Rozhkov 1970; Zhang et al. 1993) and *Abies* also serving as an occasional host (Novák 1976). *M. sutor* adults measure 15–26 mm in length (Cherepanov 1990). They have a strong spine on each side of the pronotum and overall are black in color with a yellow spot on the scutellum and several yellowish to whitish colored spots on the elytra that form three transverse dark-colored bands (Novák 1976; Bense 1995; Figure 11.27). Antennae in males are one-and-a-half to two times the body length (Rozhkov 1970).

Depending on local conditions, *M. sutor* requires one to three years to complete a single generation. Adults generally are active from late spring through summer. Adults conduct maturation feeding on twig bark for several days (Novák 1976). Females construct oviposition pits in the bark and then insert their ovipositor at the base and deposit one or a few eggs into the cambial region (Rozhkov 1970; Novák 1976). On average, females lay about 50 eggs in their lifetime (Duffy 1953). Early larval instars tunnel in the cambial region, scoring the sapwood deeply, and expel frass through the bark (Novák 1976). Late instars enter the sapwood and heartwood. Individuals overwinter as larvae. Some larval galleries extend from one side of the tree to the other side, while others are U-shaped in which the gallery starts and ends on the same side of the tree (Trägårdh 1930). Pupation occurs at the end of the gallery in spring or summer with the new adults emerging through a circular exit hole.

Monochamus sutor oviposits primarily on recently cut trees, especially along the lower trunk, and therefore greatly reduces their value (Duffy 1953; Novák 1976). At times, *M. sutor* has reached outbreak levels following large-scale defoliation events and fire (Rozhkov 1970; Zhang et al. 1993; Yuan et al. 2008). In addition, *M. sutor* is a potential vector of pinewood nematodes (Akbulut and Stamps 2012). Given the nature of *M. sutor* to rapidly infest recently cut trees, harvesting operations need to swiftly transport logs to sawmills and then process them, or if not possible, logs should be debarked (Rozhkov 1970).

FIGURE 11.27   *Monochamus sutor* adult from Austria. (Courtesy of Siga at commons.wikimedia.org.)

11.3.7 *Monochamus titillator* (F.)

Monochamus titillator is native to the eastern United States, Canada (Ontario), and the Bahamas (Browne et al. 1993; Evans 2014; Bezark 2015a) and was apparently introduced to Cuba (Peck 2005). The primary larval hosts include species of *Pinus* (Webb 1909; Baker 1972; Drooz 1985), but some authors also list *Abies* and *Picea* as potential hosts (Duffy 1953; Linsley and Chemsak 1984). Adults of *M. titillator* measure 16–31 mm in length and have a mottled appearance with a reddish-brown background color with small brown and gray patches (Webb 1909; Linsley and Chemsak 1984; Figure 11.28). The antennae are more than twice the body length in males and about the same length as the body in females (Linsley and Chemsak 1984).

The life cycle of *M. titillator* generally is univoltine with a partial second generation occurring in the southern United States (Webb 1909). Akbulut and Stamps (2012), however, state that *M. titillator* can complete up to three generations per year. Generations are overlapping, with most adults being present in the summer months, but likely year-round in the southern United States (Webb 1909; Baker 1972). Newly emerged adults feed on the bark of pine twigs until sexually mature (Luzzi et al. 1984; Akbulut and Stamps 2012). Typically, adults mate and oviposit on the trunks and major branches of severely weakened host trees, such as wind-thrown trees, bark beetle-infested trees, or on freshly cut logs (Webb 1909; Coulson et al. 1976; Akbulut and Stamps 2012). Females usually chew a funnel-shaped pit through the outer bark and then insert their ovipositor and lay one to nine eggs in a radial pattern (Webb 1909; Dodds et al. 2002). The eggs hatch in about a week with the new larvae first feeding in the cambial region and then entering the sapwood—usually constructing a deep U-shaped gallery—and then back to near the wood surface (Webb 1909). Pupation occurs at the end of the gallery with the new adult chewing a circular exit hole through the sapwood when it emerges (Webb 1909).

Most economic losses attributed to *M. titillator* result from the larval galleries that reduce wood quality and value. For example, Webb (1909) reported that, after a major wind event in the southern United States, nearly all storm-damaged trees were infested by *M. titillator* within a few months, and their galleries affected about 25% of the lumber cut from each log. *M. titillator* is also a major vector

UGA5205031

FIGURE 11.28 *Monochamus titillator* adult from Louisiana (United States). (Courtesy of Natasha Wright [Bugwood image 5205031].)

of the pinewood nematode in North America (Luzzi et al. 1984) and therefore poses a risk to Asian and European pines if introduced elsewhere in pine logs, lumber, or wood packaging (Evans et al. 1996; Sathyapala 2004). Logs should be utilized quickly to minimize losses, especially from late-instar larvae that enter the sapwood. When rapid utilization is not possible, logs can be stored under water sprinklers, which limit oviposition (Syme and Saucier 1995).

11.3.8 *Plectrodera scalator* (Fabricius)

Plectrodera scalator is native to the eastern United States where it primarily utilizes species of *Populus* as larval hosts but will also infest *Salix* (Linsley and Chemsak 1984; Solomon 1995; Bezark 2015a). Adults of *P. scalator* measure 25–40 mm in length and generally are black in color with large white areas on the head and around the eyes, a series of white transverse bands on the thorax and elytra that produce a checkered appearance, and strong lateral prothoracic black spines (Linsley and Chemsak 1984; Solomon 1995; Figure 11.29).

The life cycle of *P. scalator* usually is completed in one to two years (Solomon 1995). Adults are most active in the summer months, first feeding on leaf petioles and twigs to become sexually mature, and then mating and ovipositing at the base of host trees. Adult females chew pits in the bark just below the soil line, insert their ovipositor, and usually lay a single egg (Solomon 1980, 1995). Eggs hatch in two to three weeks, with the new larvae first tunneling downward within the cambial region of the taproot, and eventually mine the sapwood (Solomon 1980). Most galleries are below ground, but some are found above ground level, especially when multiple larvae are found in the same host (Solomon 1980). Pupation occurs within the gallery in spring or early summer, lasting about three to four weeks (Solomon 1980). New adults chew exit holes through the bark near the groundline (Solomon 1980).

Plectrodera scalator infests nursery stock as well as young trees in plantations and natural stands (Solomon 1980). In nurseries, one- to two-year-old seedlings commonly are infested. A single larva seldom kills the seedling outright but can weaken it and make it susceptible to breakage. However, when

FIGURE 11.29 *Plectrodera scalator* adult from Oklahoma (United States). (Courtesy of Ben Sale at commons.wikimedia. org.)

multiple larvae occur in a single seedling or young tree, stem breakage is common (Solomon 1980). In one survey, Solomon (1980) reported that, on average, 4.6 larvae per tree were found in three-year-old *Populus* trees. In nursery beds, *Populus* trees can be protected with insecticides (Solomon 1980). When selecting new nursery sites, managers should choose sites far from known active *P. scalator* infestations, and they should use uninfested cuttings to establish the nursery. Moreover, it is recommended to rogue old rootstocks every three years and start with new cuttings to keep *P. scalator* populations low (Solomon 1980).

11.3.9 *Saperda calcarata* Say

Saperda calcarata is native to both Canada and the United States, being found throughout most of North America (Linsley and Chemsak 1984). The principal larval hosts of *S. calcarata* include species of *Populus*, with species of *Salix* being reported as occasional hosts. *S. calcarata* adults are 18–31 mm in length and have an overall grayish color with brown dots and a faint yellowish pattern on the head, prothorax, and elytra (Linsley and Chemsak 1984; Solomon 1995; Figure 11.30). The antennae are about the length of the body.

Saperda calcarata generally has a two-year life cycle in the southern United States but three to five years in the northern United States and Canada (Drouin and Wong 1975; Solomon 1995). Adults are active from late spring through summer, first feeding on foliage and twigs for about a week before mating and ovipositing. Females chew niches in the bark along the trunk and major branches and usually lay one egg per niche (Craighead 1923; Drouin and Wong 1975). Females usually oviposit on living trees that are at least 8 cm in diameter, but trees as small as 4 cm in diameter have been infested (Solomon 1995). Females oviposit preferentially on open-grown trees (Solomon 1995). Eggs hatch in about two weeks, and the larvae first feed in the cambial region and then in the sapwood and heartwood (Craighead 1923; Drouin and Wong 1975). Completed larval galleries are 15–25 cm in length (Solomon 1995). Individuals overwinter as larvae in each year of development. Larvae expel frass and wood shavings from enlarged holes constructed where they first entered the tree (Drouin and Wong 1975). Trees respond to the larval feeding by exuding sap that flows from the entry holes (Solomon 1995). Multiple larvae often are found feeding in the same region of a tree (Solomon 1995). Pupation occurs at the end of the gallery in spring or early summer, with the new adults emerging from the original entry hole (Solomon 1995).

FIGURE 11.30 *Saperda calcarata* adult in United States. (Courtesy of Whitney Cranshaw [Bugwood image 1325084].)

Small trees can be girdled and killed by *S. calcarata* larvae, but infestations seldom kill larger trees (Solomon 1995). However, larval feeding, especially when multiple larvae are present, weakens trees and makes them prone to breakage from wind and snow (Drouin and Wong 1975; Solomon 1995). Solomon (1995) provided data from several surveys conducted in Canada and the United States in which an average of 20–63% of the *Populus* trees inspected had evidence of current and past *S. calcarata* infestation. In addition, larval galleries can serve as infection courts for various canker fungi (Anderson and Martin 1981). When infested trees are used for lumber, their value is reduced because of the larval galleries and staining (Solomon 1995). Silvicultural control options include proper site selection, maintaining well-stocked stands, and harvesting stands at maturity (Solomon 1995). Several systemic insecticides were tested against larvae by Drouin and Wong (1975).

11.3.10 *Saperda carcharias* (L.)

Saperda carcharias is native to much of Palearctic Eurasia from Spain and the United Kingdom eastward to Siberia, China, and Korea (Novák 1976; Bílý and Mehl 1989; Bense 1995). The principal larval hosts are species of *Populus*, with *Salix* species serving as occasional hosts (Duffy 1953; Bense 1995). *S. carcharias* adults are 18–30 mm long and have an overall dark body with yellowish-gray to yellowish-brown pubescence and numerous black spots on the head, prothorax, and elytra (Ritchie 1920; Novák 1976; Bense 1995; Figure 11.31). In general, the antennae of males are slightly longer than the body, whereas in females they are shorter than the body length (Novák 1976).

The life cycle of *S. carcharias* usually is completed in two to four years (Ritchie 1920; Bílý and Mehl 1989). Adults are active during the summer months. After emergence, adults conduct maturation feeding on foliage and twig bark of host trees (Ritchie 1920; Novák 1976). Adults usually mate on twigs and then the females oviposit on the trunks of both young and mature live trees, especially those growing in open settings (Novák 1976). Females use their mandibles to cut a vertical slit in the bark and then turn and deposit a single egg under the bark (Ritchie 1920). On average, each female lays 40–50 eggs (Novák 1976). Larvae do not eclose from the eggs until the following spring at which time they first feed in the cambial region and then enter the sapwood, tunneling downward at first and then reversing

FIGURE 11.31 *Saperda carcharias* adult on *Populus* in Latvia. (Courtesy of AfroBrazilian at commons.wikimedia.org.)

direction and tunneling upward (Ritchie 1920; Novák 1976). Larvae eject frass from their galleries, which can extend 25–50 cm overall (Novák 1976). Several larvae can be found in the same trees, often tunneling near each other (Ritchie 1920). Trees often become swollen around the feeding sites (Novák 1976). Pupation occurs at the end of the larval gallery in spring and lasts for about one month (Novák 1976). Adults emerge through an oval-shaped exit hole (Ritchie 1920).

Infestation by *S. carcharias* can cause severe economic losses given that otherwise healthy trees often are infested and that the larval galleries extend to the heartwood. Small-diameter trees can be killed by just a few larvae. Larger-diameter trees often suffer growth loss, and the galleries can serve as infection courts for fungi. Lumber from infested trees is greatly reduced in value because of the extensive tunneling (Ritchie 1920; Duffy 1953; Novák 1976; de Tillesse et al. 2007). Insecticides have been used to protect high-value trees, with products applied to the trunks to kill early larval instars, and systemic insecticides have been used for late-instar larvae that are deeper in the wood (CABI 2010). Various barriers have also been tested to restrict adult beetles from climbing on trunks (Allegro 1990).

11.4 Parandrinae

11.4.1 *Neandra brunnea* (Fabricius)

Neandra (= *Parandra*) *brunnea* is native to Canada and the United States, being most common in eastern North America (Brooks 1915; Linsley 1962a). In addition, *N. brunnea* has been considered established in the area of Dresden, Germany, since the early 1900s, likely having been moved in association with World War I (Cocquempot and Lindelöw 2010). Most hardwoods and some conifers serve as larval hosts to *N. brunnea*. Solomon (1995) listed the following hardwood genera as common North American hosts: *Acer, Carya, Castanea, Fagus, Fraxinus, Juglans, Liquidambar, Liriodendron, Malus, Populus, Prunus, Pyrus, Quercus, Robinia, Salix, Ulmus,* and *Tilia*. In Germany, *Populus* and *Tilia* have been reported as larval hosts (Cocquempot and Lindelöw 2010). Adults measure 8–21 mm in length, are in various shades of brown color, and are very similar in appearance to some stag beetles (Lucanidae) (Linsley 1962a; Baker 1972; Solomon 1995; Figure 11.32).

The typical life cycle of *N. brunnea* is thought to span two to four years (Gahan 1911; Brooks 1915; Craighead 1950; Baker 1972; Solomon 1995). Adults are active in summer, often laying eggs in exposed

FIGURE 11.32 *Neandra brunnea* adult from United States. (Courtesy of Joseph Berger [Bugwood image 5385937].)

wood near the ground, such as near bark wounds or on wood structures that are in contact with moist soil (Craighead 1950). Apparently some adults do not emerge from the wood when mature but rather stay within the wood, mate, and lay eggs (Craighead 1950). The larvae tunnel in both sapwood and heartwood, often gregariously, which tends to honeycomb the wood. The individual galleries are packed with granular frass. The mature larva constructs an oval pupal cell at the end of the gallery and also plugs the gallery with a wad of wood fibers. The pupal period usually spans about a month (Brooks 1915).

Neandra brunnea has been reported as a destructive pest in the United States since the late 1800s (Brooks 1915). Most reports have referred to stem breakage of mature shade trees and fruit trees especially in the lower meter of trunk, where most larval tunneling occurs (Brooks 1915). During the late 1800s and early 1900s, *N. brunnea* was considered a major pest of wood poles used to support telegraph and telephone lines in the eastern United States, especially those made from *Castanea* trees (Snyder 1911; Gahan 1911). Given that oviposition usually occurs near exposed dead wood, wounding of trees should be avoided, and when wounds do occur, they should be covered if possible (Brooks 1915; Solomon 1995).

11.5 Prioninae

11.5.1 *Mallodon downesii* Hope

Mallodon downesii is found throughout much of central and southern Africa, from Senegal to Kenya to South Africa, and also in Madagascar (Delahaye and Tavakilian 2009). This species is highly polyphagous, infesting a wide variety of woody plants that include dozens of crop trees (e.g., cacao, ceara rubber, coffee, and kapok) as well as timber and ornamental trees (e.g., species of *Acacia, Albizia, Ceiba, Cleistopholis, Cordia, Daniellia, Delonix, Ekebergia, Ficus, Khaya, Lophira, Parinari, Pterocarpus, Tamarindus, Tectona,* and many more) (Duffy 1957, 1980; Roberts 1969; Rathore 1995; Delahaye and Tavakilian 2009). *M. downesii* is found in many environments from rain forest to savanna (Roberts 1969). *M. downesii* adults measure 27–70 mm in length and are reddish brown in color with very large mandibles, antennae about half the length of the body, and elytra that are lustrous with reflexed margins (Duffy 1957; Delahaye and Tavakilian 2009; Figure 11.33).

The life cycle of *M. downesii* generally is thought to be completed in a year or less (Duffy 1957). Adults can be found nearly year-round in many locations but from March to October is most common (Roberts 1969). Eggs are laid in bark cracks or under bark of recently cut trees, logs, stumps, and exposed

FIGURE 11.33   *Mallodon downesii* adult male from Cameroon. (Courtesy of Ben Sale at commons.wikimedia.org.)

roots (Duffy 1957). Females will oviposit on both decayed and sound wood. The larvae first tunnel in the sapwood and later in the heartwood, penetrating to depths of more than 20 cm (Duffy 1957). The larval galleries are filled with coarse wood shavings. Pupation occurs within the wood often at a depth of 10–15 cm (Duffy 1957).

In many situations, *M. downesii* is considered a minor pest given that oviposition often occurs in well-decayed wood that has little commercial value (Duffy 1957). However, at times, *M. downesii* infests dying trees and recently cut logs that have sound wood; in these situations, the commercial value of the lumber is reduced (Duffy 1957; Roberts 1969). Given this beetle's life history traits, infestations could be lowered by utilizing logs soon after harvesting and by chemically treating or possibly chipping or grinding logging debris and stumps (Mayné 1914).

11.5.2 *Paroplites australis* (Erichson)

Paroplites australis is native to Australia, being found primarily in the coastal areas of New South Wales, Queensland, Tasmania, and Victoria (McKeown 1947; Duffy 1963). The principal host plants are members of the genus *Banksia*, especially *Banksia serrata* L.f., but *P. australis* has also been reported to infest native and introduced trees in the genera *Angophora, Betula, Casuarina, Eucalyptus, Quercus*, and *Salix* (Froggatt 1923; McKeown 1947; Duffy 1963; Fearn 1989; Hawkeswood 1992; Hawkeswood and Turner 2003). *P. australis* adults are 27–53 mm long, dark reddish brown, with slightly flattened elytra, antennae shorter than its body length, and a flattened prothorax that is finely toothed along the outer margin (Froggatt 1923; Elliott et al. 1998; Figure 11.34).

The duration of the life cycle has not been studied in detail but is likely annual based on observations given in Fearn (1989) and Hawkeswood (1992). Adults are present during the summer months, mating and laying eggs at night, while being mostly inactive on the bark of host trees during the day (Froggatt 1923; Hawkeswood 1992; Hawkeswood and Turner 2003). Adults lay eggs on both living and dead host trees (Hawkeswood 1992). The larvae tunnel deep within the sapwood and heartwood of the main trunk and major branches, creating wide galleries that are packed with frass and wood shavings (Froggatt 1923; Hawkeswood 1992). Several larvae can be present within the same tree. For example, Froggatt (1923) mentioned that more than 30 adults were reared from the trunk section of a single *B. serrata* tree. Pupation occurs at the end of the gallery in a large oval chamber, which is constructed close to the wood surface (Froggatt 1923; Hawkeswood 1992).

Trees in both natural and urban settings are infested by *P. australis* (Froggatt 1923; Hawkeswood 1992). Heavily infested trees are easily broken during wind storms (Froggatt 1923). *P. australis* potentially is a major pest given that it infests living trees in several plant families. Froggatt (1923) recommends destruction of infested live trees as well as recently cut or dead host trees.

FIGURE 11.34 *Paroplites australis* adult from Australia. (Courtesy of CSIRO at scienceimage.csiro.au.)

11.6 Spondylidinae

11.6.1 *Tetropium castaneum* (L.)

Tetropium castaneum is native to Palearctic Eurasia from Spain through Russia to China, and Mongolia, Korea, and Japan (Novák 1976; Bílý and Mehl 1989; Bense 1995). *T. castaneum* is not known to be established in North America, although it has been intercepted and trapped on occasion (LaBonte et al. 2005). Species of *Picea* and *Pinus* are the preferred larval hosts; however, *T. castaneum* will also infest species of *Abies* and *Larix* (Rozhkov 1970; Novák 1976). *T. castaneum* adults measure 9–18 mm in length and usually have a black head and pronotum with brown or black elytra (Novák 1976; Figure 11.35). The adult body is flattened dorsoventrally, their eyes are separated into two halves, and they have a pronounced longitudinal groove between the antennae (Novák 1976).

Tetropium castaneum usually completes its life cycle in one year, but two years may be needed occasionally. Adults are active throughout the summer months, especially during June–July, mating and laying eggs on the bark of host trees. Eggs usually are laid singly or in small batches in bark crevices or under bark flakes, hatching in 10–14 days (Novák 1976; Bílý and Mehl 1989). Larvae feed initially in the cambial region, constructing irregular galleries with granular frass, and then enter the sapwood, tunneling perpendicular to the wood surface at first to a depth of 2–4 cm and then boring parallel to the surface of the wood for another 3–4 cm (Trägårdh 1930; Novák 1976). Individuals overwinter in the larval stage at the end of their galleries, pupating in spring or summer. Adults emerge through oval-shaped exit holes.

Tetropium castaneum typically infests recently cut logs and stumps but can infest living trees, especially those stressed by drought, air pollution, disease, and defoliation. Larval galleries cause structural damage and can reduce the value of lumber (Novák 1976). Living trees can be killed by *T. castaneum* especially after multiple years of infestation. Logs should be utilized quickly to minimize infestation and larval tunneling (Novák 1976).

FIGURE 11.35 *Tetropium castaneum* adult from Russia. (Courtesy of Karill Makarov at cerambycidae.org.)

11.6.2 *Tetropium fuscum* (F.)

Tetropium fuscum is native to Eurasia, occurring from central and northern Europe to Siberia and Japan (Novák 1976; Bílý and Mehl 1989; Bense 1995). Outside its native range, *T. fuscum* has become established in Canada, being first reported in Nova Scotia in 2000 (Smith and Hurley 2000) and showing very limited spread as of 2010 (Rhainds et al. 2011). In Europe, the most common larval hosts of *T. fuscum* include species of *Picea* and *Pinus* (Bílý and Mehl 1989; Bense 1995), but *Abies* and *Larix* may also be infested (Novák 1976). In Canada, only species of *Picea* have been reported as larval hosts so far (Smith and Humble 2000; Flaherty et al. 2011). *T. fuscum* adults are 8–18 mm in length, appear dorsoventrally flattened, and their head and pronotum are dark brown to black, while the elytra are various lighter shades of brown (Novák 1976; Smith and Humble 2000; Figure 11.36). The eyes are divided (Smith and Humble 2000).

The biology of *T. fuscum* is very similar to that described earlier for *T. castaneum*. The life cycle of *T. fuscum* usually is univoltine in both Eurasia (Bílý and Mehl, 1989) and Canada (Smith and Humble 2000). Adults are active from late spring through summer, with females ovipositing in bark cracks and crevices (Smith and Humble 2000). Eggs usually are laid singly or in small clusters in bark cracks along the trunk (Smith and Hurley 2000). Resin often exudes from infested trees at sites where the larvae are feeding (Smith and Humble 2000). Larvae first tunnel in the cambial region and then form a hook-like gallery in the sapwood by first boring perpendicular to the wood surface for 2–4 cm and then boring parallel to the stem surface for about 3–4 cm (Novák 1976). Overwintering takes place in the larval stage at the end of the gallery, with pupation occurring the following spring (Bílý and Mehl 1989). Adults construct oval-shaped exit holes as they emerge (Novák 1976).

In Eurasia, *T. fuscum* generally infests weakened and recently dead host trees, but in Canada, apparently healthy spruce trees have been infested and killed (Smith and Humble 2000). In Canada, an active eradication effort was attempted from 2000 to 2006, but since then a containment program has been in place in Nova Scotia to reduce the likelihood of human-assisted transport of infested host material (CFIA 2015). Extensive tunneling of the sapwood by larvae reduces wood quality and value (Novák 1976).

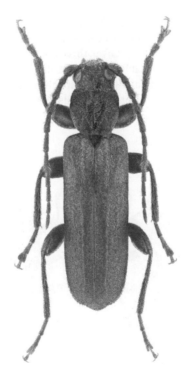

FIGURE 11.36 *Tetropium fuscum* adult from Germany. (Courtesy of Udo Schmidt at www.kaefer-der-welt.com.)

Logs should be utilized quickly to minimize infestation and wood quality as a result of larval tunneling (Novák 1976). Heat treatment has been shown to be effective in sanitizing wood cut from infested trees (Mushrow et al. 2004).

11.7 Summary and Future Outlook

The cerambycid species discussed in this chapter represent only a few of the thousands of tree-infesting species found throughout the world. Many more species could have been included but were lacking basic life history data. For example, for the large African Prioninae species *Tithoes confinis* (Castelnau) (Figure 11.37), many references were found describing its geographic range, larval host range, size, and color, but few papers provided information on its life cycle and economic impact. There is therefore a great need for basic research on the biology and ecology of most of the world's cerambycids. Not only would such basic information be useful to people within the native range of each insect but also to others throughout the world if these species become established elsewhere. This is especially important for species that may cause little economic impact in their native range but that become major pests when introduced elsewhere or when new host plants are grown within an insect's native range. Such relationships have been well documented throughout the world for many bark- and wood-infesting borers in addition to cerambycids (Yan et al. 2005; Haack 2006; Poland and McCullough 2006; Cocquempot and Lindelöw 2010; Haack et al. 2010a; Nielsen et al. 2011; Haack and Rabaglia 2013; Montecchio and Faccoli 2013; Haack et al. 2015).

Most of the tree-infesting cerambycids that have become established in areas outside their native range were likely moved inadvertently by humans in logs, lumber, firewood, wood packaging, and in live plants (Haack 2006; Haack et al. 2010a, 2010b, 2014; Liebhold et al. 2012; Rassati et al. 2016, see Chapter 13). As an example of the diversity of cerambycid species moved in international trade, data are presented in Table 11.3 on the number of cerambycid interceptions that were made at U.S. ports of entry on wood packaging and wood products during 1984–2008 and identified to at least the genus level. Overall, there were 2,655 cerambycid interceptions identified to at least the genus level, representing 76% of all 3,483 wood-associated cerambycid interceptions made during this period. Overall, the 2,655 interceptions represent 84 genera in five subfamilies, including 46 genera of Cerambycinae, 28 Lamiinae, 2 Lepturinae, 4 Prioninae, and 4 Spondylidinae (Table 11.3). Given this evidence, there is a great likelihood that many more cerambycid species will become established outside their native ranges in the future as a result of world trade and travel (Brockerhoff et al. 2014), although international standards to regulate wood packaging such as International Standards for Phytosanitary Measures (ISPM) No. 15 will help lower the arrival rate of potential pests (Haack et al. 2014).

FIGURE 11.37 *Tithoes confinis* adult male from Tanzania. (Courtesy of Ben Sale at commons.wikimedia.org.)

TABLE 11.3

Number of Cerambycid Interceptions on Wood Packaging and
Wood Products at U.S. Ports of Entry from 1984 to 2008 that
Were Identified to at Least the Genus Level (N = 2,655)

Subfamily Genus	Number of Interceptions	Major World Region Where Infested Imports Originated
Cerambycinae	807	–
Acyphoderes	1	South America
Aeolesthes	2	Asia
Aromia	1	Asia
Callidiellum	44	Asia
Callidium	64	Eurasia
Cerambyx	2	South America
Ceresium	140	Eurasia
Chlorida	2	South America
Chlorophorus	17	Eurasia
Chydarteres	1	South America
Clytus	5	Europe
Demonax	1	Asia
Dere	1	Asia
Diorthus	1	Asia
Eburia	2	Central America
Elaphidion	5	Asia, Central America
Epipedocera	1	Asia
Euryscelis	1	South America
Gnaphalodes	2	Mexico
Gracilia	1	Europe
Hesperophanes	78	Eurasia
Hylotrupes	21	Eurasia
Icosium	5	Europe
Knulliana	1	Asia
Lissonotus	1	South America
Megacyllene	4	Central America, South America
Molorchus	39	Europe
Nathrius	2	South America
Neoclytus	5	Central America, South America
Odontocera	2	South America
Pachydissus	2	Asia
Palaeocallidium	2	Asia
Perissus	1	Asia
Phoracantha	11	Africa, Oceania, South America
Phymatodes	70	Eurasia, Mexico
Placosternus	1	South America
Plagionotus	25	Eurasia
Pyrrhidium	30	Eurasia
Semanotus	3	Asia
Smodicum	1	South America
Stizocera	1	Central America
Stromatium	11	Eurasia
Trachyderes	19	South America
Trichoferus	6	Asia

(Continued)

TABLE 11.3 *(Continued)*

Number of Cerambycid Interceptions on Wood Packaging and Wood Products at U.S. Ports of Entry from 1984 to 2008 that Were Identified to at Least the Genus Level (N = 2,655)

Subfamily Genus	Number of Interceptions	Major World Region Where Infested Imports Originated
Xylotrechus	159	Eurasia
Xystrocera	13	Eurasia, Africa
Lamiinae	691	–
Acanthocinus	14	Europe
Agapanthia	2	Europe
Anelaphus	7	Central America
Anoplophora	41	Asia
Apriona	1	Asia
Ataxia	1	Mexico
Batocera	12	Asia
Coptops	5	Asia
Desmiphora	1	Mexico
Dihammus	1	Oceania
Glenea	1	Asia
Lagocheirus	2	Central America
Lamia	10	Europe
Leiopus	1	Europe
Leptostylus	4	Central America
Lepturges	1	Central America
Monochamus	540	Eurasia, Mexico
Nyssodrysina	2	Asia
Olenecamptus	4	Asia
Oncideres	2	Mexico
Petrognatha	1	Africa
Pogonocherus	8	Europe
Prosoplus	1	Asia
Pterolophia	2	Asia
Ropica	1	Asia
Saperda	22	Eurasia
Steirastoma	1	South America
Urgleptes	3	Central America, South America
Lepturinae	43	–
Leptura	17	Asia
Rhagium	26	Eurasia
Prioninae	9	–
Derobrachus	1	Mexico
Mallodon	2	Mexico
Prionus	3	Eurasia
Stenodontes	3	Central America South America
Spondylidinae	1,105	–
Arhopalus	637	Eurasia, Central America
Asemum	11	Eurasia
Oxypleurus	7	Europe
Tetropium	450	Eurasia, Canada

Source: USDA APHIS Pest ID database (see Haack et al. 2014).
Data are presented alphabetically by subfamily and genus. The principal world regions that were the origin of the infested imports are listed.

ACKNOWLEDGMENTS

The authors thank Therese M. Poland for comments on an earlier draft of this chapter as well as the many photographers who supplied photos for this chapter.

REFERENCES

Akanbi, M. O., and M. O. Ashiru. 2002. *A handbook of forest and wood insects of Nigeria*. Ibadan: Forestry Research Institute of Nigeria and Agbo Areo.

Akbulut, S., and W. T. Stamps. 2012. Insect vectors of the pinewood nematode: A review of the biology and ecology of *Monochamus* species. *Forest Pathology* 42: 89–99.

Allegro, G. 1990. Lotta meccanica contro i principali insetti xilofagi del pioppo mediante impiego di sbarramenti sui tronchi. *L' Informatore Agrario* 46: 91–95.

Allen, E. A., and L. M. Humble. 2002. Nonindigenous species introductions: A threat to Canada's forests and forest economy. *Canadian Journal of Plant Pathology* 24: 103–110.

Anderson, G. W., and M. P. Martin. 1981. Factors related to incidence of *Hypoxylon* cankers in aspen and survival of cankered trees. *Forest Science* 27: 461–476.

Bahillo de la Puebla, P., and J. C. Iturrondobeitia Bilbao. 1995. Primera cita de *Callidiellum rufipenne* (Motschulsky, 1860) para la Península Ibérica (Coleoptera: Cerambycidae). *Boletín de la Asociación española de Entomología* 19: 204.

Bain, J. 2009. *Hylotrupes bajulus*—Setting the record straight. Forest Health News 196, Rotorua, New Zealand: Scion. http://www.scionresearch.com/_data/assets/pdf_file/0008/6884/FHNews-196_July.pdf (accessed December 30, 2015).

Bain, J. 2012. *Phoracantha* longhorn beetles in New Zealand. Forest Health News 222, Rotorua, New Zealand, Scion. http://www.scionresearch.com/_data/assets/pdf_file/0010/37729/FHNews-222_Feb2012.pdf (accessed December 30, 2015).

Baker, W. L. 1972. *Eastern forest insects*. Washington, DC: USDA Forest Service, Miscellaneous Publication No. 1175.

Barr, W. F., and H. C. Manis. 1954. The red-headed ash borer in Idaho. *Journal of Economic Entomology* 47: 1150.

Bayford, E. G. 1938. Retarded development: *Hylotrupes bajulus* L. *The Naturalist, London* 980: 254–256.

Beeson, C. F. C. 1941. *The ecology and control of the forest insects of India and the neighbouring countries*. Dehra Dun: Bishen Singh Mahendra Pal Singh.

Beeson, C. F. C., and B. M. Bhatia. 1939. On the biology of the Cerambycidae (Coleopt.). *Indian Forest Records (new series) Entomology* 5: 1–235.

Bense, U. 1995. *Longhorn beetles. Illustrated key to the Cerambycidae and Vesperidae of Europe*. Weikersheim, Germany: Margraf Verlag.

Bezark, L. G. 2015a. New World Cerambycidae catalog. https://apps2.cdfa.ca.gov/publicApps/plant/bycidDB/wdefault.asp?w=n (accessed December 30, 2015).

Bezark, L. G. 2015b. Checklist of the Oxypeltidae, Vesperidae, Disteniidae and Cerambycidae, (Coleoptera) of the Western Hemisphere. (2015 ed.). https://apps2.cdfa.ca.gov/publicApps/plant/bycidDB/checklists/WestHemiCerambycidae2015.pdf (accessed December 30, 2015).

Bílý, S., and O. Mehl. 1989. Longhorn beetles (Coleoptera, Cerambycidae) of Fennoscandia and Denmark. *Fauna Entomologica Scandinavica* 22: 1–204.

Blackman, M. W., and H. H. Stage. 1924. On the succession of insects living in the bark and wood of dying, dead, and decaying hickory. *New York State College of Forestry, Syracuse, New York, Technical Publication* 17: 3–269.

Brockerhoff, E. G., M. Kimberley, A. M. Liebhold, R. A. Haack, and J. F. Cavey. 2014. Predicting how altering propagule pressure changes establishment rates of biological invaders across species pools. *Ecology* 95: 594–601.

Brooks, F. E. 1915. The parandra borer as an orchard enemy. *USDA Bureau of Entomology Bulletin* 262: 1–7.

Browne, F. G. 1968. *Pests and diseases of forest plantation trees: An annotated list of the principal species occurring in the British Commonwealth*. Oxford: Clarendon Press.

Browne, D. J., S. B. Peck, and M. A. Ivie. 1993. The longhorn beetles (Coleoptera Cerambycidae) of the Bahama Islands with an analysis of species-area relationships, distribution patterns, origin of the fauna and an annotated species list. *Tropical Zoology* 6: 27–53.

Bullas-Appleton, E., T. Kimoto, and J. J. Turgeon. 2014. Discovery of *Trichoferus campestris* (Coleoptera: Cerambycidae) in Ontario, Canada and first host record in North America. *The Canadian Entomologist* 146: 111–116.

Burfitt, C. E., K. Watson, C. A. Pratt, and J. Caputo. 2015. Total records of velvet longhorn beetle *Trichoferus campestris* Faldermann (Coleoptera: Cerambycidae) from Utah. In *USDA Interagency research forum on invasive species, 2015*, eds. K. A. McManus, and K. W. Gottschalk, 57. Ft. Collins, CO: USDA Forest Service, Forest Health Technology Enterprise Team, FHTET-2015-09.

CABI. 2010. *Saperda carcharias* (large poplar borer). Invasive species compendium. http://www.cabi.org/isc/datasheet/48316 (accessed February 9, 2016).

CABI. 2015a. *Monochamus alternatus* (Japanese pine sawyer). Invasive species compendium. http://www.cabi.org/isc/datasheet/34719 (accessed February 9, 2016).

CABI. 2015b. *Phoracantha recurva* (eucalyptus longhorned borer). Invasive species compendium. http://www.cabi.org/isc/datasheet/40371 (accessed December 30, 2015).

CABI. 2015c. *Phoracantha semipunctata* (eucalyptus longhorned borer). Invasive species compendium. http://www.cabi.org/isc/datasheet/40372 (accessed December 30, 2015).

Campadelli, G., and G. Sama. 1989. First record in Italy of a Japanese cerambycid: *Callidiellum rufipenne* Motschulsky. *Bollettino dell'Instituto di Entomologia "Guido Grandi" della Universita degli Studi di Bologna* 43: 69–73.

CFIA (Canadian Food Inspection Agency). 2015. Brown spruce longhorn beetle—*Tetropium fuscum*. http://www.inspection.gc.ca/plants/plant-pests-invasive-species/insects/brown-spruce-longhorn-beetle/eng/1330656129493/1330656721978 (accessed December 30, 2015).

Cherepanov, A. I. 1988a. *Cerambycidae of Northern Asia, Vol. 2, Cerambycinae, Part I*. New Delhi: Amerind.

Cherepanov, A. I. 1988b. *Cerambycidae of Northern Asia. Vol. 2: Cerambycinae. Part 2*. New Delhi: Amerind.

Cherepanov, A. I. 1990. *Cerambycidae of Northern Asia. Vol. 3: Lamiinae. Part 1*. New Delhi: Amerind.

Cibrián Tovar, D., J. T. Méndez Montiel, R. Campos Bolaños, H. O. Yates III, and J. E. Flores Lara. 1995. *Forest insects of Mexico*. Chapingo, Mexico: Universidad Autónoma Chapingo.

Cillie, J. J., and G. D. Tribe. 1991. A method for monitoring egg production by the *Eucalyptus* borers *Phoracantha* spp. (Cerambycidae). *South African Forestry Journal* 157: 24–26.

Cocquempot, C. 2006. Alien longhorned beetles (Coleoptera: Cerambycidae): Original interceptions and introductions in Europe, mainly France, and notes about recently imported species. *Redia* 89: 35–50.

Cocquempot, C., and A. Lindelöw. 2010. Longhorn beetles (Coleoptera, Cerambycidae). *BioRisk* 4: 193–218.

Cocquempot, C., and D. Mifsud. 2013. First European interception of the brown fir longhorn beetle, *Callidiellum villosulum* (Fairmaire, 1900) (Coleoptera, Cerambycidae). *Bulletin of the Entomological Society of Malta* 6: 143–147.

Cocquempot, C., A. Drumont, D. Brosens, and H. V. Ghate. 2014. First interception of the cerambycid beetle *Stromatium longicorne* (Newman, 1842) in Belgium and distribution notes on other species of *Stromatium* (Coleoptera: Cerambycidae: Cerambycinae). *Bulletin de la Société royale belge d'Entomologie* 150: 201–206.

Coulson, R. N., A. M. Mayyasi, J. L. Foltz, and F. P. Hain. 1976. Interspecific competition between *Monochamus titillator* and *Dendroctonus frontalis*. *Environmental Entomology* 5: 235–247.

Craighead, F. C. 1923. North American cerambycid larvae. *Bulletin of the Canada Department of Agriculture* 27: 1–239.

Craighead, F. C. 1950. *Insect enemies of eastern forests*. Washington, DC: USDA Forest Service, Miscellaneous Publication No. 657.

Curry, S. J. 1965. The biology and control of *Oemida gahani* Distant, (Cerambycidae), in Kenya. *East African Agricultural and Forestry Journal* 31: 224–235.

Danilevsky, M. L. 2015. Systematic list of longicorn beetles (Cerambycoidea) of the territory of the former USSR. http://www.cerambycidae.net/ussr.pdf (accessed December 30, 2015).

Dascalu, M-M; R Serafim, Å Lindelöw. 2013. Range expansion of *Trichoferus campestris* (Faldermann) (Coleoptera: Cerambycidae) in Europe with the confirmation of its presence in Romania. *Entomologica Fennica* 24: 142–146.

Daubenmire, R. 1978. *Plant geography: With special reference to North America*. New York, Academic Press.

Delahaye, N., and G. L. Tavakilian. 2009. Note sur *Mallodon downesii* Hope, 1843, et mise en synonymie de *M. plagiatum* Thomson, 1867 (Coleoptera, Cerambycidae). *Bulletin de la Société entomologique de France* 114: 39–45.

de Tillesse, V., L. Nef, J. G. Charles, A. Hopkin, and S. Augustin. 2007. *Damaging poplar insects—Internationally important species*. Rome: Food and Agricultural Organization. http://www.fao.org/forestry/foris/pdf/ipc/ damaging_poplar_insects_eBook.pdf (accessed December 30, 2015).

Dhakal, L. P., P. K. Jha, and E. D. Kjaer. 2005. Mortality in *Dalbergia sissoo* (Roxb.) following heavy infection by *Aristobia horridula* (Hope) beetles. Will genetic variation in susceptibility play a role in combating declining health? *Forest Ecology and Management* 218: 270–276.

Di Iorio, O. R. 1995. The genus *Neoclytus* Thomson, 1860 (Coleoptera: Cerambycidae) in Argentina. *Insecta Mundi* 9: 335–345.

Di Iorio, O. R. 2004a. Exotic species of Cerambycidae (Coleoptera) introduced in Argentina. Part 1: The genus *Phoracantha* Newman, 1840. *Agrociencia* 38: 503–515.

Di Iorio, O. R. 2004b. Exotic species of Cerambycidae (Coleoptera) introduced in Argentina. Part 2: New records, host plants, emergence periods, and current status. *Agrociencia* 38: 663–678.

Dodds, K. J., C. Graber, and F. M. Stephen. 2002. Oviposition biology of *Acanthocinus nodosus* (Coleoptera: Cerambycidae) in *Pinus taeda*. *Florida Entomologist* 85: 452–457.

Donley, D. E. 1978a. Oviposition by the red oak borer, *Enaphalodes rufulus* Coleoptera: Cerambycidae. *Annals of the Entomological Society of America* 71: 496–498.

Donley, D. E. 1978b. Distribution of the white oak borer *Goes tigrinus* DeGeer (Coleoptera: Cerambycidae) in a mixed oak stand. In *Proceedings, Central Hardwood Forest Conference*, 14–16 November 1978, 529–539. West Lafayette, Indiana: Purdue University, Department of Forestry and Natural Resources.

Donley, D. E., and Acciavatti, R. E. 1980. Red oak borer. *USDA Forest Service, Forest Insect & Disease Leaflet* 163.

Donley, D. E., and E. Rast. 1984. Vertical distribution of the red oak borer, *Enaphalodes rufulus* (Coleoptera: Cerambycidae), in red oak. *Environmental Entomology* 13: 41–44.

Drinkwater, T. W. 1975. The present pest status of eucalyptus borers *Phoracantha* spp. in South Africa. In *Proceedings of the First Congress of the Entomological Society of Southern Africa*, ed. H. J. R. Durr, J. H. Giliomee, and S. Neser, 119–129. Pretoria: Entomological Society of Southern Africa.

Drooz, A. T., ed. 1985. *Insects of eastern forests*. Washington, DC: USDA Forest Service Miscellaneous Publication 1426.

Drouin, J. A., and H. R. Wong. 1975. Biology, damage, and chemical control of the poplar borer (*Saperda calcarata*) in the junction of the root and stem of balsam poplar in western Canada. *Canadian Journal of Forest Research* 5: 433–439.

Drumont A., K. Smets, K. Scheers, A. Thomaes, R. Vandenhoudt, and M. Lodewyckx. 2014. *Callidiellum rufipenne* (Motschulsky, 1861) en Belgique: bilan de sa présence et de son installation sur notre territoire (Coleoptera: Cerambycidae: Cerambycinae). *Bulletin de la Société royale belge d'Entomologie* 150: 239–249.

Duan Y. 2001. Study on *Apriona swainsoni* Hope of devastating pest in *Sophora japonica* Linn. *Journal of Anhui Agricultural Sciences* 29: 375–377.

Duffy E. A. J. 1953. *A monograph of the immature stages of British and imported timber beetles (Cerambycidae)*. London: British Museum (Natural History).

Duffy E. A. J. 1957. *A monograph of the immature stages of African timber beetles (Cerambycidae)*. London: British Museum (Natural History).

Duffy E. A. J. 1960. *A monograph of the immature stages of neotropical timber beetles (Cerambycidae)*. London: British Museum (Natural History).

Duffy E. A. J. 1963. *A monograph of the immature stages of Australasian timber beetles (Cerambycidae)*. London: British Museum (Natural History).

Duffy E. A. J. 1968. *A monograph of the immature stages of oriental timber beetles (Cerambycidae)*. London: British Museum (Natural History).

Duffy E. A. J. 1980. *A monograph of the immature stages of African timber beetles (Cerambycidae) supplement*. Slough, England: Commonwealth Institute of Entomology.

Eldridgea, R. H., and J. A. Simpson. 1987. Development of contingency plans for use against exotic pests and diseases of trees and timber. 3. Histories of control measures against some introduced pests and diseases of forests and forest products in Australia. *Australian Forestry* 50: 24–36.

Elliot, H. J., and D. W. de Little 1984. *Insect pests of trees and timber in Tasmania*. Hobart, Tasmania: Forestry Commission.

Elliot, H. J., C. P. Ohmart, and F. R. Wylie. 1998. *Insect pests of Australian forests: Ecology and management*. Melbourne: Butterworth-Heinemann.

Endang, A. H., and N. F. Haneda. 2010. Infestation of *Xystrocera festiva* in *Paraserianthes falcataria* plantation in East Java, Indonesia. *Journal of Tropical Forest Science* 22: 397–402.

EPPO (European and Mediterranean Plant Protection Organization). 2005. *Xylotrechus altaicus. EPPO Bulletin* 35: 406–408.

EPPO (European and Mediterranean Plant Protection Organization). 2013. Pest risk analysis for *Apriona germari, A. japonica, A. cinerea*. Paris: EPPO. http://www.eppo.int/QUARANTINE/Pest_Risk_Analysis/ PRA_intro.htm (accessed December 30, 2015).

EPPO (European and Mediterranean Plant Protection Organization). 2015. *Callidiellum rufipenne*. EPPO Global Database. https://gd.eppo.int/taxon/CLLLRU (accessed December 30, 2015).

Evans, A. V. 2014. *Beetles of eastern North America*. Princeton, NJ: Princeton University Press.

Evans, H. F., D. G. McNamara, H. Braasch, J. Chadoeuf, and C. Magnusson. 1996. Pest risk analysis (PRA) for the territories of the European Union (as PRA area) on *Bursaphelenchus xylophilus* and its vectors in the genus *Monochamus. EPPO Bulletin* 26: 199–249.

Faccoli, M., R. Favaro, M. T. Smith, and J. Wu. 2015. Life history of the Asian longhorn beetle *Anoplophora glabripennis* (Coleoptera: Cerambycidae) in southern Europe. *Agricultural and Forest Entomology* 17: 188–196.

FAO (Food and Agriculture Organization). 2007. *Overview of forest pests Thailand*. Rome: Food and Agriculture Organization, Forest Resources Development Service Working Paper FBS/32E.

FAO (Food and Agriculture Organization). 2009. *Global review of forest pests and diseases*. Rome: Food and Agriculture Organization, FAO Forestry Paper 156.

Fearn, S. 1989. Some observations on the habits of *Paroplites australis* (Erichson) (Coleoptera: Cerambycidae, Prioninae) and its damaging effects on the food plant *Banksia marginata* Cav. in Tasmania. *Australian Entomological Magazine* 16(4): 81–84.

Flaherty, L., J. D. Sweeney, D. Pureswaran, and D. T. Quiring. 2011. Influence of host tree condition on the performance of *Tetropium fuscum* (Coleoptera: Cerambycidae). *Environmental Entomology* 40: 1200–1209.

French, J. R. J. 1969. Occurrence and control of European house borer in New South Wales. *Australian Forestry* 33: 13–18.

Froggatt, W. W. 1919. The native lime-tree borer (*Citriphaga mixta* Lea). *Agricultural Gazette of New South Wales* 30: 261–267.

Froggatt, W. W. 1923. *Forest insects of Australia*. Sydney: Government Printer.

Froggatt, W. W. 1925. Forest insects. No. 6. The hoop pine longicorn (*Diotima undulata* Pascoe). *The Australian Forestry Journal* 8: 6–8.

Froggatt, W. W. 1927. *Forest insects and timber borers*. Sydney: Government Printer.

Furniss, R. L. and V. M. Carolin. 1977. *Western forest insects*. Washington, DC: USDA, Forest Service, Miscellaneous Publication No. 1339.

Gahan, A. 1911. Some notes on *Parandra brunnea. Journal of Economic Entomology* 4: 299–301.

Galford, J. R. 1984. The locust borer. *USDA Forest Service, Forest Insect & Disease Leaflet* 71.

Gao, S. I., Z. C. Xu, and X. C. Gong. 2007. Progress in research on *Semanotus bifasciatus. Forest Pest and Disease* 26(3): 19–22, 38.

Gardner, J. C. M. 1957. An annotated list of East African forest insects. *East African Agriculture and Forestry Research Organisation, Forestry Technical Note* 7.

Gardner, J. C. M., and J. O. Evans. 1957. Notes on *Oemida gahani* Distant (Cerambycidae). Part II: *East African Agricultural Journal* 22: 224–230.

GBIF (Global Biodiversity Information Facility). 2014. *Cordylomera spinicornis* (Fabricius, 1775). http:// www.gbif.org/species/4738460/ (accessed December 30, 2015).

Giesbert, E. F. 1989. A new species and new record in the genus *Neoclytus* Thomson (Coleoptera: Cerambycidae) for Panama. *The Coleopterists Bulletin* 43: 269–273.

Gonzalez, O. E. and O. R. Di Iorio. 1996. Plantas hospedadoras de Cerambycidae (Coleoptera) en el noreste de Argentina. *Revista de Biologia Tropical* 44: 167–175.

Gosling, D. C. L. 1978. Observations on the biology of the oak twig pruner, *Elaphidionoides parallelus*, (Coleoptera: Cerambycidae) in Michigan. *The Great Lakes Entomologist* 11: 1–10.

Gosling, D. C. L. 1981. Correct identity of the oak twig pruner (Coleoptera: Cerambycidae). *The Great Lakes Entomologist* 14: 179–180.

Graham, S. A. 1925. The felled tree trunk as an ecological unit. *Ecology* 6: 397–411.

Gray, B. 1968. Forest tree and timber insect pests in the territory of Papua and New Guinea. *Pacific Insects* 10: 301–323.

Gray, B. 1972. Economic tropical forest entomology. *Annual Review of Entomology* 17: 313–352.

Gray, B., and F. R. Wylie. 1974. Forest tree and timber insect pests in Papua New Guinea. II: *Pacific Insects* 16: 67–115.

Grebennikov, V. V., B. D. Gill, and R. Vigneault. 2010. *Trichoferus campestris* (Falmermann) (Coleoptera: Cerambycidae), an Asian wood-boring beetle recorded in North America. *The Coleopterists Bulletin* 64: 13–20.

Gressitt, J. L., J. A. Rondon, and S. von Breuning. 1970. Cerambycid beetles of Laos (Longicornes du Laos). *Pacific Insects Monograph* 24: 1–651.

Haack, R. A. 2006. Exotic bark- and wood-boring Coleoptera in the United States: Recent establishments and interceptions. *Canadian Journal of Forest Research* 36: 269–288.

Haack, R. A. 2012. Seasonality of oak twig pruner shoot fall: A long-term dog walking study. *Newsletter of the Michigan Entomological Society* 57: 24.

Haack, R. A., and F. Slansky. 1987. Nutritional ecology of wood-feeding Coleoptera, Lepidoptera, and Hymenoptera. In *Nutritional ecology of insects, mites, spiders and related invertebrates*, eds. F. Slansky Jr., and J. G. Rodriguez, 449–486. New York: Wiley.

Haack, R. A., and R. J. Rabaglia. 2013. Exotic bark and ambrosia beetles in the USA: Potential and current invaders. In *Potential invasive pests of agricultural crop species*, ed. J. E. Peña. 48–74. Wallingford: CABI International.

Haack, R. A., D. M. Benjamin, and K. D. Haack. 1983. Buprestidae, Cerambycidae, and Scolytidae associated with successive stages of *Agrilus bilineatus* (Coleoptera: Buprestidae) infestations of oaks in Wisconsin. *The Great Lakes Entomologist* 16: 47–55.

Haack, R. A., L. S. Bauer, R.-T. Gao, J. J. McCarthy, D. L. Miller, T. R. Petrice, and T. M. Poland. 2006. *Anoplophora glabripennis* within-tree distribution, seasonal development, and host suitability in China and Chicago. *The Great Lakes Entomologist* 39: 169–183.

Haack, R. A., F. Hérard, J. Sun, and J. J. Turgeon. 2010a. Managing invasive populations of Asian long-horned beetle and citrus longhorned beetle: A worldwide perspective. *Annual Review of Entomology* 55: 521–546.

Haack, R. A., T. R. Petrice, and A. C. Wiedenhoeft. 2010b. Incidence of bark- and wood-boring insects in firewood: A survey at Michigan's Mackinac Bridge. *Journal of Economic Entomology* 103: 1682–1692.

Haack, R. A., K. O. Britton, E. G. Brockerhoff, et al. 2014. Effectiveness of the international phytosanitary standard ISPM No. 15 on reducing wood borer infestation rates in wood packaging material entering the United States. *PLoS One* 9(5): e96611. doi:10.1371/journal.pone.0096611.

Haack, R. A., Y. Baranchikov, L. S. Bauer, and T. M. Poland. 2015. Emerald ash borer biology and invasion history. In *Biology and control of emerald ash borer*, eds. R. G. Van Driesche, and R. C. Reardon, 1–13. Morgantown, WV: USDA Forest Service, Forest Health Technology Enterprise Team, FHTET-2014-09.

Hanks, L. M. 1999. Influence of the larval host plant on reproductive strategies of cerambycid beetles. *Annual Review of Entomology* 44: 483–505.

Hanula, J. L. 1996. Relationship of wood-feeding insects and coarse woody debris. In *Biodiversity and coarse woody debris in southern forests, proceedings of the workshop on coarse woody debris in southern forests: Effects on biodiversity*, eds. J. W. McMinn, and D. A. Crossley Jr., 55–81. Asheville, NC: USDA Forest Service, Southern Research Station, General Technical Report SE-94.

Harman, D. M., and A. L. Harman. 1990. Height distribution of locust borer attacks (Coleoptera: Cerambycidae) in black locust. *Environmental Entomology* 19: 501–504.

Harmon, M. E., J. F. Franklin, F. J. Swanson, et al. 1986. Ecology of coarse woody debris in temperate ecosystems. *Advances in Ecological Research* 15: 133–302.

Hart, C. J., J. S. Cope, and M. A. Ivie. 2013. A checklist of the Cerambycidae (Coleoptera) of Montana, USA, with distribution maps. *The Coleopterists Bulletin* 67: 133–148.

Hawkeswood, T. J. 1992. Review of the biology, host plants and immature stages of the Australian Cerambycidae (Coleoptera). Part 1: Parandrinae and Prioninae. *Giornale Italiano di Entomologia* 6: 207–224.

Hawkeswood, T. J. 1993. Review of the biology, host plants and immature stages of the Australian Cerambycidae. Part 2: Cerambycinae (tribes Oemini, Cerambycini, Hesperophanini, Callidiopini, Neostenini, Aphanasiini, Phlyctaenodini, Tessarommatini and Piesarthrini). *Giornale Italiano di Entomologia* 6: 313–355.

Hawkeswood, T. J. and J. R. Turner. 2003. Notes on the Australian longicorn beetle *Paroplites australis* (Erichson, 1842) (Coleoptera: Cerambycidae) with a new record from the Sydney District, New South Wales, Australia. *Entomologische Zeitschrift Stuttgart* 113(9): 270–271.

Hay, C. J. 1974. Survival and mortality of red oak borer larvae on black, scarlet, and northern red oak in eastern Kentucky. *Annals of the Entomological Society of America* 67: 981–986.

Hérard, F., M. Ciampitti, M. Maspero et al. 2006. *Anoplophora* spp. in Europe: Infestations and management process. *EPPO Bulletin* 36: 470–474.

Hesterberg, G. A., C. J. Wright, and D. J. Frederick. 1976. Decay risk for sugar maple borer scars. *Journal of Forestry* 74: 443–445.

Hickin, N. E. 1951. Delayed emergence of *Eburia quadrigeminata* Say (Col., Cerambycidae). *Entomologists Monthly Magazine* 87: 51.

Hoebeke, E. R. 1999. Japanese cedar longhorned beetle in the eastern United States. *USDA Animal and Plant Health Inspection Service, Pest Alert* 81–35–004.

Hoffard, W. H., and P. T. Marshall. 1978. *How to identify and control the sugar maple borer.* Delaware, OH: USDA Forest Service, Northeastern Area, State and Private Forestry, NA-GR-1. http://www.na.fs.fed.us/spfo/pubs/howtos/ht_mapleborer/mapleborer.htm (accessed December 30, 2015).

Humphreys, N., and E. Allen. 2000. Lesser cedar longicorn beetle—*Callidiellum rufipenne.* Exotic Forest Pest Advisory—Pacific Forestry Centre, Canadian Forest Service, No. 4. http://cfs.nrcan.gc.ca/pubwarehouse/pdfs/5507.pdf (accessed December 30, 2015).

Hutacharern, C., and S. E. Panya. 1996. Biology and control of *Aristobia horridula* (Hope) (Coleoptera: Cerambycidae), a pest of *Pterocarpus macrocarpus.* In *Impact of diseases and insect pests in tropical forests,* eds. K. S. S. Nair, J. K. Sharma, and R. V. Varma, 392–397. Proceedings of the IUFRO Symposium, Peechi, India, 23–26 November 1993.

Ivory, M. H. 1977. Preliminary investigations of the pests of exotic forest trees in Zambia. *Commonwealth Forestry Review* 56: 47–56.

Iwata, R., and F. Yamada. 1990. Notes on the biology of *Hesperophanes campestris* (Faldermann) (Col., Cerambycidae), a drywood borer in Japan. *Material und Organismen* 25: 305–313.

Iwata, R., T. Maro, Y. Yonezawa, T. Yahagi, and Y. Fujikawa. 2007. Period of adult activity and response to wood moisture content as major segregating factors in the coexistence of two conifer longhorn beetles, *Callidiellum rufipenne* and *Semanotus bifasciatus* (Coleoptera: Cerambycidae). *European Journal of Entomology* 104: 341–345.

Jaques, H. E. 1918. A long-lifed woodboring beetle. *Proceedings of the Iowa Academy of Science* 25: 175.

Johki, Y., and T. Hidaka. 1987. Group feeding in larvae of the albizia borer, *Xystrocera festiva* (Coleoptera: Cerambycidae). *Journal of Ethology* 5: 89–91.

Jones, T., and S. J Curry. 1964. *Oemida gahani* Distant (Cerambycidae), its host plants, host range and distribution. *East African Agricultural and Forestry Journal* 30: 149–161.

Khan, T. N. 1985. Community and succession of the round-head borers (Coleoptera: Cerambycidae) infesting the felled logs of White Dhup, *Canarium euphyllum* Kurz. *Proceedings of the Indian Academy of Sciences (Animal Sciences)* 94: 435–441.

Kim, K. C., and J. D. Park. 1984. Studies on ecology and injury characteristics of Japanese *Juniperus* bark borer, *Semanotus bifasciatus* Motschulsky. *Korean Journal of Plant Protection* 23: 109–115.

Kliejunas, J. T., B. M. Tkacz, H. H. Burdsall, et al. 2001. *Pest risk assessment of the importation into the United States of unprocessed Eucalyptus logs and chips from South America.* Madison, WI: USDA Forest Service, Forest Products Laboratory, General Technical Report FPL-GTR-124.

Kliejunas, J. T., H. H. Burdsall, G. A. DeNitto, et al. 2003. *Pest risk assessment of the importation into the United States of unprocessed logs and chips of eighteen eucalypt species from Australia.* Madison, WI: USDA Forest Service, Forest Products Laboratory, General Technical Report FPL-GTR-137.

Kobayashi, F., A. Yamane, and T. Ikeda. 1984. The Japanese pine sawyer beetle as the vector of pine wilt disease. *Annual Review of Entomology* 29: 115–135.

LaBonte, J. R., A. D. Mudge, and K. J. R. Johnson. 2005. Nonindigenous woodboring Coleoptera (Cerambycidae, Curculionidae: Scolytinae) new to Oregon and Washington, 1999–2002: Consequences of the intracontinental movement of raw wood products and solid wood packing materials. *Proceedings of the Entomological Society of Washington* 107: 554–564.

LeConte, J. L. 1850. An attempt to classify the longicorn Coleoptera of the part of America north of Mexico. *Journal of the Academy of Natural Sciences of Philadelphia* 2(Series 2): 2–38.

Lee, S.-I., J. R. Spence, and D. W. Langor. 2014. Succession of saproxylic beetles associated with decomposition of boreal white spruce logs. *Agricultural and Forest Entomology* 16: 391–405.

Liebhold, A. M., E. G. Brockerhoff, L. J. Garrett, J. L. Parke, and K. O. Britton. 2012. Live plant imports: The major pathway for forest insect and pathogen invasions of the US. *Frontiers in Ecology and the Environment* 10: 135–143.

Lim, J., S.-Y. Jung, J.-S. Lim, et al. 2014. A review of host plants of Cerambycidae (Coleoptera: Chrysomeloidea) with new host records for fourteen cerambycids, including the Asian longhorn beetle (*Anoplophora glabripennis* Motschulsky), in Korea. *Korean Journal of Applied Entomology* 53: 111–133.

Linsley, E. G. 1959. Ecology of Cerambycidae. *Annual Review of Entomology* 4: 99–138.

Linsley, E. G. 1961. The Cerambycidae of North America. Part I: Introduction. *University of California Publications in Entomology* 18: 1–135.

Linsley, E. G. 1962a. The Cerambycidae of North America. Part II: Taxonomy and classification of the Parandrinae, Prioninae, Spondylinae and Aseminae. *University of California Publications in Entomology* 19: 1–102.

Linsley, E. G. 1962b. The Cerambycidae of North America. Part III: Taxonomy and classification of the subfamily Cerambycinae, tribes Opsimini through Megaderini. *University of California Publications in Entomology* 20: 1–188.

Linsley, E. G. 1963. The Cerambycidae of North America. Part IV: Taxonomy and classification of the subfamily Cerambycinae, tribes Elaphidionini through Rhinotragini. *University of California Publications in Entomology* 21: 1–165.

Linsley, E. G. 1964. The Cerambycidae of North America. Part V: Taxonomy and classification of the subfamily Cerambycinae, tribes Callichromini through Ancylocerini. *University of California Publications in Entomology* 22: 1–197.

Linsley, E. G., and J. A. Chemsak. 1972. Cerambycidae of North America. Part VI, No. 1: Taxonomy and classification of the subfamily Lepturinae. *University of California Publications in Entomology* 69: 1–138.

Linsley, E. G., and J. A. Chemsak. 1976. Cerambycidae of North America. Part VI, No. 2: Taxonomy and classification of the subfamily Lepturinae. *University of California Publications in Entomology* 80: 1–186.

Linsley, E. G., and J. A. Chemsak. 1984. The Cerambycidae of North America. Part VII, No. 1: Taxonomy and classification of the subfamily Lamiinae, tribes Parmenini through Acanthoderini. *University of California Publications in Entomology* 102: 1–258.

Linsley, E. G., and J. A. Chemsak. 1995. The Cerambycidae of North America. Part VII, No. 2: Taxonomy and classification of the subfamily Lamiinae, tribes Acanthocinini through Hemilophini. *University of California Publications in Entomology* 114: 1–292.

Liu, G., and Y. Tang. 2002. The relationships between *Apriona swainsoni* and its host trees. *Scientia Silvae Sinicae* 38(3): 106–113.

Liu, H., Y. Luo, J. Wen, Z. Zhang, J. Feng, and W. Tao. 2006. Pest risk assessment of *Dendroctonus valens, Hyphantria cunea* and *Apriona swainsoni* in Beijing. *Frontiers of Forestry in China* 1: 328–335.

Łoś, K. and R. Plewa. 2011. *Callidiellum rufipenne* (motschulsky, 1862) (Coleoptera: Cerambycidae)—New to the fauna of Croatia with remarks of its biology. *Nature Journal (Opole Scientific Society)* 44: 141–144.

Löyttyniemi, K. 1983. Flight pattern and voltinism of *Phoracantha* beetles (Coleoptera: Cerambycidae) in a semihumid tropical climate in Zambia. *Annales Entomologici Fennici* 49: 49–53.

Lu, X., Z. Yang, X. Sun, L. Qiao, X. Wang, and J. Wei. 2011. Biological control of *Apriona swainsoni* (Coleoptera: Cerambycidae) by releasing the parasitic beetle *Dastarcus helophoroides* (Coleoptera: Bothrideridae). *Scientia Silvae Sinicae* 47(10): 16–121.

Luzzi, M. A., R. C. Wilkinson, and A. C. Tarjan. 1984. Transmission of the pinewood nematode, *Bursaphelenchus xylophilus*, to slash pine trees and log bolts by a cerambycid beetle, *Monochamus titillator*, in Florida. *Journal of Nematology* 16: 37–40.

Ma, L.-Q., Y.-F. Zhu, C.-J. Cao, et al. 2010. Utilization of *Pyemotes* sp. and *Scleroderma guani* Xiao et Wu to control the larvae of *Semanotus bifasciatus*. *Forest Research* 23: 313–317.

MacAloney, H. J. 1968. The sugar maple borer. *USDA Forest Service, Forest Pest Leaflet* 108.

Maier, C. T. 2007. Distribution and hosts of *Callidiellum rufipenne* (Coleoptera: Cerambycidae), an Asian cedar borer established in the eastern United States. *Journal of Economic Entomology* 100: 1291–1297.

Maier, C. T., and Lemmon, C. R. 2000. Discovery of the small Japanese cedar longhorned beetle, *Callidiellum rufipenne* (Motschulsky) (Coleoptera: Cerambycidae), in live arborvitae in Connecticut. *Proceedings of the Entomological Society of Washington* 102: 747–754.

Mattson, W. J., and R. A. Haack. 1987. The role of drought in outbreaks of plant-eating insects. *BioScience* 37: 110–118.

Mayné, R. 1914. Les ennemis de l'Hévéa au Congo Belge. *Bulletin Agricole du Congo Belge* 5: 577–596.

McKeown, K. C. 1947. Catalogue of the Cerambycidae (Coleoptera) of Australia. *Australian Museum Memoir* 10: 1–190.

McNeil, J. 1886. A remarkable case of longevity in a longicorn beetle (*Eburia quadrigeminata*). *American Naturalist* 20: 1055–1057.

Mishra, S. C., V. Veer, and A. Chandra. 1985. *Aristobia horridula* Hope (Coleoptera: Lamiidae) a new pest of shisham (*Dalbergia sissoo* Roxb.) in West Bengal. *Indian Forester* 111: 738–741.

Monné, M. A., E. H. Nearns, S. C. Carbonel Carril, I. P. Swift, and M. L. Monné. 2012. Preliminary checklist of the Cerambycidae, Disteniidae, and Vesperidae (Coleoptera) of Peru. *Insecta Mundi* 0213: 1–48.

Montecchio, L., and M. Faccoli. 2013. First record of thousand cankers disease *Geosmithia morbida* and walnut twig beetle *Pityophthorus juglandis* on *Juglans nigra* in Europe. *Plant Disease* 98: 696.

Moore, K. M. 1966. Observations on some Australian forest insects. 21. *Hesthesis cingulata* (Kirby) (Coleoptera: Cerambycidae), attacking young plants of *Eucalyptus pilularis* Smith. *Australian Zoologist* 13: 299–301.

Mushrow, L., A. Morrison, J. D. Sweeney, and D. T. Quiring, 2004. Heat as a phytosanitary treatment for the brown spruce longhorn beetle. *Forestry Chronicle* 80: 224–228.

Nair, K. S. S. 2007. *Tropical forest insect pests: Ecology, impact, and management*. Cambridge: Cambridge University Press.

Negi, S. and V. D. Joshi. 2009. Role of moisture content in rendering the sal tree component susceptible to the borer (*Hoplocerambyx spinicornis*) attack. *Asian Journal of Animal Science* 3: 190–192.

Nielsen, D. G., V. L. Muilenburg, and D. A. Herms. 2011. Interspecific variation in resistance of Asian, European, and North American birches (*Betula* spp.) to bronze birch borer (Coleoptera: Buprestidae). *Environmental Entomology* 40: 648–653.

Nose, M., and S. Shiraishi. 2008. Breeding for resistance to pine wilt disease. In *Pine Wilt Disease*, eds. B. G. Zhao, K. Futai, J. R. Sutherland, and Y. Takeuchi, 334–350. Tokyo: Springer.

Novak, V. 1976. *Atlas of insects harmful to forest trees, Vol. 1*. Amsterdam: Elsevier.

O'Connor, J. P., and R. Nash. 1984. Insects imported into Ireland. 6. Records of Orthoptera, Dermaptera, Lepidoptera and Coleoptera. *The Irish Naturalists' Journal* 21: 351–353.

Paine, T. D, and J. G. Millar. 2002. Insect pests of eucalypts in California: Implications of managing invasive species. *Bulletin of Entomological Research* 92: 147–151.

Paine, T. D., M. J. Steinbauer, and S. A. Lawson. 2011. Native and exotic pests of *Eucalyptus*: A worldwide perspective. *Annual Review of Entomology* 56: 181–201.

Pan, H. Y. 2005. *Review of the Asian longhorned beetle: Research, biology, distribution and management in China*. Rome: Food and Agriculture Organization, Forest Resources Development Service Working Paper FBS/6E.

Peck, S. B. 2005. *A checklist of the beetles of Cuba with data on distributions and bionomics (Insecta: Coleoptera)*. Arthropods of Florida and Neighboring Land 18: 1–241. Gainesville, FL: Florida Department of Agriculture and Consumer Services.

Poland, T. M., and D. G. McCullough. 2006. Emerald ash borer: Invasion of the urban forest and the threat to North America's ash resource. *Journal of Forestry* 104: 118–124.

Qiu, L. 1999. Control of wood borer larvae by releasing *Scleroderma guani* in Chinese fir forests. *Chinese Journal of Biological Control* 15: 8–11.

Quentin, R. M., and A. Villiers. 1969. Révision des Plectogasterini, nov. trib. (Col. Cerambycidae Cerambycinae). *Annales de la Société Entomologique de France (Nouvelle Série)* 5: 613–646.

Rassati, D., M. Faccoli, L. Marini, R. A. Haack, A. Battisti, and E. P. Toffolo. 2015. Exploring the role of wood waste landfills in early detection of non-native alien wood-boring beetles. *Journal of Pest Science* 88: 563–572.

Rassati, D., F. Lieutier, and M. Faccoli. 2016. Alien wood-boring beetles in Mediterranean regions. In *Insects and diseases of Mediterranean forest systems*, eds. T. D. Paine, and F. Lieutier, 293–327. Cham, Switzerland: Springer International.

Rathore, M. P. S. 1995. *Insect pests in agroforestry*. Nairobi: ICRAF Working Paper No. 70.

Rhainds, M., W. E. Mackinnon, K. B. Porter, J. D. Sweeney, and P. J. Silk. 2011. Evidence for limited spatial spread in an exotic longhorn beetle, *Tetropium fuscum* (Coleoptera: Cerambycidae). *Journal of Economic Entomology* 104: 1928–1933.

Riggins, J. J., L. D. Galligan, and F. M. Stephen. 2009. Rise and fall of red oak borer (Coleoptera: Cerambycidae) in the Ozark Mountains of Arkansas, USA. *Florida Entomologist* 92: 426–433.

Ritchie, W. 1920. The structure, bionomics, and economic importance of *Saperda carcharias* Linn., "the large poplar longhorn." *Annals of Applied Biology* 7: 299–343.

Rivas, C. 1992. *Forest pests in Central America: Handbook. Technical manual No. 3.* Turrialba, Costa Rica: CATIE.

Roberts, H. 1969. Forest insects of Nigeria with notes on their biology and distribution. *Commonwealth Forestry Institute Paper* 44: 1–206.

Robinson, W. H., and K. F. Cannon. 1979. The life history and habits of the old house borer *Hylotrupes bajulus* (L.) and its distribution in Pennsylvania. *Melsheimer Entomological Series* 27: 30–34.

Rozhkov, A. S. 1970. *Pests of Siberian larch.* Jerusalem: Israel Program for Scientific Translations.

Saint-Germain, M., P. Drapeau, and C. M. Buddle. 2007. Host-use patterns of saproxylic phloeophagous and xylophagous Coleoptera adults and larvae along the decay gradient in standing dead black spruce and aspen. *Ecography* 30: 737–748.

Salini, S., and D. S. Yadav. 2011. Occurrence of *Stromatium barbatum* (Fabr.) (Coleoptera: Cerambycidae) on grapevine in Maharashtra, India. *Pest Management in Horticultural Ecosystems* 17: 48–50.

Sama, G., J. Buse, E. Orbach, A. L. L. Friedman, O. Rittner, and V. Chikatunov. 2010. A new catalogue of the Cerambycidae (Coleoptera) of Israel with notes on their distribution and host plants. *Munis Entomology & Zoology* 5: 1–51.

Sathyapala, S. 2004. *Pest risk analysis, biosecurity risk to New Zealand of pinewood nematode (Bursaphelenchus xylophilus).* Wellington: New Zealand, Ministry of Agriculture and Forestry.

Savely, H. E. 1939. Ecological relations of certain animals in dead pine and oak logs. *Ecological Monographs* 9: 323–385.

Schabel, H. G. 2006 *Forest entomology in East Africa.* Dordrecht: Springer.

Scriven, G. T., E. L. Reeves, and R. F. Luck. 1986. Beetle from Australia threatens eucalyptus. *California Agriculture* 40(7–8): 4–6.

Shibata, E. 1994. Population studies of *Callidiellum rufipenne* (Coleoptera: Cerambycidae) on Japanese cedar logs. *Annals of the Entomological Society of America* 87: 836–841.

Shigo, A. L., W. B. Leak, and S. M. Filip. 1973. Sugar maple borer injury in four hardwood stands in New Hampshire. *Canadian Journal of Forest Research* 3: 512–515.

Shin, S. C., K. S. Choi, W. I. Choi, Y. J. Chung, S. G. Lee, and C. S. Kim. 2008. *A new illustrated book of forest insect pests.* Seoul: Upgo MunHwa.

Singh, P., and R. M. Misra. 1981. Distance response of sal heartwood borer, *Hoplocerambyx spinicornis* Newman (Cerambycidae: Coleoptera) to freshly felled sal trees. *Indian Forester* 107: 305–308.

Sjöman, H.; J. Östberg, and J. Nilsson. 2014. Review of host trees for the wood-boring pests *Anoplophora glabripennis* and *Anoplophora chinensis*: An urban forest perspective. *Arboriculture & Urban Forestry* 40: 143–164.

Smith, I. M. 2009. *Hesperophanes campestris. EPPO Bulletin* 39: 51–54.

Smith, G. A., and L. M. Humble. 2000. The brown spruce longhorn beetle. Victoria, BC: Natural Resources Canada, Canadian Forest Service, Pacific Forestry Centre, Exotic Forest Pest Advisory 5. http://cfs.nrcan.gc.ca/pubwarehouse/pdfs/5529.pdf (accessed December 30, 2015).

Smith G., and J. E. Hurley. 2000. First North American record of the Palearctic species *Tetropium fuscum* (Fabricius) (Coleoptera: Cerambycidae). *The Coleopterists Bulletin* 54: 540.

Snyder, T. E. 1911. Damage to telephone and telegraph poles by wood-boring insects. *USDA Bureau of Entomology Circular* 134: 1–6.

Solomon, J. D. 1977. Frass characteristics for identifying insect borers (Lepidoptera: Cossidae and Sesiidae; Coleoptera: Cerambycidae) in living hardwoods. *Canadian Entomologist* 109: 295–303.

Solomon, J. D. 1980. *Cottonwood borer (Plectrodera scalator)—A guide to its biology, damage, and control.* New Orleans, LA: USDA Forest Service, Southern Forest Experiment Station, Research Paper SO-157.

Solomon, J. D. 1995. *Guide to insect borers in North American broadleaf trees and shrubs.* Washington, DC: USDA Forest Service, Agriculture Handbook AH-706.

Song, S. H., L. Q. Zhang, H. H. Huang, and X. M. Cui. 1991. Preliminary study of biology of *Monochamus alternatus* Hope. *Forest Science and Technology* 6: 9–13.

Stephen, F. M., V. B. Salisbury, and F. L. Oliveria. 2001. Red oak borer, *Enaphalodes rufulus* (Coleoptera: Cerambycidae), in the Ozark Mountains of Arkansas, U.S.A: An unexpected and remarkable forest disturbance. *Integrated Pest Management Reviews* 6: 247–252.

Straw, N. A., N. J. Fielding, C. Tilbury, D. T. Williams, and D. Inward. 2014. Host plant selection and resource utilisation by Asian longhorn beetle *Anoplophora glabripennis* (Coleoptera: Cerambycidae) in southern England. *Forestry* 88: 84–95.

Sun, X.-S. 2000. The application of nematodes for control *Semanotus bifasciatus sinoauster* Gressitt. *Journal of Fujian College of Forestry* 20: 49–51.

Švácha, P., and M. L. Danilevsky. 1988. Cerambycoid larvae of Europe and Soviet Union (Coleoptera, Cerambycoidea). Part II: *Acta Universitatis Carolinae—Biologica* 31: 121–284.

Swift, I. P., and A. M. Ray. 2010. Nomenclatural changes in North American *Phymatodes* Mulsant (Coleoptera: Cerambycidae) *Zootaxa* 2448: 35–52.

Syme, J. H. and J. R. Saucier. 1995. Effects of long-term storage of southern pine sawlogs under water sprinklers. *Forest Products Journal* 45: 47–50.

Tang, Y. P., and G. H. Liu. 2000. Study on forecast of ovipositing occurrence time of *Apriona swainsoni*. *Scientia Silvae Sinicae* 36(6): 86–89.

Tavakilian, G., and H. Chevillotte. 2013. Titan database about longhorns or timber-beetles (Cerambycidae). http://lully.snv.jussieu.fr/titan/index.html (accessed December 30, 2015).

Tavakilian, G., A. Berkov, B. Meurer-Grimesi, and S. Mori. 1997. Neotropical tree species and their faunas of xylophagous longicorns (Coleoptera: Cerambycidae) in French Guiana. *The Botanical Review* 63: 304–355.

Thomas, M. C. 1999. The genus *Eburia* Audinet-Serville in Florida (Coleoptera: Cerambycidae). *Florida Department of Agriculture, Entomology Circular* 396: 1–4.

Togashi, K. 2013. Nematode-vector beetle relation and the regulatory mechanism of vector's life history in pine wilt disease. *Formosan Entomologist* 33: 189–205.

Togashi, K., and H. Magira. 1981. Age-specific survival rate and fecundity of the adult Japanese pine sawyer, *Monochamus alternatus* Hope (Coleoptera: Cerambycidae), at different emergence times. *Applied Entomology and Zoology* 16: 351–361.

Trägårdh, I. 1930. Some aspects in the biology of longicorn beetles. *Bulletin of Entomological Research.* 21: 1–8.

Troop, J. 1915. Cerambycid in bedstead (Col.). *Entomological News* 26: 281.

Turienzo, P. 2006. Definitive incorporation of *Callidiellum rufipenne* (Motschulsky, 1860) to the Argentinian fauna of Cerambycidae (Coleoptera). *Boletín de Sanidad Vegetal Plagas* 32: 155–156.

USDA Forest Service. 1991. *Pest risk assessment of the importation of larch from Siberia and the Soviet Far East.* Washington, DC: USDA Forest Service, Miscellaneous Publication No. 1495.

Van Meer, C., and C. Cocquempot. 2013. Découverte d'un foyer de *Callidiellum rufipenne* (Motschulsky, 1861) dans les Pyrénées Atlantiques (France) et correction nomenclaturale (Cerambycidae: Cerambycinae: Callidiini). *L'Entomologiste* 69: 87–95.

Vitali, F. 2015. Cerambycoidea.com: The first web-site about the world-wide Cerambycoidea. www. Cerambycoidea.com (accessed December 30, 2015).

Wagner, M. R., S. K. N. Atuahene, and J. R. Cobbinah. 1991. *Forest entomology in West Tropical Africa: Forest insects of Ghana.* Dordrecht, Netherlands, Kluwer Academic.

Wang, Q. 1995. A taxonomic revision of the Australian genus *Phoracantha* Newman (Coleoptera: Cerambycidae). *Invertebrate Taxonomy* 9: 865–958.

Wang, X.-H. 2011. *Studies on biology, ecology and biocontrol techniques of Apriona swainsoni Hope (Coleoptera: Cerambycidae).* Beijing: PhD. Thesis, Chinese Academy of Forestry.

Wang, X., Z. Yang, X. Wang, Y. Tang, and Y. Zhang, 2014. Biological control of *Apriona swainsoni* (Coleoptera: Cerambycidae) by applying three parasitoid species. *Scientia Silvae Sinicae* 50: 103–108.

Webb, J. L. 1909. *Some insects injurious to forests: The southern pine sawyer.* Washington, DC, US Government Printing Office: USDA Bureau of Entomology Bulletin 58(Part IV): 41–56.

Webster, F. M. 1889. Notes upon the longevity of the early stages of *Eburia quadrimaculata* (Say). *Insect Life* 1: 339.

Wylie, F. R. 1982. Insect problems of *Araucaria* plantations in Papua New Guinea and Australia. *Australian Forestry* 45: 125–131.

Wylie, F. R., and M. R. Speight. 2012. *Insect pests in tropical forestry*, 2nd edition Wallingford: CABI.

Xiao, G.-R., ed., 1992. *Forest insects of China* (2nd ed.). Beijing: China Forestry.

Yan, W. B. 2003. Researches on biological characters of *Semanotus bifasciatus*. *Journal of Anhui Agricultural College* 30: 57–60.

Yan, Z., J. Sun, D. Owen, Z. Zhang. 2005. The red turpentine beetle, *Dendroctonus valens* LeConte (Scolytidae): An exotic invasive pest of pine in China. *Biodiversity and Conservation* 14: 1735–1760.

Yuan, F., Y.-Q. Luo, J. Shi, et al. 2008. Invasive sequence and ecological niche of main insect borers of *Larix gmelinii* forest in Aershan, Inner Mongolia. *Forestry Studies in China* 10: 9–13.

Zhang, Q.-H., J. A. Byers, and X.-D. Zhang. 1993. Influence of bark thickness, trunk diameter and height on reproduction of the longhorned beetle, *Monochamus sutor* (Col., Cerambycidae) in burned larch and pine. *Journal of Applied Entomology* 115: 145–154.

Zhao, T. H., W. X. Zhao, R. T. Gao, Q. W. Zhang, G. H. Li, and X. X. Liu. 2007. Induced outbreaks of indigenous insect species by exotic tree species. *Acta Entomologica Sinica.* 20: 826–833.

Zhuravlev, I. I., and G. E. Osmolovskii. 1964. *Pests and Diseases of Shade Trees.* Jerusalem, Israel Program for Scientific Translations.

12

Cerambycid Pests in Agricultural and Horticultural Crops

Qiao Wang
Massey University
Palmerston North, New Zealand

CONTENTS

12.1 Introduction

Cerambycids as economic pests are not only important with respect to forest trees but also for various agricultural and horticultural crops. Damage to crops is usually caused by larval feeding and, occasionally, by adult feeding or oviposition. Larvae of most cerambycid species are borers, feeding on living, dying, dead, or rotten plant stems, branches, or twigs. Most wood-boring species feed on subcortical tissues—at least initially. Later, they may burrow further into sapwood and—even hardwood—to continue feeding. Herbaceous feeders usually bore in host stems. Root feeders may bore in the roots, hollowing out and killing the roots of the host plants, or they may live in the soil and feed on the roots. In some species, larvae damage fruit; adults cause economic damage to leaves and male flowers by feeding and to stems or branches by girdling.

Worldwide, more than 100 cerambycid species have been or currently are economic pests of crops. However, the pest status or identity of several species has not been confirmed, and several others are no longer considered economically important. Consequently, only 90 species are discussed in this chapter, of which 46 are from the Lamiinae, 34 are from the Cerambycinae, and 10 are from the Prioninae. Geographically, 48 species are considered endemic to Asia, 11 to South and Central America, 10 to Europe, 7 to North America, 6 to Africa, and 8 to Australasia.

Table 12.1 provides scientific names, common names, crop hosts, feeding sites, life cycle duration, overwintering stages, economic importance, and regions/countries where damage is considered to have an economic impact for these 90 species. Most common names are in general use in various parts of the world, and some are created here based on crops attacked, geographical distribution, morphological features, or combined information. Many species are extremely polyphagous, using a wide range of woody and herb plant species as their larval hosts; for example, the larvae of *Anoplophora chinensis* (Forster) feed on more than 100 species of living trees from 36 families (Lingafelter and Hoebeke 2002; Haack et al. 2010) and those of *Oemona hirta* (F.) on more than 200 host species from 81 families (EPPO 2013c). Therefore, the hosts listed for each cerambycid species in Table 12.1 are limited to major crops damaged rather than the full host record, and the pest status in plants other than crops is not discussed in this chapter. Feeding sites refer to the sites of plants where cerambycid larvae feed, including roots (R), main stems or trunks (M), branches (B), twigs (T), and shoots (S).

Most cerambycid species included in this chapter take at least one year to complete their life cycle. It is important to bear in mind that the information on life cycles provided in Table 12.1 is in summary form. The duration for completion of a generation may be dependent on the prevailing climatic conditions. For example, the South American species *Steirastoma breve* (Sulzer) has only one generation a year in southern Brazil but up to four generations a year in tropical areas (Kliejunas et al. 2001).

The assessment of economic importance mainly is based on literature and, to a lesser extent, on personal experience and personal communications. The countries or regions listed in Table 12.1 are limited to where damage to crops is reported for each species rather than complete geographical distribution, information about which is given in the text. Control measures, including those accepted by the growers as well as in laboratory and field trials, are provided for each species. Many synthetic insecticides used for control are not registered for cerambycid pests or have already been banned for crop pest control. Therefore, in describing chemical control available for each species, only the application method (such as spray, painting, injection, place-and-plug, and soil treatment) and chemical classes (such as organophosphate and carbamate) are given, with literature provided. Readers can make up their own minds in terms of what insecticides should be used and how to use them depending on pest species and local governmental regulations.

In this chapter, the major crops cerambycids damage are divided into eight categories on the basis of agricultural and horticultural traditions and for the convenience of workers in these sectors: (1) pome and stone fruit trees, (2) citrus trees, (3) tropical and subtropical fruit trees, (4) nut trees, (5) mulberry trees, (6) cacao, coffee, and tea crops, (7) vine crops, and (8) field crops. Many cerambycid species attack crops in more than one category; thus, the species included in each category are those that cause the most damage to crops in that category. The biology and management of cerambycid pests are discussed in individual crop categories. Under each category, the pest species are clustered according to their origins for readers' convenience.

12.2 Cerambycid Pests of Pome and Stone Fruit Trees

Pome and stone fruit trees refer to two major groups of deciduous species originating from the temperate zone of the Northern Hemisphere (Ebert 2009). The most abundant species in the first group comprise apple, pear, and quince, while those in the second group include cherries, peaches, plums, and apricots. Economically, apples, pears, peaches, and plums are ranked among the top 10 in terms of worldwide fruit production (Ebert 2009).

Dozens of cerambycid species have been recorded that attack pome and stone fruit trees, but only 20 species in the world are recognized as economic pests of these tree crops. Their biology and management measures are discussed in this section. Among these species, seven are endemic to Asia, six to Europe, four to North America, two to South America, and one to Africa.

TABLE 12.1

Cerambycid Pests of Crops in the World

Species[a]	Common Name	Crop Hosts[b]	Feeding Site[c]	Generation(s) per Year[d]	Overwintering Stage[e]	Economic Importance[f]	Countries Where the Species Is Recorded as a Pest[g]
Acalolepta cervina (Hope)	Coffee brown beetle	Coffee especially arabica coffee	M	1	L	++	China and Thailand
Acalolepta mixta (Hope)	Passion vine longicorn	Grape vine, passion vine, fig, mango, and citrus	R&M&B	1	L&P	+++	Australia
Aeolesthes holosericea (Fabricius)	Cherry stem borer	Apple, apricot, anhili, cherry, plum, peach, mulberry, and walnut	M&B	2 years	L	++	Northern India and southern China
Aeolesthes induta (Newman)	Silky gray-brown longhorn	Tea and oil tea	M&R	2–3	L&A	++	Southeastern China
Aeolesthes sarta Solsky	Apple stem borer	Apple, apricot, and walnut	M&B	2 years	L	+++	India and Iran
Agapanthia dahli (Richter)	Sunflower longicorn beetle	Sunflower	M	1–2 years	L	+	Hungary and western Russia
Analeptes trifasciata (Fabricius)	Cashew stem girdler	Cashew	M&B	2	L&P	+++	Nigeria
Anoplophora chinensis (Forster)	Asian starry citrus longicorn or Citrus longhorned beetle	Citrus particularly lime, lemon, and orange; stone and pome fruit trees	M&R	1–2 years	L	+++	Asia: China, Japan, Korea, Indonesia (Sumatra), Myanmar, Malaysia, Philippines, and Vietnam; EUROPE (introduced): France, Germany, Italy, The Netherlands, Switzerland, and UK
Anoplophora macularia (Thomson)	Macularian whitespotted longicorn	Citrus and lychee	M	1	L	++	China (Jiangsu, Taiwan) and southern Japan
Aphrodisium gibbicolle (White)	Striate-necked green longicorn	Citrus	M&B	1–2 years	L&P	+	Southern China
Apomecyna saltator (Fabricius)	Cucurbit vine borer	Vines of ridge gourd (*Luffa acutangula*), pointed gourd (*Trichosanthes dioica*) and melons, pumpkins, and vegetable marrow	M	2–3	L	++	India and China

(Continued)

TABLE 12.1 (Continued)

Cerambycid Pests of Crops in the World

Species[a]	Common Name	Crop Hosts[b]	Feeding Site[c]	Generation(s) per Year[d]	Overwintering Stage[e]	Economic Importance[f]	Countries Where the Species Is Recorded as a Pest[g]
Apriona germari (Hope)	Brown mulberry longicorn	Mulberry, apple, walnut, fig	M&B	1–3 years	L	+++	India, southern China, and Southeast Asia
Apriona rugicollis Chevrolat	Mulberry tree borer	Mulberry and teyaki	M&B	2–3 years	L	++	China, Japan, and Korea
Aristobia testudo (Voet)	Lychee longicorn	Lychee, longan, cocoa, jackfruit, guava and citrus	B&T	1	L	++	Southern China and northern India
Aromia bungii (Faldermann)	Peach rednecked longicorn	Peach, apricot, plum, and cherry	M&B	2–3 years	L	++	China, Mongolia, Korea, Vietnam; Italy and Germany (introduced)
Bacchisa atritarsis (Pic)	Black-tarsused tea longicorn	Tea and oil tea	M&B	2 years	L	++	Southern China
Bacchisa fortunei (Thomson)	Blue-backed pear longicorn	Pear, apple, and peach	B&T	2 years	L	+	China
Batocera horsfieldi Hope	Large gray and white longicorn	Walnut	M&B	2–3 years	L&P	++	China and India
Batocera lineolata Chevrolat	Cloud-marking longicorn	Walnut	M&B	2–3 years	L&P&A	++	China
Batocera rubus (Linnaeus)	Lateral-banded mango longhorn	Mango and fig	M&B	1–3	L or P	++	China and India
Batocera rufomaculata (DeGeer)	Mango tree stem borer	Mango, fig, durian, jackfruit, apple, guava, pomegranates, mulberry, and cashew	M&B	1–2	L	+++	India, Thailand, Egypt, and Israel
Bixadus sierricola (White)	West African coffee borer	Coffee particularly robusta coffee	M&R	1–2	L	+++	Western African countries
Calchaenesthes pistacivora Holzschuh	Pistachio longicorn	Pistachio	B&T	2 years	L&A	++	Iran
Celosterna scabrator (Fabricius)	Grape stem borer	Grapevine	M&B	1	L	++	India
Cerambyx carinatus (Küster)	Ring-pediceled longicorn	Cherry	M	2–3 years	L	+	Turkey

(Continued)

TABLE 12.1 (Continued)

Cerambycid Pests of Crops in the World

Species[a]	Common Name	Crop Hosts[b]	Feeding Site[c]	Generation(s) per Year[d]	Overwintering Stage[e]	Economic Importance[f]	Countries Where the Species Is Recorded as a Pest[g]
Cerambyx cerdo Linnaeus	Great capricorn beetle	Apple, pear, and cherry	M	3–5 years	L&A	+	Most European countries
Cerambyx dux (Faldermann)	Giant-eyed longicorn	Apple, pear, plum, peach, almond, cherry, plum, and almond	M&B	2–3 years	L,P&A	++	Malta, Middle East, and Mediterranean countries
Cerambyx nodulosus Germar	Nodulated longicorn	Cherry	M	3–4 years	L	+	Turkey
Cerambyx scopolii Fuessly	Capricorn beetle	Cherry, apple, plum, and walnut	M&B	2–3 years	L,P&A	+	France and Switzerland
Chelidonium argentatum (Dalman)	Greenish lemon longicorn	Citrus	M&B	1	L	++	Southern China and India
Chelidonium cinctum (Guérin-Méneville)	Lime tree borer	Citrus	B&T	1	L	+	Southern China and India
Chelidonium citri Gressitt	Green citrus longicorn	Citrus	B&T	1	L	+	Southern China
Coptops aedificator (Fabricius)	Albizia longicorn beetle	Mango and coffee	M&B	1	L	++	Middle East, Africa, and South Asia
Dectes texanus LeConte	Dectes stem borer	Sunflower and soybean	M	1	L	+++	USA (mainly southern Great Plains)
Diploschema rotundicolle (Audinet-Serville)	Brown-headed citrus borer	Citrus and peach	M&B	1	L&P	++	Brazil
Disterna plumifera (Pascoe)	Speckled longicorn	Citrus	M&B	1	L	+	Australia
Dorcadion pseudopreissi Breuning	Ryegrass root longicorn	Turf and pasture grasses	R	2 years	L&A	++	Turkey
Dorysthenes buqueti (Guérin-Méneville)	Sugarcane stem longicorn	Sugarcane	M	1–2 years	L	++	Thailand and Indonesia
Dorysthenes granulosus (Thomson)	Sugarcane root and stem longicorn	Sugarcane, citrus, mango, cassava, and longan	M&R	2 years	L	+++	Thailand and southern China
Dorysthenes hugelii (Redtenbacher)	Apple root borer	Apple but occasionally pear, peach, cherry, and plum	R	3–4 years	L&P	++	India

(Continued)

TABLE 12.1 *(Continued)*

Cerambycid Pests of Crops in the World

Species[a]	Common Name	Crop Hosts[b]	Feeding Site[c]	Generation(s) per Year[d]	Overwintering Stage[e]	Economic Importance[f]	Countries Where the Species Is Recorded as a Pest[g]
Dorysthenes hydropicus (Pascoe)	Sugarcane longicorn beetle	Sugarcane, pomelo, and reed	M&R	1–2 years	L	++	Mainland China and Taiwan
Dorysthenes paradoxus (Faldermann)	Bluegrass longicorn	Turf grass	R	3 years	L	++	Northern China
Dorysthenes walkeri (Waterhouse)	Long-toothed longicorn	Sugarcane	M&R	2 years	L	++	Southern China
Elaphidion cayamae Fisher	Spined citrus borer	Citrus (orange and mandrine)	B&T	1	L	++	Cuba
Hylettus seniculus (German)	Hylettus citrus borer	Citrus	M&B	1	L	++	Brazil
Linda nigroscutata (Fairmaire)	Apple twig borer	Apple	T&S	1	L	+	India and China
Macropophora accentifer (Olivier)	Harlequin citrus borer	Citrus and figs	M	1	L&P	++	Brazil
Macrotoma palmata (Fabricius)	Large brown longicorn	Apricot and citrus	M&B	3–4 years	L	++	Egypt
Monochamus leuconotus (Pascoe)	African coffee stem borer	Coffee especially arabica coffee	M&R	1–2 years	L	+++	Africa especially South Africa, Tanzania, and Uganda
Nadezhdiella cantori (Hope)	Gray-black citrus longicorn	Citrus	M&B	2–3 years	L&P&A	+++	China
Neoplocaederus ferrugineus (Linnaeus)	Cashew stem and root borer	Cashew	M&R	1–2	L	+++	India (particularly east and west coasts) and Nigeria
Neoplocaederus obesus (Gahan)	Cashew stem borer	Cashew and chironji	M&R	1	L	++	India, Vietnam, and southern China
Nitakeris nigricornis (Olivier)	Yellow-headed borer	Coffee especially arabica coffee	M&R&B&T	1–2 years	L	++	Eastern Africa
Nupserha vexator (Pascoe)	Sunflower stem borer	Sunflower	M	1	L	++	India
Nupserha nitidior Pic	Soybean girdle beetle	Soybean	M&R	1	L	+	India
Oberea posticata Gahan	Citrus shoot borer	Citrus	T&S	1	L&P	++	India
Oberea lateapicalis Pic	Orange shoot borer	Citrus	T&S	1–2 years	L	+	India

(Continued)

TABLE 12.1 (Continued)

Cerambycid Pests of Crops in the World

Species[a]	Common Name	Crop Hosts[b]	Feeding Site[c]	Generation(s) per Year[d]	Overwintering Stage[e]	Economic Importance[f]	Countries Where the Species Is Recorded as a Pest[g]
Oberopsis brevis (Gahan)	Girdle beetle	Soybean, rice bean, and sunflower	M&B	2	L	+++	India
Oemona hirta (Fabricius)	Lemon tree borer	Citrus, apple, persimmon, and grapevine	M&B&T	2 years	L	++	New Zealand
Oncideres cingulata (Say)	Pecan twig girdler	Pecan and hickory	B&T	1–2 years	L	++	USA
Oncideres dejeanii Thomson	Pear twig girdler	Pear	B&T	1	L	+	Brazil
Osphranteria coerulescens Redtenbacher	Rosaceae branch borer	Cherry, apricot, peach, plum, apple, pear, quince, almond, pistachio, and fig	B&T	1	L	++	Turkey and Iran
Plagiohammus colombiensis Constantino, Benavides and Esteban	Colombian coffee stem and root borer	Coffee	M&R	2 years	L&P	++	Colombia
Plagiohammus maculosus (Bates)	Speckled coffee stem and root borer	Coffee	M&R	2–3 years	L&P	++	Mexico and Guatemala
Plagiohammus mexicanus Breuning	Mexican coffee stem and root borer	Coffee	M&R	2–3 years	L&P	+	Mexico
Plagiohammus spinipennis (Thomson)	Lantana stem and root borer	Coffee	M&R	2–3 years	L&P	+	Mexico
Plagionotus floralis (Pallas)	Alfalfa root longicorn	Alfalfa and yellow sweet clover	M&R	1–2 years	L	++	Eastern Europe
Platyomopsis pedicornis (Fabricius)	Australian mango longicorn	Mango and shrubby stylo	M&B	1	L&P	+	Northern Australia
Praxithea derourei (Chabrillac)	Taladro longicorn	Apple	M&B	1	L	+	Argentina and Uruguay
Prionus californicus Motschulsky	California prionus	Cherry, peach, apricot, and citrus	R	3–5 years	L	+++	Western North America
Prionus imbricornis (Linnaeus)	Tilehorned prionus	Apple	R	3–5 years	L	++	Eastern North America

(Continued)

TABLE 12.1 (*Continued*)

Cerambycid Pests of Crops in the World

Species[a]	Common Name	Crop Hosts[b]	Feeding Site[c]	Generation(s) per Year[d]	Overwintering Stage[e]	Economic Importance[f]	Countries Where the Species Is Recorded as a Pest[g]
Prionus laticollis (Drury)	Broad-necked root borer	Apple	R	3–4 years	L	++	Eastern North America including USA and Canada
Psacothea hilaris (Pascoe)	Yellow-spotted longicorn beetle	Mulberry and figs	M	1–2	E&L	+++	China and Japan (adults also damage leaves by feeding)
Pseudonemorphus versteegi (Ritsema)	Citrus trunk borer	Citrus	M	1	L	+++	India and Myanmar
Psyrassa unicolor (Randall)	Pecan branch pruner	Pecan and hickory	B&T	1–2 years	L	+	USA
Rhytidodera bowringii White	Yellow-spotted ridge-necked longicorn	Mango	M&B&T	1–2 years	L&P	++	India and southern China
Rhytidodera simulans (White)	Mango trunk and branch borer	Mango, cashew, and rose apple	M&B&T	1	L&P	++	Indonesia and Malaysia
Rhytiphora diva (Thomson)	Alfalfa crown borer	Soybean and alfalfa	M	1	L	++	Northern Australia particularly Queensland and Western Australia
Saperda candida Fabricius	Roundheaded apple tree borer	Apple, pear, cherry, and plum	M&B	2–3 years	L&P	++	Canada (Quebec) and Eastern USA
Skeletodes tetrops Newman	Australian citrus longicorn	Citrus	B&T	1	L	+	Australia
Steirastoma breve (Sulzer)	Cacao longicorn beetle	Cacao (*Theobroma cacao*)	M&B	1–4	L	+++	South America and some Caribbean islands
Sthenias grisator (Fabricius)	Grapevine stem girdler	Mulberry and grapevine	T&S	1	L	++	India and Sri Lanka
Strongylurus thoracicus (Pascoe)	Pittosporum longicorn	Citrus	M&B&T	1	L	+	Australia
Tetrops praeustus (Linnaeus)	Plum longicorn	Apple and pear	T&S	1–2 years	L	+	European countries; introduced to northeastern USA, and northern Africa
Trachylophus sinensis Gahan	Four-ridged tea longicorn	Tea and oil tea	R&M&B	1	L	+	China

(Continued)

TABLE 12.1 (Continued)

Cerambycid Pests of Crops in the World

Species[a]	Common Name	Crop Hosts[b]	Feeding Site[c]	Generation(s) per Year[d]	Overwintering Stage[e]	Economic Importance[f]	Countries Where the Species Is Recorded as a Pest[g]
Trichoferus griseus (Fabricius)	Fig longicorn	Fig	M&B	2–3 years	L	+	Egypt
Uracanthus cryptophagus Olliff	Australian citrus branch borer	Citrus	B&T	1	L	+	Australia
Xylotrechus arvicola (Olivier)	European grapevine borer	Grapevine	M&R&B	2 years	L	+++	Spain
Xylotrechus javanicus (Castelnau & Gory)	Asian coffee stem borer	Coffee particularly arabica coffee	R&M&B	1–3	L&P	+++	India, China, Thailand, and Vietnam
Xylotrechus pyrrhoderus Bates	Grape tiger longicorn	Grapevine	M&B	1	L	++	Japan, Korea, and China

a   Synonyms are given in the text.

b   The host plants listed here are important crops the pests damage rather than the complete range of host plants.

c   Roots (R), main trunks or stems (M), branches (B), twigs (T), and shoots (S).

d   Estimated number of generations/year: 1, 2, or 3 refer to 1, 2, and 3 generations per year; however, 2 years, 3 years, or 4 years mean that one generation takes 2, 3, or 4 years.

e   Eggs (E), larvae (L), pupae (P), or adults (A).

f   +, ++, and +++ represent occasional (or minor) pest, important (or moderate) pest, and key (or serious) pest, respectively.

g   These are not exhausted distribution records.

12.2.1 Asia

12.2.1.1 Aeolesthes sarta Solsky

Common name: Apple stem borer

The adults of the cerambycine stem borer are 28–47 mm long and dark gray-brown, with the elytra covered with short silvery hairs; shiny silvery spots form two irregular bands crossing the elytra (Figure 12.1). This species is widely distributed in Central and South Asia including Afghanistan, Iran, India, Kyrgyzstan, Pakistan, Tajikistan, Turkmenistan, and Uzbekistan, and highly polyphagous with a host range of many tree species from Ulmaceae, Salicaceae, Platanaceae, Rosaceae, Juglandaceae, Sapindaceae, Betulaceae, Elaeagnaceae, Oleaceae, Fabaceae, Moraceae, and Fagaceae (Duffy 1968; EPPO 2005).

The longicorn beetle is one of the most important pests of many forest and ornamental trees such as *Populus, Platanus, Ulmus,* and other amenity trees in cities and parks in the region of its present distribution (EPPO 2005). Because this pest causes significant damage to trees in urban areas, it is also called the city longicorn beetle. In horticulture, this is a serious pest of apple and walnut trees in Asia and also impacts cherry, apricot, peach, and plum trees (Duffy 1968; Sengupta and Sengupta 1981; Shiekh 1985; Farashiani et al. 2001; Mir and Wani 2005; Bhat et al. 2010a; Khan et al. 2013). For example, *A. sarta* was first reported as a serious pest of apple and other temperate fruit trees in Jammu and Kashmir (northern India) in 1980s (Shiekh 1985). According to Bhat et al. (2010a), depending on apple cultivars and tree age, infestation rate ranges from 6.5% in American Apirogue to 23.2% in Red Delicious and from 5.5% in 10- to 15-year-old trees to 27% in those more than 30 years old in northern India. Khan and Qadri (2006) reported a 50% infestation rate in Indian apple and apricot orchards. The pest causes enormous damage to walnut trees, with a 30–40% infestation rate in northern India (Khan et al. 2013). In Iran, 40–100% of apple and walnut trees can be infested (Farashiani et al. 2001). Larvae bore into the large branches and trunks where sizable emergence holes and borings at the base of infested trees are indications of this pest's presence. Infested trees may not die for many years, but their vitality and

FIGURE 12.1 *Aeolesthes sarta* adult. (Courtesy of Xavier Gouverneur.)

productivity are impaired and leaves become wilted and dry. Usually, several generations develop on the same tree until it is eventually killed. So far, the distribution of this species still appears to be limited to Asia. However, if introduced, they can potentially become established in other parts of the world such as Europe where pome and stone fruit trees are widely grown (Vanhanen et al. 2008).

Biology of this pest is extracted from Duffy (1968), EPPO (2005), and Mazaheri et al. (2007). Pupation usually occurs in October and November, and the adults emerge from March to June. The eggs are usually laid in small batches of 5–10 on living trees, particularly in wounds in the bark, on broken ends of branches, and in pits gnawed down to the living bark. A female can lay about 50 (Duffy 1968), 123 (Mazaheri et al. 2007), or 240–270 eggs (EPPO 2005) in her lifetime. The eggs hatch in 10–14 days, and neonate larvae bore into the bark and sapwood. Frass is ejected through the entry hole. The grown larvae enter the wood and, at the end of the first season of development, make a long (about 25 cm) tunnel that first runs parallel to the axis of the trunk or branch and then turns to form a downward gallery about 15 cm long. At the bottom of this gallery, the larva overwinters, protected by a double plug made from frass. Next in the spring, the larvae continue to feed, making tunnels deep into the wood. At the end of July, they prepare pupation cells protected by double plugs made from borings. Pupation occurs in these cells, and the pupal stage lasts about two weeks. After eclosion, the adult remains in the pupation cells over the winter and then emerges in the spring. The life cycle lasts two years.

Cultural measures have been used to control the pest in orchards and gardens (Duffy 1968). For example, during outbreaks, adults can be trapped and killed using freshly cut logs of hosts. Heavily infested trees should be felled and removed before November. In a field trial, two entomopathogenic fungal isolates *Beauveria bassiana* and *Metarhizium anisopliae* gave some control (Mohi-Uddin et al. 2009). Bhat et al. (2010b) have tested the pathogenicity of three fungal isolates, *B. bassiana*, *B. brongniartii*, and *M. anisopliae*, in the laboratory and indicated that *B. bassiana* is most virulent against *A. sarta* larvae. Insertion of organophosphate insecticide-soaked small pieces of sponge into borer holes in the trunk achieved >60% larval mortality in the field (Khan and Qadri 2006). The similar application of organophosphate or aluminum phosphide can kill 70% of larvae within three weeks of application (Mohi-Uddin et al. 2009; Bhat et al. 2010c).

12.2.1.2 *Aeolesthes holosericea* (Fabricius)

Synonyms: *Pachydissus similis* Gahan, *P. velutinus* Thomson, *Ceramryx holosericeus* Fabricius
Common name: Cherry stem borer

The adults of this cerambycine species are 20–36 mm long and similar to *A. sarta* in morphology; the body is dark or reddish brown, covered with dense grayish or light brown silky pubescence; the elytra have a sheen of contrasting dull and bright areas that vary according to incidence of light (Figure 12.2). It is widely distributed in India (particularly northern India), Pakistan, Sri Lanka, southern China, Myanmar, Vietnam, Thailand, Malaysia, and Laos. So far, this species still appears to be limited to Asia. This longicorn is also highly polyphagous, attacking many tree species from Fabaceae, Rutaceae, Betulaceae, Combretaceae, Malvaceae, Phyllanthaceae, Meliaceae, Sonneratiaceae, Myrtaceae, Lythraceae, Euphorbiaceae, Anacardiaceae, Annonaceae, Moraceae, Myristicaceae, Diptocarpaceae, Pinaceae, Rosaceae, Fagaceae, and Tamiaceae (Beeson 1941; Duffy 1968; Regupathy et al. 1995; Mamlayya et al. 2009; Prakash et al. 2010; Bhawane and Mamlayya 2013; Salve 2014).

Duffy (1968) suggested that this species only attacks dead or felled trees. However, it also attacks living trees. This stem borer has become an increasingly important pest of apple trees in northern India (Tara et al. 2009; Gupta and Tara 2013, 2014) and a minor pest of mango, guava, peach, pear, and plum trees in China (Li at el. 1997). It also attacks cherry, apricot, guava, mango, mulberry, peach, pear, plum, and walnut trees in India (Tara et al. 2009). Similar to *A. sarta*, the larvae cause the damage, making horizontal zigzag tunnels into the trunk and main branches and thereby reducing the longevity and fruit yield of trees (Gupta and Tara 2014). Trees can be killed in successive years of damage.

The biology of this pest is similar to that of *A. sarta* (Duffy 1968; Tara et al. 2009; Gupta and Tara 2013). Pupation occurs in October in northern India, and adults emerge in the following March–May with

FIGURE 12.2   *Aeolesthes holosericea* adult. (Courtesy of Xavier Gouverneur.)

a peak in April. The females lay eggs in batches of four to six in the cracks or crevices under the bark during May–October. A female can lay about 200 eggs in her lifetime but, in confinement, can lay up to 92 eggs. The larvae bore into the bark, sapwood, and finally into heartwood. Their life cycle lasts about two years; the first winter is spent in the larval stage and the second in the larval or pupal stage in galleries.

So far, only chemical control appears to be effective for this pest. The application of organophosphate or aluminum phosphide in a way similar to that used for *A. sarta* can kill up to 100% of larvae (Gupta and Tara 2014).

12.2.1.3 *Aromia bungii* (Faldermann)

Synonyms: *Aromia cyanicornis* Guérin-Méneville, *Cerambyx bungii* Faldermann
Common name: Peach rednecked longicorn

The adults of this cerambycine species are 20–40 mm long and blue-black and shining—except for pronotum, which is bright red (Figure 12.3). It is widely distributed in eastern Asia including China, Japan, Korea, Mongolia, and Vietnam (Duffy 1968; Li et al. 1997; Anderson et al. 2013; EPPO 2013a, 2015). The larvae of this longicorn are reported to feed on a number of plant species from Rosaceae, Meliaceae, Poaceae, Ebenaceae, Oleaceae, Salicaceae, Juglandaceae, Lythraceae, and Theaceae with Rosaceae appearing most preferred (Anderson et al. 2013; EPPO 2013a, 2015). *A. bungii* has recently been introduced to Europe, including Germany and Italy (Anderson et al. 2013; EPPO 2013a), and may have become established in Italy (Garonna et al. 2013). Based on its known host range and distribution, this pest probably has the potential to establish in most parts of the Europe and presents a high risk (EPPO 2013a).

The peach rednecked longicorn is considered an important pest of stone fruit trees, such as peach, apricot, plum, and cherry in eastern Asia, particularly China (Duffy 1968; Li et al. 1997; Anderson et al. 2013; EPPO 2013a). Earlier references show that the larvae bore in both sapwood and heartwood of main trunks and large branches, weakening trees and occasionally killing them (Gressitt 1942; Duffy 1968).

FIGURE 12.3 *Aromia bungii* adult. (Courtesy of Antonio P. Garonna.)

More recent literature reports that the larvae mainly tunnel in the subcortical area beneath the bark and the sapwood and less commonly in the heartwood, leading to loss of fruit production and weakening of the trees (EPPO 2013a).

Pupation takes place at the end of June; the adults emerge from late June until early August, and winter is passed in the larval stage (Gressitt 1942; Liu et al. 1999). The females lay eggs in bark crevices on the trunk and main branches mainly in July (EPPO 2013a). Lifetime fecundity is unknown. This species has a life cycle of two to three years depending on climate (Liu et al. 1999; Ostojá-Starzewski and Baker 2012).

Gressitt (1942) recommended that the larvae be extracted by means of hook wire and adults collected during the daytime on branches, trunks, and flowers during July. Heavily infested trees should be felled and destroyed. Hong and Yang (2010) have tested the application of raw extracts and culture homogenates of a poisonous mushroom, *Lepiota helveola* Bresadola, to control larvae and achieved up to 80% larval mortality. There are also reports of trials using several entomophagous nematode species, *Steinernema* spp., as biological control agents against *A. bungii* (Liu et al. 1993, 1998b). However, it is not known how useful these treatments can be in the field.

12.2.1.4 *Linda nigroscutata* (Fairmaire)

Synonym: *Miocris nigroscutatus* Fairmaire
Common name: Apple twig borer

The adults of the lamiine longicorn are 15–20 mm long and bright orange-red; the pronotum has four small, round, black spots in the form of a trapeze on the disc; the elytra have an elongate scutum-shaped black mark behind the scutellum, reaching to the end of the basal quarter or basal third of suture, and the humeri generally are also marked with a black spot (Figure 12.4). This species occurs in southwest China and northeast India (Gressitt 1947; Duffy 1968; Sachan and Gangwar 1980; Cao 1981). So far, there is no report on occurrence of this lamiine outside Asia.

FIGURE 12.4 *Linda nigroscutata* adults, dorsal (left) and lateral (right) views. (Courtesy of Xavier Gouverneur.)

The apple twig borer is an apple tree pest of some economic importance in southwest China and northeast India (Fletcher 1919; Gressitt 1947; Duffy 1968; Sachan and Gangwar 1980; Cao 1981). The adult females girdle twigs and tender shoots and lay eggs in the wound subcortically, and the larvae bore in the twigs and shoots and eventually hollow them, reducing growth and yield (Beeson and Bhatta 1939; Duffy 1968; Sachan and Gangwar 1980; Cao 1981). Frass is ejected from holes made by larvae in twigs.

According to Cao (1981), the adults emerge from May to July, peaking during May. The eggs are laid singly in the girdled wounds of the current year's shoots. The larvae bore into the shoots and then continue to bore into one- and two-year-old twigs. Lifetime fecundity is unknown. This pest has one generation a year and overwinters as mature larvae that pupate in infested twigs.

Pruning and burning of infested twigs are probably the only effective measures for the control of this pest (Fletcher 1919; Cao 1981). These practices should be carried out during June–August. If the larger twigs are infested, a pesticide solution may be injected into larval tunnels.

12.2.1.5 Bacchisa fortunei (Thomson)

Synonyms: *Chreonoma fortunei* (Thomson), *C. fortunei obscuricollis* Pic, *C. fortunei flavicornis* Kano, *Plaxomicrus fortunei* Thomson
Common name: Blue-backed pear longicorn

The adults of this lamiine are 8–11 mm long; the body is orange-yellow with metallic blue elytra (Figure 12.5). *B. fortunei* (Thomson) is widely distributed in China, Japan, and Korea, and feeds on most cultivated pome and stone fruit species (Duffy 1968; Li et al. 1997). So far, this species only occurs in Asia.

This longicorn is considered a minor pest of apple, pear, and peach trees in China (Chang 1973; Qian 1987; Wei 1990; She et al. 2005). However, up to 45% infestation on apple trees has been reported in northwestern China (Xu et al. 2007). The larvae bore in twigs or small branches, making them vulnerable to wind breakage (She et al. 2005). The infested parts break open, exposing frass (Guan 1999). Growth and fruiting of damaged twigs and branches are significantly reduced.

FIGURE 12.5 *Bacchisa fortunei* adult. (Courtesy of Larry Bezark.)

Biology of this pest is summarized from Guan (1999) and She et al. (2005). Adults emerge during April–July with a peak in June. The adult females prefer twigs and branches 15–25 mm in diameter for oviposition. They gnaw oviposition slits on selected parts and lay eggs in the slits singly. A female can lay 10–30 eggs in her lifetime. Neonate larvae bore into the bark and then the sapwood. This pest takes two years to complete its life cycle and overwinters as larvae.

Cultural measures are recommended to control the pest, such as removal of eggs and neonate larvae using knives, killing larvae in tunnels using wire hooks (Guan 1999), and pruning damaged twigs and branches (She et al. 2005). Injection of organophosphate insecticides into larval tunnels is recommended for heavily infested twigs and branches (Guan 1999; She et al. 2005). Trials using a wasp (*Scleroderma guani* Xiao and Wu) parasitizing larvae and pupae have been carried out in northwestern China's Gansu Province, one of the major apple growing areas, and achieved a parasitism rate of 55.5–60.8% (Wei 1990).

12.2.1.6 *Osphranteria coerulescens* Redtenbacher

Common name: Rosaceae branch borer

The adults of this cerambycine borer are 16–22 mm long; body is elongate with a metallic violet color; the elytra are clothed with dense appressed golden pubescence (Figure 12.6). It is widely distributed in Iran, Iraq, Turkey, Pakistan, and Afghanistan, and larvae feed on various species from Rosaceae and other families (Sharifi et al. 1970; Kaplan 2013; Özdikmen 2014).

As an important pest in Iran (Sharifi et al. 1970; Abivardi 2001) and Turkey (Kaplan 2013), *O. coerulescens* damages the twigs and branches of many pome and stone fruit trees such as cherry, apricot, peach, plum, apple, pear, and quince, and several other crop species such as almond, pistachio, and fig. According to Sharifi et al. (1970), the neonate larvae immediately bore into the twigs, which start wilting with discolored leaves in two weeks. During the short period of movement up the twigs, the larvae cut the vascular bundles, reducing the amount of sap at the top of the twig. This activity apparently produces an environment

FIGURE 12.6 *Osphranteria coerulescens* adult. (Courtesy of Jiří Mička.)

favorable for the development of the larvae. After three weeks, when the larvae turn and proceed toward the base of the twig, the damaged tip of the twig falls off. The badly infested trees are quite obvious after 20–25 days because of the fallen twigs. About four weeks after penetration, the larval entrance is marked by a 10- to 15-mm-long, glutinous filament extruded from the branch at this point. On the trees, the tips of the broken branches are brownish and have irregularly cut surfaces. In addition to the central larval gallery, the branches have small holes. Fine frass is ejected through these openings. From the young branches, the larvae may continue boring into the trunks of six- to seven-year-old trees. These branches break off at the slightest pressure and, in heavily infested orchards, many may be seen on the ground in late summer, fall, and early spring. This breakage can result in severe crop reduction.

Biology of this longicorn is extracted from Sharifi et al. (1970) and Kaplan (2013). The adults emerge in May and June and prefer spring twigs for oviposition. The favored site for egg deposition is the angle between the petiole and the stem. The eggs are also laid on the twig surface—usually singly but occasionally in groups of two or three. A female can lay 30–75 eggs in her lifetime. This pest has one generation a year and overwinters as mature larvae.

Although larval parasitoids and chemical sprays may contribute to its control to some extent, the simplest, safest, and cheapest method of control is cutting the infested twigs during the summer (Sharifi and Javadi 1971).

12.2.1.7 *Dorysthenes hugelii* (Redtenbacher)

Synonyms: *Lophosternus falco* Gahan, *L. hügelii* Gahan, *L. palpalis* Gahan, *Cyrthognathus falco* Thomson, *C. hügelii* Redtenbacher
Common name: Apple root borer

The adults of this prionine root borer are very large beetles, about 29–53 mm long; the body generally is castaneous in color, with the head and prothorax slightly darker than the elytra. The species is recorded in India, Pakistan, and Nepal. There is no report on expansion of this species outside its native region. The larvae feed on roots of apple, pear, peach, cherry, and plum trees—with apple being the preferred host; adults do not feed (Singh 1941; Duffy 1968).

The apple root borer is a serious pest of apple trees in India (Singh 1941; Verma and Singh 1986) with an infestation level ranging from 2 to 16 larvae per tree. The pest infests all commercial cultivars and groups of apple plants (Sharma and Khajuria 2005; Singh et al. 2010). According to Singh (1941), the larvae bore into or girdle the roots and occasionally the stem below the surface of the ground. The infested trees generally are weakened and yield reduced. Heavily infested trees can die in a few years.

Biology of this pest has been studied by various authors (e.g., Singh 1941; Duffy 1968; Sharma and Khajuria 2005; Singh et al. 2010). The adult emergence peaks with the onset of monsoons in late June and early July. They live for a few days to about three months and are attracted to light. The female prefers sandy and sandy/loam soils for oviposition and, in her lifetime, lays 300–600 eggs, usually singly, at a depth of about 8 mm in the soil. The egg, larval, prepupal, and pupal stages occupy about 1, 42, 3 and 2–3 months, respectively. The neonate larvae move down and start feeding upon reaching the roots. The life cycle lasts three to four years, and overwintering occurs at the larval and pupal stages.

Control measures for killing adults using light trapping (Sharma and Khajuria 2005) or insecticides (Rana et al. 2004) should be directed during late June to mid-July (Singh et al. 2010). According to Sharma and Khajuria (2005), most larvae are present within a 90-cm radius and 30-cm depth of apple tree basins, where control measures should be directed.

12.2.2 Europe

12.2.2.1 *Cerambyx carinatus* (Küster)

Synonym: *Hammaticherus carinatus* Küster
Common name: Ring-pediceled longicorn

The adults of this cerambycine borer are 30–45 mm long and dark brown in color; the second antennal segment is in the shape of a ring, nearly three times as wide as long; the elytron has a spine at suture (Figure 12.7). It is widely distributed in Eastern Europe and the Middle East—such as in Bulgaria, Croatia, Iran, Macedonia, Malta, Serbia, and West Turkey, and its larvae feed on large stressed or sickly trees of the Rosaceae family (Bense 1995; Özdikmen and Turgut 2009; Hoskovec and Rejzek, 2014). So far, it has not been recorded in other parts of the world.

The ring-pediceled longicorn is a minor pest of cherry trees in Eastern Europe and the Middle East, particularly in western Turkey (Tezcan and Rejzek 2002). The larvae bore in the heartwood of the trunks, weakening the vigor of the infested trees.

The adults emerge during June–August and lay eggs in bark wounds or crevices along the trunks. The lifetime fecundity is unknown. The larvae bore in the superficial regions of the trunk in the first few months and then enter the sapwood and heartwood. The winter is passed in the larval stage. One generation takes two to three years.

Maintenance of tree health probably is the best control measure for this pest. Whitewash along the trunk may be useful in preventing oviposition.

12.2.2.2 *Cerambyx nodulosus* Germar

Common name: Nodulated longicorn

The adults are 26–46 mm long and dark brown to blackish in color; the pronotal disc bears irregular wrinkles, and the elytral apex is rounded (Figure 12.8). This species mainly occurs in southeastern Europe, Turkey, and the Near East, and its larvae feed on sickly trees of the Rosaceae family, particularly *Prunus* (Bense 1995; Hoskovec and Rejzek, 2014). So far, it has not been recorded in other parts of the world.

FIGURE 12.7   *Cerambyx carinatus* adult. (Courtesy of Jiří Mička.)

FIGURE 12.8   *Cerambyx nodulosus* adult. (Courtesy of Zoran Božović [image from Srećko Ćurčić].)

The nodulated longicorn is also a minor pest of cherry trees in Eastern Europe and the Middle East, particularly in western Turkey (Tezcan and Rejzek 2002). It also attacks *Pyrus, Malus*, and *Crataegus* trees (Bense 1995). The larvae bore in the trunks and expel large quantities of easily visible sawdust. In southeastern Bulgaria, even very small, stunted *Crataegus* shrubs growing on sandy soils of the Black Sea coastal region can serve as hosts (Tăuşan and Bucşa 2010).

The adults emerge during May–August and are active during the day. One generation lasts three to four years. The larval biology is similar to that of *C. carinatus*. The lifetime fecundity is unknown.

Control measures are similar to those for *C. carinatus*.

12.2.2.3 *Cerambyx scopolii* Fuessly

Synonyms: *C. piceus* Geoffroy, *C. niger gallicus* Voet
Common name: Capricorn beetle

The adults are 17–28 mm long and totally black in color; the elytra are rounded at the apex and covered with a fine gray pubescence (Figure 12.9). The longicorn is widely distributed in Europe, Caucasus, Transcaucasia, North Africa, and in the Near East (Bense 1995; Hoskovec and Rejzek 2014). It is polyphagous, with its larvae feeding on tree trunks and branches from the Rosaceae, Fagaceae, Juglandaceae, Betulaceae, Salicaceae, Ulmaceae, Grossulariaceae, Sapindaceae, and Oleaceae families (Duffy 1953; Hoskovec and Rejzek 2014). So far, it has not been recorded in other parts of the world.

This species is a minor pest of apple, cherry, plum, and walnut trees in Europe (Stahel and Holenstein 1948; Brauns 1952; Duffy 1953; Ferrero 1985). The larvae make large, broad tunnels deep in the heartwood of mature and often healthy trees (Figure 12.10), weakening them. When the damage occurs to the trunks, fruit drop prematurely, and the infested trees may eventually die. In the early stages of infestation, gumming and cracks and swellings in the bark can occur.

According to Duffy (1953), the adults emerge during April–July and lay eggs in the crevices in the bark of the branches and trunks of mature trees. The lifetime fecundity is unknown. The larvae initially

FIGURE 12.9 *Cerambyx scopolii* adult. (Courtesy of Boženka Hric [image from Srećko Ćurčić].)

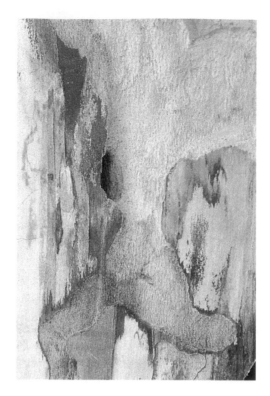

FIGURE 12.10   Galleries made by *Cerambyx scopolii* larvae. (Courtesy of György Csóka.)

feed under the bark and soon enter sapwood and then heartwood. Pupation occurs in the pupal cell in the heartwood either in August–September or in the spring. The life cycle lasts two to three years. Overwintering occurs at the larval stage, and depending on when pupation occurs, the winter can also be passed at the pupal or adult stage.

In orchards, maintenance of tree health is still probably the best control measure for this pest (Stahel and Holenstein 1948; Duffy 1953). Removal of infested branches and the use of trap-pans baited with molasses to catch adults may help reduce the pest population. When the trunks are infested, the introduction of cotton wool soaked in organophosphate insecticides into galleries and the subsequent sealing off of holes may be effective.

12.2.2.4 *Cerambyx cerdo* Linnaeus

Synonyms: *C. luguber* Voet, *C. heros* Scopoli
Common name: Great capricorn beetle

The adults are 24–55 mm long and blackish, with the elytra reddish brown toward the apex; the second antennal segment at the inner edge is as long as its width, not in the shape of a ring; the elytral apex has a spine at the suture (Figure 12.11). This species occurs in the Mediterranean region and is highly polyphagous on deciduous trees from the Rosaceae, Ulmaceae, Fagaceae, Juglandaceae, Salicaceae, Fabaceae, and Betulaceae families (Duffy 1953; Bense 1995; Hoskovec and Rejzek 2014). So far, it has not been recorded in other parts of the world.

The great capricorn beetle is a minor pest of apple, pear, cherry, and oak trees in Europe (Bruans 1952; Duffy 1953). The larvae make sizable tunnels in the sapwood and heartwood of large, mature tree trunks (Figure 12.12) and sometimes cause considerable damage. Reinfestation may continue for many years, weakening the trees that are affected.

The biology is summarized from Bruans (1952), Duffy (1953), and Hoskovec and Rejzek (2014). The adults emerge during May–August and lay eggs in the crevices of the bark. The larvae bore under the bark

FIGURE 12.11 *Cerambyx cerdo* adult. (Courtesy of György Csóka.)

FIGURE 12.12 Galleries made by *Cerambyx cerdo* larvae. (Courtesy of György Csóka.)

for a few weeks, then enter the sapwood, and finally bore into the heartwood. They spend three to four years in the trunk. Pupation occurs in the pupal cell made in the heartwood in August. The adults overwinter in the pupal cells and emerge from May onward the next year. The life cycle lasts three to five years.

Control measures are similar to those for *C. scopolii*.

12.2.2.5 *Cerambyx dux* (Faldermann)

Synonym: *Hammaticherus dux* Faldermann
Common name: Giant-eyed longicorn

The adults are 25–45 mm long and dark brown to blackish, with the elytra being black only at the base and becoming paler, reddish brown toward the apex; the eyes are large with the lower edge nearly reaching the underside of the head; the elytral apex is more or less rounded (Figure 12.13). This species is widely distributed in Eastern Europe, the Middle East, and in the Near East, particularly in Bulgaria, Macedonia, Malta, Turkey, Greece, Crimea, Iran, and Jordan (Bense 1995; Sharaf 2010; Hoskovec and Rejzek 2014). So far, it has not been recorded in other parts of the world. Its larvae are polyphagous in fruit and ornamental trees or brushes (Hoskovec and Rejzek 2014).

The giant-eyed longicorn is an important pest of apple and pear trees in the Mediterranean and Middle East regions, and it also occasionally attacks apricot, plum, peach, and almond trees (Jolles 1932; Lozovoi 1954; Saliba 1972, 1974, 1977). In Jordan, it is a serious pest of stone fruit trees with about 24% of cultivated trees being infested; plum trees are more susceptible than peach trees, and almond trees are the least susceptible host (Sharaf 2010). The larvae bore in the branches and trunks, expelling a large quantity of sawdust from tunnels. Mature trees more than five years of age are usually attacked because younger trees do not have trunks thick enough to accommodate the fully grown cerambycine larvae.

The biology of *C. dux* has been relatively well studied (Jolles 1932; Saliba 1977; Sharaf 2010). The adults emerge during March–June and may be stimulated by rain. In Jordan, the adults emerge

FIGURE 12.13   *Cerambyx dux* adult. (Courtesy of Larry Bezark.)

at the time their hosts bloom. They live for about a month and lay eggs singly or in small patches of 10 arranged in two vertical rows, with each row of five eggs placed in the bark crevices or wounded lesions of the upper trunk or the shady side of the main branch. According to Saliba (1977), the average number of eggs laid per female is approximately 13. However, the fecundity appears to be much higher than 13 eggs (Sharaf 2010). The females are attracted to black-colored, previously infested, or mechanically injured trunks and branches. The larvae bore under the bark for a few months and then enter the sapwood and heartwood. The egg stage lasts one to three months; the larval stage, 15–16 months; the pupal stage, one to two months; and the dormant adult stage in pupal cell, seven months. *C. dux* overwinters as the partially grown larvae, pupae, and adults. The life cycle lasts about two to three years.

Several control measures have been practiced (Jolles 1932; Lozovoi 1954; Saliba 1972; Sharaf 2010). Removal of heavily infested and dead trees can effectively reduce the population size in the orchards. Plugging all borer holes with cement can kill a certain number of larvae but may not be effective enough. To reduce the infestation rate, mechanical injuries to the trunks and branches should be avoided. Sprays of insecticides in May and June may kill adults and young larvae before they enter the trees. The most effective control method appears to be the application of a white latex paint, by painting or spraying, to the trunks and branches to prevent oviposition and larvae from entering the trees. The paint can be mixed with insecticides to kill adults and neonate larvae.

12.2.2.6 *Tetrops praeustus* (Linnaeus)

Synonym: *T. praeusta* Linnaeus
Common name: Plum longicorn

The adults of this lamiine borer are 3–6 mm long; the head and pronotum are black; the elytra are yellow to yellowish brown; each elytron has a black spot at the apex but sometimes the elytra are entirely black (Figure 12.14). This species is widely distributed in Europe and has become established in

FIGURE 12.14 *Tetrops praeustus* adult. (Courtesy of Eugenio Nearns, Longicorn ID, USDA APHIS ITP, Bugwood.org [5516884].)

the northeastern United States and North Africa (Yanega 1996; Howden and Howden 2000). The larvae primarily feed on various species from Rosaceae but also attack trees from the Fagaceae, Malvaceae, Ulmaceae, Salicaceae, Celastraceae, and Rhamnaceae families (Duffy 1953; Bense 1995).

The plum longicorn is a minor pest of apple and pear trees. The larvae bore inside the twigs and shoots. Leaves on primary shoots later become dry and fall off, a symptom similar to fire blight caused by *Erwinia amylovora* (Burrill) (Jerinic-Prodanovic et al. 2012).

The biology is briefly described in Duffy (1953), Bense (1995), and Jerinic-Prodanovic et al. (2012). The adults emerge in May–July and lay eggs in oviposition slits made by females in the bark of the twigs and shoots. The lifetime fecundity is unknown. The larvae bore in and under the bark and pupate in the sapwood. In Europe, this species completes its life cycle in one to two years depending on climate, and overwinters at the larval stage.

Pruning of the infested twigs can reduce the pest population to some extent.

12.2.3 North America

12.2.3.1 *Prionus californicus* Motschulsky

Common name: California prionus

There are about 27 synonyms for this species (Linsley 1962), which are not listed here. The adults of this prionine root borer are 24–55 mm long, exclusive of mandibles, and reddish brown to piceous brown in color; the antennae are 12-segmented with segments 3–11 of males being distinctly produced externally at the apex; the pronotum is short (Figure 12.15). It is widely distributed in western North America from Mexico to Alaska and highly polyphagous, feeding on species from the Fagaceae, Juglandaceae, Rosaceae, Vitaceae, Salicaceae, Myrtaceae, Ericaceae, Rutaceae, Anacardiaceae, and Pinaceae families (Linsley 1962; Bishop et al. 1984; Steffan and Alston 2005; Alston et al. 2007, 2014).

FIGURE 12.15   *Prionus californicus* adult. (Courtesy of Steven Valley, Oregon Department of Agriculture and Bugwood. org [5449551].)

In the past two decades, *P. californicus* has become a serious pest of stone fruit trees, particularly sweet cherry, peach, and apricot, in the Intermountain West region (Alston et al. 2007, 2014), and also damaged citrus trees in Phoenix, Arizona (Jeppson 1989). The larvae spend three to five years underground, causing direct or indirect death of fruit trees due to girdling of the root cambium and introduction of secondary pathogens that lead to decay (Alston et al. 2014). The damage is more severe when the trees are grown in light, well-drained soil; sometimes, the trunks of the trees are completely girdled just below the ground level (Linsley 1962). In citrus trees, larvae bore in the trunks just below the ground surface (Jeppson 1989). Canopy death or sudden loss of tree vigor can be symptoms of *P. californicus* infestation.

The biology of this pest is summarized from Linsley (1962) and Alston et al. (2014). The adults emerge in early to mid-summer and are active crepuscularly. The females lay eggs in the soil. A female can lay 200–300 eggs in her lifetime; but, in some cases, up to 600 eggs are laid. The neonate larvae tunnel in the soil to seek out tree roots. The younger larvae begin feeding on smaller roots and gradually reach the tree root crown (stump, usually at or near the soil level) as mature larvae. In cherries, a greater proportion of the larvae found at the root crown are mature larvae, while younger larvae are found in smaller roots. Following three to five years of root and root-crown feeding while moving upward in the soil, the insect pupates in a pupal cell close to the soil surface. However, early reports show that pupation can occur in the soil 8–16 cm below the surface. The life cycle lasts three to five years and overwintering occurs in the larval stage.

Once an orchard is infested by this pest, it is difficult to control its population increase and spread to nearby trees, and the application of insecticides may only suppress the pest to some extent (Alston et al. 2014). Several insecticides registered for stone fruit may provide incidental suppression of the adults. Systemic insecticides such as imidacloprid may be effective on the younger larvae on the roots but not against the older larvae in the root crown or lower trunk. Imidacloprid may suppress the local population if used annually over several years. Heavily infested trees should be removed. Maintaining tree health and preventing stress are the best options for the control of prionus infestations. Fallowing an infested field for two or more years, planting annual crops that will be tilled under each year, and avoiding planting stone fruit trees in infested sites are also effective for managing the prionus in infested soils.

12.2.3.2 *Prionus imbricornis* (Linnaeus)

Synonyms: *P. robustus* Casey, *P. diversus* Casey, *Cerambyx imbricornis* Linnaeus
Common name: Tilehorned prionus

The adults are 24–50 mm long, exclusive of mandibles, and dark brown to reddish brown in color; there are 15–20 segmented antennae (Figure 12.16). This species is widely distributed in eastern North America along the Atlantic Coast from Florida to southern Canada, and from Florida west to the Great Plains, Nebraska, and Texas (Linsley 1962; Agnello 2013). Its larvae are polyphagous, feeding on roots of many species from Fagaceae, Rosaceae, Vitaceae, Salicaceae, Juglandaceae, Sapindaceae, Cornaceae, Tiliaceae, and Poaceae (Linsley 1962; Agnello 2013).

The tilehorned prionus has been recorded as a destructive pest of the living roots of oak, chestnut, pear, apple, grape, and several herbaceous crops (Linsley 1962). It has become an important pest of apple trees in recent years and, to a lesser extent, it also attacks cherry, peach, plum, quince, pear, and pecan trees (Agnello 2013). The aboveground symptoms of infestation are a gradual thinning and yellowing of foliage and limb-by-limb mortality. Young trees may be chewed off just below the soil surface and their root systems devoured. Mature trees may have only one or two functional roots left near the surface, keeping them alive until they are blown over by wind. The larvae are reported to also damage roots of maize (reviewed by Linsley 1962).

The biology of this pest is summarized from Craighead (1915), Linsley (1962), and Agnello (2013). The adults emerge from the soil in early to mid-summer but mostly in July. They fly at dusk and night and are attracted to lights. During the day, they hide beneath the loose bark or debris at base of trees. The females lay 100–200 eggs in groups in the soil around tree bases (Linsley 1962). Agnello (2013) stated that the females live for one to two weeks, during which time they each lay hundreds of eggs. The eggs hatch in two to three weeks. The neonate larvae dig down to the roots and begin feeding on the root bark. They move through the soil from one root to another, feeding on the surfaces of small roots.

FIGURE 12.16   *Prionus imbricornis* adult. (Courtesy of Nathan P. Lord, USDA APHIS ITP, Bugwood.org [5492072].)

Eventually, they enter the wood of larger roots, which they hollow, girdle, or sever. During the summer, the larvae feed on the roots in the upper 15–45 cm of soil but in the winter, are often located at nearly twice these depths. The feeding and developmental period lasts three to five years. When ready to pupate in early spring, larvae leave the roots and construct large oval pupal cells within 6–12 cm of the soil surface. Overwintering occurs at the larval stage, and the life cycle lasts three to five years.

Several control measures are recommended by Agnello (2013). Heavily infested trees that are beyond recovery should be removed and burned before the following spring to prevent developing borers inside from completing their life cycle. Keeping trees in a healthy, vigorous condition is one of the best preventive measures against attack. Foliage sprays of broad-spectrum insecticides during peak adult activity in July may be partially effective. Although the adults are attracted to light, no trials have been conducted to test whether light traps could be useful for controlling adults.

12.2.3.3 *Prionus laticollis* (Drury)

Synonyms: *P. oblongus* Casey, *P. frosti* Casey, *P. nigrescens* Casey, *P. densus* Casey, *P. beauvoisi* Lameere, *P. kempi* Casey, *P. parvus* Casey, *P. brevicornis* Fabricius, *Cerambyx laticollis* Drury
Common name: Broad-necked root borer

The adults are 22–44 mm long, exclusive of mandibles, and black or dark brown with an obscurely reddish cast in color; the antennae are 12-segmented (Figure 12.17). This species is widely distributed in eastern North America along the Atlantic Coast from southern Canada and New York and inland to Oklahoma (Linsley 1962; Agnello 2013). Its larvae are polyphagous, feeding on roots of many species from Fagaceae, Rosaceae, Salicaceae, Tiliaceae, Vitaceae, and Pinaceae (Linsley 1962; Agnello 2013).

The broad-necked root borer has recently become an important pest of apple trees, and to a lesser extent, it also attacks cherry, peach, plum, quince, pear, and pecan in eastern North America (reviewed by Agnello 2013). It is exclusively a root feeder. The aboveground symptoms of infestation are

FIGURE 12.17 *Prionus laticollis* adult. (Courtesy of Jon M. Yuschock, Bugwood.org.)

a gradual thinning and yellowing of foliage and limb mortality. Young trees are sometimes chewed off just below the soil surface and their root systems completely consumed. Established trees may be alive after infestation with one or two roots left near the soil surface but are easily blown over by wind. It is not unusual to find from several to 20 or more borers in one tree. In the past, this species was known to attack living roots of trees and shrubs, damaging fruit trees and grapevines in the central and northeastern United States, and also to attack lead telephone cables in subterranean wooden ducts, railway ties, and other timbers in contact with the ground (Linsley 1962).

According to Laurent (1905), Linsley (1962), and Agnello (2013), the adults emerge mainly in July and live for one to two weeks. Although Linsley (1962) reported a flight period of July–August, females have not been observed flying (Agnello 2013). The adults are active during dusk and night and remain inactive under the loose bark or debris at the base of the tree during the day. The females lay eggs in the soil near the tree base and produce an average of 363 eggs each. The eggs hatch in two to three weeks. Young larvae penetrate the soil and feed on surfaces of small roots and move through the soil from one root to another. They then bore into the wood of larger roots. Similar to the tilehorned prionus, the larvae feed on the roots in the upper 15–45 cm of the soil during the summer but are often located at nearly twice these depths in winter. In the early spring, mature larvae rise to within 6–12 cm of the soil surface to pupate. Overwintering occurs in the larval stage, and the life cycle lasts three to four years.

Control measures are similar to those for the tilehorned prionus (Agnello 2013).

12.2.3.4 *Saperda candida* Fabricius

Synonyms: *S. bipunctata* Hopping and *S. bivittata* Say
Common name: Roundheaded apple tree borer

The adults of this lamiine species are 13–21 mm long and brownish to black in color; the head, pronotum, and elytra have two longitudinal white stripes running the length of the body; the antennae and legs are usually reddish brown, and the entire body is covered with gray pubescence with the ventral

FIGURE 12.18 *Saperda candida* adult. (Courtesy of Dawn Dailey O'Brien, Cornell University, Bugwood.org [5458614].)

side silvery white (Figure 12.18). This borer is native to North America and distributed across the United States (particularly the eastern part) and southern Canada (Linsley and Chemsak 1995; EPPO 2008; Agnello 2013). It was accidentally introduced to the island of Fegmarn in Germany in 2008, causing serious concern to Europe (Nolte and Krieger 2008; EPPO 2008; Kehlenbeck et al. 2009; Baufeld et al. 2010; Eyre et al. 2013) and China (Yao et al. 2009). However, it is still not clear whether this species has become established and spread in Europe or elsewhere. *S. candida* has a broad range of hosts in Rosaceae including apple, cherry, peach, plum, quince, and pear, with the preferred host being apple (EPPO 2008).

In the eastern United States and southeastern Canada, it is considered an important pest of fruit trees, particularly apple, with trees of all sizes being attacked—but with those from three- to ten-years-old suffering the most (Agnello 2013). The adults feed on leaves, twigs, and fruit of host plants, but their injury to plants is not significant. The larvae bore in the main trunks and large branches and may completely girdle young trees. Infested trees look sickly, producing sparse, pale-colored foliage. The larval attacks can kill the trees or weaken them so that they are broken by wind. Sawdust-like frass can be found at the base of infested trees.

The biology of this pest is summarized from Linsley and Chemsak (1995) and Agnello (2013). The adults fly during May–August with an emergence peak of two to three weeks in June. Most emergences occur at night through round, pencil-sized holes from the bases of infested trees. The female lives about 40 days on average, normally hiding during day and laying eggs at night. The female makes a longitudinal cut in the bark with her mandibles near the base of the tree, inserting a single egg between the bark and xylem and sealing it in place with a gummy secretion. Oviposition mainly takes place in June and July but fecundity is not known in the field. The eggs hatch in 15–20 days. The larvae bore into the bark, moving upward or downward in the trunk depending on the year and stage of growth, feeding on the inner bark layer, and widening their tunnels as they feed. By the end of their third season of feeding, the larvae have bored straight up in the trunk and constructed pupal chambers just beneath the bark surface, within which they pass their final winter. This species requires two to three years to complete its development, depending on climate.

Agnello (2013) summarized the available control measures for this pest as follows: (1) tree destruction—remove and burn heavily infested trees that are beyond recovery before the following spring to prevent developing borers inside from completing their life cycle. (2) Larval killing by hand—during bloom and again in September, inspect the bark surface of the trunk between the lower 60 cm to just below the soil surface for small pinholes with sawdust exuding from them. Insert a stiff wire that is slightly hooked at the end to reach and impale the borer. (3) Injection—inject a mixture of pyrethrum in ethanol or *para*-dichlorobenzene moth flakes in cottonseed oil into the gallery using a grease gun to kill the borers that cannot be reached with the wire. (4) Foliage sprays—apply broad-spectrum insecticides at the beginning and end of June to kill adults. (5) Oviposition barriers—in early May, wrap protective coverings of various materials around the bottom 60 cm of the trunks to prevent the female beetles from reaching their preferred oviposition sites. (6) Surface deterrents—apply a deterrent wash (alkaline mixture of insecticidal soap plus caustic potash [lye] mixed to the consistency of thick paint) on the uninfested trunk surfaces using a paintbrush every two to four weeks, depending on rainfall, from late May through July to deter egg laying.

12.2.4 South America

12.2.4.1 *Oncideres dejeanii* Thomson

Synonym: *O. pustulata* Thomson
Common name: Pear twig girdler

The adults of this lamiine longicorn are 15–26 mm long and dark brown in color; the elytra have yellowish-brown pubescence at the base and a number of more or less round pubescent spots from near the base to the apex (Figure 12.19). This species is widely distributed in Brazil (from Maranhão, Ceará to Rio Grande do Sul),

FIGURE 12.19 *Oncideres dejeanii* adult. (Courtesy of Juan Pablo Botero [image from Marcela L. Monné].)

Paraguay, Argentina, and Uruguay (Duffy 1960; Monné 2002). Its larvae are highly polyphagous, feeding on a number of plant families including Anacardiaceae, Annonaceae, Bignoniaceae, Bombacaceae, Boraginaceae, Caesalpiniaceae, Casuarinaceae, Cecropiaceae, Cupressaceae, Euphorbiaceae, Fabaceae, Flacourtiaceae, Lauraceae, Meliaceae, Mimosaceae, Moraceae, Myrtaceae, Polygonaceae, Proteaceae, Rosaceae, Rutaceae, Salicaceae, and Tiliaceae (Link et al. 1994, 1996; Duffy 1960; Monné 2002).

In recent years, the pear twig girdler has become a minor pest of pear trees in southeastern Brazil, with a potential to become an important pest (Cordeiro et al. 2010). The adults girdle healthy tree branches before oviposition, which occurs under the green bark of the girdled parts. The larvae bore and complete their development in the girdled branches. The girdled twigs and branches eventually die. The average size of infested twigs/branches is 3.5 cm in diameter with a range from 1 to 3.6 cm.

According to Paro et al. (2012), the adults are present between November and January with a peak in February in southeastern Brazil. The larvae bore in the twigs and branches. The adults feed on the flowers and leaves of host plants, but adult feeding injury is not important. The winter is passed in the larval stage. The life cycle lasts about a year but can be longer depending on climate.

Pruning of the infested branches and twigs may reduce the pest population. Insecticide sprays while adults are active may contribute to a reduction in damage.

12.2.4.2 *Praxithea derourei* (Chabrillac)

Synonyms: *Elaphidium collare* Burmeister, *Xestia derourei* Chabrillac
Common name: Taladro longicorn

The adults of this cerambycine borer are 24–34 mm long, moderately slender, and reddish brown in color with sides being more or less parallel; the vertex and pronotum are covered with golden hairs; the elytra have fairly dense punctures with the apex bearing two sharp spines (Figure 12.20). The beetle occurs throughout South America, including Brazil, Bolivia, Paraguay, Argentina, and Uruguay (Duffy 1960;

FIGURE 12.20   *Praxithea derourei* adult. (Courtesy of Juan Pablo Botero [image from Marcela L. Monné].)

Monné 2001a). Its larvae feed on a number of plant species, including Betulaceae, Fagaceae, Meliaceae, Myrtaceae, Rosaceae, Salicaceae, and Tamaricaceae (Duffy 1960; Monné 2001a).

Although this species attacks a number of ornamental and forest trees in South America (Duffy 1960), it is considered a minor to important pest of apple trees in Argentina and Uruguay (Carbonell Bruhn and Briozzo Beltrame 1980; Di Iorio 1998). This species also attacks pear trees, but its significance to pear orchards is not clear (Carbonell Bruhn and Briozzo Beltrame 1980). The larvae make tunnels down from the branches to the bases of tree trunks (Duffy 1960; Di Iorio 1998). The infested parts are weakened and, at times, killed.

The biology is summarized from Bosq (1942), Duffy (1960), and Di Iorio (1998). The adults emerge during December and January. The females lay eggs among the terminal buds, particularly on the higher branches. The lifetime fecundity is unknown in the field. Young larvae bore into the branches and tunnel down to the trunks. The life cycle lasts about a year and the larvae overwinter.

According to Bosq (1942) and Duffy (1960), the infested branches should be cut off and burned or the borer holes injected with insecticide before summer.

12.2.5 Africa

12.2.5.1 *Macrotoma palmata* (Fabricius)

Synonyms: *M. valida* Thomson, *M. coelaspis* White, *M. humeralis* White, *M. böhmi* Reitter, *M. palmata rugulosa* Kolbe, *Prionus senegalensis* Olivier, *P. spinipes* Illiger, *P. palmatus* Fabricius
Common name: Large brown longicorn

The adults of this prionine are 25–65 mm long and brown to blackish brown in color; the antennae are 11-segmented with the inner side of several basal antennal segments bearing numerous granular outgrowths or spines; the sides of the pronotum have a number of small spines and granular processes (Figure 12.21). This longicorn occurs in many African countries including Egypt, Saudi Arabia,

FIGURE 12.21 *Macrotoma palmata* adult. Specimen from MNHN-Paris. (Courtesy of Jiří Pirkl, copyright MNHN-Paris, used with permission of T. Deuve and A. Taghavian.)

Morocco, Libya, Algeria Ethiopia, Kenya, Namibia, Nigeria, Rhodesia, Senegal, Sierra Leone, Ivory Coast, Cameroon, Burundi, Rwanda, Gambia, Guinea, Uganda, Mali, Mauritania, Niger, Guinea-Bissau, Benin, Mauritius, Mozambique, Botswana, Congo, and Uganda. Its larvae are highly polyphagous, feeding on many tree species from Aceraceae, Anacardiaceae, Casuarinaceae, Lauraceae, Mimosaceae, Moraceae, Myrtaceae, Piperaceae, Platanaceae, Rhizophoraceae, Rosaceae, Rutaceae, Salicaceae, and Tamaricaceae (Duffy 1957; Tadros et al. 1993; Sama et al. 2005; Delahaye et al. 2006).

The large brown longicorn attacks forest and ornamental trees and various fruit tree species including fig, citrus, and apricot in Northern Africa (Duffy 1957; Delahaye et al. 2006; Tawfik et al. 2014). It has become an important pest of apricot (Tadros et al. 1993) and citrus (Shehata et al. 2001) orchards in Egypt. The larvae bore in the trunks and main branches, weakening the trees. Older trees such as those more than 10 years old are more susceptible to this pest (Shehata et al. 2001; Tawfik et al. 2014) and can be infested by two to eight borers per tree. The highest level of the borer population (adult exit holes) occurs on the eastern side of the orchard (Shehata et al. 2001).

The biology of this beetle is summarized from Duffy (1957), El-Sebay (1984), Tadros et al. (1993), Shehata et al. (2001), and Tawfik et al. (2014). The adults can be found between April and October, with an emergence peak in July and August. The females lay their eggs in the cracks and crevices in the bark of the trunks and main branches of healthy trees. The lifetime fecundity in the field is not known. However, in the laboratory, a female can lay an average of 50.9 eggs during the 12–18 days of the oviposition period. The larvae bore into the bark and then into the wood. The total larval period lasts two to four years, depending on temperature. The species takes three to four years to complete its life cycle and overwinters as mature larvae.

Control measures are still developing. Some laboratory tests on the effectiveness of fungus and synthetic insecticides have been made, but how useful these agents are for the control of the pest in the field is unknown.

12.3 Cerambycid Pests of Citrus Trees

The genus *Citrus* in the family Rutaceae contains some of the most commercially grown fruit species in the world—such as oranges, mandarins, and lemons. Oranges and other citrus fruit species are ranked among the top five in worldwide fruit production (Ebert 2009).

It has been widely believed that the origin of the *Citrus* genus was in the part of Southeast Asia bordered by northeast India, Myanmar, and southwest China (Scora 1975; Gmitter and Hu 1990). The fact that about 50% of cerambycid pests of citrus are endemic to Southeast Asia appears to support this hypothesis. However, some recent studies suggest that the genus could have originated in Australia, New Caledonia, and New Guinea (Liu et al. 2012a), even though relatively few cerambycids are economic pests of citrus in this region.

Jeppson (1989) listed and briefed 13 cerambycid species of some economic importance to the citrus industry; among them, *Stromatium barbatum* (Fabricius) in India and *Chloridolum alcmene* Thomson in Southeast Asia are no longer considered economically important and thus are not treated in this book; *Prionus californicus* and *Macrotoma palmata* are more important as stone fruit than as citrus trees and thus were discussed in the previous section of this chapter. In addition, *Aneflomorpha citrana* Chemsak was recorded as attacking orange twigs and small branches in Arizona in the 1950s (Gerhaedt 1961) but is not currently considered a pest of economic importance.

Although many cerambycids are reported to attack citrus trees, only 19 are considered economic pests in the world. Among these, 10 species are endemic to Asia, one of the world's largest citrus producing regions (USDA 2015). Australasia is one of the world's citrus-growing regions, but its production is relatively small compared to Asia, North and South America, Europe, and Africa (USDA 2015). Smith et al. (1997) listed five species that are distributed only in Australia and are considered economic pests of citrus in the eastern and southeastern regions of that continent. Among these species, *Acalolepta mixta* (Hope) [= *A. vastator* (Newman)] is an important pest of grapevines and thus is discussed in the section on vine crops later in the chapter; the remaining four species are detailed here. Furthermore, *Oemona hirta* (Fabricius), a New Zealand native, is a moderate pest of citrus and thus is discussed in this section—although it also attacks many plant species, including a number of fruit tree and grapevine species. South America, particularly Brazil, has become one of the major citrus growing regions in the world (USDA 2015). Many cerambycid species endemic to South America have broadened their host ranges to exotic citrus trees. For example, *Retrachydes thoracicus* (Olivier) (Penteado Dias 1980; Nascimento do et al. 1986), *Hexoplon ctenostomoides* Thomson, and *Phoebe phoebe* (Lep. & Serv.) (D'Araujo e Silva 1955) have been reported as infesting citrus and other fruit trees. However, these species are not currently considered economic pests of citrus. There are only four cerambycid species of economic importance in citrus orchards in South America, and they are discussed here. Cerambycids rarely cause any significant damage to citrus trees in Europe, North America, and Africa.

12.3.1 Asia

12.3.1.1 *Anoplophora chinensis* (Forster)

Synonyms: *A. malasiaca* (Thomson), *Melanauster chinensis* (Forster), *Cerambyx punctator* Olivier, *C. farinosus* Houttuyn, *C. chinensis* Forster
Common name: Asian starry citrus longicorn or citrus longhorned beetle

The lamiine adults are 20–40 mm long; the body is entirely shining black with the elytra marked with small, scattered, irregular, white pubescent batches; the basal halves of most antennal segments and tarsi have powdery blue pubescence; the ventral side has white pubescence; the elytra have numerous rounded tubercles on the basal fifth but lack abundant, long, suberect hairs on the surface (Figure 12.22). In rare cases, specimens do not appear to have any white patches on the elytra (Lingafelter and Hoebeke 2002). This species is primarily distributed in China, Korea, and Japan, with occasional records from Indonesia, Malaysia, Myanmar, the Philippines, and Vietnam (Duffy 1968; Lingafelter and Hoebeke 2002; CABI 2008; Haack et al. 2010). *A. chinensis* has been intercepted in many countries in Europe and North America and confirmed to have established in Europe, causing serious concern (Eyre et al. 2010; Haack et al. 2010; Peverieri et al. 2012; Schroder et al. 2012; Strangi et al. 2013). Its larvae feed on more than 100 tree species from at least 26 families including Aceraceae, Anacardiaceae, Araliaceae, Betulaceae, Elaeagnaceae, Fagaceae, Lauraceae, Oleaceae, Polygonaceae, Styracaceae, Rutaceae, Rosaceae, Salicaceae, Ulmaceae, Moraceae, Meliaceae, Leguminosae, Juglandaceae, Aquifoliaceae, Platanaceae, Euphorbiaceae, Casuarinaceae, Verbenaceae, Sapindaceae, Theaceae, and Taxodiaceae (Lingafelter and Hoebeke 2002; Haack et al. 2010; EPPO 2013b).

FIGURE 12.22 *Anoplophora chinensis* adult. (Courtesy of Steven Valley, Oregon Department of Agriculture and Bugwood.org [5445438].)

Although this longicorn attacks many tree species, it is considered a serious pest of citrus trees in China, being extremely abundant in all lowland orchards (Lieu 1945; Duffy 1968; Wang et al. 1996; You and Wu 2007; Haack et al. 2010; EPPO 2013b). It is also an important pest of many stone and pome fruit and mulberry trees there (Li et al. 1997). The adults feed on bark and leaves of host trees, but their injury is not significant. The larvae feed under the bark for about two months, after which time they bore into the woody tissues of the lowest portions of the trunks and the roots. Symptoms of infestation include frass and wood pulp extruding from frass holes, discoloration and deformation of the bark, and round emergence holes. A single larva can cause the death of a tree five- to six-years-old or younger. Mature trees can be significantly weakened by several larvae and eventually killed in successive years of infestation.

The biology of this pest is summarized from Lieu (1945), Duffy (1968), Adachi (1994), Wang et al. (1996), You and Wu (2007), Haack et al. (2010), and EPPO (2013b). The adults are present from late May to early October, with an emergence peak occurring in June and July. The females make oviposition slits with their mandibles in the living bark of lower trunks and exposed roots, lay eggs singly in the slits, and then seal the oviposition sites with secretions from the ovipositor. A female can lay an average of 70 eggs in her one- to three-month lifetime. The eggs hatch in one to three weeks, depending on temperature. The larval developmental period ranges from 10 to 15 months depending on climate. Pupation and adult development occur in the wood, and the adults exit the tree through a round hole on the bark surface. *A. chinensis* takes one year to complete its life cycle in southern China, one to two years in northern China and Japan, and probably up to three years in Northern Europe. Overwintering takes place at the larval stage.

Lieu (1945) recommended two effective control methods. First, the exposed roots and lower trunks up to 50 cm (particularly the lower 25 cm) from the ground are examined for egg slits and exuding frass at intervals of 10–14 days from early July to early October. The egg slits should be pressed with the thumb to crush the eggs, and the bark under the frass should be pared off so that the gallery can be uncovered and the larva killed. Second, whitewashing the exposed roots and lower trunks up to 50 cm from the ground with water-based pastes of quick-lime alone or mixture of quick-lime and sublimed sulphur proves to be very effective in preventing oviposition and killing the adults. The adults do not lay any eggs on treated trees, and those that visit the treated sites die in two to six days. The coating will remain for about two months. If the paste is washed away by heavy rain during the oviposition period, the trees should be retreated. Whitewashing is still the most practical and effective control method in China. Adachi (1990) has trialed the application of wire netting to prevent oviposition in a mature grove in Japan, achieving effective control. Heavily infested trees should be removed and destroyed. Chemical control is usually not very effective for this pest. However, Cavalieri (2013) trialed chemical control in Italy and achieved about 90% success. Briefly, in late May to early June, the basal trunks of trees should be sprayed or brushed with pyrethroid solutions, and then the trunks should be treated again 20 days later with a mixture of pyrethroid and neonicotinoid. Brabbs et al. (2015) summarized the literature on the biological control of this pest and discussed the potential of various biological control agents including entomopathogenic fungi *Beauveria brongniartii* Petch, parasitic nematodes *Steinernema feltiae* Filipjev and *S. carpocapsae* Weiser, and insect parasitoids *Aprostocetus anoplophorae* Delvare and *Spathius erythrocephalus* Wesmael. In China, three native parasitoids have been recorded that attack this pest, *A. prolixus* LaSalle & Huang, *Brulleia shibuensis* Matsumura, and *Scleroderma* sp. (Niu et al. 2014). However, to date, none of these agents has been used for the control of *A. chinensis* in commercial orchards. Because of this pest's invasive nature, EPPO (2013b) recommended various procedures for containing and eradicating the pest once intercepted and introduced.

12.3.1.2 *Anoplophora macularia* (Thomson)

Synonyms: *A. oshimana* Ohbayashi and *Calloplophora macularia* Thomson
Common name: Macularian whitespotted longicorn

The lamiine adults are 23–40 mm long. This species is very similar to *A. chinensis* in appearance, and they are not easily distinguished by the elytral maculation (Lingafelter and Hoebeke 2001). It differs from *A. chinensis* in having the ventral side covered with blue pubescence and the elytral surface with abundant, long, suberect hairs (Lingafelter and Hoebeke 2002). Unlike *A. chinensis*, the distribution range of *A. macularia* is limited to southern Japan, Taiwan, and southeast coast of China (Lee and Lo 1996; Lingafelter and Hoebeke 2002). So far, there has been no report of its invasion or expansion outside its primary region. It is also polyphagous with a host range of about 70 tree species from at least 19 plant families (Chang 1975).

This longicorn is an important pest of citrus and lychee (litchi) in Taiwan and is particularly abundant in abandoned citrus and lychee orchards and in areas where casuarina and melia trees are present (Chang 1970, 1975; Hwang and Ho 1994; Ho et al. 1995; Lee and Lo 1996). Larval damage can weaken or kill trees. Damage symptoms are similar to those from *A. chinensis*.

The biology of *A. macularia* is summarized from Chang (1970, 1975), Ho et al. (1995), and Lee and Lo (1996, 1998). The adults emerge between early April and October with a peak in June, bite oviposition slits in the bark near the base of the trunks, and lay eggs singly in the slits. The females live for an average of 115 days and, in their lifetime, lay an average of 204 eggs each. Larval feeding usually occurs on the phloem and sapwood under the bark of the trunk. This species has one generation a year; overwintering occurs at the larval stage, and pupation often takes place in a tunnel made into the solid wood at the upper part of the feeding area.

Several control methods recommended by Ho et al. (1995) include the white-brushing (with lime and sulfur) of the trunks before adult emergence to prevent oviposition, chemical spraying on the tree base or canopy to kill adults during April–October, biological control using egg and larval parasitic wasps and fungus *Beauveria* spp., and cultural practices such as tree base sanitation, appropriate irrigation, fertilization, and the use of resistant cultivars. Earlier, Hou (1979) suggested that the application of fumigants through the borer's tunnel openings can kill larvae. However, the effectiveness of the above control measures needs further investigation.

12.3.1.3 *Pseudonemorphus versteegi* (Ritsema)

Synonyms: *Anoplophora versteegi siamensis* Breuning, *Monochamus glabronotatus* Pic, *M. albescens* v. *subuniformis* Pic, *M. albescens* Pic, *M. versteegii* Ritsema
Common name: Citrus trunk borer

Pseudonemorphus versteegi has been placed in the genera *Monochamus* and *Anoplophora* but transferred to the genus *Pseudonemorphus* by Lingafelter and Hoebeke (2002) in their major revision of *Anoplophora*. The adults of this lamiine species are 18–33 mm long and covered with bluish pubescence; the elytra are marked with black spots (Figure 12.23). The longicorn is distributed in India, Myanmar, China, Laos, Vietnam, and Indonesia, and its host range appears to be narrow (Duffy 1968). So far, the species has not been reported to have expanded outside its primary distribution range.

This species is a serious pest of citrus trees in northern India and Myanmar (Lingafelter and Hoebeke 2002), particularly the northeastern Himalayan region of India (Saikia et al. 2012). For example, about 68% of citrus trees were infested during 1992–1995 in Meghalaya, India (Shylesha et al. 1996). *P. versteegi* is only considered a minor pest of citrus trees in southern and southwestern China (Li et al. 1997). In several other southeastern Asian countries, it is not reported as a pest. The longicorn attacks almost all commercially grown citrus species but prefers mandarin and orange (Shukla and Gangwar 1989; Singh and Singh 2012). Although several tree species other than citrus have been recorded as hosts, such as *Aglaia spectabilis* (Miquel) (Beeson and Bhatta 1939), *P. versteegi* is recognized as a pest of citrus trees only (Phukam et al. 1993).

Most adults emerge in May with some variations in different reports: April and May (Beeson and Bhatta 1939), April to June (Banerjee and Nath 1971), and March to September with a peak in May (Shylesha et al. 1996). The adults feed on citrus leaves and young bark. Oviposition occurs on the main trunks (Beeson and Bhatta 1939). Females make oviposition slits in the living bark with their mandibles and then lay eggs singly in the slits. The lifetime fecundity ranges from 35–85 eggs (Banerjee and Nath 1971) and 40–50 eggs (Shylesha et al. 1996) to an average of 69 eggs (Chatterjee

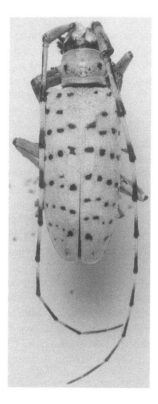

FIGURE 12.23   *Pseudonemorphus versteegi* adult. (Courtesy of Larry Bezark.)

and Ghosh 2001). Young larvae feed on the outer sapwood, forming a horizontal gallery before tunneling toward the center of the trunk (Banerjee and Nath 1971). This cerambycid has one generation a year and overwinters at the larval stage (Beeson and Bhatta 1939; Banerjee and Nath 1971; Shylesha et al. 1996).

Whitewashing the base of trees about 80 cm from the ground can prevent most ovipositions, and a tie of straw or a cotton band can prevent the beetles from ascending the trunks (Mitra and Khongwir 1928). The larvae can be killed in their tunnels by the introduction of a flexible wire (Duffy 1968). Banerjee and Nath (1971) and Shylesha et al. (1996) recommended destruction of eggs and young larvae before penetration into wood using mechanical methods such as pressing oviposition slits to crush the eggs and neonate larvae. Injection of gasoline into larval holes (5–10 mL per hole), followed by the sealing of the hole with mud or cotton plugging twice at 15-day intervals, proves to be highly effective in killing the larvae, achieving up to 82% success (Shukla and Gangwar 1989; Chatterjee and Ghosh 2001; Kalita et al. 2003).

12.3.1.4 *Aphrodisium gibbicolle* (White)

Synonyms: *Chelidonium gibbicolle rubrofemoralis* Pic, *C. gibbicolle subgibbicolle* Pic, *C. gibbicolle* White
Common name: Striate-necked green longicorn

The adults of this cerambycine borer are moderately slender and 28–34 mm long; the body is bluish green and metallic with the head and thorax being more shining; the pronotum has a number of transversely arranged wrinkles; the ventral side is grayish-green or bluish-green, covered with silvery gray pubescence (Figure 12.24). The species is distributed throughout China and in several other Asian countries including Bangladesh, India, Cambodia, Vietnam, Laos, and Thailand (Tu et al. 2006). There is no

FIGURE 12.24 *Aphrodisium gibbicolle* adult. (Courtesy of Larry Bezark.)

report of its occurrence outside its primary distribution range. This is a polyphagous cerambycid with a host range of at least 11 plant species in Rutaceae, Juglandaceae, Pinaceae, Fagaceae, Umbelliferae, and Euphorbiaceae, and it recently has been considered an important pest of oak forests in southern China (Tu et al. 2006; Zhang et al. 2010).

Aphrodisium gibbicolle is a minor to moderate pest of citrus in southern China (Hoffmann 1934; Chien 1981; Li et al. 1997). In Guangdong, the adults prefer to lay eggs on citrus branches; the larvae bore under the bark of the branches and then into the wood of the main trunks, weakening the trees (Hoffmann 1934).

Based on observations by Hoffmann (1934), the adults emerge in April to August in southern China and the life cycle lasts one year. The lifetime fecundity is 15–54 eggs (Zhang et al. 2010). Tu et al. (2006) showed that the adults emerge during May and June with an emergence peak in mid-June; this cerambycid needs two years to complete its life cycle, with the third and fourth instar larvae overwintering in the first year and prepupae and pupae overwintering in the second year. Several other *Aphrodisium* species in southern China also have a life cycle of two years (Zhang et al. 2010). On oak trees such as *Cyclobalanopsis oxyodon* and *Lithocarpus glabra*, the eggs are laid in the crevices or wounds in the bark of the branches or in trunks of 6–12 cm in diameter (Tu et al. 2006).

There has been no control method specifically developed for this pest. However, the control methods utilized for other *Aphrodisium* species could be useful for the development of control measures for *A. gibbicolle*. Zhang et al. (2010) summarized several effective ways to suppress *Aphrodisium* populations: removing and burning infested branches and trunks during the winter, whitewashing main trunks in the spring, injection of insecticides into or plugging gallery openings with cotton buds soaked with insecticide in the winter, and spraying insecticides on oviposition sites in the late spring to early summer to kill adults.

12.3.1.5 *Chelidonium argentatum* (Dalman)

Synonym: *Cerambyx argentatus* Dalman
Common name: Greenish lemon longicorn

The adults of this cerambycine species are 24–27 mm long; the body is dark green and shining; the ventral side is green and covered with silvery gray pubescence, and the antennae and legs are dark blue to blackish purple (Figure 12.25). It is distributed in China, India, Myanmar, and Vietnam, and its host range appears narrow, feeding on plants from Rutaceae, particularly citrus species (Duffy 1968).

The greenish lemon longicorn is an important pest of citrus trees in southern China (Gressitt 1942; Chang 1958; Chen et al. 1959; Chien 1981; Li et al. 1997) and southern India (Beeson and Bhatta 1939; Ramachandran 1953; Singh et al. 1983). The larvae bore into the branches, which wither and become vulnerable to wind breakage, and continue to tunnel down to the trunks. Fine, dry, and yellowish-white frass is obvious on infested branches and trunks as well as on the ground. Young trees can be hollowed to the root and killed.

The biology of this pest is summarized from Gressitt (1942), Chang (1958), and Chen et al. (1959). The adults are diurnal and present between May and August, with an emergence peak during May–June. The females lay eggs on slender branches. The lifetime fecundity is about 22 eggs. The larvae bore into the tender bark and then the wood of main branches and the trunk. They become mature in six months, overwinter in the tunnels, and pupate in April. The life cycle lasts one year.

Several methods are recommended for the control of this borer. For example, the pruning and burning of all the branches containing larvae in August (before they enter the main trunk) can significantly reduce the damage caused by *C. argentatum* (Gressitt 1942; Singh et al. 1983). In addition, Singh et al. (1983) injected 15 mL of gasoline or insecticide solutions in larval tunnels to kill those *C. argentatum* larvae that have entered the trunk. Wang et al. (1999) tested the effectiveness of an entomopathogenic nematode, *Steinernema glaseri*, for the control of *C. argentatum* larvae in the field and achieved 70% larval mortality.

FIGURE 12.25   *Chelidonium argentatum* adult. (Courtesy of Larry Bezark.)

12.3.1.6 *Chelidonium cinctum* (Guérin-Méneville)

Synonym: *Callichroma cincta* Guérin-Méneville
Common name: Lime tree borer

The cerambycine adults are 21–32 mm long and metallic bluish green in color; the antennal segments 1–3 are metallic blue and the rest are blackish purple; the elytra are metallic bluish-green with an irregular, yellow band near the basal-middle region. The species occurs in southern China, India, Laos, Myanmar, and Cambodia, and so far, only citrus species are recorded as hosts of this borer (Duffy 1968; Chiang et al. 1985).

The lime tree borer is a minor to moderate pest of orange and lime trees in southern China (Gressitt 1942; Chien 1981; Li et al. 1997) and southern India (Kunhi Kannan 1928; Beeson and Bhatta 1939; Ramachandran 1953; Singh et al. 1983). Younger branches suffer the most damage, wilting and dying. When trees are three to four years old, a single larva can kill a tree. In older trees, the borers can significantly reduce the yield and may eventually kill the trees in several successive years of infestation. Frass from frass-ejection holes is also obvious.

Callichroma cinctum adults emerge in April–June (Beeson and Bhatta 1939) and lay eggs in the axils of young living twigs; larvae bore into the twigs, causing death, and eventually bore into the main branches (Fletcher 1919). Lifetime fecundity is unknown. Young larvae bore into the center of the twigs, turn upward for about 3 cm, and cut a complete spiral around the twigs. The life cycle lasts one year, and overwintering occurs at the larval stage.

Pruning and burning of all the branches containing larvae in November and January can be effective for the control of *C. cinctum* (Fletcher 1919; Kunhi Kannan 1928; Beeson 1941). The girdled twigs turn black and are easily recognizable (Beeson 1941). Other control measures used for *C. argentatum* may also be effective for this species.

12.3.1.7 *Chelidonium citri* Gressitt

Common name: Green citrus longicorn

The cerambycine adults are 22–25.5 mm long and dark green above and golden green on the head, extremities, and sides of the prothorax and ventral surfaces; the antennae are largely violet to purplish black; the legs are bluish green to violet; the ventral surface is covered with thin silvery pubescence. This species is recorded in southwest China and feeds on citrus (Gressitt 1942) and *Murraya* (Liu and Huang 2003) in Rutaceae.

The green citrus longicorn is a minor pest of citrus trees (Chien 1981; Zhang et al. 2002) and the ornamental plant, *Murraya paniculata* (L.) (Liu and Huang 2003) in southern China. Damage symptoms are similar to those from *C. cinctum*. Infestation often reduces productivity and fruit quality (Zhang et al. 2002).

The adults emerge during April–May and lay eggs in the axils of young living twigs or petioles (Liu and Huang 2003). The lifetime fecundity in the field is unknown. The larvae bore into twigs and then branches. This species has one generation a year and overwinters at the larval stage.

Liu and Huang (2003) recommended that insecticides be sprayed on the axils of young living twigs or petioles between late June and early July to kill newly hatched larvae of *C. citri*, before they enter twigs and branches, or injected into the branches to destroy those larvae that have entered the plant. Control measures used for the other *Chelidonium* species described earlier may also be effective for the control of this pest.

12.3.1.8 *Oberea posticata* Gahan

Common name: Citrus shoot borer

The adults of this lamiine borer are 15–21 mm long and slender; the antennae are shorter or slightly longer than the body, and the body is dull brown (Figure 12.26). This species occurs in India, Myanmar, Nepal, and Taiwan. The larvae feed on tree species from Rutaceae and Moringaceae (Duffy 1968; Sasanka Goswami and Isahaque 2001a).

FIGURE 12.26   *Oberea posticata* adult. (Courtesy of Larry Bezark.)

The citrus shoot borer is considered a moderate pest of citrus, particularly mandarin and orange, in India; its larvae bore into the twigs and growing shoots, causing death and substantially reducing yield (Isahaque 1978; Ghosh 1998; Sasanka Goswami and Isahaque 2001a, 2001b). Fresh frass ejected from the twigs is evidence of infestation. Infested seedlings can be killed.

According to Sasanka Goswami and Isahaque (2001a, 2001b), the adults are present from mid-March to early September. The females make oviposition slits in the tender bark of the shoots and twigs and lay eggs singly in the slits. The lifetime fecundity varies from 14 to 23 eggs per female. The oviposition period lasts from early April to mid-July, and the larval feeding inside the shoots (as indicated by fresh frass) takes place from mid-April to early October. The developmental period from eggs to adults ranges from 87 to 105 days. This borer has one generation a year and overwinters as the mature larvae and pupae inside the shoots or twigs.

Spraying insecticides such as pyrethroid on the shoots and twigs during the oviposition period may be effective for adult control. Pruning and burning the infested twigs and shoots before the spring may also be effective.

12.3.1.9 *Oberea lateapicalis* Pic

Synonym: *O. mangalorensis* Gardner
Common name: Orange shoot borer

The lamiine adults are 12–20 mm long and slender; the body is yellow with the antennae, eyes, and apical half of the elytra being blackish brown to black; the antennae are shorter than the body. It is only recorded in India and feeds on plants from Rutaceae (Duffy 1968; Bhumannavar and Singh 1983).

The orange shoot borer is a minor pest of citrus, particularly mandarin and orange, but in neglected orchards it can be a serious pest (Bhumannavar and Singh 1983; Singh et al. 1983). The larvae bore into shoots and twigs, weakening and killing them. Fresh frass ejected from shoots and twigs is obvious.

The life cycle is summarized from Bhumannavar and Singh (1983). Emergence occurs in April and May, and the females girdle the tender shoots before laying eggs in the bark of the girdled shoots. The lifetime fecundity is unknown. The larval stage lasts 304–385 days including 120–135 days in diapause, and pupation occurs at the onset of premonsoon rains in March. *O. lateapicalis* overwinters at the larval stage in diapause. The life cycle lasts one to two years.

Spraying insecticides on the shoots and twigs during the oviposition period may effectively kill the adults (Singh et al. 1983). Pruning and burning the infested twigs and shoots before the spring may also be effective.

12.3.1.10 *Nadezhdiella cantori* (Hope)

Synonyms: *Hammaticherus scabricollis* Chevrolat, *Hamaticherus cantori* Hope
Common name: Gray-black citrus longicorn

The cerambycine beetle is 41–52 mm long, grayish-brown to brownish-black, and covered with golden pubescence; the pronotum is deeply vermiculately strigose, and each side has a stout median tubercle or spine; the elytral apex has a small tooth at the suture (Figure 12.27). The species occurs in most parts of China—particularly in the south—and in Thailand, and its larvae are recorded from Rutaceae and Vitaceae (Duffy 1968; Hua 1982). The beetle has not been recorded outside Southeast Asia to date, making it potentially important for quarantine (Wang et al. 2002b).

The gray-black citrus longicorn is a serious pest of citrus in all citrus-growing areas of China, particularly southern China and Taiwan (Gressitt 1942; Li et al. 1997). It is also reported to attack species other than citrus, such as grapevines (Hua 1982). However, *N. cantori* is not considered an important pest in Thailand and in vineyards. The larvae bore into the main branches and trunks. The infested trees are significantly weakened and subject to wind breakage. The infestation significantly reduces the yield and increases secondary pest attack. When several larvae infest a single tree, which is common, the tree ultimately dies (Duffy, 1968).

FIGURE 12.27 *Nadezhdiella cantori* adult. (Courtesy of Xavier Gouverneur.)

Several independent studies in the past 70 years report similar biological characteristics of this species. Most adults emerge in June or July and a few appear in May and August; the females lay eggs singly between late May and late July in the crevices on the bark of the trunk and main branches (Lieu 1947) or on the trunk within one meter above the ground (Chen 1942). The reproductive and flight activities of this beetle mainly occur between 17:00 and 24:00 hours, and each activity has two peaks—with the first one taking place within the three hours before sunset (21:00 hours) and the second occurring within the three hours after sunset (Wang et al. 2002b). A female lays up to 57 eggs in the laboratory, but field-collected females carry 100–150 eggs each (Duffy 1968). The larvae feed on the inner bark of the trunk and main branches for about six weeks before entering the wood (Lieu 1947). Both Chen (1942) and Lieu (1947) reported that the life cycle of this pest lasts about two years, and the winter is passed at the larval stage. However, Lieu (1947) indicated that, in some cases, the life cycle can last three years—with the third winter being passed in the pre-emergence stage. Based on the work by Chang et al. (1963), the duration of a generation depends on the date of hatching. The larvae that hatch before mid-July pupate between August and October of the following year and give rise to adults that emerge in November but may remain in the pupal chambers until the end of the following April. The larvae that hatch after August pupate in August–September of the next year, and the adults do not leave the pupal chamber until the fourth year.

Gressitt (1942) suggested that the use of light traps in the late spring and the collection of mating pairs on trees during the day could be useful control measures. Because many adults hide in big holes (usually old damage holes resulting from larval boring) in the lower portions of the trunks between late May and early August, they should be collected from these holes and destroyed during this period (Lieu 1947). Chen (1942) reported that the best method of control is to prevent oviposition by covering the trunk with flat straw ropes or filling the holes and cracks with concrete or an adhesive substance. The trunks should be examined to a height of about 2.5 m for the eggs and exudations of fine wood dust five times in June and July, the eggs removed with a hard brush or by hand, and the young larvae killed by striking the bark at points where frass is seen (Lieu 1947). The trunks can also be brushed clean and then whitewashed with a freshly made mixture of quick-lime and water (1:6) during the hot sunny days in early June to discourage oviposition. However, Chen (1942) suggested that whitewashing the trunk is not very effective. The last resource is to cut down badly damaged trees and kill the insects. Furthermore, three parasitoids that attack *A. chinensis* also parasitize this pest in China: *Aprostocetus prolixus, Brulleia shibuensis*, and *Scleroderma* sp. (Niu et al. 2014), but their application for control of this pest in commercial orchards is not practiced.

12.3.2 Australasia

12.3.2.1 *Uracanthus cryptophagus* Olliff

Common name: Australian citrus branch borer

The cerambycine beetle is narrow in shape and 24–44 mm long; the body is reddish brown to blackish brown with the head, prothorax, and legs often darker; the pronotal disc and each elytron are covered with very dense, long, golden pubescence arranged in four longitudinal lines (Figure 12.28). It occurs in eastern and northeastern New South Wales, southeastern Queensland, and southwestern Western Australia, and its larvae feed on small branches of citrus and lychee trees (Thongphak and Wang 2007).

The Australian citrus branch borer is a minor to moderate pest of all commercially grown citrus species, particularly oranges (Smith et al. 1997). The typical symptom is made by larval ringbarking the twigs and branches, which become wilted and die beyond the ring point; frass is expelled from frass-ejection holes (Duffy 1963; Smith et al. 1997). Weakened branches may fall to the ground.

The life cycle is summarized from Duffy (1963), Smith et al. (1997), and Thongphak and Wang (2007). The adults are present from July to January and are attracted to mercury vapor (MV) light. The females lay their eggs in the cracks or crevices of twigs and small branches. The lifetime fecundity is not known. The eggs hatch in about 10 days, and the larvae immediately bore into the wood

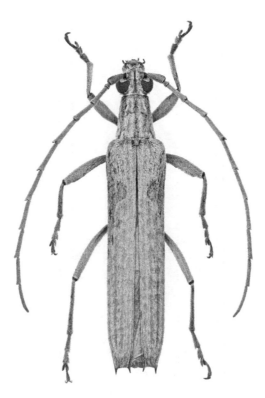

FIGURE 12.28 *Uracanthus cryptophagus* adult. (Courtesy of CSIRO, copyright CSIRO, used with permission of Adam Slipinski.)

and continue feeding in the hosts through the summer, fall, and winter. There is one generation per year. Overwintering occurs at the larval stage.

Duffy (1963) and Smith et al. (1997) provided some recommendations for the control of this pest. Trees should be inspected regularly for early signs of wilting. Infested twigs and branches should be cut as soon as possible and burned. Pruning wounds on larger branches are then sealed with wax or paint. Citrus orchards should not be established in former rain forest areas that are prone to attack by the pest. Good cultivation management to maintain the vigor of trees may be helpful. Evidence from other countries suggests that excessive use of fungicides may result in increased damage because these kill fungi that attack the borers. Infested trees may be treated by injecting insecticides into the uppermost frass-ejection holes, but the effectiveness of this chemical treatment is not clear. Furthermore, there is no insecticide that is registered for the control of this longicorn.

12.3.2.2 *Strongylurus thoracicus* (Pascoe)

Synonym: *Didymocantha thoracica* Pascoe
Common name: Pittosporum longicorn

Adults of the cerambycine borer are about 30 mm long; the body is light brown with a characteristic row of white spots on each side of the pronotum (Figure 12.29). It is found in eastern Queensland, New South Wales, and Victoria (Duffy 1963; Smith et al. 1997). The host range of this beetle is wider than that of the Australian citrus branch borer, and it feeds on tree species from the plant families Anacardiaceae, Malvaceae, Meliaceae, Pittosporaceae, Rosaceae, and Rutaceae (Duffy 1963; Smith et al. 1997).

The pittosporum longicorn is a minor pest of citrus trees, particularly orange trees in Australia (Smith et al. 1997). This species is attracted to older trees lacking vigor or to trees infected with disease. Damage is caused by larval boring inside the twigs and small branches, causing wilting. Frass ejected from plants is also evidence of infestation. Attacked twigs and branches die and snap off.

FIGURE 12.29   *Strongylurus thoracicus* adult. (Courtesy of CSIRO, copyright CSIRO, used with permission of Adam Slipinski.)

Duffy (1963) and Smith et al. (1997) described the general biology of this species. The adults emerge during October–February and lay eggs in the crevices of twigs and small branches throughout the summer. The larvae bore into, girdle, and amputate the twigs and small branches, and then they tunnel down the center of the branches for up to 1 m into the main trunk. During feeding, the larvae often come to the surface and eject frass. There is one generation per year. Overwintering occurs at the larval stage. Control measures are similar to those used for the Australian citrus branch borer.

12.3.2.3 *Skeletodes tetrops* Newman

Common name: Australian citrus longicorn

This is a cerambycine borer. The adults are 7–15 mm long; the body is straw yellow; the prothorax has two to three blackish to reddish-brown longitudinal stripes on each side, with two median stripes enclosing a short central straw-yellow stripe on the disc; each elytron has complicated longitudinal blackish to reddish-brown stripes (Figure 12.30). This longicorn occurs in southwestern Western Australia, northwestern Victoria, eastern New South Wales, eastern Queensland, and southeastern Papua New Guinea (Wang 1995). About 10 tree species from Araucariaceae, Fabaceae, Malvaceae, Meliaceae, and Rutaceae are recorded as its hosts (Duffy 1963; Webb 1987, 1988; Webb et al. 1988; Hockey and De Baar 1988; Wang 1995; Smith et al. 1997).

The Australian citrus longicorn is a minor pest of citrus trees, particularly orange, and prefers to attack older or weakened trees (Smith et al. 1997). Larvae feed under the bark of twigs and branches and make frass-ejection holes. Infested parts wilt and die.

Adults are present all year except in May and are attracted to light (Wang 1995). According to Smith et al. (1997), females lay eggs in the crevices of the bark during the summer. Lifetime fecundity is unknown. The life cycle lasts one year, and overwintering occurs at the larval stage. Control measures are similar to those used for the Australian citrus branch borer.

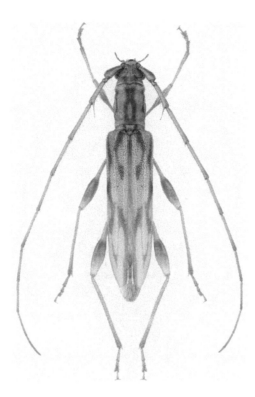

FIGURE 12.30 *Skeletodes tetrops* adult. (Courtesy of CSIRO, copyright CSIRO, used with permission of Adam Slipinski.)

12.3.2.4 *Disterna plumifera* (Pascoe)

Synonyms: *Zygocera plumifera* Pascoe, *Paradisterna plumifera* (Pascoe), *Parahybolasius fuscomaculatus*
        Breuning
Common name: Speckled longicorn

The adults of the lamiine borer are about 12 mm long; the body is robust and light or dark gray, speckled with brown marks (Figure 12.31). It is distributed in southeastern Australia, and its larvae feed on trees from Apocynaceae, Malvaceae, Meliaceae, Pinaceae, Rosaceae, and Rutaceae (Duffy 1963).

The speckled longicorn is a minor pest of citrus in eastern Queensland and northeastern New South Wales (Smith et al. 1997). All commercially grown citrus species and varieties are attacked, but the navel orange appears to be the preferred host. The larvae bore under the bark, producing a large amount of sawdust-like frass that falls to the bottom of the trunk. The infested parts are weakened, resulting in secondary rot caused by fungi. Trees more than 15 years old are usually affected the worst by this pest.

According to Smith et al. (1997), the adults start to emerge in mid-spring and continue to emerge throughout the summer. The eggs are laid on the bark of the trunk and lower sections of the main branches of citrus trees that are not vigorous due to either age or disease or both. This pest has one generation a year and overwinters as mature larvae.

Smith et al. (1997) recommended good orchard management that can maintain the vigor of trees as an ideal preventative measure for lessening the attack by this pest that prefers weakened trees. Old blocks of trees infested by this borer should be replaced. If the application of insecticides and fungicides is attempted, diseased and eaten bark is removed first and then the affected areas are sprayed with solutions containing both insecticide and fungicide. Spraying foliage with chemicals is not effective for the control of this beetle.

FIGURE 12.31   *Disterna plumifera* adult. (Courtesy of CSIRO, copyright CSIRO, used with permission of Adam Slipinski.)

12.3.2.5 Oemona hirta (Fabricius)

Synonyms: *O. humilis* Newman, *Saperda hirta* Fabricius, *S. villosa* Fabricius
Common name: Lemon tree borer

The cerambycine beetle is 11–31 mm long; the body is reddish brown to blackish brown, and the elytra are brown to reddish brown; the pronotal disc has distinct, long, transverse rugae and pale yellow hairs; the elytral surface is covered with pale yellow hairs and coarse and rugose punctures (Figure 12.32). This borer is distributed throughout New Zealand and, so far, only occurs there (Duffy 1963; Lu and Wang 2005). It was intercepted in the UK in the 1980s; in 2010, a number of *Wisteria* rootstocks imported to the UK from New Zealand were infested with larvae, which were immediately destroyed (Ostojá-Starzewski et al. 2010). The latter incidence has led to the performance of a detailed pest risk analysis organized by the European and Mediterranean Plant Protection Organisation (EPPO 2013c). *O. hirta* is extremely polyphagous, utilizing more than 200 tree and shrub species as its larval hosts—both endemic and exotic to New Zealand—from at least 63 families: Acanthaceae, Adoxaceae, Anacardiaceae, Apocynaceae, Araliaceae, Araucariaceae, Argophyllaceae, Asparagaceae, Betulaceae, Cannabaceae, Casuarinaceae, Celastraceae, Compositae, Coriariaceae, Cornaceae, Corynocarpaceae, Cruciferae, Cunoniaceae, Cupressaceae, Cyperaceae, Ebenaceae, Elaeocarpaceae, Euphorbiaceae, Fabaceae, Fagaceae, Gramineae, Grossulariaceae, Juglandaceae, Labiatae, Lauraceae, Loganiaceae, Lythraceae, Magnoliaceae, Malvaceae, Monimiaceae, Myrtaceae, Nothofagaceae, Nyssaceae, Oleaceae, Pandanaceae, Pennantiaceae, Pinaceae, Plantaginaceae, Piperaceae, Pittosporaceae, Platanaceae, Proteaceae, Phytolaccaceae, Ripogonaceae, Rhamnaceae, Rosaceae, Rubiaceae, Rutaceae, Salicaceae, Sapindaceae, Scrophulariaceae, Solanaceae, Tamaricaceae, Theaceae, Thymelaeaceae, Ulmaceae, Violaceae, and Vitaceae (Lu and Wang 2005; EPPO 2013c).

FIGURE 12.32 *Oemona hirta* adult. (Courtesy of Phil Bendle, Taranaki Educational Resource Research Analysis and Information Network.)

The lemon tree borer is an important pest of citrus, persimmon, and apple trees as well as grapevines, with citrus trees being the preferred hosts (Taylor 1957; Hosking 1978; Clearwater 1981; Clearwater & Muggleston 1985; Wang et al. 1998; Lu and Wang 2005). Almost all commercially grown citrus species in New Zealand are attacked. Dieback of the infested twigs and small branches may occur in the first year of infestation. In the second year, the larvae move downward and damage the main branches and trunk, weakening the tree. Sawdust-like frass is obvious at or around excretion holes.

The adults emerge between early September and early February with a peak occurring during October–November (Lu and Wang 2005). The eggs are usually laid singly at the leaf–twig junctions and in the crevices and wounds (particularly pruning wounds) of the twigs, branches, and trunk—with the twigs being preferred oviposition sites (Taylor 1957; Hosking 1978; Clearwater 1981; Clearwater & Muggleston 1985; Wang et al. 1998; Lu and Wang 2005). Each female produces an average of 82 eggs but only lays an average of 51 eggs in her lifetime (Wang et al. 1998). The larvae bore directly into the wood, first into the sapwood and then into the heartwood, normally along the twigs and branches toward the trunk (Hosking 1978; Wang et al. 1998). In rare cases, the larvae can enter the roots. The life cycle normally lasts two years in the field (Clearwater 1981; Wang et al. 1998; Lu and Wang 2005) but, at a constant temperature of 23°C in the laboratory, this insect can complete its life cycle within one year (Wang et al. 2002a). *O. hirta* normally overwinters as larvae.

Clearwater and Wouts (1980) injected a suspension of the nematode *Steinernema feltiae* Filipjev into frass holes and achieved more than 90% larval mortality. Pruning the infested twigs and branches and then covering the pruning cut surface with a paint containing insecticides are probably the most effective control measures for this pest (Clearwater and Muggleston 1985). Three species of parasitic wasps are found to attack the larvae of *O. hirta*, but the overall parasitism rate by these parasitoids is low (<5%), making their usefulness for the control of this pest questionable (Wang and Shi 1999, 2001).

12.3.3 South America

12.3.3.1 *Hylettus seniculus* (Germar)

Synonyms: *Nyssodrys ophthalmica* Lameere, *Lamia seniculus* Germar
Common name: Hylettus citrus borer

The lamiine adults are 18–45 mm long and robust; the antennae are 1.5 times as long as the body, which is reddish brown to dark brown and covered with grayish pubescence; each elytron has a sub-triangular brown mark near the basal-sutural area (some specimens do not have this brown mark) and a larger triangular brown mark on the apical half (Figure 12.33). The borer is widely distributed in Costa Rica, Argentina, Brazil, and Bolivia, and it feeds on tree species in Anacardiaceae, Bombacaceae, Burseraceae, Flacourtiaceae, Pinaceae, and Rutaceae (Duffy 1960; Tavakilian et al. 1997; Monné 2001b, 2005b; Di Iorio 2006; Paz et al. 2008; Ferreira-Filho et al. 2014).

The hylettus citrus borer is an important pest of citrus trees in northern Brazil, particularly in old and neglected orchards (Moreira et al. 2003). It is also collected in large numbers in northern Brazilian mango orchards, but its pest status in mango is not clear (Paz et al. 2008). The larvae bore under the bark of the main branches and trunks of citrus trees, causing leaf wilting, drying, breaking of branches, and tree death (Moreira et al. 2003). Frass is packed in tunnels behind the boring larvae; the bark of infested tree parts turns dark and can be easily removed by hand, exposing sawdust-like frass (Moreira et al. 2003).

The life history of this borer is summarized from Duffy (1960), Puzzi (1966), and Moreira et al. (2003). The adults are present almost throughout the year. The females make oviposition slits with their mandibles in the bark of the branches and trunks and lay their eggs singly in the slits. They prefer thinner branches for laying eggs. The lifetime fecundity is not known. The larvae bore into the bark and tunnel down underneath to the main trunk. Pupation occurs in shallow pupal cells located in the inner side of the bark. Overwintering appears to occur at the larval stage, but larval feeding takes place throughout the year. The life cycle probably lasts one year.

FIGURE 12.33 *Hylettus seniculus* adult. (Courtesy of Juan Pablo Botero [image from Marcela L. Monné].)

Moreira et al. (2003) described several control measures for this pest: partially scraping the bark of infested branches and trunks to kill the larvae and then pasting fungicide and hydrated lime on the wounds to prevent diseases; pruning and burning the heavily infested branches and replacement of badly infested trees to prevent reinfestation, and spraying of organophosphate insecticides in late afternoon to kill adults and neonate larvae. For small orchards, injecting organophosphate insecticides into borer tunnels through the bark is recommended.

12.3.3.2 *Diploschema rotundicolle* (Audinet-Serville)

Synonyms: *D. flavipennis* Thomson, *Phoenicocerus costicollis* Audinet-Serville, *P. rotundicollis*
        Audinet-Serville
Common name: Brown-headed citrus borer

The cerambycine adults are 30–40 mm long; the antennae are shorter than the body; the head and prothorax are reddish to blackish brown and covered with coarse light brown pubescence; the elytra are glabrous and pale yellowish brown with a short sutural spine at the apex (Figure 12.34). This species is widely distributed in central to southern Brazil, Argentina, Paraguay, and Uruguay, and its larvae are polyphagous, feeding on at least 25 host species (both endemic and exotic to South America) from 10 plant families, Rutaceae, Bignoniaceae, Cupressaceae, Erythroxylaceae, Euphorbiaceae, Meliaceae, Myrtaceae, Rhamnaceae, Rosaceae, and Sapindaceae (Duffy 1960; Monné 2001a, 2005a).

In Brazil, the brown-headed citrus borer is a moderate to serious pest of citrus trees, particularly orange and lemon (e.g., Bondar 1913; Pinto da Fonseca and Autuori 1932; Puzzi and Orlando 1959; Nascimento do et al. 1986; Machado 1992; Machado et al. 1998; Machado and Berti Filho 1999, 2006), and a minor pest of peach trees (Bondar 1913). The damage is caused by the larval tunneling in the branches and trunks. Early in the infestation, young larvae cause wilting of the buds and young leaves. Once the larvae grow larger and enter the trunk, the tree can be severely damaged. For example, Bondar (1913) reported that a lemon tree trunk was almost completely destroyed by 16 larvae.

FIGURE 12.34 *Diploschema rotundicolle* adult. (Courtesy of Juan Pablo Botero [image from Marcela L. Monné].)

Although various authors have described some aspects of its biology (e.g., Duffy 1960; Ioneda et al. 1986; Link and Costa 1994; Di Iorio 1998), the most comprehensive appraisal of the biological features of *D. rotundicolle* is given by Bondar (1913). The adults are present between late spring (November) and late fall (May), and they lay eggs in the crevices near the apex of new and tender branches. The lifetime fecundity is not known. The larvae bore in the small branches downward through the larger branches to the trunk. After about eight months' feeding, the mature larvae turn and bore upward, enlarging a portion of the tunnel to make a pupal cell for pupation and an opening about 12 mm in diameter for adult emergence. In the field (Bondar 1913) and laboratory (Machado and Berti Filho 1999), this species has one generation a year. Overwintering occurs at the larval or pupal stage.

Several control methods are recommended by various authors. The most effective control method appears to be examination of the branches in May and June for any sign of frass and removal of the infested branches before the larvae enter the main branches and the trunk. Faria et al. (1987) trialed several methods, including the application of light traps and insecticides, but concluded that none of these measures could effectively suppress this pest's population. Machado et al. (1998) tested the application of the fungus *Metarhizium anisopliae* (Metsch.) in borer galleries by using larvae of pyralid moth *Galleria mellonella* L. as vectors, which resulted in effective control of this pest. Machado and Berti Filho (2006) evaluated the effectiveness of two measures, pruning the infested branches and, following the pruning, introduction of the previously mentioned fungus into the borer galleries using the moth vector or by direct powdering into the galleries. These control practices achieved greater than 90% larval mortality.

12.3.3.3 *Macropophora accentifer* (Olivier)

Synonyms: *Prionus accentifer* Olivier, *Acrocinus accentifer* (Olivier), *Macropus accentifer* (Olivier)
Common name: Harlequin citrus borer

Adults of this lamiine borer are 18–31 mm long; the body is reddish brown and covered with a mixture of dark brown, brown, gray, pale, and white pubescence, showing harlequin patterns; the antennae are almost twice as long as the body; the pronotum has a robust spine on each side; and the elytral apex has a sharp spine at the margin (Figure 12.35). The borer is widely distributed in South America, including Brazil, Paraguay, Argentina, Uruguay, Bolivia (Monné 2001b, 2005b), and Peru (Carrasco 1978). This species is polyphagous, with at least 15 tree species from nine families being recorded as hosts: Anacardiaceae, Boraginaceae, Euphorbiaceae, Lauraceae, Meliaceae, Moraceae, Myrtaceae, Rosaceae, and Rutaceae (Monné 2001b; Machado et al. 2012).

The harlequin citrus borer has been considered an important pest of citrus trees over the past 100 years in South America, particularly in Brazil (Bondar 1913, 1923; Pinto da Fonseca and Autuori 1932; Puzzi and Orlando 1959; Duffy 1960; Zajciw 1975; Carrasco 1978; Nascimento do et al. 1986; Machado and Raga 1999). The damage is caused by larval boring inside the main trunks. If not controlled, the pest can destroy a citrus orchard in three to four years of infestation. It is also reported as a pest of fig trees in Peru (Carrasco 1978).

Although the biology of this pest is mentioned by many authors, the most comprehensive information is given by Bondar (1913, 1923) and Duffy (1960). The adults mainly emerge during August–October and lay eggs between September and January. They make oviposition slits in the bark using their mandibles before laying their eggs. The females prefer the lower portion of the trunk for oviposition. The lifetime fecundity is unknown. The larvae bore in subcortical tissues before entering the wood for pupation between July and September. The life cycle lasts about a year, and the overwintering occurs at the larval and pupal stages.

Bondar (1913, 1923) and Duffy (1960) described the effective measures for killing the larvae that should be performed during May–June: carefully examine the lower parts of the trunks of the trees, inject gasoline or insecticides into the borer holes, and block the holes with plugs. Machado and Raga (1999) have trialed applications of the fungus *M. anisopliae* and two chemicals. They show that the fungus or imidacloprid can kill about 90% of larvae in 15 days after treatment.

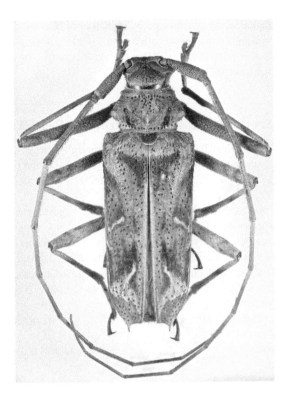

FIGURE 12.35   *Macropophora accentifer* adult. (Courtesy of Juan Pablo Botero [image from Marcela L. Monné].)

12.3.3.4 *Elaphidion cayamae* Fisher

Common name: Spined citrus borer

The cerambycine adults are 12–17 mm long; the body is dark reddish brown and irregularly covered with long, recumbent, gray pubescence; the fourth antennal segment is very short with a very long spine; the elytral apex has a very long spine at the margin and a short spine at the suture. This longicorn is widely distributed in Cuba, and its larvae feed on citrus trees (Castellanos et al. 1990; Grillo-Ravelo and Valdivies 1990, 1992; Castellanos and Jimenez 1991; Dominguez and Dominguez 1991; Vicente 1991).

Since the 1980s, the spined citrus borer has been considered an important pest of citrus trees—particularly orange and mandarin—in Cuba, resulting in yield loss (Castellanos and Jimenez. 1991). For example, in a citrus orchard, the infestation rate can be as high as 90% (Grillo-Ravelo and Valdivies. 1990). Up to 1,053 larvae/ha are found in August (Vicente 1991). The larvae feed on the branches and twigs of citrus trees, which wilt and become dry after the infestation (Grillo-Ravelo and Valdivies 1990).

The life cycle is summarized from Grillo-Ravelo and Valdivies (1990, 1992). The adults emerge from March to May and are active nocturnally. However, an attempt to use light traps for monitoring failed to catch any adults in the field. The adults can live for more than 90 days. Females lay their eggs on the twigs and branches. The lifetime fecundity in the field is not known. Larvae bore into twigs and branches and feed inside for about 10 months. The life cycle lasts one year, and overwintering occurs at the larval stage.

Pruning infested branches appears to the most effective measure for the control of this pest (Castellanos et al. 1990; Grillo-Ravelo and Valdivies 1990; Dominguez and Dominguez. 1991). To achieve the best result, pruning should take place during September–October. Pruned branches and twigs should be burned.

In recent years, a new *Elaphidion* species (unnamed) has been found causing a similar level of damage to citrus branches and twigs in Cuba (González-Risco et al. 2011, 2013). Although the harm caused by the new pest is similar to that of *E. cayamae*, the larval biology of these two species appears to be different. The new pest makes a ring around a green branch or twig in the area of the cambium, which prevents the movement of nutrients and causes symptoms of wilting, drying leaves and branches above the ring. This brings the fall of flowers and/or fruit that they hold. Control measures used for *E. cayamae* could also be effective for the new pest.

12.4 Cerambycid Pests of Tropical and Subtropical Fruit Trees

A number of tropical and subtropical fruit crops are commercially grown around the world, such as banana, mango, pineapple, lychee, figs, longan, durian, jackfruit, guava, and pomegranate. Among these crops, banana, mango, and pineapple are ranked number one, six, and nine, respectively, in worldwide fruit production (Ebert 2009).

A very large number of insect species are recorded as pests of tropical and subtropical fruit trees, but relatively few cerambycid species are of economic importance to these crops (Li et al. 1997; Hill 2008). There is no confirmed economic cerambycid pest in major herbaceous fruit crops such as bananas and pineapples. Six cerambycid species are of economic importance to tropical and subtropical fruit tree crops in Asia, one in Australasia, and one in Africa. Although some cerambycid species have been recorded attacking tropical and subtropical fruit tree crops in the New World, none is considered an economic pest of these crops in that region.

Some cerambycid species that attack tropical and subtropical fruit trees as very minor pests are considered serious pests of other crops. For example, a minor mango pest in Australia, *Acalolepta mixta* (Hope) [= *A. vastator* (Newman)], is an important vineyard pest, and thus is discussed in the Cerambycid Pests in Vine Crops section.

12.4.1 Asia

12.4.1.1 *Aristobia testudo* (Voet)

Synonyms: *Aristobia clathrator* (Thomson), *Celosterna clathrator* Thomson, *Lamia reticulator* Fabricius, *Cerambyx testudo* Voet
Common name: Lychee longicorn

The lamiine adults are 20–35 mm long and black in color; the vertex and pronotal disc are covered with yellow pubescence and two longitudinal black pubescent stripes; the elytra are covered with black pubescence and a number of yellow pubescent spots; the antennal segments 1–2 are black and segments 3–11 yellow, with segments 3–4 or 3–5 bearing black tuft hairs at the apex (Figure 12.36). The beetle is distributed in northern India, Myanmar, Thailand, Laos, Vietnam, and southern China, and its larvae feed on tree species from Annonaceae, Lythraceae, Myrtaceae, and Sapindaceae (Chen et al. 1959; Duffy 1968; Hua 2002; Shylesha et al. 2000). The species has not been recorded outside its primary distribution area.

The lychee longicorn is a moderate to serious pest of lychee, longan, cocoa, and jackfruit (Li et al. 1997)—particularly lychee (Ho et al. 1990; Li et al. 1997; He 2001; Waite and Hwang 2002) in southern China where it is also recognized as a minor pest of citrus (Li et al. 1997). The damage is caused by the adult ring-barking and larval boring. The infested twigs and branches wilt and die. In severe infestation, the tree can be killed. The twig ring-barking by ovipositing adults causes the extremities to die and snap off. In addition, this pest can cause minor to serious damage to guava in India. For example, in northeastern India, the infestation rate ranges from 8% to 76% in different guava orchards; although the pest prefers older trees (10 years old), young trees may be killed within one year of infestation (Shylesha et al. 2000).

According to Ho et al. (1990) and Waite and Hwang (2002), the adults emerge during June–August. They girdle the branches and twigs by chewing off strips of bark and then lay eggs singly in the wound, covering the wound and eggs with exudate. The lifetime fecundity in the field is unknown. The eggs hatch in August and September. The larvae initially bore under the bark and then into the xylem in January, making tunnels in the wood up to 60 cm long. These tunnels have openings packed with frass at regular intervals for aeration. The tunnels are blocked with wood fiber and frass just before pupation in June. This pest has one generation a year and overwinters at the larval stage.

FIGURE 12.36 *Aristobia testudo* adult. (Courtesy of Larry Bezark.)

Shapiro-Ilan et al. (2005) discussed the possible application of entomophagous nematodes to control cerambycid larvae, including *A. testudo*. In practice, Xu et al. (1995) trialed the injection of the nematode *Steinernema carpocapsae* (Weiser) (Agriotes strain) into larval tunnels and achieved 73–100% larval mortality under experimental conditions. Ho et al. (1990) suggested that, apart from injecting tunnels with dichlorvos (DDVP), chemical control is generally unsuccessful. The most effective control measures for this pest (Ho et al. 1990; Waite and Hwang 2002) include: (1) regular inspection of orchards during the adult activity period and removal of the beetles manually when they "play dead," (2) removal of eggs and young larvae from branches and twigs during July–September, and (3) identification of larval location based on the presence of tunnel openings packed with frass and removal of larvae with wire hooks or a knife. A skilled worker can kill 112 larvae in 2 hours.

12.4.1.2 *Batocera rubus* (L.)

Synonyms: *B. sarawakensis* Thomson, *B. sabina* Thomson, *B. mniszechii* Thomson, *B. formosana* Kriesche, *B. bipunctata* Kriesche, *B. albofasciata* Stebbing (nec DeGeer), *Lamia downesii* Hope, *L. 8maculata* Fabricius, *Cerambyx albofasciata* DeGeer, *C. albomaculatus* Retzius, *C. rubus* Linnaeus

Common name: Lateral-banded mango longicorn

The lamiine beetle is 30–60 mm long; the body is grayish brown with two orange or red marks on the pronotum ranged longitudinally; each elytron has about four whitish or yellowish spots with the median third spot being the largest; sometimes, an additional one or two much smaller spots may be present on the elytron (Figure 12.37). This species is widely distributed in southern and southeastern Asia (Duffy 1968) and has recently been intercepted in France and Italy (EPPO 2011, 2012). The borer is polyphagous, attacking many tree species from Anacardiaceae, Apocynaceae, Burseraceae, Fabaceae, Malvaceae, Moraceae, and Phyllanthaceae (Duffy 1968) in tropical and subtropical regions, but its main hosts relevant to the scope of this book are mango, fig, jackfruit (EPPO 2011), and mulberry (Butani 1978).

FIGURE 12.37   *Batocera rubus* adult. (Courtesy of Xavier Gouverneur.)

This beetle is considered a minor to moderate pest of mango and fig, particularly mango, in India (Atwal 1963; Hill 1975, 2008) and China (Li et al. 1997). The larvae bore into the trunk and main branches and mainly pupate in the branches. On the damaged branches, the foliage may die and fruit set may be impaired (EPPO 2011); severely infested trees may die.

In India, the adults emerge in May–June and lay eggs singly in the crevices or under the loose bark of the trunk and branches. The fecundity is unknown. This beetle has one to three (EPPO 2011) generations a year depending on climate. Overwintering occurs at the larval or pupal stage.

Pruning and burning all infested branches appear to be the most effective control measures for this pest (Butani 1978). Injection of insecticides or petroleum products into tunnel openings and sealing them also may be effective.

12.4.1.3 *Batocera rufomaculata* (DeGeer)

Synonyms: *B. diana* Nonfried, *B. polli* Gahan, *B. thysbe* Thomson, *B. chlorinda* Thomson, *Cerambyx cruentatus* Gmelin, *C. rufomaculatus* DeGeer

Common name: Mango tree stem borer

Adults of this lamiine borer are 26–52 mm long and similar to *B. rubus* in appearance but differ in each elytron with six orange–yellow or pale yellow spots; sometimes an additional one or two much smaller spots may be present on the elytron (Figure 12.38). This beetle is distributed throughout its primary regions, including India, Bangladesh, Nepal, Pakistan, Myanmar, Thailand, Laos, Vietnam, Malaysia, Indonesia, Sri Lanka, Andman, and southern China, and has been introduced into Lebanon, Israel, Iran, Iraq, Turkey, Egypt, Madagascar, Comoros, Mauritius, the Virgin Islands, and Puerto Rico (Duffy 1968; CABI 1994; Ben-Yehuda et al. 2000; Tozlu-Goksel 2000; Batt 2004; Potting et al. 2008). It is highly polyphagous, attacking at least 50 plant species from Anacardiaceae, Annonaceae, Apocynaceae, Arecaceae, Burseraceae, Caricaceae, Dipterocarpaceae, Euphorbiaceae, Fabaceae,

FIGURE 12.38 *Batocera rufomaculata* adult. (Courtesy of Jiří Mička.)

Lauraceae, Lecythidaceae, Malvaceae, Moraceae, Moringaceae, Musaceae, Myrtaceae, Platanaceae, Rosaceae, and Rubiaceae, but its main hosts relevant to the scope of this book are mango, fig, durian, guava, jackfruit, pomegranate, cashew, apple, walnut, and mulberry (Haq and Akhtar 1960; Atwal 1963; Duffy 1968; Palaniswami et al. 1977; Butani 1978; Sharma and Tara 1984, 1986, 1995; CABI 1994; Ben-Yehuda et al. 2000; Godse 2002; Sundararaju 2002; Batt 2004; Sudhi-Aromna et al. 2008; Swapna and Divya 2010; Maruthadurai 2012; ZetaBoards 2012; Ahmed et al. 2014).

Over the past five decades, this species has become a serious pest of a number of tropical and subtropical fruit tree crops—particularly mango, fig, durian, and jackfruit—causing significant production loss in India, Bangladesh, Thailand, Nepal, Egypt, and Israel (Palaniswami et al. 1977; Ben-Yehuda et al. 2000; Batt 2004; Sudhi-Aromna et al. 2008; Upadhyay et al. 2013; Ahmed et al. 2014). Considering its invasive ability, this species should be listed as an important quarantine pest in all subtropical and tropical regions where it has not become established. Attack by *B. rufomaculata* often leads to the death of trees.

The life cycle is summarized from Browne and Foenander (1937), Beeson (1941), Duffy (1968), Sudhi-Aromna et al. (2008), and Maruthadurai et al. (2012). The adults can be found throughout the summer when they chew small oviposition slits in the bark of branches and trunk and lay eggs singly in the slits. Most oviposition occurs during July–August. The lifetime fecundity is up to 200 eggs. The larvae initially feed under the bark, then bore into the phloem tissue, and finally bore into the heartwood to pupate. In the tropical regions, the larval feeding occurs throughout the year; and in subtropical regions, the larvae overwinter. The life cycle lasts from six months to one year depending on climate.

Various mechanical and chemical measures are recommended for the control of this pest. Pruning infested branches and covering the cuts with pruning paste may be highly effective before the larvae enter the trunk (Butani 1978; Ben-Yehuda et al. 2000). However, heavily infested trees should be felled and burned (Haq and Akhtar 1960; Duffy 1968). Spraying insecticides on the main branches and trunk (Haq and Akhtar 1960), or even entire trees (Upadhyay et al. 2013), during the oviposition period can also be effective. The injection or insertion of insecticides into the tunnels made by the larvae can lead to satisfactory control by contact or fumigant or by both modes of action (Haq and Akhtar 1960; Palaniswami et al. 1977; Butani 1978; Sharma and Tara 1986; Ben-Yehuda et al. 2000).

12.4.1.4 *Coptops aedificator* (Fabricius)

Synonyms: *C. quadristigma* Fåhraeus, *Lachnia parallela* Audinet-Serville, *Cerambyx fuscus* Olivier, *C. villica* Olivier, *Lamia aedificator* Fabricius
Common name: Albizia longicorn beetle

The lamiine adults are 13–19 mm long; the body is brown, speckled with black and gray spots, and has angular black and dark brown markings on the elytra (Figure 12.39). It is widely distributed in southern, southeastern, and western Asia and throughout Africa, and has been introduced to Taiwan and Hawaii (Duffy 1957, 1968; Dawah et al. 2013; Saha et al. 2013). Its host range is very wide, attacking about 90 plant species from Anacardiaceae, Annonaceae, Apocynaceae, Burseraceae, Caesalpiniaceae, Cannabaceae, Combretaceae, Dipterocarpaceae, Euphorbiaceae, Fabaceae, Lecythidaceae, Malvaceae, Meliaceae, Moraceae Rubiaceae, Olacaceae, Rutaceae, and Ulmaceae (Duffy 1957, 1968; Dawah et al. 2013).

Over the past century, this species has hardly been considered an important pest of any tree crops but has been treated as a quarantine pest in regions where this species does not occur, such as in the New World (Dawah et al. 2013). Recently, Dawah et al. (2013) reported that *C. aedificator* has become a serious pest of mango trees in Saudi Arabia. The damage mainly is made by larval feeding on branches and trunks. Frass expulsion holes are made at intervals, and sometimes sap oozes out of the holes, making the symptoms of the infestation obvious (Dawah et al. 2013). The pest also has caused minor damage to coffee plantations in Africa and South Asia.

In India, the adults appear almost throughout the entire year, but the peak activity period is in June (Duffy 1968). The females lay their eggs singly in oviposition slits made with their mandibles in main branches and trunks. The lifetime fecundity in the field is unknown. Larval feeding occurs throughout

FIGURE 12.39   *Coptops aedificator* adult. (Courtesy of Larry Bezark.)

the year. Larvae bore subcortically or in the inner bark (Beeson and Bhatia 1939) or into the sapwood under the bark (Dawah et al. 2013). The life cycle lasts one year, and overwintering occurs at the larval stage (Beeson and Bhatia 1939; Duffy 1968).

Careful inspection and early detection of damage may allow preventive measures to be taken and may reduce the likelihood of its spread to other regions (Dawah et al. 2013). So far, removal and destruction of infested branches or entire trees are probably the only effective control methods for this pest (Chandy 1986). The application of insecticides recommended for *Batocera* spp. may not be effective for the control of *C. aedificator* (Dawah et al. 2013).

12.4.1.5 *Rhytidodera bowringii* White

Common name: Yellow-spotted ridge-necked longicorn

This cerambycine beetle is 25–40 mm long and generally dark brown in color; the antennae are slender and slightly shorter than the body; the pronotum has regular, longitudinal ridges; the elytra are covered with gray pubescence and a number of yellow pubescent markings arranged longitudinally; the elytral apex is truncated obliquely (Figure 12.40). The species is distributed in southern China, northern India, Nepal, and Myanmar (Duffy 1968; Hua 2002). The record in Java (Duffy 1968) is not confirmed. It was intercepted in Florida in 2006 (NAPPO 2006). However, there is no evidence of its establishment outside its primary distribution region. The host range of this longicorn appears to be narrow, with only several species recorded as its hosts—including mango and cashew (Fletcher 1930; Gressitt 1942; Duffy 1968; Hua 2002; NAPPO 2006; Wang 2006).

The yellow-spotted ridge-necked longicorn is a moderate to serious pest of mango in northern India and southern China (Fletcher 1930; Gressitt 1942; Duffy 1968; Luo et al. 1990; Wang 2006; Ding et al. 2014). According to Luo et al. (1990), about 25 mango orchards were completely destroyed by this pest in 1988 in China's Hainan Province. The larvae bore into the center of shoots, branches, and trunks

FIGURE 12.40 *Rhytidodera bowringii* adult. (Courtesy of Larry Bezark.)

(Duffy 1968; Luo et al. 1990), causing death of the infested parts (Gressitt 1942; Luo et al. 1990; Wang 2006). They keep their tunnels clear by means of frass-ejection holes. However, early infestation is difficult to detect because the bored branches continue to look healthy, bear green leaves, and develop side shoots until they die back and drop off the tree.

The adults emerge during May–June. The females lay their eggs on the living shoots and branches of eight- to ten-year-old trees, and the larvae bore right down their center, making clear circular tunnels (Beeson and Bhatia 1939; Duffy 1968). The fecundity in the field is unknown. Pupation occurs in the tunnel. The life cycle lasts one year in India (Beeson and Bhatia 1939) and China's Hainan Province (Luo et al. 1990) and two years in Yunnan Province (Zhou et al. 1995). Overwintering occurs at the larval or pupal stage.

Because the adults are nocturnal and attracted to light, Gressitt (1942) suggested that the most effective measure for the control of this pest probably is the application of light traps in the orchards. The pruning and destruction of infested branches and shoots can be highly effective in preventing further damage by this pest (Luo et al. 1990). Insertion of cotton wool saturated with insecticides into frass-ejection holes and then sealing them results in effective control (Luo et al. 1990; Ding et al. 2014). Ding et al. (2014) indicated that two applications of this method during the season achieve 100% control, but injection of insecticides into the holes is less effective. Pan et al. (1997) described an integrated approach to the management of this pest including cultural measures to increase the vigor of trees, pruning infested parts to remove the larvae, and application of insecticides to kill the larvae. Shen and Han (1985) tested the effectiveness of a nematode, *Steinernema carpocapsae* (Weiser), for the control of the larvae in the laboratory, achieving 80–100% larval mortality. However, these authors suggested that the method may not be practical for large-scale control in the field. In addition, Zhang (2002) has evaluated the resistance of various mango varieties in China and suggested growing potentially resistant varieties in regions where *R. bowringii* is abundant.

12.4.1.6 *Rhytidodera simulans* White

Common name: Mango branch and trunk borer

This species is similar to *R. bowringii*. Its adults are about 30 mm long and dark brown in color; the antennae are more or less flattened and substantially shorter than the body; the pronotum is irregularly wrinkled; the elytra are covered with gray pubescence and a number of orange and gray pubescent markings arranged longitudinally; the elytral apex is toothed at the margin (Figure 12.41). This cerambycine is distributed in Indonesia, Malaysia, Thailand, and Myanmar (Duffy 1968; Anonymous 1986; Muniappan et al. 2012). To date, it has not been recorded outside Southeast Asia. Similar to *R. bowringi*, the host range of this species is narrow, with fewer than 10 species recorded as its hosts and mango, cashew, rose apple, and star fruit being the most common (Muniappan et al. 2012).

The mango branch and trunk borer is considered a moderate pest of mango trees in Indonesia (Franssen 1949, 1950) and Malaysia (Ithnin and Shamsudin 1996; Abdullah and Shamsulaman 2008). The damage is caused by the larvae boring inside twigs, branches, and trunks, usually girdling the twigs or branches just beneath the bark (Franssen 1949; Muniappan et al. 2012). Infested tree parts may die or break off.

The adults and all immature stages are present throughout the year, and the life cycle lasts about one year (Franssen 1949; Kondo and Razak 1993; Muniappan et al. 2012). The eggs are laid singly on or near the tips of the twigs of mango trees; the larvae enter the twigs and bore through the branches toward the trunk. The fecundity in the field is unknown. Pupation occurs in the tunnels after six months' larval development. The females prefer to oviposit on older trees that are weakened by diseases or other insect pests, and trees under six years old are rarely attacked.

Franssen (1949) recorded three parasitic wasp species belonging to three families—Encyrtidae, Eupelmidae, and Pteromalidae—attacking eggs of *R. simulans* but indicated that their parasitism rates are not high. There is no recommended chemical control measure for this pest, but the methods used for the control of *R. bowringi* might be effective. Because the infestation by *R. simulans* usually is secondary to the primary damage by other insect pests or diseases, control of the primary pests that precede *R. simulans*, cultural measures to promote vigorous growth of the trees, and regular inspection and

FIGURE 12.41 *Rhytidodera simulans* adult. (Courtesy of Xavier Gouverneur.)

removal of damaged branches are recommended by Franssen (1949, 1950). Kondo and Razak (1993) tested the effectiveness of a nematode, *S. carpocapsae*, in killing *R. simulans* larvae under laboratory conditions. However, the use of *S. carpocapsae* to control *R. simulans* does not appear to be practical.

12.4.2 Australasia

12.4.2.1 *Platyomopsis pedicornis* (Fabricius)

Synonyms: *Saperdopsis obscura* Breuning, *S. bispinosa* Breuning, *Rhytophora tuberculata* Hope, *Lamia pedicornis* Fabricius
Common name: Australian mango longicorn

The adults of the lamiine borer are 15–20 mm long; the body is brown and covered with grayish to brown pubescence; the elytra have a number of long erect hairs and a submedian, zigzag, grayish, pubescent band; the elytral apex is bitoothed. The beetle is distributed throughout the Australian continent except for South Australia, Victoria, and Tasmania (McKeown 1947; Hall 1980; Smith 1996), and its larvae feed on mango (Smith 1996) and shrubby stylo [*Stylosanthes scabra* (Vog.)], a legume pasture crop introduced from South America (Hall 1980).

The Australian mango longicorn is a minor pest of mango trees (Smith 1996) and shrubby stylo (Hall 1980) in northern Australia. The larvae bore into the bark and later penetrate deeper into the sapwood of the branches and trunks of mango trees, weakening the trees and reducing the yield. In shrubby stylo, the larvae tunnel into the lower main stems and tap roots and eventually kill the plants, usually after they have been weakened by fire.

Little is known about the biology of this species. The adult beetles probably lay their eggs singly into oviposition slits made by mandibles in the bark of the trunk or main branches of mango trees. It is thought that the life cycle of this species is spread over 12 months but that the larval damage is most likely to be noticed in the late wet to early dry season from March to May (Smith 1996).

Smith (1996) described a control measure for this pest on mango trees in northern Australia. Injection of a solution of organophosphate insecticides mixed with petroleum oil into the holes of larval tunnels may be effective. For small infestations involving only one or two trees on the property, the insecticide mix should be liberally painted over the suspected tunnels with a paint brush after scrubbing the area with a wire brush. Larger infestations may be treated with the same mixture but applied with a hand-operated backpack-style sprayer. The nozzle of the sprayer should be placed over one of the breather holes and a quick squirt of insecticide injected under pressure into the tunnel. Follow-up treatments may be necessary for several months.

12.4.3 Africa

12.4.3.1 *Trichoferus griseus* (Fabricius)

Synonyms: *Hesperophanes griseus* (Fabricius), *H. tomentosus* Lucas, *Callidium griseum* Fabricius
Common name: Fig longicorn

The fig longicorn is a cerambycine borer. Its adults are 10–20 mm long; the body is reddish brown and covered with a mixture of grayish and brownish pubescence (Figure 12.42). The beetle has been recorded in Algeria, Morocco, Tunisia (Duffy 1957), Egypt (El-Nahal et al. 1978; Ismail et al. 2009), and Turkey (Aksit et al. 2005). It probably is widely distributed in the Mediterranean region. Its larvae feed on trees from Anacardiaceae, Apocynaceae, Betulaceae, Fabaceae, Fagaceae, Lythraceae, Moraceae, Sapindaceae, Myrtaceae, Rosaceae, Salicaceae, and Taxaceae (Duffy 1957; Aksit et al. 2005).

This species is a minor to moderate pest of fig trees in Egypt (El-Nahal et al. 1978; Ismail et al. 2009; Imam and Sawaby 2013). The damage is caused by the larval boring into large branches and trunks. Most frass remains in the larval tunnels, but some is ejected through the holes in the bark made by the larvae. Infested trees are weakened and even killed.

The life cycle is summarized from Duffy (1957), El-Nahal et al. (1978), and Imam and Sawaby (2013). The adults emerge during June–August, with a peak in July. They do not feed and can live for only six

FIGURE 12.42   *Trichoferus griseus* adult. (Courtesy of György Csóka.)

or seven days. The females lay their eggs singly or in groups of 2–18 eggs in bark cracks, under lichens on the bark, or under loose bark on large branches and trunks. Each female lays an average of 94 eggs in her lifetime. The hatch rate is highest (98%) at 30°C as compared to 70% at 20°C. The larvae bore into the bark and later enter the wood. This borer takes two to three years to complete its life cycle in the field but can complete a generation in one year in the laboratory. Overwintering occurs at the larval stage.

Rubbing the bark to remove lichens and cutting off all infested and dead branches may suppress the pest population to a great extent (Duffy 1957). Pruning infested branches and killing larvae in the tunnels using wires may reduce the pest population by about 55% (Ismail et al. 2009). Civelek and Çolak (2008) and Ismail et al. (2009) have tested the application of aqueous extracts of several plant species to kill the larvae and have obtained some positive results. However, it is not clear whether it is practical to use these extracts for controlling this pest in the field.

12.5 Cerambycid Pests of Nut Trees

Nut trees in this chapter do not meet the botanical definition but are those that produce nuts in the culinary sense—including walnut, pistachio, almond, cashew, hickory, pecan, chestnut, sheanut, and hazelnut. Major tree nut producers are concentrated in Asia, North America, Europe, and Africa, although most countries in the world produce some nuts. For example, China and Iran are among the largest producers of walnuts, the United States and Spain of almonds, Vietnam and Nigeria of cashew nuts, Iran and the United States of pistachios, and the United States and Mexico of hickory nuts and pecans.

A large number of cerambycid species are recorded that attack nut trees around the world. However, only eight species are considered economically important to walnut, pistachio, almond, cashew, hickory, and pecan trees in Asia, North America, and Africa. There is no record of economically important cerambycid pests of chestnuts, sheanuts, and hazelnuts.

12.5.1 Asia

12.5.1.1 *Batocera horsfieldi* Hope

Synonyms: *B. adelpha* Thomson, *B. kuntzeni* Kriesche
Common name: Large gray and white longicorn

This lamiine borer is one of the largest longicorns, and its adults are 32–65 mm long. The body is dark brown to black and covered with gray and grayish brown pubescence; the pronotum has a pair of kidney-shaped white to yellowish pubescent markings; each elytron has two to three lines of irregularly shaped and arranged white pubescent markings; the basal quarter of the elytra has granular processes; the elytral apex is rounded. The large gray and white longicorn is widely distributed in China, Japan, Korea, northern India, Nepal, Vietnam, and Myanmar, and attacks at least 66 plant species from 25 families (Duffy 1968; Hua 1982, 2002; Anonymous 2013). This species has not been reported to occur or been intercepted outside its primary distribution range.

It is a moderate pest of walnut and several forest and fruit tree species in China (Li et al. 1997; Jia et al. 2001; Wang et al. 2004; Li 2009) and India (Rhaman and Khan 1942; Duffy 1968). In some Chinese walnut orchards, the infestation rate can be as high as 70% (Wang et al. 2004). Infested walnut trees are significantly weakened but rarely killed. Infested branches can often be killed.

The biology and control of this longhorn have been studied by various individuals (e.g., Beeson and Bhartia 1939; Gressitt 1942; Rhaman and Khan 1942; Duffy 1968; Wang et al. 2004). The adults emerge between May and August but can be found as late as November, depending on climate. The females make transverse oviposition slits in the bark of the living trunk and main branches and lay their eggs singly in the slits. Each female lays 55–60 eggs in her lifetime. The larvae initially bore into the bark and sapwood and later into the heartwood. Pupation occurs in the larval tunnels. The life cycle lasts two to three years, depending on climate. Overwintering is passed at the mature larval or pupal stage.

Careful inspection of individual trees for symptoms caused by oviposition and young larval feeding is highly recommended. These symptoms are transverse cuts and black patches of liquid and frass on the bark. In these early damage stages, destruction of eggs and young larvae can be achieved by cutting into or probing the affected bark using a knife or a wire. Heavily infested or dying branches and trunks should be removed and burned. In the later infestation stage, insertion of cotton-wool soaked in insecticides or a solution of the nematode *S. carpocapsae* into the larval tunnels via frass holes and then closure of these holes with clay mud or other material can be effective. Furthermore, a polyphagous parasitoid, *Dastarcus helophoroides* (Coleoptera: Bothrideridae), that attacks many cerambycid species—including *Anoplophora glabripennis* (Motschulsky) and *Monochamus alternatus* Hope (Urano 2006; Ren et al. 2012)—has been used for the control of *B. horsfieldi* on walnut and other trees, achieving between 40% and 80% control (Li 2009; Li et al. 2009a, 2009b). This parasitoid can be mass-reared artificially (Ogura et al. 1999).

12.5.1.2 *Batocera lineolata* Chevrolat

Synonyms: *B. catenata* Vollenhoven, *B. chinensis* Thomson, *B. flachi* Schwarzer, *B. hauseri* Schwarzer
Common name: Cloud-marking longicorn

This species is very similar to *B. horsfieldi* in morphology and size and once was considered a synonym of the latter. However, using external morphological and genitalia characteristics, Liu et al. (2012) confirmed that *B. lineolata* and *B. horsfieldi* are two valid species. Briefly, this beetle differs from *B. horsfieldi* in its pronotal markings (which are white) and in the wider distance between the pair of the pronotal markings. Also, this beetle's elytral apex is truncated with a small tooth at suture (Figure 12.43).

The cloud-marking longicorn is widely distributed in China, Japan, Korea, India, Myanmar, and Laos (Duffy 1968; Hua 1982, 2002; Anonymous 2013). It has been intercepted in Hawaii (Nishida 2002), Australia (Biosecurity Australia 2006), France (Menier 1992), Germany, Belgium, Austria, and the Netherlands (Anonymous 2013), causing quarantine concerns in many countries. However, there is no evidence of the establishment of this beetle outside Asia. This lamiine is also highly polyphagous,

FIGURE 12.43 *Batocera lineolata* adult. (Courtesy of Jiří Mička.)

attacking at least 47 species of trees from 15 families (Duffy 1968; Hua 1982, 2002; Anonymous 2013; Yang et al. 2013), including some important forest trees, such as oaks (Xu et al. 2010b) and poplars (Yu et al. 2007a). In recent years, it has become a moderate pest of walnut trees in China, causing damage to walnut orchards similar to that caused by *B. horsfieldi* (Yang et al. 2012, 2013).

The biology of this species is also very similar to that of *B. horsfieldi* (Kojima, 1929; Duffy 1968). The adults start to emerge at the end of April (Xia et al. 2005). The life cycle lasts two to three years, and overwintering occurs at the larval, pupal, or adult stage (Duffy 1968; Yu et al. 2007a).

Control measures used for *B. horsfieldi* could be attempted for this species. The release of *D. helophoroides* can achieve some control but appears to be less effective compared to *B. horsfieldi* (Li et al. 2013). No other tested method, including the application of insecticides, is confirmed to be effective.

12.5.1.3 *Calchaenesthes pistacivora* Holzschuh

Common name: Pistachio longicorn

The cerambycine adults are 9–12 mm long; the antennae, head, and legs are black; the pronotal disc is yellowish to reddish, with a large black marking that may be enlarged to cover the entire disc; the elytra are yellowish to reddish with two pairs of black spots (basal pair is larger) (Figure 12.44). The pistachio longicorn was first collected from Iranian pistachio orchards in 1999 and later described as a new species by Holzschuh (2003). So far, *C. pistacivora* is only found in Iran and on pistachio trees (both cultivated and wild) (Rad 2005).

Iran is the largest pistachio producer in the world. This species has recently become an increasingly important pest of pistachio trees in Iran (Rad 2006; Mehrnejad 2014), threatening production. The pest prefers trees weakened by water stress and by extreme salinity in water and soil (Achterberg and Mehrnejad 2011). The damage is caused by larval boring inside the twigs and branches. The infested twigs and branches eventually die.

FIGURE 12.44 *Calchaenesthes pistacivora* adult on pistachio flowers. (Courtesy of M. R. Mehrnejad.)

According to Rad's (2006) study in the field, the adults start to emerge in early April and feed on pistachio leaves. The females usually lay their eggs on pruned sites of branches in mid-April. Each female lays 40–45 eggs in her lifetime. The larvae bore under the bark and then into the wood of the twigs and branches. Larval development takes 16 to 18 months. Pupation occurs in the larval tunnels. The life cycle lasts two years. This species overwinters at the larval and adult stage. It passes the first winter as a larva that develops into an adult in the second fall, but the adult remains in the feeding tunnel between October and late March before emergence (Achterberg and Mehrnejad 2011).

Due to the nature of host preference by this pest, Mehrnejad (2014) suggested that maintenance of tree vigor through appropriate and regular irrigation and fertilization regimes is considered the most effective control measure. In general, removal of infested parts of trees would also be a useful measure to consider. Achterberg and Mehrnejad (2011) described a new parasitoid species, *Megalommum pistacivorae* (Hymenoptera: Braconidae), that attacks the final instar of *C. pistacivora* larvae and achieves about a 23% parasitism rate. Rad et al. (2005) tested the effectiveness of spraying insecticides to kill adults in the field during the adult activity period and suggested that, if these insecticides are used during April, the adult beetles could be effectively controlled in the field.

12.5.1.4 *Neoplocaederus ferrugineus* (Linnaeus)

Synonyms: *Plocaederus ferrugineus* (Linnaeus) and *Cerambyx ferrugineus* Linnaeus
Common name: Cashew stem and root borer

The cerambycine beetle is 25–40 mm long; the head and prothorax are blackish brown and most of the remaining parts of the body are reddish brown; the pronotal disc has irregular ridges or wrinkles; the pronotum has a pair of sharp spines at sides (Figure 12.45). It is widely distributed in southern and southeastern Asia, particularly in India, Sri Lanka, Myanmar, Thailand, Vietnam, and Cambodia (e.g., Duffy 1968; Devi and Murthy 1983; Punnaiah and Devaprasad 1995; Senguttuvan and Mahadevan 1997a, 1997b;

FIGURE 12.45 *Neoplocaederus ferrugineus* adult. (Courtesy of Narasa Reddy, SRF, NBAIR, used with permission of NBAIR director.)

Chaikiattiyos 1998; ChauIn 1998; Lay 1998; Surendra 1998). It also occurs in Nigeria (Anikwe et al. 2007), where it probably was introduced from Asia. Cashew trees were first recorded as hosts of this longhorn by Bhasin and Roonwal (1954). So far, about nine plant species are recorded as its hosts, of which four (including cashews) belong to Anacardiaceae (Duffy 1968; Mohapatra and Jena 2007; Asogwa et al. 2009a). Among five plant species tested in the laboratory, the cashew clearly is the favorite host of *N. ferrugineus* (Mohapatra and Jena 2007, 2008a).

In the past few decades, this species has become a serious pest of cashew trees in India and in several southeastern Asian countries, evoking massive studies on its biology and control. In India, this pest is particularly damaging on the east and west coast, with the infestation rate being up to 40%, and severely infested trees die within two years, causing substantial losses (Bhaskara Rao 1998; Sahu and Sharma 2008; Maruthadurai et al. 2012; Naik et al. 2012; Sahu et al. 2012; Vasanthi and Raviprasad 2013a, 2013b). Furthermore, *N. ferrugineus* has recently emerged as a serious pest of cashew trees in Nigeria, causing the sudden death of mature trees within a few weeks (Anikwe et al. 2007; Hammed et al. 2008; Asogwa et al. 2009a, 2009b). The infestation rate in Nigeria ranges from 13% to 36% (Anikwe et al. 2007; Asogwa et al. 2009a, 2009b). The major damage symptoms include extrusion of frass through the holes at the root collar region, oozing of gum at the base of the tree trunk, leaves turning yellow and falling off, and death of the tree. Affected trees may also tilt on one side due to loss of anchorage if the injury to anchoring roots is severe (Maruthadurai et al. 2012).

Emergence and oviposition occur almost all year round in the field (Mohapatra and Mohapatra 2004); there are two peaks, one in March–May and another in December–February (Senguttuvan and Mahadevan 1997b, 1999a; Mohapatra and Jena 2009). However, Raviprasad and Bhat (2010) reported that most eggs are laid during December–June. The life cycle lasts six months to one year, depending on climate. For example, this longhorn has two generations a year with some overlaps in southeastern India (Senguttuvan and Mahadevan 1997b; Mohapatra and Mohapatra 2004), while it has only one generation a year in southwestern India (Maruthadurai et al. 2012). In the laboratory, this beetle can complete one generation a year on a natural or semisynthetic diet (Rai 1983; Senguttuvan and Mahadevan 1998; Vasanthi and Raviprasad 2013b). The larvae continue to feed and develop throughout the year. According to Maruthadurai et al. (2012), the females prefer to lay their eggs in the bark crevices or physical wounds caused by previous borer damage or heavy pruning on the trunk and roots exposed above the soil; the larvae bore into the fresh tissue of the bark, feeding on the subepidermal tissues, and then into phloem and xylem. Each female lays 60–90 eggs in her lifetime. Portions of the trunk below 50 cm are most frequently attacked (Mohapatra and Mohapatra 2004). Furthermore, the females prefer trees older than four to five years (Maruthadurai et al. 2012) or more than 10 years old (Kumar et al. 1996; Senguttuvan and Mahadevan 1999a) for oviposition. Pupation occurs in the heartwood (Beeson and Bhatia 1939).

Many studies have been carried out for the development of control measures for this pest including chemical, cultural, and biological control. Control actions taken at the early stage of infestation are essential if they are to work at all, and trees with moderate and severe infestation cannot be saved (Devi and Murthy 1983; Sundararaju 1985; Senguttuvan and Mahadevan 1997a, 1997b; Mohapatra and Jena 2008b; Asogwa et al. 2009b). Cultural measures recommended include (1) removing eggs during oviposition peaks and also subsequent early instar larvae (Rao et al. 1985; Bhaskara Rao 1998; Chaikiattiyos 1998), (2) removing heavily infested and dead trees (Chaikiattiyos 1998; Lay 1998; Anikwe et al. 2007; Mohapatra and Jena 2008b), and (3) cleaning the bases of infested trees (Mohapatra and Jena 2008b). Among these cultural practices, destruction of eggs and early instar larvae appears to be the most effective. In addition, evidence shows that the level of resistance and tolerance against the pest differs substantially among cashew cultivars (Kumar et al. 1996; Sahu et al. 2012), providing opportunities for further investigation into plant breeding programs for the control of *N. ferrugineus*. Chemical methods consist of (1) swabbing, painting, or spraying the exposed roots and trunk up to a height of 1–2 m (depending on trunk length) with synthetic (such as chlorpyriphos), botanical (such as neem oil), or fossil fuel (such as kerosene) insecticides and repellents two to four times a year (Rao et al. 1985; Senguttuvan and Mahadevan 1997a, 1997b; Bhaskara Rao 1998; Lay 1998; Mohapatra and Satapathy 1998; Senguttuvan and Mahadevan 1999b; Mohapatra et al. 2000, 2007; Chakraborti 2011; Anikwe et al. 2007; Mohapatra and Jena 2008b; Raviprasad et al. 2009); (2) application of chemical dust or granules around the tree base and incorporation into the soil (Devi and Murthy 1983; Rao et al. 1985), and (3) insertion of chemicals in cotton wool under the bark (Sundararaju 1985) or injection of chemicals into the larval tunnels (Chaikiattiyos 1998). The first method is the most effective, and the synthetic insecticides are superior to others. So far, no effective predators or parasitoids have been found for *N. ferrugineus*. Biological control only involves painting the trunk with a saturated conidial-mud slurry of two fungus species, *Beauveria bassiana* (Bals.-Criv.) Vuill. and *Metarhizium anisopliae* (Metchnikoff) Sorokin (Mohapatra et al. 2007; Sahu and Sharma 2008; Mohapatra and Jena 2008b; Raviprasad et al. 2009; Ambethgar 2010), or pouring a saturated aqueous suspension of conidia through larval entry holes and soil incorporation of fungal spawn (Ambethgar 2010). However, biological control using these fungi is not very effective for the control of this pest (Mohapatra and Jena 2008b; Raviprasad et al. 2009). In conclusion, a combination of multiple tactics, particularly cultural measures followed by chemical treatment, is more effective than the use of only one tactic (Rao et al. 1985; Senguttuvan and Mahadevan 1997a, 1997b; Bhaskara Rao 1998; Lay 1998; Anikwe et al. 2007; Mohapatra et al. 2007; Mohapatra and Jena 2008b).

12.5.1.5 *Neoplocaederus obesus* (Gahan)

Synonyms: *Plocaederus pedestris* Cotes, *P. obesus* Gahan
Common name: Cashew stem borer

This cerambycine species is similar to *N. ferrugineus* in appearance. The adults are 27–45 mm long; the body is pale brown to light reddish-chestnut or testaceous, clothed with short pubescence (Figure 12.46). The borer is also widely distributed in southern and southeastern Asia, including India, the Andaman Islands, southern China, Thailand, Myanmar, Vietnam, Sri Lanka, and Bangladesh (e.g., Duffy 1968; ChauIn 1998; Liu et al. 1998; Hua 2002). So far, it has not been recorded outside its primary distribution region. *N. obesus* is more polyphagous than *N. ferrugineus*. At least 28 plant species from 12 families are recorded as its hosts, of which 7 species, including cashew, mango, and chironji, belong to Anacardiaceae (Duffy 1968; Hua 1982; Meshram 2009; Maruthadurai et al. 2012; Vasanthi and Raviprasad 2013a, 2013b).

FIGURE 12.46 *Neoplocaederus obesus* adult. (Courtesy of Larry Bezark.)

The cashew stem borer is considered an important pest of cashew trees in India (Khan 1993; Maruthadurai et al. 2012; Vasanthi and Raviprasad 2012, 2013a, 2013b), Vietnam (ChauIn 1998), and southern China (Pan and van der Geest 1990; Li et al. 1997; Liu et al. 1998). It is also an important pest of chironji trees in India (Meshram 2009; Meshram and Soni 2011a, 2011b). Infestation eventually kills the trees if no action is taken.

The biology of this species is very similar to that of *N. ferrugineus*, with damage caused to trunks and roots. It has only one generation a year in the field (Meshram 2009; Maruthadurai et al. 2012) and the laboratory (Vasanthi and Raviprasad 2013b). Each female lays 40–50 eggs in her lifetime.

Several studies have evaluated the effectiveness of cultural, biological, and chemical control methods for this pest. Gressitt (1942) recommended trimming and destroying infested branches as well as the removal of bark, frass, and larvae from attacked trunks followed by an application of tar. Pan and van der Geest (1990) suggested that manual removal of the beetle larvae from the infestation sites is probably the most effective control measure. Vasanthi and Raviprasad (2012) evaluated the effectiveness of three entomopathogenic nematodes, *Heterorhabditis indica* Poinar, *Steinernema abbasi* Elawad, and *S. bicornutum* Tallosi, against the borer and concluded that all these tested nematodes can kill the larvae about two weeks after exposure. However, it is not known whether these nematodes can successfully control the pest in the field. Meshram and Soni (2011a, 2011b) tested the effectiveness of two fungus species, *Metarhizium anisopliae* (Metsch.) Sorok. and *Beauvaria bassiana* (Bals) Vuill., and insecticides for the recovery of trees infested by the borer in the field. After removing the larvae and cleaning the frass materials, they evaluated three application methods: swabbing saturated conidial-mud slurry over the tree trunk, pouring a saturated aqueous suspension of conidia through borer entry holes, and soil incorporation of fungal spawn. Results show that pouring the conidial suspension has achieved a recovery rate of 16.0–25.0%, followed by swabbing with a conidial slurry (17.0–20.0%), and soil application (13.0–14.0%). Using the same delivery methods, conventional insecticides achieve the highest recovery rate (38.0–43.0%). However, implementation of fungal application in the integrated control of *N. obesus* should be considered because the fungi would not only be safer to nontarget organisms but also more effective in a long-term pest control program. Finally, Liu et al. (1998) suggested that cut-end coating, trunk whitewashing with a lime solution, and injection of insecticides into borer tunnels also can give effective control.

12.5.2 North America

12.5.2.1 *Psyrassa unicolor* (Randall)

Synonym: *Stenocorus unicolor* Randall
Common name: Pecan branch pruner

The cerambycine adults are slender and 7.5–13 mm long; the body is light to reddish brown, covered with short pubescence (Figure 12.47). This species occurs throughout the eastern United States from Minnesota south to Alabama and west to Texas (Linsley 1963). This branch pruner attacks pecan and hickory trees and, to a lesser extent, the oaks and a few other species (Linsley 1963; Solomon 1985).

The pecan branch pruner is only considered a minor pest of pecan trees in the United States. Linsley (1963) and Solomon (1985) described the damage made by this pest. The larvae are twig and branch pruners, usually girdling at or near a node. They tunnel in pecan twigs 2–4 mm in diameter for a distance of 12–36 cm until intersecting a larger branch, which they girdle. Branches 10–50 mm in diameter and 0.6–3.6 m long are severed with a smooth concentric cut and fall to the ground from January to May. Twenty-nine percent of the larvae are in the severed portion of the branch, 13% in the branch stub, and the remaining larvae are dislodged and lost during the break.

According to Solomon (1985), the adults emerge from late April to early June. The eggs are deposited on small twigs that arise from a larger branch 10–50 mm in diameter. The young larva tunnels down the center of the twig toward its base but does not hollow out the small twig completely. Upon reaching the larger branch, it bores into the branch and begins to girdle it. The girdle is completed during late winter and spring when the larva makes a smooth, uniform concentric circular cut, often completely severing the wood but leaving the bark intact. Pupation takes place within the gallery just beneath the bark.

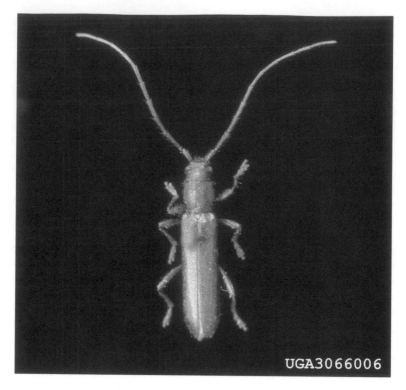

FIGURE 12.47 *Psyrassa unicolor* adult. (Courtesy of James Solomon, USDA Forest Service, Bugwood.org [3066006].)

The adult chews an irregularly shaped hole through the bark to emerge. The life cycle lasts one to two years, and overwintering occurs at the larval stage.

Severed branches on the ground under trees in orchards and ornamental plantings should be picked up in early spring and destroyed before the adults emerge in late spring and early summer. To be most effective, the pick-up and destruction practice should be undertaken for the entire orchard, woodlot, or neighborhood. Direct control in natural forest stands rarely is needed. Two icheumonid parasites, *Labena grallator* Say and *Agronocryptus discoidaloides* Viereck, may help reduce infestations (Linsley 1963).

12.5.2.2 *Oncideres cingulata* (Say)

Synonyms: *O. cingulatus pallescens* Casey, *O. proecidens* Casey, *and Saperda cingulata* Say
Common name: Pecan twig girdler

The adults of this lamiine girdler are 12–16 mm long; the body is cylindrical and generally grayish brown in color with a broad, ashy-gray band across the middle of the elytra (Figure 12.48). This longicorn is distributed throughout the eastern United States, from New England to Florida and as far west as Kansas and Arizona, and is highly polyphagous, feeding on trees from Anacardiaceae, Betulaceae, Caesalpiniaceae, Combretaceae, Cornaceae, Corylaceae, Ebenaceae, Fabaceae, Fagaceae, Juglandaceae, Lauraceae, Myricaceae, Myrtaceae, Nyssaceae, Rosaceae, Rutaceae, Salicaceae, Tiliaceae, and Ulmaceae (Bilsing 1916; Rogers 1977a; Linsley and Chemsak 1984; Rice 1995; Coppedge 2011; Day 2015).

The pecan twig girdler is a moderate pest of pecan and hickory trees in the United States (e.g., Bilsing 1916; Forcella 1984; Linsley and Chemsak 1984; Baker and Bambara 2001; Day 2015). The damage is caused by adult girdling. The females girdle the twigs and small branches using mandibles before oviposition to create proper conditions for the development of their larvae. The girdled twigs and branches break off or hang loosely on the trees. Usually, the ground under infested trees is covered with twigs

FIGURE 12.48 *Oncideres cingulata* adult. (Courtesy of Larry Bezark.)

and branches cut off by the longicorn. In pecan orchards, the fruiting twigs of heavily infested trees are often reduced, resulting in lower nut yields in the following year or years. This type of injury causes the development of many offshoots that adversely affect the symmetry of the trees. Pecan nurseries located close to heavily infested woodlots may suffer considerable losses from girdled seedlings. Severed twigs lodged in the tree canopy or on the ground often retain leaves even after the tree sheds its foliage in fall. Large trees usually sustain the most girdling, but young trees are sometimes heavily damaged.

The biology of this pest is summarized from Bilsing (1916), Rogers (1977a), Linsley and Chemsak (1984), Rice (1995), and Baker and Bambara (2001). The adults emerge from late August to early October. They feed on the tender bark near branch ends. The females girdle living twigs and small branches 0.6–2.0 cm in diameter and 25–75 cm long with their mandibles, make oviposition slits in the bark of girdled hosts, and then lay their eggs singly in the slits. Each female lays 50–200 eggs in her lifetime. The larvae bore into the host immediately after hatching. After the first instar, the larvae develop rapidly in the girdled portions of the host, boring toward the severed end of the twigs or branches by feeding only on the woody portion and leaving the bark intact. Pupation occurs during August and September, and the pupal stage lasts 12–14 days. Most larvae complete development within one year and pass the winter, but a small percentage of mature larvae pass a second winter in the galleries of girdled branches.

Baker and Bambara (2001) and Day (2015) summarized available control measures for this pest. In orchards, nurseries, and ornamental plantings, severed twigs and branches on the ground as well as those lodged on the trees should be gathered and destroyed during the fall, winter, and spring. The same practice should be applied in nearby woodlots containing oak and hickory trees if they have a history of serious damage from this pest. This practice can substantially reduce the pest population in one or two seasons. Insecticides only may be necessary to prevent damage from heavy infestations in some hardwood tree nurseries. These may be applied once or twice in September and again in October as a protectant to reduce the majority of the twig girdlers. The parasites, *Eurytoma magdalidis* Ashmead, *Iphiaulax agrili* Ashmead, and *Horismenus* sp., and a checkered flower beetle, *Cymatodera undulata* Say, may help reduce the girdler populations naturally.

12.5.3 Africa

12.5.3.1 *Analeptes trifasciata* (Fabricius)

Synonyms: *Lamia obesa* Voet, *L. trifasciata* Fabricius
Common name: Cashew stem girdler

The lamiine adults are 25–42 mm long and black in color; the elytra have three orange bands that may enlarge to cover most areas of the elytra in some specimens (Figure 12.49). This insect is widely distributed in Angola, Benin, Cameroon, the Central African Republic, the Democratic Republic of Congo, Ethiopia, Ghana, Ivory Coast, Kenya, Liberia, Niger, Nigeria, Senegal, Sierra Leone, and Togo, and attacks trees from Anacardiaceae, Lamiaceae, Malvaceae, and Myrtaceae (Duffy 1957; Wagner et al. 1991; Asogwa et al. 2011).

In the past decade or so, the cashew stem borer has become a serious pest of cashew trees in Nigerian orchards, particularly in the south (Hammed et al. 2008; Asogwa et al. 2011; Adeigbe et al. 2015). The infestation rate ranges from 40% to 80%, causing up to 55% cashew nut yield loss. Damage is caused by adult feeding on and girdling branches and young trunks, with the latter causing the most destruction. Both sexes feed by scraping the bark of cashew branches and young trunks, making V-shaped grooves, and then girdling them to provide dying sites for larval development. The stumps of the affected trees usually regenerate and produce many side shoots, giving an untidy appearance and producing no nuts. In some cases, the girdled branches and trunks recover by producing a lot of gum exudates around the girdled portion, followed by an outgrowth to heal the wound.

Information on the biology is extracted from Wagner et al. (1991) and Asogwa et al. (2011). The adults emerge in January–March and September–December. The females lay their eggs in the bark above the girdled portion that will hang on the tree or fall to the ground. The lifetime fecundity in the field is unknown. Dying and dead wood tissue provides a breeding site for the larvae, which bore inside the wood and develop to pupae and adults. This pest has two generations a year and overwinters at the larval and pupal stages.

FIGURE 12.49 *Analeptes trifasciata* adult. (Courtesy of Larry Bezark.)

The control measures are summarized from Asogwa et al. (2011) and Adeigbe et al. (2015). Cultural control appears to be the only effective measure at present. All the hanging girdled branches and those that have fallen to the ground are removed from the cashew plots on a monthly basis and burned. However, the large number of the girdled branches hanging on treetops and beyond reach by hand makes it very difficult to eradicate this pest from cashew orchards. It may be difficult for growers to afford the extra costs required to remove hanging branches from the treetops. The regenerated side shoots due to girdling attract foliar pests, which damage the apical portion and reduce photosynthetic activities of the tree. Such side shoots should be pruned immediately. Finally, removal of alternative hosts, such as *Adansonia digitata* and *Ficus mucosa*, in addition to burning the infested twigs may be necessary to effectively control the spread and damage by this pest.

12.6 Cerambycid Pests of Mulberry Trees

Mulberry trees (*Morus* spp.) are native to warm regions of Asia, Africa, and the Americas, with most species native to Asia. The fruits are red to dark purple in color, edible, and sweet with a good flavor. The leaves of several mulberry species are used as food for silkworms [*Bombyx mori* (L.)]. Silk is extracted from the cocoons of silkworms for making garments and other textile products.

Sericulture, the rearing of silkworms using mulberry leaves for the production of silk, has become one of the most important industries in many countries, particularly China, Japan, India, Korea, Brazil, Russia, Italy, and France (Sánchez 2000). At present, China and India are the two main sericultural countries, together producing more than 60% of the world's silk each year.

Mulberry trees are subject to attacks by a number of cerambycid species. For example, more than 30 longicorn species are recorded that infest mulberry trees in China (Gao et al. 1988). However, only about four species are considered economic pests of mulberry trees in the world, and all occur in Asia.

Batocera rufomaculata (DeGeer) is also an important pest of mulberry trees, but it is more severe in mango trees; thus it is treated in Section 12.4 on tropical and subtropical fruit trees. A North American species, *Dorcaschema wildii* Uhier, distributed from the eastern United States to Texas, causes very minor damage to mulberry and several ornamental trees (Solomon 1968; Linsley and Chemsak 1984), and thus is not discussed further here.

12.6.1 Asia

12.6.1.1 *Psacothea hilaris* (Pascoe)

Synonym: *Monohammus hilaris* Pascoe
Common name: Yellow-spotted longicorn beetle

The adults of this lamiine borer are 13–31 mm long; the body is dark brown with a longitudinal yellow stripe on the vertex, a longitudinal yellow stripe on each side of the pronotal disc, and a number of yellow spots on the elytra; the antennae are extremely long, about 2.5 times as long as the body in males and twice as long as the body in females (Figure 12.50). The primary distribution range of this beetle includes Japan, Korea, and China (Duffy 1968; CABI 2012). It has been intercepted in Europe and North America in warehouses, on wood and wooden spools imported from Asia, and has been confirmed as established in Italy (Lupi et al. 2013). Its host range appears to be narrow with only several plant species from Araliaceae, Asteraceae, and Moraceae being recorded as hosts (Duffy 1968; Hua 1982; Li et al. 1997).

The yellow-spotted longicorn beetle is one of the most serious pests of mulberry trees in Japan and China (Emori 1976; Masaki et al. 1976; Kawakami 1987; Gao et al. 1988; Sakakibara and Kawakami 1992; Sakakibara 1995; Shintani and Ishikawa 1999a, 1999b; CABI 2012). It can also seriously damage fig trees (Sakakibara and Kawakami 1992; Sakakibara 1995; CABI 2012). The damage is caused by larval boring in tree trunks. Severely attacked trees become weakened and eventually die (Lupi et al. 2013).

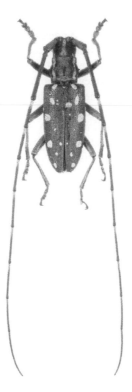

FIGURE 12.50 *Psacothea hilaris* adult. (Courtesy of Junsuke Yamasako.)

Both mulberries and figs belong to the family Moraceae. This cerambycid is also recorded as a minor pest of citrus trees in southern China (Li et al. 1997).

Studies show that there are at least two ecotypes in this species (Sakakibara and Kawakami 1992; Shintani and Ishikawa 1999a), allowing it to adapt to a wide range of environmental conditions through diapause or quiescence at the egg or larval stage under unfavorable conditions such as cold winters (Sakakibara 1995; Shintani and Ishikawa 1999a; Watari et al. 2002). This feature makes *P. hilaris* a potentially important invasive pest, and its recent establishment in Italy (Lupi et al. 2013) is a testament.

The adults can be found from May to November (Masaki et al. 1976) or from June to October (Duffy 1968; Lupi et al. 2013). Like many other lamiines, the females gnaw the bark of trees, making oviposition slits before laying eggs. They lay most of their eggs singly in oviposition slits in the basal portion of the trunk. The larvae feed on the cambial region, making irregular tunnels and ejecting dust-like frass from the crevices in the bark (Duffy 1968). Depending on climate and ecotypes, this species may have one or two generations a year (Watari et al. 2002). On an artificial diet in the laboratory, it can complete one to two generations a year depending on environmental conditions (Emori 1976). Overwintering occurs at the egg and/or larval stage.

Kawakami and Shimane (1986) described the process of producing a biopesticide containing conidia of an entomogenous fungus, *Beauveria tenella*. They have conducted field trials to determine its effectiveness for the control of adults, achieving cumulative adult mortality of 60–90% 30 days after application. Another entomogenous fungus, *Beauveria brongniartii*, was tested by Kawakami (1987) with similar killing efficiency. However, it is not known whether these fungus-based insecticides are effective in large-scale control programs. Gao et al. (1988) sprayed or brushed the trunk with synthetic insecticides, achieving between 80% and 90% control. Caution should be taken to avoid a negative impact on silkworms when synthetic chemicals are applied to mulberry trees. Although not reported for *P. hilaris*, cultural measures used for other cerambycids (see Chapter 9) could be considered and modified for the control of this species.

12.6.1.2 *Apriona germari* (Hope)

Synonyms: *A. plicicollis* Motschulsky, *A. deyrollei* Kaup, *A. cribrata* Thomson, *Lamia germari* Hope
Common name: Brown mulberry borer

The lamiine longicorn beetle is 26–50 mm long; the body is black and almost entirely covered with dense tawny brown, slightly greenish pubescence; the antennae are blackish brown, with bases of most segments ringed with grayish white pubescence; the prothorax has a pair of spines at the sides; the elytra have coarse granular processes on the basal fifth (Figure 12.51). The primary distribution range of this species includes China, Korea, India, Pakistan, Nepal, Laos, Myanmar, Thailand, and Vietnam (Duffy 1968; EPPO 2013d). Although Japan is mentioned in several publications on *A. germari* (Ibáñez Justicia et al. 2010), this pest probably is not present in Japan (EPPO 2013d). In the past two decades, *A. germari* has been intercepted in several European countries and probably also in the United States from goods imported from China, but there is no evidence that it has established outside its primary distribution range (Ibáñez Justicia et al. 2010; EPPO 2013d). This species is polyphagous, feeding on more than 30 species from Betulaceae, Bombacaceae, Fabaceae, Fagaceae, Juglandaceae, Lauraceae, Meliaceae, Moraceae, Rosaceae, Rutaceae, Salicaceae, Theaceae, and Ulmaceae (Duffy 1968; Huang et al. 2009; EPPO 2013d). Most host plant species are from Moraceae. The list of host plants of this longicorn given by Duffy (1968) and repeated by Huang et al. (2009) should be treated with some caution because Duffy (1968) did not distinguish between *A. germarii* and *A. rugicollis*, and some records appear to relate to *A. rugicollis*.

The brown mulberry borer is a serious pest of mulberry trees used for silkworm production in China (Gao et al. 1988; Zhao et al. 1994; Zhu et al. 1995) and northern India (Hussain et al. 2007, 2009; Hussain and Chishti 2012), causing significant economic loss. The infested trees are weakened and even killed. It is also considered an important pest of apple (Yang et al. 2005), fig (Qin et al. (1997), paper mulberry (Jin et al. 2011), walnut (Zhang et al. 1999), and poplar (Li 1996; Shui et al. 2009) trees. Furthermore, this species is recorded as a minor pest of many other fruit and forest trees (Li et al. 1997;

FIGURE 12.51 *Apriona germari* adult. (Courtesy of Eric Jiroux.)

Ibáñez Justicia et al. 2010; EPPO 2013d). Like many other lamiine species, *A. germari* adults need complemental nutrition for ovary development. Qin et al. (1994) showed that adults feeding on mulberry tree tissues produce normal ova, but those on apple or poplar tree tissues do not, suggesting that this species may not be able to establish in an environment where appropriate adult hosts are not present.

The adults emerge from May to October with a peak during June–August in China (Gressitt 1942). In India, they emerge from June to September, with a peak in mid-July in the subtropical region, while in the temperate region, the adult population peaks in the second fortnight of August (Hussain et al. 2007). The females use their mandibles to make oviposition slits and lay their eggs in the bark 20 cm above ground level (Duffy 1968), with a preference for primary branches 0.8–1.1 cm (Hussain and Buhroo 2012) to 1.3–1.9 cm (Yoon et al. 1997a) in diameter. They lay an average of 117 eggs each during the mean oviposition period of 43 days (Hussain and Buhroo 2012). The larvae bore into the sapwood first and then into the heartwood. Most larvae bore and move downward through the branches to the trunk (Yoon et al. 1997a), damaging the trunk and even the roots (Gressitt 1942). As the larvae tunnel downward, the larvae eject fine, wet, and sawdust-like frass at about 10 cm intervals (Duffy 1968). This species has one generation a year in southern China (Gressitt 1942; Chen et al. 1959) and northern India (Hussain and Buhroo 2012); in northern China, a life cycle lasts two to three years depending on climate (Yang et al. 2005; Huang et al. 2009). In the laboratory, this beetle can complete one generation a year on an artificial diet (Yoon et al. 1997b; Yoon and Mah 1999). Overwintering occurs at the larval stage (Ibáñez Justicia et al. 2010).

As a preventative measure, culture control can significantly reduce the damage by this borer. For example, summer pruning of the infested branches can stop further damage to the trunks, and removal of heavily infested trees can reduce infestation on mulberry farms by about 60% (Hussain and Chishti 2012). Studies showed that the level of resistance against *A. germari* varies among tree species (Zhang et al. 1999) and mulberry cultivars (Hussain and Chishti 2012), providing foundations for the development of resistant plant breeding programs for the control of this borer. An egg parasitoid, *Aprostocetus fukutai* Miwa & Sonan, is found to be an important natural enemy of *A. germari,* with a parasitism rate of about 50% in China (Yan et al. 1996). However, there is no evidence that this parasitoid is useful for large-scale biological control programs. Li et al. (2011) have tested the potential of a fungus, *Beauveria bassiana*, for biological control of *A. germari* in China. They showed that more than 95% larval mortality is achieved 10 days after inoculation in the laboratory. In the field, almost 70% of the larvae are killed 20 days after application. They suggested that this fungus has great potential to be used for the control of this pest. However, chemical control measures are still the most effective for the control of *A. germari*. For example, Gao et al. (1988) sprayed or brushed the trunk with synthetic insecticides, achieving between 80% and 90% control. In a trial on the spray effectiveness of different insecticide formulations for the control of this borer, Zhu et al. (1995) found that the controlled release microcapsule is more effective, persists longer, and has lower environmental toxicity than the emulsifiable concentrate. Yang et al. (2005) demonstrated that spraying insecticides in a microcapsule solution on the trunk and large branches during peak adult activity and then again in 20 days gives successful control. Furthermore, the injection of insecticides into borer holes can achieve >90% control (Pan, 1999), and inserting one fumigation rod into a borer tunnel that is then sealed with mud results in >97% larval mortality (Zhao et al. 1994). More recently, Shui et al. (2009) demonstrated that inserting zinc phosphide sticks into borer tunnels is a feasible method for controlling *A. germari*.

It is worth noting that, in the most recent taxonomic revision on *Apriona* (Jiroux 2011), *A. germari* is only reported from southern China (Guangdong, Hainan, Yunnan, and Tibet), India, Vietnam, Laos, Myanmar, Nepal, Bangladesh, Cambodia, and Thailand. Therefore, much of the aforementioned biology for *A. germari* outside these areas may well refer to the next species, *Apriona rugicollis* Chevrolat.

12.6.1.3 *Apriona rugicollis* Chevrolat

Synonym: *A. japonica* Thomson
Common name: Mulberry tree borer

The adults of the mulberry tree borer are 40–60 mm long; the body is dark gray to greenish in color; the basal fifth to sixth of the elytra has blackish tubercles (Figure 12.52). This species is widely

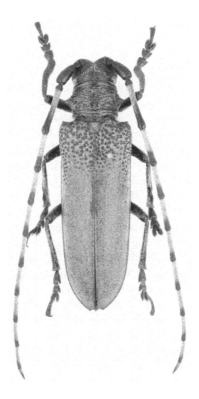

FIGURE 12.52 *Apriona rugicollis* adult. (Courtesy of Junsuke Yamasako.)

distributed in China, Korea, and Japan (Jiroux 2011). In 2009, it was intercepted in a Dutch port in a consignment of *Enkianthus* trees from Japan but did not establish (Ibáñez Justicia et al. 2010). *A. rugicollis* infests more than 26 plant species, including mulberry, fig, poplar, willow, beech, and keyaki (Japanese zelkova) trees from Cornaceae, Ebenaceae, Ericaceae, Fabaceae, Fagaceae, Juglandaceae, Lauraceae, Lythraceae, Malvaceae, Moraceae, Ulmaceae, Platanaceae, Rhamnaceae, Rosaceae, Rutaceae, and Salicaceae (Yamanobe and Hosoda 2002; Ibáñez Justicia et al. 2010; EPPO 2013d). *Apriona rugicollis* is a moderate pest of mulberry (Kikuchi 1976; Esaki 1995) and keyaki trees (Esaki 1995; Esaki 2006, 2007a), with the latter species being widely used for ornamental purposes in Japan. However, as mentioned earlier, much of the described biology and pest status for *A. germari* may refer to this species.

The adults are present from early July to early September with an oviposition peak in August (Esaki 2007a). The females prefer the main branches and trunks >10 mm in diameter for oviposition (Esaki 2007a), with most eggs being laid in the oviposition slits made on tree trunks 40–50 mm in diameter (Esaki 1995). The female lays an average of 60 eggs in her lifetime (reviewed by Yamanobe and Hosoda 2002). The adults feed on leaves and tender bark. The larvae bore under the bark and then into the heartwood. The life cycle lasts two to three years in the field (Esaki 2007a) and about one year in the laboratory (Esaki 2001). Overwintering appears to occur at the larval stage.

Esaki (2006) suggested that the removal of weeds around trees can reduce oviposition by *A. rugicollis* because the females prefer to oviposit on sites covered by weeds. Pruning the infested branches may also be effective for the control of this species. Esaki and Higuchi (2006) have tested the effectiveness of a fungus, *Beauveria brongniartii* (Saccardo) Petch, for the control of this borer in the field, demonstrating that the application of fungus-containing sheets on trees where the adults congregate for feeding can kill adults effectively. Spraying the branches with synthetic insecticides twice at an interval of three weeks during the early stage of the adult activity period can achieve 100% adult mortality over a period of nine weeks (Esaki 2007b).

12.6.1.4 *Sthenias grisator* (Fabricius)

Synonym: *Lamia grisator* Fabricius
Common name: Grapevine stem girdler

The adults of the lamiine grapevine stem girdler are 15–22 mm long; the body is stout in shape and blackish brown in color and is covered with a mix of gray and brown pubescence; the elytra have a grayish white spot in the middle at the suture and an oblique grayish white band across at the apical third margined with a thinner blackish-brown stripe at the apical side; in some specimens, the blackish stripe can expand to cover most of the apical area (Figure 12.53). This species occurs in India and Sri Lanka (Duffy 1968; Sengupta and Sengupt 1981; Bhaskar and Thomas 2010). The host range of this longicorn is quite broad, with more than 30 tree species recorded as hosts from Anacardiaceae, Apocynaceae, Casuarinaceae, Euphorbiaceae, Fabaceae, Menispermaceae, Moraceae, Moringaceae, Nyctaginaceae, Rosaceae, Rutaceae, and Vitaceae (Duffy 1968; Satyanarayana et al. 2013).

The grapevine stem girdler is considered a minor to moderate pest of mulberry, grapevine, jatropha (a crop grown for producing biodiesel), mango, almond, jackfruit, and several ornamental plant species such as roses (Singh and Singh 1962; Duffy 1968; Butani 1978; Sengupta and Sengupt 1981; Rahmathulla and Kumar 2011; Satyanarayana et al. 2013; Mani et al. 2014). In rare cases, it can be a serious pest of mulberry plantations (Bhaskar and Thomas 2010). The major damage is done by the adult girdling for egg laying (Sanjeeva Raj 1959; Duffy 1968). Portions above the girdle wilt and die.

The adults emerge almost all year round but mainly in July and August (Beeson and Bhatia 1939). The females completely girdle the shoots, tree branches up to 2 cm in diameter, vines, or bushes and then lay eggs singly in the wound (Sanjeeva Raj 1959; Duffy 1968). The lifetime fecundity in the field is unknown. The larvae bore in the portions above the girdle to avoid contact with ascending flow of sap. The life cycle may be completed within five months but, under dry conditions, a generation may take a year.

The most commonly used practice for the control of this pest is pruning and destruction of girdled apical portions from the point just below the girdle as soon as girdles are found or shoot withering is noticed (Duffy 1968; Bhaskar and Thomas. 2010). Swabbing the base of the trunks or branches with insecticides may also be effective (Butani 1978).

FIGURE 12.53   *Sthenias grisator* adult and girdle. (Courtesy of Sangamesh Hiremath.)

12.7 Cerambycid Pests of Tea, Coffee, and Cacao Crops

Tea, coffee, and cacao are the three most important crops for the production of beverages in the world. Cacao is also the key component of chocolate. Leaves of tea trees and seeds of coffee and cacao trees are harvested and consumed. Trees producing these products are small trees or bush-like woody plants. Tea is thought to originate from Asia, coffee from Africa, and cacao from Central and South America. These crops are now grown both inside and outside the regions of their origins.

Tea probably is the most popular drink in the world in terms of consumption, equaling all other manufactured drinks in the world put together—including coffee, chocolate, soft drinks, and alcohol (Macfarlane and Macfarlane 2004). China and India are the biggest tea producers in the world. Coffee is grown in most tropical regions, with Brazil and Vietnam being the top two producers. Cacao is also grown in many tropical countries, with the top two producers being Ivory Coast and Ghana in Africa.

In various books and catalogs, many cerambycid species are listed as pests of these beverage crops. For example, about 90 cerambycid species reportedly attack commercially grown coffee trees alone (Waller et al. 2007). However, only 13 species are considered pests with economic importance for these crops: three infesting tea and two impacting coffee trees in Asia, three damaging coffee trees in Africa, and one injuring cacao trees and four attacking coffee trees in Central and South America.

12.7.1 Asia

12.7.1.1 *Bacchisa atritarsis* (Pic)

Synonym: *Chreonoma atritarsis* Pic
Common name: Black-tarsused tea longicorn

The adults of the lamiine borer are 9–11 mm long; the head, pronotal disc, and ventral side are reddish to orange brown; the antennae are black with the basal two-thirds of segments 1–4 being reddish to yellowish brown; the elytra are metallic blue or purple; the legs are reddish brown to orange brown with the tarsi and apical third of tibia being blackish (Figure 12.54). This longicorn is only distributed in Taiwan and southern parts of China (Hua 2002). The number of plant species recorded as its hosts appears to be small: tea (*Camellia sinensis*), oil tea (*C. oleifera*), willow, Chinese wingnut, walnut, apple, pear, peach, and plum (Chen et al. 2002; Hua 2002; Tan et al. 2003). Of the infested species, *C. sinensis* is used in the production of tea for drinks, while seeds of *C. oleifera* are harvested for edible tea oil.

The black-tarsused tea longicorn is a moderate pest of tea and oil tea crops in southern and southeastern China and Taiwan (Chen 1958; Chang 1973; Qian 1984; Chen et al. 2002; Tan et al. 2003; Lv et al. 2011). In Zhejiang Province, it appears be a serious pest of oil tea trees (Hua et al. 2012). The damage is caused by larval tunneling in branches and trunks. The infested parts are weakened and may be broken off in the wind. The yield can be significantly reduced by this pest.

According to Qian (1984) and Chen et al. (2002), adults emerge from May to July. The females make oviposition slits in the bark of the main branches and trunks where they lay their eggs singly. The lifetime fecundity is unknown. Most oviposition occurs on branches and trunks 10–20 mm in diameter where larvae first bore under the bark, making spiral-shaped swollen marks, and then continue into the heartwood. The life cycle lasts two years, and overwintering occurs at the larval stage.

Chen et al. (2002) have practiced cultural and chemical control measures in the field and made recommendations for control methods. Careful examination of plants for characteristic symptoms (spiral-shaped swollen marks) and then pruning and destroying the infested portions can reduce further damage and infestation. However, when the main trunks are infested, chemical control probably is the only option. Swabbing the trunk below the damage marks with a systemic organophosphorous insecticide during August kills almost all larvae 20–30 days after application.

FIGURE 12.54   *Bacchisa atritarsis* adult. (Courtesy of Larry Bezark.)

According to Chen et al. (2002), *Dolichomitus mclanomcrus tinctipcnnis* (Camcron) is reported to be a larval parasitoid of this longicorn, but whether this wasp species can be used for the control of this pest is unknown.

12.7.1.2 *Trachylophus sinensis* Gahan

Synonym: *T. piyananensis* Kano
Common name: Four-ridged tea longicorn

The cerambycine adults are 25–38 mm long; the body is blackish brown and covered with dense, shining, golden pubescence; the pronotal disc has irregular wrinkles and four longitudinal and oblique ridges (Figure 12.55). This longicorn mainly occurs in southern and southeastern China and in Myanmar where *Camellia* spp. have been reported as hosts (Duffy 1968; Liau 1968; Hua 1982; Qian 1984).

This longicorn attacks tea and oil tea, causing minor to moderate damage to these economic crops (Liau 1968; Qian 1984; Tan et al. 2003; Hua et al. 2012). The damage is caused by larval tunneling in branches, trunks, and roots. The infested plants are weakened and their yield reduced.

Adults emerge between March and June and lay their eggs on the trunk surface just above the soil level, on mosses and lichens attached to the bark (Liau 1968), or on the bark of the trunk and branches (Tan et al. 2003; Hua et al. 2012). Each female lays 29–30 eggs in their lifetime. The larvae bore into the trunk and roots (Liau 1968) or into the trunk and branches (Tan et al. 2003; Hua et al. 2012). The life cycle lasts about one year, and overwintering occurs at the larval stage (Liau 1968).

Because the main damage occurs in the trunk and roots, chemical control is the only option that has been considered to date. Injecting insecticides into larval tunnels in October or spraying insecticides on the lower trunk between the ground and upward to 30 cm during June–July has is effective for the control of this pest (Liau 1968). In addition, the chemical control method for *B. atritarsis* (Chen et al. 2002) may also be effective for this pest.

FIGURE 12.55   *Trachylophus sinensis* adult. (Courtesy of Xavier Gouverneur.)

12.7.1.3 *Aeolesthes induta* (Newman)

Synonym: *Hammaticherus indutus* Newman
Common name: Silky gray-brown longhorn

The cerambycine adults are 23–38 mm long; the body is brown to blackish brown and covered with dense, shining, light brown pubescence, giving a satin look; the ventral side is covered with grayish brown pubescence; the pronotal disc has irregular wrinkles on the sides and a rectangular area in the middle that is arranged longitudinally (Figure 12.56). This longicorn is widely distributed in southern and southeastern Asia, including India, Sri Lanka, China, the Philippines, Indonesia, Myanmar, Thailand, and Laos, and it attacks a number of woody species in Anacardiaceae, Dipterocarpaceae, Euphorbiaceae, Fabaceae, Malvaceae, Meliaceae, Myrtaceae, Pinaceae, Rubiaceae, Rutaceae, and Theaceae (Chen 1958; Duffy 1968; Hua 1982; Qian 1984; Tan et al. 2003; Lv et al. 2011). Duffy (1968) cited a 1920s' report on serious damage caused by this species to cocoa in the Philippines. However, further details about its current pest status for this crop are unknown.

The silky gray-brown longhorn is a moderate (occasionally serious) pest of tea and oil tea crops in southeastern China (Chen 1958; Qian 1984; Tan et al. 2003; Liu et al. 2010; Lv et al. 2011). The infestation rate varies depending on the distance of trees from the road, the landscape where tea plantations are located, and whether roots are exposed to the air. For example, about 30% of trees near the roadside are infested, while less than 10% of trees 70–100 m away from the road are attacked (Liu et al. 2010; Yang and Liu 2010; Yang et al. 2011). Liu et al. (2010) reported an infestation rate ranging from 7% to 50% in a tea orchard in Zhejiang Province, with more serious damage occurring near a hilltop rather than midway along the hillside. According to Yang and Liu (2010), higher infestation rates (51%) occur in trees planted in shallow soil with their roots partially exposed to the air, while infestation rates are lower (17%) in those planted in thick soil with their roots completely covered. Infestation can cause trunks to decay (Chen 1958). The infested trees are generally weakened, and heavily damaged ones die.

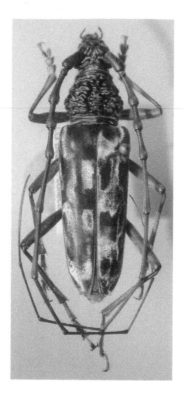

FIGURE 12.56 *Aeolesthes induta* adult. (Courtesy of Larry Bezark.)

The adults emerge between April and July and lay their eggs in crevices or joints on trunks or branches that are 2–3.5 cm in diameter and 7–35 cm above the ground. Most oviposition occurs in late May. The lifetime fecundity in the field is unknown. The larvae first bore under the bark and then into the heart-wood. They bore downward and make tunnels extending to at least 35 cm below the ground's surface, leaving holes 2–3 cm aboveground for frass expulsion. The life cycle lasts two to three years depending on climate. Overwintering occurs at the larval stage in the first year or first and second years and at the adult stage in the pupal cells in the second or third year.

Yang et al. (2011) indicated that the use of light traps during the adult activity period can substantially reduce the infestation on tea trees by this pest in Zhejiang Province. Covering exposed roots with soil can increase the vigor of trees and prevent oviposition, thereby reducing infestation rate (Yang and Liu 2010; Lv et al. 2011). Insertion of cotton wool saturated with organophosphate insecticides into borer holes near the ground and sealing those holes with mud can also effectively control this pest (Yang and Liu 2010; Lv et al. 2011).

12.7.1.4 *Acalolepta cervina* (Hope)

Synonyms: *Dihammus cervinus* (Hope), *Monohammus cervinus* Hope
Common name: Coffee brown borer

The lamiine adults are 15–27 mm long; the body is uniformly covered with dense, grayish brown pubescence; pubescence on the basal portions of most antennal segments is grayish; the scutellum is covered with golden pubescence; the pronotum has a spine on each side (Figure 12.57). The coffee brown borer is widely distributed in Asia including China, India, Myanmar, Thailand, Vietnam, Nepal, Laos, Japan, and Korea (Duffy 1968; Hua 1982; Waller et al. 2007). Its larvae feed on at least 16 forest tree species in Adoxaceae, Bignoniaceae, Combretaceae, Daphniphyllaceae, Lamiaceae, Lythraceae, Moraceae, Rubiaceae, Salicaceae, Scrophulariaceae, and Theaceae (Duffy 1968). Chen et al. (1959) listed coffee (family Rubiaceae) as a host that is not listed in Duffy (1968).

FIGURE 12.57 *Acalolepta cervina* adult. (Courtesy of Larry Bezark.)

In the past two decades, the coffee brown borer has been recognized as a moderate (Loh and Zhang 1988; Kuang et al. 1997; Li et al. 1997; Yu and Kuang 1997) to serious pest (Wei and Yu 1998; Rhainds et al. 2002; Waller et al. 2007; Lan and Wintgens 2012) of Arabica coffee in China and Thailand. The infested trees are seriously weakened with leaves turning yellow and falling off and the trunk and main branches breaking up. In severe cases, trees die. Coffee plants of different ages, ranging from three to seven years, have a similar level of infestation (Rhainds et al. 2002). In southwestern China, this pest is much more prevalent than *Xylotrechus javanicus* (Castelnau & Gory), another coffee pest in Asia (see the following section), although the latter is a more serious pest (Kuang et al. 1997; Rhainds et al. 2002).

The adults emerge between April and October in India (Duffy 1968) and between March and June in China (Kuang et al. 1997; Yu and Kuang 1997). The females cut transverse oviposition slits in the bark of the trunk where they lay their eggs singly during April–June. A female lays 40–60 eggs in her lifetime. The larvae bore into the bark and then the sapwood (Rhainds et al. 2002). Larval tunnels in the trunk usually are spiral upward (Lan and Wintgens 2012). This longicorn has one generation a year and overwinters at the mature larval stage (Loh and Zhang 1988; Waller et al. 2007; Lan and Wintgens 2012).

Lan and Wintgens (2012) summarized several commonly used control measures. Swabbing the trunk with contact insecticides from the base to 1 m aboveground during the oviposition period can reduce egg laying and, during May–July, may kill neonate larvae and prevent them from boring into the trunk. These authors also recommend cultural practices including removing heavily infested trees in August and September, planting shade trees, avoiding moving infested trees to noninfested areas, and proper pruning and fertilizer application. Wei and Kuang (2002) have tested the efficiency of biological control using a fungus, *Beauveria bassiana,* against the pest in both the laboratory and the field. In the field trials, they pushed "fungus mud" containing fungal spores into larval tunnels and achieved more than 90% larval mortality 20 days after application. However, it is not known whether this biological control method can be used successfully on a large scale.

12.7.1.5 Xylotrechus javanicus (Castelnau & Gory)

Synonyms: *X. lyratus* Pascoe, *X. quadripes* Chevrolat, *Clytus sappho* Pascoe, *C. javanicus* Castelnau & Gory
Common name: Asian coffee stem borer

The adults of this cerambycine borer are 13–17 mm long and black in color; the head, prothorax, and bands on the elytra are covered with grayish or yellowish pubescence; the pronotal disc has a round, sub-asperate, medial black spot and a pair of smaller sublateral black spots; the elytra have a basal transverse band, an oblique band on the shoulder, a band beginning immediately behind the scutellum—running close to the suture and then diverging posteriorly and turning outward, a transverse or slightly oblique band behind the middle that gradually widens toward the suture, and an apical band with an oblique front margin (Figure 12.58). *Xylotrechus javanicus* (as *X. quadripes*) was erroneously reported by several authors as occurring in Africa (Venkatesha and Dinesh 2012), but it is only distributed in southern and southeastern Asia—including China, India, Sri Lanka, Bangladesh, Cambodia, Indonesia, Laos, Myanmar, the Philippines, Thailand, and Vietnam (Duffy 1968; CABI 1998). So far, about 20 plant species from Anacardiaceae, Bignoniaceae, Burseraceae, Cannabaceae, Fabaceae, Lamiaceae, Moraceae, Oleaceae, and Rubiaceae have been reported as its hosts with a preference for Arabica coffee trees as hosts (Duffy 1968; Waller et al. 2007; Lan and Wintgens 2012; Venkatesha and Dinesh 2012). This pest's enormous economic importance has triggered massive studies over the past 100 years, particularly in India and China. Venkatesha and Dinesh (2012) provided an excellent review of those studies with more than 100 references cited.

The Asian coffee stem borer is a serious pest of Arabica coffee trees in India, China, Thailand, and Vietnam, causing substantial losses in production (Duffy 1968; Visitpanich 1994; Li et al. 1997; Yu and Kuang 1997; Rhainds et al. 2002; Waller et al. 2007; Lan and Wintgens 2012; Venkatesha and Dinesh 2012). About two months after infestation, an early external symptom may appear—ridges on the surface of the trunk or main branches, each caused by a larva tunneling under the bark. With continued larval tunneling, the damaged tree withers, showing yellowing and wilting symptoms. Severely damaged trees die.

FIGURE 12.58   *Xylotrechus javanicus* adult. (Courtesy of Kwan Han.)

This pest prefers to attack older trees. For example, in southwestern China, five- to seven-year-old plants are 10 times more heavily infested than three- to four-year-old plants (Rhainds et al. 2002).

The emergence period varies in different countries probably because of different climatic conditions (Venkatesha and Dinesh 2012). For example, emergence occurs in April–May and September–December in India; most beetles emerge during May–July and some during February–March and November–December in Vietnam; and three emergence peaks take place in May, October, and December in China. The females lay their eggs singly or in small batches of several eggs in the cracks and crevices of bark on the trunk and main branches. They prefer to lay eggs in bright sunlight. The lifetime fecundity is 50–103 eggs. The larvae bore into the bark and then into the sapwood and heartwood, making tunnels downward with some extending into the roots. About 80% of adults emerge from the trunk and the rest from the roots. In the warm coffee-growing regions, larval development continues all year round, and overwintering occurs at the larval or pupal stage. This pest may have one generation a year in India (Venkatesha and Dinesh 2012) and two or three generations a year in China (Kung 1977; Loh and Zhang 1988; Lan and Wintgens 2012), depending on climate. The borer can be reared successfully in the laboratory using cut trunks and branches of Arabica coffee trees (Gowda et al. 1992) or with an artificial diet (Raphael et al. 2005). It can complete development from eggs to adults in 71–90 days on cut Arabica coffee trunks (Seetharama et al. 2005).

Many tactics have been developed and practiced for the control of this pest (reviewed by Duffy 1968; Waller et al. 2007; Lan and Wintgens 2012; Venkatesha and Dinesh 2012). Venkatesha and Dinesh (2012) listed 33 species of natural enemies that attack *X. javanicus,* of which 29 are parasitic wasps, two are avian predators, and two are fungal pathogens. Although various trials in the laboratory and field using some of these natural enemies demonstrate promising outcomes in terms of killing the borer, there appears to be no large-scale application of any of these agents for the control of the pest. Venkatesha and Dinesh (2012) suggested that because the larvae and pupae are well protected in tunnels, application of their natural enemies may not be highly successful; but further investigation into the application of biological control agents that attack eggs may be useful. A number of field trials using a synthetic male-produced sex pheromone have also been carried out for the control of this pest (Venkatesha and Dinesh 2012). While Hall et al. (2006) and Jayarama et al. (2007) suggested that pheromone traps may be able to contribute to successful control of the female beetles, the application of pheromone traps does not seem to be a viable strategy in China (Rhainds et al. 2001). Further studies on the application of semiochemicals are warranted.

The commonly used control measures (Waller et al. 2007; Lan and Wintgens 2012; Venkatesha and Dinesh 2012) are summarized as follows:

1. *Cultural control:* Before the beginning of adult emergence, scrubbing the scaly bark of the trunk and main branches with a brush, a piece of metal, or with gloves made from thick fabric (to make the bark's surface smooth) can prevent females from laying eggs and can destroy eggs and young larvae. This measure is highly effective but labor-intensive. Examination for the characteristic borer ridges and then removal of the infested parts or uprooting of entire trees can reduce further infestation of healthy trees. The removed trees or branches should be burned. If the uprooted trees are used for firewood, they should be submerged in water for 7–10 days to kill the borers, which can otherwise complete development in the cut trunks and branches. Because the females prefer to lay eggs in bright sunlight, planting appropriate shade trees alongside coffee trees can reduce infestation.

2. *Chemical control:* Swabbing or painting the trunk and main branches with a coal tar distillate, a lime solution, or an insecticide solution before the beginning of adult emergence can reduce oviposition to some extent and kill eggs and young larvae. However, swabbing systemic insecticides on the plants is not effective for the control of larvae that have already bored into the plant tissue.

3. *Resistant cultivars:* A number of attempts have been made to develop cultivars resistant to this borer. For example, a new Arabica cultivar in India, Chandragiri, has good yield potential and a high tolerance to the pest because the drooping branches of these plants cover the entire trunk and act as a barrier against borer attack (Jayarama 2007).

12.7.2 Africa

12.7.2.1 *Monochamus leuconotus* (Pascoe)

Synonyms: *Phygas fasciatus* Fåhraeus, *Anthores leuconotus* Pascoe
Common name: African coffee stem borer

The lamiine adults are 17–35 mm and dark brown in color, with yellowish-brown patterns; the elytra have an extensive area of white pubescence that forms a very wide band across the subbasal and medial regions and a narrower band in the apical region; these two white bands are connected along the suture (Figure 12.59). This borer occurs throughout Central, South, and East Africa—including Angola, Burundi, Cameroon, Congo, Congo Democratic Republic, Ethiopia, Kenya, Malawi, Mozambique, Namibia, Rwanda, South Africa, Sudan, Tanzania, Uganda, Zambia, and Zimbabwe. To date, it has not established in West Africa or any continents other than Africa (Duffy 1957; CABI 2005a). However, a recent study (Kutywayo et al. 2013) suggests that the areas suitable for *M. leuconotus* infestation will increase significantly in African countries due to climate change. About a dozen woody plant species in Erythroxylaceae and Rubiaceae are listed as its hosts by Duffy (1957), including all coffee species (Rubiaceae), with Arabica coffee being the preferred host (Duffy 1957; CABI 2005a; Rutherford and Phiri 2006).

The African coffee stem borer has been a serious pest of coffee trees in South and East Africa since the 1930s (Duffy 1957; Hill 1975; Hillocks et al. 1999; Waller et al. 2007; Crowe 2012). In southern Africa, yield losses as high as 25% have been reported with more than 80% of coffee farms affected (Rutherford and Phiri 2006). Ringbarking in the trunks by the early instar larvae in the first two to three months after hatch causes severe damage to trees, which become weakened and display yellowing foliage, shoot dieback, and defoliation (Duffy 1957; Waller et al. 2007; Crowe 2012). Other signs include frass-ejection holes, from which frass is extruded. Young trees with small trunk diameters can be killed by these larvae, although infestation of older trees reduces production but rarely kills them. When the larvae become older, they bore into the heartwood, but their damage to trees is not as serious as that caused by younger larvae.

FIGURE 12.59 *Monochamus leuconotus* adult on a coffee tree. (Courtesy of Dumisani Kutywayo.)

Adults usually emerge during November–January with a peak in mid-December, and most oviposition occurs in mid-January (Tapley 1960; Schoeman et al. 1998). The females use their mandibles to make oviposition slits (usually horizontally) in the bark of the lower trunk (usually the lower 50 cm) and deposit eggs singly in these slits (Duffy 1957). According to Schoeman et al. (1998), a female lays an average of 80 eggs during her life span; the larvae bore into the bark and feed on phloem and cambium tissue for three to four months before entering the heartwood, feeding on xylem tissue for about 12 months. The larvae usually bore spirally downward to the roots (Duffy 1957). At warmer, lower altitudes, larvae are found mostly at the base of the trunks and main root systems, while at higher, cooler altitudes and also in heavy shade, more larvae develop well above the ground level (Tapley 1960). The adult beetles usually do not cause significant damage, although shoots and twigs are occasionally ringbarked (Schoeman et al. 1998). The life cycle lasts 12–25 months, with most individuals requiring 16–20 months (Tapley 1960). More recently, Waller et al. (2007) suggested that it takes between 18 and 24 months depending on the latitude and altitude. The larvae appear to continue developing all year round when the temperature is suitable, and overwintering usually occurs at the larval stage.

Although a number of control measures have been developed and practiced over the past century, this pest is still difficult to control (Rutherford and Phiri 2006; Waller et al. 2007). The most effective control method has been the brush or spray application of chemical pesticides to the trunk as a band around the lower 50 cm section and the root collar region. In past years, organochlorides achieved outstanding success in the control of this pest (Tapley 1960; Da Ponte 1965; Crowe 1966; Bigger 1967; McNutt 1967; De Villiers 1970; Schoeman 1991, 1995; Schoeman and Pasques 1993a, 1993b). However, these pesticides are no longer used due to their hazardous effects on the environment and on nontarget organisms and also their risk to the users (Rutherford and Phiri 2006). Alternative pesticides including organophosphates (Schoeman and Pasques 1993; Waller et al. 2007), pyrethroids (Schoeman 1995), phenylpyrazoles, and natural lime solution (Rutherford and Phiri 2006) can also provide effective control. The timing of the applications is extremely important, that is, the insecticides should be applied just before the start of the monsoon rains (Rutherford and Phiri 2006). Schoeman (1990, 1991, 1995) and Rutherford and Phiri (2006) summarized cultural control methods currently practiced. For example, well-maintained coffee trees are less likely to be attacked by this stem borer because it usually attacks weakened or debilitated plants. Planting resistant species or cultivars may suppress the pest population to some extent. Destruction of heavily infested plants (uprooting) before the onset of rains, and hence before the adults emerge, can reduce further infestation. Although labor-intensive, the removal of loose bark at the base of trees and wrapping tree trunks with banana fiber reduce oviposition and thus infestation. Jonsson et al. (2015) suggested that lowering coffee tree density and removing shade trees in or around the orchards can reduce damage because this pest avoids sun-explored trees. Several wasp species parasitizing *M. leuconotus* (including an egg parasitoid, *Aprostocetus* sp.) may contribute to high mortality rates for this pest in the field (Tapley 1960). Although there is no report on the commercial use of these parasitoids for pest control, further investigations into their potential and application may be warranted. Massive application of chemical pesticides could be one of the reasons why these natural enemies do not control the pest (Rutherford and Phiri 2006). Preliminary testing of the fungus *Beauvaria bassiana* for biological control has been investigated in South Africa with some promising results (Schoeman and Schoeman 1997). However, no commercial biological control products appear to be available (Rutherford and Phiri 2006).

12.7.2.2 *Bixadus sierricola* (White)

Synonym: *Monochamus sierricola* White
Common name: West African coffee borer

The lamiine beetle is 20–30 mm long, chestnut brown in color, and covered with grayish white pubescence; the elytra have a more or less V-shaped brown mark near the middle. It occurs throughout Western and Central Africa, including Angola, Benin, Cameroon, Central African Republic, Ivory Coast, Equatorial Guinea, Gabon, Gambia, Ghana, Guinea, Guinea-Bissau, Nigeria, the Republic of the Congo, Sierra Leone, and Uganda (Duffy 1957; CABI 2005b; Waller et al. 2007). To date, it has not become established on any other continents. Duffy (1957) listed all coffee species and two other plant species as the hosts of *B. sierricola* but mentioned that some earlier authors have suggested that this longicorn beetle could be polyphagous.

The West African coffee borer is a moderate to serious pest of coffee trees in West and Central Africa (Padi and Adu-Acheampong 2001; Crowe 2012) depending on latitude and altitude. In Uganda, up to 80% of trees can be infested by this pest (Padi and Adu-Acheampong 2001). *B. sierricola* attacks all commercially grown coffee species (Crowe 2012). However, there is still controversy as to which coffee species is preferred as a host by this pest. For example, Fonseca Ferrão (1965) reported that it prefers to attack Robusta coffee, although Waller et al. (2007) suggested that Arabica coffee suffers the most damage. Ringbarking in the trunks by early instar larvae during the first two to four weeks after hatch causes serious damage to coffee trees, leading to loss of branches in older trees and the death of young trees (Duffy 1957; Fonseca Ferrão 1965; Waller et al. 2007). Although older trees often survive attacks, they are frequently broken by wind or attacked by secondary pests such as termites or by fungal diseases (Padi and Adu-Acheampong 2001; Waller et al. 2007). According to Waller et al. (2007), this pest prefers to attack trees four to five years old. Following ringbarking, the older larvae bore into the heartwood, producing large amounts of frass that drops at the base of the tree. Borer holes occur exclusively on the trunk up to 3 m from the ground level (Padi and Adu-Acheampong 2001).

This species has one or two generations a year depending on latitude and altitude (Duffy 1957; Fonseca Ferrão 1965). For the beetles with two generations a year, most adults of the first generation emerge between late August and early September, and most adults of the second generation emerge between late January and early February. For those with only one generation a year, the adults emerge in greatest numbers at the beginning of the long rains (Fonseca Ferrão 1965). The adults are active at dusk, searching for mates and laying eggs, and can live for two to three weeks. The eggs are laid singly in crevices or wounds in the bark of the trunk, usually 15–20 cm above the root collar. One female can lay 50–60 eggs in her life span. The larvae bore into the bark and ring it for a few weeks before entering the heartwood. They tunnel downward in the trunk through the root collar to the level of the second bundle of secondary roots. The larvae appear to continue developing all year round, and overwintering usually occurs at the larval stage.

Crowe (2012) suggested that the chemical control measures used for *M. leuconotus* can also be effective for the control of *B. sierricola* but does not give any details. Several cultural control measures are summarized by Duffy (1957) and Waller et al. (2007): (1) uproot and burn infested trees, (2) keep the base of the tree free of lichen and other covers to facilitate early detection of the pest, and (3) separate the coffee trees from the edge of the forest with a 30 m belt of food crops. The use of light traps is also suggested for catching adults but their practicality and effectiveness are not known. Two parasitoid species, one wasp (*Gabunia ruficoxis* Kriechbaumer) and one fly (*Phorostoma* sp.), were reported in earlier references (cited by Duffy 1957) as attacking the pest, but information about their usefulness in the control of this species is unknown. In the 1960s, the same organochlorides used to control *M. leuconotus* were used to paint lower sections of trunk (up to 50 cm) and the root collar region during adult emergence peaks; they provided excellent control for six months (Fonseca Ferrão 1965). More recently, an aluminum phosphide paste has been tested for its effectiveness in the control of *B. sierricola* larvae (Padi and Adu-Acheampong 2001). The paste is squeezed into fresh borer holes, and then those holes are sealed off with plasticine. Results show 100% larval mortality within 15 days after treatment, with no subsequent reinfestation by the beetle recorded within nine months.

12.7.2.3 *Nitakeris nigricornis* (Olivier)

Synonym: *Dirphya nigricornis* Olivier
Common name: Yellow-headed borer

This species was called *Dirphya nigricornis* Olivier until the recent revisional work by Teocchi et al. (2014). The adults of this lamiine borer are slender, 20–31 mm long, and yellowish to orange brown in color; the antennae, eyes, at least three-quarters of the elytra (from the basal one-quarter to the apex), and the tibiae and tarsi are dark brown to black (Figure 12.60). The yellow-headed borer is distributed only in eastern and southern Africa including Kenya, Mozambique, South Africa, Tanzania, and Malawi

FIGURE 12.60 *Nitakeris nigricornis* adult. (Courtesy of Larry Bezark.)

(Rutherford and Phiri 2006; Waller et al. 2007). Earlier, this species was recorded in both East and West Africa (Duffy 1957). However, more recent studies show that the borer in West Africa is *Neonitocris princeps* (Jordan) rather than *N. nigricornis* (Waller et al. 2007). *N. princeps* is a very minor pest of coffee trees and thus is not treated further in this chapter. The yellow-headed borer only attacks plants in Rubiaceae, including *Rytigynia schumanii* in Tanzania and *Vangueria rotundata* in Kenya in addition to *Coffea* spp. in many African countries (Duffy 1957; Waller et al. 2007). However, *D. nigricornis* prefers Arabica coffee as a host (Crowe 1963, 2012; Rutherford and Phiri 2006; Waller et al. 2007).

The larvae of this species cause minor to moderate damage to Arabica coffee in eastern Africa (Crowe 2012). However, serious damage to coffee plantations has been recorded in Kenya and in northern Malawi (Crowe 1963; Wanjala 1988; Wanjala and Khaemba 1988; Rutherford and Phiri 2006; Waller et al. 2007). The oviposition activity at the tips of the fruiting branches can be destructive, causing them to wilt and die. The larvae bore inward into the branches and then downward to the trunk and main roots. On the way downward, the larvae leave a series of flute-like holes along the branches and down the side of the trunk, where sawdust-like frass is ejected. Affected branches are weakened and may break under the weight of developing coffee berries; green shoots may wilt and die.

Adults may be collected almost all year round, with two emergence peaks—one in April–May and another in November–December (Crowe 1963). The females gnaw oviposition slits in the bark with their mandibles and lay their eggs singly inside the slits, mainly on the green distal and immature internodes, with the tertiary or primary shoots being preferred (Wanjala 1990). Each female can lay 20–30 eggs in her lifetime (Wanjala 1988). The larvae enter the wood, tunnel toward the trunk, and may eventually reach the main roots. This pest prefers weakened trees to healthy ones for oviposition, and the larval development can be completed in dead wood (Crowe 1963). The life cycle lasts one year (Wanjala 1988; Crowe 2012) to 20 months (Wanjala 1988), depending on local climates. The borer continues to develop all year around and overwinters at the larval stage.

Cultural control appears to be the most effective measure for the control of this pest. At the first signs of infestation (branch tip wilting and flute holes at the top of the branches), the affected branches should be removed and destroyed before the larvae reach the trunk (Crowe 1963, 2012; Rutherford and Phiri 2006; Waller et al. 2007). According to Crowe (2012), reinfestations only occur in areas where there are numerous shrubs of the coffee family Rubiaceae growing near coffee plantations, and a campaign to eradicate these shrubs may be necessary if the pest is particularly troublesome. Piercing the larvae with wire is also suggested (Crowe 1963). Wanjala (1988) listed 12 species of parasitoids and predators that attack this pest, but the role these natural enemies may play in its control is not known. If the larvae reach the trunk, chemical control may be undertaken. For example, the lowest frass-ejection holes can be enlarged using a knife, and then a persistent insecticide can be injected with an oil can or other suitable applicator (Waller et al. 2007; Crowe 2012).

12.7.3 South America

12.7.3.1 *Steirastoma breve* (Sulzer)

Synonyms: *S. depressum* Hart, *Cerambyx depressus* Fabricius, *C. brevis* Sulzer
Common name: Cacao longicorn beetle

The lamiine adults are 12–30 mm long and stout; the body is reddish or blackish brown and covered with grayish pubescence; the pronotum has three longitudinal, raised ridges on the disc and five short, blunt tubercles at each side; each elytron has a longitudinal carina from the base to near the apex (Figure 12.61). The longicorn is widely distributed in Latin America from the West Indies and Guatemala to Argentina. Its larvae feed on about 25 tree species in Arecaceae, Bombacaceae, Cecropiaceae, Fabaceae, Lecythidaceae, Malvaceae, Myrtaceae, Salicaceae, and Sterculiaceae, with a preference for Malvaceae—the species to which cacao (*Theobroma cacao* L.) belongs (Duffy 1960; Monné 2001b, 2005b).

The cacao longicorn beetle has been recorded as the most serious cerambycid pest of cacao trees in the New World since the nineteenth century (Guppy 1911; Duffy 1960; Entwistle 1972; Sánchez and

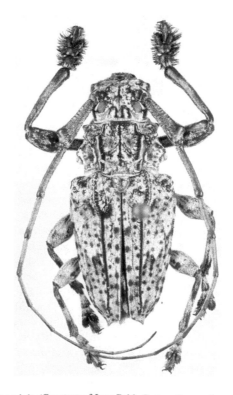

FIGURE 12.61   *Steirastoma breve* adult. (Courtesy of Juan Pablo Botero [image from Marcela L. Monné].)

Capriles 1979; Mendes and Garcia 1984; Liendo-Barandiaran et al. 2010). The adults feed on the thin bark and underlying soft tissues of young branches, leaving scarring that can be readily seen. However, the main damage is caused by larval tunneling in the branches and trunks. The larvae feed subcortically, making irregular spiral-like galleries and resulting in a ringed branch or trunk. Depending on the age and location of the damage, these events can kill the apical area. If the trunk is attacked, the entire plant can be killed. The larvae eject frass from oval holes. In dry weather, the ejected frass, together with gummy, gelatinous sap from the wound, can be found attached to the bark of trunks and branches, but this may be washed away by rain. When the larvae bore into the heartwood, they weaken the branches and trunks, making them liable to break off in windy conditions.

The life history is summarized from Guppy (1911), Duffy (1960), Mendes and García (1984), Kliejunas et al. (2001), and Liendo-Barandiaran et al. (2005, 2010). The adults are present all year round. The females make oviposition slits with their mandibles in the bark around the preferred root collar or lower trunk of the host trees and then lay their eggs singly in the slits. The lifetime fecundity in the field is unknown. The larvae bore into the bark and feed subcortically most of the time and then enter the wood. The adults are attracted to semiochemicals released from males and cacao trees. The number of generations this borer has depends on the local climate. For example, the species has one generation a year in southern Brazil but up to four generations a year in the tropical areas of northern Brazil. Overwintering, if any, takes place at the larval stage.

Preventative cultural control probably still is the most effective measure for this pest (Guppy 1911; Duffy 1960; Entwistle 1972). Traps composed of fresh branches (3–12 cm in diameter and 90–120 cm long) of *Pachira aquatica* Aublet, one of the borer's preferred hosts, are hung on cacao tree branches 150–250 cm above the ground to attract adults to lay eggs. The attracted adults can then be collected and destroyed. The traps should be replaced with newly cut branches every 2–3 weeks in the rainy season and every 10–12 days in the dry season. The used trap branches should be burned. Larvae found underneath tree bark can be cut out with a knife or killed with a piece of wire. Because unshaded plants are preferred for oviposition, adequate shade cover is important. Several parasitoid and predator species reportedly attack the cacao longicorn (reviewed by Duffy 1960), but their application for pest control has not been practiced in commercial cacao orchards. At present, it is not known whether the semiochemicals associated with this species can be used to control adults. Chemical control has achieved good results (Garcia et al. 1985). For example, endosulfan had been sprayed onto trunks and branches but that organochlorine insecticide is now banned. However, because the adults are active all year round in tropical regions, chemical sprays may provide only limited control.

12.7.3.2 *Plagiohammus maculosus* (Bates)

Synonym: *Hammoderus maculosus* Bates
Common name: Speckled coffee stem and root borer

The lamiine adults are 25–34 long; the body is reddish to blackish brown and covered with yellowish-green pubescence; the pronotal disc has a wide, longitudinal, yellowish-white stripe at each side; each elytron has six regular yellowish-white maculae that may be divided into eight or more smaller spots (Figure 12.62). The longicorn occurs from Mexico to Nicaragua (Monné 2005b) and feeds on Arabica coffee trees in Mexico, Guatemala, El Salvador, and Honduras (Barrera et al. 2004; Avila 2005; Barrera 2008).

The speckled coffee stem and root borer has been a moderate coffee pest since 1935 in Guatemala, with the infestation levels ranging between 5% and 25%. In Mexico, 34.8% of 23 sampled highland coffee plantations were infested, with levels ranging from 0.8% to 24.5%. Producers consider this borer one of the most important pests of coffee trees in Mexico (Constantino et al. 2014). According to Barrera (2008), the damage is caused by larval tunneling in the trunk and main roots. Borer attack reduces plant growth and eventually causes death either directly by root damage or indirectly by facilitating trunk breakage by wind. White to yellowish sawdust-like frass is present at the base of the infested coffee plants, which may become wilted, yellow, and decayed. The abundance of this longicorn is greater in high-altitude coffee plantations (>1,000 m above the sea level) and in places with long summers or low rainfall. Abandoned coffee plantations are more severely impacted.

FIGURE 12.62 *Plagiohammus maculosus* adult. (Courtesy of José Rafael Esteban [image from Luis M. Constantino].)

Barrera (2008) summarized the general biology of the coffee pests *P. maculosus, P. spinipennis*, and *P. mexicanus*. The adults are more visible at the beginning of the rainy season (April–June), which is the period when egg laying occurs. The females lay their eggs on the bark of the coffee trunks between the ground and 30 cm above. The lifetime fecundity is unknown. The larvae penetrate the trunk and bore all the way down to the roots. Pupation occurs in a chamber in the trunk near the ground where an adult exit opening is made. The larval period lasts from two to three years. Overwintering occurs at the larval and pupal stages.

Control measures (Espinosa 2000; Barrera 2008) are summarized as follows. The coffee plants with frass at the base of their trunks should be located. If damage is recent (the frass is white or pale yellow), the infested trunks should be removed. Weeds should be managed by shading, mulching, ground cover, and mechanical removal. In places where the pest appears yearly, a preventive insecticide applied with a brush or manual pump is recommended, treating from the trunk base up to 60 cm. The application may be repeated once or twice every 20 days. In order to kill the larvae within the trunk, a cotton ball soaked in an insecticide can be inserted through the respiration and excretion openings made by the larvae, or an insecticide solution can be injected into the openings with a syringe. With this manner of treatment, the orifice is enlarged, the insecticide applied, and the orifice sealed with mud or clay.

12.7.3.3 *Plagiohammus spinipennis* (Thomson)

Synonyms: *Hammoderus spinipennis* Thomson, *Hammatoderus jacoby-i* Nonfried
Common name: Lantana stem and root borer

This species is similar to *P. maculosus* but differs in that each elytron has three to four white maculae (Constantino et al. 2014) (Figure 12.63). It is distributed from Mexico to Panama and southward

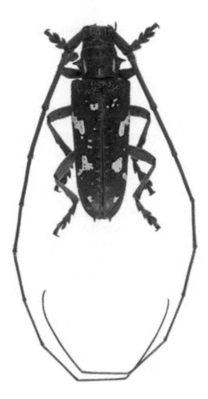

FIGURE 12.63 *Plagiohammus spinipennis* adult. (Courtesy of José Rafael Esteban [image from Luis M. Constantino].)

to Colombia, Venezuela, and Peru (Monné 2005b; Esteban-Durán et al. 2010; Constantino et al. 2014). The larvae feed on several plant species from Asteraceae, Lamiaceae, Rubiaceae, and Verbenaceae (Maes et al. 1994; Arguedas and Chaverri 1997; Barrera 2008). Because some of its preferred hosts, *Lantana* spp. (Verbenaceae), have become important weeds in Hawaii and Australia, this longicorn has been introduced to these two areas as a biological control agent for weed control (Harley 1969, 1973).

The lantana stem and root borer is a minor to moderate pest of coffee trees in the highland coffee growing region (≥1,000 m above sea level) of the Mexican Pacific Coast (Barrera et al. 2004; Barrera 2008). The damage symptoms, biology, and control measures are similar to those of *P. maculosus*. When attacking *Lantana*, the larvae girdle the trunks, causing a swollen area to form, and then bore downward into the trunks and roots, killing the plants. In *Tectona grandis* L.f., the larvae feed on the phloem, resulting in trunk bulges at the point of attack and giving rise to new buds below that point (Arguedas and Chaverri 1997).

12.7.3.4 *Plagiohammus mexicanus* Breuning

Common name: Mexican coffee stem and root borer

The adult body is about 25–30 mm long and blackish brown in color; the elytra have very dense white dots throughout. This longicorn is only recorded in Mexico (Monné 2005b; Barrera 2008). The preferred host is Arabica coffee (Barrera 2008).

The Mexican coffee stem and root borer is a minor pest of coffee trees in the highland coffee growing region (≥1,000 m above sea level) of the Mexican Pacific Coast (Barrera et al. 2004; Barrera 2008). The damage symptoms, biology, and control measures are similar to those of *P. maculosus* (Barrera 2008).

12.7.3.5 *Plagiohammus colombiensis* Constantino, Benavides, and Esteban

Common name: Colombian coffee stem and root borer

Adults are 25–26 mm long and grayish brown in color; the pronotal disc has dense golden yellow pubescence except for two longitudinal light brown lines; each elytron has six large golden yellow spots and 50–55 small and very small golden yellow spots scattered over the surface (Figure 12.64). It occurs in Colombia and attacks Arabica coffee (Constantino et al. 2014). The primary host plant of *P. colombiensis* is presumed to be a native forest species because the main borer attacks have occurred in Colombia in the coffee plantations that used to be forests. The destruction and removal of native forest species may be the main cause of adaptation of this borer to coffee plantations (Constantino et al. 2014).

According to Constantino et al. (2014) and Constantino and Benavides (2015), the Colombian coffee stem and root borer is an important pest of coffee plantations in that country. The adults emerge in May and lay their eggs on the bark of the trunk near the base. The lifetime fecundity is not known. The larva (Figure 12.65) bores into the central portion of the trunk and tunnels down toward the main roots (Figure 12.66). Affected trees are recognized by piles of white sawdust-like frass that have accumulated at the base of the trunk. When the larvae reach the main roots, they tunnel upward to a height of about 10–30 cm above the soil where pupation takes place in a chamber. The life cycle takes 18 months from egg to adult. The adults are nocturnal.

Constantino and Benavides (2015) recommended the following measures for the control of this pest: (1) inspection of trees—looking for sawdust at the base of the trunks; (2) removal of dead, withered, or unproductive trees with adult emergence holes (Figure 12.67), including stumps; (3) injection of organophosphorus or pyrethroid insecticide solutions into the base of infested but not dying trees through the larval holes, which are then immediately sealed with clay, mud, or wax.

FIGURE 12.64 *Plagiohammus colombiensis* adult. (Courtesy of Luis M. Constantino.)

FIGURE 12.65 *Plagiohammus colombiensis* larva. (Courtesy of Luis M. Constantino.)

FIGURE 12.66 Gallery made by a *Plagiohammus colombiensis* larva in a trunk. (Courtesy of Luis M. Constantino.)

FIGURE 12.67 *Plagiohammus colombiensis* adult emergence hole. (Courtesy of Luis M. Constantino.)

12.8 Cerambycid Pests in Vine Crops

In general, the term vine refers to any plant with a growth habit of trailing and climbing stems or runners. Many vine species currently are grown as crops in gardens and commercial farms or orchards around the world for fresh fruit, vegetables, and wine making. There is little information on cerambycid pests in kiwifruit and legume vines. Therefore, vine crops in this chapter refer to grape and cucurbit vines where cerambycid stem borers are recorded as pests.

A number of cerambycid species use these plants as hosts, and some may be pests. For example, about seven species are recorded as attacking grapevines (Mani et al. 2014). Most species listed currently are not considered economically important, such as *Chlorophorus varius* Müller in North Africa (El-Minshawy 1976), while some species not listed have become highly important pests of grapevines such as *Xylotrechus arvicola* (Olivier) in European vineyards. Some cerambycid species that also attack grapevines are considered more important pests of other crops, such as *Oemona hirta* in citrus and *Sthenias grisator* in mulberry trees, both of which are discussed in the earlier sections on citrus and mulberry crops, respectively. A South American species, *Adetus analis* (Haldeman), that was recorded as a minor pest of chayote vine (*Sechium edule*) in Brazil (Pigatti et al. 1979; de Souza Filho et al. 2001) is not significant enough to be treated in this chapter. In this section, five species are considered economic pests of vine crops throughout the world.

12.8.1 Asia

12.8.1.1 *Xylotrechus pyrrhoderus* Bates

Common name: Grape tiger longicorn

The cerambycine adults are 9–14 mm long; the prothorax is dark brown to orange, the elytra are black with thick yellowish to whitish bands on the basal and subapical areas; the basal band is connected

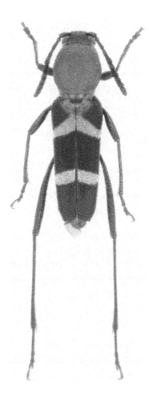

FIGURE 12.68 *Xylotrechus pyrrhoderus* adult. (Courtesy of Junsuke Yamasako.)

by a sharp angle with the subbasal band along the elytral suture; the subapical area has a transverse whitish pubescent band (Figure 12.68). This longicorn is distributed in East Asia including Japan, Korea, and China (Hua 2002; Han and Lyu 2010). To date, there is no record of its occurrence in other regions. The main host plants are from the family Vitaceae, including *Vitis vinifera* L. and *Ampelopsis brevipedunculata* (Maxim.) (Han and Lyu 2010).

The grape tiger longicorn is considered a moderate to serious pest of grapevines in Japan (Matsumoto 1920; Ashihara 1982; Sakai et al. 1982; Kiyota et al. 2009), Korea (Kim et al. 1991), and China (Miao 1994; Guo 1999; Li 2001; Huang and Yang 2002). The newly hatched larvae bore into the young stems and bud base and make tunnels transversely inside. Frass is packed in the tunnels and not ejected, making detection of larvae difficult. The infested areas turn dark or black late in the season. The infested stems may become wilted and easily broken by wind after the larvae enter the heartwood. In some Chinese vineyards, between 20% and 90% of grape plants are infested by this pest, depending on management level and plant age, with older and poorly managed plants suffering more severe damage than well-managed and younger plants (Huang and Yang 2002). However, according to Kim et al. (1991), the injury level caused by the larvae is higher in vigorous shoots than in weak ones. Most larval damage occurs during May–June. The adults feed on young shoots, buds, and leaves, but their damage is not significant (Guo 1999).

Adult emergence occurs between July and October, with a peak in August (Matsumoto 1920; Ashihara 1982; Kim et al. 1991; Huang and Yang 2002). The adults live for 7–40 days and, between August and early November, lay eggs on the buds and on the young shoots that are 8–10 mm in diameter. Each female lays 17–90 eggs in her lifetime; these hatch in a week or two depending on temperature. The larvae stop developing when they reach about 4 mm in length (usually in late November) and then overwinter. The larvae become active again in the spring. Pupation occurs in the stems during July–August, and the pupal stage lasts 10–14 days; the newly eclosed adults remain in the pupal cells for 9–12 days before chewing an exit hole and emerging. The life cycle lasts one year.

The most common management measures for this pest are cultural and chemical control (Kim et al. 1991; Miao 1994; Guo 1999; Huang and Yang 2002). The main cultural control method is pruning infested branches/stems after the harvest season and burning pruned materials before June. Buds are carefully examined in summer, and any larvae found are killed using a knife. Spraying insecticides during August–October can be effective in killing adults and newly hatched larvae. In addition, a parasitoid wasp, *Odontobracon bicolor* Ashmead, has been identified as one of the natural enemies of *X. pyrrhoderus*, and its emergence period is from late July to mid-August, which is earlier than that of the grape tiger longicorn (Kim et al. 1991). However, the effectiveness of this parasitoid in the control of the grape tiger longicorn is unknown.

12.8.1.2 *Celosterna scabrator* (Fabricius)

Synonyms: *C. scabrator griseator* Aurivillius, *Aristobia murina* Nonfried, *Psaromaia renei* Pascoe,
       Lamia gladiator Fabricius, *L. spinator* Fabricius, *L. scabrator* Fabricius
Common name: Grape stem borer

The adult of this lamiine borer is 30–40 mm long and dull, yellowish brown in color; the sides of the body and legs are bluish; the elytra are yellowish gray with a large number of black spots varying in size from pinheads to minute specks (Figure 12.69). This south Asian species occurs in India, Pakistan, Sri Lanka, and Nepal (Beeson, 1941; Duffy 1968). It has been introduced into Madagascar and Réunion. The borer attacks a wide range of host plants, including genera *Eucalyptus, Acacia, Cassia, Casuarina, Pithecellobium, Prosopis, Tectona, Shorea, Pyrus, Mangifera, Zizyphus* (Beeson 1931, 1941; Chatterjee and Singh 1968; Duffy 1968), and *Vitis* (Upasani and Phadnis 1968).

This species has been recognized as a pest of forest trees; the eggs are laid on young living plants, and the larvae bore into the trunks and the roots, which are hollowed out. The infested tree stops growing and frequently dies (Beeson 1931, 1941; Chatterjee and Singh 1968; Duffy 1968). This borer generally attacks living trees, but its larvae can develop inside cut and stacked timber. The larval stage lasts about

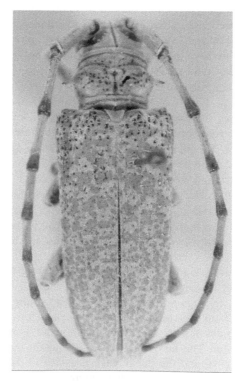

FIGURE 12.69   *Celosterna scabrator* adult. (Courtesy of Larry Bezark.)

nine months. According to Chatterjee and Singh (1968), the adults feed on the bark of young eucalypt trees, causing death or breakage of the shoots, lay eggs under the bark of one- to four-year-old trees, and the larvae tunnel downward in the pith to the roots, causing cessation of growth or death. This species is also reported to attack citrus, apple, and mango trees, but its significance for these tree crops is not known (Mani et al. 2014).

Since the late 1960s, this borer has become a moderate to serious pest of grapevines in India (Upasani and Phadnis 1968; Azam 1979; Ranga Rao et al. 1979; Jagginavar et al. 2006, Mani et al. 2014). For example, up to 30% damage was reported in a vineyard in Karnataka, India (Jagginavar et al. 2006). The larvae bore into the vines, tunneling upward and downward until maturation and pupation. The damage caused by this borer to grapevines comprises holes in the main stem and branches, discoloration, withering and premature leaf drop, shriveling of the grape clusters before maturity, and injuries to the vascular bundles, eventually resulting in the withering of the vines above the site of attack (Upasani and Phadnis 1968; Mani et al. 2014). Previously, the borer was considered problematic only in old and neglected vineyards but, in recent years, severe damage has also occurred in one-year-old vineyards (Mani et al. 2014). The adults may also cause some damage to the tender shoots by scraping. Extrusion of frass from holes on the trunk and branches is a common sign with sawdust-like frass on the ground. Gummosis (oozing of a resinous substance from the hole) is also observed on damaged areas.

The adults emerge from early July to late October, with a peak between mid-July and late August (Beeson 1931, 1941). According to Mani et al. (2014), the females make slits on the bark of trunks and branches of vines, lay their eggs singly in the slits, and cover them with a gummy substance. The eggs hatch in 10 days. A female can lay 12–15 eggs in her lifetime, but this fecundity could be underestimated. The larval period lasts six to eight months and overwintering takes place during the larval stage. The mature larvae remain in the vine until May when pupation occurs. The pupal period is 25–35 days, and adults live for 20–25 days. A complete life cycle lasts from 10 months (Mani et al. 2014) to 1 year (Beeson 1941).

Chemical control is by far the most commonly used measure for the management of this pest (Azam 1979; Ranga Rao et al. 1979; Gandhale et al. 1983; Chandrasekaran et al. 1990; Jagginavar et al. 2008). Spraying insecticides on the crops during July–August can be effective in the control of adults (Azam 1979; Ranga Rao et al. 1979). The application of insecticides by immersing one or two roots per plant can control larvae (Chandrasekaran et al. 1990). Furthermore, Jagginavar et al. (2008) reported that injection of insecticides into stems results in more than 90% larval mortality. During August–September, the trunks and branches should be carefully examined, and the eggs and young larvae in the oviposition slits in the bark can be mechanically removed using a knife. This cultural measure can be used to effectively control this pest (Azam 1979; Ranga Rao et al. 1979).

12.8.1.3 *Apomecyna saltator* (Fabricius)

Synonyms: *A. subuniformis* Pic, *A. excavaticeps* Pic, *A. multinotata* Pic, *A. niveosparsa tonkinea* Pic, *A. pertigera* Thomson, *A. neglecta* Pascoe, *Lamia saltator* Fabricius
Common name: Cucurbit vine borer

The lamiine beetle is 10–12 mm long and slate-gray in color, with several white spots arranged in three evenly shaped "V" markings across the elytra (Figure 12.70). It is widely distributed in tropical and subtropical China (Qian 1985; Li et al. 1997), India, Bangladesh, and Sri Lanka (Duffy 1968). This species also occurs in Hawaii (Tavakilian and Chevillotte 2013), where it probably was introduced. The borer attacks a range of species in the family Cucurbitaceae (Khan 2012). The record of this species in Australia (May 1946) is likely another *Apomecyna* species, *A. histrio* (F.), which also infests cucurbits but is not considered an economic pest.

The cucurbit vine borer is a minor to moderate pest infesting vines of gourds, cucumbers, pumpkins, squashes, and watermelons in China (Qian 1985; Li et al. 1997) and India (Shah and Vora 1973, 1974; Samalo 1985; Sontakke and Satpathy 1993; Singh et al. 2008; Khan 2012). According to Singh et al. (2008), neonate larvae bore into the vines at the nodal joints and damage nodal regions; late instars enter the main vines, causing swelling and oozing of sap. The tunnels in the stems are filled with glutinous waste material. Young plants die due to severe infestation.

FIGURE 12.70   *Apomecyna saltator* adult. (Courtesy of Larry Bezark.)

The adults emerge two to three times in a year, depending on climate. The females make oviposition slits on the internodes into which they lay eggs (Shah and Vora 1974; Sontakke 2002). A female lays 38–52 eggs in her lifetime. The eggs are laid singly and hatch in five to eight days. There are six distinct larval instars with a total larval period of 31–42 days. The pupal period is 8–10 days with 2–3 days as prepupae. The adults live for 35–44 days. The total life cycle of this pest ranges from 80–98 days (Shah and Vora 1974) to 85–107 days (Sontakke 2002). Overwintering occurs during the larval stage. There are two to three generations a year.

Chemical control measures have been evaluated in India. For example, insecticide sprays can effectively control this pest (Samalo 1985; Sontakke and Satpathy1993), but granular or dust insecticides applied to the soil can only achieve 60–76% control (Samalo 1985). Singh et al. (2008) tested the potential resistance of various pointed gourd cultivars against the pest but found that none of them is sufficiently resistant. Both Samalo (1985) and Qian (1985) recommended that seedlings be carefully examined, infested stems removed, and postharvest residues destroyed.

12.8.2 Australasia

12.8.2.1 *Acalolepta mixta* (Hope)

Synonyms: *A. vastator* (Newman), *Monohammus vastator* Newman, *M. mixtus* Hope
Common name: Passion vine longicorn

The adults of this lamiine longicorn are about 30 mm long and speckled gray-brown; the prothorax has a prominent spine on each side, and the antennae are about twice as long as the body (Figure 12.71). It is widely distributed in mainland Australia, Lord Howe Island, Samoa (Duffy 1963; Slipinski and Escalona 2013), and Solomon Islands (Vitali and Casadio 2007; Chin and Brown 2008). A previous distributional record in the Philippines (Duffy 1963) is questionable. This borer has a wide range of host plants, including the genera *Mangifera* (mango) (Smith 1996), *Citrus* (Smith et al. 1997), *Agathis, Araucaria,*

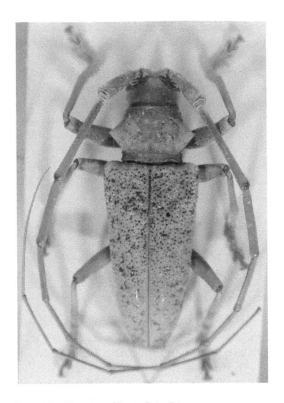

FIGURE 12.71   *Acalolepta mixta* adult. (Courtesy of Larry Bezark.)

Cassinia, Cymbidium, Diospyros, Ficus (fig), *Flindersia, Laportea, Ozothamnus, Passiflora* (passion vine), *Pinus, Vitis* (grapevine), *Toona,* and *Wisteria* with preferred hosts being fig trees, passion vines, and grapevines (Slipinski and Escalona 2013).

Acalolepta mixta is a serious pest of grapevines in southeastern Australia (Dunn and Zurbo 2014), infesting up to 70% of vines in some vineyards in New South Wales (Goodwin et al. 1994). It is also considered an important pest of fig trees and passion vines (Duffy 1963) and a minor pest of mango (Smith 1996) and citrus trees (Smith et al. 1997) in Australia. In vineyards (Dunn and Zurbo 2014), the larvae bore into the vine wood and tunnel throughout the trunk and into the roots. Sawdust-like frass is often visible in tunnels and around the infested vine trunk. Extensive damage to the vine trunk can lead to dieback and significant crop losses. Damage symptoms in passion vines are similar to those of grapevines (Duffy 1963). In the case of fig trees, the larvae penetrate the bark and then the sapwood; circular pieces of the bark die and fall out, leaving round pits and exposing the sapwood (Duffy 1963).

Goodwin and Pettit (1994) provided a detailed description of the life history of this pest in vineyards. The adults emerge between October and March with an emergence peak in January–February and an oviposition peak in January–March. The adults can live for up to six months. The females prefer to lay their eggs singly on the vine trunks and the base of young vine canes on the main or secondary arms. On average, a female lays 2.25 eggs per vine but only 0.8 larvae per vine survive to maturation. However, the total number of eggs a female can lay in her lifetime is unknown. The larvae develop under the surface of the bark during the first four instars and then bore into the wood in the later five instars. About 75% of mature larvae occur in the trunks, with more than 60% of all larvae completing their development in the area proximal to the fork in the grapevine. They complete their development in about 38 weeks. The pupal developmental period is about 20 days. The winter is passed during the mature larval and pupal stages. This borer has one generation a year. The adults can disperse up to 150 m by flight and by walking along the vine rows (rather than randomly) in the vineyards (Goodwin et al. 1994).

No parasites or predators have been recorded for this pest, but *Beauvaria bassiana* infects a small number of larvae during wet conditions in the field (Goodwin and Pettit 1994). Several control measures used on fig trees and passion vines are summarized by Duffy (1963). For example, the infested branches are cut off and burned, infested or loose bark is scraped away and the surface painted with bluestone paint or a lime/sulphur wash, and when necessary, chemical pesticides are injected into the vines. In vineyards, careful pruning and destruction of the infested vines may remove many larvae, but retraining of vines may be necessary following the pruning of those vines with serious infestations (Dunn and Zurbo 2014). Seasonal spraying in vineyards of organophosphate insecticides at two- to four-week intervals to control adults and young larvae is also widely practiced, but insecticide residues detected in grapes and wines are a concern (Goodwin and Ahmad 1998). Furthermore, the immediate reduction in the pest population made by a seasonal spraying scheme is followed by pest resurgence (Goodwin 2005a). To reduce the number of sprays and increase control effectiveness, Goodwin (2005b) recommended a single application of insecticide at dormancy. Two insecticides, fipronil (a phenylpyrazole) and bifenthrin (a pyrethroid) are registered for this purpose.

12.8.3 Europe

12.8.3.1 *Xylotrechus arvicola* (Olivier)

Synonyms: *Clytus heydeni* Stierlin, *C. arvicola* (Olivier), and *Callidium arvicola* Olivier
Common name: European grapevine borer

The adults of this cerambycine borer are 8–20 mm long and reddish brown to dark brown in color; the antennae and legs are reddish yellow, and the forehead bears distinct yellowish pubescence; the pronotum has yellow frontal and basal angles; the elytra are reddish brown with yellow transverse stripes (Figure 12.72). It is widely distributed in the Holomediterranean region including Europe, North Africa, and West and Central Asia (Bense 1995; Ocete et al. 2002b; Tavakilian and Chevillotte 2013; Ilic and Curcic 2013). This species is polyphagous, with its larvae feeding on at least 15 broadleaf plant genera

FIGURE 12.72   *Xylotrechus arvicola* adult. (Courtesy of Boženka Hric [image from Srećko Ćurčić].)

in Betulaceae, Fagaceae, Malvaceae, Moraceae, Rosaceae, Ulmaceae, Salicaceae, and Vitaceae (Bense 1995; Ocete et al. 2002a, 2002b, 2004, 2008, 2010; Soria et al. 2013).

This species was considered a minor pest of various tree species. However, it has become an increasingly serious pest of grapevines in Spain since the 1990s (Ocete et al. 2002a, 2002b, 2004, 2008, 2010; Soria et al. 2013). For example, in the vineyards of northern and central Spain, the infestation rate increased from 51% in 2004 to 96% in 2008; the vines started to die in 2005; and the death rate increased to 17% in 2008 (Ocete et al. 2010). The continuous use of nonselective chemical controls and the consequent changes in entomofauna could explain the origin and increase of this pest in Spanish vineyards (Soria et al. 2013). Garcia Calleja (2004a, 2004b) detailed the significant increase in damage and economic loss caused by this borer. Ocete et al. (2008) reported that, due to the damage by this borer, the inflorescence size and number of flowers are considerably reduced and that, during harvest, the grape clusters are smaller and looser, weighing on average five times less than those collected from the sound branches. Moreover, the wine made from the grapes on the infested branches has a significantly lower alcoholic percentage and a higher organic acid concentration.

The larvae bore into the vine branches, trunks, and roots. They excavate large and numerous galleries inside the wood, and infestations are difficult to detect and control (Ocete et al. 2008). Furthermore, Ocete et al. (2002a) indicated that the borer may be associated with pathogenic fungal species affecting the vine wood. More recent studies suggest that this borer, in addition to causing a progressive decay of the branches, could facilitate the transmission of plant fungus diseases—further threatening the vineyards by spreading grapevine decline pathogens (Benavides et al. 2013; Soria et al. 2013). The destruction by the borer and then the fungal diseases progressively weaken the affected vine as indicated by drastic reduction in yield and premature death. During the dying period, the shoots are without vigor; they look quite similar to those infected by the Eutypa dieback complex (Ocete et al. 2002b).

The adults emerge during May–August (Bense 1995; Benavides et al. 2013), and the females lay most of their eggs in the first two weeks after emergence (García-Ruiz et al. 2012). The female that emerges from grape wood collected from the field lays about 196 eggs in her lifetime (García-Ruiz et al. 2012). In the field, the life cycle lasts two years and the winter is passed at the larval stage (Bense 1995). However, in the laboratory, the developmental period from larvae to adults ranges from 269 to 321 days, depending on diet (García-Ruiz et al. 2012). The optimal temperature for the development of this borer is estimated to be from 31.7°C to 32.9°C (García-Ruiz et al. 2011).

So far, no effective control measures have been developed. Ocete et al. (2002b) suggested that to eliminate the sources of the infestation and infection, it is necessary to cut back the trunk of vines to enable regrowth and to cover the wounds with sealing paste. In vineyards, it is also recommended that this borer should be managed using an integrated pest management program (Ocete et al. 2008). For example, Soria et al. (2013) suggested that integrated pest management measures should be performed in mid-June and the end of July to reduce adult populations. Urgent investigations into the development of cultural and chemical control measures are warranted.

12.9 Cerambycid Pests in Field Crops

Field crops refer to any of the herbaceous plants grown on a large scale in cultivated fields—primarily grain, forage, sugar, oil, or fiber crops. Field crops are of ultimate importance for human survival. Future trajectories of food prices, food security, and crop land expansion are closely linked to future average crop yields in the major agricultural regions of the world.

Hundreds of insect species damage these crops, causing significant loss in yield and income and threatening food security. However, cerambycids that damage field crops are relatively rare. To date, only about 17 cerambycid species have been reported to be of economic importance in sunflower, soybean, sugarcane, and pasture and grain crops throughout the world. Among these species, *Oxymerus aculeatus* Dupont adults damage corn plants by feeding on the male flowers in Brazil, reducing seed production (Pires et al. 2011). However, this species is not considered further in this chapter because little is known about its larval biology. *Rhytiphora diva* (Thomson) [= *Zygrita diva* (Thomson)] and *R. stigmatica* (Pascoe) [= *Corrhenes stigmatica* (Pascoe)] (Slipinski and Escalona 2013) attack legume crops including soybean and lucerne (alfalfa) in northern Australia, but *R. diva* is far more common than *R. stigmatica* (Anonymous 2010). Therefore, only *R. diva* is discussed here. In North America, *Dectes texanus* LeConte, *Ataxia hubbardi* Fisher, and *Saperda inornata* Say were recorded as pests of sunflowers by Rogers (1977b). However, the latter two species are no longer considered pests of economic importance to field crops today. Therefore, only *D. texanus*, which is also a major pest of soybeans, is detailed in this chapter.

In this section, 13 cerambycid species that infest bean, sunflower, sugarcane, rice, sorghum, and pasture crops are discussed, including 1 in North America, 3 in Europe, and 9 in Asia. There is no record of any cerambycid that causes economic damage to wheat, tomato, and potato crops.

12.9.1 Asia

12.9.1.1 Obereopsis brevis (Gahan)

Synonyms: *Obereopsis subbrevis* Breuning, *Oberea brevis* Gahan
Common name: Girdle beetle

The girdle beetle is a lamiine widely distributed in India and Myanmar. The body is 9–9.5 mm long; the head, prothorax, and basal one-half to four-fifths elytra are yellowish; the antennae and apical one-half to one-fifth elytra are dark blue to black (Figure 12.73). This is a highly polyphagous species, feeding on numerous herbaceous species from Fabaceae (or Leguminosae), Asteraceae (or Compositae), Euphorbiaceae, Phyllanthaceae, and Malvaceae (Shrivastava et al. 1989).

Quite some time ago, this beetle reportedly attacked soybeans in India but was not a concern until the late 1960s and early 1970s, when serious damage to soybeans was reported (Srivastava et al. 1972; Gangrade and Singh 1975). Since then, it has become an increasingly serious pest of soybeans in India, causing substantial losses (Singh et al. 1976; Singh and Dhamdhere 1983; Kundu and Trimohan 1986;

FIGURE 12.73 *Obereopsis brevis* adult. (Courtesy of Prakash Kumar, CAD Chambal Agriculture Research Station, NBAIR, used with permission of NBAIR director.)

Rai and Patel 1990; Singh and Singh 1996; Upadhyay et al. 1999; Kumar et al. 2012; Netam et al. 2013; Jayanti et al. 2014; Sharma et al. 2014). It is also a moderate pest of garden beans *Lablab purpureus* (L.) Sweet, cowpeas *Vigna unguiculata* (L.) Walp. (Singh and Singh 1966), and rice beans *V. umbellata* (Thunb.) Ohwi & H. Ohashi (Verma and Singh 1991). The damage to these legume crops is caused by both adults and larvae. Adult females girdle the stems, petioles, petiolets, and branches before laying eggs. The parts above the girdle wilt and die. The larvae tunnel upward and downward in the stem, which weakens or even kills the whole plant. According to Jayanti et al. (2014), the girdle formation by the adult females results in a significant reduction in the number of pods and seeds and in seed weight. Losses caused by this pest are also associated with crop stage and different infestation levels. For example, up to a 10% infestation level does not reduce yield at all crop stages, but between 37 and 44 days after germination, the crop appears to be most vulnerable to infestation. Severe yield reduction occurs in those plants infested in the last week of August (40.0%) followed by the first and second weeks of September (37.6% and 32.3%, respectively). No further yield reduction takes place in the plants infested during the fourth week of September and the first week of October. Singh and Singh (1996) have examined the plant parts the beetle prefers to attack and the relationship between the parts attacked and yield loss. They note that the beetle inflicts maximum girdle formation on the petiole (43.1%), followed by the branch (22.6%), stem (21.5%), and petiolet (12.8%). Plants with girdles on the main stem suffer 15.74% premature mortality and the highest grain weight loss (47.2%), followed by girdles on the petiole, branch, and petiolet, which account for a grain weight loss of 40.3%, 39.9%, and 30.2%, respectively.

Furthermore, *O. brevis* has expanded its host range to attack sunflowers, with the infestation rate ranging from 60% (Pachori and Sharma 1997) to about 95% (Veda and Shaw 1994) in different regions of India, causing significant losses. The damage to sunflowers is made by girdling and tunneling the stems, damaged plants wilt, turn black, and look sickly (Veda and Shaw 1994).

The adults emerge twice a year, once between late July and early September and again between early September and mid-October, with an activity peak occurring in late August (Netam et al. 2013). In an earlier study, Rai and Patel (1990) examined the incidence of *O. brevis* in soybean plots sown on July 15, 1988, in India, and showed that the pest first appeared on August 10, and activity continued until October 12. A female girdles the plant at two levels, cuts a slit in between using her mandibles, and lays an egg in the slit (Dutt and Pal 1982; Pal 1983; Gautam 1988). A female produces 31–60 eggs but appears to lay only 7–13 eggs in her lifetime (Singh and Singh 1966). On soybean plants in the laboratory, the egg stage lasts 3–4 days, the larval stage 34–47 days, and the pupal stage 8–11 days (Gangrade et al. 1971). The developmental periods for all stages may be longer in the field, especially in the fall when some of the larvae enter hibernation. In another laboratory study (Dutt and Pal 1984), the egg stage is 3–4 days and the larval stage 24–44 days at 29–30°C. The beetle has two generations a year, and the larvae of both generations can enter diapause (Gangrade and Singh 1975). According to Singh and Singh (1966), the pest overwinters as mature and diapausing larvae in the feeding tunnel; the larvae become active again after the first rains in late June or early July and pupate soon afterward; the pupal stage lasts up to a month, and the adults emerge in time to breed on the monsoon crop. Dutt and Pal (1988) reported that the termination of larval diapause and emergence of adults can be forecasted on the basis of premonsoon precipitation and temperature. For example, exposure of larvae to 90–100% relative humidity and 30°C in the laboratory can terminate diapause in the following summer but not in the preceding winter. A premonsoon precipitation of 200 mm in the rainy season but not in the winter at 30°C secures the release of larvae from diapause and leads to adult emergence.

Because of its economic significance, a lot of effort has been made to control this pest, and many measures have been developed in the past 50 years. To date, the use of insecticides appears to be the most effective control measure for this pest (Chaudhary and Tripathi 2011). The classes of insecticides used for the control of this pest range from organochloride, organophosphate, organothiophosphate, organophosphorus, carbamate, benzamide, and pyrethroid to anthranilic diamide. Some of these insecticides have now been banned for use in agriculture (such as organochlorides). These insecticides can be sprayed once at the crop age of 30–45 days, depending on the infestation rate (Singh and Singh 1966; Singh 1986; Pande et al. 2000; Upadhyay and Sharma 2000; Yadav et al. 2001). In general, the spray should be made within 7–10 days of the presence of girdling symptoms (Sharma et al. 2014). If the infestation continues two weeks after the spray, a second spray is recommended (Parsai et al. 1990;

Purwar and Yadav 2004; Gupta 2008; Chaudhary and Tripathi 2011; Jain and Sharma 2011; Sharma et al. 2014). If the infestation is observed 75–80 days after sowing, spraying will not improve yield (Sharma et al. 2014). Organophosphate (Singh 1986; Singh and Singh 1989; Patil et al. 2002; Dahiphale et al. 2007) or carbamate (Singh 1986) insecticides in granular formulations can be used to treat the soil at the time of sowing or about 25 days afterward.

Crop resistance against the beetle has been investigated over the past 30 years, with attempts made to find highly resistant cultivars and to reduce insecticide applications (Bhattacharjee and Goswami 1984; Sharma 1994, 1995; Gupta et al. 1995; Salunke et al. 2002; Gupta et al. 2004; Manoj et al. 2005; Patil et al. 2006; Sinha and Netam 2013; Sharma et al. 2014). However, these attempts have only achieved limited success. Further investigations are warranted.

Several cultural control measures have been practiced and been demonstrated to be effective to a large extent. For example, soybeans should not be sown before the arrival of the monsoon period (July). Research shows that sowing in July results in the lowest infestation rate and the highest yield (Singh and Gangrade 1977; Singh and Dhamdhere 1983; Parsai and Shrivastava 1993; Meena and Sharma 2006). Soybean fields should be closely monitored, and the infested parts of the plants that have wilted as a result of attack should be collected at least once every 10 days and destroyed to get rid of the eggs and young larvae (Singh and Singh 1966; Singh and Dhamdhere 1983; Sharma et al. 2014). Intercropping soybean with maize or sorghum should be avoided because such an arrangement increases the infestation of soybeans by the beetle (Singh et al. 1990; Sharma et al. 2014). Sharma et al. (2014) suggested the use of *Dhaincha* as the trap crop for the girdle beetle, but the effectiveness of this measure is not known. Finally, deep summer plowing and crop rotation may also reduce the pest population.

Attempts have been made to explore biological control of this pest. Initially, a hymenopteran parasitoid (*Dinarmus* sp.) was found to parasitize *O. brevis* larvae (Singh and Singh 1966; Gangrade et al. 1971). However, no follow-up work has been undertaken, probably due to the low parasitism rate. Also used for the control of several other cerambycid pests mentioned earlier, the fungus pesticide *Beauveria bassiana* has been sprayed twice at 35 and 64 days after sowing, achieving only moderate success (Purwar and Yadav 2004). In an experiment where cultural and biological control measures were integrated for the management of this beetle and a lepidopteran pest on soybeans, Chaudhary et al. (2012) reported that the program involving removal of infested plant parts plus release of *Trichogramma* sp. at 30 days after sowing plus *Bacillus thuringiensis* (Bt) sprays at 50 days after sowing was highly effective.

Despite these efforts, the pest is still causing significant losses. This might well be due to the overuse of pesticides. More research is badly needed.

12.9.1.2 *Nupserha nitidior* Pic

Synonyms: *N. malabarensis* var. *nitidior* Pic, *N. nitidior* m. *atripennis* Breuning, *N. nitidior*
          m. *atroreductipennis* Breuning
Common name: Soybean girdle beetle

The lamiine adults are 6–7 mm long and dark yellowish in color without any specific patterns. This species appears to occur only in India. The host range is not very clear but appears to include herbaceous species from Asteraceae and Fabaceae (Duffy 1968; Anonymous 1973; Kashyap et al. 1978).

The soybean girdle beetle is considered a minor pest of soybeans in India (Thakur and Bhalla 1980; Chandel et al. 1995). Similar to *O. brevis*, this species damages the crops by adult girdling of the stems and larval tunneling in the stems and roots, causing similar symptoms (Anonymous 1973).

The adults emerge in July and August, girdle plants, and lay a single egg in an oviposition slit. The larvae bore into the stem and roots and enter diapause in the roots or soil between September and November, depending on climate. Diapaused larvae overwinter. This species has one generation a year.

Control measures are similar to those developed for *O. brevis*, including cultural and chemical control and use of resistant cultivars (Anonymous 1973; Kashyap et al. 1978; Thakur and Bhalla 1982; Chandel et al. 1995).

12.9.1.3 *Nupserha vexator* (Pascoe)

Synonym: *Glenea vexator* Pascoe
Common name: Sunflower stem borer

Because the identification of this lamiine species is not 100% certain, (Patil and Jadhav (2009) and Patil et al. (2009) who reported the occurrence of the damage tentatively used the name *Nupserha* sp. near *vexator* (Pascoe) (Figure 12.74). *Nupserha vexator* is distributed in India, Sri Lanka, and Myanmar. The species, formerly referred to as *N. antennata* Gahan (Duffy 1968; Slipinski and Escalona 2013), was introduced to Australia to control the noxious weed *Xanthium pungens* Wallr. in Queensland (Wapshere 1974) but apparently did not become established (Slipinski and Escalona 2013). Known hosts include species of Asteraceae.

In the early 1990s, the beetle was first noticed to damage sunflowers in central-western India. Since 1998, it has become a moderate to serious pest of sunflower crops in this region (Patil and Jadhav 2009; Patil et al. 2009). The infestation rate ranges from 10% to 70%, depending on year and sampling area, causing up to 17–31% yield loss (Patil et al. 2009). Patil et al. (2009) also reported that the infestation rate in July-sown sunflower crops is twice as high as that in August-sown sunflowers. The damage is caused by larval tunneling in the stem. The infested plants break easily, look sickly, and produce few or no seeds.

The information on biology mainly is extracted from Duffy (1968), Wapshere (1974), and Patil et al. (2009). The adults emerge during the first monsoon rains in June–July and lay their eggs singly in leaf axils. The adult males live for <1 month while females survive for up to 2.5 months. A female lays 30–260 eggs in her lifetime. The larvae enter the stem 7.5–9.0 cm above the root collar region and bore upward through the internal pith tissue. The first five larval instars last about 1.5 months, and the final two instars require several months. Before becoming mature, the larvae move downward, make an exit hole near the ground, and enter the soil for pupation. Some mature larvae remain in the stem and pupate there. Overwintering occurs at the mature larval stage. There is one generation a year.

FIGURE 12.74 *Nupserha vexator* adult. (Courtesy of Larry Bezark.)

Measures for the control of this pest are still being developed. The screening of cultivars resistant to the pest is promising. For example, Patil and Jadhav (2009) reported that the stem borer infestation rates on sunflowers varied from 0% to >80% in various germplasms under investigation over a five-year period, suggesting that it is possible to select for resistant cultivars for the effective control of this borer. Three chemical insecticides have been tested for control of the exposed stages of the pest—the eggs, newly hatched larvae, and the adults (Patil et al. 2009). Results show that spraying an organothiophosphate insecticide 40 days after the adult emergence achieves effective and economic control of the borer, followed by the organophosphate and organochlorine insecticides. However, organochlorines are being phased out globally.

12.9.1.4 *Dorysthenes granulosus* (Thomson)

Synonyms: *Paraphrus granulosus* (Thomson) and *Cyrthognathus granulosus* Thomson
Common name: Sugarcane root and stem longicorn

The prionine adults are 25–62 mm long; the body is reddish brown with a blackish-brown head and antennal segments 1–3, and each side of pronotum has three saw-like processes (Figure 12.75). It resembles *D. buqueti* (Guérin-Méneville) (see the next entry) but differs in having a shorter distance between the antennal tubercules; the lateral processes (spines) of pronotum are longer and sharper, and the lower side of antennal segments 3–5 have numerous granular processes. It is widely distributed in southern and southeastern Asia including southern China, Thailand, Vietnam, Laos, Myanmar, and India (Duffy 1968; Pu 1980; Li et al. 1997; Hua 2002; Thongphak and Hanlaoedrit 2012). *D. granulosus* and *D. buqueti* are often confused in reports from Thailand where both species occur (Thongphak and Hanlaoedrit 2012; D. Thongphak 2014, personal communication; W. Lu 2014, personal communication). According to Thongphak (2014, per. comm.), *D. granulosus* is a more abundant and more important pest of sugarcane than *D. buqueti* in northern Thailand. Based on photos in Thongphak and Hanlaoedrit (2012),

FIGURE 12.75 *Dorysthenes granulosus* adult. (Courtesy of Nathan P. Lord, USDA APHIS ITP, Bugwood.org [5491724].)

Anonymous (Fig. 5 beetle on the left, 2014), and personal communications with Thongphak and Lu in 2015, the Thai species mentioned in Sommartya and Suasa-ard (2006), Pliansinchai et al. (2007), Sommartya et al. (2007), and Suasa-ard et al. (2008, 2012) should be *D. granulosus* rather than *D. buqueti*. The host range of *D. granulosus* appears to be very wide, including numerous species from Poaceae, Sapindaceae, Anacardiaceae, Euphorbiaceae, Rutaceae, Myrtaceae, Fagaceae, Pinaceae, Arecaceae, and Meliaceae.

Since the 1970s, *D. granulosus* has become an increasingly more serious pest of sugarcane in southern China, particularly in Guangxi, Guangdong, Hainan, and Yunnan (Plant Protection Group of Guangxi Sugarcane Research Institute 1975; Zeng and Huang 1981; Yang et al. 1989; Liao et al. 2006; Yu et al. 2006, 2007, 2008, 2009, 2010; Long and Wei 2007; Gong et al. 2008; Xiong et al. 2010), and in northern Thailand (Sommartya and Suasa-ard 2006; Pliansinchai et al. 2007; Sommartya et al. 2007; Suasa-ard et al. 2008, 2012; Thongphak and Hanlaoedrit 2012). In the late 1970s, the infestation rate reached 5–40% in southern Guangxi, Hainan, and Yunnan (Zeng and Huang 1981). In the past decade, frequent outbreaks have occurred in the major sugarcane growing regions of Guangxi, with infestation rates of 15% to more than 60% (Liao et al. 2006; Yu et al. 2006, 2007; Long and Wei 2007; Gong et al. 2008), causing 10–50% yield loss (Liao et al. 2006). Young larvae feed on tender roots and then bore into main roots, tunnel upward in the stems up to 30–60 cm above the ground, and hollow out the stems (Zeng and Huang 1981; Liao et al. 2006; Long and Wei. 2007). As a result of larval feeding, sugarcane leaves turn yellow to brown, stems are easily broken by wind, and the whole plant eventually dies. Various authors reported that the damage by this pest is more severe to sugarcane planted in sandy soil and/ or grown from the ratooning of the stubble for successive years (Zeng and Huang 1981; Long and Wei 2007; Yu et al. 2007). In northern Thailand, *D. granulosus* is considered a major pest of sugarcane, but much research remains to be done for better understanding of its pest status and biology in that region (Thongphak and Hanlaoedrit 2012).

This pest also damages roots and stumps of various other economic plants in southern China. For example, about 10% of less-than-one-year-old mango seedlings are infested, becoming wilted and dying (Zhang 1992). Longan seedlings and young citrus trees are heavily damaged, with up to 12% mortality in longan seedlings and about 30% of citrus trees losing economic value (Zhu and Xu 1996; Zhu 2012). More recently, it caused economic damage to cassava (*Manihot esculenta* Crantz) (Chen et al. 2012). The infestations on those crops occur mainly because they are planted on previous sugarcane farms without proper treatment of cane residues and soil. Since the 1990s, many farmers have replaced their sugarcane crops with fruit trees and other crops for high economic gain, but they have suffered huge economic losses due to this pest (Zhu and Xu 1996). Studies recommend that deep plowing, destroying cane residues, and soil treatment using pesticides be performed before planting other crops. *D. granulosus* larvae feed on roots and stumps of young citrus and *Eucalyptus citriodora* (Hook.) K.D. Hill & L.A.S. Johnson trees—weakening and even killing the trees (Zhu 1995)—and on stumps of harvested pine trees, causing economic losses (Wang 1994). However, these trees are not planted on previous sugarcane farms. In addition, Qin et al. (2010) reported that *D. granulosus* larvae are found to feed on rhizomes of coconut trees in southern China.

Long and Wei (2007) and Gong et al. (2008) have suggested possible reasons for the recent outbreaks of *D. granulosus* on sugarcane in southern China. These include (1) increase in sugarcane growing area, (2) increase in ratoon cropping, (3) decrease in rotation with rice, (4) planting susceptible varieties for high yield and sugar content, (5) increase in pesticide resistance, (6) lower rainfall in the past decade, and (7) increase in eucalypt growing area causing drought and providing more food for the pest.

Zeng and Huang (1981) gave an excellent description of the general biology of this longicorn. The adults emerge from April to June, with a peak in May. They mate on the emergence day or the next day and then lay eggs in soil 10–30 mm below the surface. The adults are nocturnal and attracted to light. A female can lay up to 783 eggs, with an average of 251 eggs in her lifetime. The eggs hatch in 10–18 days—most during early to mid-June. The larvae have 15–18 instars and last almost two years. The larvae go through 18 instars (Yu et al. 2012). The mature larvae leave the sugarcane stumps and overwinter in the soil. During March–April, they make large pupal cells near the stumps or 10–20 cm away from stumps and pupate in the cells 20–30 cm below the surface. The life cycle lasts two years. Yu et al. (2007) demonstrated that a female lays an average of 300 eggs in her lifetime and the pupal period lasts 15–31 days. In Thailand, the eggs are laid singly around the root zone at a depth of 6–35 mm;

each female lays approximately 160–310 eggs, with an incubation period of 14–23 days (Pliansinchai et al. 2007). Furthermore, Long and Wei (2007) tested larval resistance against starvation and indicated that the young larvae survive for more than two months without feeding and that the mature longicorns survive for more than four months with no food. They start feeding and damaging immediately after the food is available. This feature, together with the wide host range, makes the pest highly likely to be transported to other areas and become established. The biology described in China is very similar to that in Thailand (Pliansinchai et al. 2007). The authors show that the larvae are found throughout the year, with the highest population occurring in October and the lowest in April.

A number of measures have been developed for the control of *D. granulosus* in the past two decades and they are summarized here. (1) Light and/or pitfall traps can be used to attract adults during April–June when most of them are active. The light traps are set up about 100 m apart and 1.5 m high. Under each light trap, a pitfall trap (40 cm by 40 cm by 60 cm) is dug in the soil and filled with water (Yu et al. 2007). Lights are turned on between 19:30 and 23:00 hours (Mai 1988) or between 19:00 and 21:00 hours (Yu et al. 2009), depending on temperature. Pitfall traps (30 cm long by 20 cm wide by 40 cm deep) of higher density (4 m apart) are made if the light traps are not used (Yu et al. 2008). The trapped adults are destroyed every day. Pliansinchai et al. (2007) also suggested that the adult trapping practice should be carried out during April–June in Thailand. (2) Cultural control includes irrigation by flooding to drown larvae, crop rotation with rice or sweet potato to reduce infestation (Yu et al. 2007), and deep plowing (Gong et al. 2008) using a rotary cultivator (Yang 1992) to destroy sugarcane stumps and to kill the larvae after harvest or before planting. (3) Resistant cultivars are planted to suppress the pest population. Pliansinchai et al. (2007) indicated that some cultivars, such as Uthong 3 and K 88-92, are resistant to pest attack and should be grown in the pest outbreak areas. (4) Chemical control involves the use of carbamate, phosphorus, or organophosphate insecticides in granular formulation for soil treatment. Yang (1992) and Yu et al. (2007) recommended that, following plowing and before planting new crops, the soil be treated with insecticides. Gong et al. (2008) suggested that the insecticides should be used twice a year, first at planting time and second at the egg hatch peak period to kill young larvae. Pliansinchai et al. (2007) also suggested that the larval populations could be controlled at the early stages in April–July in Thailand. (5) Biological control involves the applications of fungus and nematode insecticides. Zeng and Huang (1981) first suggested that the fungus *M. anisopliae* may be used for the control of this pest. Yu et al. (2010) have then screened and obtained a highly virulent strain against *D. granulosus* in China. Sommartya et al. (2007) and Suasa-ard et al. (2008, 2012) also demonstrated that this fungus is effective for the control of *D. granulosus* larvae in Thailand. Wang et al. (1999) have tested the effectiveness of the nematode *Steinernema glaseri* for the control of *D. granulosus* in the laboratory but found this nematode caused little mortality of the beetle.

12.9.1.5 *Dorysthenes buqueti* (Guérin-Méneville)

Synonym: *Lophosternus buqueti* Guérin-Méneville
Common name: Sugarcane stem longicorn

This prionine species (Figure 12.76) is very similar to *D. granulosus* (see earlier) but differs in having a wider distance between the antennal tubercules, the lateral processes (spines) of the pronotum are shorter and blunter, and the lower side of antennal segments 3–5 is smooth without granular processes. This species is recorded in Thailand, Indonesia (Java), India (Assam), Myanmar, Malaysia, Nepal, Laos, and China (Yunnan) (Duffy 1968; Chiang et al. 1985; Pitaksa 1993; Pramono et al. 2001a, 2001b; Thongphak and Hanlaoedrit 2012). The main hosts are recorded from Poaceae (Duffy 1968; Pitaksa 1993; Pramono et al. 2001a). Pramono et al. (2001a) have tested host preference of this species among 11 plant species from Anacardiaceae, Euphorbiaceae, Fabaceae, Malvaceae, and Poaceae, and found that sugarcane from Poaceae is the preferred host.

In the past, *D. buqueti* was only considered a minor pest of bamboo, feeding on rhizomes of bamboo clumps in Southeast Asia (Duffy 1968; Chiang et al. 1985). However, it has become an increasingly important pest of sugarcane in Thailand since the mid-1980s (Prachuabmoh 1986; Pitaksa 1993; Charernsom and Suasa-ard 1994; Leslie 2004) and in West Java, Indonesia, since the 1990s (Pramono et al. 2001a,

FIGURE 12.76 *Dorysthenes buqueti* adult. Specimen from Natural History Museum (London). (Courtesy of Jiří Pirkl, copyright Natural History Museum [London], used with permission of M.V.L. Barclay.)

2001b; Anonymous 2014). *D. buqueti* appears to be less damaging than *D. granulosus* in Thailand (Thongphak and Hanlaoedrit 2012; Thongphak 2014, per. comm.) although sugarcane weight losses to the pest by up to 43% (Prachuabmoh 1986) and 51% (Pitaksa 1993) have been reported. In the laboratory, the larval feeding can reduce sugarcane stem weight by 41.4% in 10–12 days (Pramono et al. 2001a). According to Pramono et al. (2001b), since 1995, when *D. buqueti* was first reported attacking 100 ha of sugarcane in Indonesia, the infestation area had expanded to 6,000 ha in 1999, suggesting that this pest has the potential to threat the entire Indonesian sugar industry. The larvae do not attack the cane roots but bore directly into the base of stems and then tunnel the stem upward to 20–30 cm above the ground (Pramono et al. 2001a). One larva can attack more than one stem. Canes of all ages are attacked, but the infestation rate on ratoon canes is much higher than on plant canes (Prachuabmoh 1986; Pitaksa 1993; Pramono et al. 2001a). Similar to *D. granulosus,* this pest causes more severe damage to sugarcanes in sandy soil (Prachuabmoh 1986). The infested crops turn yellow and suffer poor growth. The plants ultimately die, and crops fail to ratoon properly in cases of severe infestation (Pramono et al. 2001a).

In Thailand, the adults emerge from the soil in April, live for 15–20 days, and each female lays 500–600 eggs that hatch in 15–18 days (Prachuabmoh 1986). Pramono et al. (2001a) have made detailed observations on the biology of *D. buqueti* in Indonesia, which is summarized here. The adults are nocturnal, attracted to light, and live for an average of 31.3 days. Each female lays 180–287 eggs in her lifetime. The eggs are laid in the soil around the root zone and hatch in 15–32 days. The total larval duration ranges from 20 to 21 months with 10 instars, and different larval instars and pupae coexist at the same time. Pupation occurs in a 5.0 by 4.2 cm mud chamber in the soil about 20 cm below the surface. The pupal stage lasts 17–31 days. The fully developed adults usually remain in the pupal chamber for about 7–10 days before emergence. In Thailand, the larval stage lasts about one year (Prachuabmoh 1986) and life cycle takes one to two years to complete (Pitaksa 1993), although in Indonesia one generation takes two years (Pramono et al. 2001a). The winter is passed at the larval stage. Like *G. granulosus*, the larvae and pupae of *D. buqueti* can survive for a long time under unfavorable conditions, making it possible for the species to spread to other areas with the aid of transportation (Pramono et al. 2001a).

Similar measures that are used for the control of *D. granulosus* could be applied to manage this pest. Prachuabmoh (1986) and Pitaksa (1993) described some chemical and cultural control methods used in Thailand. Pramono et al. (2001a) have experimentally tested an integrated set of control measures in Indonesia, which includes the intensive cultivation, hand-collection of larvae and adults, cane sanitation in the stubble, and application of an entomopathogenic fungus (*Metarhizium flavoviride* Gams & Roszypal [Pramono et al. 2001b]). These authors show that the integrated pest management strategy can reduce the pest population by 60% and increase millable cane yield by 25.8%.

12.9.1.6 *Dorysthenes hydropicus* (Pascoe)

Synonyms: *Cyrtognathus breviceps* Fairmaire, *C. chinensis* Thomson, *Prionus hydropicus* Pascoe
Common name: Sugarcane longicorn beetle

The adults of this prionine longicorn are 25–47 mm long; the body is shiny and chestnut brown to blackish brown with a pair of large and long, knife-shaped mandibles which are curved backward and crossed against each other at about an apical third when resting (Figure 12.77). This species is widely distributed in China (from Jilin to Guangdong, Taiwan) and Korea (Chen et al. 1959; Toepfer et al. 2014). Earlier, only sugarcane from Poaceae was recorded as its host (Chen et al. 1959; Tseng 1962; Duffy 1968). With further studies, the host range of *D. hydropicus* is in fact very wide, with its larvae feeding on plants from Poaceae, Fabaceae, Rutaceae, Salicaceae, Thymelaeaceae, Santalaceae, Cupressaceae, Pinaceae, Ulmaceae, Rosaceae, Theaceae, and Myrtaceae (Fan and Liu 1987; Chiang 1989; Qiao et al. 2011; Toepfer et al. 2014). Qiao et al. (2011) have investigated the suitability of three host plant species from Rutaceae, Thymelaeaceae, and Fabaceae, respectively, and found that the larvae feeding on pomelo [*Citrus maxima* (Burm. f.) Merr. = *C. grandis* (L.) Osbeck] from Rutaceae perform the best.

FIGURE 12.77   *Dorysthenes hydropicus* adult. (Courtesy of Meiying Lin.)

It is a moderate pest of sugarcane in Taiwan, Jiangxi, Hunan, and Guangxi (Chen et al. 1959; Tseng 1962, 1964; Pemberon 1963; Jiang 1991; Li et al. 1997). However, its damage to sugarcane appears to be severe in central Taiwan (Tseng 1962, 1964), with a yield loss of 30–50% cited by Fei and Wang (2007). In the late 1950s and early 1960s, yield loss to *D. hydropicus* damage was reported to be as high as 40% in Taiwan (Tseng 1962). According to Jiang's (1991) description, the larvae feed on the roots first, then bore into the stumps, and tunnel the main stems upward up to the ground level. The stumps and stems are hollowed up, resulting in wilting and death of entire plants. The infested canes are easily pulled out and broken. The damage to the ratoon cane is more severe than to the plant cane, and the infestation rate is higher in the unirrigated fields (Tseng 1962).

In the 1980s, *D. hydropicus* was reported to seriously damage common reeds [*Phragmites australis* (Cav.) Trin. ex Steud.] in Hubei Province, China, with its larvae feeding on the reed stems below the surface (Fan and Liu 1987). The infestation rate usually is 20–40% and could reach 70% in some areas. The infestation weakens and eventually kills the plants. Reed is widely used for paper making and building materials, and its root is a common ingredient in Chinese medicine. Since 2000, thousands of pomelo (*C. maxima*) trees have been killed by *D. hydropicus* in orchards in Guangdong Province (Qin et al. 2008). Authors reported that about 25% of the 40,000 trees in one pomelo orchard were killed by this pest. Trees usually die one to two years after infestation. Almost 100% of young trees (one to two years old) are killed. One to seven larvae are extracted from the stump of each dead tree. The pomelo plants are widely used for altar decoration and the fruits for eating and Chinese medicine.

In sugarcane fields, the adults emerge and mate after spring–summer rains between April and June, depending on when a reasonable rain (20–30 mm) occurs (Chen et al. 1959; Jiang 1991). The adults are nocturnal and attracted to light. According to Jiang (1991), the females lay their eggs in the soil about 3–4 cm below the surface, usually near the sugarcane plants, with each female ovipositing about 200 eggs. The larvae initially feed on roots and then bore into the stumps and stems. The mature larvae leave the canes, make pupation chambers in the soil 4–6 cm below the surface, and pupate in March–April. Overwintering occurs at the larval stage. There is one generation a year in Taiwan.

In reed fields in Hubei (Fan and Liu 1987), the adults emerge in May–June and lay their eggs in the soil 2–3 cm below the surface; each female lays an average of 180 eggs. The larvae feed on subterranean stems or bore into stems up to the ground surface and cut the stems at the ground level. They may move in the soil to feed on other stems when one stem is completely consumed but do not feed on stems above the surface. The mature larvae make large pupal chambers 3–7 cm below the surface and pupate there in April–May. The life cycle takes two years, and overwintering occurs at the larval stage in Hubei. In the pomelo orchards in Guangdong (Qin et al. 2008), the adults emerge between late April and late June, with an emergence peak during late May. Oviposition occurs between May and July with a peak during early to mid-June. The eggs are laid in the soil 1–3 cm below the surface. The female egg load ranges from 73 to 543 eggs depending on a female's body size, with an average of 296 eggs per female. Most eggs hatch during late June to early July. The damage by larvae takes place almost all year round. The adults are attracted to light, with one lamp being able to trap more than 2,000 adults per night. The larvae feed on small and tender roots and then bore into the main roots. After they consume all internal contents of the main roots, they move to another plant and continue boring. The larvae only feed on the roots between 15 and 60 cm below the surface. The mature larvae make pupation chambers in the soil 20–30 cm below the surface and pupate there during late March to early April. The pupal stage lasts 15–25 days. The larvae can survive for several months without food. Both Qin et al. (2008) and Qiao et al. (2011) showed that one generation takes one to two years, and overwintering is passed at the larval stage.

Control measures are still being developed, particularly for infestation by this pest in citrus orchards. Several tactics are recommended: (1) using light or pitfall traps to attract adults during the peak adult activity period (Fan and Liu 1987; Qin et al. 2008)—this tactic should be suitable for all types of crops that this pest infests. (2) Chemical control involves the use of organophosphate insecticides in granular formulation for soil treatment. The method has only been recommended for sugarcane fields so far. In the 1960s, the chemicals used were organochlorines (Tseng 1964, 1966), which have now been phased out. Before planting, one treatment is sufficient for the planted cane fields, but two treatments (second application in January–February) may be necessary for the ratoon cane fields (Fei and Wang 2007). For citrus orchards, once damage symptoms occur, any control measure cannot save the infested trees (Qin et al. 2008). (3) Biological control involves applications of fungus and nematode insecticides. In Guangdong, Ma et al. (2013) have screened the virulence of 36 strains of entomopathogenic fungi *Metarhizium* spp. against *D. hydropicus* larvae feeding on *C. grandis* and found that two strains are highly virulent to the pest. However, there has been no report on the field application of these fungi for control of this pest. Xu et al. (2010) have tested the effectiveness of three entomopathogenic nematodes, *Heterorhabditis bacteriophora* Poinar, *Steinernema scapterisci* Nguyen & Smart, and *S. carpocapsae* (Weiser) for killing *D. hydropicus* larvae feeding on *C. grandis*. They showed some promising results, but whether these nematodes could be used in the field is unknown.

12.9.1.7 *Dorysthenes walkeri* (Waterhouse)

Synonyms: *Cyrtognathus siamensis* Nonfried, *Baladeva walkeri* Waterhouse
Common name: Long-toothed longicorn

The adults of this prionine beetle are very large in size, ranging from 37 to 65 mm long; the body is blackish brown to black and shiny; the mandibles are very long and curved downward and backward (Figure 12.78). This longicorn is distributed in Myanmar, Vietnam, Laos, Thailand, and in southern China (Fujiang, Guangdong, Yunnan, and Hainan) (Gressitt 1940; Chien 1982; Hua 1982) and were introduced into Iran (Özdikmen et al. 2009). The larvae of this species feed on various species from Arecaceae, Marantaceae, and Poaceae (Chien 1982; Qin et al. 2010).

The long-toothed longicorn was first found to attack sugarcane in the late 1970s and became an important sugarcane pest in Hainan and southwest Guangdong, causing substantial losses (Chien 1982; Mai 1983). The larvae feed on the roots in early instars and the stems in late instars. The damaged canes turn brown and are easily broken by wind. The damage to the ratoon canes usually is more severe than to the plant canes. Sugarcane grown in sandy soil is subject to more severe damage. Between 32% and 60%

FIGURE 12.78 *Dorysthenes walkeri* adult. (Courtesy of Jiří Mička.)

yield loss has been recorded in some ratoon cane growing regions in Hainan (Mai 1983). Qin et al. (2010) reported that this pest also attacks coconut in southern China.

Both Chien (1982) and Mai (1983) have made observations on the general biology of *D. walker*. The adults emerge between May and September, with a peak in June–August depending on when a reasonable rain occurs. Drought or lack of rain during this period of the year may delay emergence. Mating occurs on the same day as emergence. Two to three days after mating, the females start laying eggs. The eggs are laid singly (or two at a time) in the soil 1.5–5 cm below the surface. The average egg load per female is 288 eggs, but many of these eggs have not been laid when the female dies. The adults live for 6–14 days, are nocturnal, and are most active between 20:00 and 01:00 hours. The young larvae move downward in the soil, approach the cane stumps or ratoons, and feed on the roots. Three to four instars later, they bore into the ratoon cane or plant cane stems below the surface and tunnel the stems upward up to 30 cm (usually 10 cm) above the surface. The larvae can also make tunnels up to 80 cm (usually 20–30 cm) long in soil 20–40 cm below the surface. The mature larvae make pupation chambers in the soil and pupate there. The pupal stage lasts 13–22 days, but emerged adults can stay in the pupation chambers for more than 10 days waiting for rain before emerging from the soil. Overwintering occurs at the larval stage. One generation takes two years.

Several control methods are recommended by Chien (1982) and Mai (1983, 1988). (1) Cultural control involves plowing the sugarcane field and rotating with soybean and peanut crops every two to three years. (2) Adult trapping can be carried out during the adult emergence season, particularly after a rain. Light traps can be set 100 m apart and 4–6 m above the ground with lights on between 19:30 and 23:00 hours. (3) Soil treatment with *Metarhizium* or organophosphate insecticides is another method. Mai (1983) showed that the fungus-based insecticide can kill 94% of larvae of all instars, but the chemical insecticides can be only effective when used for the early instar larvae.

12.9.1.8 *Dorysthenes paradoxus* (Faldermann)

Synonyms: *Cyrthognathus aquilinus* Thomson, *Prionus paradoxus* Faldermann
Common name: Bluegrass longicorn

The prionine adults are 33–41 mm long; the body is chestnut brown to blackish brown (Figure 12.79). It resembles *D. hydropicus* but differs in having the external angles of antennal segments 3–10 sharper and the anterior process at the pronotal side smaller and closer to the mid process. The species is widely distributed from Heilongjiang in northern China to Guizhou in southwestern China (Chen et al. 1959; Hua 1982; Wang et al. 2009) and also in Mongolia and eastern Russia (Tavakilian and Chevillotte 2013). The bluegrass longicorn larvae feed on plants from Poaceae, Fagaceae, and Ulmaceae (Chen et al. 1959; Yan et al. 1997; Wang et al. 2009).

The crop species recorded as host plants of *D. paradoxus* include sugarcane and rice in southern China, and maize, sorghum, and turfgrass (*Poa annua* L.) in northern China (Chen et al. 1959; Yan et al. 1997). More than half a century ago, Chen et al. (1959) stated that the larvae of this species cut the stems of maize and sorghum at the ground level. However, little information on its current pest status with regard to sugarcane, rice, maize, and sorghum is available, suggesting that it is not a concern for those crops at present. Yan et al. (1997) and Hao et al. (1999) reported that *D. paradoxus* has recently caused serious damage to *P. annua* turfgrass in Shanxi Province, resulting in patched death. Larvae cause the damage by feeding on grass roots.

According to Yan et al. (1997), adults emerge from the soil during early July to early September with an emergence peak in late July to late August. The newly eclosed adults stay in pupation chambers for days and wait for rain before emerging from the soil. Mating occurs within a few minutes after emerging from the soil. The females lay eggs in the soil 2–3 cm below the surface. The average egg load per female is 121 eggs but only about 36 eggs are laid before death. The eggs hatch 7–10 days after being laid. The adults live for five to seven days. The larvae make tunnels in the soil and move upward and downward. The early instar larvae overwinter among the roots and late instars in vertical tunnels 45–65 cm below the surface. The overwintered late instar larvae move upward to 3–5 cm below the surface during April–May and start feeding on the roots again. The mature larvae make pupation chambers in the tunnels and pupate there during June and July. The pupal stage lasts 20–25 days. It takes three years to complete a life cycle.

FIGURE 12.79 *Dorysthenes paradoxus* adult. Specimen from Natural History Museum (London). (Courtesy of Jiří Pirkl, copyright Natural History Museum [London], used with permission of M.V.L. Barclay.)

Yan et al. (1997) recommended several control measures for this pest during late July to late August: (1) scattering mixtures of soil, organophosphate insecticides, and fertilizers on the lawn surface and then watering to kill the neonate larvae or spraying organophosphate insecticides in the rain to kill adults and (2) hand-collection of adults. The adults should also be nocturnal and attracted to light. Therefore, light trapping could also be used.

12.9.2 Australasia

12.9.2.1 *Rhytiphora diva* (Thomson)

Synonym: *Zygrita diva* Thomson
Common name: Alfalfa crown borer, or Lucerne crown borer

The lamiine adults are about 15 mm long and bright orange in color, with black or dark brownish legs and antennae; the elytra have two black spots near the apex but may have no spots or more spots (Figure 12.80). This species occurs in all Australian states except South Australia, Victoria, and Tasmania (Duffy 1963), and in Papua New Guinea (Tavakilian and Chevillotte 2013). Several legume genera are recorded as its host plants (Jarvis and Smith 1946).

This species has been a minor to serious pest of soybeans and alfalfa in northern Queensland since the 1940s (Jarvis and Smith 1946; Williamson 1976; Romano and Kerr 1977) and in northern Western Australia since the 1970s (Strickland 1981; Heap 1991). It is also of some economic importance to soybeans grown in northern New South Wales (Goodyer 1972). Brier (2007) and Bailey and Goodyer (2007) summarize the damage caused by this pest. In some soybean fields in northern Queensland, more than 80% of plants may be infested. The major damage occurs when the mature larvae ringbark the plants internally before making pupal chambers. This activity causes plant death above the girdle. In southern Queensland, this usually occurs after seeds are fully developed with no yield loss. In tropical regions, however, the larval

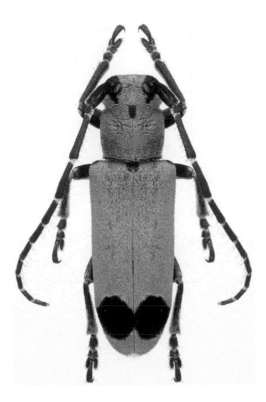

FIGURE 12.80 *Rhytiphora diva* adult. (Courtesy of CSIRO, copyright CSIRO, used with permission of Adam Slipinski.)

development is more rapid, and there can be considerable crop losses. In addition, *R. diva* is very damaging to edamae (soybeans) where green immature pods are harvested by mechanical pod pluckers. The stems of infested plants are weakened and snapped off, contaminating the harvested product.

The infestations usually occur from flowering onward. The adults lay their eggs in the stems of young soybeans. The larvae tunnel up and down through the pith in the stems and usually pupate in the tap roots (Brier 2007). According to Jarvis and Smith (1946), the larval development lasts three or more months, and pupation takes place in the end of the larval tunnel, from which the adult beetle cuts its way to the surface. There probably is only one generation a year, although the larvae of different instars occur together in the same field. The winter usually is passed at the larval stage.

Cultural control appears to be the most commonly used control measure for this pest. Jarvis and Smith (1946) recommended deep plowing of infested alfalfa fields before sowing. It is best to plough the infested legume field and to sow another crop in the farm rotation (Duffy 1963). Brier (2007) and Bailey and Goodyer (2007) also recommended cultural measures. For example, planting susceptible crops close to alfalfa should be avoided. In at-risk regions, later soybean plantings should be considered to shorten crop development, and in tropical regions, winter plantings would be an option to reduce loss. Shepard et al. (1981) have screened soybean lines for the alfalfa crown borer-resistant genotypes and obtained some promising results. So far, there are no pesticides registered for the alfalfa crown borer in soybeans.

12.9.3 Europe

12.9.3.1 *Agapanthia dahli* (Richter)

Synonyms: *A. dahli theryi* Pic, *A. dalhi erivanica* Pic, *Saperda dahli* Richter, *S. nigricornis* Fabricius
Common name: Sunflower longicorn beetle

The adults of this lamiine borer are 9.5–22 mm long; the body dorsum has greenish yellow pubescence of an unclear pattern and rusty erected hairs; the antennal segments 1–2 are black and the rest yellowish

FIGURE 12.81   *Agapanthia dahli* adult. (Courtesy of Milan Djurić [image from Srećko Ćurčić].)

or reddish yellow in the basal part and black in the apical part, and the pronotum has three yellowish longitudinal bands (Figure 12.81). It is widely distributed in Southern and Eastern Europe (including western Russia and Turkey), northern Iran, Kazakhstan, and Mongolia, and its host range appears to be limited to a number of herbaceous species from Asteraceae and Apiaceae (Horvath and Bukaki 1988; Bense 1995; Horvath and Hatvani 2001; David'yan 2008).

The sunflower longicorn beetle is a minor pest of sunflower crops in Eastern Europe, particularly in Hungary and western Russia (Horvath and Bukaki 1988; Horvath and Hatvani 2001; David'yan 2008). It damages sunflowers by the larval tunneling in the stems. Late sown crops usually suffer more severe damage. The infested stems are often cracked, resulting in lower seed yield and seed-oil content.

The adults are present from April to July and are active during the day (Bense 1995; David'yan 2008). The females gnaw sunflower stems up to 90 cm above the ground to make oviposition slits and lay their eggs in the slits singly. A female can lay about 50 eggs in her lifetime. The larvae bore into the stems and tunnel downward. Most larvae overwinter in the stems 8–80 cm above the ground but may overwinter in the stem below the ground surface. Some larvae pupate in the following spring but others overwinter twice. One generation lasts one to two years. For the control of this pest, David'yan (2008) recommended removal of sunflower stems from the field after harvest and deep plowing the field during the fall.

12.9.3.2 *Dorcadion pseudopreissi* Breuning

Common name: Ryegrass root longicorn

The adults of the lamiine longicorn (Figure 12.82) are robust beetles and brown to dark brown in color; the vertex and pronotal disc have a white to gray longitudinal pubescent stripe in the middle; each elytron has a white to gray pubescent stripe along the suture and one near the margin;

FIGURE 12.82   *Dorcadion pseudopreissi* adult. (Courtesy of I. A. Susurluk.)

the antennae reach the apical fifth of the elytra in males and the middle of the elytra in females. This species appears to occur only in central and western Turkey (Özdikmen 2010). Several turf and pasture grass species from Poaceae are recorded as the host plants of this longicorn beetle (Kumral et al. 2010, 2012).

The ryegrass root longicorn has become a major turf and pasture grass pest in western Turkey since 1996, causing up to 30% loss (Kumral et al. 2010). The pest has spread into urban landscapes, home lawns, and football fields (Kumral et al. 2012). The main economic damage is caused by larval feeding on the roots, and the damaged grasses turn yellow and die (Kumral et al. 2010; Susurluk et al. 2011). Kumral et al. (2012) have tested the susceptibility of five common grass species (*Lolium perenne* L., *Poa pratensis* L., *Festuca rubra* L., *F. arundinacea* Schreb, and *Agrostis stolonifera* L.) to this pest in western Turkey. Their results show that *F. arundinacea* is a turf species that is tolerant to *D. pseudopreissi* larvae, and *A. stolonifera* is not a suitable host of this pest.

Kumral et al. (2012) showed that the females lay their eggs in the soil in May, the eggs hatch in early June, and the larvae feed on grass roots for more than one year. The lifetime fecundity is unknown. One generation takes two years, and overwintering occurs at the mature larval and adult stages. Due to the nature of the two-year life cycle, the adults are present all year round except June–July, the eggs are present in May, the immature larvae in May–July, the mature larvae all year round except June, and the pupae appear in May–August.

There is no report on cultural and chemical control measures for this new pest. However, Susurluk et al. (2009) have tested the potential of an entomophagous nematode species for the control of *D. pseudopreissi* in the laboratory and have obtained some promising outcomes. A field test of the nematode *Heterorhabditis bacteriophora* Poinar against *D. pseudopreissi* larvae shows that the application of this nematode at 0.5 million infective juveniles per m^2 to the turf after *D. pseudopreissi* adults lay eggs significantly reduces the damage caused by this pest (Susurluk et al. 2011). However, it is still unknown whether this control measure is practical for large-scale use.

12.9.3.3 *Plagionotus floralis* (Pallas)

Synonym: *Echinocerus floralis* Pallas
Common name: Alfalfa root longicorn

The adults of this cerambycine longicorn are 6–20 mm long and black in color; the pronotum has a yellow pubescent front margin and a transverse yellow pubescent stripe behind the middle; the elytra have five transverse yellow pubescent stripes (Figure 12.83). It is widely distributed in Middle and Eastern Europe to the Caucasus Mountains, Iran, Syria, and Middle Siberia (Kaszab, 1971). *Plagionotus floralis* is polyphagous, with the larval host range being a number of species from Fabaceae, Asteraceae, Amaranthaceae, Theaceae, and Euphorbiaceae (Bense 1995; Toshova et al. 2010).

This species is a moderate pest of alfalfa in Eastern Europe (Mészáros 1990; Toshova et al. 2010; Bozsik 2013) and has become increasingly important in recent years (Zhekova and Petkova 2010). It also damages yellow sweet clover *Melilotus officinalis* (L.) (Toshova et al. 2010). The damage is caused by larval feeding on the main roots and stumps, resulting in yellow shoots and leaves and even death, especially in dry conditions. Toshova et al. (2010) suggested that the larvae may also bore into the lower section of the stems. According to Jermy and Balázs (1990), the damage by this longicorn is more severe in older alfalfa plants (three to four years old or more), causing up to 70–90% plant mortality in the European part of Russia. In eastern Azerbaijan of Iran, most damage also occurs in greater than three-year-old alfalfa and nonirrigated fields (Noushad and Kazemi 1995).

Using CSALOMON® ARb3z fluorescent yellow traps with a floral attractant in Bulgarian alfalfa fields, Toshova et al. (2010) reported that the adults are present from late May to late July. More recently, Imrei et al. (2014) have used the traps modified from CSALOMON® to monitor the adult activity period in Bulgaria and Hungary and have found that adults are active from May to August. The eggs are laid singly on the base of stems. The lifetime fecundity in the field is not known. The larvae bore into the main roots and tunnel in the middle. According to Toshova et al. (2010), this longicorn has one generation a year and overwinters at the larval stage in Bulgaria, but the life cycle may last two years in

FIGURE 12.83   *Plagionotus floralis* adult. (Courtesy of Boženka Hric [image from Srećko Ćurčić].)

the southern Urals and in western Siberia. The phenology of this pest in Iran is similar to that in Bulgaria (Noushad and Kazemi 1995).

Mitrjuskin (1940) reported that a parasitic wasp (*Bracon lautus* Szepl.) may be an effective natural enemy of *P. floralis*, which could be used for the control of this pest. The author also suggests that the removal of infested plants may be an alternative control measure. Plowing older fields and crop rotation with nonhost crops currently are considered the most effective measures (Imrei et al. 2014). Petkova and Zhekova (2012) have recently tested the resistance and tolerance of several alfalfa hybrids against the pest and concluded that those with the highest forage and seed productivity are attacked the least. Finally, the adult traps mentioned earlier are reasonably effective in adult monitoring and trap-and-kill programs (Toshova et al. 2010; Imrei et al. 2014).

12.9.4 North America

12.9.4.1 *Dectes texanus* LeConte

Common name: Dectes stem borer

The adults of this lamiine species are 5–16 mm long and pale gray in color, with long gray and black banded antennae (Figure 12.84). It is widely distributed from the eastern United States to Montana and Arizona, and southward to central Mexico and the Cape region of Baja California (Linsley and Chemsak 1995). So far, it has not been recorded outside its native range in North America. The larvae of *D. texanus* feed on a number of herbaceous species from Asteraceae, Brassicaceae, Cucurbitaceae, Fabaceae, Malvaceae, Solanaceae, and Zygophyllaceae (Linsley and Chemsak 1995).

This species is now a serious pest of both soybeans and sunflowers in the United States, particularly in the Midwest (Michaud and Grant 2010; Sloderbeck and Buschman 2011). It infested soybeans in Missouri, Arkansas, Louisiana, and Tennessee in the early 1970s but was not considered important at

UGA5024026

FIGURE 12.84 *Dectes texanus* adult. (Courtesy of Whitney Cranshaw, Colorado State University, Bugwood.org [UGA5024026].)

the time (Patrick 1973). A few years later, this species started causing economic losses to soybean crops by larval girdling of the stem prior to harvest (Campbell and Vanduyn 1977). In the meantime, the stem borer has become a serious pest of sunflower crops in the southern Great Plains (Rogers 1977b) by larval burrowing within and internal girdling of the stalks (Rogers 1985). Nearly 40% of soybean plants are infested in Tennessee and Arkansas, with the infestation rates from individual fields varying greatly (0–100%) within states (Tindall et al. 2010). This pest has recently been receiving increased attention as a pest of soybeans in the North American Great Plains and is now well established as a pest of soybeans in at least 14 states across eastern North America (Buschman and Sloderbeck (2010). More recently, serious damage to soybeans was recorded in Kansas (Harris and McCornack 2014) and Nebraska (Rystrom and Wright 2013, 2014). Economic injury occurs when soybean plants become susceptible to late season lodging due to the larval girdling. About 80–90% of sunflower plants in cultivated varieties are infested (Michaud and Grant 2009). The mature larvae girdle the stalks at the base in preparation for overwintering, a behavior that reduces stalk breakage force by 34–40% (Michaud et al. 2007a), leading to significant yield losses through lodging. In Kansas, >70% of sunflower plants are infested, and planting date has no effect on infestation rate (Michaud et al. 2009). The pest also has caused damage to sunflowers in North and South Dakota in recent years (Knodel et al. 2010). According to Michaud and Grant (2005, 2010), the stem borer prefers sunflower to soybean plants if given a choice, and the adults from sunflower plants are heavier and live longer than those from soybean plants.

Biological information is extracted from several studies of soybean and sunflower fields (Patrick 1973; Hatchett et al. 1975; Rogers 1985; Knodel et al. 2010; Crook et al. 2004; Michaud and Grant 2005; Rystrom and Wright 2013). Depending on regions, the adults emerge from June to August and live for about a month. The females start to lay their eggs four to eight days after mating. The eggs are laid singly in the pith of the leaf petioles, side branches, and primary stems of soybeans and in the leaf petioles of sunflowers. Each female lays about 50 eggs in her lifetime. The eggs hatch in 6–10 days. The larvae bore into the pith and tunnel throughout the stem. The larvae are cannibalistic, and only one survives in a single plant. Before overwintering and at about the time when the plants are mature and ready for harvest, the mature larvae girdle the stems near ground level, causing the plants to break off at the slightest pressure. The overwintered larvae pupate from late May to early August. The stem borer has one generation a year.

A number of measures have been developed or tested for the control of this important pest, which are summarized here.

1. *Cultural control:* For the control of this pest on soybeans, winter burial of stems with deep plowing or row bedding methods has proved to be effective in reducing larval survival and adult emergence. Burial of stems at a depth of about 5 cm is sufficient to decrease the larval survival and adult emergence (Campbell and Vanduyn 1977). Because the females prefer sunflowers over soybeans for feeding and oviposition, sunflowers can be planted as the trap crops in the soybean fields to control the pest on soybeans (Michaud et al. 2007b). Michaud et al. (2007b) demonstrated that the females do not avoid ovipositing in plants already containing their own eggs or those of the conspecific females. As a result, sunflower plants can accumulate multiple eggs and subsequent larval combat typically results in the survival of only one. For example, planting sunflowers in the nonirrigated corners of a center pivot irrigated soybean field in 2005 reduced infestation of soybeans by 65% compared with a control field without adjacent sunflowers. In 2006, surrounding a 0.33 ha soybean field with six rows of sunflowers reduced soybean infestation to <5% of plants, compared with 96% of sunflower plants.

   Several cultural measures for the control of this pest in sunflowers are recommended. For example, planting sunflowers before mid-April or after late May in Texas significantly reduces stalk infestation (Rogers 1985). It is also indicated that delayed planting can be a reliable and effective management tool for growers in the central (Charlet et al. 2007a) and southern Great Plains (Knodel et al. 2010). However, planting date has no effect on the pest population in Kansas (Michaud et al. 2009). Ensuring adequate moisture by irrigation during the growing season can assist in reducing stem-infesting insect densities (Charlet et al. 2007b) and irrigated sunflowers have thick stalks, which may reduce lodging and yield losses from the stem borer

(Knodel et al. 2010). Rogers (1985) showed that a single disk and sweep tillage of the sunflower stubble in October and January can cause up to 73.5% and 39.7% mortality of the overwintering larvae, respectively. Finally, reduced plant spacing potentially can delay the onset of girdling behavior by the borer larvae and thus mitigate losses that otherwise result from the lodging of girdled plants (Michaud et al. 2009). In response to doubts as to whether the stem borers attacking soybeans and sunflowers are the same species, Niide et al. (2006) reconfirmed that these borers are the same species and that farmers should not plant soybeans and sunflowers in rotation.

2. *Chemical control:* Chemical control of this pest is difficult because the larvae spend about 10 months inside the stems or in stubbles (Campbell and Vanduyn 1977; Rystrom and Wright 2013). In soybean fields, insecticide sprays and granules prove to be ineffective against the larvae (Campbell and Vanduyn 1977), leaving the potential of chemical control measures to adults. However, sprays of an IGR and a pyrethroid, applied at two-week intervals, are not effective (Andrews and Williams 1988), probably due to a prolonged window of adult activity (Rystrom and Wright 2013). Sloderbeck and Buschman (2011) carried out three-year large-scale field trials in central Kansas to determine whether the stem borer can be controlled at its adult stage. They concluded that the aerial applications of a pyrethroid on July 6, 12, and 15 were successful in significantly reducing adults but that the applications on July 1, 20, and 24 were less successful. These data suggest that, for central Kansas, two aerial applications may be required to control the stem borer in soybeans. Due to the lack of effective sampling methods, treatment thresholds and timing have not yet been established for this pest. Rystrom and Wright (2014) attempted to establish the annual trends in the adult emergence and activity using field cages and regular sweep net sampling in Nebraska for 2013 and 2014. However, these measures can be time-consuming and unreliable. To establish a more reliable monitoring system, Harris and McCornack (2014) have shown some promise in the use of aerial imagery to track and monitor plant health in soybean fields. This may eventually lead to better understanding of *D. texanus* biology, which will aid in developing and implementing site-specific management strategies.

   Chemical control of the stem borer in sunflower fields also proves to be difficult due to the similar reasons described for soybean fields (Knodel et al. 2010), although Charlet et al. (2007a) suggested that chemical control is often reliable in protecting the sunflower crop from the stem pest and relatively insensitive to application timing.

3. *Application of plant resistance:* Richardson (1976) found some evidence of soybean plant resistance against the stem borer over a three-year study. In a more recent study, Niide et al. (2012) have found that no soybean cultivars have resistance to the larval damage. However, these authors show that one of the genotypes tested has plant antibiosis to the larvae and two tested have oviposition antixenosis.

   Michaud and Grant (2009) have investigated why the infestation of wild sunflowers by *D. texanus* is remarkably rare, whereas 80–90% of plants may be infested in the cultivated varieties. They conclude that a combination of physical and chemical properties of wild sunflowers confers substantial resistance to oviposition by *D. texanus* and that this natural antixenosis has been inadvertently diminished in the course of breeding cultivated varieties. In another study, Charlet et al. (2009) also found some potential for developing resistant genotypes in cultivated varieties for this pest in the central Great Plains. Furthermore, in the southern Great Plains, the perennial sunflower species are resistant to stalk infestation by the stem borer, indicating the possibility of breeding cultivars resistant to the pest (Knodel et al. 2010). Further studies are warranted before the plant resistance—including both antixenosis and antibiosis—can be used in the field.

4. *Biological control:* Hatchett et al. (1975) recorded seven species of hymenopteran parasites (representing the families Braconidae, Pteromalidae, and Ichneumonidae) that were reared from *D. texanus* larvae. More than three decades later, a dipteran parasitoid, *Zelia tricolor* (Coquillett) (Tachinidae) (Tindall and Fothergill 2010), and a hymenopteran wasp, *Dolichomitus irritator* (Fabricus) (Ichneumonidae) (Tindall and Fothergill 2012), were reported as attacking *D. texanus* in soybeans. However, to date, none of these natural enemies has been commercially applied for the control of this stem borer in the field.

12.10 Concluding Remarks

This chapter probably is the most comprehensive treatment of cerambycid pests on crops from a global view. It covers 90 cerambycid species of economic importance in Africa, Eurasia, North and South America, and Australasia. Species are ordered based on the main crops they damage, which are divided into eight categories: pome and stone fruit trees; citrus trees; tropical and subtropical fruit trees; nut trees; mulberry trees; cacao, coffee, and tea crops; vine crops; and field crops. Under each species, known distribution, biology, and control measures are provided. Effective measures for management of insect pests usually are developed and designed on the basis of our understanding of their biology. However, essential knowledge of biology of many cerambycid pests is still lacking. Future work should focus on biology of cerambycid pests of crops.

ACKNOWLEDGMENTS

I would like to thank W. Lu, D. Thongphak, M. R. Mehrnejad, J. Yamasako, L. Bezark, M. Y. Lin, L. M. Constantino, and A. Slipinski for providing references or comments on local pest or taxonomic states; R. A. Haack for constructive comments on an earlier manuscript; numerous photographers (acknowledged in figure captions) for providing images; and the University of Georgia for maintaining the Bugwood website.

REFERENCES

Abdullah, F., and K. Shamsulaman. 2008. Insect pests of *Mangifera indica* plantation in Chuping, Perlis, Malaysia. *Journal of Entomology* 5:239–251.

Abivardi, C. 2001. *Iranian entomology—Applied entomology*. Berlin: Springer.

Achterberg, K. van, and M. R. Mehrnejad. 2011. A new species of *Megalommum* Szepligeti (Hymenoptera, Braconidae, Braconinae), a parasitoid of the pistachio longhorn beetle (*Calchaenesthes pistacivora* Holzschuh; Coleoptera, Cerambycidae) in Iran. *Zookeys* 112:21–38.

Adachi, I. 1990. Control methods for *Anoplophora malasiaca* (Thomson) (Coleoptera: Cerambycidae) in citrus groves. 2. Application of wire netting for preventing oviposition in a mature grove. *Applied Entomology and Zoology* 25:79–83.

Adachi, I. 1994. Development and life cycle of *Anoplophora malasiaca* (Thomson) (Coleoptera, Cerambycidae) on citrus trees under fluctuating and constant temperature regimes. *Applied Entomology and Zoology* 29:485–497.

Adeigbe, O. O., F. O. Olasupo, B. D. Adewale, and A. A. Muyiwa. 2015. A review of cashew research and production in Nigeria in the last four decades. *Scientific Research and Essays* 10:196–209.

Agnello, A. M. 2013. Apple-boring beetles. New York State Integrated Pest Management Program, Cornell University. http://nysipm.cornell.edu/factsheets/treefruit/pests/ab/ab.asp (accessed April 30, 2015).

Ahmed, K. U., M. M. Rahman, M. Z. Alam, M. M. Hossain, and M. G. Miah. 2014. Evaluation of relative host preference of *Batocera rufomaculata* De Geer on different age stages of jackfruit trees. *Asian Journal of Scientific Research*. 1–6. doi: 10.3923/ajsr.2014

Aksit, T., I. Cakmak, and F. Özsemerci. 2005. Some new xylophagous species on fig trees (*Ficus carica* cv. Calymirna L.) in Aydin, Turkey. *Turkish Journal of Zoology*. 29:211–215.

Alston, D. G., J. D. Barbour, and S. A. Steffan. 2007. California prionus *Prionus californicus* Motschulsky (Coleoptera: Cerambycidae). Washington State University Tree Fruit Research and Extension Center, Orchard Pest Management Online. http://jenny.tfrec.wsu.edu/opm/displayspecies.php?pn=643 (accessed March 31, 2015).

Alston, D. G., S. A. Steffan, and M. Pace. 2014. Prionus root borer (*Prionus californicus*). Utah Pests Fact Sheet: http://utahpests.usu.edu/ipm/htm/fruits/fruit-insect-disease/prionus-borers10 (accessed March 31, 2015).

Ambethgar, V. 2010. Field assessment of delivery methods for fungal pathogens and insecticides against cashew stem and root borer, *Plocaederus ferrugineus* L. (Cerambycidae: Coleoptera). *Journal of Biopesticides* 3:121–125.

Anderson, H., A. Korycinska, D. Collins, S. Matthews-Berry, and R. Baker. 2013. Rapid pest risk analysis for *Aromia bungii*. UK: The Food and Environment Research Agency. http://www.fera.defra.gov.uk/plants/plantHealth/pestsDiseases/documents/aromiaBungii.pdf (accessed March 29, 2015).

Andrews, G. L., and R. L. Williams. 1988. An estimate of the effect of soybean stem borer on yields. *Technical Bulletin, Mississippi Agricultural and Forestry Experiment Station* 153:1–5.

Anikwe, J. C., F. A. Okelana, H. A. Otuonye, L. A. Hammed, and O. M. Aliyu. 2007. The integrated management of an emerging insect pest of cashew: A case study of the cashew root and stem borer, *Plocaederus ferrugineus* in Ibadan, Nigeria. *Journal of Agriculture, Forestry and Social Sciences* 5(1): unpaginated.

Anonymous. 1973. Biology and control of soybean girdler *Nupserha nitidior* Pic var *atripennis* Brng (Coleoptera: Cerambycidae) in Himachal Pradesh. India: Himachal Pradesh University.

Anonymous. 1986. Burma—New record of mango pest. *Quarterly Newsletter, Asia and Pacific Plant Protection Commission, FAO, Thailand* 29:37.

Anonymous. 2010. Lucerne crown borer. Brisbane: Queensland Department of Agriculture, Fisheries and Forestry. https://www.daff.qld.gov.au/plants/field-crops-and-pastures/broadacre-field-crops/integrated-pest-management/a-z-insect-pest-list/lucerne-crownborers (accessed January 20, 2015).

Anonymous. 2013. *Batocera* species (longhorn beetles). QS. *Entomology* 2013(5), 10. National Plant Protection Organization, the Netherlands.

Anonymous. 2014. Dossier on *Dorysthenes buqueti* as a pest of sugarcane. Sugar Research Australia: Indooroopilly, Queensland. http://www.sugarresearch.com.au/icms_docs/163515_Sugarcane_longhorn_stemborer_Dorysthenes_buqueti_Dossier.pdf (accessed December 24, 2014).

Arguedas, M., and P. Chaverri. 1997. *Abejones barrenadores* (Cerambycidae). Cartago, Costa Rica, ITCR-CIT. (Serie Plagas y Enfermedades Forestales N° 20). 8 pp.

Ashihara, W. 1982. Seasonal life history of the grape tiger borer *Xylotrechus pyrrhoderus*. *Bulletin of the Fruit Tree Research Station Series E* 4:91–113.

Asogwa, E. U., J. C. Anikwe, T. C. N. Ndubuaku, and F. A. Okelana. 2009a. Distribution and damage characteristics of an emerging insect pest of cashew, *Plocaederus ferrugineus* L. (Coleoptera: Cerambycidae) in Nigeria: A preliminary report. *African Journal of Biotechnology* 8:53–58.

Asogwa, E. U., J. C. Anikwe, T. C. N. Ndubuaku, F. A. Okelana, and L. A. Hammed. 2009b. Host plant range and morphometrics descriptions of an emerging insect pest of cashew, *Plocaederus ferrugineus* L. (Coleoptera: Cerambycidae) in Nigeria: A preliminary report. *International Journal of Sustainable Crop Production* 4:27–32.

Asogwa, E. U., T. C. N. Ndubuaku, and A. T. Hassan. 2011. Distribution and damage characteristics of *Analeptes trifasciata* Fabricius 1775 (Coleoptera: Cerambycidae) on cashew (*Anacardium occidentale* Linnaeus 1753) in Nigeria. *Agriculture and Biology Journal of North America* 2:421–431.

Atwal, A. S. 1963. Insect pests of mango and their control. *Punjab Horticultural Journal* 3:235–58.

Avila, L. A. 2005. Impacto agronómico del daño causado por el gusano barrenador del tallo del cafeto (*Plagiohammus maculosus* Bates) en el municipio de Santa Cruz Naranjo del departamento de Santa Rosa, Guatemala. Tesis. Fitotecnia. Universidad de San Carlos de Guatemala.

Azam, K. M. 1979. A cerambycid beetle pest of grapevines (*Coelosterna scabrator*). [Foreign Title: Un coleoptere cerambycide parasite de la vigne (Coelosterna scabrator F.)]. *Progres Agricole et Viticole* 96:433–434.

Bailey, P. T., and G. Goodyer. 2007. Lucerne. In *Pests of field crops and pastures: Identification and control*, ed. P. T. Bailey, pp. 425–440. Collingwood: CSIRO Publishing.

Baker, J. R., and S. Bambara. 2001. Twig girdler *Oncideres cingulata* (Say) (Cerambycidae, Coleoptera). Publication No. Ent/ort-96, North Carolina Cooperative Extension Service, North Carolina State University. http://www.ces.ncsu.edu/depts/ent/notes/O&T/trees/note96/note96.html (accessed July 17, 2015).

Banerjee, S. N., and D. K. Nath. 1971. Life-history, habits and control of the trunk borer of orange, *Anoplophora versteegi* (Ritsema) (Cerambycidae: Coleoptera). *Indian Journal of Agricultural Sciences* 41:765–771.

Barrera, J. F. 2008. Coffee pests and their management. In *Encyclopedia of entomology*, ed. J. L. Capinera, 961–998 (2nd ed.). Dordrecht: Springer.

Barrera, J. F., J. Herrera, J. Villalobos, and B. Gómez. 2004. El barrenador del tallo y la raíz del café. Una plaga silenciosa. Proyecto Manejo Integrado de Plagas, folleto técnico No. 9. El Colegio de la Frontera Sur ECOSUR, Tapachula, Chiapas, México. 8 p.

Batt, A. E. M. 2004. Field and laboratory observations on some tree borers and their hosts in North Sinai Governorate, Egypt. *Egyptian Journal of Agricultural Research* 82:559–572.

Baufeld, P., E. Pfeilstetter, and G. Schrader. 2010. Ergebnisse einer Risikobewertung des Rundkopfigen Apfelbaumbohrers (*Saperda candida*). *Julius-Kuhn-Archiv* 428:234–235.

Beeson, C. F. C. 1931. The life history and control of *Celosterna scabrator* F. Col., Cerambyeidae. *Indian Forest Record, Calcutta* 16:279–294.

Beeson, C. F. C. 1941. The ecology and control of the forest insects of India and the neighbouring countries. Delhi: J. Singh at the Vasant Press.

Beeson, C. F. C., and B. M. Bhatia. 1939. On the biology of the Cerambycidae (Coleoptera). *Indian Forest Record* 5:1–235.

Benavides, P. G., P. M. Zamorano, C. A. O. Perez, L. Maistrello, and R. O. Rubio. 2013. Biodiversity of pathogenic wood fungi isolated *Xylotrechus arvicola* (Olivier) galleries in vine shoots. *Journal International des Sciences de la Vigne et du Vin* 47:73–81.

Bense, U. 1995. *Longicorn beetles: Illustrated key to the Cerambycidae and Vesperidae of Europe.* Weikersheim: Margraf.

Ben-Yehuda, S., Y. Dorchin, and Z. Mendel. 2000. Outbreaks of the fig borer *Batocera rufomaculata* and other cerambycids in fruit plantations in Israel. *Alon Hanotea* 54:23–29.

Bhasin, D. G., and M. L. Roonwal. 1954. A list of insect pests of forest plants in India and the adjacent countries. *Indian Forest Bulletin (Entomology)* 171:1–93.

Bhaskar, H., and J. Thomas. 2010. Incidence of the stem girdler, *Sthenias grisator* in mulberry. *Insect Environment* 16:20–20.

Bhaskara Rao, E. V. V. 1998. Integrated production practices of cashew in India. In *Integrated production practices of cashew in Asia*, eds. M. K. Papademetriou, and E. M. Herath. Bangkok: Food and Agriculture Organization of the United Nations Regional Office for Asia and Pacific. http://www.fao.org/docrep/005/ ac451e/ac451e04.htm#bm04 (accessed October 4, 2015).

Bhat, J. A., N. A. Wani, G. M. Lone, and M. S. Pukhta. 2010a. Influence of apple cultivars, age, and edaphic factors on infestation by the stem borer, *Aeolesthes sarta* Solsky. *Indian Journal of Applied Entomology* 24:89–92.

Bhat, J. A., N. A. Wani, M. S. Pukhta, and S. Mohi-ud-Din. 2010b. Pathogenicity of *Beauveria bassiana* in combination with synthetic insecticide to apple stem borer *Aeolesthes sarta*. *Journal of Mycology and Plant Pathology* 40:304–305

Bhat, M. A., M. A. Mantoo, and F. A. Zaki. 2010c. Management of tree trunk borer, *Aeolesthes sarta* (Solsky) infesting apple in Kashmir. *Journal of Insect Science* (Ludhiana) 23:124–128.

Bhattacharjee, N. S., and K. Goswami. 1984. Preliminary studies on the incidence of girdle beetle, *Oberea brevis* (Swedenbord) (Lamiidae, Coleoptera) in soybean. *Indian Journal of Entomology* 46:126–128.

Bhawane, G. P., and A. B. Mamlayya. 2013. *Artocarpus hirsutus* (Rosales: Moraceae): A new larval food plant of *Aeolesthes holosericea* (Coleoptera: Cerambycidae). *Florida Entomologist* 96:274–277.

Bhumannavar, B. S., and S. P., Singh. 1983. Some observations on the biology and habits of orange shoot borer, *Oberea lateapicalis* Pic. Coleoptera: Lamiidae. *Entomon* 8:331–336.

Bigger, M. 1967. Pest profiles III-white borer. *Tanganyika Coffee News* 8:14–16.

Bilsing, S. W. 1916. Life history of the pecan twig girdler. *Journal of Economic Entomology* 9:110–115.

Biosecurity Australia. 2006. Technical justification for Australia's requirement for wood packaging material to be bark free. Canberra: Biosecurity Australia. http://www.daff.gov.au/_data/assets/pdf_ file/0013/12361/2006-13a.pdf (accessed January 15, 2015).

Bishop, G. W., J. L. Blackmer, and C. R. Baird. 1984. Observations on the biology of *Prionus californicus* Mots. on hops, *Humulus lupulus* L., in Idaho. *Journal of Entomological Society of British Columbia* 81:20–24.

Bondar, G. 1913. Brocas das laranjeiras e outras Auranciaceas. *Boletim do Ministerio da Agricultura, Rio de Janeiro* 2:81–93.

Bondar, G. 1923. Pragas dos pomares no Brasil. Pragas das laranjeiras e outras auranciáceas. *Boletim de Agricultura, Comercio e Industria, Salvador* 1–6:5–53.

Bosq, J. M. 1942. Un taladro dañino para nuestros frutales y forestales, *Praxithea derourei* (Chabrill.). *Publicaciones Misceláneas del Almanaque del Ministerio de Agricultura y Ganadería, Buenos Aires* 121:425–430.

Bozsik, A. 2013. The occurrence of and damage caused by alfalfa root longhorn beetle (*Plagionotus floralis* Pallas, 1773) in alfalfa stands of various ages in Hungary. *Növényvédelem* 49:361–365 [in Hungarian].

Brabbs, T., D. Collins, F. Hérard., M. Maspero, and D. Eyre. 2015. Prospects for the use of biological control agents against *Anoplophora* in Europe. *Pest Management Science* 71:7–14.

Brauns, A. 1952. Capricorn beetles damaging fruit trees. *Nachrichtenblatt des Deutschen Pflanzenschutzdienstes* 4:66–67.

Brier, H. 2007. Pulses—Summer (including peanuts). In *Pests of field crops and pastures: Identification and control,* ed. P. T. Bailey, 169–258. Collingwood: CSIRO Publishing.

Browne, F. G., and E. C. Foenander. 1937. An entomological survey of tapped jelutong trees. *Malaysian Forester* 6:240–254.

Buschman, L. L., and P. E. Sloderbeck. 2010. Pest status and distribution of the stem borer, *Dectes texanus,* in Kansas. *Journal of Insect Science* 10: Article 198.

Butani, D. K. 1978. Insect pests of fruit crops and their control: 25—Mulberry. *Pesticides* 12:53–59.

CABI. 1994. *Distribution map of Batocera rufomaculata.* Distribution Maps of Plant Pests Issue June, Map 542.

CABI. 1998. *Xylotrechus quadripes.* Distribution Maps of Plant Pests Issue December: Map 589.

CABI. 2005a. *Monochamus leuconotus.* Distribution Maps of Plant Pests December (1st rev.): Map196.

CABI. 2005b. *Bixadus sierricola.* Distribution Maps of Plant Pests December Issue: Map 670.

CABI. 2008. *Anoplophora chinensis.* Distribution Maps of Plant Pests Map 595 (1st rev.)

CABI. 2012. *Psacothea hilaris.* Distribution Maps of Plant Pests June Issue: Map 762.

Campbell, W. V., and J. W. Vanduyn. 1977. Cultural and chemical control of *Dectes texanus texanus* (Coleoptera: Cerambycidae) on soybeans. *Journal of Economic Entomology* 70:256–258.

Cao, C. Y. 1981. Biology and control of *Linda nigroscutata. Acta Phytophylacica Sinica* 8:193–196 [in Chinese with English abstract].

Carbonell Bruhn, J., and J. Briozzo Beltrame. 1980. Los taladros *Praxithea derourei* Chabrillac, *Trachyderes thoracicus* Olivier and *T. striatus* Fabricius (Coleoptera: Cerambycidae) y su relacion con los cultivos de manzanos en Uruguay. *Investigaciones Agronomicas, Centro de Investigaciones Agricolas 'Alberto Boerger'* 1:11–14.

Carrasco, Z. F. 1978. Cerambicidos (Insecta: Coleoptera) sur-peruanos. *Revista Peruana de Entomologia* 21:75–78.

Castellanos, G. L., and C. R. Jimenez. 1991. Comportamiento del barrenador de los citricos, *Elaphidion cayamae* Fischer, en arboles de citricos en la empresa Horquita. *Centro Agricola* 18:19–27.

Castellanos, G. L., C. R. Jimenez, S. J. Hurtado, and H. O. Homen. 1990. Efectividad de diferentes programas de podas para el control del barrenador de la rama de los citricos. *Centro Agricola* 17:60–66.

Cavalieri, G. 2013. Summary of 2008–2011 trials on the possibility of control of *Anoplophora chinensis* with pesticides. *Journal of Entomological and Acarological Research* 45:19–20.

Chaikiattiyos, S. 1998. Integrated production practices of cashew in Thailand. In *Integrated production practices of cashew in Asia,* eds. M. K. Papademetriou, and E. M. Herath. Bangkok: Food and Agriculture Organization of the United Nations Regional Office for Asia and Pacific. http://www.fao.org/docrep/005/ac451e/ac451e09.htm#fn9 (accessed October 4, 2015).

Chakraborti, S. 2011. Management of the early stages of cashew stem and root borer, *Plocaederus ferrugineus* L. (Cerambycidae, Coleoptera). *Journal of Entomological Research* 35:203–207.

Chandel, Y. S., R. K. Gupta, and O. P. Sood. 1995. Screening of soybean varieties/germplasm to girdle beetle *Nupserha nitidior* Pic var *atripennis* Brig. *Indian Journal of Plant Protection* 23:79–81.

Chandrasekaran, J., R. S. Azhakiamanavalan, and I. H. Louis. 1990. Control of grapevine borer, *Coelosterna scabrator* F. (Cerambycidae: Coleoptera). *South Indian Horticulture* 38:108–108.

Chandy, K. T. 1986. *Drumstick booklet No 398: Vegetable production.* New Delhi: New Delhi Indian Institute.

Chang, K. P. 1970. A study of *Anoplophora macularia,* a serious insect pest of litchi. *Taiwan Agriculture Quarterly* 6:133–143.

Chang, S. C. 1973. Cerambycids of the genus *Chreonoma* in Taiwan. *Journal of Agriculture and Forestry* 22:183–189 [in Chinese with English abstract].

Chang, S. C. 1975. The host plants, egg laying and larval feeding habits of Macularia white spotted longicorn beetles. *Journal of Agriculture and Forestry* 24:13–20 [in Chinese with English abstract].

Chang, S. M., Y. W. Shen, and Y. F. Si. 1963. Studies on the citrus-trunk cerambycid, *Nadezhdiella cantori* Hope. *Acta Phytophylacica Sinica* 2:167–172 [in Chinese with English abstract].

Chang, Y. G. 1958. Preliminary study on a species of Citrus tree borer, *Chelidonium argentatum* (Dalman) (Coleoptera, Cerambycidae). *Acta Entomologica Sinica* 8:281–289 [in Chinese with English abstract].

Charernsom, K., and D. W. Suasa-ard. 1994. Pest status, biology and control measures for soil pest of sugar cane in South East Asia. In *Biology, pest status and control measure relationships of sugarcane insect pests,* eds. A. J. M. Carnegie and D. E. Conlong, 53–60. Durban: Proceedings of the International Society of Sugar Cane Technologists Second Entomology Workshop.

Charlet, L. D., R. M. Aiken, R. F. Meyer, and A. Gebre-Amlak. 2007a. Impact of combining planting date and chemical control to reduce larval densities of stem-infesting pests of sunflower in the central plains. *Journal of Economic Entomology* 100:1248–1257.

Charlet, L. D., R. M. Aiken, R. F. Meyer, and A. Gebre-Amlak. 2007b. Impact of irrigation on larval density of stem-infesting pests of cultivated sunflower in Kansas. *Journal of Economic Entomology* 100:1555–1559.

Charlet, L. D., R. M. Aiken, J. F. Miller, and G. J. Seiler. 2009. Resistance among cultivated sunflower germplasm to stem-infesting pests in the Central Great Plains. *Journal of Economic Entomology* 102:1281–1290.

Chatterjee, H., and J. Ghosh. 2001. Life table study of citrus trunk borer *Anoplophora versteegi* (Ritsema) (Cerambycidae: Coleoptera) and evaluation of some insecticidal formulations against it in the Himalayan region of West Bengal. *Journal of Interacademicia* 5:206–211.

Chatterjee, P. N., and P. Singh. 1968. *Celosterna scabrator* Fabricius (Lamiidae: Coleoptera), new pest of Eucalyptus and its control. *Indian Forester* 94:826–830.

Chaudhary, H. R., B. Ram, H. P. Meghwal, and J. Chaman. 2012. Evaluation of organic modules against insect pests of soybean. *Indian Journal of Entomology* 74:163–166.

Chaudhary, H. R., and N. N. Tripathi. 2011. Evaluation of some insecticides against major insect pests infesting soybean. *Indian Journal of Applied Entomology* 25:53–55.

ChauIn, N. M. 1998. Integrated production practices of cashew in Vietnam. In *Integrated production practices of cashew in Asia,* eds. M. K. Papademetriou, and E. M. Herath. Bangkok: Food and Agriculture Organization of the United Nations Regional Office for Asia and Pacific. http://www.fao.org/docrep/005/ac451e/ac451e0a.htm#fn10 (accessed October 4, 2015).

Chen, F. G. 1942. *A preliminary study on citrus cerambycid, Nadezhdiella cantori (Hope).* Chengtu: Szechwan Agricultural Research Station.

Chen, H. L., L. Y. Dong, C. L. Zhou, X. L. Liu, and F. Lian. 2002. Studies on the biology and control of *Bacchisa atritarsis* Pic. *Jiangxi Plant Protection* (3):65–67 [in Chinese with English abstract].

Chen, J. T. 1958. Morphology and habit of six main borers of the tea bush. *Acta Entomologica Sinica* 8:272–280 [in Chinese with English abstract].

Chen, S. X., Y. Z. Xie, and G. F. Deng. 1959. *Economic insect fauna of China, Fasc. 1—Coleoptera: Cerambycidae (I).* Beijing: Science Press [in Chinese].

Chen, Q., F. P. Lu, H. Lu, et al. 2012. Phenology and control of *Dorysthenes granulosus* and *Anomala corpulenta* on cassava. *Chinese Journal of Tropical Crops* 33:332–337 [in Chinese].

Chiang (Jiang), S. N. 1989. *Cerambycid larvae of China.* Chongqing: Chongqing Press [in Chinese].

Chiang, S. N., F. J. Pu, and L. Z. Hua. 1985. *Economic insect fauna of China, Fasc. 35—Coleoptera: Cerambycidae (III).* Beijing: Science Press [in Chinese].

Chien, T. Y. 1981. Records on the larvae of the citrus stem-borers. *Entomotaxonomia* 3:239–242 [in Chinese with English abstract].

Chien, T. Y. 1982. Studies on *Dorysthenes (Baladeva) walkeri* Waterhouse. *Acta Entomologica Sinica* 25: 31–34 [in Chinese with English abstract].

Chin, D., and H. Brown. 2008. Longicorn borer in fruit trees (*Acalolepta mixtus*). *Northern Territory Government Department of Primary Industry, Fisheries and Mines Agnote* 129:1–2

Civelek, H. S., and A. M. Çolak. 2008. Effects of some plant extracts and bensultap on *Trichoferus griseus* (Fabricius, 1792) (Coleoptera: Cerambycidae). *World Journal of Agricultural Sciences* 4:721–725.

Clearwater, J. R. 1981. Lemon tree borer, *Oemona hirta* (Fabricius), life cycle. *New Zealand Department of Science, Industry and Research Information Series* 105:1–3.

Clearwater, J. R., and S. J. Muggleston. 1985. Protection of pruning wounds: Favoured oviposition sites for the lemon tree borer. *Proceedings of New Zealand Weed and Pest Control Conference* 38:199–202.

Clearwater, J. R., and W. M. Wouts. 1980. Preliminary trials on the control of lemon tree borer with nematodes. *Proceedings of New Zealand Weed and Pest Control Conference* 33:133–135.

Constantino, C. L. M., and M. P. Benavides. 2015. El barrenador del tallo y la raíz del café, *Plagiohammus colombiensis*. *Revista Cenicafé* 66:17–24.

Constantino, C. L. M., P. M. Benavides, and J. R. Esteban. 2014. Description of a new species of coffee stem and root borer of the genus *Plagiohammus* Dillon and Dillon from Colombia (Coleoptera: Cerambycidae: Lamiinae), with a key to the Neotropical species. *Insecta Mundi* 337:1–21.

Coppedge, B. R. 2011. Twig morphology and host effects on reproductive success of the twig girdler *Oncideres cingulata* (Say) (Coleoptera: Cerambycidae). *The Coleopterists Bulletin* 65:405–410.

Cordeiro, G., N. Anjos, P. G. Lemes, and C. A. R. Matrangolo. 2010. Ocorrencia de *Oncideres dejeanii* Thomson (Cerambycidae) em *Pyrus pyrifolia* (Rosaceae), em Minas Gerais. *Pesquisa Florestal Brasileira* 30:153–156.

Craighead, F. C. 1915. Contributions toward a classification and biology of North American Cerambycidae. *USDA Office of Secretary Report* 107:1–24.

Crook, D. J., J. A. Hopper, S. B. Ramaswamy, and R. A. Higgins. 2004. Courtship behavior of the soybean stem borer *Dectes texanus texanus* (Coleoptera: Cerambycidae): Evidence for a female contact sex pheromone. *Annals of the Entomological Society of America* 97:600–604.

Crowe, T. J. 1963. The biology and control of *Dirphya nigricornis* (Olivier), a pest of coffee in Kenya (Coleoptera: Cerambycidae). *Journal of the Entomological Society of Southern Africa* 25:304–312.

Crowe, T. J. 1966. Recent advances in the use of insecticides on coffee in East Africa. *Agricultural and Veterinary Chemicals* 7:33–35.

Crowe, T. J. 2012. Coffee pests in Africa. In *Coffee: Growing, processing, sustainable production—A guidebook for growers, processors, traders and researchers*, ed. J. N Wintgens, 425–462, 2nd revised edition. Hoboken, NJ: Wiley, John & Sons Incorporated.

Dahiphale, K. D., D. S. Suryawanshi, S. K. Kamble, S. P. Pole, and I. A. Madrap. 2007. Effect of new insecticides against the control of major insect pests and yield of soybean [*Glycine max* (L.) Merrill]. *Soybean Research* 5:87–90.

Da Ponte, A. M. 1965. An experiment on the control of the trunk borer, *Anthores leuconotus*, of arabica coffee. *Publicações Ocacional, Instituto de Investigação Agronómica de Angola*:1–10.

D'Araujo e Silva, A. G. 1955. Seis noyas brocas da laranjeira. I. Broca das pontas. *Bol Fitossanitar* 6:35–44.

David'yan, G. E. 2008. *Agapanthia dahli* (Richt.)—Long-horned beetle, sunflower long-horned beetle. In *Interactive agricultural ecological atlas of Russia and neighboring countries: Economic plants and their diseases, pests and weeds* [online], eds. A. N. Afonin, S. L. Greene, N. I. Dzyubenko, and A. N. Frolov. Available at http://www.agroatlas.ru/en/content/pests/Agapanthia_dahli/ (accessed March 16, 2015).

Dawah, H. A., S. A. Alkahtani, A. H. Hobani, and S. N. Sahloli. 2013. The first occurrence of *Coptops aedificator* (Fabricius) (Coleoptera: Cerambycidae) a pest of cultivated mango in South-Western Saudi Arabia. *Journal of Jazan University-Applied Sciences Branch* 2:1–10.

Day, E. 2015. Twig girdler/twig pruner. Publication No. 2911-1423 (ENTO-124NP), Virginia Cooperative Extension, Virginia Polytechnic Institute and State University.

Delahaye, N., A. Drumont, and J. Sudre. 2006. Catalogue des Prioninae du Gabon (Coleoptera, Cerambycidae). *Lambillionea Supplément* 106:1–32.

de Souza Filho, M. F., D. Gabriel, and J. A. de Azevedo Filho. 2001. Caracterizacao morfologica de tres especies de broca-da-haste em chuchuzeiro (Coleoptera: Cerambycidae, Lamiinae). *Neotropical Entomology* 30:475–477

Devi, M. R., and P. R. K. Murthy. 1983. Protection of cashew from tree-borer pest. *Pesticides* 17:37–37.

de Villiers, E. A. 1970. The white coffee borer. *Farming in South Africa* 46:28–29.

Di Iorio, O. R. 1998. Torneutini (Coleoptera: Cerambycidae) of Argentina. Part II: Biology of *Coccoderus novempunctatus* (Germar, 1824), *Diploschema rotundicolle* (Serville, 1834) and *Praxithea derourei* (Chabrillac, 1857). *Giornale Italiano di Entomologia* 9:3–25.

Di Iorio, O. R. 2006. New records, remarks and corrections to host plants of Cerambycidae (Coleoptera) from Argentina. *Giornale Italiano di Entomologia* 11:159–178.

Ding, L. F., J. Y. Ma, H. B. Du, Y. N. Chen, and H. M. Chen. 2014. Comparison of several pesticides in the effectiveness for the control of mango stem borer (*Rhytidodera bowringi* White). *China Tropical Agriculture* 2014(1):49–50 [in Chinese].

Dominguez, J. E., and J. R. Dominguez. 1991. Danos, perdidas causadas y control de *Elaphidion cayamae* en naranja 'Valencia' de la Empresa de Citricos Victoria de Giron. *Proteccion de Plantas* 1:23–31.

Duffy, E. A. J. 1953. *A monograph of the immature stages of British and imported timber beetles (Cerambycidae)*. London: British Museum (Natural History).

Duffy, E. A. J. 1957. *A monograph of the immature stages of African timber beetles (Cerambycidae).* London: British Museum (Natural History).

Duffy, E. A. J. 1960. *A monograph of the immature stages of Neotropical timber beetles (Cerambycidae).* London: British Museum (Natural History).

Duffy, E. A. J. 1963. *A monograph of the immature stages of Australasian timber beetles (Cerambycidae).* London: British Museum (Natural History).

Duffy, E. A. J. 1968. *A monograph of the immature stages of Oriental timber beetles (Cerambycidae).* London: British Museum (Natural History).

Dunn, G., and B. Zurbo. 2014. Grapevine pests and their management. PRIMEFACT 511 (2nd ed.) NSW Department of Primary Industries. http://www.dpi.nsw.gov.au/_data/assets/pdf_file/0010/110998/Grapevine-pests-and-their-management.pdf (accessed January 15, 2015).

Dutt, N., and P. K. Pal. 1982. Role of mandibular length of *Obereopsis brevis* Swed. in selecting the site of oviposition and damage in soybean, *Glycine max* (L.) Merr. *Journal of Entomological Research* 6:37–47.

Dutt, N., and P. K. Pal. 1984. Morphology and bioecology of immature forms of soybean girdler, *Obereopsis brevis* Swed. (Col., Lamiidae). *Annals of Entomology* 2:97–106.

Dutt, N., and P. K. Pal. 1988. Termination of larval diapause and its importance in forecasting incidence of soybean girdler, *Obereopsis brevis* (Coleoptera, Lamiidae). *Indian Journal of Agricultural Sciences* 58:38–42.

Ebert, G. 2009. Fertilising for high yield and quality: Pome and stone fruits of the temperate zone. *Bulletin of International Potash Institute, Horgen, Switzerland* 19:1–74.

El-Minshawy, A. M. 1976. On the control of the cerambicid beetle *Chlorophorus varius* Mull. (Coleoptera, Cerambicidae) in grape orchards, with some biological observations. *Agricultural Research Review* 54:167–169.

El-Nahal, A. K. M., A. M. Awadallah, N. A. Abou Zeid, S. F. Zaklama, and M. M. Kinawy. 1978. On the biology of the longicorn beetle *Hesperophanes griseus* F. (Coleoptera, Cerambycidae). *Agricultural Research Review* 56:49–56

El-Sebay, Y. 1984. Biological, ecological and control studies on the wood borers, *Bostrychopsis reichei* Mars. and *Dinoderus bifoveolatus* Woll (Bostrychidae) and *Macrotoma palmata* F. (Cerambycidae). Ph.D. Thesis, Faculty of Agriculture, Al-Azhar University, Egypt.

Emori, T. 1976. Ecological study on the occurrence of the yellow-spotted longicorn beetle, *Psacothea hilaris.* Part I: Effects of temperature and photoperiod conditions on their development. *Japanese Journal of Applied Entomology and Zoology* 20:129–132 [in Japanese with English abstract].

Entwistle, P. 1972. *Pests of Cocoa.* London: Longman.

EPPO. 2005. Data sheets on quarantine pests: *Aeolesthes sarta. EPPO Bulletin* 35:387–389. http://www.eppo.int/QUARANTINE/insects/Aeolesthes_sarta/DS_Aolesthes_sarta.pdf (accessed December 25, 2014).

EPPO. 2008. *Saperda candida* (Coleoptera: Cerambycidae), round-headed apple tree borer. EPPO RS 2008/139. http://www.eppo.int/QUARANTINE/Alert_List/insects/saperda_candida.htm (accessed November 25, 2014).

EPPO. 2011. *Batocera rubus* detected in a bonsai plant in France. *EPPO Reporting Service* 6:8–8.

EPPO. 2012. EPPO database on quarantine pests. http://www.scoop.it/t/almanac-pests/p/1143030444/mango-longhorn-beetle-batocera-rubus-basic-data 2011 (accessed April 16, 2015).

EPPO. 2013a. *Aromia bungii* (Coleoptera: Cerambycidae), redneck longhorned beetle. EPPO, Paris. http://www.eppo.int/QUARANTINE/Alert_List/insects/Aromia_bungii.htm (accessed March 16, 2015).

EPPO. 2013b. PM 9/16 (1) *Anoplophora chinensis*: Procedures for official control. *EPPO Bulletin* 43:518–526.

EPPO. 2013c. Pest risk analysis for *Oemona hirta.* Paris: EPPO. Available at http://www.eppo.int/QUARANTINE/Pest_Risk_Analysis/PRA_intro.htm (accessed March 22, 2015).

EPPO. 2013d. *Pest risk analysis for Apriona germari, A. japonica, A. cinerea.* EPPO, Paris. Available at http://www.eppo.int/QUARANTINE/Pest_Risk_Analysis/PRA_intro.htm (accessed January 20, 2015).

EPPO. 2015. EPPO datasheet on pests recommended for regulation—*Aromia bungii. EPPO Bulletin* 45:4–8.

Esaki, K. 1995. Ovipositional characteristics of *Apriona japonica* Thomson (Coleoptera: Cerambycidae) in a *Zelkova serrata* Makino plantation. *Journal of the Japanese Forestry Society* 77:596–598 [in Japanese with English abstract].

Esaki, K. 2001. Artificial diet rearing and termination of larval diapause in the mulberry longicorn beetle, *Apriona japonica* Thomson (Coleoptera: Cerambycidae). *Japanese Journal of Applied Entomology and Zoology* 45:149–151 [in Japanese with English abstract].

Esaki, K. 2006. Deterrent effect of weed removal in *Zelkova serrata* on oviposition of *Apriona japonica* Thomson (Coleoptera, Cerambycidae). *Applied Entomology and Zoology* 41:83–86.

Esaki, K. 2007a. Life cycle, damage analysis and control of *Apriona japonica* Thomson (Coleoptera, Cerambycidae) in young *Zelkova serrata* plantation. *Bulletin of the Ishikawa-ken Forest Experimental Station* 39:1–44 [in Japanese with English abstract].

Esaki, K. 2007b. Management of *Apriona japonica* Thomson (Coleoptera: Cerambycidae) adults by spraying feeding trees with fenitrothion. *Journal of the Japanese Forest Society* 89:61–65 [in Japanese with English abstract].

Esaki, K., and T. Higuchi. 2006. Control of *Apriona japonica* (Coleoptera: Cerambycidae) adults using non-woven fabric sheet-formulations of an entomogenous fungus, *Beauveria brongniartii*, hung on feeding trees. *Journal of the Japanese Forest Society* 88:441–445 [in Japanese with English abstract].

Espinosa, A. E. 2000. Control of the coffee shoot borer *Plagiohammus maculosus* in the region of Santa Clara. *Instituto Interamericano de Ciencias Agricolas—II Simposio Latinoamericano sobre caficultura*:81–83.

Esteban-Durán, J. R., A. H. Salazar, M. González, P. del Estal-Padillo, and L. Castresana-Estrada. 2010. Species of the Tribe Lamiini (Coleoptera: Cerambycidae) from the Reserva Biológica Alberto Manuel Brenes, San Ramón, Costa Rica. *Spanish Journal of Agriculture Research* 8:1024–1032.

Eyre, D., H. Anderson, R. Baker, and R. Cannon. 2013. Insect pests of trees arriving and spreading in Europe. *Outlooks on Pest Management* 24:176–180.

Eyre, D., R. Cannon, D. McCann, and R. Whittaker. 2010. Citrus longhorn beetle, *Anoplophora chinensis*: An invasive pest in Europe. *Outlooks on Pest Management* 21:195–198.

Fan, N, X., and J. H. Liu. 1987. A preliminary study on the damage to common reed (*Phragmites australis*) by *Dorysthenes hydropicus*. *Kunchong Zhishi (Entomological Knowledge)* 24:278–279 [in Chinese].

Farashiani, M. E., S. E. Sadeghi, and M. Abaii. 2001. Geographic distribution, and hosts of sart longhorn beetle, *Aeolesthes sarta* Solsky (Col.: Cerambycidae) in Iran. *Journal of Entomological Society of Iran* 20:81–96.

Faria, A. M., S. C. S. Fernandes, J. C. C. Santos, et al. 1987. Estudos sobre controle da broca dos ramos e do tronco dos citros *Diploschema rotundicolle* (Serville, 1834) (Coleoptera–Cerambycidae). *Biologico* 53:41–43.

Fei, W. Q., and Y. Q. Wang. 2007. *Plant protection manual.* Taichung (Taiwan): Agricultural Chemicals and Toxic Substances Research Institute. For Subterranean pests: http://www.tactri.gov.tw/wSite/htdocs/ppmtable/sci-04.pdf (accessed March 25, 2015) [in Chinese].

Ferreira-Filho, P. J., C. F. Wilcken, J. C. Guerreiro, A. C. V. Lima, J. B. Carmo, and J. C. Zanuncio. 2014. First record of the wood-borer *Hylettus seniculus* (Coleoptera: Cerambycidae) in *Pinus caribaea* var. *hondrurensis* plantations in Brazil. *Florida Entomologist* 97:1838–1841.

Ferrero, F. 1985. Fruit crops—Two little-known pests on cherry. *Phytoma* (364):36–36.

Fletcher, T. B. 1919. Report of the imperial entomologist. *Scientific Reports to Agricultural Research Institute, Pusa 1918–1919*:86–103.

Fletcher, T. B. 1930. Report of the imperial entomologist. *Scientific Reports to Agricultural Research Institute, Pusa 1928–1929*:66–77.

Fonseca Ferrão, A. P. S. 1965. *Brocas do tronco do café.* Chianga: Instituto de Investigação Agronómica de Angola.

Forcella, F. 1984. Tree size and density affect twig girdling intensity of *Oncideres cingulata* (Coleoptera: Cerambycidae). *The Coleopterists Bulletin* 38:37–42.

Franssen, C. J. H. 1949. Levenswijze en bestrijding van de mangga-boktor (*Rhytidodera simulans* White). *Mededelingen van het Algemeen Proefstation voor den Landbouw* 95:1–40.

Franssen, C. J. H. 1950. Levenswijze en bestrijding van de manggaboktor *Rhytidodera simulans* White. *Landbouw* 22:1–38.

Gandhale, D. N., B. G. Awate, and R. N. Pokharkar. 1983. Chemical control of grapevine stem borer *Celosterna scabrator* Fbr. (Lamiidae: Coleoptera) in Maharashtra. *Entomon* 8:307–308.

Gangrade, G. A., K. N. Kapoor, and P. J. Gujrati. 1971. Biology, behaviour, diapause and control of *Oberea brevis* Swed. (Col.: Cerambycidae-Lamiinae) on soybeans in Madhya Pradesh, India. *The Entomologist* 104:260–264.

Gangrade, G. A., and O. P. Singh. 1975. Soybean plant response to the attack of *Oberea brevis* Swed. (Col., Cerambycidae). *Journal of Applied Entomology (Zeitschrift für Angewandte Entomologie)* 79:285–290.

Gao, Z. S., J. Chen, C. S. Zhu, D. Y. Zhou, and P. C. Lou. 1988. Control of cerambycids on mulberry trees. *Acta Sericologica Sinica* 14:188–190 [in Chinese with English abstract].

Garcia, J. J. S., O. Trevizan, and A. C. B. Mendes. 1985. Eficiência de inseticidas no controle de *Steirastoma breve* (Sulzer, 1776) (Coleoptera, Cerambycidae), broca do cacaueiro na Amazônia Brasileira. *Anais da Sociedade Entomológica do Brasil* 14:237–241.

Garcia Calleja, A. 2004a. Estudio de los danos que causa *Xylotrechus arvicola* Olivier. *Boletin de Sanidad Vegetal, Plagas* 30:25–31.

Garcia Calleja, A. 2004b. Datos sobre la dinamica de poblacion de *Xylotrechus arvicola* Olivier en Valladolid. 2002. *Boletin de Sanidad Vegetal, Plagas* 30(1):33–40.

García-Ruiz, E., V. Marco, and I. Pérez-Moreno. 2011. Effects of variable and constant temperatures on the embryonic development and survival of a new grape pest, *Xylotrechus arvicola* (Coleoptera: Cerambycidae). *Environmental Entomology* 40:939–947.

García-Ruiz, E., V. Marco, and I. Pérez-Moreno. 2012. Laboratory rearing and life history of an emerging grape pest, *Xylotrechus arvicola* (Coleoptera: Cerambycidae). *Bulletin of Entomological Research* 102:89–96.

Garonna, A. P., F. Nugnes, B. Espinosa, R. Griffo, and D. Benchi. 2013. *Aromia bungii*, nuovo tarlo asiatico ritrovato in Campania. *Informatore Agrario* 69:60–62.

Gautam, R. D. 1988. Some observations on the biology of soybean girdle beetle, *Obereopsis (Oberea) brevis* (Swedenbord) (Lamiinae: Cerambycidae: Coleoptera). *Bulletin of Entomology* 29:112–114.

Gerhaedt, P. D. 1961. Notes on a cerambycid *Aneflomorpha citrana* Chemsak, causing injury to orange trees in Arizona (Coleoptera). *Pan-Pacific Entomologist* 37:131–135.

Ghosh, PK. S. Ray, P. Tamang, and M. C. Diwakar. 1998. New record of *Oberea posticata* Gahan Coleoptera: Cerambycidae on Sikkim mandarin *Citrus reticulata* Blanco. *Plant Protection Bulletin* 47:38–38.

Gmitter, F., and X. L. Hu. 1990. The possible role of Yunnan, China, in the origin of contemporary *Citrus* species (Rutaceae). *Economic Botany* 44:267–277.

Godse, S. K. 2002. An annotated list of pests infesting cashew in Konkan Region of Maharashtra. *Cashew* 16:15–20.

Gong, H. L., Y. X. An, C. X. Guan, et al. 2008. Pest status and control strategies of *Dorysthenes granulosus* (Thomson) on sugarcane in China. *Sugarcane and Canesugar* (5):1–5 and 38 [in Chinese].

González-Risco, L., H. Grillo-Ravelo, and L. Valero-González. 2011. Caracterización de daños provocados por *Elaphidion* sp. (Coleoptera: Cerambycidae), nueva plaga en plantaciones citrìcolas de Jagüey Grande. *Centro Agrícola* 38:29–34.

González-Risco, L., H. Grillo-Ravelo, and L. Valero-González. 2013. Dinámica de aparición de ramas afectadas for *Elaphidion* sp. n. (Coleoptera: Cerambycidae) nueva plaga de los cétricos en EN Jagüey Grande. *CitriFrut* 30:17–21.

Goodwin, S. 2005a. Chemical control of fig longicorn, *Acalolepta vastator* (Newman) (Coleoptera: Cerambycidae), infesting grapevines. *Australian Journal of Entomology* 44:71–76.

Goodwin, S. 2005b. A new strategy for the chemical control of fig longicorn, *Acalolepta vastator* (Newman) (Coleoptera: Cerambycidae), infesting grapevines. *Australian Journal of Entomology* 44:170–174.

Goodwin, S., and N. Ahmad. 1998. Relationship between azinphos-methyl usage and residues on grapes and in wine in Australia. *Pesticide Science* 53:96–100.

Goodwin, S., and M. A. Pettit. 1994. *Acalolepta vastator* (Newman) (Coleoptera, Cerambycidae) infesting grapevines in the Hunter Valley, New South Wales. 2. Biology and ecology. *Journal of the Australian Entomological Society* 33:391–397.

Goodwin, S., M. A. Pettit, and L. J. Spohr. 1994. *Acalolepta vastator* (Newman) (Coleoptera, Cerambycidae) infesting grapevines in the Hunter Valley, New South Wales. 1. Distribution and dispersion. *Journal of the Australian Entomological Society* 33:385–390.

Goodyer, G. J. 1972. Insect pests of soybeans. *Agricultural Gazette of New South Wales* 83:342–344.

Gowda, D. K. S., M. G. Venkatesha, and P. K. Bhat. 1992. Rearing of white stemborer, *Xylotrechus quadripes* Chev. (Coleoptera: Cerambycidae) on freshly cut stems of *Coffea arabica* L. *Journal of Coffee Research* 22:69–72.

Gressitt, J. L. 1940. The longicorn beetles of Hainan Islands (Coleoptera: Cerambycidae). *Philippine Journal of Science* 72:1–239.

Gressitt, J. L. 1942. Destructive long-horned beetle borers at Canton, China. *Special Publications of Ningnan Natural History Survey Museum* 1:1–60.

Gressitt, J. L. 1947. Chinese longicorn beetles of the genus *Linda* (Coleoptera: Cerambycidae). *Annals of the Entomological Society of America* 40:545–555.

Grillo-Ravelo, H., and Y. Valdivies. 1990. Estudio bioecologico de *Elaphidion cayamae* Fisher (Coleoptera: Cerambycidae), nueva playa de los citricos en Jaguey Grande (I). *Centro Agricola* 17:41–51.

Grillo-Ravelo, H., and Y. Valdivies. 1992. Estudios bioecologicos de *Elaphidion cayamae* Fisher (Coleoptera: Cerambycidae) (II). *Centro Agricola* 19:23–32.

Guan, Y. Q. 1999. Occurrence and control of *Chreonoma fortunei*. *Jiangxi Fruit Trees* 1999(1):36–36 [in Chinese].

Guo, J. F. 1999. Control of grape tiger longicorn. *China Rural Science & Technology* (12):16–16 [in Chinese].

Guppy, P. L. 1911. The life-history and control of the cacao beetle. *Circular of the Board of Agriculture in Trinidad* 1:1–35.

Gupta, A., D. Sharma, and A. Bagmare. 1995. Screening soybean germplasms for resistance to *Obereopsis brevis* (Swed.) and *Ophiomyia phaseoli* (Tryon). *Crop Research* 10:338–343

Gupta, M. P. 2008. Efficacy and economics of biorationals and their admixtures against incidence of major insect pests of soybean. *Annals of Plant Protection Sciences* 16:282–288.

Gupta, M. P., S. K. Chourasia, and H. S. Rai. 2004. Field resistance of soybean genotypes against incidence of major insect pest. *Annals of Plant Protection Sciences* 12:63–66.

Gupta, R., and J. S. Tara. 2013. First record on the biology of *Aeolesthes holosericea* Fabricius, 1787 (Coleoptera: Cerambycidae), an important pest on apple plantations (*Malus domestica* Borkh.) in India. *Munis Entomology & Zoology* 8:243–251.

Gupta, R., and J. S. Tara. 2014. Management of apple tree borer, *Aeolesthes holosericea* Fabricius on apple trees (*Malus Domestica* Borkh.) in Jammu Province, Jammu and Kashmir State, India. *Journal of Entomology and Zoology Studies* 2:96–98.

Haack, R. A., F. Hérard, J. H. Sun, and J. J. Turgeon. 2010. Managing invasive populations of Asian longhorned beetle and citrus longhorned beetle: A worldwide perspective. *Annual Review of Entomology* 55:521–46.

Hall, D. R., A. Cork, S. J. Phythian, et al. 2006. Identification of components of male-produced pheromone of coffee white stemborer, *Xylotrechus quadripes*. *Journal of Chemical Ecology* 32:195–219.

Hall, T. J. 1980. Attack on the legume *Stylosanthes scabra* (Vog.) by *Platyomopsis pedicornis* (F.) (Coleoptera: Cerambycidae). *Journal of the Australian Entomological Society* 19:277–279.

Hammed, L. A., J. C. Anikwe, and A. R. Adedeji. 2008. Cashew nuts and production development in Nigeria. *American-Eurasian Journal of Scientific Research* 3:54–61.

Han, Y., and D. Lyu. 2010. Taxonomic review of the genus *Xylotrechus* (Coleoptera: Cerambycidae: Cerambycinae) in Korea with a newly recorded species. *Korean Journal of Applied Entomology* 49:69–82.

Hao, C., C. X. Yan, G. H. Zhang, and G. F. Yu. 1999. Morphological anatomy of alimentary canal and reproductive system of *Dorysthenes paradoxus* Faldermann. *Journal of Shanxi Agricultural University* 19:7–13 [in Chinese with English abstract].

Haq, K. A., and M. Akhtar. 1960. The mango borer and its control. *Punjab Fruit Journal* 23:203–204.

Harley, K. L. S. 1969. Assessment of the suitability of *Plagiohammus spinipennis* (Thoms.) (Col., Cerambycidae) as an agent for control of weeds of the genus *Lantana* (Verbenaceae). I: Life—History and capacity to damage *L. camara* in Hawaii. *Bulletin of Entomological Research* 58:567–574.

Harley, K. L. S. 1973. Biological control of *Lantana* in Australia. *Proceedings of the 3rd. International Symposium on the Biological Control of Weeds*. Memoirs. Montpellier, France: 23–29.

Harris, A., and B. McCornack. 2014. Using spectral response properties to identify and characterize infestations of *Dectes texanus* in soybean (abstract). The 62nd Entomological Society of America Annual Meeting, 16–19 November 2014, Portland, OR.

Hatchett, J. H., D. M. Daugherty, J. C. Robbins, R. M. Barry, and E. C. Houser. 1975. Biology in Missouri of Dectes texanus, a new pest of soybean. *Annals of the Entomological Society of America* 68:209–213.

He, D. P. 2001. On the integrated management of litchi pests in Guangdong province. *South China Fruits* 30: 23–26 [in Chinese].

Heap, M. A. 1991. *The Lucerne crown borer, Zygrita diva Thomson (Coleoptera: Cerambycidae), a pest of soybean*. Perth: University of Western Australia.

Hill, D. S. 1975. *Agricultural insect pests of the tropics and their control*. 2nd edition. Great Britain: Cambridge University Press.

Hill, D. S. 2008. *Pests of crops in warmer climates and their control*. UK: Springer.

Hillocks, R. J., N. A. Phiri, and D. Overfield. 1999. Coffee pest and disease management options for smallholders in Malawi. *Crop Protection* 18:199–206.

Ho, D. P., H. W. Liang, Z. W. Feng, and X. D. Zhao. 1990. A study of the biology and control methods of the long horn beetle *Aristobia testudo* (Voet). *Natural Enemies of Insects* 12:123–128 [in Chinese with English abstract].

Ho, K. Y., K. C. Lo., C. Y. Lee., A. S. Hwang. 1995. Ecology and control of the white-spotted longicorn beetle on citrus. *Taiwan Agricultural Research Institute Special Publication* 51:263–278.

Hockey, M. J., and M. De Baar. 1988. New larval food plants and notes for some Australian Cerambycidae (Coleoptera). *Australian Entomological Magazine* 15:59–66.

Hoffmann, W. E. 1934. Tree borers and their control in Kwangtung. *Lingnan Agricultural Journal* 1:37–59.

Holzschuh, C. 2003. Beschreibung von 72 neuen Bockkäfern aus Asien, vorwiegend aus China, Indien, Laos und Thailand (Coleoptera, Cerambycidae). *Entomologica Basiliensia* 25:147–241.

Hong, N., and S. B. Yang. 2010. Biological control of *Aromia bungii* by *Lepiota helveola* spent culture broth and culture homogenates. *Acta Edulis Fungi* 17:67–69 [in Chinese with English abstract].

Horvath, Z., and G. Bukaki. 1988. Ujabb adatok a napraforgon karosito *Agapanthia dahli* Richt. (Coleoptera: Cerambycidae) biologiajahoz. *Novenyvedelem* 24:298–302.

Horvath, Z., and A. Hatvani. 2001. Uj karte, vo a napraforgon: A sargagyurus bogancscincer (*Agapanthia dahli* Richt.). In *6 Tiszantuli novenyvedelmi forum, 6–8 November 2001, Debrecen, Hungary*, ed. G. K. Kovics, 118–120.

Hosking, G. P. 1978. *Oemona hirta* (Fabricius) (Coleoptera: Cerambycidae)—Lemon tree borer. *Forest and Timber Insects in New Zealand* (31):1–4.

Hoskovec, M., and M. Rejzek. 2014. *Cerambycidae—Longhorn beetles of the West Palaearctic Region*. http://www.cerambyx.uochb.cz (accessed March 25, 2015).

Hou, R. F. 1979. Studies on control of the white-spotted longicorn beetle, *Anoplophora macularia* (Thomson), in the coastal windbreaks. II. Tunnel fumigation and factors affecting infestation. *Bulletin of the Society of Entomology, National Chung-Hsing University* 14:35–41.

Howden, H., and A. Howden. 2000. *Tetrops praeusta* (L.) (Coleoptera: Cerambycidae), a potential pest? *Insecta Mundi* 14:220–220.

Hua, L. Z. 1982. *A checklist of the longicorn beetles of China (Coleoptera: Cerambycidae)*. Guangzhou: Zhongshan (Sun Yat-sen) University Press [in Chinese].

Hua, L. Z. 2002. *List of Chinese insects* (Vol. 2). Guangzhou: Sun Yat-sen University Press.

Hua, Z. Y., J. T. Wang, J. Liu, H. J. Wang, J. P. Shu, and T. S. Xu. 2012. Insect pests of *Camellia oleifera* and their natural enemies in Guizhou, China. *Journal of Zhejiang A&F University* 29:232–243 [in Chinese with English abstract].

Huang, G. T., and B. Yang. 2002. Phenology and control of grape tiger longicorn. *Agriculture of Henan* 2002(5): 17–17 [in Chinese].

Huang, J., W. Wang, S. Zhou, and S. Wang. 2009. Review of the Chinese species of *Apriona* Chevrolat, 1852, with proposal of new synonyms (Coleoptera, Cerambicidae, Lamiinae, Batocerini). *Les Cahiers Magallanes* 94:1–23.

Hussain, A., and A. A. Buhroo. 2012. On the biology of *Apriona germari* Hope(Coleoptera: Cerambycidae) infesting mulberry plants in Jammu and Kashmir, India. *Nature and Science* 10:24–35.

Hussain, A., and M. Z. Chishti. 2012. Cultural control of *Apriona germari* Hope (Coleoptera: Cerambycidae: Lamiinae) infesting mulberry plants in Jammu and Kashmir (India). *Journal of Entomological Research* 36:207–210.

Hussain, A., M. Z. Chishti, A. A. Buhroo, and M. A. Khan. 2007. Adult population of *Apriona germari* Hope (Coleoptera: Cerambycidae) in mulberry farms of Jammu and Kashmir State. *Pakistan Entomologist* 29:15–17.

Hussain, A., M. A. Khan, M. Z. Chishti, and A. A. Buhroo. 2009. Cerambycid borers of mulberry (*Morus* spp.) in Jammu and Kashmir, India. *Indian Journal of Applied & Pure Biology* 24:101–103.

Hwang, A. S., and K. Y. Ho. 1994. Relationship between growth vigor of citrus trees and damage by white-spotted longicorn beetles. *Chinese Journal of Entomology* 14:271–275.

Ibáñez Justicia, A., R. Potting, and D. J. van der Gaag. 2010. Pest risk assessment—*Apriona* spp. Plant Protection Service, Ministry of Agriculture, Nature and Food Quality, the Netherlands.

Ilic, N., and S. Curcic. 2013. The longicorn beetles (Coleoptera: Cerambycidae) of Rtanj Mountain (Serbia). *Acta Entomologica Serbica* 18:69–94.

Imam, A. I., and R. F. Sawaby. 2013. Arthropod diversity associated with infestation spots of fig tree borer under rain-fed conditions of Maged Valley, Matrouh, Egypt. *Egyptian Academic Journal of Biological Sciences* 6:11–19.

Imrei, Z., Z. Kováts, T. B. Toshova, et al. 2014. Development of a trap combining visual and chemical cues for the alfalfa longhorn beetle, *Plagionotus floralis*. *Bulletin of Insectology* 67:161–166.

Ioneda, T., K. Watanabe, and G. I. Imai. 1986. Aspectos morfologicos e biologicos de *Diploschema rotundicolle* (Serville, 1834) (Coleoptera: Cerambycidae). *Biologico* 51:281–283.

Isahaque, N. M. M. 1978. A new record of *Oberea posticata* Gah. (Coleoptera: Cerambycidae) on citrus plants in Assam. *Science and Culture* 44:130–130.

Ismail, A. I., F. M. El-Hawary, and A. S. H. Abdel-Moniem. 2009. Control of the fig longicorn beetle, *Hesperophanes griseus* (Fabricius) (Coleoptera: Cerambycidae) on the fig trees, *Ficus carica* L. *Egyptian Journal of Biological Pest Control* 19:89–91.

Ithnin, B., and O. M. Shamsudin. 1996. Insect pests—Mango planting guide. *Malaysian Agricultural Research and Development Institute Publication* 6:32–42.

Jagginavar, S. B., N. D. Sunitha, and D. R. Patil. 2006. Seasonal incidence, injury and integrated management of *Celosterna scabrator* Fabr (Coleoptera: Cerambycidae) in grape vine ecosystem. *Proceedings of International Symposium on Grape Production and Processing, Baramati, India, 6–11 Feb 2006*: 120–121.

Jagginavar, S. B., N. D. Sunitha, and D. R. Paitl. 2008. Management strategies for grape stem borer *Celosterna scrabrator* Fabr. (Coleoptera: Cerambycidae). *Indian Journal of Agricultural Research* 42:307–309.

Jain, N., and D. Sharma. 2011. Evaluation of yield losses by girdle beetle and its management in soybean crop in Kota region, Rajasthan, India. *Life Science Bulletin* 8:123–125.

Jarvis, H., and J. H. Smith. 1946. Legume pests. *Queensland Agricultural Journal* 62:79–89.

Jayanti, L., R. G. Shali, and K. Nanda. 2014. Studies on seasonal incidence, nature of damage and assessment of losses caused by girdle beetle on *Glycine max. Annals of Plant Protection Sciences* 22:320–323.

Jayarama, B. K. 2007. Chandragiri—Farmer friendly new arabica plant variety. *Indian Coffee* 71:21–25.

Jayarama, B. K., M. V. D'Souza, and D. R. Hall. 2007. (S)-2-Hydroxy-3-decanone as sex pheromone for monitoring and control of coffee white stemborer (*Xylotrechus quadripes*). *Proceedings of the 21st International Conference on Coffee Science, 11–15 September, Montpellier, France* (Publication No. 2-900212-20-0):1291–1300.

Jeppson, L. R. 1989. Biology of citrus insects, mites, and mollusks, In *The citrus industry. Vol. 5,* eds. W. Reuther, E. C. Calavan, and G. E. Carman, 1–81. Oakland, CA: Division of Agriculture and Natural Resources, University of California.

Jerinic-Prodanovic, D., R. Spasic, and D. Smiljanic. 2012. *Janus compressus* F. and *Tetrops praeusta* L., little known fruit pests in Serbia. *Biljni Lekar* 40:291–295.

Jermy, T., and K. Balázs. 1990. *Handbook of plant protection Zoology.* Budapest: Akadémiai Kiadó.

Jia, X. Y., G. Y. Ma, L. G. Wang, W. Liang, and H. Wen. 2001. Integrated control of walnut pests. *China Fruits* (1):39–40 [in Chinese].

Jiang, B. H. 1991. *Atlas of insect and rodent pests of sugarcane in Taiwan.* Tainan: Taiwan Sugar Research Institute [in Chinese].

Jin, F., B. Z. Ji, S. W. Liu, L. Tian, and J. Gao. 2011. Influence of oviposition secretion of *Apriona germari* Hope (Coleoptera: Cerambycidae) on water content, pH level and microbial quantity in its incisions on paper mulberry (*Broussonetia papyrifera*). *Acta Entomologica Sinica* 54:477–482 [in Chinese with English abstract].

Jiroux, E. 2011. Révision du genre *Apriona* Chevrolat, 1852 (Coleoptera, Cerambycidae, Lamiinae, Batocerini). *Les Cahiers Magellanes* 5:1–103.

Jolles, P. 1932. A study of the life-history and control of *Cerambyx dux* Fald, a pest of certain stone-fruit trees in Palestine. *Bulletin of Entomological Research* 23:251–256.

Jonsson, M., I. A. Raphael, B. Ekbom, and S. Kyamanywa, J. Karungi. 2015. Contrasting effects of shade level and altitude on two important coffee pests. *Journal of Pest Science* 88:281–287.

Kalita, H., A. C. Barbora, S. C. Borah, and P. Handique. 2003. Chemical control of citrus trunk borer, *Anoplophora versteegi* (Ritsema) in khasi mandarin. *Journal of Applied Zoological Researches* 14:181–183.

Kaplan, C. 2013. Guneydogu Anadolu Bolgesi meyve agaclarinda zararli Osphranteria coerulescens Redtenbacher, 1850 (Coleoptera: Cerambycidae)'in yayililisi, konukculari, zarar sekli ve bazi biyolojik ozellikleri. *Bitki Koruma Bulteni* 53:1–6.

Kashyap, N. P., S. F. Hameed, and D. N. Vaidya. 1978. Preliminary study on the varietal susceptibility of soybean to *Nupserha nitidor* var. *atripennis* Pic. (Lamidae Coleoptera). *Tropical Grain Legume Bulletin* (13/14):50–52.

Kaszab, Z. 1971. *Cincérek—Cerambycidae. Fauna Hungariae 106.* Budapest: Akadémiai Kiadó.

Kawakami, K. 1987. The use of an entomogenous fungus, *Beauveria brongniartii*, to control the yellow-spotted longicorn beetle, *Psacothea hilaris. Extension Bulletin, ASPAC Food and Fertilizer Technology Center for the Asian and Pacific Region, Taiwan* 257:38–39.

Kawakami, K., and T. Shimane. 1986. Microbial control of the yellow-spotted longicorn beetle, *Psacothea hilaris* (Coleoptera, Cerambycidae) by an entomogenous fungus *Beauveria tenella. The Journal of Sericultural Science of Japan* 55:227–234.

Kehlenbeck, H. P. Baufeld, and G. Schrader. 2009. Neuer Schadorganismus an Apfel und anderen Geholzen in Deutschland: Risikobewertung zu *Saperda candida. Journal fur Kulturpflanzen* 61:417–421.

Khan, M. F., and S. S. Qadri. 2006. Ecologically safe control of apple and apricot tree trunk borer *Aeolesthes sorta* using the insecticide implantation technique. *International Pest Control* 48:86–87

Khan, M. M. H. 2012. Morphometrics of cucurbit longicorn (*Apomecyna saltator* F.) Coleoptera: Cerambycidae reared on cucurbit vines. *Bangladesh Journal of Agricultural Research* 37:543–546.

Khan, S. A., S. Bhatia, and N. Tripathi. 2013. Entomological investigation on *Aeolesthes sarta* (Solsky), a major pest on walnut trees (*Juglans Regia* L.) in Kashmir Valley. *Journal of Academia and Industrial Research* 2:325–330.

Khan, T. N. 1993. Biology and ecology of *Plocaederus obesus* Gahan (Coleoptera: Cerambycidae), a comparative study. *Proceedings of the Zoological Society* 46:39–49.

Kikuchi, M. 1976. Control of insect pests of mulberry in Japan. *Japan Pesticide Information* 29:9–11.

Kim, S. B., M. S., Yiem, H. I. Jang, W. S. Kim, and J. H. Oh. 1991. Studies on the ecological characteristics and chemical control of grape tiger longicorn *Xylotrechus pyrrhoderus* Bates in grapevines. *Research Reports of the Rural Development Administration* 33:38–47.

Kiyota, R., R. Yamakawa, K. Iwabuchi, K. Hoshino, and T. Ando. 2009. Synthesis of the deuterated sex pheromone components of the grape borer, *Xylotrechus pyrrhoderus. Bioscience, Biotechnology and Biochemistry* 73:2252–2256.

Kliejunas, J. T., B. M. Tkacz, H. H. Burdsall Jr., et al. 2001. Pest risk assessment of the importation into the United States of unprocessed Eucalyptus logs and chips from South America. *USDA Forest Service General Technical Report FPL-GTR-124*:1–144.

Knodel, J. J., L. D. Charlet, and J. Gavloski. 2010. Integrated pest management of sunflower insect pests in the Northern Great Plains. *NDSU Extension Service, Fargo, ND, E-1457*:1–20.

Kojima, T. 1929. The habits of *Batocera lineolata*, Chevr. *Journal of Applied Zoology* 1:43–45.

Kondo, E., and A. R. Razak. 1993. Infectivity of entomopathogenic nematodes, *Steinernema carpocapsae*, on the mango shoot borer, *Rhytidodera simulans. Japanese Journal of Nematology* 23:28–36.

Kuang, R. P., X. W. Yu, and N. Zhong. 1997. A study on the structure and temporal-spatial attacking traits of coffee stem-borer population in Simao region. *Zoological Research* 18:33–38 [in Chinese with English abstract].

Kumar, D. P., P. S. Rai, and M. Hedge. 1996. Life cycle and impact of the stem borer *Plocaederus ferrugineus* in cashew plantation in Karnataka, India, with notes on resistance to the borer. *Proceedings of the IUFRO Symposium, Peechi, India, 23–26 November 1993*:324–327.

Kumar, U., S. P. Kumar, and S. Surabhi. 2012. Spectrum of insect pest complex of soybean (*Glycine max* (L) Merrill) at Lambapeepal village in Kota region, India. *International Research Journal of Biological Sciences* 1:80–82.

Kumral, N. A., U. Bilgili, and E. Açikgöz. 2010. *Dorcadion pseudopreissi*, a new turf pest in Turkey, the biology and damage on different turf species. *The 9th European Congress of Entomology, Budapest, 22–27 August 2010* (abstract).

Kumral, N. A., U. Bilgili, and E. Açikgöz. 2012. *Dorcadion pseudopreissi* (Coleoptera: Cerambycidae), a new turf pest in Turkey, the bio-ecology, population fluctuation and damage on different turf species. *Turkish Journal of Entomology* 36:123–133 [in Turkish with English abstract].

Kundu, G. G., and Trimohan. 1986. Effect of infestation by the girdle beetle *Obereopsis bevis* Swed. Coleoptera Lamiidae on plant parameters contributing to yield in soybean crop. *Journal of Entomological Research* 10:57–62.

Kung, P. C. 1977. Studies on two long-horned beetles infesting coffee trees in Kwangsi Autonomous Region. *Acta Entomologica Sinica* 20:49–56 [in Chinese with English abstract].

Kunhi Kannan, K. 1928. The large citrus borer of south India, *Chelidonium cinctum* Guer. *Bulletin of Department of Agriculture* 8:1–24.

Kutywayo, D., A. Chemura, W. Kusena, P. Chidoko, and C. Mahoya. 2013. The impact of climate change on the potential distribution of agricultural pests: The case of the coffee white stem borer (*Monochamus leuconotus* P.) in Zimbabwe. *Plos One* 8(8): Article Number: e73432. doi: 10.1371/journal.pone.0073432

Lan, C. C., and J. N. Wintgens. 2012. Major pests of coffee in the Asia-Pacific region. In *Coffee: Growing, processing, sustainable production—A guidebook for growers, processors, traders and researchers* (2nd rev. ed.), ed. J. N. Wintgens, 463–477. Hoboken, NJ: Wiley, John & Sons Incorporated.

Laurent, P. 1905. Egg potential of *Prionus laticollis*. *Entomological News* 16:62.

Lay, M. M. 1998. Integrated production practices of cashew in Myanmar. In *Integrated production practices of cashew in Asia*, eds. M. K. Papademetriou, and E. M. Herath. Bangkok: Food and Agriculture Organization of the United Nations Regional Office for Asia and Pacific. http://www.fao.org/docrep/005/ac451e/ac451e06.htm#fn6 (accessed March 15, 2015).

Lee, C. Y., and K. C. Lo. 1996. Reproductive biology of the white-spotted longicorn beetle, *Anoplophora macularia* (Thomson) (Coleoptera: Cerambycidae), on sweet oranges. *Journal of Agricultural Research of China* 45:297–304 [in Chinese with English abstract].

Lee, C. Y., and K. C. Lo. 1998. Rearing of *Anoplophora macularia* (Thomson) (Coleoptera: Cerambycidae) on artificial diets. *Applied Entomology and Zoology* 33:105–109.

Leslie, G. 2004. Pests of sugarcane. In *Sugarcane*, eds. G. James, 78–100, second edition. Oxford: Blackwell Science.

Li, H. P., D. Z. Huang, and Z. G. Wang. 2011. Potential of *Beauveria bassiana* for biological control of *Apriona germari*. *Frontiers of Agriculture in China* 5:666–670.

Li, J. Q. 2009. Biocontrol of *Batocera horsfieldi* (Coleoptera: Cerambycidae) with parasitoid *Dastarcus helophoroides* (Coleoptera: Bothrideridae). Ph.D. Thesis, Northwest A&F University of Science and Technology, Yanglin, China [in Chinese with English abstract].

Li, J. Q., Z. Q. Yang, Z. X. Mei, B. Feng, P. Wang, and X. Y. Wang. 2013. Control effects on *Batocera lineolata* (Coleoptera: Cerambycidae) attacking walnut trees by releasing parasitoid *Dastarcus helophoroides* (Coleoptera: Bothrideridae). *Chinese Journal of Biological Control* 29:194–199 [in Chinese with English abstract].

Li, J. Q., Z. Q. Yang, Z. X. Mei, and Y. L. Zhang. 2009a. Pest risk analysis and control countermeasure of *Batocera horsfieldi*. *Forest Research* 22:148–153 [in Chinese with English abstract].

Li, J. Q., Z. Q. Yang, Y. L. Zhang, Z. X. Mei, Y. R. Zhang, and X. Y. Wang. 2009b. Biological control of *Batocera horsfieldi* (Coleoptera: Cerambycidae) by releasing its parasitoid *Dastarcus helophoroides* (Coleoptera: Bothrideridae). *Scientia Silvae Sinicae* 45:94–100 [in Chinese with English abstract].

Li, K. Z. 1996. Poplar stem-boring pests and their control. *Journal of Northeast Forestry University* 7:51–61 [in Chinese with English abstract].

Li, L. Y., R. Wang, and D. F. Waterhouse. 1997. *The distribution and importance of arthropod pests and weeds of agriculture and forestry plantations in southern China*. Canberra: ACIAR.

Li, S. H. 2001. Grape production in China. In *Grape production in the Asia-Pacific region*, eds. M. K. Papademetriou, and F. J. Dent, 19–27. Bangkok: FAO Regional Office.

Liao, H. H., M. Lu, L. Y. Bu, H. H. Lin, and Z. X. Mo. 2006. *Dorysthenes granulosus* (Thomson) causes serious damage to sugarcanes in Beihai City, Guangxi. *Guangxi Plant Protection* 19:22–24 [in Chinese].

Liau, T. L. 1968. *Trachylophus sinensis* Gahan and its control. *Plant Protection Bulletin, Taiwan* 10:71–78 [in Chinese with English abstract].

Liendo-Barandiaran, C. V., B. Herrera-Malaver, F. Morillo, P. Sánchez, and J. V. Hernández. 2010. Behavioral responses of *Steirastoma breve* (Sulzer) (Coleoptera: Cerambycidae) to host plant *Theobroma cacao* L., brushwood piles, under field conditions. *Applied Entomology and Zoology* 45:489–496.

Liendo-Barandiaran C. V., F. Morillo, P. Sánchez, et al. 2005. Olfactory behavior and electroantennographic responses of the cocoa beetle, *Steirastoma breve* (Coleoptera: Cerambycidae). *Florida Entomologist* 88:117–122.

Lieu, K. O. V. 1945. The study of wood borers in China. I. Biology and control of the citrus-root-cerambycids, *Melanauster chinensis*, Forster (Coleoptera). *Florida Entomologist* 27:62–101.

Lieu, K. O. V. 1947. The study of wood borers in China-II: Biology and control of the citrus-trunk-cerambycld, *Nadezhdiella cantori* (Hope) (Coleoptera). *Notes d'Entomologie Chinoise* 11:69–119.

Lingafelter, S. W., and E. R. Hoebeke. 2001. Variation and homology in elytral maculation in the *Anoplophora malasiaca/macularia* species complex (Coleoptera: Cerambycidae) of Japan and Taiwan. *Proceedings of the Entomological Society of Washington* 103:757–769.

Lingafelter, S. W., and E. R. Hoebeke. 2002. *Revision of Anoplophora (Coleoptera: Cerambycidae)*. Washington, DC: The Entomological Society of Washington.

Link, D., and E. C. Costa. 1994. Nivel de infestacao da broca dos citros, *Diploschema rotundicolle* (Serville, 1834) em cinamomo e plantas citricas, em Santa Maria, RS. *Ciencia Rural* 24:7–10.

Link, D., E. C. Costa, and A. B. Thum. 1994. Bionomia comparada dos serradores *Oncideres saga saga* (Dalman, 1923) e *Oncideres dejeani* (Thomson, 1868) (Coleoptera: Cerambycidae) in *Parapiptadenia rigida*. *Ciência Florestal* 4:137–144.

Link, D., E. C. Costa, and A. B. Thum. 1996. Alguns aspectos da biologia do serrador, *Oncideres dejeani* Thomson, 1868 (Coleoptera: Cerambycidae). *Ciência Florestal* 6:21–25.

Linsley, E. G. 1962. *The Cerambycidae of North America. Part II: Taxonomy and classification of the Parandrinae, Prioninae, Spondylinae and Aseminae.* Berkeley, CA: University of California Press.

Linsley, E. G. 1963. *The Cerambycidae of North America. Part IV: Taxonomy and classification of the subfamily Cerambycinae, tribes Elaphidionini through Rhinotragini.* Berkeley, CA: University of California Press.

Linsley, E. G., and J. A. Chemsak. 1984. *The Cerambycidae of North America Part VII, No 1: Taxonomy and classification of the subfamily Lamiinae, tribes Parmenini through Acanthoderini.* Berkeley, CA: University of California Press.

Linsley, E. G., and J. A. Chemsak. 1995. *The Cerambycidae of North America. Part VII, No 2: Taxonomy and classification of the subfamily Lamiinae, Tribes Acanthocinini through Hemilophini.* Berkeley, CA: University of California Press.

Liu, K. D., S. B. Liang, and S. S. Deng. 1998a. Integrated production practices of cashew in China. In *Integrated production practices of cashew in Asia*, eds. M. K. Papademetriou, and E. M. Herath Bangkok: Food and Agriculture Organization of the United Nations Regional Office for Asia and Pacific. http://www.fao.org/docrep/005/ac451e/ac451e03.htm#fn3 (accessed January 16, 2015).

Liu, Q. Z., Y. Z. Wang, F. Q. Tong, W. Zhang, and L. N. Xu. 1998b. Study on application techniques of *Steinernema* nematodes against RNL. *Journal of China Agricultural University* 3:17–21 [in Chinese with English abstract].

Liu, Q. Z., Y. Z. Wang, and J. L. Zhou. 1999. Biology of RNL's boring trunk and expelling frass. *Journal of China Agricultural University* 4:87–91 [in Chinese with English abstract].

Liu, Q. Z., D. W. Yang, and L. L. Liang. 2010. Analysis of damage characteristics of tea longhorn in organic tea garden in Cangnan County of Zhejiang Province. *Acta Agriculturae Zhejiangensis* 22:220–223 [in Chinese with English abstract].

Liu, Y., E. Heying, and S. Tanumihardjo. 2012a. History, global distribution, and nutritional importance of citrus fruits. *Comprehensive Reviews in Food Science and Food Safety* 11:565–576.

Liu, Y., S. Xiong, J. Q. Ren, X. X. Zhang, and L. Chen. 2012b. Comparative morphological study on genus *Batocera* (Coleoptera, Cerambycidae, Lamiinae, Batocerini). *Acta Zootaxonomia Sinica* 37:701–711 [in Chinese with English abstract].

Liu, Y. L., and S. C. Huang. 2003. Harm of *Chelidonium citri* Gressitt to *Murraya paniculata* (L.) and economic damage appraisal. *Guangxi Sciences* 10:142–145 [in Chinese].

Liu, Z., G. L. Zhang, Y. Li, and J. Zong. 1993. Biological control of peach rednecked longicorn *Aromia bungii* with entomopathogenic nematodes. *Chinese Journal of Biological Control* 9:186–186 [in Chinese with English abstract].

Loh, Y. M., and J. Zhang. 1988. Pests of coffee. In *The manual of pests and diseases of tropical crops in China*, ed. H. K. Zhang, 25–42. Beijing: Agricultural Publisher [in Chinese].

Long, H. N., and Q. X. Wei. 2007. Occurrence and control of *Dorysthenes granulosus* in Liuzhou. *Guangxi Plant Protection* 20:41–42 [in Chinese].

Lozovoi, D. J. 1954. Control of *Cerambyx dux* Fald on stone fruits. *Sad i Ogorod* (6):79–79.

Lu, W., and Q. Wang. 2005. Systematics of the New Zealand longicorn beetle genus *Oemona* Newman with discussion of the taxonomic position of the Australian species, *O. simplex* White (Coleoptera: Cerambycidae: Cerambycinae). *Zootaxa* 971:1–31.

Luo, Y. M., S. M. Cai, and Q. A. Jin. 1990. *Rhytidodera bowringii* White in Hainan Island. *Chinese Journal of Tropical Crops* 11:107–112 [in Chinese with English abstract].

Lupi, D., C. Jucker, and M. Colombo. 2013. Distribution and biology of the yellow-spotted longicorn beetle *Psacothea hilaris hilaris* (Pascoe) in Italy. *EPPO Bulletin* 43(2):316–322.

Lv, K. S., X. M. Lu, Z. P. Huang, Y. L. Liu, and Z. H. Pang. 2011. The preliminary survey on the pest and diseases of *Camellia oleifera* branches in Liuzhou City. *Guangdong Forestry Science and Technology* 27: 58–62 [in Chinese with English abstract].

Ma, W. S., H. L. Qiao, X. Q. Nong, et al. 2013. Screening for virulence strains of *Metarhizium* against *Dorysthenes hydropicus* Pascoes. *Zhongguo Zhongyao Zazhi (China Journal of Chinese Materia Medica)* 38:3438–3441 [in Chinese with English abstract].

Macfarlane, A., and I. Macfarlane. 2004. *The empire of tea.* New York: The Overlook Press.

Machado, L. A. 1992. Ocorrencia de *Diploschema rotundicolle* (Serville, 1834) (Coleoptera: Cerambycidae) em laranjeiras novas, no Estado de Sao Paulo. *Revista de Agricultura* 67:81–82.

Machado, L. A., and E. B. Berti Filho. 1999. Criacao artificial da broca-dos-citros *Diploschema rotundicolle* (Serville, 1834) (Col.: Cerambycidae). *O Biologico* 61:5–11.

Machado, L. A., and E. B. Berti Filho. 2006. Biological control using *Metarhizium anisopliae* against the citrus borer *Diploschema rotundicolle*, in two methods of application, associated with control by pruning. *Arqivos do Instituto Biologico Sao Paulo* 73:439–445.

Machado, L. A., E. Berti Filho, L. G. Leite, and E. M. Silva da. 1998. Efeito da pratica cultural (catacao manual de ramos atacados) associada ao controle biologico com o fungo *Metarhizium anisopliae* no combate a broca-dos-citros *Diploschema rotundicolle* (Serville, 1834) (Col.: Cerambycidae). *Arquivos do Instituto Biologico* 65:35–42.

Machado, L. A., and A. Raga. 1999. Eficiencia de inseticidas quimico e biologico sobre uma broca-do-tronco dos citros. *Revista de Agricultura* 74(2):229–234.

Machado, V. S., J. P. Botero, A. Carelli, M. Cupello, H. Y. Quintino, and M. V. P. Simões. 2012. Host plants of Cerambycidae and Vesperidae (Coleoptera, Chrysomeloidea) from South America. *Revista Brasileira de Entomologia* 56:186–198.

Maes, J. M., A. Allen., M. A. Monné, and F. T. Hovore. 1994. Catálogo de los Cerambycidae (Coleoptera) de Nicaragua. *Revista Nicaraguense de Entomología* 27:1–58.

Mai, Y. S. 1983. Preliminary study on *Dorysthenes walker. Sugarcane and Canesugar* (12):31–33 [in Chinese].

Mai, Y. S. 1988. Application of light traps to control *Dorysthenes walkeri* and *D. granulosus. Sugarcane and Canesugar* (3):45–48 [in Chinese].

Mamlayya, A. B., S. R. Aland, S. M. Gaikwad, and G. P. Bhawane. 2009. Incidence of beetle *Aeolesthes holosericea* Fab. on *Samanea saman* and *Albizia lebbeck* trees at Kolhapur Maharashtra. *Bionotes* 11:133–133.

Mani, M., C. Shivaraju, and S. Narendra Kulkami. 2014. *The grape entomology.* New Delhi: Springer.

Manoj, A., A. N. Sharma, and R. N. Singh. 2005. Screening of soybean genotypes for resistance against three major insect-pests. *Soybean Genetics Newsletter* 32:1–8.

Maruthadurai, R., A. R. Desai, H. R. C. Prabhu, and N. P. Singh. 2012. Insect pests of cashew and their management. *Technical Bulletin of ICAR (Research Complex for Goa, India)* 28:1–16.

Masaki, I. B. A., S. Inoue, and M. Kikuchi. 1976. Ecological studies on the yellow-spotted longicorn beetle, *Psacothea hilaris* Pascoe. *Journal of Sericultural Science of Japan* 45:156–160.

Matsumoto, S. 1920. Studies on injurious insects of the vine. *Supplementary Report of Rinji-Hokoku Prefectural Agricultural Experimental Station, Okayama* 21:1–28.

May, A. W. S. 1946. Pests of cucurbit crops. *Queensland Agricultural Journal* 62:137–150.

Mazaheri, A., B. Hatami, J. Khajehali, and S. E. Sadeghi. 2007. Reproductive parameters of *Aeolesthes sarta* Solsky (Col., Cerambycidae) on *Ulmus carpinifolia* Borkh. under laboratory conditions. *Journal of Science and Technology of Agriculture and Natural Resources* 11:333–343.

McKeown, K. C. 1947. Catalogue of the Cerambycidae (Coleoptera) of Australia. *Australian Museum Memoir* 10:1–190.

McNutt, D. N. 1967. The white coffee-borer *Anthores leuconotus* Pasc. (Col., Lamiidae): Its identification, control and occurrence in Uganda. *East African Agricultural and Forestry Journal* 32:469–473.

Meena, N. L., and U. S. Sharma. 2006. Effect of sowing date and row spacing on incidence of major insect-pests of soybean, *Glycine max* (L.) Merrill. *Soybean Research* 4:73–76.

Mehrnejad, M. R. 2014. Pest problems in pistachio producing areas of the world and their current means of control. *Acta Horticulturae (ISHS)* 1028:163–169.

Mendes, A., and J. García. 1984. Biología do besouro do cacao *Steirastoma breve* (Sulzer) (Coleoptera: Cerambycidae). *Revista Theobroma* 14:61–68.

Menier, J. J. 1992. Capture insolite d'un *Batocera lineolata* en région parisienne (Col. Cerambycidae). *Entomologiste* 48:221–223.

Meshram, P. B. 2009. Stem borer *Plocaederus obesus* Gahn (Coleoptera: Cerambycidae) as a pest of *Buchanania lanzan* (Spreng). *World Journal of Zoology* 4:305–307.

Meshram, P. B., and K. K. Soni. 2011a. Application of delivery methods for fungal pathogens and insecticides against chironji (*Buchanania lanzan*) stem borer, *Plocaederus obesus* Gahn. *Journal of Plant Protection Research* 51:337–341.

Meshram, P. B., and K. K. Soni. 2011b. Application of delivery methods for fungal pathogens and insecticides against chironji (*Buchanania lanzan*) stem borer, *Plocaederus obesus* Gahn. *Asian Journal of Experimental Biological Sciences* 2:53–57.

Mészáros, Z. 1990. Cincérek—Cerambycidae. In *A növényvédelmi állattan kézikönyve 3/A (Handbook of plant protection zoology)*, eds. T. Jermy, and K. Balázs, 215–234. Budapest: Akadémiai Kiadó [in Hungarian].

Miao, W. R. 1994. Biological characteristics and control of grape tiger longicorn. *China Forestry Science and Technology* (1):33–33 [in Chinese with English abstract].

Michaud, J. P., and A. K. Grant. 2005. The biology and behavior of the longhorned beetle, *Dectes texanus* on sunflower and soybean. *Journal of Insect Science* 5: Article 25.

Michaud, J. P., and A. K. Grant. 2009. The nature of resistance to *Dectes texanus* (Col., Cerambycidae) in wild sunflower, *Helianthus annuus*. *Journal of Applied Entomology* 133:518–523.

Michaud, J. P., and A. K. Grant. 2010. Variation in fitness of the longhorned beetle, *Dectes texanus*, as a function of host plant. *Journal of Insect Science* 10: Article 206.

Michaud, J. P., A. K. Grant, and J. L. Jyoti. 2007a. Impact of the stem borer, *Dectes texanus*, on yield of the cultivated sunflower, *Helianthus annuus*. *Journal of Insect Science* 7: Article 21.

Michaud, J. P., J. A. Qureshi, and A. K. Grant. 2007b. Sunflowers as a trap crop for reducing soybean losses to the stalk borer *Dectes texanus* (Coleoptera: Cerambycidae). *Pest Management Science* 63:903–909.

Michaud, J. P., P. W. Stahlman, J. L. Jyoti, and A. K. Grant. 2009. Plant spacing and weed control affect sunflower stalk insects and the girdling behavior of *Dectes texanus* (Coleoptera: Cerambycidae). *Journal of Economic Entomology* 102:1044–1053.

Mir, G. M., and M. A. Wani. 2005. Severity of infestation and damage to walnut plantation by important insect pests in Kashmir. *Indian Journal of Plant Protection* 33:188–193.

Mitra, S. K., and P. C. Khongwir. 1928. Orange cultivation in Assam. *Bulletin of Agricultural Department of Assam* 2:1–19.

Mitrjuskin, K. P. 1940. Control of *Plagionotus floralis* Pallas to protect lucerne crops. *Sotsialisticheskoe Zernovoe Khozyaistvo* (5):117–20 [in Russian].

Mohapatra, L. N., and R. N. Mohapatra. 2004. Distribution, intensity and damage of cashew stem and root borer, *Plocaederus ferrugineus* in Orissa. *Indian Journal of Entomology* 66:4–7.

Mohapatra, L. N., and C. R. Satapathy. 1998. Effect of prophylactic measures in management of cashew stem and root borer *Plocaederus ferrugineus*. *Indian Journal of Entomology* 60:257–261.

Mohapatra, R. N., and B. C. Jena. 2007. Biology of cashew stem and root borer, *Plocaederus ferrugineus* L. on different hosts. *Journal of Entomological Research* 31:149–154.

Mohapatra, R. N., and B. C. Jena. 2008a. Determination of feeding potentiality of cashew stem and root borer *Plocaederus ferrugineus* L. grubs under laboratory condition. *Journal of Plant Protection and Environment* 5:76–79.

Mohapatra, R. N., and B. C. Jena. 2008b. Effect of prophylactic measures in management of cashew stem and root borer (*Plocaederus ferrugineus* L.). *Journal of Plantation Crops* 36:140–141.

Mohapatra, R. N., and B. C. Jena. 2009. Seasonal incidence of cashew stem and root borer, *Plocaederus ferrugineus* Linnaeus in Orissa. *Journal of Insect Science* 22:393–397.

Mohapatra, R. N., B. C. Jena, and P. C. Lenka. 2007. Effect of some IPM components in management of cashew stem and root borer. *Journal of Plant Protection and Environment* 4:21–25.

Mohapatra, R. N., L. K. Rath, and L. N. Mohapatra. 2000. Some prophylactic measures against cashew stem and root borer (*Plocaederus ferrugineus* L). *Journal of Applied Biology* 10:88–90.

Mohi-Uddin, S., Y. Munazah, M. D. J. Ahmed, and S. B. Ahmed. 2009. Management of apple stem borer, *Aeolesthes sarta* Solsky (Coleoptera: Cerambycidae) in Kashmir. *Environment and Ecology* 27:931–933.

Monné, M. A. 2001a. Catalogue of the Neotropical Cerambycidae (Coleoptera) with known host plant—Part II: Subfamily Cerambycinae, Tribes Graciliiini to Trachyderini. *Publicações Avulsas do Museu Nacional* 90:1–119.

Monné, M. A. 2001b. Catalogue of the Neotropical Cerambycidae (Coleoptera) with known host plant—Part III: Subfamily Lamiinae, Tribes Acanthocinini to Apomecynini. *Publicações Avulsas do Museu Nacional* 92:1–94.

Monné, M. A. 2002. Catalogue of the Neotropical Cerambycidae (Coleoptera) with knownhost plant—Part IV: Subfamily Lamiinae, Tribes Batocerini to Xenofreini. *Publicações Avulsas do Museu Nacional* 94:1–92.

Monné, M. A. 2005a. Catalogue of the Cerambycidae (Coleoptera) of the Neotropical Region. Part I: Subfamily Cerambycinae. *Zootaxa* 946:1–765.

Monné, M. A. 2005b. Catalogue of the Cerambycidae (Coleoptera) of the Neotropical Region. Part II: Subfamily Lamiinae. *Zootaxa* 1023:1–760.

Moreira, M. A. B., J. O. L. de Jr Oliveira, and M. A. Monné. 2003. Ocorrência de *Hylettus seniculus* (Germar, 1824), em pomares cítricos de Roraima, Brasil, e alternativas de controle. *Acta Amazonica* 33:607–611.

Muniappan, R., B. M. Shepard, G. R. Carner, and P. A. C. Ooi. 2012. *Arthropod pests of horticultural crops in tropical Asia.* Oxfordshire: CABI.

Naik, C. M., A. K. Chakravarthy, and B. Doddabasappa. 2012. Seasonal distribution of insect pests associated with cashew (Anacardium occidentale L.) in Karnataka. *Environment and Ecology* 30:1321–1323.

NAPPO. 2006. Pest of mango, *Rhytidodera bowringii* White, intercepted at Florida Port. North American Plant Protection Organization's (NAPPO) Phytosanitary Alert System http://www.pestalert.org/viewNewsAlert.cfm?naid=13

Nascimento do, A. S., A. L. M. Mesquita, and H. V. Sampaio. 1986. Planta-armadilha no controle de coleobroca em citros. *Informe Agropecuario* 12:13–18.

Netam, H. K., R. Gupta, and S. Soni. 2013. Seasonal incidence of insect pests and their biocontrol agents on soybean. *IOSR Journal of Agriculture and Veterinary Science* 2:7–11.

Niide, T., R. D. Bowling, and B. B. Pendleton. 2006. Morphometric and mating compatibility of *Dectes texanus texanus* (Coleoptera: Cerambycidae) from soybean and sunflower. *Journal of Economic Entomology* 99:48–53.

Niide, T., R. A. Higgins, R. J. Whitworth, W. T. Schapaugh, C. M. Smith, and L. L. Buschman. 2012. Antibiosis resistance in soybean plant introductions to *Dectes texanus* (Coleoptera: Cerambycidae). *Journal of Economic Entomology* 105:598–607.

Nishida, G. M. 2002. *Hawaiian terrestrial arthropod checklist*, 4th edition. Honolulu: Bishop Museum.

Niu, J. Z., H. Hull-Sanders, Y. X. Zhang, J. Z. Lin, W. Dou, and J. J. Wang. 2014. Biological control of arthropod pests in citrus orchards in China. *Biological Control* 68:15–22.

Nolte, O., and D. Krieger. 2008. Nachweis von *Saperda candida* Fabricius 1787 auf Fehmarn eine weitere, bereits in Ansiedlung befindliche, eingeschleppte Käferart in Mitteleuropa. *DGaaE Nachrichten* 22:133–136.

Noushad, S. A., and M. H. Kazemi. 1995. Biological study on alfalfa root longhorn beetle *Plagionotus floralis* (Pall.) (Col; Cerambycidae) in East-Azarbaidjan. *Journal of Entomological Society of Iran* 15:5–13.

Ocete, R., M. Lara, L. Maistrello, A. Gallardo, and M. A. Lopez. 2008. Effect of *Xylotrechus arvicola* (Olivier) (Coleoptera, Cerambycidae) infestations on flowering and harvest in Spanish vineyards. *American Journal of Enology and Viticulture* 59:88–91.

Ocete, R., M. A. Lopez, A. Gallardo, M. A. Perez, and I. M. Rubio. 2004. Efecto de los danos de *Xylotrechus arvicola* (Olivier) (Coleoptera, Cerambycidae) sobre las caracteristicas de los racimos de la variedad de vid Tempranillo en La Rioja. *Boletin de Sanidad Vegetal, Plagas* 30:311–316.

Ocete, R., M. A. Lopez, C. Prendes, C. D. Lorenzo, and J. L. Gonzalez-Andujar. 2002a. Relacion entre la infestacion de *Xylotrechus arvicola* (Coleoptera, Cerambycidae) (Olivier) y la presencia de hongos patogenos en un vinedo de la Denominacion de Origen "La Mancha". *Boletin de Sanidad Vegetal, Plagas* 28:97–102.

Ocete, R., M. A. Lopez, C. Prendes, C. D. Lorenzo, J. L. Gonzalez-Andujar, and M. Lara. 2002b. *Xylotrechus arvicola* (Olivier) (Coleoptera, Cerambycidae), a new impacting pest on Spanish vineyards. *Vitis* 41:211–212.

Ocete, R., J. M. Valle, K. Artano, et al. 2010. Evolution of the spatio-temporal distribution of *Xylotrechus arvicola* (Olivier) (Coleoptera, Cerambycidae) in La Rioja vineyard (Spain). *Vitis* 49:67–70.

Ogura, N., K. Tabata, and W. Wang. 1999. Rearing of the colydiid beetle predator, *Dastarcus helophoroides*, on artificial diet. *BioControl* 44:291–299.

Ostojá-Starzewski, J. O., and R. H. A. Baker. 2012. *Red-necked longhorn: Aromia bungii.* UK: FERA Plant Pest Factsheet.

Ostojá-Starzewski, J., A. MacLeod, and D. Eyre. 2010. *Lemon tree bore Oemona hirta.* UK: FERA Plant Pest Factsheet. http://www.fera.defra.gov.uk/plants/publications/documents/factsheets/lemonTreeBorer.pdf (accessed March 15, 2015).

Özdikmen, H. 2010. The Turkish Dorcadiini with zoogeographical remarks (Coleoptera: Cerambycidae: Lamiinae). *Munis Entomology & Zoology* 5:380–498.

Özdikmen, H. 2014. A synopsis of Turkish Callichromatini (Coleoptera: Cerambycidae: Cerambycinae). *Munis Entomology & Zoology* 9:554–563.

Özdikmen, H., H. Ghahari, and S. Turgut. 2009. New records for Palaeactic Cerambycidae from Iran with zoogeographical remarks (Col.: Cerambycoidea: Cerambycidae). *Munis Entomology & Zoology* 4:1–18.

Özdikmen, H., and S. Turgut. 2009. On Turkish *Cerambyx* Linnaeus, 1758 with zoogeographical remarks (Coleoptera: Cerambycidae: Cerambycinae). *Munis Entomology & Zoology* 4:301–319.

Pachori, R., and D. Sharma. 1997. Changing host preference of *Obereopsis brevis* (Swed.) as a new pest of sunflower in India. *Journal of Entomological Research* 21:195–196.

Padi, B., and R. Adu-Acheampong. 2001. An integrated pest management strategy for the control of the coffee stem borer, *Bixadus sierricola* White (Coleoptera: Lamiidae). *19eme Colloque Scientifique International sur le Cafe*, Trieste, Italy, 14–18 mai 2001, pp. 1–5.

Pal, P. K. 1983. Role of mandibular length of *Obereopsis brevis* in selecting the site of oviposition and damage in soybean *Glycine max*. *Journal of Entomological Research* 6:37–47.

Palaniswami, M. S., T. R. Subramaniam, and P. C. S. Babu. 1977. Studies on the nature of damage and chemical control of the mango stem borer *Batocera rufomaculata* De Geer in Tamil Nadu. *Pesticides* 11:11–13.

Pan, C. R. 1999. Experiment on prevention and cure of *Apriona germarii* in poplar by injection method. *Journal of Zhejiang Forestry Science and Technology* 19:56–57 [in Chinese with English abstract].

Pan, X. L., J. D. Shen, Y. L. Yan, and R. M. Fu. 1997. Integrated management of mango stem borer (*Rhytidodera bowringii* White). *Chinese Journal of Tropical Crops* 18:79–83 [in Chinese with English abstract].

Pan, X. L., and L. P. S. van der Geest. 1990. Insect pests of cashew in Hainan, China, and their control. *Journal of Applied Entomology* 110:370–377.

Pande, A. K., S. V. Mahajan, A. T. Munde, and S. M. Telang. 2000. Evaluation of insecticides against major pests of soybean (*Glycine max* L.). *Journal of Soils and Crops* 10:244–247.

Paro, C. M., A. Arab, and J. Vasconcellos-Neto. 2012. Population dynamics, seasonality and sex ratio of twig-girdling beetles (Coleoptera: Cerambycidae: Lamiinae: Onciderini) of an Atlantic rain forest in south-eastern Brazil. *Journal of Natural History* 46:1249–1261.

Parsai, S. K., and S. K. Shrivastava. 1993. Effect of dates of sowing and different soyabean varieties on infestation by the girdle beetle, *Oberiopsis brevis* Swed. *Bhartiya Krishi Anusandhan Patrika* 8:1–4.

Parsai, S. K., S. K. Shrivastava, and S. S. Shaw. 1990. Comparative efficacy and economics of synthetic pyrethroids in the management of soybean girdle beetle. *Current Research—University of Agricultural Sciences (Bangalore)* 19:134–136.

Patil, A. K., S. R. Kulkarni, R. V. Kulkarni, and G. L. Chunale. 2006. Field screening of soybean genotypes for resistance to girdle beetle. *Research on Crops* 7:321–323.

Patil, A. K., V. S. Teli, and B. R. Patil. 2002. Evaluation of insecticides against girdle beetle infesting soybean. *Journal of Maharashtra Agricultural Universities* 27:321–322.

Patil, B. V., G. G. Bilapate, and R. N. Jadhav. 2009. Control of stemborer [*Nupserha* sp. near *vexator* (Pascoe)], a new pest of sunflower (*Helianthus annuus* L.), by conventional insecticides. *Helia* 32:91–98.

Patil, B. V., and R. N. Jadhav. 2009. Evaluation of sunflower germplasms for stem borer [*Nupserha* sp. near *vexator* (Pascoe)] resistance. *Helia* 32:99–106.

Patrick, C. R. 1973. Observations on the biology of *Dectes texanus texanus* (Coleoptera: Cerambycidae) in Tennessee. *Journal of the Georgia Entomological Society* 8:277–279.

Paz, J. K. S., P. R. R. Silva, L. E. M. Pádua, S. Ide, E. M. S. Carvalho, and S. S. Feitosa. 2008. Monitoramento de coleobrocas associadas à mangueira no município de José de Freitas, Estado do Piauí. *Revista Brailesira de Fruticultura* 30:348–355.

Pemberon, C. E. 1963. Important Pacific insect pests of sugar cane. *Pacific Science* 17:251–252.

Penteado Dias, A. M. 1980. Biology and ontogeny of *Trachyderes thoracicus* (Coleoptera: Cerambycidae). *Revista Brasileira de Entomologia* 24:131–136.

Petkova, D. S., and E. D. Zhekova. 2012. Forage and seed productivity in alfalfa, attacked by *Plagionotus floralis* Pall. (Coleoptera: Cerambycidae). *Bulgarian Journal of Agricultural Science* 18:708–712.

Peverieri, G. S., G. Bertini, P. Furlan, G. Cortini, and P. F. Roversi. 2012. *Anoplophora chinensis* (Forster) (Coleoptera: Cerambycidae) in the outbreak site in Rome (Italy): Experiences in dating exit holes. *Redia-Giornale di Zoologia* 95:89–92.

Phukam, E., J. N. Khound, and S. K. Dutta. 1993. Survey for host range of Citrus trunk borer, *Anoplophora versteegi* (Ritsema) (Coleoptera: Cerambycidae) in Assam. *Indian Journal of Entomology* 55:34–37.

Pigatti, A., A. P. Takematsu, and P. R. de Almeida. 1979. Ensaio preliminar de campo para controle da 'broca da haste' do chuchuzeiro—*Adetus muticus*, Thomson, 1857—(Coleoptera–Cerambycidae–Lamiinae) e observacoes gerais sobre seus habitos. *Biologico* 45:309–312.

Pinto da Fonseca, J., and M. Autuori. 1932. Lista dos prineipaes insectos que atacam plantas citricas no Brasil. *Revue d'Entomologie* 2:202–216.

Pires, E. M., I. Moreira, M. A. Soares, J. A. Marinho, R. Pinto, and J. C. Zanuncio. 2011. *Oxymerus aculeatus* (Coleoptera: Cerambycidae) causing damage on corn plants (*Zea mays*) in Brazil. *Revista Colombiana de Entomologia* 37:82–83.

Pitaksa, C. 1993. *Ecological studies, crop loss assessment and potential control of sugarcane stem boring grub, Dorysthenes buqueti guerin (Coleoptera: Cerambycidae).* MSc. Thesis. Bangkok: Kasetsart University [in Thai with English abstract].

Plant Protection Group of Guangxi Sugarcane Research Institute. 1975. Preliminary observations on biology of *Dorysthenes granulosus*. *Sugarcane and Canesugar* (8):30–32 [in Chinese].

Pliansinchai, U., V. Jarnkoon, S. Siengsri, C. Kaenkong, S. Pangma, and P. Weerathaworn. 2007. Ecology and destructive behaviour of cane boring grub (*Dorysthenes buqueti* Guerin) in north eastern Thailand. *XXVI Congress, International Society of Sugar Cane Technologists, ICC, Durban, South Africa, 29 July–2 August, 2007*:863–867.

Potting, R., D. J. van der Gaag, and B. Wessels-Berk. 2008. *Short pest risk analysis—Batocera rufomaculata, mango tree stem borer.* Wageningen: The Netherlands Plant Protection Service. http://www.vwa.nl/onderwerpen/english/dossier/pest-risk-analysis/evaluation-of-pest-risks

Prachuabmoh, O. 1986. Pest and disease records in Thailand: Sugarcane stem boring grub (*Dorystenes buqueti* Guerin) (Fam. Cerambycidae Order Coleoptera). *Quarterly Newsletter, FAO Asia and Pacific Plant Protection Commission* 29:11–12.

Prakash, P. J., J. R. S. Singh, B. V. Sanjeeva Roa, G. P. Mahobia, M. Vihaya Kumar, and B. C. Prasad. 2010. Population dynamics of round headed stem borer, *Aeolesthes holosericea* Fabr. (Coleoptera: Cerambycidae) in arjun ecosystem of Andhra Pradesh and its correlation with abiotic factors. *Entomon* 35:143–146.

Pramono, D., A. Rival, and D. P. Putranto. 2001a. *Dorysthenes* sp (Coleoptera: Cerambycidae), a new pest of sugarcane plantations in Indonesia—Biology and integrated control. *Proceedings of International Society of Sugar Cane Technologists* 2:398–400.

Pramono, D., A. Thoharisman, D. P. Putranto, D. Juliadi, and E. M. Achadian. 2001b. Effectiveness of an indigenous entomopathogenic fungus as a biocontrol agent of *Dorysthenes* sp (Coleoptera: Cerambycidae). *Proceedings of International Society of Sugar Cane Technologists* 2:401–403.

Pu, F. J. 1980. *Economic insect fauna of China, Fasc. 19—Coleoptera: Cerambycidae (II).* Science Press: Beijing [in Chinese].

Punnaiah, K. C., and V. Devaprasad. 1995. Management of cashew stem and root borer *Plocaederus ferrugineus* L. *Cashew* 9:17–23.

Purwar, J. P., and S. R. Yadav. 2004. Effect of bio-rational and chemical insecticides on stem borers and yield of soybean (*Glycine max* (L.) Merrill). *Soybean Research* 2:54–60.

Puzzi, D. 1966. Pragas de pomares de citros seu combate. *Instituto Biológico, publicação no* 116:1–55.

Puzzi, D., and A. Orlando. 1959. Principais pragas dos pomares citricos. Recomendacoes para o controle. *Biologico* 25:1–20.

Qian, T. Y. 1984. Descriptions of the larvae of four species of tea tree borers (Coleoptera: Cerambycidae). *Wuyi Science Journal* 4:189–193 [in Chinese with English abstract].

Qian, T. Y. 1985. A note on gourd stem borer *Apomecyna saltator neveosparsa* infesting Cucurbitaceae (Coleoptera: Cerambycidae). *Wuyi Science Journal* 5:47–50 [in Chinese with English abstract].

Qian, T. Y. 1987. Descriptions of the larvae of five species of peach tree borers [*Aromia bungii, Xystrocera globosa, Neocerambyx oenochrousx, Linda fraternal* and *Bacchisa fortune* (Coleoptera: Cerambycidae)], important pests in Fujian Province, China. *Wuyi Science Journal* 7:221–226 [in Chinese with English abstract].

Qiao, H. L., J. Chen, C. Q. Xu, et al. 2011. The larval growth and development of *Dorysthenes hydropicus* and its relationship with host plants. *Journal of Biology* 28:35–37, 42 [in Chinese with English abstract)].

Qin, D. R., T. B. Guo, F. R. Jiang, Y. Zhou, H. Z. Zhao, and K. F. Wang. 1994. A study on the relationship between ovary development and complemental nutrition of *Apriona* (Hope). *Journal of Nanjing Forestry University* 18:46–50 [in Chinese with English abstract].

Qin, H. Z., Y. M. Ju, and W. Z. Wang. 1997. A study on control of damage of *Apriona germari* (Hope) on *Ficus carica*. *Journal of Shanghai Agricultural College* 15:239–242 [in Chinese with English abstract].

Qin, R. M., J. Chen, C. Q. Xu, et al. 2008. Preliminary studies on biological characteristics of *Dorysthenes hydropicus*. *Zhongguo Zhongyao Zazhi (China Journal of Chinese Materia Medica)* 33:2887–2891 [in Chinese with English abstract].

Qin, W. Q., C. J. Lv, C. X. Li, S. C. Huang, and Z. Q. Peng. 2010. Research on pests of coconut in China. *Chinese Agricultural Science Bulletin* 26:200–204 [in Chinese with English abstract].

Rad, H. H. 2005. Study on the biology and distribution of the long-horned beetle *Calchaenesthes pistacivora*, a new cultivated and wild pistachio pest in Kerman Province. *Final project report of the Pistachio Research Institute of Iran*:1–23.

Rad, H. H. 2006. Study on the biology and distribution of long-horned beetles *Calchaenesthes pistacivora* n. sp (Col.: Cerambycidae): A new pistachio and wild pistachio pest in Kerman Province. *Acta Horticulturae (ISHS)* 726:425–430.

Rad, H. H., R. Zeydabadi, A. Rajabi, H. Bahreyney, and M. Morovati. 2005. Study of some insecticide effects on the long-horned beetle *Calchaenesthes pistacivora*. AGRIS, FAO. http://agris.fao.org/agris-search/search.do?recordID=IR2008000725 (accessed January 28, 2015).

Rahmathulla, V. K., and C. M. K. Kumar. 2011. Pest incidence of stem girdle beetle, *Sthenias grisator* (Cerambycidae: Coleoptera) on mulberry plantation. *Insect Environment* 17:111–112.

Rai, P. S. 1983. Bionomics of cashew stem and root borer *Plocaederus ferrugineus*. *Journal of Maharashtra Agricultural Universities* 8:247–249.

Rai, R. K., and R. K. Patel. 1990. Girdle beetle, *Obereopsis brevis* Swed. incidence in kharif soybean. *Orissa Journal of Agricultural Research* 3:163–165.

Ramachandran, S. 1953. A note on the identity of the cerambycid borer of oranges in South India. *Indian Journal of Entomology* 14:214–214.

Rana, V. K., S. P. Bhardwaj, and R. Kumar. 2004. Evaluation of insecticides against apple root-borer (*Dorysthenes hugelii*) (Cerambycidae: Coleoptera). *Indian Journal of Agricultural Sciences* 74:287–288.

Ranga Rao, P. V., K. M. Azam, K. Laxminarayana, and E. L. Eshbaugh. 1979. A new record of *Coelosterna scabrator* F. (Cerambycidae: Coleoptera) on grapevines in Andhra Pradesh. *Indian Journal of Entomology* 41:289–290.

Rao, B. H. K., R. Ayyanna, and K. L. Narayana. 1985. Integrated control of cashew stem and root borer *Plocaederus ferrugineus* L. *Acta Horticulturae* 108:136–138.

Raphael, P. K., K. Surekha, and R. Naidu. 2005. Laboratory rearing of coffee white stemborer *Xylotrechus quadripes* Chevrolet in artificial diet. *Proceedings of the ASIC 2004 20th International Conference on Coffee Science, 11–15 October, Bangalore, India (Publication No. 2-900212-19-7)*:1277–1281.

Raviprasad, T. N., and P. S. Bhat. 2010. Age estimation technique for field collected grubs of cashew stem and root borer (*Plocaederus ferrugineus* Linn.). *Journal of Plantation Crops* 38:36–41.

Raviprasad, T. N., P. S. Bhat, and D. Sundararaju. 2009. Integrated pest management approaches to minimize incidence of cashew stem and root borers (*Plocaederus* spp.). *Journal of Plantation Crops* 37:185–189.

Regupathy, A., Chandrashekharan, T. Manoharan, and S. Kuttalam. 1995. *Guide to forest entomology*. Coimbatore: Sooriya Desktop Publisher.

Ren, L. L., J. Shi, Y. N. Zhang, and Y. Q. Luo. 2012. Antennal morphology and sensillar ultrastructure of *Dastarcus helophoroides* (Fairmaire) (Coleoptera: Bothrideridae). *Micron* 43:921–928.

Rhainds, M., C. C. Lan, S. King, R. Gries, L. Z. Mo, and G. Gries. 2001. Pheromone communication and mating behaviour of coffee white stemborer, *Xylotrechus quadripes* Chevrolat (Coleoptera: Cerambycidae). *Applied Entomology and Zoology* 36:299–309.

Rhainds, M., C. C. Lan, M. L. Zhen, and G. Gries. 2002. Incidence, symptoms, and intensity of damage by three coffee stem borers (Coleoptera: Cerambycidae) in South Yunnan, China. *Journal of Economic Entomology* 95:106–112.

Rhaman, K. A., and A. W. Khan. 1942. A study of the life-history and control of *Batocera horsfieldi* Hope (Lamiidae: Coleoptera)—A borer pest of walnut tree in the Punjab. *Proceedings of Indian Academy of Sciences Section B* 15:202–205.

Rice, M. E. 1995. Branch girdling by *Oncideres cingulata* (Coleoptera; Cerambycidae) and relative host quality of persimmon, hickory, and elm. *Annals of the Entomological Society of America* 88:451–455.

Richardson, L. G. 1976. Resistance of soybeans to a stem borer *Dectes texanus texanus* Leconte. *Dissertation Abstracts International B* 36:5960B.

Rogers, C. E. 1977a. Bionomics of *Oncideres cingulate* (Coleoptera: Cerambycidae) on mesquite. *Journal of the Kansas Entomological Society* 50:222–228.

Rogers, C. E. 1977b. Cerambycid pests of sunflower: Distribution and behavior in the Southern Plains. *Environmental Entomology* 6:833–838.

Rogers, C. E. 1985. Cultural management of *Dectes texanus* (Coleoptera, Cerambycidae) in sunflower. *Journal of Economic Entomology* 78:1145–1148.

Romano, I., and J. C. Kerr. 1977. Soybeans in the central and upper Burnett. *Queensland Agricultural Journal* 103:64–67.

Rutherford, M. A., and N. Phiri. 2006. *Pests and diseases of coffee in Eastern Africa: A technical and advisory manual.* UK: CAB International.

Rystrom, Z. D., and R. Wright. 2013. Dectes stem-borer, *Dectes texanus* LeConte, in Nebraska soybeans. *The 61st Entomological Society of America Annual Meeting*, 10–13 November 2013, Austin, TX.

Rystrom, Z. D., and R. Wright. 2014. Studying the dectes stem borer (*Dectes texanus*) in Nebraska. *The 62nd Entomological Society of America Annual Meeting*, 16–19 November 2014, Portland, Oregon.

Sachan, J. N., and S. K. Gangwar. 1980. Insect pests of apple in Meghalaya. *Bulletin of Entomology* 21:113–121.

Saha, S., H. Özdikmen, M. K. Biswas, and D. Raychaudhuri. 2013. Exploring flat faced longhorn beetles (Cerambycidae: Lamiinae) from the Reserve Forests of Dooars, West Bengal, India. *ISRN Entomology* 2013: Article 737193, 8. doi: http://dx.doi.org/10.1155/2013/737193

Sahu, K. R., and D. Sharma. 2008. Management of cashew stem and root borer, *Plocaederus ferrugineus* L. by microbial and plant products. *Journal of Biopesticides* 1:121–123.

Sahu, K. R., B. C. Shukla, B. S. Thakur, S. K. Patil, and R. R. Saxena. 2012. Relative tolerance of cashew cultivars against cashew stems and root borer (CSRB), *Plocaederus ferrugineus* L. in Chhattisgarh. *Uttar Pradesh Journal of Zoology* 32:311–316.

Sakai, T., Y. Nakagawa, J. Takahashi, K. Iwabuchi, and K. Ishii. 1982. Isolation and identification of the male sex pheromone of the grape borer *Xylotrechus pyrrhoderus* Bates (Coleoptera: Cerambycidae). *Chemistry Letters* 13:263–264.

Saikia, K., N. S. A. Thakur, A. Ao, and S. Gautam. 2012. Sexual dimorphism in *Pseudonemorphus versteegi* (Ritsema) (Coleoptera: Cerambycidae), citrus trunk borer. *Florida Entomologist* 95:625–629.

Sakakibara, M. 1995. Egg periods in several populations of yellow-spotted longicorn beetle, *Psacothea hilaris* Pascoe (Coleoptera, Cerambycidae). *Japanese Journal of Applied Entomology and Zoology* 39:59–64 [in Japanese with English abstract].

Sakakibara, M., and K. Kawakami. 1992. Larval diapause inheritance mode in 2 ecotypes of the yellow-spotted longicorn beetle, *Psacothea hilaris* (Pascoe) (Coleoptera, Cerambycidae). *Applied Entomology and Zoology* 27:47–56.

Saliba, L. J. 1972. Toxicity studies on *Cerambyx dux* (Coleoptera: Cerambycidae) in Malta. *Journal of Economic Entomology* 65:424–424.

Saliba, L. J. 1974. Adult behaviour of *Cerambyx due* Faldermann. *Annals of the Entomological Society of America* 67:47–50.

Saliba, L. J. 1977. Observations on biology of *Cerambyx due* Faldermann in Maltese islands. *Bulletin of Entomological Research* 67:107–117.

Salunke, S. G., U. S. Bidgire, D. G. More, and S. S. Keshbhat. 2002. Field evaluation of soybean cultivars for their major pests. *Journal of Soils and Crops* 12:49–55.

Salve, U. S. 2014. Neem, *Azadirachta indica* Juss. a new host plant of insect *Aeolesthes holosericea* Fabricius from Maharashtra, India. *Online International Interdisciplinary Research Journal* 4:199–203.

Sama, G., J. C. Ringenbach, and M. Rejzek. 2005. A preliminary survey of the Cerambycidae of Libya (Coleoptera). *Bulletin de la Societe Entomologique de France* 110:439–454.

Samalo, A. P. 1985. Chemical control of pointed gourd vine borer (*Apomecyna saltator*) Fabr. *Madras Agricultural Journal* 72:325–329.

Sánchez, M. M. 2000. World distribution and utilization of mulberry and its potential for animal feeding. In *Mulberry for animal production*, ed. M. M. Sánchez, 1–9. FAO Animal Production and Health Paper No. 147.

Sánchez, P., and L. Capriles de Reyes. 1979. Insectos asociados al cultivo del cacao en Venezuela. *Boletín Técnico de Estación Experimental de Caucagua Centro Nacional de Investigaciones Agropecuarias* 11:1–56.

Sanjeeva Raj, P. J. 1959. The bionomics of the stem girdler *Sthenias grisator* Fab. (Cerambycidae: Coleoptera) from Tambaram, South India. *Indian Journal of Entomology* 21:163–166.

Sasanka Goswami, and N. M. M. Isahaque. 2001a. Biology of *O. posticata*: Biology and host plants of citrus shoot borer, *Oberea posticata* Gahan in Assam. *Journal of the Agricultural Science Society of North-East India* 14:57–61.

Sasanka Goswami, and N. M. M. Isahaque. 2001b. Seasonal cycle of citrus shoot borer, *Oberea posticata* Gah. in Assam. *Journal of the Agricultural Science Society of North-East India* 14:133–137.

Satyanarayana, J., R. Sudhakar, G. Sreenivas, D. V. V. Reddy, and T. V. K. Singh. 2013. Succession of potential insect and mite pests and known insect predators and parasitoids on *Jatropha curcas* L. in Andhra Pradesh, India. *International Journal of Bio-resource and Stress Management* 4:565–570.

Schoeman, P. S. 1990. Beheer van die witkoffiestamboorder. *Inligtingsbulletin—Navorsingsinstituut vir Sitrus en Subtropiese Vrugte* (220):3–4.

Schoeman, P. S. 1991. Biology and control of white coffee stemborer. *Inligtingsbulletin—Navorsingsinstituut vir Sitrus en Subtropiese Vrugte* (233):1–3.

Schoeman, P. S. 1995. Integrated control of arabica coffee pests. Part II: An overview of measures to control the white coffee stemborer, *Monochamus leuconotus* (Pascoe) (Coleoptera: Cerambycidae). *Inligtingsbulletin—Instituut vir Tropiese en Subtropiese Gewasse* (280):11–20.

Schoeman, P. S., and B. Pasques. 1993a. Chlordane stem treatments to combat white coffee stem borer. *Inligtingsbulletin—Instituut vir Tropiese en Subtropiese Gewasse* (249):17–19.

Schoeman, P. S., and B. Pasques. 1993b. Chemicals tested against the white coffee stemborer. *Inligtingsbulletin—Instituut vir Tropiese en Subtropiese Gewasse* (252):16–18.

Schoeman, P. S., and M. H. Schoeman. 1997. Fungus could be used to control white coffee stem borer. *Neltropika Bulletin* (296):36–42.

Schoeman, P. S., H. van Hamburg, and B. P. Pasques. 1998. The morphology and phenology of the white coffee stem borer, *Monochamus leuconotus* (Pascoe) (Coleoptera: Cerambycidae), a pest of Arabica coffee. *African Entomology* 6:83–89.

Schroder, T., E. Pfeilstetter, and K. Kaminski. 2012. Zum Sachstand des Citrus-Bockkafers, *Anoplophora chinensis*, in der EU und den in der Kommissionsentscheidung 2008/840/EG festgelegten Bekampfungsstrategien unter besonderer Berucksichtigung des Monitorings. *Journal fur Kulturpflanzen* 64:86–90.

Scora, R. W. 1975. On the history and origin of citrus. *Bulletin of the Torrey Botanical Club* 102:369–375.

Seetharama, H. G., V. Vasudev, P. K. V. Kumar, and K. Sreedharan. 2005. Biology of coffee white stem borer *Xylotrechus quadripes* Chev. (Coleoptera: Cerambycidae). *Journal of Coffee Research* 33:98–107.

Sengupta, C. K., and T. Sengupta. 1981. Cerambycidae (Coleoptera) of Arunachal Pradesh. *Records of the Zoological Survey of India* 78:133–154.

Senguttuvan, T., and N. R. Mahadevan. 1997a. Integrated management of cashew stem borer, *Plocaederus ferrugineus* L. *Pest Management in Horticultural Ecosystems* 3:79–84.

Senguttuvan, T., and N. R. Mahadevan. 1997b. Studies on the population fluctuation and management of cashew stem and root borer, *Plocaederus ferrugineus* L. *Pest Management in Horticultural Ecosystems* 3:85–94.

Senguttuvan, T., and N. R. Mahadevan. 1998. Comparative biology of cashew stem and root borer, *Plocaederus ferrugineus* L. on natural host and semi-synthetic diet. *Journal of Plantation Crops* 26:133–138.

Senguttuvan, T., and N. R. Mahadevan. 1999a. Influence of tree characteristics on the incidence of cashew stem and root borer *Plocaederus ferrugineus* L. *Journal of Plantation Crops* 27:59–62.

Senguttuvan, T., and N. R. Mahadevan. 1999b. Prophylactic control of stem and root borer (*Plocaederus ferrugineus*) in cashew (Anacardium occidentale). *Indian Journal of Agricultural Sciences* 69:163–165.

Shah, A. H., and V. J. Vora. 1973. A new record of *Apomecyna neglecta* Pasc. as a pest of parwal in Gujarat. *Indian Journal of Entomology* 35:343–343.

Shah, A. H., and V. J. Vora. 1974. Biology of the pointed gourd vine borer, *Apomecyna neglecta* Pasc. (Cerambycidae: Coleoptera) in South Gujarat. *Indian Journal of Entomology* 36:308–311.

Shapiro-Ilan, D. I., L. W. Duncan, L. A. Lacey, and R. Han. 2005. Orchard applications. In *Nematodes as biocontrol agents*, eds. P. S. Grewal, R. U. Ehlers, and D. I. Shapiro-Ilan, 215–229. Wallingford (UK): CABI Publishing.

Sharaf, N. S. 2010. Colonization of *Cerambyx dux* Faldermann (Coleoptera: Cerambycidae) in stone-fruit tree orchards in Fohais Directorate, Jordan. *Jordan Journal of Agricultural Sciences* 6:560–578.

Sharifi, S., and I. Javadi. 1971. Control of Rosaceae branch borer in Iran. *Journal of Economic Entomology* 64:484–486.

Sharifi, S., I. Javadi, and J. A. Chemsak. 1970. Biology of the Rosaceae branch borer, *Osphranteria coerulescens* (Coleoptera: Cerambycidae). *Annals of the Entomological Society of America* 63:1515–1520.

Sharma, A. N. 1994. Quantifying girdle beetle resistance in soybean. *Soybean Genetics Newsletter* 21:124–127.

Sharma, A. N. 1995. Determining appropriate screening parameters for evaluating soybean genotypes for tolerance to major insect-pests. *Journal of Insect Science* 8:167–170.

Sharma, A. N., G. K. Gupta, R. K. Verma, et al. 2014. *Integrated pest management for soybean*. New Delhi: National Centre for Integrated Pest Management.

Sharma, B., and J. S. Tara. 1984. Leaf yield studies of three mulberry varieties infested with *Batocera rufomaculata* (Coleoptera: Cerambycidae) in Jammu Division, Jammu and Kashmir State, India. *Zoologica Orientalis* 1:5–8.

Sharma, B., and J. S. Tara. 1986. Studies on the chemical control of *Batocera rufomaculata* De Geer (Coleoptera: Cerambycidae), a serious pest of mulberry in Jammu and Kashmir State, India. *Indian Journal of Sericulture* 25:84–87.

Sharma, B., and J. S. Tara. 1995. Infestation by fig borer, *Batocera rufomaculata* De Geer (Coleoptera: Cerambycidae) in various varieties and age groups of mulberry plants. *Journal of Insect Science* 8:104–105.

Sharma, J. P., and D. R. Khajuria. 2005. Distribution and activity of grubs and adults of apple root borer *Dorysthenes hugelii* Redt. *Acta Horticulturae* 696:387–393.

She, D. S., F. J. Fu, F. Chen, C. L. Zhou, and G. S. Ma. 2005. The occurrence of diseases and pests on the trunk and branches of Yunhexueli pear cultivar and their control. *South China Fruits* (4):66–68 [in Chinese].

Shehata, W. A., A. M. Okil, and Y. El-Sebay. 2001. Effect of some ecological factors on population level of *Macrotoma palmata* L. (Col., Cerambycidae). *Egyptian Journal of Agricultural Research* 79:105–115.

Shen, J. D., and Q. Y. Han. 1985. A preliminary study of controlling *Rhytidodera bowrigii* with DD-136 nematode. *Natural Enemies of Insects* 7:28–29 [in Chinese].

Shepard, M., M. Keeratikasikorn, P. R. B. Blood, and D. Morgan. 1981. Soybean resistance to adults of Lucerne crown borer, *Zygrita diva* Thomson (Coleoptera, Cerambycidae). *Journal of the Georgia Entomological Society* 16:34–40.

Shiekh, A. G. 1985. Insect pests of temperate fruits and their management. *Proceedings of National Workshop-cum-seminar on Temperate Fruits, SKUAST, Malangpora, Kashmir*:95–98.

Shintani, Y., and Y. Ishikawa. 1999a. Transition of diapause attributes in the hybrid zone of the two morphological types of *Psacothea hilaris* (Coleoptera: Cerambycidae). *Environmental Entomology* 28:690–695.

Shintani, Y., and Y. Ishikawa. 1999b. Geographic variation in cold hardiness of eggs and neonate larvae of the yellow-spotted longicorn beetle *Psacothea hilaris*. *Physiological Entomology* 24:158–164.

Shrivastava, S. K., S. K. Parsai, and R. K. Choudhary. 1989. Record of new alternate host plants of girdle beetle, *Oberea brevis* Swed. (Coleoptera: Lamiinae). *Journal of Entomological Research* 13:147–148.

Shui, S. Y., J. B. Wen, M. Chen, X. I. Hu, F. Liu, and J. Li. 2009. Chemical control of *Apriona germari* (Hope) larvae with zinc phosphide sticks. *Forestry Studies in China* 11:9–13 [in Chinese with English abstract].

Shukla, R. P., and S. K. Gangwar. 1989. Management of citrus trunk borer, *Anoplophora (Monohammus) versteegi* Rits. (Coleoptera: Cerambycidae) in Meghalaya. *Indian Journal of Hill Farming* 2:95–96.

Shylesha, A. N., N. S. A. Thakur, and Mahesh Kumar. 1996. Bioecology and management of citrus trunk borer *Anoplophora (Monohammus) versteegi* Ritsema (Coleoptera: Cerambycidae): A major pest of citrus in North Eastern India. *Pest Management in Horticultural Ecosystems* 2:65–70.

Shylesha, A. N., N. S. A. Thakur, and Ramchandra. 2000. Incidence of litchi trunk borer, *Aristobia testudo* Voet (Coleoptera: Lamiidae) on guava in Meghalaya. *Pest Management in Horticultural Ecosystems* 6:156–157.

Singh, D. P., and K. M. Singh. 1966. Certain aspects of bionomics and control of *Oberea brevis* S., a new pest of beans and cow peas. *Labdev Journal of Science and Technology* 4:174–177.

Singh, H. S., L. K. Bharathi, B. Sahoo, and G. Naik. 2008. New record of vine borer (*Apomecyna saltator*) and its differential damage to pointed gourd (*Trichosanthes dioica*) varieties/accessions in Orissa. *Indian Journal of Agricultural Sciences* 78:813–814.

Singh, K. M., and T. K. Singh. 2012. Life cycle and host preference of citrus trunk borer, *Anoplophora versteegi* Ritsema (Cerambycidae: Coleoptera). *Indian Journal of Entomology* 74:120–124.

Singh, M., J. P. Sharma, and D. R. Khajuria. 2010. Impact of meteorological factors on the population dynamics of the apple root borer, *Dorysthenes hugelii* (Redt.), adults in Kullu valley of Himachal Pradesh. *Pest Management and Economic Zoology* 18:134–139.

Singh, O. P. 1986. Toxicity of insecticides on the eggs and grubs of girdle beetle *Oberea brevis* Swed. a pest of soybean in Madhya Pradesh. *Legume Research* 9:11–15.

Singh, O. P., and S. V. Dhamdhere. 1983. Girdle beetle a major pest of soybean. *Indian Farming* 32:35–36.

Singh, O. P., and G. A. Gangrade. 1977. Note on girdle-beetle infestation in relation to dates of planting of soybean. *Indian Journal of Agricultural Sciences* 47:425–426.

Singh, O. P., S. K. Mehta, S. M. Sharma, and G. A. Gangrade. 1976. Response of new strains of soybean to girdle beetle *Oberea brevis* Swed. (Coleoptera: Lamiidae). *JNKVV Research Journal* 10:267–268.

Singh, O. P., and K. J. Singh. 1989. Efficacy and economics of granular insecticides in the management of the stem borers of soybean in Madhya Pradesh, India. *Pestology* 13:8–12.

Singh, O. P., and K. J. Singh. 1996. Yield response to girdling by the girdle beetle *Obereopsis brevis* (Swed) on different plant parts of soya bean. *Tropical Agriculture* 73:77–79.

Singh, O. P., P. P. Singh, and K. J. Singh. 1990. Influence of intercropping of soybean varieties with sorghum on the incidence of major pests of soybean. *Legume Research* 13:21–24.

Singh, R. N. 1941. The life-history, biology and ecology of the apple root borer, *Lophosternus hugelii* Redtenbacher in Kumaun. *Indian Journal of Agricultural Science* 11:925–940.

Singh, S. M., and S. Singh. 1962. On the occurrence of stem-girdler *Sthenias grisator* Fabr., Cerambycidae; Coleoptera on grapes *Vitis* spp. in Uttar Pradesh. *Horticultural Research Institute, Saharanpur*:142–145.

Singh, S. P., N. S. Rao, and K. K. Kumar. 1983. Studies on the borer pests of citrus. *Indian Journal of Entomology* 45:286–294.

Sinha, D., and H. K. Netam. 2013. Screening of soybean varieties against girdle beetle. *Journal of Plant Development Sciences* 5:73–76.

Slipinski, A., and H. E. Escalona. 2013. *Australian longhorn beetles (Coleoptera: Cerambycidae). Vol. 1: Introduction and subfamily Lamiinae*. Melbourne: CSIRO Publishing.

Sloderbeck, P. E., and L. L. Buschman. 2011. Aerial insecticide treatments for management of Dectes stem borer, *Dectes texanus*, in soybean. *Journal of Insect Science* 11: Article 49, doi: 10.1673/031.011.4901

Smith, D., G. A. C. Beattie, and R. Broadley. 1997. *Citrus pests and their natural enemies: Integrated pest management in Australia*. Brisbane: Department of Primary Industries and Horticultural Research and Development Corporation.

Smith, E. 1996. Control of longicorn beetles in mango trees. *Agnote* (695):1–2.

Solomon, J. D. 1968. Cerambycid borer in mulberry. *Journal of Economic Entomology* 61:1023–1025.

Solomon, J. D. 1985. Observations on a branch pruner, *Psyrassa unicolor* Randall Coleoptera Cerambycidae, in pecan trees. *Journal of Entomological Science* 202:258–262.

Sommartya, P., and W. Suasa-ard. 2006. Sugarcane longhorn stem borer *Dorysthenes buqueti* (Coleoptera: Cerambycidae) and its natural enemies in Thailand (abstract). *ISSCT Entomology workshop,* 14–24 May 2006, Cairns, Australia.

Sommartya, P., W. Suasa-ard, and A. Puntongcum. 2007. Natural enemies of sugarcane longhorn stem borer, *Dorysthenes buqueti* Guerin (Coleoptera: Cerambycidae), in Thailand. *XXVI Congress, International Society of Sugar Cane Technologists, ICC, Durban, South Africa*, 29 July–2 August, 2007:858–862.

Sontakke, B. K. 2002. Biology of the pointed gourd vine borer, *Apomecyna saltator* Fab. (Coleoptera: Cerambycidae) in Western Orissa. *Journal of Applied Zoological Researches* 13:197–198.

Sontakke, B. K., and C. R. Satpathy. 1993. Field evaluation of synthetic pyrethroids against vine borer, *Apomecyna saltator* Fabr. infesting pointed gourd. *Indian Journal of Entomology* 55:390–392.

Soria, F. J., M. A. Lopez, M. A. Perez, L. Maistrello, I. Armendariz, and R. Ocete. 2013. Predictive model for the emergence of *Xylotrechus arvicola* (Coleoptera: Cerambycidae) in La Rioja vineyards (Spain). *Vitis* 52:91–96.

Srivastava, A. S., K. M. Srivastava, B. K. Awasthi, and P. M. Nigam. 1972. Damage of stem borer *Oberia brevis* S. (Cerambycidae: Coleoptera), a new pest of soybean crop in Uttar Pradesh. *Labdev Journal of Science and Technology* 10-B:53–54.

Stahel, M., and R. Holenstein. 1948. Der Buchenbock, ein Kirschbaumschadling. *Schweizer Zeitschrift für Obst- und Weinbau* 57:413–416.

Steffan, S. A., and D. G. Alston. 2005. Prionus root borer (*Prionus californicus*). Utah State University Extension Fact Sheet HG/Orchard/2005-01. http://extension.usu.edu/files/publications/publication/HG_ Orchard_2005-01.pdf (accessed March 31, 2015).

Strangi, A., G. S. Peverieri, and P. F. Roversi. 2013. Managing outbreaks of the citrus long-horned beetle *Anoplophora chinensis* (Forster) in Europe: Molecular diagnosis of plant infestation. *Pest Management Science* 69:627–634.

Strickland, G. R. 1981. Integrating insect control for Ord soybean production. *Journal of Agriculture, Western Australia* 22:81–82.

Suasa-ard, W., P. Sommartya, P. Buchatian, A. Puntongcum, and R. Chiangsin. 2008. Effect of *Metarhizium anisopliae* on infection of sugarcane stems borer, *Dorysthenes buqueti* Guerin (Coleoptera: Cerambycidae) in laboratory. *Proceedings of the 46th Kasetsart University Annual Conference, Kasetsart, 29 January– 1 February, 2008 (Subject: Plants)*:155–160.

Suasa-ard, W., K. Suksen, and O. Kernasa. 2012. Utilisation of the green muscardine, *Metarhizium anisopliae*, to control the sugarcane longhorn stem borer *Dorysthenes buqueti* guerin (Coleoptera: Cerambycidae). *International Sugar Journal* 114:37–40.

Sudhi-Aromna, S., K. Jumroenma, P. Chaowattanawong, W. Plodkornburee, and Y. Sangchote. 2008. Studies on the biology and infestation of stem borer, *Batocera rufomaculata*, in durian. *ISHS Acta Horticulturae* 787: CD-rom format CD only.

Sundararaju, D. 1985. Chemical control of cashew stem and root borer, *Plocaederus ferrugineus* L. at Goa. *Journal of Plantation Crops* 13:63–66.

Sundararaju, D. 2002. Pest and disease management of cashew in India. *Cashew* 16(4):32–38.

Susurluk, I. A., N. A. Kumral, U. Bilgili, and E. Acikgoz. 2011. Control of a new turf pest, *Dorcadion pseudopreissi* (Coleoptera: Cerambycidae), with the entomopathogenic nematode *Heterorhabditis bacteriophora*. *Journal of Pest Science* 84:321–326 [in Turkish with English abstract]

Susurluk, I. A., N. A. Kumral, A. Peters, U. Bilgili, and E. Açikgöz. 2009. Pathogenicity, reproduction and foraging behaviours of some entomopathogenic nematodes on a new turf pest, *Dorcadion pseudopreissi* (Coleoptera: Cerambycidae). *Biocontrol Science and Technology* 19:585–594.

Swapna, S., and S. K. Divya. 2010. Survey on the incidence of the stem borer, *Batocera rufomaculata* in mulberry plantations in Palakkad district, Kerala. *Journal of Entomological Research* 34:147–149.

Tadros, A. W. M. M. Kinawy, and F. F. Abdallah. 1993. Population dynamics and host range of *Macrotoma palmata* F. (Coleoptera: Cerambycidae). *Insect Science and its Application* 14:713–718.

Tan, J. C., J. W. Zhang, N. W. Xiao, T. Z. Yuan, and X. Deng. 2003. A list of tea pest insects and mites in Hunan Province. *Journal of Hunan Agricultural University (Natural Sciences)* 29:296–307 [in Chinese with English abstract].

Tapley, R. G. 1960. The white coffee borer, *Anthores leuconotus* Pase., and its control. *Bulletin of Entomological Research* 51:279–301.

Tara, J. S., R. Gupta, and M. Chhetry. 2009. A study on *Aeolesthes holosericea* Fabricius, an important pest of apple plantations (*Malus domestica* Borkh.) in Jammu region. *Asian Journal of Animal Science* 3:222–224.

Tăusan, I., and C. Bucsa. 2010. Genus *Cerambyx* L. 1758 (Coleoptera: Cerambycidae) in the Natural History Museum collections of Sibiu (Romania). *Brukenthal Acta Musei* 3:607–612.

Tavakilian, G., and H. Chevillotte. 2013. TITAN: Cerambycidae database (version 12, Oct 2011). In: *Species 2000 & ITIS Catalogue of Life, 11th March 2013*, eds. Y. Roskov, T. Kunze, L. Paglinawan, et al., Digital resource at www.catalogueoflife.org/col/ (accessed January 15, 2015). Reading: Species 2000.

Tavakilian, G. L., A. Berkov, B. Meurer-Grimes, and S. Mori. 1997. Neotropical tree species and their faunas of xylophagous longicorns (Coleoptera: Cerambycidae) in French Guiana. *The Botanical Review* 63:304–355.

Tawfik, H. M., W. A. Shehata, F. N. Nasr, and F. F. Abd-Allah. 2014. Population dynamics of *Macrotoma palmata* F. (Col.: Cerambycidae) on casuarina trees in Alexandria, Egypt. Alex. *Journal of Agricultural Research* 59:141–146.

Taylor, H. S. 1957. Citrus borer. *New Zealand Journal of Agriculture* 94:357–358.

Teocchi, P., J. Sudre, and E. Jiroux. 2014. Synonymies, diagnoses et bionomie de quelques Lamiaires africains (20e note) (Coleoptera, Cerambycidae, Lamiinae). *Les Cahiers Magellanes* 15:51–95.

Tezcan, S., and M. Rejzek. 2002. Longhorn beetles (Coleoptera: Cerambycidae) recorded in cherry orchards in western Turkey. *Zoology in the Middle East* 27:91–100.

Thakur, A. K., and O. P. Bhalla. 1980. Effect of soil moisture and soil depth on larval diapause of soybean girdler *Nupserha nitidior* Pic var. *atripennis* (Brng) (Col.: Cerambycidae). *Tropical Grain Legume Bulletin* (20):14–16.

Thakur, A. K., and O. P. Bhalla. 1982. Studies on the response of insecticides against soybean girdler *Nupserha nitidior* Pic var *atripennis* Breuning. *Pesticides* 16:21–22.

Thongphak, D., and C. Hanlaoedrit. 2012. Taxonomy of sugarcane longhorn beetle *Dorysthenes* spp. derived from North and Northeast of Thailand. *KKU Research Journal* 17:1–8 [in Thai].

Thongphak, D., and Q. Wang. 2007. Taxonomic revision of the longicorn beetle genus *Uracanthus* Hope (Coleoptera: Cerambycidae: Cerambycinae) from Australia. *Zootaxa* 1569:1–139.

Tindall, K. V., and K. Fothergill. 2010. *Zelia tricolor* (Diptera: Tachinidae): First host record from *Dectes texanus* (Coleoptera: Cerambycidae). *Florida Entomologist* 93:635–636.

Tindall, K. V., and K. Fothergill. 2012. *Dolichomitus irritator* (Hymenoptera: Ichneumonidae): A new parasite of *Dectes texanus* (Coleoptera: Cerambycidae) in soybeans. *Florida Entomologist* 95:238–240.

Tindall, K. V., S. Stewart, F. Musser, et al. 2010. Distribution of the long-horned beetle, *Dectes texanus*, in soybeans of Missouri, Western Tennessee, Mississippi, and Arkansas. *Journal of Insect Science* 10: Article 178.

Toepfer, S., H. M. Li, S. G. Pak, et al. 2014. Soil insect pests of cold temperate zones of East Asia, including DPR Korea: A review. *Journal of Pest Science* 87:567–595.

Toshova, T. B., D. I. Atanasova, M. Tóth, and M. A. Subchev. 2010. Seasonal activity of *Plagionotus (Echinocerus) floralis* (Pallas) (Coleoptera: Cerambycidae: Cerambycinae) adults in Bulgaria established by attractant baited fluorescent yellow funnel traps. *Acta Phytopathologica et Entomologica Hungarica* 45:391–399.

Tozlu-Goksel. 2000. The tropical fig borer, *Batocera rufomaculata,* new for Turkey. *Zoology in the Middle East* 20:121–124.

Tseng, H. T. 1962. Injury of *Dorysthenes (Cyrtognathus) hydropicus* Pascoe to sugarcanes in Taichung District. *Report of Taiwan Sugar Experimental Station* 27:111–117 [in Chinese with English abstract].

Tseng, H. T. 1964. A study on chemical control of the sugar cane long horn beetle, *Dorysthenes hydropicus* Pescoe. *Report of the Taiwan Sugar Experiment Station* 35:141–147 [in Chinese with English abstract].

Tseng, H. T. 1966. A further study on chemical control of the sugar cane longhorn beetle *Dorysthenes hydropicus. Report of the Taiwan Sugar Experiment Station* 42:47–50 [in Chinese with English abstract].

Tu, K. L., G. Y. Li, Z. N. Xue, G. P. Li, D. B. Jiang, and D. B. Liao. 2006. Research on biological characteristics of *Aphrodisium gibbicolle. Guangxi Forestry Science* 35:61–65 [in Chinese with English abstract].

Upadhyay, S., and S. Sharma. 2000. Efficacy of some insecticides against eggs and grubs of soybean girdler (*Oberea brevis*) in the field. *Indian Journal of Agricultural Sciences* 70:57–58.

Upadhyay, S., S. Sharma, and R. C. Mishra. 1999. Effect of girdle beetle (*Oberea brevis*) on quantitative characters of soybean (*Glycine max*) cultivars. *Indian Journal of Agricultural Sciences* 69:806–807.

Upadhyay, S. K., B. Chaudhary, and B. Sapkota. 2013. Integrated management of mango stem borer *(Batocera rufomaculata* Dejan*)* in Nepal. *Global Journal of Biology, Agriculture and Health Sciences* 2:132–135.

Upasani, E. R., and N. A. Phadnis. 1968. A new record of borer *Celosterna scabrator* Fbr. (Lamiidae, Coleoptera) as a pest of grape vine, *Vitis vinifera* L. in Maharashtra. *Indian Journal of Entomology* 30:177–177.

Urano, T. 2006. Experimental release of adult *Dastarcus helophoroides* (Coleoptera: *alternatus* (Coleoptera: Cerambycidae). *Bulletin of FFPRI* 5:257–263.

USDA. 2015. *Citrus: World markets and trade.* Washington DC: USDA Foreign Agricultural Service.

Vanhanen, H., T. O. Veteli, and P. Niemelä. 2008. Potential distribution ranges in Europe for *Aeolesthes sarta, Tetropium gracilicorne* and *Xylotrechus altaicus,* a CLIMEX analysis. *EPPO Bulletin* 38:239–248.

Vasanthi, P., and T. N. Raviprasad. 2012. Relative susceptibility of cashew stem and root borers (csrb), *Plocaederus* spp. and *Batocera rufomaculata* (de Geer) (Coleoptera: Cerambycidae) to entomopathogenic nematodes. *Journal of Biological Control* 26:23–28.

Vasanthi, P., and T. N. Raviprasad. 2013a. Antennal sensilla of cashew stem and root borers *Plocaederus ferrugenius* and *P. obesus* (Coleoptera: Cerambycidae). *International Journal of Science and Research* 2:62–69.

Vasanthi, P., and T. N. Raviprasad. 2013b. Biology and morphometrics of cashew stem and root borers (CSRB) *Plocaederus ferrugenius* and *Plocaederus obesus* (Coleoptera: Cerambycidae) reared on cashew bark. *International Journal of Scientific and Research Publications* 3:1–7.

Veda, O. P., and S. S. Shaw. 1994. Record of stem boring beetle on sunflower in Jhabua Hills of M. P. *Current Research—University of Agricultural Sciences (Bangalore)* 23:118–119.

Venkatesha, M. G., and A. S. Dinesh. 2012. The coffee white stemborer *Xylotrechus quadripes* (Coleoptera: Cerambycidae): Bioecology, status and management. *International Journal of Tropical Insect Science* 32:177–188.

Verma, K. L., and M. Singh. 1986. Population distribution of apple root borer, *Dorysthenes hugleii* Redtenbacher in Himachal Pradesh. *Conference: Advances in research on temperate fruits, 15–18 March 1984, Solan, India*:355–357.

Verma, S. N., and O. P. Singh. 1991. Rice bean, *Vigna umbellate*, a new host of soybean girdle beetle (*Obereopsis brevis* S.). *Indian Journal of Plant Protection* 19:104–104.

Vicente, E. A. 1991. Dinamica de danos provocados por *Elaphidion cayamae* Fisher en Contramaestre. *Centro Agricola* 18:53–57.

Visitpanich, J. 1994. The biology and survival rate of the coffee stem borer, *Xylotrechus quadripes* Chevrolat (Coleoptera, Cerambycidae) in Northern Thailand. *Japanese Journal of Entomology* 62:731–745.

Vitali, F., and C. A. Casadio. 2007. Contribution to the cerambycid fauna of the Solomon Islands (Coleoptera Cerambycidae). *Entomapeiron Neoentomology, Genova* 1:1–36.

Wagner, M. R., S. K. N. Atuahene, and J. R. Cobbinah. 1991. *Forest entomology in west tropical Africa: Forest insects of Ghana*. The Netherlands: Kluwer Academic Publishers.

Waite, G. K., and J. S. Hwang. 2002. Pests of lichi and longan. In *Tropical fruit pests and pollinators—Biology, economic importance, natural enemies and control*, eds. J. E. Pena, J. L. Sharp, and M. Wysoki, 331–360. Wallingford (UK): CABI Publishing.

Waller, J. M., M. Bigger, and R. J. Hillocks. 2007. Stem- and branch-borers. In *Coffee pests, diseases and their management*, eds. J. M. Waller, M. Bigger, and R. J. Hillocks, 41–67. Oxfordshire (UK): CABI Publishing.

Wang, J. J. 1994. Damage to pine stumps by *Dorysthenes granulosus*. *Guangxi Forestry* (4):30–30 [in Chinese].

Wang, L. Z., Q. L. Lang, X. H. Xia, et al. 2009. Checklist of oak (*Quercus*) pests in Cerambycidae. *Journal of Shenyang Agricultural University* 40:546–557 [in Chinese with English abstract].

Wang, Q. 1995. The Australian longicorn beetle genus, *Skeletodes* (Coleoptera: Cerambycidae). *The Coleopterists Bulletin* 49:109–118.

Wang, Q., L. Y. Chen, W. Y. Zeng, and J. S. Li. 1996. Reproductive behavior of *Anoplophora chinensis* (Forster) (Coleoptera: Cerambycidae), a serious pest of citrus trees. *The Entomologist* 115:40–49.

Wang, Q., and G. L. Shi. 1999. Parasitic natural enemies of lemon tree borer. *Proceedings of New Zealand Plant Protection Conference* 52:60–64.

Wang, Q., and G. L. Shi. 2001. Host preference and sex allocation of three hymenopteran parasitoids of a longicorn pest, *Oemona hirta* (Fabricius) (Coleoptera: Cerambycidae). *Journal of Applied Entomology* 125:463–468.

Wang, Q., G. Shi, and L. K. Davis. 1998. Reproductive potential and daily reproductive rhythms of *Oemona hirta* (Coleoptera: Cerambycidae). *Journal of Economic Entomology* 91:1360–1365.

Wang, Q., G. Shi, D. Song, D. J. Rogers, L. K. Davis, and X. Chen. 2002a. Development, survival, body weight, longevity, and reproductive potential of *Oemona hirta* (Coleoptera: Cerambycidae) under different rearing conditions. *Journal of Economic Entomology* 95:563–569.

Wang, Q., W. Y. Zeng, L. Y. Chen, J. S. Li, and X. M. Yin. 2002b. Circadian reproductive rhythms, pair-bonding, and evidence for sex-specific pheromones in *Nadezhdiella cantori* (Coleoptera: Cerambycidae). *Journal of Insect Behavior* 15:527–539.

Wang, S. L., A. Q., Wang, M. H., Xia, and S. H. Dong. 2004. Study on the occurrence of *Batocera horsfieldi* on walnut and its control. *China Fruits* (2):11–13 [in Chinese].

Wang, S. W. 2006. Control methods for mango stem borer. *Zhi Fu Tian Di* 2006(6):29–29 [in Chinese].

Wang, Z. W., Y. M. Liao, G. Y. Zhang, C. Z. Yang, and Z. M. Li. 1999. Culture of an entomogenous nematode and determination of its pathogenicity to two longicorn species. *Journal of Guangxi Agricultural and Biological Science* 18:109–113 [in Chinese with English abstract].

Wanjala, F. M. E. 1988. *Studies on population dynamics of the yellow headed borer, Dirphya nigricornis Olivier (Coleoptera: Cerambycidae), a pest of coffee in Kenya*. Ph.D. Thesis, University of Nairobi, Kenya.

Wanjala, F. M. E. 1990. Potential for integrated control of the yellow headed borer, *Dirphya nigricornis* Olivier (Coleoptera: Cerambycidae) in Kenya. *Integrated Pest Management in Tropical and Subtropical Cropping Systems '89, held on 8–15 February 1989, Bad Durkheim, Germany* (2):323–330.

Wanjala, F. M. E., and B. M. Khaemba. 1988. Reproductive characteristics that influence birth rates of *Dirphya nigricornis* Olivier (Coleoptera: Cerambycidae) in Kenya. *Environmental Entomology* 17:542–545.

Wapshere, A. J. 1974. An ecological study of an attempt at biological control of noogoora burr (*Xanthium strumarium*). *Australian Journal of Agricultural Research* 25:275–292.

Watari, Y., T. Yamanaka, W. Asano, and Y. Ishikawa. 2002. Prediction of the life cycle of the west Japan type yellow-spotted longicorn beetle, *Psacothea hilaris* (Coleoptera: Cerambycidae) by numerical simulations. *Applied Entomology and Zoology* 37:559–569.

Webb, G. A. 1987. Larval host plants of Australian Cerambycidae (Coleoptera) held in some Australian insect collections. *Forestry Commission of New South Wales Technical Paper* 8:1–19.

Webb, G. A. 1988. Notes on the biology of *Skeletodes tetrops* Newman (Coleoptera: Cerambycidae). *Australian Entomological Magazine* 15:73–75.

Webb, G. A., G. A. Williams, and P. De Keyzer. 1988. Some new and additional larval host records for Australian Cerambycidae (Coleoptera). *Australian Entomological Magazine* 15:95–104.

Wei, J. N., and R. P. Kuang. 2002. Biological control of coffee stem borers, *Xylotrechus quardripes* and *Acalolepta cervinus*, by *Beauveria bassiana* preparation. *Insect Science* 9:43–50.

Wei, J. N., and X. W. Yu. 1998. Study on living strategies of coffee stem borers, *Xylotrechus quardripes* and *Acalolepta cervinus*. *Zoological Research* 19:218–224 [in Chinese with English abstract].

Wei, X. D. 1990. Control of pear borer, *Chreonoma fortunei* (Col.: Cerambycidae) with *Scleroderma guani* (Hym.: Bethylidae). *Chinese Journal of Biological Control* 6:116–117 [in Chinese with English abstract].

Williamson, A. J. P. 1976. Soybeans in Queensland. *Queensland Agricultural Journal* 102:573–582.

Xia, J. P., J. H. Dai, L. D. Liu, and X. Y. Hu. 2005. Progress in research on *Batocera horsfieldi*. *Journal of Hubei Forestry Science and Technology* 132:42–44 [in Chinese with English abstract].

Xiong, G. R., Z. P. Li, G. L. Feng, et al. 2010. Occurrence and control of sugarcane pests in Hainan Province. *Chinese Journal of Tropical Crops* 31:2243–2249 [in Chinese with English abstract].

Xu, C., J. Xu, J. Chen, J. Yu, R. Chen, R. Qin, et al. 2010a. Lethal effects of entomopathogenic nematodes on larvae of *Dorysthenes hydropicus* in laboratory experiment. *Zhongguo Zhongyao Zazhi (China Journal of Chinese Materia Medica)* 35:1239–1241 [in Chinese with English abstract].

Xu, C., Z. D. Yang, and Y. Q. He. 2007. Biological characteristics and control of pear eye longicorn beetle. *Journal of Northwest Forestry University* 22:109–110 [in Chinese with English abstract].

Xu, J. L., R. H. Han, X. L. Liu, L. Cao, and P. Yang. 1995. The application of codling moth nematode against the larvae of the lichi longicorn beetle. *Acta Phytophylacica Sinica* 22:12–16 [in Chinese with English abstract].

Xu, T., B. Z. Ji, Q. D. Zhang, H. J. Wang, Y. W. Ju, and Z. F. Hua. 2010b. Investigations on the damages of *Batocera lineolata* in high altitude oak forests of Sanqingshan Mountain. *Forest Pest and Disease* 29:23–25 [in Chinese with English abstract].

Yadav, M. K., S. M. Matkar, A. N. Sharma, M. Billore, K. N. Kapoor, and G. L. Patidar. 2001. Efficacy and economics of some new insecticides against defoliators and stem borers of soybean (*Glycine max* (L.) Merrill). *Crop Research* 21:88–92.

Yamanobe, T., and H. Hosoda. 2002. High survival rates for the longicorn beetle, *Apriona japonica* Thomson (Coleoptera, Cerambycidae) in beech trees (*Fagus crenata* Blume) planted in lowlands. *Japanese Journal of Applied Entomology and Zoology* 46:256–258 [in Japanese with English abstract].

Yan, C. X., C. Hao, and M. Q. Wang. 1997. A preliminary study on *Dorysthenes paradoxus*. *Journal of Shanxi Agricultural University* 17:342–345 [in Chinese with English abstract].

Yan, Y. H., D. Z. Huang, Z. G. Wang, D. R. Ji, and J. J. Yan. 1996. Preliminary study on the biological characteristics of *Aprostocetus futukai* Miwa et Sonan. *Journal of Hebei Agricultural University* 19:41–46 [in Chinese with English abstract].

Yanega, D. 1996. *Field guide to Northeastern longhorned beetles (Coleoptera: Cerambycidae)*. Champaign: Illinois Natural History Survey

Yang, C. 1992. Integrated management of *Dorysthenes granulosus* Thomson. *Sugarcane and Canesugar* (1):11–15 [in Chinese].

Yang, C., H. Y. Chen, Z. Q. Huang, and X. Y. Yang. 1989. Damage status and chemical control of *Dorysthenes granulosus* on sugarcane at Qianjin Farm. *Sugarcane and Canesugar* (6):13–17 [in Chinese].

Yang, D. W., and Q. Z. Liu. 2010. Tea longhorn beetle damage and its control emphasis in organic tea gardens. *Journal of Anhui Agricultural Sciences* 38(3):1305 and1312 [in Chinese with English abstract].

Yang, D. W., Q. Z. Liu, X. Y. Li, et al. 2011. Relationship between the damage degree of tea longhorn beetle larvae and the distances of tea clumps to trap lamp and roadside. *Acta Agriculturae Zhejiangensis* 23(1):97–100 [in Chinese with English abstract].

Yang, H., W. Yang, C. P. Yang, et al. 2012. Sexual recognition and mating behaviour of the populous longicorn beetle, *Batocera lineolata* (Coleoptera: Cerambycidae). *Munis Entomology & Zoology* 7:191–199.

Yang, H., W. Yang, C. P. Yang, et al. 2013. Electrophysiological and behavioral responses of the whitestriped longhorned beetle, *Batocera lineolata*, to the diurnal rhythm of host plant volatiles of holly, *Viburnum awabuki*. *Journal of Insect Science* 13: Article 85.

Yang, Q. L., M. C. Zhang, G. M. Wang, and H. P. Zhang. 2005. Effectiveness of using 8% cypermethrin for control of *Apriona germarii*. *China Fruits* (1):54–54 [in Chinese].

Yao, Y. X., W. X. Zhao, and W. X. Huai. 2009. Be cautious of a serious pest on orchard-*Saperda candida*. *Chinese Bulletin of Entomology* 46:819–822 [in Chinese with English abstract].

Yoon, H. J., and Y. I. Mah. 1999. Life cycle of the mulberry longicorn beetle, *Apriona germari* Hope on an artificial diet. *Journal of Asia-Pacific Entomology* 2:169–173.

Yoon, H. J., I. G. Park, Y. I. Mah, S. B. Lee, and S. Y. Yang. 1997a. Ecological characteristics of mulberry longicorn beetle, *Apriona germari* Hope, at the hibernation stage in mulberry fields. *Korean Journal of Applied Entomology* 36:67–72 [in Korean with English abstract].

Yoon, H. J., I. G. Park, Y. I. Mah, and K. Y. Seol. 1997b. Larval development of mulberry longicorn beetle, *Apriona germari* Hope, on the artificial diet. *Korean Journal of Applied Entomology* 36:317–322 [in Korean with English abstract].

You, D. K., and Y. Q. Wu. 2007. The occurrence, damage characters and control of citrus borer. *South China Fruits* (4):23–23 [in Chinese].

Yu, J. X., Y. R. Zhang, and W. H. Zhong. 2007a. Study on the occurrence and non-pollution control of *Batocera horsfieldi* on poplar. *Journal of Huinan Forestry Science and Technology* 34:30–31.

Yu, X. W., and R. P. Kuang. 1997. Spatial distribution and its application of population of coffee stem-borer. *Zoological Research* 18:39–44 [in Chinese with English abstract].

Yu, Y. H., G. J. Chen, D. W. Wei, X. R. Zeng, and T. Zeng. 2012. Division of larval instars of *Dorysthenes granulosus* based on Crosby growth rule. *Journal of Southern Agriculture* 43:1485–1489 [in Chinese with English abstract].

Yu, Y. H., R. Z. Chen, D. W. Wei, X. F. Chen, T. Zeng, and Z. Y. Wang. 2009. Trapping and killing of *Dorysthenes granulosus* adults using frequency vibration-killing lamps. *Guangxi Agricultural Sciences* 40:1552–1554 [in Chinese with English abstract].

Yu, Y. H., D. W. Wei, T. Zeng, Z. Y. Wang, R. Z. Chen, and X. F. Chen. 2008. Preliminary report on controlling *Dorysthenes granulosus* (Thomson) by pitfall traps. *Journal of Anhui Agricultural Science* 36:11840–11841 [in Chinese with English abstract].

Yu, Y. H., T. Zeng, D. W. Wei, X. R. Zeng, L. F. Li, and Z. Y. Wang. 2006. Damage status and control strategies of *Dorysthenes granulosus* (Thomson) in Guangxi sugarcane region. *Guangxi Agricultural Sciences* 37: 545–547 [in Chinese with English abstract].

Yu, Y. H., T. Zeng, D. W. Wei, X. R. Zeng, L. F. Li, and Z. Y. Wang. 2007b. Sugarcanes in Guangxi are seriously damaged by *Dorysthenes granulosus* in recent years. *Plant Protection* 33:136–137 [in Chinese].

Yu, Y. H., T. Zeng, X. R. Zeng, D. W. Wei, Z. Y. Wang, and S. X. Ren. 2010. Virulence of *Metarhizium* strains to the larvae of *Dorysthenes granulosus* Thomson. *Chinese Journal of Biological Control* 26(Suppl.):8–13 [in Chinese with English abstract].

Zajciw, D. 1975. Descriptions of larva and pupa of *Macropophora accentifer* Oliver, 1795 Coleoptera, Cerambycidae, Lamiinae. *Anais da Academia Brasileira de Ciências* 472:347–350.

Zeng, C. F., and Q. S. Huang. 1981. Preliminary study on occurrence and control of *Dorysthenes granulosus*. *Entomological Knowledge (Kunchong Zhishi)* (1):18–20 [in Chinese].

ZetaBoards. 2012. Mango stem borer—*Batocera rufomaculata*. http://carnivoraforum.com/topic/9703715/1/ (accessed June 12, 2015).

Zhang, B. Q. 2002. *Preliminary studies on mango resistant mechanism to Rhytidodera bowringii White*. Master's Thesis. Hainan: South China University of Tropical Agriculture.

Zhang, C. 1992. Damage to mango seedlings by *Dorysthenes granulosus*. *Guangxi Agricultural Sciences* (6): 259–259 [in Chinese with English abstract].

Zhang, Q. B, H. D. Lei, B. M. Lin, et al. 2002. Several citrus trunk diseases and pests. *South China Fruits* 31: 12–14 [in Chinese].

Zhang, Q. D., B. Z. Ji, T. Xu, et al. 2010. Progress and perspectives on the genus *Aphrodisium*. *Scientia Silvae Sinicae* 46:144–151 [in Chinese with English abstract].

Zhang, S. P., D. Z. Huang, and J. J. Yan. 1999. The sequence choice of the main tree species to *Apriona germarii* in northern part of China. *Journal of Northeast Forestry University* 27:17–21 [in Chinese with English abstract].

Zhao, K. J., Y. M. Zhou, H. B. Zhu, K. Z. Xu, and H. X. Sun. 1994. Study on the control of the mulberry borer (*Apriona germari*) using a fumigating rod. *Plant Protection* 20:10–11 [in Chinese].

Zhekova, E. D., and D. S. Petkova. 2010. Productivity of alfalfa germplasms—New data about alfalfa root longhorn beetle (*Plagionotus floralis* Pall.). *Banat's Journal of Biotechnology* 1:56–60.

Zhou, Y. S., F. R. Shen, H. P. Zhao, and Q. Shi. 1995. The biology and control of *Rhytidodera bowringi* White. *Journal of Southwest Agricultural University* 17:451–455 [in Chinese with English abstract].

Zhu, J. W. 1995. Damage to *Eucalyptus citriodora* by *Dorysthenes granulosus*. *Guangxi Forestry* (1):19–19 [in Chinese].

Zhu, J. W. 2012. Preliminary study on damage to *Citrus sinensis* (L.) cv. Hong Jiang Cheng by *Dorysthenes granulosus*. *Agriculture and Technology* (10):109–109 [in Chinese].

Zhu, Z. C., X. B. Wang, J. G. Tang, and S. F. Min. 1995. Studies on the relationship between insecticide formulation and its toxicity and persistence in controlling forest insect pests. *Scientia Silvae Sinicae* 31:330–337 [in Chinese with English abstract].

Zhu, J. H., and M. Y. Xu. 1996. Damage to longan and citrus by *Dorysthenes granulosus*. *Plant Protection* 22:38–38 [in Chinese].

13

Invasive Cerambycid Pests and Biosecurity Measures

Dominic Eyre
Defra, Department for Environment, Food & Rural Affairs
Sand Hutton, York, UK

Robert A. Haack
USDA Forest Service
Lansing, Michigan

CONTENTS

13.1 Introduction

Longhorn beetles (Cerambycidae) in general, and some species in particular, have increased in importance for national and regional plant protection agencies over recent decades. Expensive eradication campaigns have been carried out in order to eliminate some longhorn beetles. For example, the cost of eradication campaigns undertaken between 1996 and 2013 against the cerambycid

Anoplophora glabripennis (Motschulsky), commonly called the Asian longhorn beetle, were estimated to have exceeded US$537 million for all U.S. infestations in Illinois, Massachusetts, New Jersey, New York, and Ohio (Rhonda Santos, USDA-APHIS, personal communication Feb 2014). Although other taxa—such as fungal pathogens [e.g., ash dieback, *Hymenoscyphus fraxineus* (T. Kowalski) Baral, Queloz, Hosoya, comb. nov., in Europe (Pautasso et al. 2013; Baral 2014)] and the buprestid emerald ash borer, *Agrilus planipennis* Fairmaire, in the United States (Kovacs et al. 2010)—have in some cases caused greater damage; the potential to counter outbreaks of some cerambycids is often larger in part due to their relatively slow rate of spread. The possibility of eradicating invasive populations of longhorn beetles has provided the justification for abatement efforts in North America and Europe against the Asian *Anoplophora* species, some of which have been successful (Haack et al. 2010).

In addition to the relatively slow rate of spread of many cerambycids, other factors relevant to their status as quarantine pests are (1) the many and diverse pathways along which they can be moved, (2) their potential for causing significant damage, and (3) the difficulties of detecting either juvenile stages within trees or wood or early stage outbreaks. Wood packaging, trees, timber, and wood products that are infested with the juvenile stages of cerambycids may have little or no external signs of infestation. As a result, visual inspection alone may be inadequate for detecting infested consignments and other more costly methods may be necessary to intercept infested materials or goods.

The estimated 36,000 described species of Cerambycidae worldwide (see Chapter 1) are divided into the xylophagous species that feed on wood and the phytophagous species that feed on herbaceous plants (Bense 1995). It is the xylophagous species that have drawn the greatest attention as invasive pests. Only a minority of longhorn beetles are significant pests; in Europe, for example, about 20% of all cerambycid species are thought to be important as forest pests or timber beetles (Hellrigl 1974). The xylophagous species tend to be specialized into those that attack different parts of their host—such as roots, stems, branches, or twigs, and there is also a spectrum of preferences for attacking either dead, dying, weakened, or healthy trees (Bense 1995; Hanks 1999). These preferences are an important indication of whether a species may become a significant invasive pest of standing trees. As is the case with other invasive species, the damage in the invaded area can be greater than the damage the pests cause in their native range due to the greater susceptibility of host trees in the invaded area, the lack of control by natural enemies, and the widespread planting of susceptible trees (Haack et al. 2010). This chapter provides a description of the impact that invasive longhorn beetles have had, the routes along which they have been moved to new areas, and the measures that have been taken to prevent their movement and establishment in new locations.

13.2 Interceptions of Cerambycids on Imported Materials

Databases of pests intercepted on imported materials have been analyzed to try to understand the most important pathways for the movement of pests. Although such databases are useful for establishing that there is a biosecurity threat associated with trade, it is more difficult to use such data to conclusively establish the lack of a threat for a particular trade or to provide a fair comparison of the relative threat of particular trades. Humble (2010) listed some of the shortcomings of most interception databases: (1) they are not based on random sampling; (2) negative inspections are not recorded; (3) once a quarantine organism has been discovered, consignments may be destroyed without further inspection, and thus other exotic organisms can be missed; (4) only a small percentage of individual shipments are inspected; and (5) organisms often are not identified as to species because many are intercepted as larvae that can be difficult to identify to species level. In addition to these shortcomings, trade volumes and sources can change rapidly, the number of consignments inspected varies from year to year in response to national and regional plant health and wider government priorities, and the method and intensity of quarantine inspections can vary within and among countries and also over time. Furthermore, different proportions of consignments from different trades can be inspected, reflecting the perceived quarantine risks of each trade. For example, plants for planting arriving into the European Union (EU) are treated in EU legislation as being high risk, and member states are required to inspect all consignments. In contrast, the movement of some genera of plants among EU member states is not regulated by plant health legislation, and other higher risk plants are controlled by the issuing of Plant Passports, which is an industry-led scheme rather than government-led scheme (EU 2000). Thus, consignments of plants being moved among EU member states are not routinely subject to official inspections despite the fact these movements have been shown to be pathways for the transplantation of longhorn beetles (Giltrap et al. 2009). This lack of official inspection means that there is a lack of knowledge concerning the extent and risk of longhorn beetles being moved among member states.

The Agricultural Quarantine Inspection Monitoring (AQIM) sampling strategy was developed by the U.S. Department of Agriculture (USDA) to try to overcome some of the difficulties relating to data gathered during routine plant health or biosecurity inspections. It is a random protocol designed to be sufficient to detect nonindigenous pests infesting greater than 10% of a shipment with 95% confidence (Venette et al. 2002). AQIM has been used to show (1) the arrival rate of insect species in the United States via foreign trade and the most important pathways (Work et al. 2005), (2) the importance of the live plant pathway for the importation of forest pests into the United States (Liebhold et al. 2012), and (3) the impact of international phytosanitary regulations such as International Standards for Phytosanitary Measures No. 15 (ISPM 15) (Haack et al. 2014).

An analysis of the 1985–2000 interception records of the USDA's Animal and Plant Health Inspection Service (APHIS) Port Information Network (now referred to as PestID) database by Haack (2006) revealed that Cerambycidae represented the second most frequently intercepted insect family associated with wood (1,642 interceptions) following Scolytidae, now considered to be a subfamily of Curculionidae (i.e., Scolytinae), of which there were 5,008 interceptions (Bright 2014; Jordal et al. 2014). However, analysis of this data set in four-year time periods showed that the number of Scolytinae interceptions declined from 2,215 during 1985–1988 to 687 during 1997–2000, whereas there was an increase in the number of interceptions of Cerambycidae from 211 during 1985–1988 to 686 during 1997–2000. The increase in the number of interceptions of Cerambycidae between 1985 and 2000 was mainly the result of an increase in the number of interceptions on material from China (from 15 interceptions during 1985–1988 to 357 during 1997–2000). The decline in the number of interceptions of Scolytinae was in part explained by a 1996 change in U.S. import regulations that required all unmanufactured solid wood items to be "totally free from bark" or else be certified by the exporting country as treated for wood pests (USDA-APHIS 1995; Haack 2006).

13.3 Pathways: Plants for Planting

The movement of plants for planting is thought to be the most important pathway for the introduction of nonnative forest pests into North America. Liebhold et al. (2012) estimated that 70% of damaging forest insects and pathogens established in the United States between 1860 and 2006 most likely entered on imported live plants. Smith et al. (2007) found that 89% of nonnative invertebrate plant pests introduced into Great Britain since 1970 had potentially been introduced with live plants, especially ornamentals.

Differences in the proportion of consignments inspected at the point of entry may have influenced the predominance of the connection between live plants and the number of forest pests that have been linked with their import. In the United States, approximately 50% of commerce is believed to be carried on wooden pallets (Haack et al. 2014) but only about 2% of cargo arriving at maritime ports, airports, and border crossings is inspected (Work et al. 2005). In contrast, although published figures are not available, the USDA is thought to inspect a portion of all consignments of plants for planting arriving in the United States. After consignments associated with wood packaging material arrive in a country, the packaging is often discarded because it will already have served its purpose to the importer. This means that any emerging pests are unlikely to be noticed. In contrast, live plants are usually imported in a dormant state and then grown on at a nursery for a period before they are sold. While the plants are held at the nursery, they will be watered or irrigated, examined for their general health and perhaps pruned, repotted, and treated with plant protection products. All these operations provide an opportunity for nursery employees to notice any diseases and emerging pests such as cerambycids and hence increase the chance of a connection being made between the trade and the pest. In the case of longhorn beetles, however, infestations may go unnoticed unless exit holes are found, frass builds up on the outside of the host, or an emerged adult is detected.

The citrus longhorn beetle, *Anoplophora chinensis* (Forster), is one of the most notable worldwide invasive longhorn beetles that has been transported with plants for planting. Some authors (e.g., Adachi 1994) refer to *A. malasiaca* (Thomson); however, *A. malasiaca* was synonymized with *A. chinensis* by Lingafelter and Hoebeke (2002). They argued that *A. chinensis* and *A. malasiaca* could not be separated on the basis of the color and size of elytral macula (spots) and the presence or absence of hair on the pronotum because there is considerable variation of such features and considerable overlap in specimens from the same locality (CABI 2014a). For the purposes of this chapter, the synonymization has been accepted, but some Japanese researchers still consider *A. chinensis* and *A. malasiaca* to be separate species on the bases of morphological characteristics and DNA from the CO1 sequence of the mitochondria (Ohbayashi et al. 2009; Iwaizumi et al. 2014).

A. chinensis is a damaging pest for a range of deciduous trees and originates in Asia, with most records being from China, Japan, and South Korea. Other Asian countries or regions where *A. chinensis* has been recorded are Malaysia, North Korea, Vietnam, Taiwan, the Philippines, Indonesia, and Myanmar, although there are many fewer records from these countries (Gressit 1951; CABI and EPPO 1997; Lingafelter and Hoebeke 2002). *A. chinensis* is a serious pest of fruit trees, especially citrus in Japan (Adachi 1994) and China (Lieu 1945). It was first discovered outside of Asia in 2000 when an outbreak (in this case meaning a breeding population in a nonnative area) was discovered in Parabiago, Lombardy, in northern Italy, and since then outbreaks have been discovered elsewhere in Italy and the rest of Europe. Table 13.1 and Figure 13.1 show that most outbreaks of *A. chinensis* in Europe have been in Italy and the Netherlands, which are the EU countries that import the greatest number of plants for planting (FAO 2014).

EUROPHYT is a phytosanitary database of interceptions of plant health pests and pathogens for the European Union and contains records that have been reported by member states or Switzerland to the European Commission. These data are published by the European and Mediterranean Plant Protection Organization (EPPO) and also by the EU (EPPO 2013a; EU 2014). Between 1998 and 2013, there were 455 Cerambycidae interception records in the EUROPHYT database. Table 13.2 shows the number of interceptions of Cerambycidae by year and product category, and Table 13.3 shows the taxa and origin of Cerambycidae intercepted on plants for planting.

Between 1998 and 2013, there were 54 interceptions of Cerambycidae on plants for planting reported on EUROPHYT. This includes plants described as "already planted," "not yet planted," "bonsais," and "seeds." This analysis has excluded 39 records from 2008 apparently relating to plant genera that were in the same consignment as an *Acer* sp. infested with *A. chinensis* but were not themselves confirmed as being infested. The only species of longhorn beetle recorded in association with plants for planting on EUROPHYT was *A. chinensis* (38 records); the other records were *Anoplophora* sp. (11, many of which

TABLE 13.1

Recorded Outbreaks of *Anoplophora chinensis* Outside the Countries Where It Is a Native Species as of February 2016

Country	Location	Outbreak Detected	Declared Eradicated	Eradication Year	References
Italy	Parabiago	2000	No	NA	Colombo and Limonota 2001, Caremi and Ciampitti 2006, Hérard et al. 2006, Tomiczek and Hoyer-Tomiczek 2007
France	Soyons	2003	Yes	2006	Hérard et al. 2006
Italy	Montichiari	2007	No	NA	Petruzziello 2009
Netherlands	Westland	2007	Yes	2010	Van der Gaag et al. 2010
Croatia	Turanj–Sveti Filip i Jakov	2007	No	NA	Van der Gaag et al. 2010, Vukadin and Hrasovec 2008, EPPO 2016
Italy	Gussago	2008	No	NA	Bazzoli and Alghisi 2013
Italy	Rome	2008	No	NA	Van der Gaag et al. 2010, Roselli et al. 2013, Peverieri et al. 2012
Netherlands	Boskoop	2009	Yes	2010	Van der Gaag et al. 2010
Italy	Prato	2014	No	NA	EPPO 2016

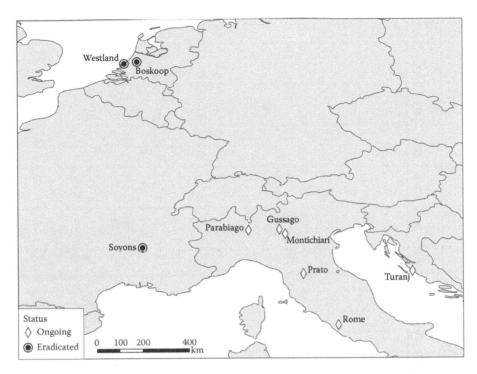

FIGURE 13.1 Locations of outbreaks of *Anoplophora chinensis*, the citrus longhorn beetle, in Europe as of February 2016.

were likely *A. chinensis*) and Cerambycidae (5). Maples were the reported hosts for *A. chinensis* for 37 of the 38 interceptions of *A. chinensis* on plants for planting, including *Acer palmatum* Thub. (20 interceptions), *Acer buergerianum* Miq. (3), *Acer rubrum* L. (1), and *Acer* sp. (13). The origin of consignments infested with either *A. chinensis* or *Anoplophora* sp. (49) was China (32), the Netherlands (8), Italy (3), Japan (3), South Korea (1), and unknown (2).

The predominance of live plant imports from China being the source of *A. chinensis* interceptions was also found in records from Europe, North America, and New Zealand between 1980 and 2008 (Haack et al. 2010). In total, there were 74 interceptions of *A. chinensis* between 1980 and 2008, with the highest number (24) in 2008; the highest number of interceptions (41) took place in the Netherlands. Most of the interceptions of *A. chinensis* were in live plants from China (85%), Japan (13%), and South Korea (2%) (Haack et al. 2010).

This data set also demonstrates the difficulties that national plant protection organizations (NPPOs) have faced when trying to detect *A. chinensis* at points of entry (Haack et al. 2010). For example, of the 74 interceptions, only 7 were at points of entry, with all the others occurring post-entry, although most (86) of the 145 interceptions of *A. glabripennis* were at points of entry. The majority (78) of the 86 interceptions at points of entry were in the United States (46) and Canada (32). In EU countries, the majority of interceptions of *A. glabripennis* were post-entry, with all 27 interceptions in the United Kingdom being post-entry (Haack et al. 2010). Most of these records refer to findings of the pest in close association with imported items, such as wood packaging materials, after they had been cleared through ports.

In 2008, there were multiple finds of *A. chinensis* in imported plants that had been planted in private gardens in the United Kingdom (Eyre et al. 2010). Plant health authorities in the United Kingdom first became aware of the situation when a member of the public from Tyne and Wear in northeast England reported finding a live adult *A. chinensis*. The beetle had emerged from an *Acer palmatum* that had been obtained via an offer of free plants in a national newspaper. The plant was traced to a mail order company in Guernsey (an island Crown dependency in the English Channel) that had imported the plants from China via the Netherlands. Approximately 60,000 *A. palmatum* had been sent to some 45,000 customers

TABLE 13.2

Number of Interceptions of Longhorn Beetles per Year in the European Union as Recorded on EUROPHYT

	Product Category	1998	1999	2000	2001	2002	2003	2004	2005	2006	2007	2008	2009	2010	2011	2012	2013	Grand Total
Plants for planting and cut branches	Intended for planting: already planted					1			1	1	1	14	3	4			1	26
	Intended for planting: bonsais	1				1				2	2	2	2			2		12
	Intended for planting: not yet planted	1	1									5	4	4				15
	Intended for planting: seeds			1														1
	Cut flowers and branches with foliage							1										1
	Cut branches without foliage								1									1
Wood packaging	Dunnage		1										2	1	3	1	2	10
	Wood packaging material			3	3	9	8	12	12	6	10	3	8	7	10	22	30	143
	Wood pallet		1	1			1	5		1	5	6	7	6	4	16	28	81
	Wooden crate			14				3	3	2	3	4	2	1	6	20	14	72
Others	Products: wood and bark	2	7	13	2	5		3	12	7		2	1	4	11	2	2	73
	Object with wooden parts								1	1	1		1				1	5
	Packing materials	1	2			1										3		7
	Products: by-products of plant origin											1						1
	Others/unknown						1				3		1			1	1	7
Grand Total		5	12	32	5	17	10	24	30	20	25	37	31	27	34	67	79	455

TABLE 13.3

Number of Interceptions of Longhorn Beetles in the European Union on Plants for Planting between 1998 and 2013 by Taxon and Country of Export Sorted by Host Plant

Cerambycid Taxon	Country of Export	Acer buergerianum	Acer palmatum	Acer rubrum	Acer sp.	Cercis sp.	Ilex sp.	Pinus sp.	Taxus cuspidata	Wisteria sp.	Grand Total
Anoplophora chinensis	China	2	10	1	9	1	0	0	0	0	23
	Italy	0	3	0	0	0	0	0	0	0	3
	Japan	0	1	0	1	0	0	0	0	0	2
	Republic of Korea	1	0	0	0	0	0	0	0	0	1
	Netherlands	0	6	0	1	0	0	0	0	0	7
	Unknown	0	0	0	2	0	0	0	0	0	2
Anoplophora sp.	China	0	4	0	4	0	1	0	0	0	9
	Japan	0	0	0	0	0	0	0	1	0	1
	Netherlands	0	1	0	0	0	0	0	0	0	1
Cerambycidae	China	0	0	0	1	0	0	0	0	0	1
	India	0	0	0	0	0	0	1	0	0	1
	Republic of Korea	0	1	0	0	0	0	0	0	0	1
	New Zealand	0	0	0	0	0	0	0	0	2	2
Grand Total		3	26	1	18	1	1	1	1	2	54

around the United Kingdom. The trees generally were small (with a maximum trunk diameter of only a few centimeters), and the probability of two or more *A. chinensis* being able to complete their development within the same tree was considered to be low. Given that *A. chinensis* reproduce sexually, an outbreak could only be initiated at locations where both a male and a female were to emerge, locate each other, and mate. Therefore, the risks of establishment were considered highest at sites where two or more of the potentially infested plants had been planted. In order to evaluate the risk, an estimated 326 trees were destructively sampled and a further 253 were visually inspected. No *A. chinensis* were found during the course of these inspections; hence, it was concluded that the risk of establishment was low.

In addition to the UK finds described here, the Netherlands reported numerous interceptions of *A. chinensis* on *Acer* trees imported from Asia in 2007 and 2008. In 2007, the Dutch Plant Health Service inspected approximately 100 sites holding *Acer* trees imported from China or Japan for the presence of *A. chinensis*, and infested consignments were found at six locations. At one of these six locations, live *A. chinensis* larvae were found in two consignments of *Acer* trees from two different locations in Japan, and a further three consignments of *Acer* (two from Japan and one from China) had exit holes that were probably caused by *A. chinensis*. Between January and March 2008, 45 consignments of *Acer* were imported into the Netherlands from China and Japan. *A. chinensis* was intercepted in two of these consignments at the point of entry and in a further 12 consignments during post-entry inspections (van der Gaag et al. 2008).

The outbreak of *A. chinensis* that was first discovered in 2000 in Parabiago, Lombardy, Italy, currently is the most significant *Anoplophora* outbreak in Europe, and the pest may have spread too far for eradication to be practical (Brockerhoff et al. 2010). Between 2001 and 2013, almost €18 (US$20) million was spent on eradicating *A. chinensis* in Lombardy, and 25,000 trees were cut down as part of the eradication campaign (Ciampitti and Cavagna 2014). Another very significant outbreak of *A. chinensis* was detected in Europe in December 2009 when Dutch inspectors detected seven exit holes in two dead stumps of *A. palmatum* and one exit hole in a *Carpinus* tree in Boskoop in the Netherlands (van der Gaag et al. 2010). The significance of this outbreak was not its scale but the importance of its location. Boskoop is an important center for the production of hardy ornamental nursery stock for Europe; therefore, an outbreak in this area meant that there was a significant risk that the pest could have been moved to other parts of the EU in infested plants.

As of February 2016, no established populations of *A. chinensis* have been reported in North America, although in 2001, five *A. chinensis* adults were thought to have emerged from *Acer* bonsai trees being held outdoors at a nursery in Washington State and possibly had become established nearby. Out of concern that the beetles could mate and lay eggs on nearby trees, an eradication program was initiated by Washington State in 2002. This involved cutting and chipping all potential host trees within 200 m of the nursery, treating all potential host trees within 200–400 m of the nursery with systemic insecticides, and conducting annual surveys of all host trees within 800 m of the nursery for five years. No evidence was ever found that *A. chinensis* had become established, and thus the quarantine was terminated in 2007 (Haack et al. 2010).

The lemon tree borer, *Oemona hirta* (Fab.), is another example of a pest that has been moved among continents in plants for planting. This beetle was intercepted in the United Kingdom once in 1983 and at two nurseries in 2010. The 2010 finds were entered on EUROPHYT as "Cerambycidae" and identified to species level at a later date. The pest is native to New Zealand and is not known to be established elsewhere. The 2010 interceptions were of larvae associated with *Wisteria* plants imported from New Zealand for growing on and then sale (EPPO 2013c).

13.4 Pathways: Wood Packaging Materials

Wood packaging materials are frequently made of raw wood that may not have undergone sufficient processing or treatment to remove or kill pests; therefore, it becomes a pathway for the introduction and spread of pests such as Cerambycidae (IPPC 2009). Of the 455 interceptions of longhorn beetles recorded on the EUROPHYT database between 1998 and 2013, the majority (306) were interceptions relating to consignments of wood packaging materials. The interceptions have been divided into those made on unspecified wood packaging material (143), wood pallets (81), wooden crates (72), and dunnage (10) (Table 13.2). When the beetles were identified to genus or species level (181 interceptions), almost half the interceptions were related to either *Anoplophora glabripennis* (57 interceptions) or *Anoplophora* sp. (22) (Table 13.4). The *Anoplophora* sp. are very likely to have been *A. glabripennis* given that there is only one record of another *Anoplophora* sp. being associated with wood packaging material in Europe— an interception of *A. chinensis* in 2007.

The second most commonly intercepted species of longhorn beetle was *Apriona germari* (Hope), with 26 interceptions being recorded on EUROPHYT and most of these being in 2012 (15) and 2013 (8). The risk of this pest, along with the risks relating to *Apriona japonica* Thomson and *Apriona cinerea* Chevrolat, has been assessed for the Netherlands (Justica et al. 2010) and the wider EPPO region (EPPO 2013b). The EPPO assessment was that *A. germari* presents a particular risk to the Mediterranean area, southeast Europe, northern Turkey, and to the oceanic areas of southwest Europe. The following hosts are those most commonly reported for these *Apriona* pests in the literature: *Morus* spp., *Populus* spp., *Salix* spp., *Malus* spp., *Ficus carica* L., *Broussonetia papyrifera* (L.) Hert. Ex Vent., *Artocarpus heterophyllus* Lamark, and *Sophora japonica* (L.) Schott (EPPO 2014).

Figure 13.2 shows the number of interceptions of Cerambycidae in association with wood packaging in the EU between 1999 and 2013. This figure indicates that there were more than double the number of interceptions in 2012 and 2013 than in all previous years, and that wood packaging from China was the source of the majority of interceptions every year.

During 1984–2008, there were 3,483 cerambycid interceptions associated with wood products or packaging at U.S. points of entry, of which 89% were identified to subfamily level or lower, 76% to genus, and 19% to species. The data set analyzed here is the same as the PestID data described in Haack et al. (2014) as well as the one described for just the Scolytinae in Haack and Rabaglia (2013). These 3,483 cerambycid interceptions represented at least 85 genera in six subfamilies and 86 distinct species. For those individuals that were identified to at least the genus level, the top 10 intercepted genera in decreasing order were *Arhopalus*, *Monochamus*, *Tetropium*, *Xylotrechus*, *Ceresium*, *Hesperophanes*, *Phymatodes*, *Callidium*, *Callidiellum*, and *Anoplophora*. Similarly, for those individuals identified to the species level, the top five intercepted species were *T. castaneum* (L.), *Arhopalus syriacus* (Reitter), *Arhopalus rusticus* (L.), *Callidiellum rufipenne* (Motschulskyi), and *Pyrrhidium sanguineum* (L.) (Table 13.5). Of the 677 interceptions that were identified to species level, there were 254 interceptions (38%) of *T. castaneum*, and the top five species accounted for 59% of all interceptions. This data set was likely influenced by the ease of identification to generic or specific level; for example, the lack of taxonomic keys or reference specimens could have made the identification of some of those intercepted Cerambycidae difficult or impossible.

Figure 13.3 shows the number of interceptions of Cerambycidae in the United States on wood packaging from various regions of the world and by the type of associated imported commodities between 1984 and 2008. Most Cerambycidae during that entire time period were intercepted on wood packaging from Europe, followed by Asia and North America (mostly Mexico). The United States has a policy of only inspecting limited amounts of wood packaging from Canada because most bark-infesting and wood-infesting insects native to Canada are also native to the United States and because of the long largely forested border between the two countries (Haack et al. 2014). Wood packaging associated with tiles and quarry products such as marble accounted for the majority of the cerambycid interceptions from Europe.

The link between heavy commodities such as stone and the interception of longhorn beetles may be due to the thickness of wood required to support the commodities, with thicker wood being both more likely to harbor insects and more difficult to heat treat or fumigate. A second explanation for the link between pest levels in wood packaging material and heavy consignments such as stone is that low-quality wood is more frequently used to make the wood packaging material (John Morgan, Forestry Commission, UK, personal communication). Wood packaging material used to transport heavy items is often damaged in transit; therefore, the wood packaging material is considered to be single use in contrast to wood packaging material that may be used to transport food or items to be stored in warehouses, which needs to be made of higher quality wood.

Figure 13.4 illustrates the number of interceptions of Cerambycidae in the United States on wood per year between 1984–2008 for the five countries which accounted for the highest number of interceptions: China, Italy, Mexico, Turkey and Spain. The highest number of cerambycid interceptions on imports

TABLE 13.4

Number of Interceptions of Longhorn Beetles in the European Union on Wood Packaging between 1998 and 2013 by Taxon and Country of Export

Cerambycid Taxon	Country of Export	1998	1999	2000	2001	2002	2003	2004	2005	2006	2007	2008	2009	2010	2011	2012	2013	Grand Total
Anoplophora chinensis	China	–	–	–	–	–	–	–	–	–	1	–	–	–	–	–	–	1
Anoplophora glabripennis	China	–	–	4	–	1	–	3	3	–	–	4	3	4	5	11	19	57
Anoplophora sp.	China	–	–	–	–	1	2	–	6	–	4	1	–	1	1	5	2	22
Apriona germarii	China	–	–	–	–	–	–	–	–	–	–	–	2	–	1	15	8	26
Batocera lineolata	China	–	–	–	–	–	–	1	1	–	–	–	–	–	–	–	2	3
Batocera sp.	China	–	–	–	–	–	–	–	–	–	–	–	–	–	–	1	1	3
Cerambycidae	Australia	–	–	–	–	–	–	–	–	–	1	–	–	–	–	–	–	1
	Belarus	–	–	–	–	–	–	–	–	–	–	–	–	–	1	–	–	1
	Brazil	–	–	–	–	–	–	–	–	–	–	–	2	–	–	–	–	2
	China	–	1	1	–	3	7	8	1	3	7	4	2	6	9	19	28	99
	Costa Rica	–	–	–	–	–	–	–	–	–	–	–	–	–	–	–	1	1
	Germany	–	–	–	–	–	–	–	–	–	–	–	–	–	–	–	1	1
	India	–	–	–	–	–	–	–	–	–	2	2	–	1	–	1	2	8
	Indonesia	–	–	–	–	–	–	–	–	–	–	–	–	–	1	–	–	1
	Japan	–	–	–	–	–	–	–	–	–	–	–	1	–	–	–	–	1
	Portugal	–	–	–	–	–	–	–	–	–	–	–	1	–	1	–	–	2
	Russian Federation	–	–	–	–	–	–	–	–	–	–	–	–	–	1	–	–	1
	Turkey	–	–	–	–	–	–	–	–	–	–	1	–	–	–	–	–	1
	Ukraine	–	–	–	–	–	–	–	–	–	–	–	–	–	–	4	–	4
	Vietnam	–	–	–	–	–	–	–	–	–	–	–	1	–	–	–	1	2
Chlorophorus sp.	China	–	–	–	–	1	–	–	–	–	1	–	–	–	–	–	1	3
Criocephalus rusticus	South Africa	–	–	–	–	–	–	–	–	–	–	–	1	–	–	–	–	1
Monochamus alternatus	China	–	–	1	–	–	–	1	–	3	–	–	2	–	–	–	1	8

(Continued)

TABLE 13.4 (Continued)

Number of Interceptions of Longhorn Beetles in the European Union on Wood Packaging between 1998 and 2013 by Taxon and Country of Export

Cerambycid Taxon	Country of Export	1998	1999	2000	2001	2002	2003	2004	2005	2006	2007	2008	2009	2010	2011	2012	2013	Grand Total
Monochamus sp.	Belarus	—	—	—	—	—	—	—	—	—	—	—	—	—	—	—	1	1
	China	—	1	9	3	2	—	4	3	—	—	1	2	2	1	3	—	31
	Kazakhstan	—	—	—	—	—	—	—	—	—	—	—	—	—	—	—	3	3
	Russian Federation	—	—	—	—	—	—	2	—	3	—	—	1	1	1	—	—	8
	Taiwan	—	—	1	—	—	—	—	—	—	—	—	—	—	—	—	—	1
	United States	—	—	1	—	—	—	1	—	—	1	—	—	—	—	—	1	4
Monochamus sutor	Latvia	—	—	1	—	—	—	—	—	—	—	—	—	—	—	—	—	1
	Russian Federation	—	—	—	—	—	—	—	—	—	—	—	—	—	2	—	—	2
Phoracantha semipunctata	Brazil	—	—	—	—	—	—	—	—	—	—	—	—	—	—	—	1	1
	China	—	—	—	—	—	—	—	—	—	—	—	—	—	—	—	1	1
Saperda sp.	China	—	—	—	—	—	—	—	1	—	—	—	—	—	—	—	1	2
Xylotrechus sp.	China	—	—	—	—	1	—	—	—	—	—	—	—	—	—	—	—	1
	India	—	—	—	—	—	—	—	—	—	1	—	—	—	—	—	—	1
Grand Total		—	2	18	3	9	9	20	15	9	18	13	19	15	23	59	74	306

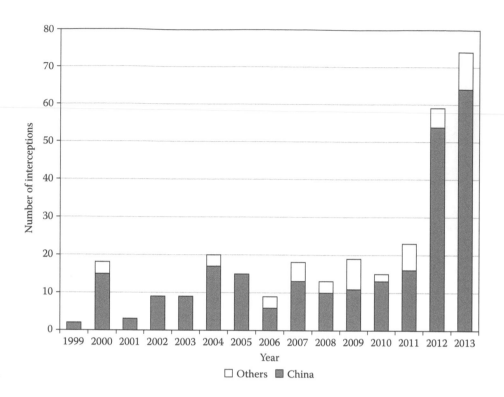

FIGURE 13.2 Annual number of interceptions of Cerambycidae in association with wood packaging material in the European Union as recorded on EUROPHYT between 1999 and 2013.

from China was in 1998, possibly reflecting more intensive inspections following the discoveries of *A. glabripennis* in New York City in 1996 and in Chicago in 1998 (Haack et al. 2010). The total number of cerambycid interceptions per year in the United States remained below 50 between 1984 and 1993 but had shown a dramatic increase between 1994 and 2008, with the highest number of interceptions in 2007 (377 interceptions).

Table 13.6 lists the top 20 combinations of imported commodity and country of origin in terms of the number of cerambycid interceptions made in association with imports to the United States between 1984 and 2008. The list is dominated by trades in either tiles (four of the top eight trades) or quarry products (8 of the top 20 trades), which include products such as marble, granite, and slate. The list also includes three metal commodity or product trades. Nine of the trades were with China and three with Italy. The number of interceptions per year relating to the top four commodities (Figure 13.5) shows that interceptions relating to tiles and quarry products significantly increased from the mid-1990s to 2008.

The prime example of a nonnative longhorn beetle that has been moved to new geographic regions in wood packaging material is *A. glabripennis*. There have been numerous outbreaks of this pest of deciduous trees in North America, Europe, and Japan. Table 13.7 lists the outbreaks of this pest that had been reported outside its native range by February 2016. Some of the listed outbreaks had multiple foci, which could indicate multiple introductions or possible natural or human-assisted spread, such as the outbreak areas in New York. Outbreaks have tended to occur in the more industrialized countries of the world, which is logical given that they are also likely to be importing the greatest volumes of products from China. In the United States, all the *A. glabripennis* outbreaks have been on the eastern side of the country—the area that has been judged to be most at risk based on host availability and climate (Peterson et al. 2004). *A. glabripennis* had been declared eradicated in 9 of the 35 outbreak areas as of February 2016. Nevertheless, it is anticipated that other currently undetected outbreaks exist in Europe and North America given that 16 *A. glabripennis* outbreaks were detected between 2012 and 2015. Figures 13.6 and 13.7 show the locations of the *A. glabripennis* outbreaks in North America and Europe, respectively.

TABLE 13.5

Cerambycidae Intercepted in the United States between 1984 and 2008 on Wood Packaging Material and Identified to Species Level

Cerambycid Taxon	No. Interceptions	Most Numerous Commodity	Second Most Numerous Commodity	Most Common Origin	Second Most Common Origin
Acanthocinus aedilis (Linnaeus)	2	Tiles		Spain	Turkey
Acanthocinus griseus (Fabricius)	2	Tiles		Spain	Turkey
Agapanthia irrorata (Fabricius)	1	Tiles		Spain	
Anaglyptus mysticus (Linnaeus)	1	Tiles		Italy	
Anelaphus moestus (Leconte)	1	Unknown		Honduras	
Anoplophora glabripennis Motschulsky	6	Iron	Machinery	China	
Arhopalus asperatus (Leconte)	1	Unknown		Unknown	
Arhopalus ferus Mulsant	3	Tiles		Turkey	
Arhopalus productus (Leconte)	1	Unknown		Honduras	
Arhopalus rusticus (Linnaeus)	33	Quarry product	Tiles	Spain	China
Arhopalus syriacus (Reitter)	56	Quarry product	Tiles	Italy	Turkey
Aromia moschata (Linnaeus)	1	Quarry product		China	
Asemum striatum (Linnaeus)	5	Quarry product	Tiles	Spain	China
Batocera rufomaculata (De Geer)	1	Unknown		India	
Callidiellum rufipenne (Motschulsky)	31	Steel	Machinery	Japan	China
Callidiellum villosulum Fairmaire	8	Woodenware	Engines	China	
Callidium aeneum (De Geer)	1	Tiles		Italy	
Callidium violaceum (Linnaeus)	6	Tiles	Quarry product	Italy	China
Chlorida festiva (Linnaeus)	2	Tiles		Colombia	Dom. Rep.
Chlorophorus annularis (Fabricius)	6	*Bambasa* sp.	Household goods	China	Indonesia
Chlorophorus pilosus glabromaculatus Forster	1	Tiles		Italy	
Chlorophorus strobilicola Champion	2	Woodenware		India	
Clytus lama Mulsant	1	Tiles		Italy	
Dere thoracica White	1	Unknown		China	
Derobrachus geminatus LeConte	1	Unknown		Mexico	
Desmiphora hirticollis (Olivier)	1	Unknown		Mexico	
Eburia mutica Leconte	1	Unknown		Mexico	
Elaphidion irroratum (Linnaeus)	1	Tiles		Dom. Rep.	
Elaphidion mucronatum (Say)	1	Unknown		China	

(Continued)

TABLE 13.5 (Continued)

Cerambycidae Intercepted in the United States between 1984 and 2008 on Wood Packaging Material and Identified to Species Level

Cerambycid Taxon	No. Interceptions	Most Numerous Commodity	Second Most Numerous Commodity	Most Common Origin	Second Most Common Origin
Euryscelis saturalis Olivier	1	Tiles		Colombia	
Gnaphalodes trachyderoides Thomson	2	Machinery		Mexico	
Gracilia minuta (Fabricius)	1	Unknown		France	
Hesperophanes campestris (Faldeman)	1	Unknown		China	
Hylotrupes bajulus (Linnaeus)	21	Tiles	Quarry product	Italy	China
Icosium tomentosum Lucas	5	Quarry product		Greece	Israel
Knulliana cincta (Drury)	1	Household goods		China	
Lagocheirus araneiformis Linnaeus	1	Tiles		Dom. Rep.	
Lagocheirus undatus (Voet)	1	Unknown		Costa Rica	
Lamia textor (Linnaeus)	4	*Quercus* sp.	Tiles	Italy	France
Lissonotus flavocinctus Dupont	1	Tiles		Colombia	
Megacyllene antennatus (White)	2	*Prosopis* sp.		Mexico	
Molorchus minor (L.)	15	Tiles	Quarry product	Italy	Spain
Monochamus alternatus Hope	17	Quarry product	Tiles	China	Taiwan
Monochamus carolinensis (Oliver)	2	Machinery		Mexico	Unknown
Monochamus clamator (Leconte)	1	Unknown		Unknown	
Monochamus galloprovincialis (Olivier)	5	Tiles	Quarry product	Spain	Turkey
Monochamus sartor (Fabricius)	5	Tiles	Aluminum	Italy	Greece
Monochamus scutellatus (Say)	2	*Pinus* sp.		Canada	China?
Monochamus sutor (Linnaeus)	9	Tiles	Quarry product	Spain	Turkey
Monochamus teserula White	1	Unknown		China	
Nathrius brevipennis (Mulsant)	2	Furniture		Chile	
Neoclytus caprea (Say)	1	Unknown		South Korea	
Neoclytus cordifer (Klug)	2	Foodstuffs		Honduras	
Neoclytus olivaceus L. & G.	1	Unknown		Brazil	
Nyssodrysina haldemani Leconte	2	Unknown		Bangladesh	
Oxypleurus nodieri Mulsant	7	Quarry product	Tiles	Turkey	Spain
Palaeocallidium rufipenne Motschulsky	1	Unknown		Japan	
Perissus delerei Tippmann	1	Unknown		Pakistan	

(Continued)

TABLE 13.5 *(Continued)*

Cerambycidae Intercepted in the United States between 1984 and 2008 on Wood Packaging Material and Identified to Species Level

Cerambycid Taxon	No. Interceptions	Most Numerous Commodity	Second Most Numerous Commodity	Most Common Origin	Second Most Common Origin
Petrognatha gigas (Fab)	1	Unknown		Africa country unknown	
Phoracantha recurva Newman	2	Parts	Tiles	Brazil	
Phoracantha semipunctata Fabricius	4	Machinery	Tools	Australia	Brazil
Phymatodes testaceus (Linnaeus)	5	Tiles	*Acer* sp.	Italy	Germany
Plagionotus christophi Kraatz	3	Ironware		China	
Plagionotus detritus Linnaeus	1	Tiles		Italy	
Pogonocherus hispidus (Linnaeus)	3	Tiles	Quarry product	Italy	
Pogonocherus perroudi Mulsant	2	Tiles		Spain	Turkey
Prionus californicus Mots	1	Unknown		Indonesia	
Pyrrhidium sanguineum (Linnaeus)	25	Tiles	Steel	Italy	Belgium
Rhagium inquisitor (Linnaeus)	8	Tiles	Quarry product	France	Italy
Rhagium mordax (Degeer)	2	Tiles		Belgium	Italy
Saperda carcharias (Linnaeus)	5	Tiles	Parts	Italy	
Semanotus ligneus (Fabricius)	1	Decorations		China	
Smodicum cucujiforme (Say)	1	Paint		Unknown	
Stromatium barbatum (Fabricius)	1	Unknown		Taiwan	
Stromatium longicorne (Newman)	3	Equipment	Furniture	India	Pakistan
Tetropium castaneum (Linnaeus)	254	Tiles	Quarry product	Italy	France
Tetropium fuscum Fabricius	9	Tiles	Quarry product	Italy	China
Tetropium gabrieli Weise	11	Tiles	Quarry product	Italy	China
Trachyderes mandibularis Serville	2	Tiles		Italy	Mexico
Xylotrechus grayi (White)	1	Iron		China	
Xylotrechus magnicollis Fairmaire	10	Tiles	Electrical parts	China	Hong Kong
Xylotrechus pyrrhoderus (Bates)	2	Unknown		South Korea	Unknown
Xylotrechus rusticus (Linnaeus)	21	Tiles	Quarry product	Greece	Italy
Xylotrechus stebbingi Gahan	1	Tiles		Italy	
Xystrocera globosa (Olivier)	5	Hardware	Machinery	India	Singapore
Total	677				

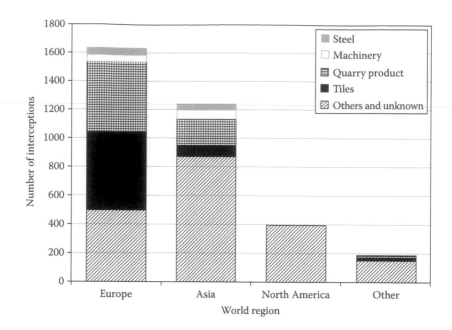

FIGURE 13.3 Number of interceptions of Cerambycidae on wood packaging material in the United States by world region for the period 1984–2008 divided by the type of imported commodity.

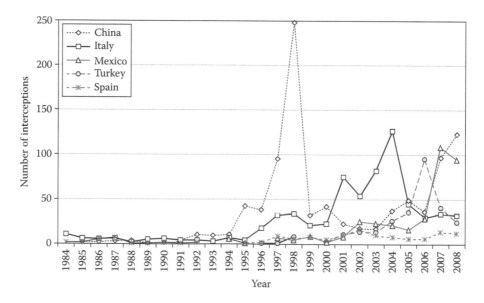

FIGURE 13.4 Annual number of interceptions of Cerambycidae on wood packaging material in the United States between 1984 and 2008 for the five countries with the highest number of interceptions. (From USDA APHIS PestID Database.)

Millions of hectares of monoculture forests have been planted in China since 1949 (Hsiao 1982), and in northern China these were mainly *Populus* species (Luo et al. 2003). The widespread planting of suscep-tible *Populus* spp. is thought to have led to the rapid spread and high outbreak levels of *A. glabripennis* within China (EPPO 1999) and the beetle has been a major pest there since the 1970s–1980s (Lingafelter and Hoebeke 2002; Haack et al. 2010). *A. glabripennis* is thought to be responsible for US$1.5 billion worth of damage to Chinese forests each year, and this represents 12% of all losses attributable to forest pests in that country (Hu et al. 2009; Meng et al. 2015). It has been estimated that, in China, hundreds

TABLE 13.6

Trades (country of origin and product combinations) Responsible for the Highest Number of Cerambycid Interceptions on Wood Packaging in the United States from 1984 through 2008

Origin	Commodity	No. of Interceptions
Italy	Tiles	347
Turkey	Quarry product	168
Italy	Quarry product	167
China	Quarry product	162
Turkey	Tiles	85
China	Iron	79
Spain	Tiles	75
China	Tiles	74
China	Machinery	51
China	Ironware	48
Italy	Metalware	46
Spain	Quarry product	45
France	Quarry product	32
China	Glass	27
Greece	Quarry product	24
China	Woodenware	20
Bulgaria	Quarry product	19
China	Equipment	17
Portugal	Quarry product	16
China	Metalware	15

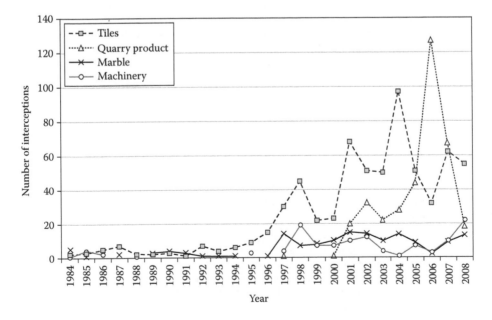

FIGURE 13.5 Annual number of interceptions of Cerambycidae on wood packaging material in the United States by commodity type between 1984 and 2008 for the top four commodity classes. (From USDA APHIS PestID Database.)

TABLE 13.7

Recorded Outbreaks of *Anoplophora glabripennis* Outside the Countries Where It Is Native

Country	Location	Outbreak Detected	Year Eradicated[a]	Source
United States	New York[b]	1996	–	Haack et al. 1997
United States	Chicago, Illinois	1998	2008	Poland et al. 1998, EPPO 2016
Austria	Braunau-am-Inn	2001	2013	Tomiczek et al. 2002, EPPO 2016
United States	Jersey City, New Jersey	2002	2013	Smith et al. 2009, EPPO 2016
Japan	Yokohama City	2002	2005	Takahashi and Ito 2005
France	Gien, Loiret, Centre	2003	–	Cocquempot and Herard 2003
Canada	Toronto, Ontario	2003	2013	Hu et al. 2009, EPPO 2016
France	Sainte-Anne-sur-Brivet	2004	–	Hérard et al. 2006
Germany	Neukirchen am Inn	2004	2016	Hérard et al. 2006, EPPO 2016
Germany	Bornheim North Rhine Westphalia	2005	–	Tomiczek and Hoyer-Tomiczek 2007
Italy	Corbetta, Lombardy	2007	–	Maspero et al. 2007
France	Strasbourg	2008	–	Hérard et al. 2009
United States	Worcester, Massachusetts	2008	–	Smith and Wu 2008
Italy	Cornuda	2009	–	Zampini et al. 2013
Netherlands	Almere	2010	2011	Morall 2011, PPS Netherlands 2010, PPS Netherlands 2011
United States	Boston, Massachusetts	2010	2014	Dodds and Orwig 2011,USDA-APHIS 2014
Germany	Weil am Rhein Baden-Wüerttemberg	2011	–	Saxony State Office for the Environment 2016
Switzerland	Brunisried (Canton of Fribourg)	2011	–	Forster and Wermelinger 2012
United States	Bethel, Ohio	2011	–	Dodds et al. 2013
Germany	Feldkirchen near Munich	2012	–	Bayerische Landesandstlt fur Landwirschaft 2012, JKI 2012
Netherlands	Winterswijk	2012	–	PPS Netherlands 2012
Switzerland	Winterthur (Canton of Zurich)	2012	–	Forster and Wemelinger 2012
United Kingdom	Paddock Wood, Kent,	2012	–	Straw et al. (2014)
Austria	Geinberg, Ried Im Innkreis	2012	–	Hoyer et al. 2013, EPPO 2016
France	Corsica	2013	–	EPPO 2016
Austria	Gallspach	2013	–	Schreck and Tomiczech 2013
Italy	Marche	2013	–	EPPO 2016
Canada	Mississauga[c]	2013	–	Turgeon et al. 2015
Switzerland	Marly	2014	–	EPPO 2016
Germany	Neubiberg	2014	–	EPPO 2016
Germany	Ziemetshausen (Schönebach)	2014	–	EPPO 2016
Germany	Magdeburg	2014	–	EPPO 2016
Turkey	Zeytinburnu[d]	2014	2015	Ayberk et al. 2014, EPPO 2016
Finland	Vantaa	2015	–	EPPO 2016
Switzerland	Berikon	2015	–	EPPO 2016

[a] En dash (–) indicates the outbreak has not been eradicated.

[b] *A. glabripennis* has been eradicated from three areas: Islip (2011), Manhattan (2013), and Staten Island (2013).

[c] Recent evidence from mitochondrial DNA haplotype studies and backdating of the two Canadian *A. glabripennis* infestations in the Toronto area suggest that the two populations could be related (Turgeon et al. 2015).

[d] The official status in Turkey is absent because the Turkish National Plant Protection Organization have not detected the pest, although investigations are ongoing.

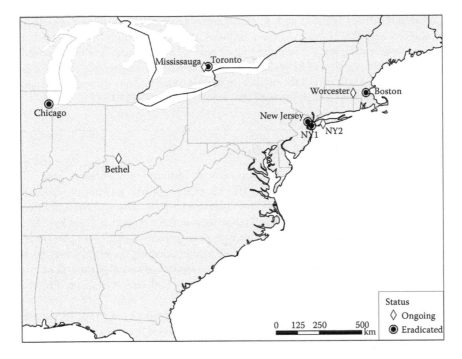

FIGURE 13.6 Locations of outbreaks of *Anoplophora glabripennis*, the Asian longhorn beetle, in North America as of January 2016. NY1 = The USDA have declared the Asian longhorn beetle as eradicated from Islip (2011), Staten Island (2013), and Manhattan (2013). NY2 = Outbreaks are ongoing in Brooklyn/Queens, central Long Island.

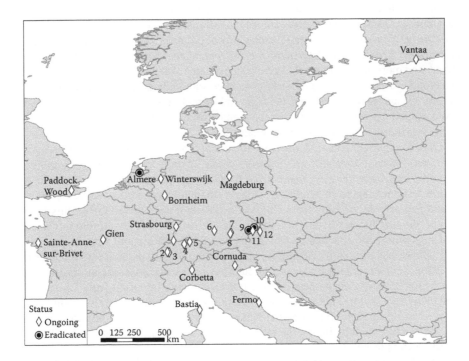

FIGURE 13.7 Locations of outbreaks of *Anoplophora glabripennis*, the Asian longhorn beetle, in Europe as of February 2016. 1 = Weil am Rhein, DE; 2 = Marly, CH; 3 = Brunisried, CH; 4 = Berikon, CH; 5 = Winterhur, CH; 6 = Ziemetshausen, DE; 7 = Feldkirchen, DE; 8 = Neubiberg, DE; 9 = Braunau am Inn, AT; 10 = Neukirchen am Inn, DE; 11 = Geinberg, AT; and 12 = Gallspach, AT (AT = Austria, CH = Switzerland, DE = Germany).

of millions of trees have been infested with *A. glabripennis* and cut down to slow the spread of the pest (Lingafelter and Hoebeke 2002). Some of these millions of *A. glabripennis*–infested trees likely were used to construct wood packaging, hence providing an easy pathway for the pest to be moved outside of Asia (Haack et al. 2010). An official EU visit to the areas of China that produce stone for Europe also found that exports were a small part of the market for many producers; thus they were not specialized in using wood packaging material complying with the requirements of ISPM 15 (EU 2013b).

13.5 Pathways: Finished Wood Products

The risks associated with finished wood products such as furniture or ornaments generally are considered low because of all the wood processing that takes place during production. Other factors that reduce the risk that finished wooden products would serve as a pathway for movement of cerambycids are: (1) these products are less likely to be stored outdoors and hence emerging pests have less chance of reaching host trees and (2) finished products are likely to be dispersed to retailers or consumers relatively quickly and not stored en masse in a single location like wood packaging material; therefore, the founder population likely would be smaller. Because Cerambycidae reproduce sexually, the risk of establishment would require that a male and female emerge near each other in both time and space. The emergence of one longhorn beetle of either sex is not a risk in itself.

Recorded interceptions of Cerambycidae in traded wooden products are rare in comparison with other pathways; for example, Cocquempot and Lindelow (2010) listed the main pathways for cerambycid introduction into Europe as timber, timber for pulp, wood packaging material, and plants for planting. However, many recent (2000–2010) interceptions of Cerambycidae have been related to manufactured wood products, including species such as *Chlorophorus annularis* (Fab.) and *Trichoferus campestris* (Faldermann) (Cocquempot and Lindelow 2010). One of the reasons for the relatively low number of cerambycid interceptions in manufactured wood products in the EU is that they are not "controlled goods," and there is no requirement to inspect such imported consignments on arrival in Europe. Given that these goods are not inspected at points of entry, many of the records associated with such goods originate from members of the public who are the eventual purchasers of products such as furniture. Between 1988 and 2008, for example, there were 83 detections of exotic organisms associated with imported furniture in New Zealand, and more than half of these occurred post-entry, with many being detected by members of the public (Froud et al. 2008). Even if the risks associated with individual consignments of manufactured wood products are judged to be very low, the large and increasing volume of wood furniture that is now being moved around the world suggests that the cumulative amount of goods moving via this pathway is sufficiently significant to deserve further investigation.

The brown fir longhorn beetle, *Callidiellum villosulum* (Fairmaire), is native to southeastern China and Taiwan, and known host plants are members of the Taxodiaceae including *Cunninghamia lanceolata* (Lamb.) and *Cryptomeria japonica* (L. f.) (Cocquempot and Mifsud 2013). In April 2013, *C. villosulum* was intercepted in Malta on saleable wooden commodities, some of which were not debarked and had arrived via Italy from China (Cocquempot and Mifsud 2013); this interception is believed to be the first for this beetle in Europe. This same pest was also found in Canada in 2012. Sections of logs with bark attached had been imported from China and used to provide the bases for artificial Christmas trees; a *C. villosulum* larva was found in association with one of these bases (Burleigh 2013). This was not the first time that such products had been associated with *C. villosulum*; there were 20 interceptions of the pest in Christmas tree bases imported from China in the United States (USDA-APHIS 2004). This provides an example of how difficult it is for plant health authorities to monitor and control the international movement of goods that may be infested with invasive pests, especially when the labeling on the product does not give any indication that there is a risk of invasive pests being present.

An analysis of the value of wood furniture imports into the United States between 1972 and 2010 demonstrates a switch from home-produced furniture to furniture imported from Asia (Luppold and Bumgardner 2011), hence increasing the risk of introduction of quarantine pests. In 1972, for example, the United States imported less than US$400 million dollars (adjusted for inflation to 1982 dollars) worth of wooden furniture, with Canada and Europe being the origin for most of the furniture. However, by 2006, imports of wood furniture into the United States reached a peak of US$6 billion, with the combined imports from China, Vietnam, Malaysia, Indonesia, Thailand, and the Philippines accounting for more than 65% (Luppold and Bumgardner 2011) (Figure 13.8).

In comparison with other manufactured goods, the risks of introducing wood-boring beetles have been demonstrated to be particularly high for upholstered furniture. Between 2000 and 2005, for example, imports of upholstered furniture from China into New Zealand increased from 962 units to 39,308 units, and similarly for Malaysia from 5,377 to 14,758 units, and for Thailand from 1,690 to 7,422 units (Froud et al. 2008). Overall, from 2000 to 2005, the total number of imports for upholstered furniture into New Zealand increased by more than fourfold from 16,895 to 65,456 units (Froud et al. 2008). In June 2006, teams from the Ministry of Agriculture and Forestry Biosecurity New Zealand purchased and destructively analyzed 49 imported couches—39 from China and 10 from Malaysia. This analysis confirmed the practice by some manufacturers of placing good quality (flawless) wood on visible surfaces and poor quality (insect damaged and bark covered) timber in internal sections of the furniture. Of the 49 couches inspected, 30 included wood with bark, 19 had insect contaminants, 32 had visible insect damage (including borer exit holes and bark beetle galleries), and fungal samples were collected from 11. Four longhorn beetles were found during the inspections, and the one that was identified to species was *Xylotrechus magnicollis* (Fairmaire) (Froud et al. 2008).

In July 2013, a member of the public in England reported an adult *Monochamus alternatus* Hope emerging from a recently purchased dining room chair (Ostojá-Starzewski 2014). The beetle had developed in the softwood framework that was covered in layers of foam and plastic. A further adult *M. alternatus* emerged from the chair, and three dead and one live larvae were found within the wood. Samples of wood in which the two beetles had developed were found to contain large numbers of pinewood nematodes, *Bursaphelenchus xylophilus* (Steiner and Buhrer). These examples demonstrate that there is a threat of invasive longhorn beetles becoming established in new environments as a result of the importation of finished wooden products; however, there have been few attempts to quantify the level of risk.

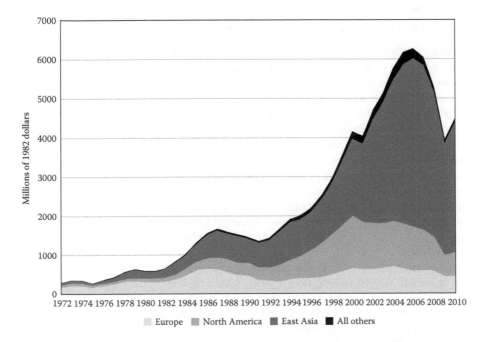

FIGURE 13.8 Value of U.S. imports of wood furniture by world region (in constant 1982 dollars), 1972–2010 (From Luppold, W. G., and M. S. Bumgardner, *Bioresources*, 6, 4895–4908, 2011.)

13.6 Pathways: Timber Other Than Wood Packaging Materials

There were 73 interceptions of Cerambycidae on "wood products or bark" between 1998 and 2013 recorded on EUROPHYT (Table 13.8). These interceptions were dominated by *Monochamus* spp. Of the 55 interceptions where the beetles were identified to at least the genus level, 51 were either *Monochamus* spp. (50 interceptions) or *M. alternatus* (1). All the *Monochamus* spp. interceptions were on commodities manufactured in either Europe (mainly non-EU countries) or Asia.

Monochamus galloprovincialis (Olivier) was intercepted in Turkey on industrial wood imported from the Ukraine in 2011 (Bozkurt et al. 2013). Three species of *Monochamus* were found in timber from the Komi Republic in northwest Russia during 1985–1989 in timber destined for Bulgaria—namely *M. sutor* (L.), *M. urussovi* (Fischer), and *M. galloprovincialis* (Olivier) (Tsankov et al. 1996). Between 2002 and 2005, Ostrauskas and Tamutis (2012) surveyed temporary storage sites for timber and wood imported into Lithuania from Russia using baited multiple-funnel traps. Seventeen species of longhorn beetles were caught, although all species were native to Europe. Similarly, four species of longhorn beetles were found in a timber mill in northern Sweden on timber imported from Siberia; these were *Tetropium gabrieli* Weise, *Tetropium aquilonium* Plavistshikov, *M. urrosovi* (Fischer & Walderheim), and *Acanthocinus griseus* (Fab.) (Lundberg and Petersson 1997). The arrival of these species was considered to be beneficial, however, because they are native to, but rare, in Sweden.

Kliejunas et al. (2001) assessed risks associated with the importation into the United States of unprocessed *Eucalyptus* logs and chips from South America. A number of species of wood-boring beetles were considered to be a high risk to U.S. *Eucalyptus* production including the following cerambycids: *Phoracantha semipunctata* (Fab.), *Chydarteres striatus* (Fab.), *Retrachyderes thoracicus* (Olivier), *Trachyderes* spp., *Steirastoma breve* Sulzer, and *Stenodontes spinibarbis* (L.). At the time of this assessment, *Eucalyptus* wood was not a regulated material in the EU; there was only one record of a pest being found on *Eucalyptus* wood, which was an unidentified Coleoptera. This unregulated pathway would have provided a potential introduction route for nonnative pests into Europe. *Eucalyptus* forestry is now extensive in Europe with more than 1.1 million ha being cultivated (Eyre 2005).

13.7 Pest Risk Assessments for Cerambycids

National and regional plant protection organizations have carried out numerous pest risk analyses (PRAs) for longhorn beetles in order to evaluate the threat they pose to countries and regions remote from the native regions of these beetles. Common triggers for carrying out a PRA are reports of a pest being moved in trade or successfully invading and damaging plants in a new environment. For example, the EPPO (2013c) carried out a PRA for *O. hirta*, a cerambycid native to New Zealand, following the discovery of the pest in the United Kingdom in association with plants from New Zealand, and MacLeod et al. (2002) performed a PRA for *A. glabripennis* following the confirmation of an outbreak of the beetle in New York and Illinois in 1996 and 1998, respectively (Haack et al. 1997; Poland et al. 1998).

The biology of some subfamilies of Cerambycidae makes them more likely to be moved in traded goods or wood packaging materials than others. For example, cerambycids in the subfamilies Prioninae, Lepturinae, and Parandrinae mostly develop in decaying wood, which is hence unlikely to be suitable for wood packaging, whereas most species of Cerambycinae and Lamiinae develop in living, dying, or recently dead plants, which are more likely to be used as wood packaging (Cocquempot and Lindelow 2010).

Polyphagous cerambycids such as the lamiine *A. chinensis* and *A. glabripennis* have been more successful at establishing in Europe than species with a narrower host range given that no strictly monophagous (feeding on a single host species) exotic longhorn beetles are known to have become established in Europe to date (Cocquempot and Lindelow 2010; Rassati et al. 2016). Table 13.9 shows the number of interceptions of longhorn beetles in wood at points of entry in the United States between 1984 and 2008 divided by subfamily. These data indicate that beetles in the subfamilies Spondylidinae (1,124 interceptions), Cerambycidae (1,081), and Lamiinae (814) accounted for 98% of

TABLE 13.8

Number of Interceptions of Longhorn Beetles in the European Union on Wood and Bark Products between 1998 and 2013 by Taxon and Country of Export

Harmful Organism	Country of Export	1998	1999	2000	2001	2002	2004	2005	2006	2008	2009	2010	2011	2012	2013	Grand Total
Cerambycidae	Cameroon										1		4			5
	China								1							1
	Congo													1	2	3
	Republic of the Congo												1			1
	Gabon									1						1
	Mongolia											1				1
	Russian Federation					2			2	1		1				6
Criocephalus rusticus									2							2
Monochamus alternatus	China		2													2
Monochamus sp.	Czech Republic		1	1												2
	Mongolia											1				1
	Romania												1			1
	Russian Federation			4	2	1	3	12	2				2			26
	Slovakia	2	3	8		2										15
	Ukraine												3	1		4
Tetropium sp.	Slovakia		1													1
Xylotrechus sp.	Tanzania											1				1
Grand Total		2	7	13	2	5	3	12	7	2	1	4	11	2	2	73

TABLE 13.9

Cerambycidae Intercepted in the United States between 1984 and 2008 on Wood Packaging Material by Subfamily

Subfamily	Number	Percent of Interceptions Identified to Subfamily (%)	Top Three Genera in Decreasing Order
Family only	396		
Cerambycinae	1,081	35	*Xylotrechus, Ceresium, Hesperophanes*
Lamiinae	814	26.4	*Monochamus, Anoplophora, Saperda*
Lepturinae	56	1.8	*Rhagium, Leptura*
Prioninae	12	0.4	*Prionus, Stenodontes, Mallodon*
Spondylidinae	1,124	36.4	*Arhopalus, Tetropium, Asemum*
Grand total	3,483	100	

all interceptions, while the Prioninae and Lepturinae comprised only about 2%. Six key elements of PRAs for longhorn beetles are discussed in the following text.

13.7.1 The Pathway of Pest Movement

Understanding the routes along which pest longhorn beetles may travel is an essential part of understanding the potential risk to an importing country. As discussed in previous sections, some pathways are inherently riskier than others due to (1) the volume of goods that are carried along the pathway, (2) the degree to which the goods are inspected on arrival in the importing country, and (3) the extent and duration that the potentially infested goods (and pests) are kept together after their arrival.

Population genetics can be used to establish the most likely source of invasive populations of insects; this could be populations in the native environment of the pest or another invasive population where the pest has become established. This can help explain the pathways along which pests may have moved. For example, Miller et al. (2005) indicated that there had been at least three separate introductions of the North American chrysomelid pest of maize, *Diabrotica virgifera virgifera* Leconte, into Europe and that some outbreaks were the result of intra-European movements. Carter et al. (2010) studied the mitochondrial sequence data of invasive populations of *A. glabripennis* in North America and Europe. They suggested that separate introductions of beetles from Asia were responsible for the appearance of *A. glabripennis* infestations in (1) New York City; (2) Carteret and Linden, New Jersey; and (3) Staten Island and Prall's Island, New York. It was indicated that local spread was responsible for moving the pest from New York City (e.g., Queens) to other areas of Long Island. Samples from outbreaks in Austria, France, Germany, and Italy indicated that there had been separate introductions to each of these locations—most likely from China rather than spread among the different outbreak areas in Europe. Carter et al. (2010) were not able to identify locations within China that might have been the source of the invasive *A. glabripennis* populations in North America or China, possibly because they were derived from areas of China where the populations of *A. glabripennis* were invasive, and therefore there was no clear link between the population genetics and locations. The outbreaks in Europe appeared to have been initiated with a low number of founders, and any genetic bottlenecks did not appear to inhibit population development.

13.7.2 Propagule Pressure

Propagule pressure is a measure of the number of individuals released into a nonnative environment and incorporates the number of introduction events and the number of individuals associated with each introduction (Reaser et al. 2008). Propagule pressure clearly plays an important role in invasion ecology (Reaser et al. 2008; Eschtruth and Battles 2011). Brockerhoff et al. (2014) proposed that propagule pressure may be the most important factor in establishment success of nonnative species of various taxa in a variety of ecosystems worldwide. Logically, this will only be the case when comparing taxa that are able to survive in both the biotic and abiotic environments being assessed. Propagule pressure is likely to be more important for sexually reproducing pests, such as Cerambycidae, than it is for pests that reproduce asexually, such as aphids. This is because when cerambycids are moved internationally as juveniles, there is a need for an individual of each sex to be introduced into locations close enough for them to be able to find each other once they have emerged as adults and mate. Brockerhoff et al. (2014) found a positive association between interception frequency and probability of establishment for bark beetles and longhorn beetles in the United States and worldwide. Their work showed that, for the most frequently intercepted species (i.e., those exerting the greatest propagule pressure), a fractional change (e.g., 50% fewer) in the numbers arriving will result in little change in establishment rates but that such fractional changes may have a strong impact on establishment success of species that arrive less frequently.

In a study of the destination of packaging materials that are frequently associated with exotic forest insects, Colunga-Garcia et al. (2009) found that 84–88% of the imported tonnage ended up at only 4–6% of the urban areas in the contiguous United States. Such studies are useful for assessing the overall risk of pests to countries and can also be used to target inspection and monitoring activities. Colunga-Garcia et al. (2010) analyzed the significance of a number of factors in determining the success rate of invasions of *A. glabripennis* and the Asian strain of the gypsy moth, *Lymantria dispar* (L.), in the United States and found (1) that propagule pressure was a more significant factor than percent tree cover in determining successful invasions and (2) that the predominant land use where most forest insect pests invaded was commercial–industrial. Such information could be used to optimize surveillance for these pests.

13.7.3 Pest Establishment in the Nonnative Environment

A range of techniques and tools can be used to evaluate the potential for a cerambycid species to survive in a nonnative environment, and several of these methods have been reviewed recently by Eyre et al. (2012) and Sutherst (2014). One of the indicators that can be used to evaluate the potential of an exotic cerambycid's survival at latitudinal extremes is the concept of degree-days. The number of degree-days needed to complete development can be measured in laboratory studies such as those by Adachi (1994) for *A. chinensis* and by Keena and Moore (2010) for *A. glabripennis*. Degree-days have been used in risk assessments by the EPPO (2013c) to illustrate the extent to which the climate that *O. hirta* experiences in its native environment is similar to the climate that it might experience in Europe were it to become established there.

Such analyses can be taken a step further by estimating the potential length of time for a species of longhorn beetle to complete its life cycle by equating the average climate conditions at a particular location to the number of degree-days above a threshold required for the organism to complete its life cycle. An example is shown in Figure 13.9 for *A. chinensis*. This is based upon a requirement for 1,900 degree-days in excess of a threshold of 11°C in line with Adachi (1994).

If risk assessors have either, or ideally both, records of where a pest insect is known to be present globally and laboratory temperature-response data, it is possible to create a model projecting where the pest may be able to survive in nonnative regions. Figure 13.10 shows the predicted environmental index (EI) for *A. glabripennis* in Europe from a CLIMEX model (Sutherst et al. 2007) that was published in MacLeod et al. (2012). Model parameter sets are developed based on the known distribution of a species in its native environment and/or the use of laboratory temperature response studies. The environmental index is a measure of how suitable the climate is for *A. chinensis* in different European locations. An EI value of zero is a prediction that the species would not be able to survive in that location for the longer term. Higher EI values are an indication of greater suitability for a species. Vanhanen et al. (2008) created CLIMEX models for three Asian species of longhorn beetles, *Aeolesthes sarta* (Solsky), *Tetropium gracilicorne* (Reitter), and *Xylotrechus altaicus* (Gebler), and predicted that all three had the potential to establish in Europe.

The Genetic Algorithm for Rule-set Prediction (GARP) ecological niche modeling technique was used to project the potential distribution of *A. glabripennis* and *A. chinensis* in the United States based on their distribution in Asia (Peterson and Vieglais 2001; Peterson et al. 2004). These analyses suggested that the area at greatest risk from *A. glabripennis* in North America

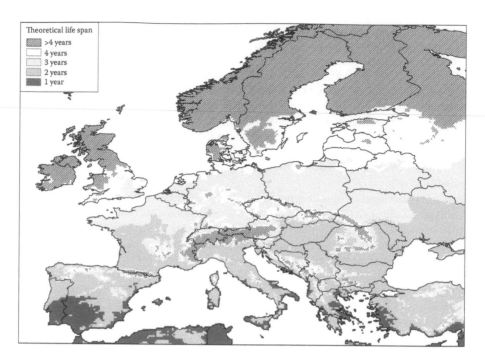

FIGURE 13.9 Predicted generation time for *Anoplophora chinensis* in Europe based on the CLIMEX model published in the supplementary material of Robinet et al. (2012).

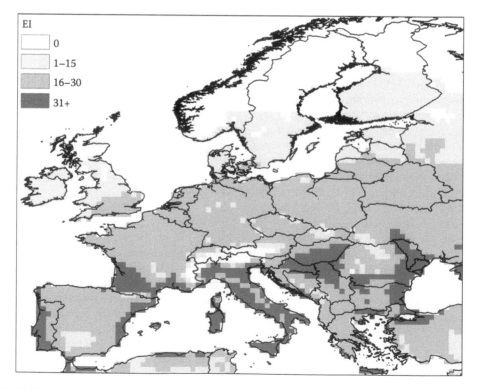

FIGURE 13.10 Environmental index for *Anoplophora glabripennis* in Europe based on a CLIMEX model published in the report of a project for the European Food Safety Authority. (From MacLeod, A., H., et al., External scientific report: Pest risk assessment for the European Community plant health: A comparative approach with case studies, Annex P2 ALB Method 3 no RROs. Available at: http://www.efsa.europa.eu/en/supporting/pub/319e.htm.)

was the eastern United States, especially the Great Lakes region. The predictions for *A. chinensis* (Peterson and Vieglais 2001) indicated that the areas at greatest risk were more southerly and dispersed across North America.

13.7.4 The Host Species

Whether host plants are native or exotic or a mixture of both is another important factor in the eventual impact of a longhorn beetle. If a longhorn beetle only attacks uncommon ornamentals, the overall impact is likely to be low. For example, *Frea marmorata* Gerstaecker is thought to have a very low chance of establishing in Europe because it is known from eastern Africa (Kenya, Malawi, Tanzania, and Zimbabwe), and the known host is coffee, *Coffea arabica*, which is not grown commercially in Europe (Cocquempot 2007). However, longhorn beetles that can damage widely planted exotic plants can be significant pests. *Phoracantha semipunctata* (Fab.) is native to Australia but has become established in all continents where eucalypts are grown, with the exception of Asia (Paine et al. 2011). This species has caused serious damage in some Mediterranean countries, including Morocco where around 170,000 ha of eucalyptus plantations were lost in 1981 (El-Yousfi 1982). The impact appears to be greatest in regions of the world with Mediterranean climates (CABI 2014c). Pests that have significant impacts on eucalyptus plantations have the potential to cause considerable economic impact because eucalypts have been planted in more than 90 countries, they account for 8% of productive planted forests worldwide, and in 2000 they provided 33% of the global wood supply (Laclau et al. 2013).

Some invasive longhorn beetles are important pests of fruit trees and have the potential to cause significant economic damage in invaded areas. *Aromia bungii* (Faldermann), the red-necked longhorn beetle, is an Asian pest of *Prunus* spp., especially *P. persica* L. (peach) and *P. armeniaca* L. (apricot) but to a lesser extent plum (e.g., *P. domestica* L.) and cherry (e.g., *P. avium* L.) and some other plants and deciduous trees such as *Populus* spp., *Punicia granatum* L. (pomegranate), *Olea europaea* L. (olive), and *Bambusa textilis* McClure (weaver's bamboo) (Liu et al. 1993; Anderson et al. 2013). In 2011, an outbreak of *A. bungii* was discovered near Kolbermoor in southern Bavaria, Germany, and in 2012, an outbreak of this pest was discovered between Napoli and Pozzuoli in southern Italy where it was infesting apricot, cherry, and plum trees (Anderson et al. 2013; EPPO 2016). *A. chinensis* is a significant pest of citrus and other fruit trees in China and Japan (Lieu 1945; Adachi 1994) and has the potential to cause severe damage in citrus groves and other fruit trees on other continents.

In addition to creating pest risk maps based on climate, it is possible to combine climate data with other risk factors in order to generate predictions of the endangered areas within a pest risk assessment area, that is, the areas where economically important loss is likely to occur (Baker et al. 2012). Magarey et al. (2011) described the North Carolina State University–APHIS Plant Pest Forecasting System (NAPPFAST), which links climatic and geographical databases with interactive templates for biological modeling. Three factors that can be used as indicators of overall plant pest risk in a particular location are host density, climate, and pathways. Decision makers may prefer to look at a single map that integrates the risks associated with these three factors, and NAPPFAST can be used to create such maps (Magarey et al. 2011). Pareto dominance is a technique for combining risk factors that does not rely on linear weighted averaging or standardization of the individual criteria (Yemshanov et al. 2010). The USDA has published a map based on a Pareto dominance analysis that shows which parts of the United States are most at risk of an outbreak of *M. alternatus* (USDA-CPHST 2011).

13.7.5 Preferences for Site of Attack and Health Status of Hosts

Some species of Cerambycidae have a preference for attacking hosts at a certain location and also for trees of a certain age or health status. These preferences will influence the nature of the damage that the beetle may cause and whether the damage is mainly economic or environmental. The eucalyptus longhorn borer, *P. semipunctata*, is an example of a pest that has a different pest status in its native environment when compared with its behavior in some invaded environments. In its native Australia, *P. semipunctata* primarily is a scavenger of dead, stressed, or dying trees, but it can cause significant mortality in invaded areas (CABI 2014c). It typically is associated with trees under moisture stress and therefore causes severe damage in semiarid regions as well as during drought periods as has occurred in southern Europe and northern Africa (Turnbull 1999; Kliejunas et al. 2001).

Trichoferus campestris was reported emerging from a cutlery drawer imported from China to the United Kingdom in June 2013 (Fera, unpublished data). An assessment of the phytosanitary risk of this beetle to the United Kingdom has been complicated because of conflicting evidence over whether it attacks live trees. For example, the Centre for Agriculture and Biosciences International (CABI 2015d) states that the economic impact of this beetle is comparatively small as with most drywood-boring pest species of Cerambycidae, whereas the EPPO (2009a) reported that *T. campestris* is able to attack healthy or slightly stressed trees of many important species. More recently, despite its establishment in Eastern Europe, apparent westward spread, and occasional finds in Canada and the United States, this pest's damage to living trees appears to have been minimal (Sabol 2009; Grebennikov et al. 2010; Dascalu et al. 2013; Bullas-Appleton et al. 2014; EPPO 2016). The current evaluation of the risks posed by this beetle to the United Kingdom can be seen on the UK plant health risk register (Baker et al. 2014). In common with many lamiine and cerambycine species, *A. glabripennis* can infest and develop in both healthy and stressed trees (EPPO 1999; Cocquempot and Lindelow 2010; Haack et al. 2010), a characteristic that has made it a highly invasive pest. Many of the outbreaks of *A. glabripennis* in North America and Europe have been in towns and cities such as New York, Chicago, Toronto, and Braunau-am Inn, Austria (Haack et al. 2010). In these locations, *A. glabripennis* attacks trees along streets and in yards and parks, resulting at first in upper trunk and branch dieback, which poses a threat to people, vehicles, or property (MacLeod et al. 2003). In China, damage by *A. glabripennis* has killed millions of trees, especially species of *Populus*, which were planted to counteract desertification in northern regions (Hu et al. 2009; Haack et al. 2010).

In contrast, *O. hirta*, which specializes in infesting small branches and twigs, rarely kills trees in its native environment, and mature trees with a large diameter trunk are unlikely to be killed. In addition, the damage to large branches of host trees does not tend to kill them but tends to degrade the condition of infested trees (EPPO 2013c).

Longhorn beetles can be vectors of highly damaging pests, for example, the pinewood nematode, *B. xylophilus,* which is native to North America. Without regulatory controls, the outbreak of this pest in Portugal and Spain has the potential to cause some US$24 billion damage in the EU (Soliman et al. 2012). There are also damaging outbreaks of the pinewood nematode in China where the pest is spreading both naturally as well as through human assistance (Hu et al. 2013). The pinewood nematode is discussed in more detail in Chapter 6.

13.7.6 Pest Spread in the Invaded Environment

The difficulty of detecting longhorn beetles makes the accurate assessment of their rate of spread in a new environment daunting, but this becomes more feasible when there is an efficient and systematic trapping system available for the pest. Species that are deemed by national or regional plant protection organizations to be significant enough pests to monitor are often subject to containment or eradication measures. Therefore, any measurements of the rate of spread can also be used to measure the effectiveness of the control methods. In general, the overall rate of natural spread of many invasive longhorn beetles is slow because adults reproduce near the host plant from which they emerge (Haack et al. 2010; Rhainds et al. 2011). In contrast, the rate of spread of other invasive wood-boring beetles can be much faster; for example, the buprestid *Agrilus planipennis* is estimated to be spreading at a rate of around 20 km a year in North America (Prasad et al. 2010), and the ambrosia beetle (*Xylosandrus germarus*) is able to spread over tens of kilometers per year (Rassati et al. 2016).

The spread of the invasive species *T. fuscum* (Fab.) has been studied in Canada by analyzing the results of a trapping program using traps baited with a male-produced pheromone and host volatiles (Rhainds et al. 2011). This species was first discovered in North America in Halifax, Nova Scotia, in 1999, where it was attacking red spruce, *Picea rubens* Sarg., but it is thought to have been established since at least 1990 (Rhainds et al. 2011). *T. fuscum* is native to Europe where it primarily infests Norway spruce, *P. abies* (L.) (Smith and Hurley 2000). There has been an eradication and containment program for this pest in Canada since May 2000 (Sweeney et al. 2007). A formal eradication program for *T. fuscum* took place in Nova Scotia from 2000 to 2006. This involved the cutting and disposal of selected high-risk trees and regulating the movement of high-risk materials. A containment program was enacted beginning in 2007 (Gregg Cunningham, Canadian Food and Inspection Agency, personal communication). Since the time of its original introduction in the early 1990s, the results of the trapping program indicated that between 1990 and 2010 the pest had spread approximately 80 km. The rate of spread of the beetle is thought to have been limited because spruce is rarely used as firewood in Canada, thus reducing the risk of the pest being spread by this pathway (Rhainds et al. 2011). The potential spread rate of longhorn beetles can also be studied using flight mills. For example, Sweeney et al. (2009) found that both sexes of *T. fuscum* flew an average of more than 1 km per day, with some individuals flying more than 9 km a day and some not flying at all.

Sawyer et al. (2011) have studied the spread of *A. glabripennis* in the United States at outbreak sites in suburban and urban areas and compared these findings with data from outbreaks in more open habitats. In urban and suburban landscapes where tree density is moderate to high and the tree diversity is high, *A. glabripennis* tends to spread slowly, developing in the initial years on just a few trees near the point of origin. In Chicago, surveys indicated that where hosts were readily available, females did not travel far from the trees from which they emerged, as evidenced by 90% of the oviposition sites being found within 140 m of the nearest exit hole and 99% of oviposition sites being within 300 m of the nearest exit hole. In contrast, at an outbreak site in Linden, New Jersey, which is a more open habitat with low tree density, *A. glabripennis* spread about 3.2 km to the south and west in five years. In line with the observations in the United States, an analysis of the trees infested with *A. glabripennis* in Mississauga, Canada, revealed that a population, possibly founded by one or a few individuals from a distant outbreak (at least 7.5 km away), remained on a single tree for about six or seven years before spreading naturally to other trees around 300 m away (Turgeon et al. 2015).

A model created using survey data from the outbreak of *A. glabripennis* in Cornuda, Italy, indicated that 80% of dispersal occurred within 300 m of the nearest infested tree-cluster, but some individuals could move farther than 2 km (Favaro et al. 2015). *A. glabripennis* dispersal distances in Cornuda were shown to lessen as the population density of the pest declined due to eradication activity (Favaro et al. 2015).

Keena and Major (2001) studied how various factors influenced the propensity of *A. glabripennis* to fly in the laboratory at 15°C, 22°C, and 30°C using fresh cut logs. Adults did not fly at 15°C, and the propensity to fly increased with temperature. Well-fed females would remain on good-quality hosts and chew oviposition pits rather than fly, but when host quality was poor, more than 50% of females flew within the first 30 minutes. Thus, if emerging female beetles detect that the tree from which they have emerged is in poor condition, perhaps as a result of successive generations of *A. glabripennis* infestation, they would be expected to be more likely to disperse than if the host was still in relatively good health.

Trotter and Hull-Sanders (2015) used data from the *A. glabripennis* outbreak in Worcester, Massachusetts, and graph theory to create four dispersal kernels for the beetle based on different assumptions. One of the assumptions of the models was that the dispersal of *A. glabripennis* could be the result of individuals dispersing a long way from heavily infested trees (those trees with >100 exit holes) or, alternatively, it could be the result of beetles dispersing shorter distances from trees with lower levels of infestation. Under the four different scenarios, the fiftieth percentile annual dispersal kernel was between 37 and 234 m and the ninety-ninth dispersal kernel was between 1,358 and 2,979 m.

The potential rate of spread of invasive longhorn beetles has been studied in three mark-recapture studies in Gansu Province, China. Smith et al. (2001) carried out a mark-recapture study of *A. glabripennis* using 16,511 adult beetles. The beetles were either captured as they emerged from poplar bolts that had been transported to the release site or were collected locally. Capture stations, consisting of about 12 poplar trees (averaging 7.2 cm diameter at breast height and 7.8 m tall) were established at varying distances from the release point. A total of 188 beetles were recaptured, and of these, the mean dispersal distance was 266 m, with 98% being recaptured within 560 m.

In a second study, Bancroft and Smith (2005) found strong evidence for density-dependent dispersal of *A. glabripennis*: the beetles were more likely to disperse when their density on a tree was high. Females and smaller beetles dispersed further than males and larger beetles. Flying *A. glabripennis* tended to head toward large, lush trees. Over the course of this study, the average daily temperature and humidity at 14:00 hours were 32°C and 51% relative humidity (RH). Beetles were captured by shaking trees in the early morning when temperatures were less than 20°C and adult beetles were less active.

The propensity of beetles to move declined with increasing temperature, but the distance traveled increased with increasing temperature. In a third, more recent mark-recapture study (Li et al. 2010), the mean dispersal distance of *A. glabripennis* was 424 m, there was no difference between male and female adults in dispersal direction or distance traveled, and the furthest dispersal distance recorded was 2,644 m. Beetles tended to fly into the wind, but there was no strong relationship found among wind speed, temperature, or relative humidity.

Previous studies show that spread of *A. glabripennis* can be very minimal at the start of outbreaks (Sawyer et al. 2011), but after this period it is likely to be in the region of hundreds of meters per year (Smith et al. 2001; Li et al. 2010; Sawyer et al. 2011). The spread rate can also be influenced by tree density and species composition (Sawyer et al. 2011), and some studies have shown that weather, including temperature and wind, can be a factor (Keena and Major 2001; Bancroft and Smith 2005). For example, the threshold for *A. glabripennis* flight is between 15°C and 22°C (Keena and Major 2001), and the optimum conditions for flight are likely to be warmer than 20°C (Bancroft and Smith 2005). In northern Europe, the number of days in the summer and the proportion of each summer day when conditions would be suboptimal for *A. glabripennis* flight would be much higher than those in southern Europe. A comparison of the actual spread rate of *A. glabripennis* in areas with very different climatic conditions—for example, comparing the rate of spread in northern Europe (the United Kingdom and the Netherlands) with the rate of spread in locations with much warmer summers such as Italy or Ohio (United States)—might provide an indication of how summer air temperatures affect rate of spread.

The studies described here for *A. glabripennis* have aimed to establish how far the beetle might be able to move without human intervention. However, in addition to the international movement of longhorn beetles in wood or plants, the risk of pests being moved by humans also needs to be assessed on a more local scale for outbreak situations. For example, following the discovery of *A. glabripennis* in Brooklyn, New York, in 1996, survey teams started to investigate areas where the pest might have been moved accidentally, which led to the discovery of a second infestation approximately 50 km east in Amityville. The Amityville infestation is thought to have been initiated by the movement of infested tree sections meant for disposal or sale as firewood (Haack et al. 1997).

13.8 Preventing the Establishment of Nonnative Longhorn Beetles

International and local movements of goods and people provide opportunities for nonnative longhorn beetles to establish in new locations. The sections that follow describe the measures that can be taken to prevent establishment of longhorn beetles at different stages before and after their arrival in a new location.

13.8.1 Pre- and Post-Export Treatments and Measures

There are a number of international advisory documents and standards that have been developed to reduce the risk of wood-boring pests such as Cerambycidae being moved in wood and wood packaging materials. The Food and Agriculture Organization (FAO) of the United Nations published a guide to the implementation of phytosanitary standards in forestry in 2011 (FAO 2011). Some of the postharvest measures recommended to reduce pest spread are: on-site treatment of freshly felled logs, inspection of logs entering sawmills, storing log piles under cover, general hygiene measures, and postharvest treatments such as fumigation.

The EPPO, the regional plant protection organization for Europe and the Mediterranean region, has developed and published a number of standards relevant to longhorn beetles in commodities. For example, EPPO PM 10/6 concerns heat treatments of wood to control insects (EPPO 2009c), EPPO 10/7 describes methyl bromide fumigation of wood to control insects and wood-borne nematodes (EPPO 2009d), and EPPO PM 10/8 covers the disinfestation of wood with ionizing radiation (EPPO 2009b). The EPPO also has developed two commodity-specific phytosanitary measures of relevance to Cerambycidae, one for Coniferae (PM 8/2) (EPPO 2009e) and another for *Quercus* and *Castanea* (PM 8/3) (EPPO 2010). These standards list the relevant pests for each plant genus and provide a description of what treatments would be appropriate for wood of the different tree genera that has been processed to different levels such as plants for planting, cut branches, or squared wood.

On an international level, phytosanitary treatments have been developed by the Technical Panel on Phytosanitary Treatments (TPPT) of the International Plant Protection Convention (IPPC). The TPPT evaluates data from national and regional plant protection organizations and develops wood treatment measures (IPPC 2014).

The IPPC is close to finalizing a standard called "Management of Phytosanitary Risks in the International Movement of Wood," which will include information on appropriate types of treatments to reduce risks associated with movement of wood other than wood packaging material (IPPC 2013). There is also a more recently commissioned international standard called "International Movement of Wood Products and Handicrafts Made from Wood." The commodities that are regulated for wood-boring pests vary among countries and regions. For example, finished wooden products such as furniture are not considered to be controlled goods within the EU (EU 2000), whereas there are treatment requirements for the import of wooden manufactured items into Australia (Australian Government 2015).

The international phytosanitary standard for wood packaging materials, ISPM 15, was published originally in 2002 (IPPC 2002) and has since been revised (IPPC 2009, 2016). The 2016 version of ISPM 15 gives the option of heat treatment using dielectric heating (e.g., microwave). Wood packaging materials composed of wood not exceeding 20 cm when measured across the smallest dimension of the piece or the stack must be heated to achieve a minimum temperature of 60°C for one continuous minute throughout the entire profile of the wood (including its surface) when using dielectric heating. The goal of ISPM 15 is to reduce the risks that wood-associated insect pests and diseases will be moved around in wood packaging through international trade (Haack and Brockerhoff 2011; Haack et al. 2014). Internationally moved wood packaging materials are required to be marked to indicate that they conform to this standard. To comply with the standard, the wood used needs to be debarked and either heat-treated or fumigated with methyl-bromide (IPPC 2016). Despite the introduction of this standard, interceptions of nonnative longhorn beetles continue with wood packaging materials (Figure 13.2). Using a compilation of interception records in North America, Europe, and Australasia, Haack et al. (2010) reported that of the interceptions of *Anoplophora* species where the country of origin was known (71%), 97% of the interceptions on wood packaging materials originated in China. Similarly, in the EUROPHYT records from 1999–2013, 257 (84%) of the 306 interceptions of Cerambycidae were in association with wood packaging materials from China.

In response to the link between wood packaging materials from China and the outbreaks of *A. glabripennis* in Europe, the EU introduced requirements to inspect set proportions of consignments of wood packaging materials associated with specified commodities (EU 2013a). This decision came into effect on March 31, 2013, requiring that the wood packaging materials associated with 90% of consignments of slate, marble, and granite should be inspected, along with 15% of consignments of two other categories of stone imports. In Austria, 451 consignments (1,374 containers) of stone imports were inspected between April 1, 2013 and April 14, 2014. Noncompliance with ISPM 15 was reported in 44 cases (9.8%), and live Cerambycidae were found in 38 consignments, including *A. glabripennis, T. campestris*, and *A. germari* (Krehan 2014).

In June 2013, the EU's Food and Veterinary Office (FVO) carried out an audit of the measures taken by China to ensure that wood packaging materials exported to Europe meet EU requirements (EU 2013b). The FVO concluded (p. 17) that "the current system of official controls in China did not adequately ensure that wood packaging materials which form part of consignments of goods exported to the EU are marked and treated according to ISPM 15." Thus, flaws related to the treatment of wood packaging materials in China have left open a pathway for the movement of *A. glabripennis* and other wood pests to other parts of the world. Although there is a strong link between wood packaging material from China and the spread of *A. glabripennis* to North America and Europe, problems with pests in wood packaging material are a worldwide problem. For example, Table 13.4 shows that, in the EU, in addition to wood packaging material from China, longhorn beetles have been found in wood packaging material from EU countries, other European countries, other Asian countries, Africa, Australasia, and from the Americas. Table 13.6 shows that the trades responsible for the greatest number of interceptions of longhorn beetles on wood packaging material in the United States includes those from Italy, Turkey, Spain, France, Greece, and Bulgaria, as well as from China.

Haack et al. (2014) compared pest interception rates in the United States from periods before and after the implementation of ISPM 15. Data from 2003 to 2009 (about two years before versus four years after implementing ISPM 15) indicated that infestation rates of wood-infesting insects declined between 36% and 52% to a level of about 0.11% following the implementation of ISPM 15. These results led the authors to conclude that ISPM 15 did result in a lowering of the level of infestation in wood packaging materials, but if infestation rates prior to 2003 had been available for making comparisons, it is likely that it would have been possible to demonstrate that the impact of ISPM 15 had been even greater. In addition, Haack et al. (2014) demonstrated the importance of an infestation rate even as low as 0.1% by estimating that this would represent 13,000 infested containers entering the United States each year based upon an annual total of 25 million shipping containers—of which 52% have wood packaging materials.

The main fumigants currently used for quarantine and biosecurity purposes are methyl bromide and phosphine, and there are problems with both. Methyl bromide is an ozone-depleting compound, and although there is an exemption from the control measures agreed under the Montreal Protocol (UNEP 2011) for quarantine and preshipment purposes, international policy requires its replacement if technically and economically feasible (Banks 2012). The use of phosphine requires long exposure times to be fully effective, which can make it inconvenient and costly (Banks 2012), and some storage pests have developed a resistance to it (Jagadeesan et al. 2012). The difficulty with the long exposure time needed for phosphine can be reduced if treatments are carried out during transit, which is the case for the export of *Pinus radiata* D. Don logs from New Zealand (Glassey et al. 2005). Disinfestation of wood and wood packaging materials is the largest global use of methyl bromide, even though heat is widely utilized as an alternative to satisfy the requirements of ISPM 15. To date, no alternative fumigation treatments have been found that provide an equivalent level of efficacy and hence allow inclusion in ISPM 15, but some of the alternatives that have been considered are sulfuryl fluoride, methyl isothiocyanate, hydrogen cyanide, and methyl iodide (Banks 2012).

Sulfuryl fluoride has been shown to be able to penetrate wood as effectively as methyl bromide (Ren et al. 2011), and its efficacy has been demonstrated against *A. glabripennis* in *Populus* logs (Barak et al. 2006), *Hylotrupes bajulus* L. in the timbers of an historic building in Istanbul (Yildirim et al. 2012), and the cerambycid-vectored pinewood nematode, *B. xylophilus*, in boards of *Pinus pinaster* Aiton (Bonifacio et al. 2014). The application of methyl isothiocyanate for 24 hours at 15°C killed all *Semanotus japonicus* Lacordaire (eggs), *C. rufipenne* (all stages), and *M. alternatus* Hope (eggs) at 20 g/m^3 and all *M. alternatus* larvae at 40 g/m^3; however, it had little effect on ambrosia beetles (Scolytinae) (Naito et al. 1999).

Stejskal et al. (2014) tested hydrogen cyanide (HCN) against three pests of trees. At a temperature of 20–21°C, all *A. glabripennis* and *H. bajulus* were killed by a dose of 20g/m^3 of HCN within one hour. Similarly, all *B. xylophilus* were killed by exposure to 20g/m^3 of HCN for 18 hours at 25°C. Methyl iodide has been shown to be effective at killing *M. alternatus* and *Arholalus rusticus* when applied for 24 hours at 84g/m^3 at 10°C, 60g/ m^3 at 15°C, or 36g/ m^3 at 25°C (Soma et al. 2006). Soma et al. (2007) found that when methyl iodide was applied at 60g/m^3 for 24 hours at around 10°C, it killed *M. alternatus* larvae and *C. rufipenne* adults.

Zahid et al. (2013) studied the efficacy of freezing naturally infested logs of tropical origin, *Acacia parramattensis* and *Acacia decurrens*, timber infested with termite (Isoptera), and timber blocks inoculated with Bostrichidae as a phytosanitary treatment. No insects were found alive after the wood had been subjected to temperatures of −18°C to −25°C for 24 hours. Dead insects recovered from the samples included Cerambycidae, Bostrichidae, and Scolytinae.

Zhan et al. (2011b) studied the impact of X-ray irradiation on *M. alternatus* mature larvae, pupae, and adults and found irradiation tolerance increased in older life stages. In order to sterilize logs infested with *M. alternatus*, the authors recommended a minimum dose of 60–80 Gy for larvae and 140 Gy for pupae and newly developed adults. Zhan et al. (2011a) obtained similar results for *Monochamus sutor*.

13.8.2 Detection of Infested Consignments

The majority of interceptions of longhorn beetles reported on the EUROPHYT database have been associated with wood packaging materials. Even though wood packaging is a known pathway for the international movement of longhorn beetles, the large volume of trade that arrives with wood packaging materials means that inspection of all consignments is not possible. In addition, shipping manifests do not currently state whether or not wood packaging materials have been used, making the targeting of inspections more difficult. However, there has been a history of longhorn beetles being found in association with certain types of cargo, for example, the top 20 trades listed in Table 13.6. These kinds of data can be used to target inspections on those consignments that are most likely to be infested as has happened for stone imports into the EU from China (EU 2013a).

In recognition of the difficulty of detecting longhorn beetle larvae in young trees, in 2012, measures were adopted in the EU to reduce the risk of *A. chinensis* being introduced in imported plants, which preceded a two-year ban on the import of *Acer* sp. from China (EU 2012). Among other requirements, the 2012 EU Commission Decision specified that 10% of *A. chinensis* host plants (or 450 plants if the consignment is >4,500 plants) imported into the EU should be destructively sampled. This practice should help overcome the difficulty of detecting early stage larvae in large consignments of young plants.

In addition to visual inspection, some new methods for detecting longhorn beetles in wood packaging materials and young plants have been researched and developed. A team in Austria has been using sniffer dogs to detect *Anoplophora* in wood packaging, imported trees, and at outbreak sites in Germany and Switzerland (Hoyer-Tomiczek and Sauseng 2013). In the United States, sniffer dogs were 80–90% effective at detecting *A. glabripennis* frass in controlled environments (Errico 2013; Meng et al. 2015).

Three types of infrared thermography camera have been evaluated for their ability to detect young larvae of insects such as *A. chinensis* within imported young trees, but currently available technology does not seem to be adequate for this use (Hoffmann et al. 2013). Mankin et al. (2008) and Schofield (2011) have investigated the use of acoustic equipment to detect and identify longhorn beetles within wood and have shown that it has some potential. In a study with bolts of wood infested with *A. glabripennis*, Zorovic and Cokl (2015) were able to demonstrate that laser vibrometry could be used to distinguish infested and noninfested bolts. In summary, although many cerambycids are intercepted, the methods and the resources available to plant protection services are not adequate to prevent the international movement of cerambycids on wood packaging materials and plants for planting.

13.8.3 Monitoring in and Around Ports and Other High-Risk Sites

Setting up traps at locations in and around ports provides an opportunity to detect beetles that have not been found in the course of inspections of imported goods and could provide an early indication of an outbreak near a port. Rassati et al. (2014) set up traps for wood-boring beetles around four international ports in northeastern Italy. In each port, four traps were baited with one of four different lures (single-lure traps) and one trap included all four lures (a multi-lure trap). Two of the lures tested were generic attractants for bark and wood-boring beetles (ethanol and [−] α-pinene), while the other attractants were specific for bark beetles (frontalin, ipsenol, and ipsdienol). Ipsenol and ipsdienol were tested together. The mean number of trapped species was significantly higher in multi-lure traps than in each individual single-lure traps. Multi-funnel traps were compared with cross-vane traps (Figure 13.11). Although there was no significant difference in their trapping performance, the multi-funnel traps were found to be more suitable than cross-vane traps around ports because they were easier and quicker to set up and more durable to adverse conditions such as high wind. Eleven species of longhorn beetles were trapped, including two nonnative species, *Neoclytus acuminatus* (Fab.) and *X. stebbingi* Gahan (Rassati et al. 2014).

A larger study was carried out at 15 Italian seaports in 2012 (Rassati et al. 2015a) using black multi-funnel traps and multi-lures composed of (-) α-pinene, ipsenol, ipsdienol, 2-methyl-3-buthen-2-ol, and ethanol. In this study, 81 species of wood-boring beetles were trapped, 49 Scolytinae, 26 Cerambycidae, and 6 Buprestidae. Three species of nonnative longhorn beetles were trapped, two of which were known to be established in Italy, *P. recurva* Newman and *X. stebbingi* Gahan, and one species *Cordylomera spinicornis* (Fabricius) that had not been recorded in Italy before but was not considered to have become established. Of the eight alien species (Scolytinae and Cerambycidae) that were considered established, one was trapped exclusively at ports, two exclusively in surrounding forests, and five in both locations. Of the six species not considered established, three were trapped only in ports and three only in forests. This result reinforced the recommendation of the pilot study (Rassati et al. 2014) that trapping should be carried out both within

FIGURE 13.11 Types of pheromone traps that can be used for surveying cerambycidae. Left: multi-funnel trap; right: cross-vane trap. (Courtesy of Toby Petrice, USDA Forest Service.)

ports and in surrounding forests. There was a significant effect of the volume of imported commodities on the alien species richness both in the ports and the surrounding forests, and broadleaf-dominated forests supported more alien species and individuals than conifer-dominated forests. Based on these conclusions, the authors recommended that extensive monitoring programs should be implemented and concentrated in ports with large volumes of imports and in the surrounding broadleaf forests.

The results of a field study in the tropical montane rain forest of southern China have provided some support to the hypothesis that pheromone traps can be used to attract a range of longhorn beetles species (Wickham et al. 2014). Ten compounds were tested for their attractiveness to Cerambycidae. In this study, 55 species were captured by traps baited with five compounds, (2R*,3S*)-2,3-hexanediol, 3-hydroxyhexan-2-one, (2R*,3R*)-2,3-octanediol, (2R*,3S*)-2,3-octanediol, and 2-(undecyloxy)ethanol, accounting for 77% of the total number of species captured in the whole study. In a study in Michigan, Graham et al. (2012) found that both cross-vane and funnel traps were effective at capturing cerambycids, although the number of cerambycids was higher in cross-vane traps. A second conclusion was that traps for cerambycids need to be placed on supporting bars (around 1.5 m above ground level) and also in tree canopies in order to catch the widest diversity of species. Cerambycid pheromones and their potential applications in pest monitoring are detailed in Chapter 5.

In addition to the use of insect traps, trap trees or "sentinel trees" can be used to monitor for the presence of invasive longhorn beetles. These could be locally important tree species or species that are known to be favored hosts of target species. They can be planted in the areas surrounding high-risk sites such as ports (Wylie et al. 2008) or used in outbreak situations (Hérard et al. 2009; Rassati et al. 2016); they must be monitored regularly for signs of infestation.

The Canneto sull'Oglio district is an important nursery production area in Lombardy, Italy, where around three million potential hosts of *Anoplophora chinensis* are grown. Plant health authorities in Lombardy have established a system of monitoring the nurseries and environment in this area for this pest. Over an area of 539 km^2, plants in every 0.5 by 0.5 km grid square are monitored annually to ensure that any outbreak of *A. chinensis* are detected at a very early stage and to minimize the possibility of infested nursery plants (Ciampitti et al. 2015).

Rassati et al. (2015b) surveyed around wood waste landfill sites in 12 Italian towns. They trapped 74 species of wood-boring beetles, including three nonnative species of longhorn beetles, and concluded that landfill sites for wood waste should be high-priority sites for the detection of nonnative species.

13.8.4 Detection of Outbreaks

One of the characteristics of *Anoplophora* outbreaks in Europe and North America has been that, in many cases, the confirmation of the outbreaks did not take place until many years after the pest became established. Table 13.10 shows examples of the estimates of when the beetles likely became established and the year when they were first detected—usually a lag of at least five years. The implication of such delays in detection are that the invasive beetles will have had a chance to build in numbers and spread geographically by natural means (flying) and anthropogenic means (human movement of felled trees, parts of trees, or plants for planting).

Host utilization data at the species level recorded from outbreak sites can be invaluable for targeting surveys toward the tree species that are most likely to be infested. Infestation records at outbreak sites in combination with data on the number of trees of different species present at a site can provide an indication of the plant species that longhorn beetles are likely to infest environments with a number of native and nonnative trees. For example, in a study of hosts in an area of Lombardy that was heavily infested with *A. chinensis*, 24.1% of *Corylus* trees in the area were infested, whereas only 2.2% of *Acer palmatum* were infested (Cavagna et al. 2013). Haack et al. (2006) analyzed the *A. glabripennis*-infested host genera present in Chicago between 1998 and 2003 and showed that most of the infested trees there were species of *Acer, Ulmus, Fraxinus, Aesculus, Betula*, and *Salix. Acer* and *Ulmus* trees were overrepresented among the infested trees when compared with the number of trees present in the area, whereas *Fraxinus, Aesculus, Betula*, and *Salix* were underrepresented.

It is important to note that there may be a difference between the tree species on which the female longhorn beetles lay the highest number of eggs and the tree species from which the highest number of adults emerge, reflecting differing survival rates or development rates within different hosts. For example, in parts of Lombardy, 81% of the infested *Acer pseudoplatanus* trees had *A. chinensis* exit holes compared with 16% of the infested *Malus* trees, that is, of the trees that had been infested by *A. chinensis*, complete development had occurred in a much higher proportion of the *Acer pseudoplatanus* than the *Malus* sp. (Cavagna et al. 2013). In general, it is necessary to find live stages of longhorn beetles in a host plant to confirm an established population of the pest. However, a method has been developed to confirm the

TABLE 13.10

Estimated Initiation and Detection Date for Selected *Anoplophora* Outbreaks

Species	Country	Location	Probable Initial Infestation	Outbreak Detected	References
Ac	Italy	Parabiago	1980–1990?	2000	Hérard et al. 2006
Ac	Italy	Rome	2000 or 2001	2008	Peverieri et al. 2012
Ac	France	Soyons	1998 or earlier	2003	Hérard et al. 2006
Ac	Netherlands	Westland	2002	2007	Van der Gaag et al. 2010
Ac	Croatia	Turanj	2007	2007	Van der Gaag et al. 2010
Ag	United States	Brooklyn, NY	1980s or early 1990s	1996	Haack et al. 1997
Ag	United States	Cataret, NY	1997	2004	Sawyer and Panagakos 2009
Ag	United States	Chicago, IL	1991–1993	1998	Poland et al. 1998
Ag	France	Strasbourg	2003	2008	EPPO 2016
Ag	United States	Worcester, MA	1998	2008	Dodds and Orwig 2011
Ag	Netherlands	Winterswijk	2007 or 2008	2012	PPS Netherlands 2012
Ag	Switzerland	Winterthur (Canton of Zurich)	2006	2012	Forster and Wermelinger 2012
Ag	Canada	Mississauga	2004 or earlier	2013	Turgeon et al. 2015

Note: Ac = *Anoplophora chinensis*; Ag = *Anoplophora glabripennis*.

presence of *A. chinensis* using PCR amplification of DNA extracted from fecal material (present in frass) collected in the field. This technique could be used to confirm the presence of the pest without having to cut and split trees to search for larvae (Strangi et al. 2013). Another novel technique being developed for identifying trees infested with *A. chinensis* is the use of an "electronic nose," but further research is needed before it can be used operationally (Villa et al. 2013).

13.8.5 Surveys

The standard method for surveying for *Anoplophora* species in detection and eradication campaigns is the visual inspection of trees for signs and symptoms of beetle damage. Surveying can be carried out from ground level, which is effective for detecting damage that is low down on host trees, and binoculars can be used to detect damage higher up in host trees. Surveying from the ground is appropriate for *A. chinensis* because this species tends to lay its eggs close to the base of trunks or on exposed roots, although this is not always the case (Hérard et al. 2006; van der Gaag et al. 2010).

Ground surveying, by naked eye for the lower trunks and branches and using binoculars for higher trunks and branches, also is a standard technique for survey of *A. glabripennis*, but it has been reported to be only around 20% effective at identifying infested trees (Nehme et al. 2010). For *A. glabripennis,* the inspection of trees using tree climbers is more effective because the female adults tend to initiate oviposition in the upper trunk and main branches (Haack et al. 2006). However, this survey technique should only be carried out by people specially trained in climbing trees. Also, it should be recognized that tree climbing is slower and hence more expensive than surveillance from the ground (Hu et al. 2009; Nehme et al. 2010).

Decisions on the size of the area to survey will be a balance between a judgment on how far the beetle may have spread or flown and the resources available to search for the pest. The EPPO has recommended that an intensive delimiting survey be carried out around infested trees to a radius of at least 1 km for *A. chinensis* (EPPO 2013e) and at least 1–2 km for *A. glabripennis* (EPPO 2013d). For *A. glabripennis,* the USDA has recommended surveys of all potential host trees within 1.5 miles (2.4 km), including a half-mile infested core area and a one-mile delimiting survey area of infested trees using a combination of ground surveyors, tree climbers (whenever possible), and inspectors in bucket trucks (cherry pickers) (USDA-APHIS 2008). Beyond the 1.5 mile radius, USDA guidelines suggest that high-risk sites be surveyed (e.g., around tree trimming companies that may have moved host material from the infested site) and an area-wide survey of a sample of host trees in each square mile within 25 miles of the center of the outbreak (USDA-APHIS 2008). In 2013, a member of the public found an adult *A. glabripennis* on their car at a location 2 km outside the regulated area covering 152 km^2 that had been established to contain the outbreak that had been discovered in Toronto, Canada, in 2003. The location was 10 km from the epicenter of the 2003 outbreak and 7.5 km from where the closest tree with an exit hole had been detected (Turgeon et al. 2015). This demonstrates that, even with a very extensive quarantine area, there is a possibility of satellite outbreaks in locations outside this area.

The results of surveys from the ground at Paddock Wood, Kent, UK, where an outbreak of *A. glabripennis* was confirmed in 2012, provide an example of how difficult it can be to spot signs of infestation from the ground. Intensive ground surveys in the area around known infested trees indicated that 24 trees were infested, but when trees within 100 m of known infested trees were felled and examined for signs of *A. glabripennis*, a further 42 infested trees were confirmed. Therefore, presuming that all infested trees were detected once the trees had been felled, surveillance from the ground was 36% effective (Straw et al. 2014).

Data relating to the ability of beetle species to disperse and the recorded rate of spread of invasive longhorn beetles can be used to plan surveys. Other considerations are the host and habitat preferences of the beetles. For example, *A. glabripennis* has been hypothesized as being specialized in colonizing trees on the edge of forests and having evolved in riparian habitats (Williams et al. 2004). Until recently, all the invasive populations of *A. glabripennis* in North America and Europe had been found in urban or open agricultural environments. However, surveys between 2008 and 2010 confirmed the presence of the pest in closed canopy forests close to Worcester, Massachusetts, a finding that has implications for planning surveys and the potential impact of the pest (Dodds and Orwig 2011). An analysis of the distribution records of *A. glabripennis* has shown that, despite the ability of the beetle to be able to colonize closed canopy forests around Worcester, the majority of locations where it had been found were within 30 m of a road (Shatz et al. 2013). This could be a product of the beetles preferred habitat and may also be linked to roads aiding the spread of the beetles. Both *A. glabripennis* and *A. chinensis* have been able to successfully colonize and spread in habitats dominated by humans (i.e., among urban trees). This ability could be one of the most significant factors in making these species two of the most important invasive longhorn beetles worldwide.

Investigations into identifying pheromones of *Anoplophora* species have been going on for more than a decade (Wang 1998; Zhang et al. 2002; see Chapter 5). Following trials in China in 2007 and 2008, traps using semiochemical baits were deployed in Worcester between 2009 and 2011. In 2009, nine *A. glabripennis* were caught, and the average distance to the closest infested host was 80 m. In 2011, a total of 500 traps (450 containing lures and 50 unbaited) were deployed, and 23 beetles were caught. The use of traps can reduce eradication costs by focusing survey work on those areas that are more likely to be infested and can speed up the removal of infested trees (Nehme et al. 2013).

13.9 Pest Management Actions and Movement Restrictions

Despite the legislation and measures that are in place to prevent the introduction of longhorn beetles, some cerambycids have been introduced into nonnative environments. Some of these species are known to be damaging pests, and state led eradication campaigns have been put in place to prevent the beetles becoming established. In other cases, the introduced cerambycid has been judged to be either not of sufficient threat or too widespread when discovered for an eradication campaign to be necessary or feasible.

13.9.1 Eradication of Invaded Pests

In eradication campaigns for *Anoplophora* species, the main eradication tool has been the felling and destruction of trees that are known to be infested as well as potentially infested or high-risk trees (i.e., known hosts in close proximity to the infested trees). The logic for felling trees that have not been confirmed as being infested is that it is currently not possible to confirm the presence or absence of larvae from the trees without felling and often splitting them. In common with making the decisions about how extensive surveys should be, decisions on which potentially infested trees need to be removed are also a balance between the risks of missing infested trees and the costs of removing more trees. When considering the costs of removing trees, in addition to the costs of the tree removal operation, the losses to landowners and environmental damage have to be weighed against the potential losses that could ensue if an invasive longhorn beetle were to become widely established. Other influences on the number of trees that should be removed are the number of trees of host species present in an outbreak area, as well as host size, nearby habitat types, and climatic conditions.

In the EU, there is now a requirement to remove all "specified plants" within 100 m of trees infested by *A. chinensis*, and emergency measures list 18 genera and 2 species of specified plants that are considered as its hosts in Europe (EU 2012). The EPPO recommends removal of all potential hosts within 100 m of infested trees for *A. glabripennis* and *A. chinensis* (EPPO 2013d, 2013e). The USDA guidelines list 10 genera and 1 species of trees that should be considered as hosts for *A. glabripennis* and state that total host removal and/or chemical treatment within a radius of half-mile (800 m) of infested trees would mean that only a low percentage of beetles would be likely to disperse beyond the treatment area (USDA-APHIS 2008). In Canada, *A. glabripennis*-infested trees and potential host trees in the genera *Acer, Betula, Populus,* and *Salix* growing within 400 m (and later changed to 800 m) of each infested tree are cut (Haack et al. 2010; Turgeon et al. 2015).

Because *A. chinensis* larvae tend to infest host trees near the base of the trunk, it is recommended that tree stumps be removed or, where this is not possible, the stumps should be ground down to at least 40 cm below the soil level (EPPO 2013e). In the eradication campaign for the outbreak of *A. chinensis* discovered in Rome in July 2008, some of the infested trees were close to Roman walls or in historic gardens, and the removal of these trees could have caused unacceptable damage. As an alternative to stump removal, some of the stumps and tree bases were covered in a fine wire mesh to prevent escape of any emerging *A. chinensis* adults in the future. A similar tactic was used for five standing *Platanus* trees in Rome, three of which were known to be infested. The base of the trees was covered in a combination of a mosquito polyethylene net and wire mesh that was buried in a ring around each tree trunk. There were several cases where beetles were trapped below this mesh screening, and thus, the method was judged to be effective (van der Gaag et al. 2010; Roselli et al. 2013).

Being the main tool of *Anoplophora* eradication, the felling of infested and potentially infested trees will help limit the extent of any outbreak by improving the possibilities of detecting all of the infested trees. It may also help reveal the history of the outbreak and hence may give indications of when and where the outbreak may have begun. For example, by analyzing the growth rings around the *A. chinensis* outbreak site in Rome, it was possible to trace the history of the outbreak back to the first emergence of a beetle in 2002 (Peverieri et al. 2012). Copini et al. (2014) have demonstrated that, in the temperate climate of northwestern Europe, it is possible to analyze the exit holes left by *Anoplophora* to determine which of three phases of the year the exit hole might have been created, namely: (1) during dormancy between October and late April/early May; (2) the period of tree-ring growth in which they would be created within the tree ring—between late April/early May and the beginning of September, and (3) during the end of the growing season, after September. Once infested trees and potentially infested trees have been examined for the pest, EPPO guidelines for *A. glabripennis* recommend that they be chipped to a size of less than 2.5 cm in any dimension or burned on site (EPPO 2013d). The USDA also permits the option of moving large trunks to approved burning sites with the condition that they are covered during transit and burned within 24 hours of arrival (USDA-APHIS 2008).

In eradication campaigns, when possible, all infested and potentially infested hosts should be cut down and destroyed before the next adult emergence period. Smith et al. (2004) studied the emergence of *A. glabripennis* in China and created a model that could be used to predict the emergence data of adults based on degree-days in excess of 10°C after January first each year. This model applied well to the emergence of *A. glabripennis* in Cornuda, Italy (Faccoli et al. 2015). Such models can be implemented using mean climatic data to estimate emergence in an average year and also with data from an individual year to get a more accurate projection of probable emergence dates.

The life cycle of longhorn beetles, with most time being spent as larvae feeding inside the host plant tissues, means that insecticide treatments only can be used to target some life stages of the pest and are unlikely to be the decisive factor in eradication campaigns. Two of the methods of using insecticides against longhorn beetles are (1) foliar sprays of the foliage and/or bark in order to target emerging adults that may be chewing out of tunnels in the trees or landing on the trees to feed or chew oviposition slits in the bark and (2) the injection of systemic insecticides into either the ground or the trunk. These systemic insecticides will move within the tree and are considered to be effective against adults feeding on leaves and young bark, ovipositing females, adults emerging from the tree, and young larvae as they first burrow into their host trees, but they are only partially effective against larvae that are already established deep in the sapwood or heartwood of the host trees (USDA-APHIS 2008).

Pyrethroid insecticides are recommended for application to bark of host trees in the United States in order to target *A. glabripennis* adults emerging from infested trees and to prevent their dispersal. They are also recommended for application to large tracts of wooded land as a quick means of suppressing populations or preventing infestation before arrangements can be made to destroy these trees (USDA-APHIS 2008). Deltamethrin, a pyrethroid insecticide, was applied to *Acer* trees within polytunnels (hoop greenhouses) at a nursery in southern England where *A. chinensis* adults had been found in order to kill any adults that had not been detected (Eyre et al. 2010). Another pyrethroid insecticide, lambda-cyhalothrin, was found to be the most effective product against *A. chinensis* adults, and neonicitinoids, including thiamethoxam, sprayed or painted onto trunks were found to be effective treatments at targeting young *A. chinensis* larvae in trials using potted trees in Italy (Cavalieri 2013).

In trials using insecticide injection into a range of deciduous trees in China that targeted *A. glabripennis*, Poland et al. (2006) found that injections with imidacloprid were more effective than injections with either emamectin benzoate or azadirachtin—although this result varied among tree genera. More recently, the impact of imidacloprid injections on adults feeding on treated trees has been studied by feeding adults with twigs collected between June and September from *Acer platanoides* L. trees that had been treated with imidacloprid as part of the *A. glabripennis* eradication campaign at Worcester,

Massachusetts (Ugine et al. 2012). The authors reported: (1) 35% mortality within 21 days of beetles feeding on twigs from trunk-injected trees; however, the concentration of imidacloprid within the twigs was <1 ppm for 68% of these twigs. (2) Mortality of 82% for those that were fed on twigs with >1 ppm imidacloprid. (3) The number of larvae produced from adults that had fed on treated twigs over a three-week period was reduced by 67% compared to those that had fed on untreated twigs. (4) In choice tests, there was no tendency for beetles to move from injected to control twigs over a four-day test period. (5) There was great variability in the concentration of imidacloprid within twigs from treated twigs, with 39% of twig samples having no quantifiable imidacloprid in the bark. The overall impact of imidacloprid injections on *A. glabripennis* populations would be expected to be higher in the field than these results indicate because Ugine et al. (2012) did not take into account impacts on ovipositing females, egg hatch, and larval survival.

Preventing the movement of potentially infested plants or wood has been shown to be essential to avoid the pest being transported to new locations; for example, as described earlier in this chapter, *A. glabripennis* was found to have been moved in wood about 50 km from the original outbreak area in New York City (Haack et al. 1997). Furthermore, when an outbreak of *A. chinensis* was discovered in Boskoop, in the Netherlands, an important site for producing trees that are distributed throughout Europe, there was a risk that the pest may have been spread to multiple locations around Europe before the outbreak was detected (van der Gaag et al. 2010). EU member states are required to establish demarcated areas of at least 2 km around trees infested by *A. chinensis* and prevent the movement of any potentially infested material outside these areas (EU 2012).

Three methods of using pheromones for mating disruption as a population suppression tool have been tested against *T. fuscum* in Nova Scotia, Canada. Two studies were conducted in 2008 and 2009, using fuscomol, the aggregation pheromone for *T. fuscum*. In the first study, fuscomol in Hercon flakes was applied to two 4-ha plots twice a year in 2008 and 2009. In 2008, the flakes were applied from the ground using modified leaf blowers and, in 2009, they were applied from a helicopter (Sweeney et al. 2011a). In a second study, high densities of pheromone traps (100 per ha in a 10 m by 10 m grid) were set up in four 1-ha plots (Sweeney et al. 2011b). In both of these studies *Picea* logs were put into the plots to evaluate the efficacy of the treatments. The percentage of logs infested with *T. fuscum* was lower in the treated plots than in the control plots, and the number of larvae was lower per square meter of surface area of the bait logs. In 2012, a pilot study showed that pheromone traps baited with the fungus *Beauveria bassiana* (Bals.-Criv.) Vuill. in collecting cups contaminated 28% of *T. fuscum*, and 67% of these beetles became infected and produced conidia (Sweeney et al. 2013).

Tree removal is the main tool for eradicating *Anoplophora* sp. in Europe, but biological control could be used to compliment other management strategies and where eradication is no longer considered possible (Brabbs et al. 2015). In 2002, the eulophid egg parasitoid *Aprostocetus anoplophorae* Delvare was discovered in Italy infesting the eggs of *Anoplophora chinensis*. It is suspected to have arrived within *A. chinensis* eggs imported from Japan. Recorded levels of parasitism of *A. chinensis* in Italy vary between 21% and 72% (Hérard et al. 2005; Maspero et al. 2008). The beetle *Dastarcus longulus* Sharp (Bothrideridae) is an effective parasite of *A. glabripennis* (Li et al. 2009a), is amenable to mass rearing (Yang et al. 2014), and has strong natural dispersal (Wei et al. 2007; Li et al. 2009b; Wei and Niu 2011); but unfortunately, it is not host specific and so could potentially damage nontarget species of longhorn beetles (Brabbs et al. 2015). Some entomopathogenic fungi have been tested for their potential use against longhorn beetles, including *Anoplophora* sp. In Japan, nonwoven fiber bands impregnated with the entomopathogenic fungus *Beauveria brongniartii* (Sacc.) Petch are sold to be attached around tree trunks for the control of orchard pests such as *Psacothea hilaris* and *Anoplophora chinensis* (Hajek et al. 2006). Such products potentially could be incorporated in eradication programs for invasive longhorn beetles.

13.9.2 Publicity

Members of the public, including scientists, amateur entomologists, nursery workers, and arborists, can provide an invaluable addition to official staff in the detection of longhorn beetles. The size and distinctive appearance of many longhorn beetles naturally attract attention and hence make them more likely to be reported. Many outbreaks have been first reported by members of the public rather than by officials; for example, the confirmation of an outbreak of *A. glabripennis* at Paddock Wood, Kent, UK, in 2012, followed the initial report of an adult beetle by a member of the public (Straw 2014) as has been the case with many of the outbreaks in the United States and Canada (Haack et al. 1997; Haack et al. 2010). Awareness raising and education of the public are not only necessary to make people aware of invasive longhorn beetles but also essential to (1) encourage people working with potentially infested plants and wood to comply with voluntary or statutory measures that have been implemented to prevent the introduction and spread of invasive longhorn beetles, including those that work with trees in urban areas, woodlands, and forests, and beneath power lines—such as arborists, tree surgeons, foresters, and garden center workers, and (2) to explain the rationale and need for any measures that are being taken to eradicate longhorn beetles to those who are being affected most by such measures—such as residents who have trees in their gardens that may need to be removed or treated.

Raising public awareness of *Anoplophora* in the case of outbreaks is recommended by the EPPO and forms part of the measures required by the EU when outbreaks of *A. chinensis* are discovered in member countries (EU 2012; EPPO 2013d, 2013e). The U.S. guidelines for outbreaks of *A. glabripennis* include recommendations on how to inform the public by hosting public meetings, setting up *A. glabripennis* telephone hotlines to respond to queries from the public, notifying landowners about required measures, and engaging with the media (USDA-APHIS 2008). The Lombardy Plant Protection Service has implemented an information campaign about *A. chinensis* in order to educate the public about the pest. This led to (1) the report of a new outbreak of *A. chinensis* in Brescia Province, Lombardy, a site 150 km east of the main outbreak area (which is close to Milan), and 2) the first record of *A. glabripennis* in Italy. The media campaign has included articles in newspapers and magazines, features on television and radio, plus eye-catching posters on the Milan underground train network (Ciampitti and Cavagna 2014).

13.10 Outbreaks and Established Populations of Nonnative Species

Tobin et al. (2014) have collected records of eradication and containment programs of arthropod plant pests and plant pathogens from 104 countries around the world. Of the 299 taxa (mainly species but some genera) that were listed in the database as of February 2014, 159 were arthropods, including 27 Coleoptera, most of which were Curculionidae (9), Cerambycidae (6), and Chrysomelidae (5). The six species of Cerambycidae listed in the database were *A. chinensis, A. glabripennis, C. rufipenne, T. fuscum, Saperda candida* Fab., and *H. bajulus*. Although this database is not exhaustive, it provides a guide to the best publicized campaigns against exotic pests (including Cerambycidae) and pathogens. As mentioned in the introduction, some of the biological and ecological properties of Cerambycidae, especially their relatively slow rate of spread, have made expensive eradication efforts a better alternative than simply allowing a damaging nonnative pest to spread. Given a realistic chance of eradication, national plant protection organizations have carried out expensive campaigns that would not be implemented for more mobile pests.

Table 13.11 shows records of longhorn beetles that have managed to become temporarily or permanently established in Europe or North America, including species that have been damaging and those that have been comparatively benign in the invaded region. Some details of eradication campaigns against some of the species listed in Table 13.11 have been provided earlier in this chapter. Following is a description of two of the other species in Table 13.11 that have been subject to eradication or containment campaigns that have been described in publications (*Saperda candida* and *Hylotrupes bajulus*) and one species that became widely established before eradication could be implemented (*Psacothea hilaris*).

Saperda candida, the round headed apple borer, is native to the eastern United States and Canada; apple (*Malus*) is its preferred host plant but this species will also feed on other plants in the family Rosaceae, including *Cotoneaster, Crataegus, Cydonia, Prunus, Pyrus*, and *Sorbus*. It is no longer a significant pest in apple orchards because insecticide applications for other pests keep it under control, but it remains a major pest of several ornamental trees (Johnson and Lyon 1991; EPPO 2011). In 2008, the first outbreak of this pest in Europe was discovered on Ferhmarn, Germany, an island on the eastern

TABLE 13.11

Nonnative Longhorn Beetles with Temporary or Permanent Populations in Europe and North America

Cerambycid	Hosts in Invaded Regions	Country/Region of Origin	Country/Region Where Invasive (Year First Reported, if Known)	Species Known to be Under Eradication or Containment Programs	References
Acanthoderes jaspidea (Germar)	*Acacia, Albizzia*	Brazil	The Azores, Portugal (1880)	No	Cocquempot and Lindelow 2010
Acrocinus longimanus L.	Moraceae, Apocynaceae	Brazil	Madeira, Portugal, and mainland Portugal (1977)	No	Cocquempot and Lindelow 2010
Anoplophora chinensis (Forster)	*Acer, Betula, Carpinus, Corylus, Platanus, Fagus,* and others	Asia	Europe (2000)	Yes	Colombo and Limonata 2001
Anoplophora glabripennis (Motschulsky)	*Acer, Aesculus, Betula, Populus, Salix, Ulmus,* and others	China and Korea	North America (1996), Japan and Europe (2001)	Yes	Haack et al. 2010
Aromia bungii (Faldermann)	*Prunus, Populus,* and other broadleaved trees	Asia	Italy	Yes	Garonna et al. 2013
Callidium violaceum (L.)	*Juniperus*	Europe/Asia	USA (1907)—this species is possibly holarctic	No	Aukema et al. 2010, Maier 2009
Callidiellum rufipenne (Motschulsky)	*Cryptomeria, Chamaecyparis, Cupressus, Juniperus, Pinaceae Thuja,* and others	Japan	North America (1997), Europe (Italy, 1989), Spain, France	No	Poland and Haack 2003, Haack 2006
Chlorophorus annularis (Fab.)	*Citrus, Gossypium, Pyrus, Malus, Tectoan, Dipterocarpus, Vitis,* and *Liquidambar* (*Bambusa, Dendrocalamus,* and *Phyllostachys*)	Temperate Asia	Spain (1991); Europe	No	Cocquempot and Lindelow 2010, Mattson et al. 2007
Cyrthognathus forficatus (Fab.)	Unknown	Africa	Malta (1872)	No	Cocquempot and Lindelow (2010)
Derolus mauritanicus Buquet, 1840	*Nerium oleander*	Northern Africa	France (1884), Spain (these records are considered uncertain)	No	Cocquempot and Lindelow 2010
Deroplia albida (Brullé, 1838)	*Pelargonium*	Canary Islands, Spain	Mainland Spain (1988)	No	Cocquempot and Lindelow 2010
Hylotrupes bajulus L.	*Pinus, Picea, Abies,* and *Larix*	Europe and Asia	Much of the world, incl. USA (1850)	No	CABI 2015b; Aukema et al. 2010

(Continued)

TABLE 13.11 *(Continued)*

Nonnative Longhorn Beetles with Temporary or Permanent Populations in Europe and North America

Cerambycid	Hosts in Invaded Regions	Country/Region of Origin	Country/Region Where Invasive (Year First Reported, if Known)	Species Known to be Under Eradication or Containment Programs	References
Neoclytus acuminatus (Fab.)	*Ulmus, Fraxinus, Juglans, Celtis, Morus,* and *Quercus ilex*	North America	Italy, Croatia, France, Slovenia, Serbia, and Montenegro (1908)	No	Jucker et al. 2006, Matosevic and Zivkovic 2013; Rassati et al. 2016
Parandra brunnea (Fab.)	*Tilia, Populus,* and other deciduous trees	North America	Germany	No	Katschak 2004, Vanhanen et al. 2008
Phoracantha recurva (Newman)	*Eucalyptus* spp.	Australia	USA (1995), Africa, South America, Asia, New Zealand, and Europe	No	Jucker et al. 2006; Grebennikov et al. 2010
Phoracantha semipunctata	*Eucalyptus* spp.	Australia	USA (1984), Africa (Egypt 1950), South America, Asia, New Zealand, and Europe (Sicily 1975)	No	Jucker et al. (2006); Grebennikov et al. 2010)
Phryneta leprosa (Fabricius)	*Morus nigra*	Tropical Africa	Malta, France (1997)	No	Mifsud & Dandria 2002, Cocquempot an Lindelow 2010
Phymatodes lividus (Rossi)	*Quercus, Ulmus*	Europe	North America	No	Mattson 1994, Swift and Ray 2010
Phymatodes testaceus (L.)	*Quercus, Tsuga, Abies, Malus, Fagus, Prunus, Castanea, Carya,* and *Salix*	Europe	USA (1903)	No	Aukema et al. 2010, LaBonte et al. 2005
Psacothea hilaris (Pascoe)	*Ficus* and *Morus*	Asia	Italy (2005)	No	Jucker et al. 2006, Lupi et al. 2013

(Continued)

TABLE 13.11 (Continued)

Nonnative Longhorn Beetles with Temporary or Permanent Populations in Europe and North America

Cerambycid	Hosts in Invaded Regions	Country/Region of Origin	Country/Region Where Invasive (Year First Reported, if Known)	Species Known to be Under Eradication or Containment Programs	References
Saperda candida Fab.	*Malus, Pyrus, Prunus*, and other Rosaceous hosts	North America	Germany (2008)	Yes	Eyre et al. 2013
Saperda populnea L.	*Populus*	Europe	North America (this species is possibly holarctic)	No	Aukema et al. 2010
Sybra alternans (Wiedemann)	*Ficus*	Asia	USA (1992)	No	Haack 2006
Tetropium fuscum (Fab.)	*Picea*	Europe	Canada (1999)	Yes	Poland and Haack 2003, Smith and Hurley 2000
Tetrops praeusta (L.)	Rosaceae, including *Crataegus* and *Malus*	Europe	North America (1996)	No	Haack 2006
Trichoferus campestris (Faldermann)	Range of broadleaved trees, also timber of *Abies, Picea*, and *Pinus*	Asia and SE European Russia	North America (2002), Central Europe	No	Grebennikov et al. 2010, Sabol 2009
Xylotrechus stebbingi Gahan	*Quercus*	Asia	Europe (1990)	No	Biscaccianti 2007, Sama and Cocquempot 1995, Teunissen 2002, Rassati et al. 2016

side of the Jutland Peninsula (Nolte and Krieger 2008). An eradication campaign is underway involving surveillance of host plants, destruction of infested plants, the application of insecticides, and preventing host plants from being moved (Eyre et al. 2013).

Hylotrupes bajulus, the European house borer, is native to North Africa but is now found in Europe, North and South America, South Africa, Asia Minor, China, and Russia (Durr 1954; Duffy 1968). In the 1950s, it was detected in New South Wales, Queensland, and Victoria in Australia, following the importation of prefabricated houses from Europe, but these populations were eradicated 20 years later. In 2004, *H. bajulus* was detected in dead pine trees near Perth in Western Australia (Gove et al. 2007), which led to a response program that has reduced populations of the pest especially in suburban and residential areas. The program has now switched from eradication to containment, with a greater focus on self-management by stakeholders (Liu et al. 2010; DAFF 2011).

Psacothea hilaris (Pascoe), the yellow spotted longhorn beetle, is an Asian species that has become established in Italy (Lupi et al. 2013). It was first detected there in 2005, but a recent survey has found that the beetle has become established over an area of more than 60 km² south of Lake Como in Lombardy (Lupi et al. 2013), where it is damaging old, young, stressed, and healthy host trees, mainly figs, *Ficus carica*, and to a lesser extent mulberry, *Morus* sp. This pest has been found in the United States and Canada in warehouses in association with wood and wooden spools from Asia. It also has been found a few times in the United Kingdom but is not known to have become established there. In 2008, it was found in a private garden in Derbyshire (EPPO 2016), and in 2014, in a private home in South Yorkshire, but the source of the beetles in the United Kingdom could not be detected. The pest also has been found in an UK warehouse associated with goods on wooden pallets and among the wood packaging for imported medical equipment (Fera, unpublished data).

13.11 Conclusions

- In common with the spread of other invasive species, the movement of nonnative cerambycids to new locations is a by-product of increasing levels of international trade.
- Wood packaging, especially the wood packaging used to transport heavy commodities such as quarry products and tiles, has been a major pathway in the spread of nonnative cerambycids. The most significant pest moved along this pathway is *Anoplophora glabripennis*, the Asian longhorn beetle. There are outbreaks of this pest in Canada, the United States, and Western Europe. In excess of US$500 million has been spent on eradicating this pest from the United States since it was first discovered in 1996. However, there are a number of infestations in North America and Europe that had yet to be eradicated as of 2016.
- Plants for planting have been another major pathway for the spread of cerambycids. One of the most significant pests moved along this pathway is *Anoplophora chinensis*, the citrus long-horn beetle. More than 25,000 trees have been cut down in order to eradicate this pest from Lombardy, Italy. Although this pest is now considered established in Europe, all outbreaks are subject to ongoing containment and eradication campaigns.
- Invasive longhorn beetles can have large economic impacts—such as damaging commercial eucalyptus plantations, environmental impacts—such as damaging trees planted to prevent desertification in northern China, and social impacts—such as damaging street trees in North America and Europe.
- Measures have been introduced to try to prevent the movement of longhorn beetles such as ISPM 15, an international standard for treating wood packaging materials used in international trade, and the measures introduced by the European Union to prevent the introduction of long-horn beetles in *Acer* trees.
- Risk assessments are carried out by national and regional plant protection organizations in order to evaluate the threat posed by particular cerambycid species and pathways.

- Outbreaks of longhorn beetles often are not detected until many years after their initiation, and pests may have become well established before an eradication campaign can begin. Therefore, close monitoring and public education are extremely important.

- Eradication campaigns against longhorn beetles generally involve extensive removal of trees and labor-intensive visual surveys to detect infested trees.

ACKNOWLEDGMENTS

The authors thank Christopher Malumphy (Fera) and Anastasia Korycinska (Defra) for comments on an earlier draft of this chapter.

REFERENCES

Adachi, I. 1994. Development and life cycle of *Anoplophora malasiaca* (Thomson) (Coleoptera: Cerambycidae) on citrus trees under fluctuating and constant temperature regimes. *Applied Entomology and Zoology* 29:485–497.

Anderson, A., A. Korycinska, D. Collins, S. Matthews-Berry, and R. Baker. 2013. Rapid pest risk analysis for *Aromia bungii*. https://secure.fera.defra.gov.uk/phiw/riskRegister/downloadExternalPra.cfm?id=3818 (accessed February 25, 2016).

Aukema, J. E., D. G. McCullough, B. von Holle, A. M. Liebhold, K. Britton, and S. J. Frankel. 2010. Historical accumulation of nonindigenous forest pests in the continental United States. *BioScience* 60:886–897.

Australian Government. 2015. Wooden articles input permit. http://www.agriculture.gov.au/import/goods/timber/low-risk (accessed February 26, 2016).

Ayberk, H., H. Ozdikmen, and H. Cebeci. 2014. A serious pest alert for Turkey: A newly introduced invasive longhorned beetle, *Anoplophora glabripennis* (Cerambycidae: Lamiinae). *Florida Entomologist* 97:1852–1855.

Baker, R. H. A., J. Benninga, J. Bremmer, et al. 2012. A decision-support scheme for mapping endangered areas in pest risk analysis. *EPPO Bulletin* 42:65–73.

Baker, R. H. A., H. Anderson, S. Bishop, A. MacLeod, N. Parkinson, and M. G. Tuffen. 2014. The UK Plant Health Risk Register: A tool for prioritizing actions. *EPPO Bulletin* 44:187–194.

Bancroft, J. S., and M. T. Smith. 2005. Dispersal and influences on movement for *Anoplophora glabripennis* calculated from individual mark-recapture. *Entomologia Experimentalis et Applicata* 116:83–92.

Banks, J. 2012. Gas processes, CA and fumigation, for quarantine and biosecurity. http://ftic.co.il/2012AntalyaPDF/SESSION%2008%20PAPER%2001.pdf (accessed February 26, 2016).

Barak, A. V., Y. Wang, G. Zhan, Y. Wu, L. Xu, and Q. Huang. 2006. Sulfuryl fluoride as a quarantine treatment for *Anoplophora glabripennis* (Coleoptera: Cerambycidae) in regulated wood packing material. *Journal of Economic Entomology* 99:1628–1635.

Baral, H.-O. 2014. *Hymenoscyphus fraxineus*, the correct scientific name for the fungus causing ash dieback in Europe. *IMA Fungus* 5:79–80.

Bayerische Landesanstalt für Landwirtschaft. 2012. Baumschädling aus Asien wurde in Feldkirchen gefunden. http://www.vaterstetten.de/city_info/display/dokument/show.cfm?region_id=276&id=355887 (accessed February 19, 2016).

Bazzoli, M., and E. Alghisi. 2013. Fighting *Anoplophora chinensis* in a wooded habitat. The Gussago case. *Journal of Entomological and Acarological Research* 45(s1):31.

Bense, U. 1995. *Longhorn beetles: Illustrated key to the Cerambycidae and Vesperidae of Europe.* Weikersheim: Margraf Verlag.

Biscaccianti, A. B. 2007. The Cerambycidae of Mt. Vesuvius (Coleoptera). In *Ricerche preliminari sugli Artropodi del Parco Nazionale del Vesuvio*, eds. G. Nardi and V. Vomero, 249–278. Verona: Conservazione Habitat Invertebrati.

Bonifacio, L. F., E. Sousa, P. Naves, et al. 2014. Efficacy of sulfuryl fluoride against the pinewood nematode, *Bursaphelenchus xylophilus* (Nematoda: Aphelenchidae), in *Pinus pinaster* boards. *Pest Management Science* 70:6–13.

Bozkurt, V., A. Ozdem, and E. Ayan. 2013. Coleopteran pests intercepted on imported forest products in Turkey. In *Fourth International Scientific Symposium "Agrosym 2013", Jahorina, Bosnia and Herzegovina, 3–6 October, 2013*, ed. D. Kovacevic, 646–652. Bosnia: University of East Sarajevo.

Brabbs, T., D. Collins, F. Hérard, M. Maspero, and D. Eyre. 2015. Prospects for the use of biological control agents against *Anoplophora* in Europe. *Pest Management Science* 71:7–14.

Bright, D. E. 2014. A Catalog of Scolytidae and Platypodidae (Coleoptera), Supplement 3 (2000–2010), with notes on subfamily and tribal reclassifications. *Insecta Mundi* 356:1–336.

Brockerhoff, E. G., M. Kimberley, A. M. Liebhold, R. A. Haack, and J. F. Cavey. 2014. Predicting how altering propagule pressure changes establishment rates of biological invaders across species pools. *Ecology* 95:594–601.

Brockerhoff, E. G., A. M. Liebhold, B. Richardson, and D. M. Suckling. 2010. Eradication of invasive forest insects: Concepts, methods, costs and benefits. *New Zealand Journal of Forestry Science* 40: S117–S136.

Bullas-Appleton, E., T. Kimoto, and J. J. Turgeon. 2014. Discovery of *Trichoferus campestris* (Coleoptera: Cerambycidae) in Ontario, Canada and first host record in North America. *The Canadian Entomologist* 146:11–116.

Burleigh, J. 2013. The insects that tried to steal Christmas. *BC Forest Professional May/June* 2013:10–26.

CABI. 2015a. Crop Protection Compendium datasheet for *Anoplophora chinensis*. http://www.cabi.org/cpc/datasheet/5556 (accessed February 28, 2016).

CABI. 2015b. Crop Protection Compendium datasheet for *Hylotrupes bajulus*. http://www.cabi.org/cpc/datasheet/28192 (accessed February 28, 2016).

CABI. 2015c. Crop Protection Compendium datasheet for *Phoracantha semipunctata*. http://www.cabi.org/cpc/datasheet/40372 (accessed February 28, 2016).

CABI. 2015d. Crop Protection Compendium datasheet for *Trichoferus campestris*. http://www.cabi.org/cpc/datasheet/27900 (accessed February 28, 2016).

CABI, and EPPO. 1997. Data sheets on quarantine pests: *Anoplophora malasiaca* and *Anoplophora chinensis*. In *Quarantine pests for Europe*, eds. I. Smith, D. McNamara, P. Scott, M. Holderness and B. Burger, 57–60. Wallingford, Oxon: CAB International.

Caremi, G., and M. Ciampitti. 2006. *Anoplophora chinensis* in Lombardia region: Spread and control strategies. *ATTI of the Giornate Fitopatologiche* 1:205–210.

Carter, M., M. Smith, and R. Harrison. 2010. Genetic analyses of the Asian longhorned beetle (Coleoptera, Cerambycidae, *Anoplophora glabripennis*), in North America, Europe and Asia. *Biological Invasions* 12:1165–1182.

Cavagna, B., M. Ciampitti, A. Bianchi, S. Rossi, and M. Luchelli. 2013. Lombardy region experience to support the prediction and detection strategies. *Journal of Entomological and Acarological Research* 45(s1):1–6.

Cavalieri, G. 2013. Summary of 2008–2011 trials on the possibility of controlling *Anoplophora chinensis* with pesticides. *Journal of Entomological and Acarological Research* 45(s1):23–24.

Ciampitti, M., and B. Cavagna. 2014. Public awareness: A useful tool for the early detection and a successful eradication of the longhorned beetles *Anoplophora chinensis* and *A. glabripennis*. *EPPO Bulletin* 44:248–250.

Ciampitti, M., B. Cavagna, A. Bianchi, V. Cappa, S. Asti, and A. Fumagalli. 2015. Stepped-up surveillance for early detection of *Anoplophora chinensis* in a nursery district. *EPPO Bulletin* 45:269–272.

Cocquempot, C. 2007. Alien longhorned beetles (Coleoptera Cerambycidae): Original interceptions and introductions in Europe, mainly in France, and notes about recently imported species. *Redia* 89:35–50.

Cocquempot, C., and F. Herard. 2003. Les *Anoplophora* un danger pour la pépinière et les espaces verts. *PHM Revue Horticole* 449:28–33.

Cocquempot, C., and A. Lindelow. 2010. Longhorn beetles (Coleoptera, Cerambycidae) *BioRisk* 4:193–218.

Cocquempot, C., and D. Mifsud. 2013. First European interception of the brown fir longhorn beetle, *Callidiellum villosulum* (Fairmaire, 1900) (Coleoptera, Cerambycidae). *Bulletin of the Entomological Society of Malta* 6:143–147.

Colombo, M., and L. Limonta. 2001. *Anoplophora malasiaca* Thomson (Coleoptera Cerambycidae Lamiinae Lamiini) in Europe. *Bollettino di Zoologia Agraria e di Bachicoltura* 33:65–68.

Colunga-Garcia, M., R. A. Haack, and A. O. Adelaja. 2009. Freight transportation and the potential for invasions of exotic insects in urban and periurban forests of the United States. *Journal of Economic Entomology* 102:237–246.

Colunga-Garcia, M., R. A. Haack, R. D. Magarey, and M. L. Margosian. 2010. Modeling spatial establishment patterns of exotic forest insects in urban areas in relation to tree cover and propagule pressure. *Journal of Economic Entomology* 103:108–118.

Copini, P., U. Sass-Klaassen, J. den Ouden, G. M. J. Mohren, and A. J. M. Loomans. 2014. Precision of dating insect outbreaks using wood anatomy: The case of *Anoplophora* in Japanese maple. *Trees—Structure and Function* 28:103–113.

DAFF. 2011. European house borer response. http://www.northam.wa.gov.au/Assets/Documents/Document-Centre/building-services/New-Dwelling-Information/European-House-Borer.pdf (accessed February 25, 2016).

Dascalu, M.-M., R. Serafim, and A. Lindelow. 2013. Range expansion of *Trichoferus campestris* (Faldermann) (Coleoptera: Cerambycidae) in Europe with the confirmation of its presence in Romania. *Entomologica Fennica* 24:142–146.

Dodds, K. J., and D. A. Orwig. 2011. An invasive urban forest pest invades natural environments—Asian longhorned beetle in northeastern US hardwood forests. *Canadian Journal of Forest Research* 41:1729–1742.

Dodds, K., H. Hull-Sanders, N. Siegert, and M. Bohne. 2013. Colonization of three maple species by Asian longhorned beetle, *Anoplophora glabripennis*, in two mixed-hardwood forest stands. *Insects* 5:105–119.

Duffy, E. A. J. 1968. *A monograph of the immature stages of oriental timber beetles (Cerambycidae).* London: British Museum (Natural History).

Durr, H. J. R. 1954. *The European House Borer Hylotrupes bajulus (L.) Serville (Coleoptera: Cerambycidae) and its control in the Western Cape Province.* Pretoria: Department of Agriculture South Africa.

El-Yousfi, M. 1982. *Phoracantha semipunctata* au Maroc. Ecologie et méthodes de lutte. In *Note Technique de la Division de Recherches et d'Experimentation Forestières.* Rabat, Morocco: Direction des Eaux et des Forêts.

EPPO. 1999. EPPO Datasheets on quarantine pests: *Anoplophora glabripennis. EPPO Bulletin* 29:497–501.

EPPO. 2009a. Datasheets on pests recommended for regulation: *Hesperophanes campestris. EPPO Bulletin* 39:51–54.

EPPO. 2009b. Disinfestation of wood with ionizing radiation. *EPPO Bulletin* 39:34–35.

EPPO. 2009c. Heat treatment of wood to control insects and wood-borne nematodes. *EPPO Bulletin* 39:31–31.

EPPO. 2009d. Methyl bromide fumigation of wood to control insects. *EPPO Bulletin* 39:32–33.

EPPO. 2009e. PM 8/2(1): Coniferae. *EPPO Bulletin* 39:420–449.

EPPO. 2010. PM 8/3 (1): *Quercus* and *Castanea. EPPO Bulletin* 40:376–386.

EPPO. 2011. Pest risk analysis for *Saperda candida.* http://www.eppo.int/QUARANTINE/Pest_Risk_Analysis/PRA_intro.htm (accessed February 26, 2016).

EPPO. 2013a. 2013/246 EPPO report on notifications of non-compliance. *EPPO Reporting Service* 2013:No. 246.

EPPO. 2013b. Pest risk analysis for *Apriona germari, A. japonica, A. cinerea.* http://www.eppo.int/QUARANTINE/Pest_Risk_Analysis/PRA_intro.htm (accessed February 27, 2016).

EPPO. 2013c. Pest risk analysis for *Oemona hirta.* http://www.eppo.int/QUARANTINE/Pest_Risk_Analysis/PRA_intro.htm (accessed February 25, 2016).

EPPO. 2013d. PM 9/15 (1) *Anoplophora glabripennis*: Procedures for official control. *EPPO Bulletin* 43:510–517.

EPPO. 2013e. PM 9/16 (1) *Anoplophora chinensis*: Procedures for official control. *EPPO Bulletin* 43:518–526.

EPPO. 2014. *Apriona germari. EPPO Bulletin* 44:155–158.

EPPO. 2016. *EPPO Global Database.* https://gd.eppo.int/ (accessed February 15, 2016).

Errico, M. 2013. Asian longhorned beetle detector dog pilot project. In *Proceedings. 23rd U.S. Department of Agriculture interagency research forum on invasive species 2012,* eds. K. A McManus, and K. W Gottschalk, 18. Newtown Square, PA: USDA Forest Service, Northern Research Station. General Technical Report NRS-P-114.

Eschtruth, A. K., and J. J. Battles. 2011. The importance of quantifying propagule pressure to understand invasion: An examination of riparian forest invasibility. *Ecology* 92:1314–1322.

EU. 2000. *Council Directive 2000/29/EC of 8 May 2000 on protective measures against the introduction into the Community of organisms harmful to plants or plant products and against their spread within the Community.* http://eur-lex.europa.eu/LexUriServ/LexUriServ.do?uri=CELEX:02000L0029-20090303:EN:NOT (accessed December 31, 2015).

EU. 2012. Commission implementing decision of 1 March 2012 as regards emergency measures to prevent the introduction into and the spread within the Union of *Anoplophora chinensis (Forster)*. http://eur-lex .europa.eu/legal-content/EN/TXT/?uri=CELEX%3A32012D0138 (accessed February 27, 2016).

EU. 2013a. Commission implementing decision of 18 February 2013 on the supervision, plant health checks and measures to be taken on wood packaging material actually in use in the transport of specified commodities originating in China 2013. http://eur-lex.europa.eu/legal-content/EN/ TXT/?uri=CELEX%3A32013D0092 (accessed February 27, 2016).

EU. 2013b. Final report of an audit carried out in China from 18 to 28 June 2013 in order to evaluate the measures taken by China to ensure that wood packaging material exported to the European Union meets EU requirements. http://ec.europa.eu/food/fvo/audit_reports/details.cfm?rep_id=3204 (accessed December 31, 2015).

EU. 2014. Europhyt interceptions 2014. http://ec.europa.eu/food/plant/plant_health_biosecurity/europhyt/ interceptions/index_en.htm (accessed February 27, 2016).

Eyre, D. 2005. Pest risk analysis for *Chrsophtharta bimaculata* (Olivier). Central Science Laboratory 2005. https://secure.fera.defra.gov.uk/phiw/riskRegister/plant-health/documents/Chrysoph.pdf (accessed February 27, 2016).

Eyre, D., R. Cannon, D. McCann, and R. Whittaker. 2010. Citrus longhorn beetle, *Anoplophora chinensis*: An invasive pest in Europe. *Outlooks on Pest Management* 21:195–198.

Eyre, D., R. H. A. Baker, S. Brunel, et al. 2012. Rating and mapping the suitability of the climate for pest risk analysis. *EPPO Bulletin* 42:48–55.

Eyre, D., H. Anderson, R. Baker, and R. Cannon. 2013. Insect pests of trees arriving and spreading in Europe. *Outlooks on Pest Management* 24:176–180.

Faccoli, M., R. Favaro, M. T. Smith, and J. Wu. 2015. Life history of the Asian longhorn beetle *Anoplophora glabripennis* (Coleoptera Cerambycidae) in southern Europe. *Agricultural and Forest Entomology* 17:188–196.

FAO. 2011. *Guide to the implementation of phytosanitary standards in forestry*. Rome: FAO.

FAO. 2014. The impact of global trade and mobility on forest health in Europe. http://www.fao.org/docrep/ meeting/030/mj554e.pdf (accessed February 27, 2016).

Favaro, R., L. Wichmann, H-P. Ravn, and M. Faccoli. 2015. Spatial spread and infestation risk assessment in the Asian longhorned beetle, *Anoplophora glabripennis*. *Entomologia Experimentalis et Applicata* 155:95–101.

Forster, B., and B. Wermelinger. 2012. First records and reproductions of the Asian longhorned beetle *Anoplophora glabripennis* (Motschulsky) (Coleoptera, Cerambycidae) in Switzerland. *Mitteilungen der Schweizerischen Entomologischen Gesellschaft* 85:267–275.

Froud, K. J., H. G. Pearson, B. J. T. McCarthy, and G. Thompson. 2008. Contaminants of upholstered furniture from China and Malaysia. In *Surveillance for Biosecurity: Pre-Border to Pest Management*, eds. K. J. Froud, I. A. Popay, and S. M. Zydenbos, 63–75. Christchurch: Wickliffe Limited.

Garonna, A. P., F. Nugnes, B. Espinosa, R. Griffo, and D. Benchi. 2013. *Aromia bungii*, a new Asian worm found in Campania. *Informatore Agrario* 69:60–62.

Giltrap, N., D. Eyre, and P. Reed. 2009. Internet sales of plants for planting—An increasing trend and threat? *EPPO Bulletin* 39:168–170.

Glassey, K. L., G. P. Hosking, and M. Goss. 2005. Phosphine as an alternative to methyl bromide for the fumigation of pine logs and sawn timber. In *Proceedings of annual international conference on methyl bromide alternatives and emissions reductions*, 63–1—63–2. Fresno, CA: Methyl Bromide Alternatives Outreach Office.

Gove, A. D., R. Bashford, and C. J. Brumley. 2007. Pheromone and volatile lures for detecting the European house borer (*Hylotrupes bajulus*) and a manual sampling method. *Australian Forestry* 70:134–136.

Graham, E. E., T. M. Poland, D. G. McCullough, and J. G. Millar. 2012. A comparison of trap type and height for capturing cerambycid beetles (Coleoptera). *Journal of Economic Entomology* 105:837–846.

Grebennikov, V. V., B. D. Gill, and R. Vigneault. 2010. *Trichoferus campestris* (Faldermann) (Coleoptera: Cerambycidae), an Asian wood-boring beetle recorded in North America. *Coleopterists Bulletin* 64:13–20.

Gressit, J. L. 1951. *Longicorn beetles of China*. Vol. 2, *Longicornia*. Paris: Paul Lechevalier.

Haack, R. A. 2006. Exotic bark- and wood-boring Coleoptera in the United States: Recent establishments and interceptions. *Canadian Journal of Forest Research* 36:269–288.

Haack, R. A., and E. G. Brockerhoff. 2011. ISPM No. 15 and the incidence of wood pests: Recent findings, policy changes, and current knowledge gaps. *Proceedings: 42nd Annual Meeting of the International Research Group on Wood Protection,* IRG-WP 11-30568. Stockholm: IRG Secretariat.

Haack, R. A., and R. J. Rabaglia. 2013. Exotic bark and ambrosia beetles in the USA: Potential and current invaders. In *Potential invasive pests of agricultural crops,* ed. J. E. Pena, 48–74. Wallingford: CAB International.

Haack, R. A., L. S. Bauer, R.-T. Gao, J. J. McCarthy, D. L. Miller, T. R. Petrice, et al. 2006. *Anoplophora glabripennis* within-tree distribution, seasonal development, and host suitability in China and Chicago. *The Great Lakes Entomologist* 39:169–183.

Haack, R. A., K. O. Britton, E. G. Brockerhoff, et al. 2014. Effectiveness of the international phytosanitary standard ISPM No. 15 on reducing wood borer infestation rates in wood packaging material entering the United States. *PLoS One* 9: e96611.

Haack, R. A., F. Herard, J. H. Sun, and J. J. Turgeon. 2010. Managing invasive populations of Asian longhorned beetle and citrus longhorned beetle: A worldwide perspective. *Annual Review of Entomology* 55:521–546.

Haack, R. A., K. R. Law, V. C. Mastro, H. S. Ossenburgen, and B. J. Raimo. 1997. New York's battle with the Asian long-horned beetle. *Journal of Forestry* 95:11–15.

Hajek, A. E., B. Huang, T. Dubois, M. T. Smith, and Z. Li. 2006. Field studies of control of *Anoplophora glabripennis* (Coleoptera: Cerambycidae) using fiber bands containing the entomopathogenic fungi *Metarhizium anisopliae* and *Beauveria brongniartii*. *Biocontrol Science and Technology* 16:329–343.

Hanks, L. M. 1999. Influence of the larval host plant on reproductive strategies of cerambycid beetles. *Annual Review of Entomology* 44:483–505.

Hellrigl, K. 1974. Cerambycidae. In *Die Forstschädlinge Europas Vol. 2,* ed. W. Schwenke, 130–202. Hamburg and Berlin: Paul Parey.

Hérard, F., M.-C. Bon, M. Maspero, C. Cocquempot, and J. Lopez. 2005. Survey and evaluation of potential natural enemies of *Anoplophora glabripennis* and *A. chinensis*. In *Proceedings. 16th U.S. Department of Agriculture interagency research forum on invasive species 2005,* ed. K. W. Gottschalk, 34. Newtown Square: USDA Forest Service, Northeastern Research Station, General Technical Report NE-337.

Hérard, F., M. Ciampitti, M. Maspero, et al. 2006. *Anoplophora* species in Europe: Infestations and management processes. *EPPO Bulletin* 36:470–474.

Hérard, F., M. Maspero, N. Ramualde, et al. 2009. *Anoplophora glabripennis* infestation (Col.: Cerambycidae) in Italy. *EPPO Bulletin* 39:146–152.

Hoffmann, N., T. Schröder, F. Schluter, and P. Meinlschmidt. 2013. Potential of infrared thermography to detect insect stages and defects in young trees. *Journal fur Kulturpflanzen* 65:337–346.

Hoyer-Tomiczek, U., H. Krehan, and G. Hoch. 2013. Wood boring insects in packaging material: Recent interceptions and a new Asian longhorned beetle outbreak in Austria. In *Proceedings. 24th U.S. Department of Agriculture interagency research forum on invasive species 2013,* eds. K. A. McManus and K. W. Gottschalk, 79. Fort Collins: USDA Forest Service, Forest Health Technology Enterprise Team, FHTET 13-01.

Hoyer-Tomiczek, U., and G. Sauseng. 2013. Sniffer dogs to find *Anoplophora* spp. infested plants. *Journal of Entomological and Acarological Research* 45(s1):10–12.

Hsiao, K.-J. 1982. Forest entomology in China—A general review. *Crop Protection* 1:359–367.

Hu, J., S. Angeli, S. Schuetz, Y. Luo, and A. E. Hajek. 2009. Ecology and management of exotic and endemic Asian longhorned beetle *Anoplophora glabripennis*. *Agricultural and Forest Entomology* 11:359–375.

Hu, S.-J., T. Ning, D.-Y. Fu, et al. 2013. Dispersal of the Japanese pine sawyer, *Monochamus alternatus* (Coleoptera: Cerambycidae), in mainland China as inferred from molecular data and associations to indices of human activity. *PLoS One* 8: e57568.

Humble, L. 2010. Pest risk analysis and invasion pathways—Insects and wood packing revisited: What have we learned? *New Zealand Journal of Forestry Science* 40: S57–S72.

IPPC. 2002. ISPM 15: Guidelines for regulating wood packaging material in international trade. In *International standards for phytosanitary measures.* Rome: FAO. http://www.fao.org/docrep/009/a0450e/a0450e00.htm (accessed February 27, 2016).

IPPC. 2009. ISPM 15: Regulation of wood packaging material in international trade. http://www.ispm15.com/ISPM15_Revised_2009.pdf (accessed February 28, 2016).

IPPC. 2013. Draft ISPM: Management of pest risks associated with the international movement of wood (2006-029). https://www.ippc.int/publications/draft-ispm-management-pest-risks-associated-international-movement-wood (accessed February 28, 2016).

IPPC. 2014. Technical Panel on Phytosanitary Treatments (TPPT). https://www.ippc.int/en/core-activities/standards-setting/expert-drafting-groups/technical-panels/technical-panel-phytosanitary-treatments/ (accessed February 28, 2016).

IPPC. 2016. ISPM 15: Regulation of wood packaging material in international trade. https://www.ippc.int/en/publications/640/ (accessed February 28, 2016).

Iwaizumi, R., M. Arimoto, and T. Kurauchi. 2014. A study on the occurrence and fecundity of white spotted longicorn, *Anoplophora malasiaca* (Coleoptera: Cerambycidae). *Research Bulletin of the Plant Protection Service Japan* 50:9–15.

Jagadeesan, R., P. J. Collins, G. J. Daglish, P. R. Ebert, and D. I. Schilipalius. 2012. Phosphine resistance in the rust red flour beetle, *Tribolium castaneum* (Coleoptera: Tenebrionidae): Inheritance, gene interactions and fitness costs. *PLoS One* 7: e31582.

Johnson, W., and H. Lyon. 1991. *Insects that feed on trees and shrubs*. Ithaca, NY: Comstock Publishing Associates.

Jordal, B. H., S. M. Smith, and A. I. Cognato. 2014. Classification of weevils as a data-driven science: Leaving opinion behind. *ZooKeys* 439:1–18.

Jucker, C., A. Tantardini, and M. Colombo. 2006. First record of *Psacothea hilaris* (Pascoe) in Europe (Coleoptera Cerambycidae Lamiinae Lamiini). *Bollettino di Zoologia Agraria e di Bachicoltura* 38:187–191.

Justica, A. I., R. Potting, and D. J. van der Gaag. 2010. Pest risk assessment for *Apriona* spp. Wageningen: Plant Protection Service of the Netherlands. https://www.vwa.nl/txmpub/files/?p_file_id=2001667 (accessed February 28, 2016).

Katschak, G. 2004. Comments on *Parandra brunnea* (F.). *Coleo* 5:17–21.

Keena, M., and W. Major. 2001. *Anoplophora glabripennis* (Coleoptera: Cerambycidae) flight propensity in the laboratory. In *Proceedings U.S. Department of Agriculture interagency research forum on invasive species 2001*, eds. S. L. C. Fosbroke and K. W. Gottschalk, 81. Newtown Square, USDA Forest Service, Northeastern Research Station, General Technical Report NE-285.

Keena, M. A., and P. M. Moore. 2010. Effects of temperature on *Anoplophora glabripennis* (Coleoptera: Cerambycidae) larvae and pupae. *Environmental Entomology* 39:1323–1335.

Kliejunas, J. T., B. M. Tkacz, M. Borys, et al. 2001. *Pest risk assessment of the importation into the United States of unprocessed Eucalyptus logs and chips from South America*. Madison, WI: USDA Forest Service, Forest Products Laboratory, General Technical Report FPL–GTR–124.

Kovacs, K. F., R. G. Haight, D. G. McCullough, R. J. Mercader, N. W. Siegert, and A. M. Liebhold. 2010. Cost of potential emerald ash borer damage in U.S. communities, 2009–2019. *Ecological Economics* 69:569–578.

Krehan, H. 2014. First experiences with wood packaging inspections in Austria according implementing decision 2013/92/EU. *Forstschutz Aktuell* 59:3–7.

LaBonte, J. R., A. D. Mudge, and K. J. R. Johnson. 2005. Nonindigenous woodboring Coleoptera (Cerambycidae, Curculionidae: Scolytinae) new to Oregon and Washington, 1999–2002: Consequences of the intracontinental movement of raw wood products and solid wood packing materials. *Proceedings of the Entomological Society of Washington* 107:554–564.

Laclau, J.-P., J. L. de Moraes Goncalves, and J. L. Stape. 2013. Perspectives for the management of eucalypt plantations under biotic and abiotic stresses. *Forest Ecology and Management* 301:1–5.

Li, G.-H., R.-T. Gao, M. T. Smith, and L.-C. Kong. 2010. Study on dispersal of *Anoplophora glabripennis* (Motsch.) (Coleoptera: Cerambycidae) population. *Forest Research* 23:678–684.

Li, M. L., Y. Z. Li, Q. O. Lei, and Z. Q. Yang. 2009a. Biocontrol of Asian longhorned beetle larva by releasing eggs of *Dastarcus helophoroides* (Coleoptera: Bothrideridae). *Scientia Silvae Sinicae* 45:78–82.

Li, Y., Z. Yang, and M. Li. 2009b. Study on the relationship between temperature and the growth and procreation of *Dastarcus helophoroides* Fairmaire. *Journal of Northwest A & F University Natural Science Edition* 37:125–129.

Liebhold, A. M., E. G. Brockerhoff, L. J. Garrett, J. L. Parke, and K. O. Britton. 2012. Live plant imports: The major pathway for forest insect and pathogen invasions of the US. *Frontiers in Ecology and the Environment* 10:135–143.

Lieu, K. O. V. 1945. The study of woodborers in China—I: Biology and control of the citrus-root-cerambycids, *Melanauster chinensis,* Forster (Coleoptera). *The Florida Entomologist* 27:61–101.

Lingafelter, S. W., and E. R. Hoebeke. 2002. *Revision of Anoplophora (Coleoptera: Cerambycidae)*. Washington, DC: The Entomological Society of Washington.

Liu, S. R., W. Proctor, and D. Cook. 2010. Using an integrated fuzzy set and deliberative multi-criteria evaluation approach to facilitate decision-making in invasive species management. *Ecological Economics* 69:2374–2382.

Liu, Z., G. L. Zhang, Y. Li, and J. Zong. 1993. Biological control of peach rednecked longicorn *Aromia bungii* with entomopathogenic nematodes. *Chinese Journal of Biological Control* 9:186–186.

Lundberg, S., and R. Petersson. 1997. Observations of the beetle fauna in Russian timber at a saw mill in Vasterbotten, northern Sweden. *Entomologisk Tidskrift* 118:49–51.

Luo, Y., J. Wen, and Z. Xu. 2003. Current situation of research and control on poplar longhorned beetle, especially for *Anoplophora glabripennis* in China. *Nachrichtenblatt des Deutschen Pflanzenschutzdienstes* 55:66–67.

Lupi, D., C. Jucker, and M. Colombo. 2013. Distribution and biology of the yellow-spotted longicorn beetle *Psacothea hilaris hilaris* (Pascoe) in Italy. *EPPO Bulletin* 43:316–322.

Luppold, W. G., and M. S. Bumgardner. 2011. Thirty-nine years of U.S. wood furniture importing: Sources and products. *Bioresources* 6:4895–4908.

MacLeod, A., H. F. Evans, and R. H. A. Baker. 2002. An analysis of pest risk from an Asian longhorn beetle (*Anoplophora glabripennis*) to hardwood trees in the European community. *Crop Protection* 21:635–645.

MacLeod, A., H. F. Evans, and R. H. A. Baker. 2003. The establishment potential of *Anoplophora glabripennis* in Europe. *Nachrichtenblatt des Deutschen Pflanzenschutzdienstes* 55:83–84.

MacLeod, A., H. Anderson, S. Follak, et al. 2012. External scientific report: Pest risk assessment for the European Community plant health: A comparative approach with case studies, Annex P2 ALB Method 3 no RROs. http://www.efsa.europa.eu/en/supporting/pub/319e.htm (accessed February 27, 2016).

Magarey, R. D., D. M. Borchert, J. S. Engle, M. Colunga-Garcia, F. H. Koch, and D. Yemshanov. 2011. Risk maps for targeting exotic plant pest detection programs in the United States. *EPPO Bulletin* 41:46–56.

Maier, C. T. 2009. Distribution and host records of Cerambycidae (Coleoptera) associated with Cupressaceae in New England, New York, and New Jersey. *Proceedings of the Entomological Society of Washington* 111:438–453.

Mankin, R. W., M. T. Smith, J. M. Tropp, E. B. Atkinson, and D. Y. Jong. 2008. Detection of *Anoplophora glabripennis* (Coleoptera: Cerambycidae) larvae in different host trees and tissues by automated analyses of sound-impulse frequency and temporal patterns. *Journal of Economic Entomology* 101:838–849.

Maspero, M., C. Jucker, and M. Colombo. 2007. First record of *Anoplophora glabripennis* (Motschulsky) (Coleoptera Cerambicidae Lamiinae Lamiini) in Italy. *Bollettino di Zoologia Agraria e di Bachicoltura* 39:161–164.

Maspero, M., C. Jucker, M. Colombo, F. Herard, M. Ciampitti, and B. Cavagna. 2008. News about CLB and ALB in Italy. *Forstschutz Aktuell* 44:25–26.

Matosevic, D., and I. P. Zivkovic. 2013. Alien phytophagous insect and mite species on woody plants in Croatia. *Sumarski List* 137:191–205.

Mattson, W. J., P. Niemela, I. Millers, and Y. Inguanzo. 1994. *Immigrant phytophagous insects on woody plants in the United States and Canada: An annotated list.* St Paul: USDA Forest Service, North Central Forest Experiment Station, General Technical Report NC-169.

Mattson, W., H. Vanhanen, T. Veteli, S. Sivonen, and P. Niemelä. 2007. Few immigrant phytophagous insects on woody plants in Europe: Legacy of the European crucible? *Biological Invasions* 9:957–974.

Meng, P. S., K. Hoover, and M. A. Keena. 2015. Asian longhorned beetle (Coleoptera: Cerambycidae), an introduced pest of maple and other hardwood trees in North America and Europe. *Journal of Integrated Pest Management* 6:1–13.

Miller, N., A. Estoup, S. Toepfer, et al. 2005. Multiple transatlantic introductions of the western corn rootworm. *Science (Washington)* 310:992–992.

Mifsud, D., and D. Dandria. 2002. Introduction and establishment of *Phryneta leprosa* (Fabricius) (Coleoptera, Cerambycidae) in Malta. *Central Mediterranean Naturalist* 3:207–210.

Morall, L. 2011. Insect pests on trees and shrubs in forests and rural areas in 2010. *Vakblad Natuur Bos Landschap* 8:23–27.

Naito, H., Y. Soma, I. Matsuoka et al. 1999. Effects of methyl isothiocyanate on forest insect pests. *Research Bulletin of the Plant Protection Service Japan* 35:1–4.

Nehme, M., M. Keena, A. Zhang, A. Sawyer, and K. Hoover. 2010. Monitoring Asian longhorned beetles in Massachusetts. In *Proceedings 21st U.S. Department of Agriculture interagency research forum on invasive species* 2010, eds. K. A. McManus and K. W. Gottschalk, 109–110. Newtown Square, PA, USDA Forest Service, Northern Research Station, NRS-P-75.

Nehme, M. E., M. A. Keena, P. Meng, et al. 2013. Development of a trapping system for Asian longhorned beetle using semiochemicals. *Journal of Entomological and Acarological Research* 45(s1):20.

Nolte, O., and D. Krieger. 2008. Record of *Saperda candida* Fabricius 1787 on Fehmarn—another already expanding invasive beetle species in Central Europe. *DGaaE Nachrichten* 22:133–136.

Ohbayashi, N., J. Ogawa, and Z.-H. Su. 2009. Phylogenetic analysis of the lamiine genus *Anoplophora* and its relatives (Coleoptera, Cerambycidae) based on the mitochondrial COI gene. *Special Bulletin of the Japanese Society of Coleopterology* 7:309–324.

Ostojá-Starzewski, J. C. 2014. Imported furniture—A pathway for the introduction of plant pests into Europe. *EPPO Bulletin* 44:34–36.

Ostrauskas, H., and V. Tamutis. 2012. Bark and longhorn beetles (Coleopetera: Curculionidae, Scolytinae et Cerambycidae) caught by multiple funnel traps at the temporary storages of timbers and wood in Lithuania. *Baltic Forestry* 18:263–269.

Paine, T. D., M. J. Steinbauer, and S. A. Lawson. 2011. Native and exotic pests of *Eucalyptus*: A worldwide perspective. *Annual Review of Entomology* 56:181–201.

Pautasso, M., G. Aas, V. Queloz, and O. Holdenrieder. 2013. European ash (*Fraxinus excelsior*) dieback—A conservation biology challenge. *Biological Conservation* 158:37–49.

Peterson, A. T., and D. A. Vieglais. 2001. Predicting species invasions using ecological niche modeling: New approaches from bioinformatics attack a pressing problem. *BioScience* 51:363–371.

Peterson, A. T., R. Schachetti-Pereira, and W. W. Hargrove. 2004. Potential geographic distribution of *Anoplophora glabripennis* (Coleoptera: Cerambycidae) in North America. *American Midland Naturalist* 151:170–178.

Petruzziello, L. 2009. Il "Tarlo Asiatico" a Montichiari -BS-, *Anoplophora chinensis:* Una vera minaccia per le nostre piante. http://www.istitutobonsignori.it/bonsignori/images/dicono_di_noi/Anoplophora%20 chinensis%20a%20Montichiari%20-BS-.pdf (accessed February 28, 2016).

Peverieri, G. S., G. Bertini, P. Furlan, G. Cortini, and P. F. Roversi. 2012. *Anoplophora chinensis* (Forster) (Coleoptera Cerambycidae) in the outbreak site in Rome (Italy): Experiences in dating exit holes. *Redia-Giornale Di Zoologia* 95:89–92.

Plant Protection Service of the Netherlands. 2010. First outbreak of *Anoplophora glabripennis* in the Netherlands. https://www.vwa.nl/onderwerpen/kennis-en-advies-plantgezondheid/dossier/pest-reporting/pest-reports (accessed February 28, 2016).

Plant Protection Service of the Netherlands. 2011. March 2011, Eradication accomplished, first outbreak of *Anoplophora glabripennis* in the Netherlands 2011. https://www.vwa.nl/onderwerpen/kennis-en-advies-plantgezondheid/dossier/pest-reporting/pest-reports (accessed February 28, 2016).

Plant Protection Service of the Netherlands. 2012. Outbreak of *Anoplophora glabripennis* in one tree of *Acer platanoides* in a residential area of a minor town (Winterswijk). https://www.vwa.nl/onderwerpen/ kennis-en-advies-plantgezondheid/dossier/pest-reporting/pest-reports (accessed February 28, 2016).

Poland, T., and R. A. Haack. 2003. Exotic forest insect pests and their impact on forest management. In *Proceedings: Society of American Foresters 2002 National Convention,* 132–141. Bethesda, Society of American Foresters.

Poland, T. M., R. A. Haack, and T. R. Petrice. 1998. Chicago joins New York in battle with the Asian longhorned beetle. *Newsletter of the Michigan Entomological Society* 43:15–17.

Poland, T. M., R. A. Haack, T. R. Petrice, D. L. Miller, L. S. Bauer, and R. Gao. 2006. Field evaluations of systemic insecticides for control of *Anoplophora glabripennis* (Coleoptera: Cerambycidae) in China. *Journal of Economic Entomology* 99:383–392.

Prasad, A. M., L. R. Iverson, M. P. Peters, et al. 2010. Modeling the invasive emerald ash borer risk of spread using a spatially explicit cellular model. *Landscape Ecology* 25:353–369.

Rassati, D., E. Petrucco Toffolo, A. Roques, A. Battisti, and M. Faccoli. 2014. Trapping wood boring beetles in Italian ports: A pilot study. *Journal of Pest Science* 87:61–69.

Rassati, D., M. Faccoli, E. P. Toffolo, A. Battisti, and L. Marini. 2015a. Improving the early detection of alien wood-boring beetles in ports and surrounding forests. *Journal of Applied Ecology* 52:50–58.

Rassati, D., M. Faccoli, L. Marini, R. A. Haack, A. Battisti, and E. P. Toffolo. 2015b. Exploring the role of wood waste landfills in early detection of non-native wood-boring beetles. *Journal of Pest Science* 88:563–572.

Rassati, D., F. Lieutier, and M. Faccoli. 2016. Alien wood-boring beetles in Mediterranean regions. In *Insects and Diseases of Mediterranean Forest Systems*, eds. D. T. Paine, and F. Lieutier, 293–327. Cham: Springer International Publishing.

Reaser, J. K., L. A. Meyerson, and B. Von Holle. 2008. Saving camels from straws: How propagule pressure-based prevention policies can reduce the risk of biological invasion. *Biological Invasions* 10:1085–1098.

Ren, Y., B. Lee, and B. Padovan. 2011. Penetration of methyl bromide, sulfuryl fluoride, ethanedinitrile and phosphine into timber blocks and the sorption rate of the fumigants. *Journal of Stored Products Research* 47:63–68.

Rhainds, M., W. E. Mackinnon, K. B. Porter, J. D. Sweeney, and P. J. Silk. 2011. Evidence for limited spatial spread in an exotic longhorn beetle, *Tetropium fuscum* (Coleoptera: Cerambycidae). *Journal of Economic Entomology* 104:1928–1933.

Robinet, C., H. Kehlenbeck, D. J. Kriticos, et al. 2012 A suite of models to support the quantitative assessment of spread in pest risk analysis. *PLoS One* 7: e43366.

Roselli, A., A. Bianchi, L. Nuccitelli, G. Perverieri Sabbatini, and P. Roversi. 2013. Control strategies of *Anoplophora chinensis* in an area of considerable artistic and archaeological value in Rome. *Journal of Entomological and Acarological Research* 45(s1):27–29.

Sabol, O. 2009. *Trichoferus campestris* (Coleoptera: Cerambycidae)—a new species of longhorn beetle for the Czech Republic and Slovakia. *Klapalekiana* 45:199–201.

Sama, G., and C. Cocquempot. 1995. Note on the European extension of *Xylotrechus stebbingi* Gahan, 1906 (Coleoptera, Cerambycidae, Clytini). *Entomologiste (Paris)* 51:71–75.

Sawyer, A., and W. Panagakos. 2009. Spatial dynamics of the Asian longhorned beetle: Cataret, NJ, to Staten Island, NY, in nine years? In *Proceedings 19th U.S. Department of Agriculture interagency research forum on invasive species* 2008, eds. K. McManus and K. W. Gottschalk, 68. Newtown Square, PA: USDA Forest Service, Northern Research Station, General Technical Report NRS-P-36.

Sawyer, A., W. Panagakos, A. Horner, and K. Freeman. 2011. Asian longhorned beetle, over the river and through the woods: Habitat-dependent population spread. In *Proceedings. 21st U.S. Department of Agriculture interagency research forum on invasive species 2010*, eds. K. A. McManus and K. W. Gottschalk, 52–54. Newtown Square, PA, USDA Forest Service, Northern Research Station, NRS-P-75.

Saxony State Office for the Environment. 2016. Asian longhorned beetle, *Anoplophora glabripennis*: New find of the dangerous tree pest in June 2015 Baden-Württemberg—risk also exists for Saxony 2016. https://www.landwirtschaft.sachsen.de/landwirtschaft/2132.htm (accessed February 28, 2016).

Schofield, J. 2011. *Real-time acoustic detection of invasive wood-boring beetles*. Ph.D. Dissertation, University of York, Department of Electronics. http://etheses.whiterose.ac.uk/1978/1/js517_thesis.pdf (accessed February 28, 2016).

Shatz, A. J., J. Rogan, F. Sangermano, Y. Ogneva-Himmelberger, and H. Chen. 2013. Characterizing the potential distribution of the invasive Asian longhorned beetle (*Anoplophora glabripennis*) in Worcester County, Massachusetts. *Applied Geography* 45:259–268.

Schreck, M., and C. Tomiczek. 2013. *Asiatischer Laubholxbock in Gallspach (oo) entdeckt*. BFW 2013. http://bfw.ac.at/rz/bfwcms.web?dok=9797 (accessed February 28, 2016).

Smith, G., and J. E. Hurley. 2000. First North American record of the Palearctic species *Tetropium fuscum* (Fabricius) (Coleoptera: Cerambycidae). *The Coleopterists Bulletin* 54:540.

Smith, M., and J. Wu. 2008. Asian longhorned beetle: Renewed threat to northeastern USA and implications worldwide. *International Pest Control* Nov/Dec:311–316.

Smith, M. T., J. Bancroft, G. H. Li, R. Gao, and S. Teale. 2001. Dispersal of *Anoplophora glabripennis* (Cerambycidae). *Environmental Entomology* 30:1036–1040.

Smith, M. T., P. C. Tobin, J. Bancroft, G. H. Li, and R. T. Gao. 2004. Dispersal and spatiotemporal dynamics of Asian longhorned beetle (Coleoptera: Cerambycidae) in China. *Environmental Entomology* 33:435–442.

Smith, M. T., J. J. Turgeon, P. De Groot, and B. Gasman. 2009. Asian longhorned beetle *Anoplophora glabripennis* (Motschulsky): Lessons learned and opportunities to improve the process of eradication and management. *American Entomologist* 55:21–25.

Smith, R. M., R. H. A. Baker, C. P. Malumphy, et al. 2007. Recent non-native invertebrate plant pest establishments in Great Britain: Origins, pathways, and trends. *Agricultural and Forest Entomology* 9:307–326.

Soliman, T., M. C. M. Mourits, W. van der Werf, G. M. Hengeveld, C. Robinet, and A. G. J. M. Oude Lansink. 2012. Framework for modelling economic impacts of invasive species, applied to pine wood nematode in Europe. *PLoS One* 7: e45505.

Soma, Y., H. Komatsu, Y. Abe, T. Itabashi, Y. Matsumoto, and F. Kawakami. 2006. Effects of some fumigants on mortality of the pine wood nematode, *Bursaphelenchus xylophilus* infesting wooden packages: 6. Mortality of pine wood nematode and longhorn beetles by methyl iodide tarpaulin fumigation. *Research Bulletin of the Plant Protection Service Japan* 42:7–13.

Soma, Y., H. Komatsu, T. Oogita, et al. 2007. Mortality of forest insect pests by methyl iodide tarpaulin fumigation. *Research Bulletin of the Plant Protection Service Japan* 43:9–15.

Stejskal, V., O. Douda, M. Zouhar, et al. 2014. Wood penetration ability of hydrogen cyanide and its efficacy for fumigation of *Anoplophora glabripennis, Hylotrupes bajulus* (Coleoptera), and *Bursaphelenchus xylophilus* (Nematoda). *International Biodeterioration & Biodegradation* 86:189–195.

Strangi, A., G. Sabbatini Peverieri, and P. F. Roversi. 2013. Managing outbreaks of the citrus long-horned beetle *Anoplophora chinensis* (Forster) in Europe: Molecular diagnosis of plant infestation. *Pest Management Science* 69:627–634.

Straw, N. A., N. J. Fielding, C. Tilbury, D. T. Williams, and D. Inward. 2014. Host plant selection and resource utilisation by Asian longhorn beetle *Anoplophora glabripennis* (Coleoptera: Cerambycidae) in southern England. *Forestry* 88:84–95.

Sutherst, R. W. 2014. Pest species distribution modelling: Origins and lessons from history. *Biological Invasions* 16:239–256.

Sutherst, R. W., G. F. Maywald, and D. J. Kriticos. 2007. Climex version 3 user's guide. *Hearne scientific software Pty Ltd 2007.* http://www.hearne.com.au/getattachment/0343c9d5-999f-4880-b9b2-1c3eea908f08/Climex-User-Guide.aspx (accessed February 28, 2016).

Sweeney, J., J. Price, W. Mackay, B. Guscott, P. de Groot and J. Gutowski. 2007. Detection of the brown spruce longhorn beetle, *Tetropium fuscum* (F.) with semiochemical-baited traps, tree bands, and visual surveys. In *Proceedings. 17th U.S. Department of Agriculture interagency research forum on invasive species 2006*, ed. K. W. Gottschalk, 95–96. Newtown Square, PA: USDA Forest Service, Northern Research Station, General Technical Report NRS-P-10.

Sweeney, J., P. Silk, D. Pureswaran and L. Flaherty. 2009. Research update on the brown spruce longhorn beetle, *Tetropium fuscum* (Fabr.). In *Proceedings. 20th U.S. Department of Agriculture interagency research forum on invasive species 2009*, eds. K. McManus and K. W. Gottschalk, 56–57. Newtown Square, PA: USDA Forest Service, Northern Research Station, General Technical Report NRS-P-51.

Sweeney, J., P. Silk, M. Rhainds, J. E. Hurley, and E. Kettela. 2011a. Population suppression of *Tetropium fuscum* (F.) by pheromone-mediated mating disruption. In *Proceedings. 22nd U.S. Department of Agriculture interagency research forum on invasive species 2011*, eds. K. McManus and K. W. Gottschalk, 91. Newtown Square, PA: USDA Forest Service, Northern Research Station, General Technical Report NRS-P-92.

Sweeney, J., P. Silk, M. Rhainds, J. E. Hurley, and W. Mackay. 2011b. Mass trapping for population suppression of an invasive longhorn beetle, *Tetropium fuscum* (F.) (Coleoptera: Cerambycidae). In *Proceedings. 22nd U.S. Department of Agriculture interagency research forum on invasive species 2011*, eds. K. McManus and K. W. Gottschalk, 92. Newtown Square, PA: USDA Forest Service, Northern Research Station, General Technical Report NRS-P-92.

Sweeney, J., P. Silk, C. Hughes, R. Lavallee, M. Blais, and C. Guertin. 2013. Auto-dissemination of *Beauveria bassiana* for control of brown spruce longhorned beetle, *Tetropium fuscum* (F.), Coleoptera: Cerambycidae. In *Proceedings. 24th U.S. Department of Agriculture interagency research forum on invasive species 2013*, eds. K. A. McManus and K. W. Gottschalk, 98. Fort Collins, CO: USDA Forest Service, Forest Health Technology Enterprise Team, FHTET 13-01.

Swift, I. P., and A. M. Ray. 2010. Nomenclatural changes in North American *Phymatodes* Mulsant (Coleoptera: Cerambycidae). *Zootaxa* 2448:35–52.

Takahashi, N., and M. Ito. 2005. Detection and eradication of the Asian longhorned beetle in Yokohama, Japan. *Research Bulletin of the Plant Protection Service Japan* 41:83–85.

Teunissen, A. P. J. A. 2002. Observations of *Xylotrechus stebbingi* in Greece. An Asian cerambycid beetle which became recently established in the Mediterranean area (Coleoptera: Cerambycidae). *Entomologische Berichten (Amsterdam)* 62:57–58.

Tobin, P., J. Kean, D. Suckling, D. McCullough, D. Herms, and L. Stringer. 2014. Determinants of successful arthropod eradication programs. *Biological Invasions* 16:401–414.

Tomiczek, C., and U. Hoyer-Tomiczek. 2007 Asian longhorned beetle (*Anoplophora glabripennis*) and citrus longhorned beetle (*Anoplophora chinensis*) in Europe—actual situation. *Forstschutz Aktuell* 38:2–5.

Tomiczek, C., H. Krehan, and P. Menschhorn. 2002. Dangerous Asiatic longicorn beetle found in Austria: New danger for our trees? *AFZ/Der Wald, Allgemeine Forst Zeitschrift fur Waldwirtschaft und Umweltvorsorge* 57:52–54.

Trotter, R. T. III, and H. M. Hull-Sanders. 2015. Quantifying dispersal of the Asian longhorned beetle (*Anoplophora glabripennis*, Coleoptera) with incomplete data and behavioral knowledge. *Biological Invasions* 17:3359–3369.

Tsankov, G., A. Maslov, B. Kovalyov, B. Ogibin, and L. Matusevich. 1996. The cerambycids of genus *Monochamus* and their development in a timber imported from the Republic of Komi. *Nauka za Gorata* 33:67–76.

Turgeon, J. J., M. Orr, C. Grant, Y. Wu, and B. Gasman. 2015. Decade-old satellite infestation of *Anoplophora glabripennis* Motschulsky (Coleoptera: Cerambycidae) found in Ontario, Canada outside regulated area of founder population. *The Coleopterists Bulletin* 69:674–678.

Turnbull, J. W. 1999. Eucalypt plantations. *New Forests* 17:37–52.

Ugine, T. A., S. Gardescu, P. A. Lewis, and A. E. Hajek. 2012. Efficacy of imidacloprid, trunk-injected into *Acer platanoides,* for control of adult Asian longhorned beetles (Coleoptera: Cerambycidae). *Journal of Economic Entomology* 105:2015–2028.

UNEP. 2011. The Montreal protocol on substances that deplete the ozone layer. http://ozone.unep.org/new_site/en/Treaties/treaties_decisions-hb.php?sec_id=5 (accessed February 28, 2016).

USDA-APHIS. 1995. 7 CFR Parts 300 and 319—importation of logs, lumber, and other unmanufactured wood articles. Federal Register, 25 May 1995. 60:27, 665–27, 682.

USDA-APHIS. 2004. Pest Datasheet for *Callidiellum villosulum* (Fairmaire) (Coleoptera: Cerambycidae). http://www.fs.fed.us/foresthealth/publications/Callidiellum_villosulum_APHIS_Fact_sheet.pdf (accessed February 28, 2016).

USDA-APHIS. 2008. New Pest Response Guidelines: Asian longhorned beetle *Anoplophora glabripennis.* https://www.aphis.usda.gov/plant_health/plant_pest_info/asian_lhb/downloads/alb_response_guidelines.pdf (accessed February 25, 2016).

USDA-APHIS. 2014. *USDA declares a Boston, Massachusetts areas free of Asian longhorned beetle.* https://www.aphis.usda.gov/aphis/newsroom/news/SA_By_Date/SA_2014/SA_05/CT_alb_boston (accessed February 28, 2016).

USDA-CPHST. 2011. Pareto risk map: *Monochamus alternatus,* Japanese pine sawyer beetle. caps.ceris.purdue.edu/dmm/1965 (accessed February 28, 2016).

van der Gaag, D., M. Ciampitti, B. Cavagna, M. Maspero, and F. Hérard. 2008. Pest risk analysis for *Anoplophora chinensis.* Wageningen: Plant Protection Service of the Netherlands. https://secure.fera.defra.gov.uk/phiw/riskRegister/plant-health/documents/Anoplop.pdf (accessed February 29, 2016).

van der Gaag, D. J., G. Sinatra, P. F. Roversi, A. Loomans, F. Herard, and A. Vukadin. 2010. Evaluation of eradication measures against *Anoplophora chinensis* in early stage infestations in Europe. *EPPO Bulletin* 40:176–187.

Vanhanen, H., T. O. Veteli, and P. Niemelä. 2008. Potential distribution ranges in Europe for *Aeolesthes sarta, Tetropium gracilicorne* and *Xylotrechus altaicus,* a CLIMEX analysis. *EPPO Bulletin* 38:239–248.

Venette, R. C., R. D. Moon, and W. D. Hutchison. 2002. Strategies and statistics of sampling for rare individuals. *Annual Review of Entomology* 47:143–174.

Villa, G., L. Bonanomi, D. Guarino, L. Pozzi, and M. Maspero. 2013. Use of the electronic nose for the detection of *Anoplophora chinensis* (Forster) on standing trees: Preliminary results. *Journal of Entomological and Acarological Research* 45(s1):13–14.

Vukadin, A., and B. Hrasovec. 2008. *Anoplophora chinensis* (Forster) in Croatia. *Forstschutz Aktuell* 44:23–24.

Wang, Q. 1998. Evidence for a contact female sex pheromone in *Anoplophora chinensis* (Forster) (Coleoptera: Cerambycidae: Lamiinae). *The Coleopterists Bulletin* 52:363–368.

Wei, J. R., Z. Q. Yang, J. H. Ma, and H. Tang. 2007. Progress on the research of *Dastarcus helophoroides. Forest Pest and Disease* 26:23–25.

Wei, J. R., and Y. L. Niu. 2011. Evaluation of biological control of *Anoplophora glabripennis* (Coleoptera: Cerambycidae) by releasing adult *Dastarcus helophoroides* (Coleoptera: Zopheridae): a case study in Xi'an city, northwestern China. *Acta Entomologica Sinica* 54:1399–1405.

Wickham, J. D., R. D. Harrison, W. Lu, et al. 2014. Generic lures attract cerambycid beetles in a tropical montane rain forest in southern China. *Journal of Economic Entomology* 107:259–267.

Williams, D. W., H. P. Lee, and I. K. Kim. 2004. Distribution and abundance of *Anoplophora glabripennis* (Coleoptera: Cerambycidae) in natural *Acer* stands in South Korea. *Environmental Entomology* 33:540–545.

Work, T. T., D. G. McCullough, J. F. Cavey, and R. Komsa. 2005. Arrival rate of nonindigenous insect species into the United States through foreign trade. *Biological Invasions* 7:323–332.

Wylie, F. R., M. Griffiths, and J. King. 2008. Development of hazard site surveillance programs for forest invasive species: A case study from Brisbane, Australia. *Australian Forestry* 71:229–235.

Yang, Z.-Q., X.-Y. Wang, and Y.-N. Zhang. 2014. Recent advances in biological control of important native and invasive forest pests in China. *Biological Control* 68:117–128.

Yemshanov, D., F. H. Koch, Y. Ben-Haim, and W. D. Smith. 2010. Detection capacity, information gaps and the design of surveillance programs for invasive forest pests. *Journal of Environmental Management* 91:2535–2546.

Yildirim, N., H. Taskin, and R. Karaman. 2012. A fumigation treatment applied in Istanbul-Beylerbeyi Palace by using sulfuryl fluoride against Coleoptera species. *Journal of the Faculty of Forestry, Istanbul University* 62:47–52.

Zahid, M. I., C. A. Grgurinovic, and T. Zaman. 2013. Eradication of insect pests of subtropical and tropical forest products with freezing storage. *Journal of Tropical Forest Science* 25:475–486.

Zampini, M., M. Vettorazzo, M. Faccoli, R. Favaro, I. D. Cin, and M. Coppe. 2013. *Anoplophora glabripennis* outbreak management in Veneto region, Italy. *Journal of Entomological and Acarological Research* 45(s1):30.

Zhan, G., B. Li, Y. Wang, M. Hu, L. Li, and H. Qin. 2011a. Primary results on X-ray (9 MeV) irradiation on larva and female adults of *Monochamus sutor*. *Plant Quarantine (Shanghai)* 25:18–20.

Zhan, G., X. Wang, Y. Wang, et al. 2011b. Irradiation of X-rays (9 Mev) on *Monochamus alternatus*. *Plant Quarantine (Shanghai)* 25:12–17.

Zhang, A. J., J. E. Oliver, J. R. Aldrich, B. D. Wang, and V. C. Mastro. 2002. Stimulatory beetle volatiles for the Asian longhorned beetle, *Anoplophora glabripennis* (Motschulsky). *Zeitschrift für Naturforschung C-a Journal of Biosciences* 57:553–558.

Zorovic, M., and A. Cokl. 2015. Laser vibrometry as a diagnostic tool for detecting wood-boring beetle larvae. *Journal of Pest Science* 88:107–112.

Index